T0181085

Communications
in Computer and Information Science 1330

Editorial Board Members

More information about this series at http://www.springer.com/series/7899

Yuqing Sun · Dongning Liu ·
Hao Liao · Hongfei Fan ·
Liping Gao (Eds.)

Computer Supported Cooperative Work and Social Computing

15th CCF Conference, ChineseCSCW 2020
Shenzhen, China, November 7–9, 2020
Revised Selected Papers

 Springer

Editors
Yuqing Sun
Shandong University
Jinan, China

Dongning Liu
Guangdong University of Technology
Guangzhou, China

Hao Liao
Shenzhen University
Shenzhen, China

Hongfei Fan
Tongji University
Shanghai, China

Liping Gao
University of Shanghai for Science and
Technology
Shanghai, China

ISSN 1865-0929   ISSN 1865-0937 (electronic)
Communications in Computer and Information Science
ISBN 978-981-16-2539-8   ISBN 978-981-16-2540-4   (eBook)
https://doi.org/10.1007/978-981-16-2540-4

This Springer imprint is published by the registered company Springer Nature Singapore Pte Ltd.
The registered company address is: 152 Beach Road, #21-01/04 Gateway East, Singapore 189721, Singapore

Preface

Welcome to ChineseCSCW 2020, the 15th CCF Conference on Computer Suppor
Cooperative Work and Social Computing.

ChineseCSCW 2020 was organized by the China Computer Federation (CC
co-hosted by the CCF Technical Committee on Cooperative Computing (CCF TCC
and the Shenzhen Artificial Intelligence and Data Science Institute (Longhua), a
supported by the CAS Institute of Automation and Shenzhen University. The ther
of the conference was *Human-centered Collaborative Intelligence*, which reflects tl
emerging trend of the combination of artificial intelligence, human-system collabor,
tion, and AI-empowered applications.

ChineseCSCW (initially named CCSCW) is a highly reputable conference series o
computer supported cooperative work (CSCW) and social computing in China with
long history. It aims at bridging Chinese and overseas CSCW researchers, practitioners
and educators, with a particular focus on innovative models, theories, techniques
algorithms, and methods, as well as domain-specific applications and systems, covering
both the technical and social aspects in CSCW and social computing. The conference
was initially held biennially, starting in 1998, and has been held annually since 2014.

This year, the conference received 137 submissions, and after a rigorous
double-blind peer review process, only 40 of them were eventually accepted as full
papers to be presented orally, resulting in an acceptance rate of 29%. The program also
included 15 short papers, which were presented as posters. In addition, the conference
featured six keynote speeches, seven high-level technical seminars, the 1st ChineseCSCW Cup Big Data Challenge, and the Award Ceremony for the 10th
Anniversary of the Establishment of CCF TCCC. We are grateful to the distinguished
keynote speakers, *Prof. Enhong Chen* from the *University of Science and Technology
of China*, *Prof. Bin Guo* from the *Northwestern Polytechnical University*, *Prof. Zhihui
Zhan* from the *South China University of Technology*, *Prof. Xin Yao* from the *Southern
University of Science and Technology*, *Prof. Yunjun Gao* from *Zhejiang University*,
and *Prof. Yong Li* from *Tsinghua University*.

We hope that you have enjoyed ChineseCSCW 2020.

November 2020

Yong Tang
Dajun Zeng

Organization

ChineseCSCW 2020 was organized by the China Computer Federation (CCF), co-hosted by the CCF Technical Committee on Cooperative Computing (CCF TCCC) and the Shenzhen Artificial Intelligence and Data Science Institute (Longhua), and supported by the CAS Institute of Automation and Shenzhen University, during November 7–9, 2020.

Steering Committee

Yong Tang	South China Normal University, China
Weiqing Tang	Chinese Academy of Sciences, China
Ning Gu	Fudan University, China
Shaozi Li	Xiamen University, China
Bin Hu	Lanzhou University, China
Yuqing Sun	Shandong University, China
Xiaoping Liu	Hefei University of Technology, China
Zhiwen Yu	Northwestern University of Technology, China
Xiangwei Zheng	Shandong Normal University, China
Tun Lu	Fudan University, China

General Chairs

Yong Tang	South China Normal University, China
Dajun Zeng	Institute of Automation, Chinese Academy of Sciences, China

Program Committee Chairs

Yuqing Sun	Shandong University, China
Dongning Liu	Guangdong University of Technology, China
Hao Liao	Shenzhen University, China

Organization Committee Chairs

Xiaoping Liu	Hefei University of Technology, China
Zhiwen Yu	Northwestern University of Technology, China
Tun Lu	Fudan University, China
Tianzi Jin Miao	Institute of Automation, Chinese Academy of Sciences, China

Publicity Chairs

Xiangwei Zheng   Shandong Normal University, China
Jianguo Li   South China Normal University, China

Publication Chairs

Bin Hu   Lanzhou University, China
Hailong Sun   Beihang University, China

Finance Chair

Xiaoyan Huang   Institute of Automation, Chinese Academy of Sciences, China

Paper Award Chairs

Shaozi Li   Xiamen University, China
Yichuan Jiang   Southeast University, China

Program Committee

Tie Bao	Jilin University, China
Hongming Cai	Shanghai Jiao Tong University, China
Zhicheng Cai	Nanjing University of Science and Technology, China
Buqing Cao	Hunan University of Science and Technology, China
Donglin Cao	Xiamen University, China
Jian Cao	Shanghai Jiao Tong University, China
Chao Chen	Chongqing University, China
Jianhui Chen	Beijing University of Technology, China
Longbiao Chen	Xiamen University, China
Liangyin Chen	Sichuan University, China
Qingkui Chen	University of Shanghai for Science and Technology, China
Ningjiang Chen	Guangxi University, China
Weineng Chen	South China University of Technology, China
Yang Chen	Fudan University, China
Shiwei Cheng	Zhejiang University of Technology, China
Xiaohui Cheng	Guilin University of Technology, China
Yuan Cheng	Wuhan University, China
Lizhen Cui	Shandong University, China
Weihui Dai	Fudan University, China
Xianghua Ding	Fudan University, China
Wanchun Dou	Nanjing University, China
Bowen Du	University of Warwick, UK
Hongfei Fan	Tongji University, China

Shanshan Feng	Shandong Normal University, China
Liping Gao	University of Shanghai for Science and Technology, China
Ning Gu	Fudan University, China
Kun Guo	Fuzhou University, China
Wei Guo	Shandong University, China
Yinzhang Guo	Taiyuan University of Science and Technology, China
Tao Han	Zhejiang Gongshang University, China
Fei Hao	Shanxi Normal University, China
Fazhi He	Wuhan University, China
Haiwu He	Chinese Academy of Sciences, China
Bin Hu	Lanzhou University, China
Daning Hu	South University of Science and Technology, China
Wenting Hu	Jiangsu Open University, China
Yanmei Hu	Chengdu University of Technology, China
Changqin Huang	South China Normal University, China
Bo Jiang	Zhejiang Gongshang University, China
Bin Jiang	Hunan University, China
Jiuchuan Jiang	Nanjing University of Finance and Economics, China
Weijin Jiang	Xiangtan University, China
Yichuan Jiang	Southeast University, China
Lu Jia	China Agricultural University, China
Yi Lai	Xi'an University of Posts and Telecommunications, China
Dongsheng Li	Microsoft Research, China
Guoliang Li	Tsinghua University, China
Hengjie Li	Lanzhou University of Arts and Science, China
Jianguo Li	South China Normal University, China
Li Li	Southwest University, China
Renfa Li	Hunan University, China
Shaozi Li	Xiamen University, China
Taoshen Li	Guangxi University, China
Xiaoping Li	Southeast University, China
Lu Liang	Guangdong University of Technology, China
Bing Lin	Fujian Normal University, China
Dazhen Lin	Xiamen University, China
Dongning Liu	Guangdong University of Technology, China
Hong Liu	Shandong Normal University, China
Li Liu	Chongqing University, China
Shijun Liu	Shandong University, China
Shufen Liu	Jilin University, China
Xiaoping Liu	Hefei University of Technology, China
Tun Lu	Fudan University, China
Huijuan Lu	China Jiliang University, China
Dianjie Lu	Shandong Normal University, China
Qiang Lu	Hefei University of Technology, China

Pin Lv	Guangxi University, China
Hui Ma	University of Electronic Science and Technology of China and Zhongshan Institute, China
KeJi Mao	Zhejiang University of Technology, China
Haiwei Pan	Harbin Engineering University, China
Li Pan	Shandong University, China
Limin Shen	Yanshan University, China
Yuliang Shi	ShanDa Dareway Company Limited, China
Yanjun Shi	Dalian University of Science and Technology, China
Xiaoxia Song	Datong University, China
Kehua Su	Wuhan University, China
Hailong Sun	Beihang University, China
Ruizhi Sun	China Agricultural University, China
Yuqing Sun	Shandong University, China
Yuling Sun	East China Normal University, China
Wen'an Tan	Nanjing University of Aeronautics and Astronautics, China
Lina Tan	Hunan University of Technology and Business, China
Yong Tang	South China Normal University, China
Shan Tang	Shanghai Polytechnic University, China
Weiqing Tang	Beijing Zhongke Fulong Computer Technology Company Limited, China
Yan Tang	Hohai University, China
Yiming Tang	HeFei University of Technology, China
Yizheng Tao	China Academy of Engineering Physics, China
Shaohua Teng	Guangdong University of Technology, China
Dakuo Wang	IBM Research – USA, USA
Hongbin Wang	Kunming University of Science and Technology, China
Hongjun Wang	Southwest Jiaotong University, China
Hongbo Wang	University of Science and Technology Beijing, China
Lei Wang	Dalian University of Technology, China
Tianbo Wang	Beihang University, China
Tong Wang	Harbin Engineering University, China
Wanyuan Wang	Southeast University, China
Yijie Wang	National University of Defense Technology, China
Zhiwen Wang	Guangxi University of Science and Technology, China
Yiping Wen	Hunan University of Science and Technology, China
Chunhe Xia	Beihang University, China
Fangxion Xiao	Jinling Institute of Technology, China
Zheng Xiao	Hunan University, China
Xiaolan Xie	Guilin University of Technology, China
Zhiqiang Xie	Harbin University of Science and Technology, China
Yu Xin	Harbin University of Science and Technology, China
Jianbo Xu	Hunan University of Science and Technology, China
Jiuyun Xu	China University of Petroleum, China
Meng Xu	Shandong Technology and Business University, China

Heyang Xu	Henan University of Technology, China
Jiaqi Yan	Nanjing University, China
Bo Yang	University of Electronic Science and Technology of China, China
Chao Yang	Hunan University, China
Dingyu Yang	Shanghai DianJi University, China
Gang Yang	Northwestern Polytechnical University, China
Jing Yang	Harbin Engineering University, China
Lin Yang	Shanghai Computer Software Technology Development Center, China
Xiaochun Yang	Northeastern University, China
Xu Yu	Qingdao University of Science and Technology, China
Zhiwen Yu	Northwestern Polytechnical University, China
Jianyong Yu	Hunan University of Science and Technology, China
Yang Yu	Zhongshan University, China
Zhengtao Yu	Kunming University of Science and Technology, China
An Zeng	Guangdong Polytechnic University, China
Dajun Zeng	Chinese Academy of Sciences, China
Changyou Zhang	Chinese Academy of Sciences, China
Jifu Zhang	Taiyuan University of Science and Technology, China
Liang Zhang	Fudan University, China
Miaohui Zhang	Jiangxi Academy of Sciences, China
Peng Zhang	Fudan University, China
Shaohua Zhang	Shanghai Software Technology Development Center, China
Wei Zhang	Guangdong University of Technology, China
Zhiqiang Zhang	Harbin Engineering University, China
Zili Zhang	Southwest University, China
Xiangwei Zheng	Shandong Normal University, China
Ning Zhong	Beijing University of Technology, China
Yifeng Zhou	Southeast University, China
Tingshao Zhu	Chinese Academy of Science, China
Xia Zhu	Southeast University, China
Xianjun Zhu	Jinling University of Science and Technology, China
Yanhua Zhu	The First Affiliated Hospital of Guangdong Pharmaceutical University, China
Jianhua Zhu	City University of Hong Kong, China
Qiaohong Zu	Wuhan University of Technology, China

Contents

Domain-Specific Collaborative Applications

Collaborative Mechanisms, Models, Approaches, Algorithms, and Systems

Social Media and Online Communities

Crowdsourcing, Crowd Intelligence, and Crowd Cooperative Computing

Selective Self-attention with BIO Embeddings for Distant Supervision Relation Extraction

Mingjie Tang and Bo Yang[✉]

School of Computer Science and Engineering,
University of Electronic Science and Technology of China,
Chengdu 611731, Sichuan, China
yangbo@uestc.edu.cn

Abstract. Distant supervision method is proposed to label instances automatically, which could operate relation extraction without human annotations. However, the training data generated in this way intrinsically include massive noise. To alleviate this problem, attention mechanism is employed by most prior works that achieves significant improvements but could be still imcompetent for one-sentence bags which means only one sentence within a bag. To this end, in this paper, we propose a novel neural relation extraction method employing BIO embeddings and a selective self-attention with fusion gate mechanism to fix the aformentioned defects in previous attention methods. First, in addition to commonly adopted embedding methods in input layer, we propose to add BIO embeddings to enrich the input representation. Second, a selective self-attention mechanism is proposed to capture context dependency information and combined with PCNN via a Fusion Gate module to enhance the representation of sentences. Experiments on the NYT dataset demonstrate the effectiveness of our proposed methods and our model achieves consistent improvements for relation extraction compared to the state-of-the-art methods.

Keywords: Distant supervision relation extraction · One-sentence bag problem · Selective self-attention · BIO embeddings

1 Introduction

Relation extraction, which aims at extracting ternary relations between entity pairs in the given sentences, is a crucial task in natural language processing (NLP) [1,2]. Conventional methods with supervised learning paradigm[4,5] require large-scale manually labeled data, which is time-consuming and labor-intensive. To this end, a distant supervision method [1] was proposed to overcome the lack of labeled training data, which automatically generates the data by aligning entities in knowledge bases (KBs). The assumption for distant supervision is that if a entity pair has a relation corresponding to the KBs, then all sentences mentioning this entity pair hold that relation.

© Springer Nature Singapore Pte Ltd. 2021
Y. Sun et al. (Eds.): ChineseCSCW 2020, CCIS 1330, pp. 3–16, 2021.
https://doi.org/10.1007/978-981-16-2540-4_1

However, intuitively, it is inevitable that distant supervision suffers from wrongly-labeled problem due to its strong assumption and the dataset created by distant supervision can be seriously noisy [6,9]. In order to alleviate this problem, a multi-instance learning (MIL) method was proposed [6] to be incorporated with distant supervision. In MIL method, the dataset generated by distant supervision is divided into bags where each bag consists of sentences involving the same entity pair.

In recent years, many neural relation extraction in MIL methods with attention mechanism have been proposed to mitigate the influence of noisy data [2,10] and achieved significant improvements. These proposed methods mainly focus on the correctly labeled sentences in one bag and utilize them to train a robust model. However, most of the bags in the distant supervision dataset (e.g. NYT dataset [6]) are one-sentence, and more importantly, the one-sentence bag is highly likely to be wrongly labeled. We investigate NYT dataset and find out that about 80% of its training examples are one-sentence bags while up to 40% of them are wrongly labeled, taking labeled sentences in NYT dataset for example in Table 1. Actually, vanilla selective attention mechanism is vulnerable to such situations and could hurt the model performance. Furthermore, we argue that sentences with wrong label in a bag also contain implicitly useful information for training model and are supposed to be well exploited.

Table 1. Two examples of one-sentence bag with correct and wrong label in distant supervision respectively.

One-sentence bag	Label	Correct
In december, **Brad Grey**, the chairman of **Paramount Pictures**, stood up and greeted the celebrity-packed crowds	/business/person/company	True
The theater has never been so popular, said the actress **Joanne Woodward**, a **Westport** resident who organized the restoration effort	/people/person/place_of_birth	False

Based on the above observations, we propose a Selective Self-Attention (SS-Att) framework for distant supervision relation extraction in this paper. The framework is composed of three ideas. First, motivated by the Neural Entity Recognition (NER) task, we employ BIO (Beginning, Inside, Outside) embedding approach to enrich the input representation. Second, a Selective Self-attention mechanism is proposed to capture rich context dependency information and combined with the widely-used piecewise CNN (PCNN) [11] via a Fusion Gate module, which is leveraged to enhance the feature representation of sentences and overcome the one-sentence bag problem. Moreover, similar to Ye et al. [9], we devise a relation-aware attention mechanism that all relations are used to calculate the relation-aware bag representations rather than the only relation of each bag and wrongly-labeled bags are utilized to obtain rich bag-level representations with this mechanism.

We evaluate our model on NYT (New York Time [6]) dataset, the experiments and extensive ablation studies demonstrate the efficiency of our ideas and experimental results show that compared with the state-of-the-art works, the proposed methods achieve improvements for relation extraction.

2   Related Work

Many conventional methods with supervised learning paradigm [4, 5] designed hand-crafted features to train feature-based and kernel-based models, which have high performance and better practicability. However, these methods require large-scale manually labeled data which is hard to get. To address this issue, distant supervision was proposed [1] by automatically aligning plain texts toward Freebase relational facts to generate relation label for entity pairs. However, it is such a strong assumption that inevitably introduce the wrongly-labeled problem. Therefore, in order to alleviate such problem, a line of works [6, 7] solve distant supervision relation extraction with multi-instance learning method that all sentences mentioning the same entity pair are taken into a bag.

In recent years, with the development of deep learning methods [8], many deep neural networks have been developed for distant supervision relation extraction. Zeng et al. [11] proposed piecewise convolutional networks (PCNNs) to extract feature representations from sentences and select the most reliable sentence for bag representation. Lin et al. [2] adopt PCNNs as sentence encoder and propose a selective attention mechanism to generate bag representation over sentences. Liu et al. [3] propose a soft-label method to denoise the sentences. Ye et al. [9] propose a intra-bag and inter-bag attention mechanism to extend the sentence-level attention to bag-level. These methods all achieve significant improvement in relation extraction, but they cannot handle the one-sentence bag problem as a result of the drawback of vanilla selective attention.

To address the one-sentence bag problem, we propose a selective self-attention mechanism with fusion gate which could capture rich context dependency information and enhance the representation of sentences. Further, a BIO embeddings method is proposed to enrich the input representation.

3   Methodology

In this section, we introduce the proposed SS-Att framework for distant supervision relation extraction. The framework of our method is shown in Fig. 1, which consists of the following neural layers: input layer, encoder layer, fusion layer and output layer.

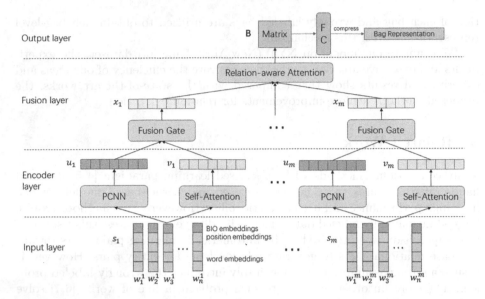

Fig. 1. The framework of our proposed SS-Att for distant supervision relation extraction

3.1   Input Layer

Besides commonly used word embeddings and position embeddings in previous methods [2,11], we add extra BIO embeddings to input representations. Let $b = \{s_1, s_2, \ldots, s_m\}$ denote a bag of sentences which include the same entity pair, and m is the number of sentences within this bag. Let $s_i = \{w_1^i, w_2^i, \ldots, w_n^i\}$ denote the i-th sentence in a bag and n is its length, as well as the number of words. Given a bag within a set of sentences, the input layer transforms all the sentences into a matrix S, which includes word embeddings, position embeddings and BIO embeddings of each word. The descriptions of the three embeddings are as follows.

- **Word Embeddings** The word embeddings are pre-trained with word2vec method [12] for each word. The dimension of word embeddings is d_w.
- **Position Embeddings** We employ position embeddings [11] to represent the relative distance between current word and target entity pair. Rich information is captured by position embeddings which are mapped into two vectors p_j^i and q_j^i of d_p dimensions.
- **BIO Embeddings** We add BIO embeddings into our framework which enhance the representation of the entity pairs. Specifically, if a word is the start of an entity, we labeled this word as B, while the rest part of the entity as I and other words as O. Then the BIO embeddings are obtained by looking up a embedding table according to its label. The dimension of BIO embeddings is d_b. Table 2 illustrates how to generate BIO labels.

These three embeddings are concatenated for each word and then a sentence is transformed into a matrix $\mathbf{S} = [\mathbf{w_1}, \mathbf{w_2}, \ldots, \mathbf{w_n}]$ as the input, where the word representation $\mathbf{w_i} \in \mathbb{R}^{d_h}$, $d_h = d_w + 2 * d_p + d_b$.

Table 2. Illustration of BIO information for a given instance.

Sentence	Possibilities	for	more	conflict	in	Iran	and	elsewhere	in	the	Middle	East	are	adding	to	the	surge	.
Label	O	O	O	O	O	B	O	O	O	O	B	I	O	O	O	O	O	O

3.2 Encoder Layer

In encoder layer, we employ two modules, PCNN and our newly added Self-Attention mechanism, to capture content feature and context dependency information respectively.

Piecewise Convolutional Neural Network. PCNN has been proved its efficiency by many baseline works [2, 9] and integrated with our model. For sentence s_i, the matrix S of input representation is input into a CNN with d_c filters. Then the output vectors of CNN are divided into three segments through a piecewise max pooling [11], which extract features and filter information. Note that the boundaries of segments are determined by the position of the entity pair. As a result, we can obtain the representation of sentences $\mathbf{U} = [\mathbf{u_1}, \mathbf{u_2}, \ldots, \mathbf{u_m}]$ through PCNN, where $\mathbf{u_i} \in \mathbb{R}^{3d_c}$.

Self-attention Mechanism. Different from the self-attention mechanism proposed by Vaswani et al. [13] which implements pairwise computation, we employ the newly-promoted self-attention mechanism [14] to capture context dependency by compressing the word-level representations into sentence-level representations, which estimates the contribution of each word for context dependency. Formally, for the sentence matrix S obtained in input layer, we calculate attention weights in the following equations:

$$\mathbf{A} = \mathbf{W}_{s_2} tanh(\mathbf{W}_{s_1} \mathbf{S} + b_{s_1}) + b_{s_2} \tag{1}$$

$$\mathbf{K} = softmax(\mathbf{A}) \tag{2}$$

where $\mathbf{W}_{s_1}, \mathbf{W}_{s_2} \in \mathbb{R}^{d_h \times d_h}$ are both learnable parameters and K is attention weights matrix. After that, we can calculate the self-attention result as

$$v = \sum \mathbf{K} \odot \mathbf{S} \tag{3}$$

where $\odot$ indicates element-wise multiplication and $\sum$ is performed along sequential dimension. As a result, $\mathbf{V} = [\mathbf{v_1}, \mathbf{v_2}, \ldots, \mathbf{v_m}]$ is the representation of sentences as well, which is employed as a complement to the corresponding PCNN-resulting one and $v_i \in \mathbb{R}^{d_h}$.

3.3   Fusion Layer

Fusion layer aims to fusion vectors into a new vector selectively, which includes two modules. One is fusion gate which dynamically select features between sentence representations of PCNN and self-attention respectively while the other is relation-aware attention which is leveraged to calculate all relation representation for one bag.

Fusion Gate. In order to combine the sentence representation together dynamically, we devise a fusion gate mechanism referred to Li et al. [15]. Formally, we input the sentence representation u_i of PCNN and v_i of self-attention into fusion gate module and proceed with the following procedure:

$$\alpha = sigmoid(\lambda \cdot (W_{g_1}U)) \tag{4}$$

$$\mathbf{X}_{(\mathbf{v})} = tanh(W_{g_2}V) \tag{5}$$

$$\mathbf{X} = \alpha \cdot U + (1 - \alpha) \cdot \mathbf{X}_{(\mathbf{v})} \tag{6}$$

where $W_{g_1} \in \mathbb{R}^{3d_c \times 3d_c}$ and $W_{g_2} \in \mathbb{R}^{3d_c \times d_h}$ are both learnable parameters, λ is used to control smoothness as a hyper-parameter, and $\mathbf{X} = [\mathbf{x_1}, \mathbf{x_2}, \ldots, \mathbf{x_m}] \in \mathbb{R}^{3d_c \times m}$ is the representation of sentences within one bag.

Relation-Aware Attention. We argue that sentences with wrong label in a bag also contain implicitly useful information for training model, thus, a relation-aware attention mechanism is leveraged to handle these problem. Given the representation of sentences within bag b and $\mathbf{R} \in \mathbb{R}^{3d_c \times r}$ is a relation embedding matrix initialized from the beginning where r is the number of relations.

To capture complete information for bag representations and address the wrong-label issue, similar to Ye et al. [9], our approach calculates bag representations $\mathbf{b_k}$ for bag b with all relations as

$$\mathbf{b_k} = \sum_{i=1}^{m} \alpha_{ki}\mathbf{x_i} \tag{7}$$

where $k \in \{1, 2, \ldots, r\}$ denotes the relation index and α_{ki} denotes the attention weight between the i-th sentence in the bag and the k-th relation. Further, α_{ki} is calculated as

$$\alpha_{ki} = \frac{exp(e_{ki})}{\sum_{j=1}^{m} exp(e_{kj})} \tag{8}$$

where $exp(e_{ki})$ denotes similarity between the i-th sentence in the bag and the k-th relation query. Simply, we employ a dot product on vectors to calculate the similarity as

$$exp(e_{ki}) = \mathbf{r_k s_i}^{\top} \tag{9}$$

where $\mathbf{r_k}$ denotes the k-th row in the relation embedding matrix $\mathbf{R}$. As a result, we can get the relation representation matrix $\mathbf{B} \in \mathbb{R}^{r \times 3d_c}$ composed by the bag representation $\mathbf{b_k}$, where each row represent a relation representation of this bag.

3.4 Output Layer

In this layer, relation representation matrix $\mathbf{B}$ is used to predict the final relation under a series of transformation operations. First, a fully-connected layer is leveraged on $\mathbf{B}$ that transform matrix $\mathbf{B}$ to a predicting matrix $\mathbf{B}' \in \mathbb{R}^{r \times r}$. Second, $\mathbf{B}'$ is compressed into a vector $\mathbf{b}' \in \mathbb{R}^r$ by extracting the main diagonal element. Then, $\mathbf{b}'$ is fed into a softmax function to predict the bag relation. Formally,

$$\mathbf{p} = softmax(C(FC(\mathbf{B}))) \in \mathbb{R}^r \tag{10}$$

where $C(\cdot)$ denotes compressing operation and $FC(\cdot)$ denotes a fully-connected layer respectively, and we also employ dropout strategy in fully-connected layer to prevent overfitting.

3.5 Objective Function and Optimization

Same as the related research [2], the objective function is negative log-likelihood loss, which is defined as

$$L_{NLL} = -\frac{1}{|T|} \sum_{l=1}^{|T|} \log p(y_i|b_i; \theta) \tag{11}$$

where $|T|$ indicates the number of bags in training set and y_i indicates the label of a bag while θ is the set of model parameters to be learned. We utilize mini-batch stochastic gradient descent (SGD) strategy to minimize the objective function and learn model parameters.

4 Experiments

In this section, firstly, we give a introduction of the dataset and evaluation metrics. Then we specify training details and list our experimental parameter settings. Further, we conduct experiments with our method regarding performance compared with previous baselines and competitive methods. Besides, ablation experiments and case study show that our SS-Att framework is effective to extract context features and address one-sentence bag problem.

4.1 Dataset and Evaluation Metrics

We conduct experiments on a widely-used [9,16] dataset New York Times (NYT) released by Riedel et al. [6] This dataset was generated by automatically aligning Freebase with the New York Times corpus. There are 52 actual relations and a special NA realation which means an entity pair with no relation. For training set, it consists of 522,611 sentences, 281,270 entity pairs and 18,152 relation facts while test set consists of 172,448 sentences, 96,678 entity pairs and 1950 relational facts.

Following the existing studies [2,9], we evaluate our model in the held-out evaluation. Specifically, Precision-recall (PR) curves, area under curve (AUC) values and Precision@N (P@N) values are adopted as metrics for experiments. Besides, in order to show the performance on processing one-sentence bag, the accuracy of classification (Acc.) on not-NA sentences is calculated for illustration.

4.2 Training Details and Hyperparameters

For a fair comparison with baselines methods, most of the hyperparameters follow the settings of prior works [2,11] and are listed in Table 3. We use 50-dimensional word embeddings and 5-dimensional position embeddings released by Lin et al. [2] as well as our 20-dimensional BIO embeddings for initialization. The filters number of CNN d_c is set to 230 and the window size of CNN is set to 3. All the weight matrices are initialized with Xavier initialization while deviations are initialized as constant zero. In output layer, we adopt dropout strategy for regularization and set the drop probability as 0.5. We adopt a grid search method to select optim learning rate λ among $b = \{0.5, 0.1, 0.05, 0.01, 0.001\}$ and learning rate is decayed to one tenth every 100,000 steps. Moreover, we adopt gradient clip in training step following the Ye et al. [9], which is set to 0.5. And the hyper-parameter λ in fusion gate is optimally set to 0.8.

Table 3. Hyperparameter Settings

Word Embedding Size	50
Position Embedding Size	5
BIO Embedding Size	20
CNN Filter Number	230
CNN Window Size	3
Dropout Probability	0.5
Learning Rate	0.1
Batch Size	50
Gradient Clip	50
λ	0.8

4.3 Overall Performance

Comparison with Baseline Methods. We compare our proposed methods with previous baselines, which are briefly summarized as follows.

– **Mintz** [1] is a multi-class logistic regression model for distant supervision which is used as a baseline for comparison.

- **MultiR** [7] is a probabilistic graphical model using multi-instance learning for distant supervision.
- **MIMLRE** [17] is a graphical model for multi-instance multi-label learning.
- **PCNN+ATT** [2] is a CNN model using sentence-level attention.
- **PCNN+ATT+SL** [3] is a CNN model adopting a soft label method to mitigate the impact of wrongly-labeled problem.
- **PCNN+BAG-ATT** [9] is a CNN model using intra-bag attention and inter-bag attention to cope with the sentence-level noise and bag-level noise respectively which achieves a state-of-the-art performance.

PR Curves. Figure 2 shows the PR curves of our proposed SS-Att compared with baselines mentioned above, where Mintz [1], MultiR [7] and MIMLRE [17] are traditional feature-based methods, and Lin et al. [2], Liu et al. [3] and Ye et al. [9] are PCNN-based ones. In order to be consistent with Ye et al. [9], we also plot the curves with only the top 2000 points. From the results, we can have the following observations: (1) All neural-based models outperform feature-based models significantly which means features designed by human are limited and cannot work well in noisy date environment. (2) Compared to the previous state-of-the-art models, our model achieve a better performance consistently.

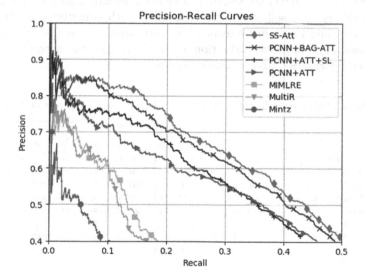

Fig. 2. PR curves of previous baselines and our model

AUC Values. Further, we report AUC values for a overall comparison in Table 4. The empirical results show that our proposed method can achieve considerable improvements compared with the state-of-the-art approaches and reach a better performance on addressing one-sentence bag problem by employing BIO embeddings and selective self-attention with fusion gate mechanism.

Table 4. AUC values of previous baselines and our model, where PCNN+HATT and PCNN+BAG-ATT are two models proposed in Han et al. [10] and Ye et al. [9] respectively.

Model	AUC
PCNN+HATT	0.42
PCNN+BAG-ATT	0.42
SS-Att	0.45

Top N Precision. Following Ye et al. [9], we compare our proposed SS-Att with previous work mentioned above for top-N precision (P@N). P@N means the the top N highest probabilities for the precision of the relation classification results in the test set. We randomly select one sentence, two sentences and all of them for each test entity pair to generate three test sets. We report the results for the P@100, P@200, P@300 values and means of them on these three test sets in Table 5. From the table we can see that: (1) Our proposed method achieves higher P@N values than any other work which means SS-Att has a more powerful capability to capture better semantic features. (2) Compare the results between *SS-Att w/o BIO* and *SS-Att w/o Self-Att*, we can draw a conclusion that BIO embeddings and self-attention mechanism both contribute to the model performance while the improvement of self-attention is more significant. One rational reason for that is self-attention mechanism is good at extracting context dependency information and semantic features.

Table 5. P@N values in entity pairs for different number of test sentences.

Method	One				Two				All			
P@N(%)	100	200	300	Mean	100	200	300	Mean	100	200	300	Mean
PCNN+ATT (Lin et al. [2])	73.3	69.2	60.8	67.8	77.2	71.6	66.1	71.6	76.2	73.1	67.4	72.2
PCNN+ATT+SL (Liu et al. [3])	84.0	75.5	68.3	75.9	86.0	77.0	73.3	78.8	87.0	84.5	77.0	82.8
PCNN+HATT (Han et al. [10])	84.0	76.0	69.7	76.6	85.0	76.0	72.7	77.9	88.0	79.5	75.3	80.9
PCNN+BAG-ATT (Ye et al. [9])	86.8	77.6	73.9	79.4	91.2	79.2	75.4	81.9	91.8	84.0	78.7	84.8
SS-Att (ours)	**86.9**	**78.1**	**74.2**	**79.7**	**91.3**	**82.5**	**76.3**	**83.3**	**93.0**	**86.5**	**78.6**	**86.0**
SS-Att w/o BIO	86.0	76.0	72.0	77.9	91.1	81.5	78.0	83.5	93.0	85.0	78.2	85.4
SS-Att w/o Self-Att	83.0	74.0	70.3	75.7	91.2	81.5	75.6	82.7	91.2	82.5	78.3	84.0
SS-Att w/o ALL	78.9	72.8	67.5	73.0	84.8	76.2	72.5	77.8	86.8	83.0	74.9	81.5

Accuracy of Classification. For a further evaluation on the performance of our model in addressing one-sentence bag problem, we conduct experiments on the bags including only one sentence from training and test set of NYT, which take up 80% of the original dataset. From the Table 6 we can know that the proposed method do achieve a improvement on one-sentence problem compared to the previous methods. Moreover, note that PCNN+ATT obtain a light decrease on performance compared to PCNN, which confirms our opinion that vanilla attention mechanism may hurt the model performance in one-sentence bag situation.

Table 6. Result of training and testing on the bags including only one sentence from NYT dataset for the AUC value and Accuracy on not-NA bags.

Model	AUC	Accuracy
PCNN	0.35	81%
PCNN+ATT	0.34	77%
SS-Att (ours)	0.42	87%

4.4 Ablation Study

In order to verify the effectiveness of each module in our SS-Att, we conduct extensive ablation study. Specifically, SS-Att w/o Self-Att denotes removing the self-attention mechanism with fusion gate, SS-Att w/o BIO denotes removing BIO embeddings, and SS-Att w/o ALL denotes removing both self-attention and BIO embeddings.

The P@N results of ablation study are listed in the bottom of Table 5 and the corresponding AUC values are listed in Table 7 while PR-curves are shown in Fig. 3. From the results, we find that BIO embeddings and self-attention both contribute to the model performance while the improvement of self-attention is more significant, which verifies the effectiveness of our proposed methods. Moreover, by removing both of the two modules, the model performance degenerate sharply but still have a capability to make a correct prediction for distant supervision relation extraction.

Table 7. Ablation study on AUC value.

Model	AUC
SS-Att w/o Self-Att	0.40
SS-Att w/o BIO	0.43
SS-Att w/o ALL	0.38
SS-Att (ours)	0.45

4.5 Case Study

In Table 8, we show some examples for a case study, which illustrate the effectiveness of BIO embeddings and self-attention mechanism with fusion gate.

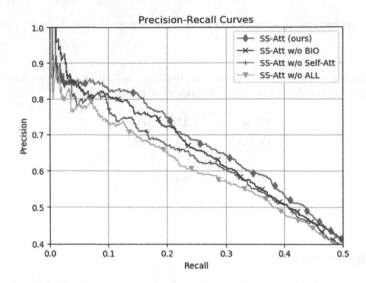

Fig. 3. Performance comparison for ablation study on PR curves

First, for Bag 1 and Bag 2, we observe that the model will classify both Bag 1 and 2 into a wrong NA label without employing self-attention mechanism with fusion gate. Further, for the result of Bag 1 we find that, even if self-attention module is employed, the model still makes a wrong prediction because of the lack of BIO embeddings, which demonstrates the effectiveness of the proposed BIO embeddings.

Then we take two error cases in Bag 3 and Bag 4 for examples to further verify the effectiveness of self-attention mechanism with fusion gate. The label

Table 8. A case study for one-sentence bags.

Bag	Sentence	Relation	SS-Att(ours)	w/o Self-Att	w/o BIO
B1	But the greatest player produced by guadeloupe was the stalwart defender **Marius Trésor**, who played for **France** in the 1978 and 1982 world cups	*/people/person/nationality*	Correct	Wrong	Wrong
B2	... at any of the sites and pilgrimage headquarters, 1105-a providence road, behind the **Maryland** presbyterian church in **Towson**	*/location/location/contains*	Correct	Wrong	Correct
B3	According to glazer, philadelphia's **Brian dawkins** and **Jacksonville**'s donovin darius have trained at a mixed martial arts gym	*/people/person/place_of_birth*	Correct	Wrong	Correct
B4	Cardiff demanded when **Von Habsburg** tracked the artist down in **Berlin**	*NA*	Correct	Wrong	Correct

of these two bags are */people/person/place_of_birth* and *NA* respectively, and experimental result shows that the model is unable to make a correct prediction without the support of self-attention. One possible reason for it is that due to the lack of self-attention, the remaining modules cannot extract enough context dependency information and obtain rich semantic features which are important to one-sentence bags.

5 Conclusion

In this paper, we propose SS-Att framework to cope with the noisy one-sentence bag problem in distant supervision relation extraction. First, we propose to add BIO embeddings to enrich the input representation. Second, rich context dependency information is captured by the proposed Selective Self-Attention, which is combined with PCNN module via a gate mechanism named Fusion Gate to enhance the feature representation of sentences and overcome the one-sentence bag problem. Further, we employ relation-aware attention that calculate weights for all relations dynamically to alleviate the noise in wrongly-labeled bags. Experimental results on a widely-used dataset show that our model consistently outperform the performance of models using only vanilla selective attentions. In the future, we would like to explore the direction about multi-label problem of relation extraction and further conduct experiments on multi-label loss functions.

Acknowledgements. This work is supported by National Natural Science Foundation of China (Project No. 61977013) and Sichuan Science and Technology Program, China (Project No. 2019YJ0164).

References

1. Mintz, M., Bills, S., Snow, R.: Distant supervision for relation extraction without labeled data. In: Proceedings of the Joint Conference of the 47th Annual Meeting of the ACL and the 4th International Joint Conference on Natural Language Processing of the AFNLP, pp. 1003–1011 (2009)
2. Lin, Y., Shen, S., Liu, Z., Luan, H.: Neural relation extraction with selective attention over instances. In: Proceedings of the 54th Annual Meeting of the Association for Computational Linguistics (ACL), pp. 2124–2133 (2016)
3. Liu, T., Wang, K., Chang, B., Sui, Z.: A soft-label method for noise-tolerant distantly supervised relation extraction. In: Proceedings of the 2017 Conference on Empirical Methods in Natural Language Processing (EMNLP), pp. 1790–1795 (2017)
4. Zelenko, D., Aone, C.: Kernel methods for relation extraction. J. Mach. Learn. Res. 1083–1106 (2003)
5. Culotta, A., Sorensen, J.: Dependency tree kernels for relation extraction. In: Proceedings of the 42nd Annual Meeting of the Association for Computational Linguistics (ACL), pp. 423–429 (2004)
6. Riedel, S., Yao, L., McCallum, A.: Modeling relations and their mentions without labeled text. In: Balcázar, J.L., Bonchi, F., Gionis, A., Sebag, M. (eds.) ECML PKDD 2010. LNCS (LNAI), vol. 6323, pp. 148–163. Springer, Heidelberg (2010). https://doi.org/10.1007/978-3-642-15939-8_10

7. Hoffmann, R., Zhang, C., Ling, X., Zettlemoyer, L., Weld, D.S.: Knowledge-based weak supervision for information extraction of overlapping relations. In: Proceedings of the 49th Annual Meeting of the Association for Computational Linguistics (ACL): Human Language Technologies, pp. 541–550 (2011)
8. LeCun, Y., Bengio, Y., Hinton, G.: Deep learning. Nature **521**(7553), 436–444 (2015)
9. Ye, Z.X., Ling, Z.H.: Distant supervision relation extraction with intra-bag and inter-bag attentions. In: Proceedings of the 2019 Conference of the North American Chapter of the Association for Computational Linguistics (NAACL): Human Language Technologies, pp. 2810–2819 (2019)
10. Han, X., Yu, P., Liu, Z., Sun, M., Li, P.: Hierarchical relation extraction with coarse-to-fine grained attention. In: Proceedings of the 2018 Conference on Empirical Methods in Natural Language Processing (EMNLP), pp. 2236–2245 (2018)
11. Zeng, D., Liu, K., Chen, Y., Zhao, J.: Distant supervision for relation extraction via piecewise convolutional neural networks. In: Proceedings of the 2015 Conference on Empirical Methods in Natural Language Processing (EMNLP), pp. 1753–1762 (2015)
12. Mikolov, T., Chen, K., Corrado, G., Dean, J.: Efficient estimation of word representations in vector space. In: Proceedings of the 1st International Conference on Learning Representations (ICLR), pp. 1–12 (2013)
13. Vaswani, A., et al.: Attention is all you need. In: Proceedings of the 31st Annual Conference on Neural Information Processing Systems (NIPS), pp. 5998–6008 (2017)
14. Lin, Z., et al.: A structured self-attentive sentence embedding. In: Proceedings of the 5th International Conference on Learning Representations (ICLR), pp. 1–15 (2017)
15. Li, Y., et al.: Self-attention enhanced selective gate with entity-aware embedding for distantly supervised relation extraction. In: Proceedings of 34th AAAI Conference on Artificial Intelligence (AAAI), pp. 8269–8276 (2020)
16. Yuan, Y., et al.: Cross-relation cross-bag attention for distantly-supervised relation extraction. In: Proceedings of the 33th AAAI Conference on Artificial Intelligence (AAAI), pp. 419–426 (2019)
17. Surdeanu, M., Tibshirani, J., Nallapati, R., Manning, C.D.: Multi-instance multi-label learning for relation extraction. In: Proceedings of the 2012 Joint Conference on Empirical Methods in Natural Language Processing and Computational Natural Language Learning, pp. 455–465 (2012)

Service Clustering Method Based on Knowledge Graph Representation Learning

Bo Jiang, Xuejun Xu$^{(\boxtimes)}$, Junchen Yang, and Tian Wang

Zhejiang Gongshang University, Hangzhou 310018, China

Abstract. With the changing of users' requirements, the number of Web services is growing rapidly. It has been a popular research field to discover the suitable service accurately and quickly in service computing research. At present, most of the Web services published on the Internet are described in natural language. This trend is becoming more and more obvious. Existing service clustering methods are not only limited to a specifically structured document but also rarely consider the relationship between services into semantic information. In response to the problems mentioned above, this paper suggests a Service Clustering method based on Knowledge Graph Representation Learning (SCKGRL). This method firstly crawled the services data from ProgrammableWeb.com, use natural language tools to process the web services description document, and obtain the service function information set. Secondly, we constructed the service knowledge graph by using the service-related information, the triples are converted into vectors and minimize the dimension of service feature vectors due to the knowledge representation learning method. Finally, the services were clustered by the Louvain algorithm. The experiments show that SCKGRL gives better performance compared with other methods, such as LDA, VSM, WordNet, and Edit Distance, which provides on-demand service more accurately.

Keywords: Service clustering · Semantic information · Knowledge graph · Representation learning

1 Introduction

Web services are applications or business logic that can be accessed using standard Internet protocols. Since the mobile Internet, cloud computing, and SOA (Service-Oriented Architecture) technologies are developing rapidly, the scale and type of services are increases quickly. At the same time, service description languages are also diversified, whether it is WSDL (Web Services Description Language), OWL-S (Ontology Web Language for Services), or the very popular RESTful service that uses natural language text to describe [1]. How to accurately locate the service that meets the user's specific business needs from the large number of service sets whose functional attribute differences are difficult to define, and accurately and efficiently discover that meeting the needs of the masses of users is a challenge in the research field of SOC.

The knowledge graph [2] is an open model of a knowledge domain for semantic processing proposed by Google in 2012 [2]. The data storage of the knowledge graph is

© Springer Nature Singapore Pte Ltd. 2021
Y. Sun et al. (Eds.): ChineseCSCW 2020, CCIS 1330, pp. 17–31, 2021.
https://doi.org/10.1007/978-981-16-2540-4_2

usually in the form of triples (h, r, t) to represent the different concepts in the physical world and the interrelationship between concepts, where h represents the head entity and r represents the relationship, t represents the tail entity. Knowledge is presented in the form of graphs, thereby achieving the role of knowledge description. Researchers have found that by combining and analyzing the relationship information between entities, they can learn rich representation features and have achieved good results in search engines [3], recommendation systems [4], and relationship mining [5]. Knowledge representation learning has received widespread attention [4–10], among which the TransX series model of knowledge representation learning based on the translation mechanism has become a representative model [5–10].

In service clustering, clustering can gather similar services together. In the service discovery stage or in the service recommendation stage, the search scope of the service is narrowed to reduce the number of service matches. Related research confirmed that by calculating functional similarity, clustering Web services into a certain fixed-function category can improve the performance of Web service search engines [11–13]. Literature [14] extracts key features that reflect service functions, quantifies key features, and finally clusters services into clusters of similar function sets for WSDL documents. Literature [15] apply the LDA model to extract the hidden service function theme information of the Web service, reduces dimensionality and encodes the topic information, and then calculates the similarity between services for service clustering. Literature [16] considers the division of services with similar functions into homogeneous service communities to classify services. However, existing work generally has the following two shortcomings:

(1) The types of service documents for clustering are relatively single: such as WSDL documents and OWL-S documents, and most of these services follow the SOAP protocol, and less attention is paid to currently popular RESTful services.
(2) In our real world, Web services do not exist independently. They usually have a certain relationship and influence each other in a certain way (such as the relationship of shared tags). The semantic relationship between these services is rarely fully utilized.

In response to the problems summarized above, this paper suggests a service clustering method based on knowledge graph representation learning--SCKGRL. This method applies knowledge graph technology to the field of service computing and enhances the semantic description between services. Firstly, extract all useful function information sets in the service description through natural language technology. Secondly, combine the functional information set of the service and other related service attributes to construct a service knowledge graph. Then according to the representation learning method, Translating the triples into vectors to minimize the dimension of service feature vectors and calculate the similarity of the service. Finally, they were clustered by the Louvain algorithm. Experiments are performed on real service data sets on ProgrammableWeb.com, and four different service similarity calculation methods are selected for assessment experiments.

2 Related Work

Service clustering can reduce the search scope of services, thereby improving the efficiency of service discovery. At present, academia has carried out a lot of work and achieved considerable research results. Literature [17] proposed a probabilistic topic model MR-LDA that considers multiple Web service relationships. Modeling through the existence of mutual combination and sharing of tags between Web services. This method increases the relevant features between services to achieve the purpose of accurate clustering. Literature [18] suggests that new attributes such as context can be extracted from WSDL structured text to achieve aggregation of similar services. Literature [19] proposed a method of clustering all words according to semantics based on the Word2vec tool, and then when training the topic information of the service, the auxiliary effect of the words belonging to the same cluster as the words in the service description document is considered. Literature [11] proposed a domain service clustering model DSCM and based on this model, clustered services subject-oriented, and organized services with similar functions in specific domains into subject clusters. Literature [20] Taking into account the unsatisfactory results of using traditional modeling methods due to the sparse service representation data, BTM can be used to learn the hidden topics in the entire service description texts set, and the topic distribution of each text can be acquired through inference, and then used K-Means algorithm performs clustering. Literature [21] uses the machine learning model to acquire the synonym table in the description texts and generate the domain feature word set to minimize the dimension of the service feature vector, and cluster services in the same domain through the S-LDA model.

Most of the existing service clustering methods are not limited to specific service document types such as WSDL documents or OWL-S. They lack attention to RESTful style types and the rich semantic relationships between services are rarely used.

Through the analysis of the above related work, this paper suggests to apply the knowledge graph related technology to maximize the description of the semantic features of the service, represented in the form of triples, and extract the functional information set of the service from the service description through natural language technology as a type of entity in the service knowledge graph, combining other related features of the service, construct the knowledge graph. At the same time, after comparing different dimensionality reduction techniques, we choose the translation model algorithm to represent the service triples in the form of vectors, which greatly reduces the dimension of service feature vectors, and the modular Louvain algorithm is used to group similar services into one category.

3 SCKGRL Method

To improve service clustering performance, the SCKGRL method needs to preprocess the collected service data set, extract the functional information set of the service from the service description through natural language technology as a type of entity in the service knowledge graph, combining other related attributes of the service, construct the service knowledge graph. Then according to the representation learning method, Translating the triples into vectors to minimize the dimension of service feature vectors, calculate the pairwise similarity between services, and apply the service similarity to the Louvain algorithm to achieve service clustering. The data set for constructing the service knowledge graph comes from real service data on PWeb. The following sections will detail the main parts of the framework. Figure 1 shows the framework of the SCKGRL method.

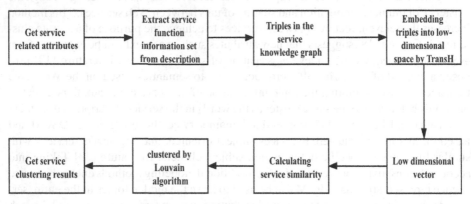

Fig. 1. The framework of SCKGRL

3.1   Service Function Information Set Extraction

We need to obtain the relevant attributes of the service as the corresponding service entity to provide the required materials for constructing the service knowledge graph. Service function information refers to the development of the description of the service function characteristics from the perspective of user intent, but the user's requirements are difficult to obtain directly, and the service description document is mainly used to explain and describe the service function so that users can quickly judge the service Whether to meet its needs.

The structure of API is shown in Fig. 2, which contains information such as service name, service tag, service category and service description. In order for this method to work effectively, the functional information that can express the service needs to be extracted from the service description as a type of entity.

Definition 1. (Service function information) SF = (Verb, Object), where Verb represents the execution action, Object represents the operation object.

For the service information described in natural language, this paper uses the open-source tool Stanford Parser to analyze the service description, extract the set of dependencies, and then directly and indirectly reason to obtain the core function information set of the service. This tool is currently one of the more popular natural language processing tools. Its functions include: (1) part-of-speech tagging; (2) Analyze grammatical relationships and generate a corresponding set of SD (Stanford Dependencies); (3) extract grammatical structure; (4) tense reduction. The latest version of Stanford Parser contains 50 SDs. Perform natural language processing on the service description, and use

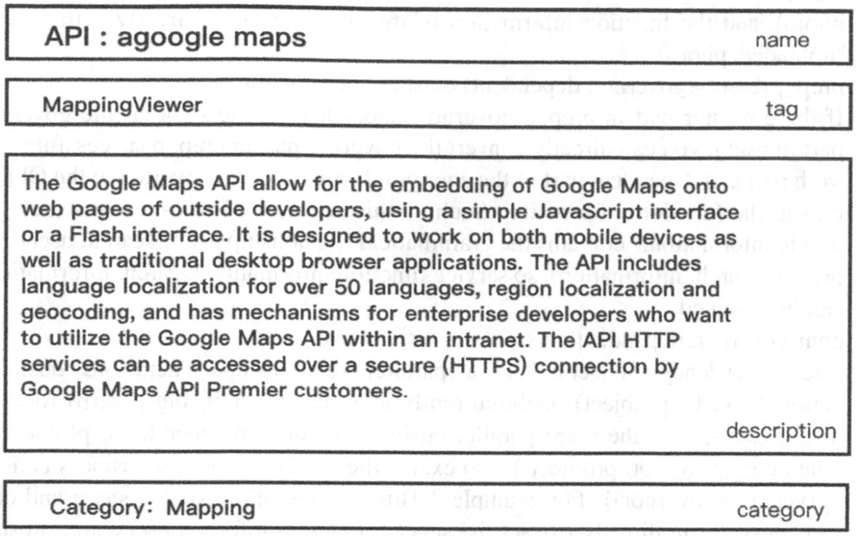

API : agoogle maps	name

MappingViewer	tag

The Google Maps API allow for the embedding of Google Maps onto web pages of outside developers, using a simple JavaScript interface or a Flash interface. It is designed to work on both mobile devices as well as traditional desktop browser applications. The API includes language localization for over 50 languages, region localization and geocoding, and has mechanisms for enterprise developers who want to utilize the Google Maps API within an intranet. The API HTTP services can be accessed over a secure (HTTPS) connection by Google Maps API Premier customers.

description

Category: Mapping	category

Fig. 2. The structure of API

Stanford Parser for grammatical analysis to obtain the SD set. Regarding SD, it can be expressed as a two-tuple: sdType(governor, dependent), where governor represents the dominant word, and dependent represents the dependent word. Each sentence will be parsed to get the corresponding SD set. The SD set represents the relationship between words in a sentence. For instance: "dobj(send, message)" represents that the dependency between "send" and "message" is "dobj", that is the verb-object relationship. "Send" is the dominant word, and "message" is the object of "send". For the obtained SD set, the service information extraction involved in this paper is summarized in 4 situations after comprehensive analysis:

(1) dobj(governor, dependent):
The structure of the dependency relationship of "dobj" refers to the verb-object phrase, the governor part is directly converted to the Verb part of the function information, and the dependent part is directly converted to the function information Centralized Object part. For example, the sentence "AddToAny allows users to share content with a single sharing button" contains the grammatical relationship dobj(share, content), and directly generates the service function information as (share, content) from the above dependency relationship.

(2) nsubjpass (governor, dependent):
The dependency of "nsubjpass" is expressed as a passive verb-object relationship, that is, the dependency will only appear when the sentence contains the passive voice. For this dependency, the governor part is directly converted to the Verb part in the function information, and the dependent part is directly converted to the Object part in the function information. For example, the sentence "Your photo will be uploaded to the server." contains the grammatical relationship nsubjpass(uploaded,

photo), and the function information of the service can be directly extracted as (uploaded, photo).

(3) prep_p&nsubj(governor, dependent) exist at the same time:

If the governor part in prep_p(governor, dependent) is the same as the governor part in nsubj, you can directly convert the governor part in prep_p successfully the Verb part and dependent part of the energy information are converted to the Object part in the function information. Such as this sentence "The user can search for movie information." contains the grammatical relationship nsubj(search, user) and prep_p(search, information), so service function information (search, information) can be obtained.

(4) conj(governor, dependent):

The dependency of "conj" is a parallel relationship. There are business actions{p(verb), p(object)} in the currently acquired SD set. If conj(p(verb), (dependent)) exists, then there are parallel business actions {p(dependent), p(object)}, when conj(governor, p(object)) also exists, there will be parallel business actions {p(verb), p(governor)}. For example, "This mashup allow you to share and collect music." can directly extract the service function information (share, music), and then use the grammatical relationship conj(share, collect) to find the potential service function information (collect, music).

The basic service function information extracted by the Stanford Parser tool only contains two-tuples similar to the verb-object structure. For texts described in complex and natural language, some service functions may be semantically incomplete. For example, "The user can search for vintage shop." we expect to extract the function information (search, vintage shop), but according to the above four scenarios, the function information (search, shop) will be generated. Therefore, the service function information must be semantically extended.

The semantic extension of service function information is mainly for the noun part. Although the modifiers of the noun part include qualifiers, adverbs, nouns, adjectives and gerunds, we have analyzed a large number of texts and found that only nouns and gerunds can affect the semantics of service function information. The nn grammatical relationship in Stanford Parser represents the relationship between two nouns in a sentence, so semantic expansion is carried out through nn grammar.

By referring to the above four types of function information set extraction types, the specific extraction service function information set is as follows:

Algorithm 1. function information extraction

Input: SD set (from service description text)

Output: Service function information collection useful in text NF

1. *init(D);* // Initialize service function information set
2. FOR *sd* ∈ *D*:// Traverse all sd in set D
3.   IF(*sd* meets meet the above 4 situations between two words):
4.   *sf←create (sd)* // Create corresponding service function information
5.   IF(*sf* Imperfect semantics):
6.   *esf←enrichSF(sd)*// Extended basic function information
7.   *NF←esf*// Save to function information collection
8.   ELSE *NF←sf*
9.   END IF
10.   END IF
11. END FOR

3.2 Construct the Service Knowledge Graph

As a domain knowledge graph, the service knowledge graph has clear domain concepts, terminology, and data, and the related concepts and scope are relatively controllable, so it is suitable for top-down construction. The service knowledge graph constructed in this section is mainly expressed in the form of triples to provide data materials for subsequent knowledge representation learning.

By processing the service description document, we extracted the functional information set of the service, combined with the relevant attributes of the service, and finally determined the four types of entities of the service knowledge graph: API, Category, Tag, and Function. The service knowledge graph is used to describe API the entity type is the service information developed at the center, and three relationships are defined, as follows:

(1) There is a belong_to relationship between API and Category, expressed as (API, belong_to, Category);
(2) There is a label relationship between API and Tag, expressed as (API, label, Tag);
(3) There is a has relationship between API and Function, expressed as (API, has, Function);

As shown in Fig. 3: Services may share common categories, tags, and functions. Through the presentation of the service knowledge graph, the relationship between services can be depicted to a great extent.

3.3 Service Knowledge Graph Representation Learning

Recently, the representation learning technology has been enthusiastically sought after in various fields. This technology can accurately calculate the semantic connection between objects in a vector space, thereby powerful solving the problem of data sparseness.

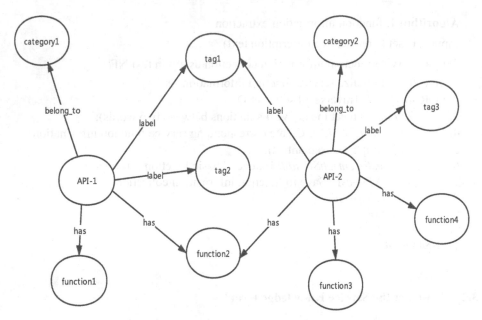

Fig. 3. Simplified diagram of service knowledge graph

Knowledge representation models are mainly divided into neural network models, tensor models and translation models. Literature [6] proposed the translation model TransE inspired by translation invariance. Since the knowledge graph based on the translation model has a small amount of calculation and can effectively represent the semantic knowledge base, this paper assists in calculating the similarity between service entities by constructing a service knowledge graph. TransE has a good effect when dealing with only a single relationship between entities. However, when there are multiple relationships between entities, the entities and relationships of TransE are embedded in the same plane, which cannot accurately represent the multi-relationship model between entities. Therefore, this paper adopts an improved translation model TransH, which can express multiple relationships between entities.

Embedding Triples in Low-Dimensional Space. We convert the triples between entities and vectors through the translation model TransH, which greatly reduces the dimension of service feature vectors. The translation model TransE maps the head vector to the tail vector through the relation vector. By setting the values of the three, the relationship between the three is true: $h + r - t \approx 0$, When the relationship between the three is not established, its value should be great. However, in real life, a pair of entities may have multiple relationships, and the translation model TransE cannot accurately reflect the multiple relationships between entities. The emergence of the translation model TransH improves the shortcomings of TransE. Figure 4 shows the TransH model.

The TransH model in knowledge representation learning introduces a mechanism that projects onto the hyperplane of a specific relationship, so that entities can play different roles in multiple-relationships. For a relation r, it is translated into a specific vector l_r

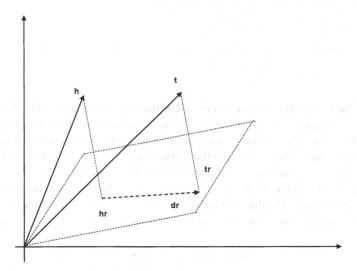

Fig. 4. TransH model

through the translation model transH, and the relation vector is placed on the specific hyperplane l_{nr} of the relation. Through this translation mechanism, the relationship and the entity are divided into two parts, avoiding errors in the vector representation of the entity due to multiple-relationships. For a triple, h and t are first mapped to l_{hr}, l_{tr}, on the hyperplane l_{nr}, Formulas (2) and (3) can connect l_{hr}, l_{tr} and l_{nr} to the hyperplane.

The score function is described in Eq. (1):

$$d(l_{hr} + l_r, l_{tr}) = \|l_{hr} + l_r - l_{tr}\|_2 \tag{1}$$

which can provide

$$l_{hr} = l_h - l_{nr}^T * l_h * l_{nr} \tag{2}$$

$$l_{tr} = l_t - l_{nr}^T * l_t * l_{nr} \tag{3}$$

$$d(l_{hr} + l_r, l_{tr}) = \left\|(l_h - l_{nr}^T * l_h * l_{nr}) + l_r - (l_t - l_{nr}^T * l_t * l_{nr})\right\|_2 \tag{4}$$

constraint condition:

$$\forall h, t \in E, \|l_h\|_2 \leq 1, \|l_t\|_2 \leq 1 \tag{5}$$

$$\forall r \in R, \left|l_{nr}^T * l_r\right|/\|l_r\|_2 \leq \varepsilon \tag{6}$$

$$\forall r \in R, \|l_r\|_2 = 1 \tag{7}$$

The loss function is described Eq. (8):

$$L = \sum_{(h,r,t)} \sum_{(h',r,t') \in k'} [(\lambda + d(l_h + l_r, l_t) - d(l_{h'} + r, l_{t'}))]_+ + c \sum_{e \in E} \|e\|^2$$
$$+ c \sum_{r \in R} \left[\frac{(l_{nr}^T * r)^2}{\|l_r\|^2} - \varepsilon^2 \right] \tag{8}$$

Calculate the Similarity Between Service Entities. In this paper, the classification standard of clustering is to judge the service distance based on the similarity between services. The greater the similarity, the more similar the services and the closer the distance. On the contrary, the less similar the services. Calculating service similarity is one of the very important steps of service clustering. Since TransH uses Euclidean distance to calculate the loss function, this paper uses Euclidean distance to calculation the distance between entities. The distance between entities in space determines the degree of similarity between entities. The entity similarity formula is described in Equation:

$$sim(A, B) = \frac{1}{\|A - B\| + 1} \tag{9}$$

Among them, A and B represent two service entity in low-dimensional space. It is easy to deduced that the $sim(A, B)$ is closer to 1, the closer the service entity A and service entity B are in space, and the higher the similarity between the two services. On the contrary, the lower the degree of similarity between the two services.

3.4 Louvain Clustering Algorithm Based on Service Similarity

The essence of service clustering is to classify services with similar properties into the same class. The services in each aggregated category have the same characteristics as much as possible, and the characteristic difference between different aggregated categories is as large as possible. The Louvain algorithm is a community discovery algorithm for social network mining. It is an algorithm based on multi-level optimization modularity. Louvain algorithm has the advantages of fast and accurate, and is considered to be one of the best performing community discovery algorithms.

We use the Louvain algorithm to cluster the calculated service pairwise similarity values to obtain the clustering results. We treat each service as a separate node, and the similarity between the two services is used as the weight of the connected edge. The calculation formula of Louvain algorithm is:

$$Q = \frac{1}{2m} * \sum \left[A_{i,j} - \frac{k_i * k_j}{2m} \right] * \delta(C_i, C_j) \tag{10}$$

Among them, m is the total number of service pairwise similarities, $A_{i,j}$ represents the edge weight between service i and j, k_i represents the sum of the weights of all connected edges that point to service i, k_j is same k_i. C_i is the cluster number of service i. The $\delta(C_i, C_j)$ function means that if the services i and j are in the same cluster, the return value is 1, otherwise it returns 0.

In the process of clustering using Louvain algorithm, for each service i, try to assign i to the cluster where each neighbor service is located and calculate the modularity increment ΔQ before and after the assignment. The simplified calculation formula is as follows:

$$\Delta Q = k_{i,in} - \frac{\sum_{tot} *k_i}{m}$$ (11)

Among them, $k_{i,in}$ represents the sum of the weights of services and service i in cluster C. $\sum tot$ represents the sum of the weights of the edges connected by the service entities of cluster C, k_i represents the sum of the weights of the connected edges of service i.

For any service i, calculate the change ΔQ of the modularity value after merging it into the adjacent cluster, and merge it into the cluster with the largest value of ΔQ. If the calculation result is negative, the cluster of i will not be changed. When all services cannot be moved, it means that the classification has reached the optimal state and the algorithm ends.

Output service clustering results.

4   Experimental Analysis

4.1   Data Set Description

In order to show the specific implementation process of the method in this paper and verify its effectiveness, in November 2018, we crawled the information of 18536 API services from the PWeb. Including service name, description information, tag, category and other information. In the crawled data, there are 903 APIs with the category "Tools", and the category "Background" contains only 1 API. Therefore, the top ten web services with the largest number of categories are selected, and a total of 5580 web services are used for the experiment. The detailed API distribution is shown in Table 1.

Table 1. API in top 10 categories

Category	Amount	Category	Amount
Tools	903	Enterprise	515
Financial	786	Social	492
Messaging	609	Mapping	428
eCommerce	557	Science	381
Payments	545	Government	364

4.2   Evaluation Criteria

In the experimental evaluation phase, we select accuracy, recall, purity and comprehensive evaluation index F1 are used for evaluation. Among them, the accuracy rate indicates the probability that all services that are classified into the same cluster should be classified into the cluster, and the recall rate indicates that all services that are clustered into the same cluster account for all the services that should be clustered into the cluster account for all the proportion of clusters that should be clustered. F1 reflects the comprehensive performance of the method. The detailed calculation formula is as follows:

$$P = \frac{s_{ucc}(c_i)}{s_{ucc}(c_i) + m_{ispl}(c_i)} \tag{12}$$

$$R = \frac{s_{ucc}(c_i)}{s_{ucc}(c_i) + m_{issed}(c_i)} \tag{13}$$

$$Pu(c_i) = \frac{1}{n_i} * \max_j (n_i^j) \tag{14}$$

$$purity = \sum_{i=1}^{k} \frac{n_i}{n} Pu(c_i) \tag{15}$$

$$F1 = \frac{2 * P * R}{P + R} \tag{16}$$

Among them: P is the accuracy rate, R is the recall rate, $s_{ucc}(c_i)$ is the number of API successfully clustered into category c_i, $m_{issed}(c_i)$ is the number of services that should have been clustered into but were incorrectly clustered into other categories, n_i is the number of API services in category c_i, n_i^j refers to the number of API successfully classified into category c_i in j categories, Pu is the clustering purity of each topic cluster, *purity* is the average cluster purity of all clusters.

4.3 Comparison Method

Choose the following method to compare with the SCKGRL method.

(1) LDA method. This method uses LDA topic analysis to calculate the similarity of the description texts, and use the Louvain algorithm to cluster the services.
(2) WordNet method. This method calculates service similarity by using WordNet tool. WordNet is a large-scale English thesaurus, which is associated with vocabulary and semantics, which is convenient for calculating similarity. Finally, use the Louvain algorithm for clustering.
(3) VSM method. The VSM is used to calculate the similarity of the description texts, and the Louvain algorithm is used to cluster.
(4) ED (Edit Distance) method. This method is a method of calculating string similarity. The ED method is used to calculate the similarity of description texts, and use the Louvain algorithm for clustering.
(5) The SCKGRL method. Firstly, construct the service knowledge graph. Then use the translation model TransH algorithm to translating the triples into vectors to minimize the dimension of service feature vectors and calculate the similarity. Finally, use the Louvain algorithm for clustering.

4.4 Clustering Performance Comparison

For verifying the effectiveness of the SCKGRL method, we selected the LDA method, the WordNet method, the VSM method and the ED method to conduct comparative experiments. The four methods were used to compare the service similarity index calculated by knowledge graph technology. The Louvain community discovery algorithm is used to cluster services.

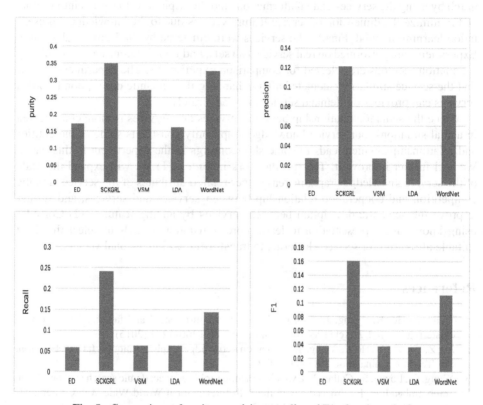

Fig. 5. Comparison of purity, precision, recall, and F1 of each method

It can be clearly seen from Fig. 5 that the purity, precision, recall and comprehensive evaluation index F1 of the SCKGRL method are the highest. Compared with the ED method, the VSM method, the LDA method and the WordNet method, the F1 obtained by this method is increased by 332%, 326%, 341%, 44.2% respectively, this method can effectively improve the service clustering effect. The LDA method, ED method and VSM method are not particularly good clustering effects on the data set in this paper. This may be due to the distribution of service similarity values being too dense, resulting in too tight edges in the entire graph network, and the degree of discrimination is not particularly large, which affects the clustering effect.

5 Conclusions and Prospects

For the existing shortcomings of service clustering methods, we presented a service clustering method SCKGRL, which combines knowledge graph technology with service computing to enhance the semantic description between services. Using natural language processing technology to process the web services description document, and obtain the service function information set. Then, we constructed the service knowledge graph by using the service-related information, and the triples are converted into vectors and minimize the dimension of service feature vectors due to the knowledge representation learning method. Finally, the services were clustered by the Louvain algorithm. Experiments are performed on real service data sets, and four different service similarity calculation methods are selected for comparison experiments. The experimental show that the service similarity calculated after enhancing the semantic description between services can provides on-demand service more accurately.

Since the semantic relationship between services and services is more complicated in actual situations, our service knowledge graph only constructs some of the entities and relationships, which leads to some shortcomings in the experiment in this paper. We will further improve it. The specific work includes: (1) We will expand the scale of the service knowledge graph, increase other related attributes of the service, and the competition and cooperation relationship between services and services, and further improve the semantic description between services by adding features. (2) Consider using knowledge representation to learn more translation models to select the most suitable algorithm for service clustering to transform low-dimensional vectors.

References

1. Tian, G., He, K., Wang, J., Sun, C., Xu, J.: Domain-oriented and tag-aided web service clustering method. J. Acta Electronica Sinica **43**(7), 1266–1274 (2015)
2. Xu, Z., Sheng, Y., He, L., Wang, Y.: Review on knowledge graph techniques. J. Univ. Electron. Sci. Technol. China **45**(4), 589–606 (2016)
3. Xiong, C., Power, R., Callan, J.: Explicit semantic ranking for academic search via knowledge graph embedding. In: Proceedings of the 26th International World Wide Web Conferences Steering Committee, pp. 1271–1279 (2017)
4. Wang, H., Zhang, F., Xie, X., Guo, M.: Deep knowledge-aware network for news recommendation. In: Proceedings of the 2018 World Wide Web Conference, pp. 1835–1844 (2018)
5. Lin, H., Liu, Y., Wang, W., Yue, Y., Lin, Z.: Learning entity and relation embeddings for knowledge resolution. J. Procedia Comput. Sci. **108**, 345–354 (2017)
6. Bordes, A., Usunier, N., Garcia-duran, A.: Translating embeddings for modeling multi-relational data. J. Adv. Neural Inf. Process. Syst. 2787–2795 (2013)
7. Wang, Z., Zhang, J.W., Feng, J.L., Chen, Z.: Knowledge graph embedding by translating on hyperplanes. In: Proceedings of the Twenty-Eighth AAAI Conference on Artificial Intelligence. AAAI (2014)
8. Ji, G., He, S., Xu, L., Liu, K., Zhao, J.: Knowledge graph embedding via dynamic mapping matrix. In: Proceedings of the 53rd Annual Meeting of the Association for Computational Linguistics and the 7th International Joint Conference on Natural Language Processing, pp. 687–696 (2015)

9. Chen, W., Wen, Y., Zhang, X.: An improved TransE-based method for knowledge graph representation. J. Comput. Eng. **46**(5), 63–69 (2020)
10. Cao, Z., Qiao, X., Jiang, S., Zhang, X.: An efficient knowledge-graph-based web service recommendation algorithm. J. Symmetry **11**(3), 392 (2019)
11. Li, Z., Wang, J., Zhang, N., He, C., He, K.: A topic-oriented clustering approach for domain services. J. Comput. Res. Dev. **51**(2), 408–419 (2014)
12. Jiang, B., Ye, L., Wang, J., Wang, Y.: A semantic-based approach to service clustering from service documents. In: IEEE International Conference on Services Computing. IEEE (2017)
13. Jiang, B., Chen, Y., Wang, Y., Liu, P.: Service discovery method for agile mashup development. In: Sun, Y., Lu, T., Yu, Z., Fan, H., Gao, L. (eds.) ChineseCSCW 2019. CCIS, vol. 1042, pp. 30–49. Springer, Singapore (2019). https://doi.org/10.1007/978-981-15-1377-0_3
14. Elgazzar, K., Hassan, A.E., Martin, P.: Clustering WSDL documents to bootstrap the discovery of web services. In: 2010 IEEE International Conference on Web Services (ICWS). IEEE Computer Society (2010)
15. Yu, Q., Wang, H., Chen, L.: Learning sparse functional factors for large scale service clustering. In: 2015 IEEE International Conference on Web Services (ICWS). IEEE Computer Society (2015)
16. Yu, Q., Rege, M.: On service community learning: a co-clustering approach. In: 2010 IEEE International Conference on Web Services (ICWS). IEEE Computer Society (2010)
17. Shi, M., Liu, J., Zhou, D., Cao, B., Wen, Y.: Multi-relational topic model-based approach for web services clustering. J. Chin. J. Comput. **042**(004), 820–836 (2019)
18. Liu, W., Wong, W.: Web service clustering using text mining techniques. J. Int. J. Agent-Oriented Softw. Eng. **3**(1), 6–26 (2009)
19. Shi, M., Liu, J., Zhou, D., Tang, M., Cao, B.: WE-LDA: a word embeddings augmented LDA model for web services clustering. In: 2017 IEEE International Conference on Web Services (ICWS). IEEE Computer Society (2017)
20. Chen, T., Liu, J., Cao, B., Li, R.: Web services clustering based on biterm topic model. J. Comput. Eng. Sci. **040**(010), 1737–1745 (2018)
21. Zhao, Y., Li, Z., Chen, P., He, J., He, K.: Web service semantic clustering method oriented domain. J. Chin. Comput. Syst. **40**(01), 83–90 (2019)

An Efficient Framework of Convention Emergence Based on Multiple-Local Information

Cui Chen, Chenxiang Luo, and Wu Chen[✉]

Southwest University, Chongqing 400700, China
{cc1026,luolatazhu8}@email.swu.edu.cn,
chenwu@swu.edu.cn

Abstract. Convention emergence is an effective strategy to solve the coordination problem in Multiagent systems (MASs). To achieve coordination of the system, it is important to investigate how rapidly synthesize an ideal convention through repeated local interactions and learning among agents. Many methods have been proposed to generate convention, but it is difficulty to deal with the effectiveness of small-world networks and the efficiency of convergence speed. In addition, there is still the limitation of local observation of agents, which affects the learning in the MASs. This paper presents a novel learning strategy to accelerate the emergence of convention based on a Multiple-Local information table (ML-table), which integrates Local information table (L-table) collected from neighbors. Extensive agent-based simulations verify that our algorithm outperforms the advanced algorithms in terms of convergence efficiency and robustness in various networks.

Keywords: Multiagent systems (MASs) · Convention emergence · Multiple-local Information

1 Introduction

MASs [14] is composed of a certain number of individuals that can perceive the environment and respond accordingly to the environment based on the received information. Simultaneously a lot of computing networks, which have a certain spatial topology in distributed environments can be regarded as MASs. In MASs, by encouraging agents to choose the same action with mostly acceptable, the coordination problem of the MASs can be effectively solved. The emergence of the same action is called a convention, and the process of emergence is called a convention emergence. Convention likes a public constraint which plays an important role in real life by regulating the action of people. People driving on the same side of the road can be considered as a convention in traffic regulations. In short, because of the existence of the convention, the interaction among agents will be coordinated smoothly, which fundamentally ensuring the coordination between agents and the operation of MASs.

Y. Sun et al. (Eds.): ChineseCSCW 2020, CCIS 1330, pp. 32–45, 2021.
https://doi.org/10.1007/978-981-16-2540-4_3

How to effectively and quickly form a convention is a key issue in MASs. Until now, there has been two main ways to address the issue of the emergence of the convention: top-down (centralization) [1,11], bottom-up (autonomous agent interaction) [16,24,27,28]. The top-down approach through manager enables global regulation of the action, which is selection of controlled agents and then build up a convention among them. The bottom-up focuses on repeated local interactive learning among agents to develop a convention. However, in the distributed multiagent interaction environment, to achieve centralization, each controlled agent needs to interact with the manager, which requires a very reliable connection. Once the manager fails, it will cause the entire MASs paralysis, which is a great challenge to the robustness of the distributed networks. In contrast, the independent interaction among agents method is only through interact with its neighbors and changes its state according to the state of the interactive agent, therefore, it provides a reliable communication link that will not affect the entire MASs due to some failures. It is important to study what mechanisms can promote the convention to form in an ideal direction through local interactions among agents, which is important to ensure effective coordination between agents.

Reinforcement learning approaches [2,8,12,16,24,26,27] have proved that it can promote the convention emergence in distributed system. Although these studies provide some simple and general insights into the emergence of convention without centralization [28], the interaction protocols and learning settings in these studies are too simplified. Firstly, in MASs, each agent usually only observes and interacts with the partially observed (in a given network, an agent usually can only observe and interact with neighbors), but the response strategies of other agents are changing along with time, which causes the environment of MASs is dynamic and unpredictable, which causes the learning process of an agent may frequently fluctuate with local observation. How to integrate the local observation information of each agent into reliable Multiple-Local information among agents becomes very important. Secondly, most of the existing advanced algorithms are not suitable for all complex networks. Many algorithms cannot form a convention in small-world networks. However, the relationship between agents is usually controlled by some complex relationships [5]. It is also considered that the topology of complex networks is the basis for determining the emergence of the convention. Based on research on the convention emergence in a particular network structure is not sufficient to represent complex realities. It is necessary to conduct research on all complex network structures and propose an algorithm that will be robust. Above all shows that reinforcement learning is still insufficient in the study of the convention emergence.

So, in this paper, to overcome this limitation, we propose a Multiple-Local table learning (Multiple-L) strategy. Firstly, we propose L-table, which represents the neighbors' information that each agent can directly observe in the current interaction episode. Then, with the observation mechanism [6], agents obtain and collect the L-tables of neighbors. Simultaneously, we propose a strategy in which the agent only interacts with the neighbors to integrate the collected

ML-tables, it increases the information of observed area and improves the estimation of available action of agents. This strategy contributes to the robustness of various networks and accelerates the learning process of coordination game. So the effectiveness and efficiency in convention emergence is well solved. This paper also investigates how some key parameters such as the number of available action, population size, and network topology affect the emergence of the convention. The results of the experiments show that the strategy proposed can quickly and effectively reduce the convention formation time under various network structures. Simultaneously, experiments prove the necessity of communicating the proposed L-table.

This paper is organized as follows. In Sect. 2, we will introduce the related work of the research problem formed by the convention. Section 3 elaborates on the algorithm proposed in this paper, explaining some basic definitions and experimental settings. Section 4 shows our results are compared with existing methods. Section 5 includes conclusions and prospects for the future.

2 Related Work

The study on the convention emergence has received a lot of attention in MASs. Shoham and Tennenholtz [17] introduced the problem of convention emergence into MASs. They proposed four basic types of policy update rules and showed the external majority (EM) strategy performed best. Subsequently, they first modeled the agent's interaction process as the coordinated game [18]. Sen and Airia proposed social learning [16] framework, it shows that convention can be formed in a network where agents are randomly connected. The local interaction between each pair of agents is modeled as the standard form of a two-player game and the convention corresponds to a uniform Nash equilibrium of the coordinated game. However, their social learning framework was based on the assumption of random interaction, without considering the underlying topology, which may not accurately reflect the actual interaction with the real world. Later, many papers [15,20,21] subsequently expanded the interaction mode between agents and evaluated heterogeneity by using more realistic and more complex networks. However, they are still slow in the speed of convention emergence.

Yu et al. [28] proposed collective learning framework imitates interpersonal relationships with a certain degree of the society, where people usually collect opinions of their friends before making final decision, based on majority vote. This framework assumes each episode of interaction, agent integrates all the actions provided by the neighbors based on the learning experience to determine the optimal action of the coordination game. Although they have studied the convention emergence in many complex networks, they have only verified in a few small size networks through experiments. When MASs has a large number of agents or its network topology is small-world (especially the reconnect probability is less than 0.1), no convention will emerge, and as described in [20], contrary to the appear a consistent convention, different sub-conventions coexist in the MASs. However, Yu et al. did not consider the relationship about each agent.

Based on this, Mihaylov et al. [10] proposed the Win-Stay Lose-probabilistic-Shift (WSLpS) based on the inspiration of cooperative games, which is applied in wireless sensor networks to solve coordination problems. It is proposed that when the agent successfully interacts with the interactive agents, then the agent maintains the current action. On the contrary, if losing, agent has a probability to decide whether to update the current action to the interactive agents' action. The biggest difficulty is that if the agent chooses to update action but due to treating them equally, and then may face choose difficulty and result in blindness. As in the paper [10], when the available action number exceeds 5, the MASs cannot form a convention, so it proposes Win-Stay Lose-Learn (WSLL), by combining the advantages of reinforcement learning and WSLpS, proposing a new judgment with win or lose. However, when the action space exceeds 20, the speed of convention emergence becomes significantly slow.

In a particular network topology, each agent using reinforcement learning to reach convention can only observe and interact with its neighbors, and other agents in the network will have a certain impact on the learning environment. Agents take actions to be based on its local observation of the global environment [4], it will lead to the instability of agent action along with the process of convention emergence. Not only various complex networks are possibly existing in the real world, but a lot of algorithms failing to a convention in the small-world networks which easily reach sub-convention. At the same time, On paper [7], when the agent collects neighbor action, through introducing extra median payoff based on the proportion of the action of a current agent, thereby accelerating the convention emergence. The proportion of each action in the neighbors can directly lead to changes in the environment of the current agent, simultaneously the selection of its current action. However, it only considered the currently action taken by the current agent in the neighbors. Based on this, the core idea of the algorithm in this paper is to introduce additional communication to reduce the locality of the agent action due to local observation of the global environment [9], communicating the Local information table proposed in this paper, at the beginning, it contains the proportion of all available actions and then updates with iteration, which expands the scope of communication with neighbors, effectively using the local observation information about the agent to accelerate the convention emergence.

3 Preliminaries and Method

Before introducing the proposed algorithm, we first give some basic concepts and describe scenario of this paper.

3.1 Coordination Problem and Basic Concepts

Definition 1. (Convention) The social law that constrains the action of the agents to a specific strategy is called the (social) convention [19].

In previous studies, social conventions (i.e., when all or at least majority) agents in a society follow the same action, it is considered that a limit has been established on the set of available actions to agents and usually modeled as a pure coordinated game (Table 1). The convention usually corresponds to a Nash equilibrium [25], which all agents choose the same action, this abstraction covers many practical scenarios. A typical example is the 'traffic rules' scenario. When driving on the road, 'walking to the right' or 'walking to the left' all can be regarded as a convention.

Table 1. Two-person pure coordination game

	Action 1	Action 2
Action 1	1, 1	−1, −1
Action 2	−1, −1	1, 1

3.2   Interaction Model

Considering a population of agents, each agent is connected based on the static network topology. In each episode t, each agent chooses all neighbors to interact. The interaction model of MASs is the undirected graph, $G = (V, E)$, where G describes the network topology, V and E respectively indicate the set of nodes and the edges between nodes. Particularly, when $(v_i, v_j) \in E$, then v_i, v_j are called neighbors. And $nei(i)$ is the set of the neighbors of agent i, i.e., $nei(i) = \{V_j | (v_i, v_j) \in E\}$. Four classic representative complex network structures are considered, respectively random networks, small-world networks [23], scale-free networks [3] and ring networks.

3.3   Multiple-L Strategy

Figure 1 shows the overall framework of our strategy. Specifically, in a complex network, Multiple-L uses the following steps to reach convention. At the beginning, each agent i creates a Local information table L^0-table by collecting its best-response action against each of its neighbors. Then, in the next rounds of interaction, each agent i using the idea of consensus algorithm thoughts to estimate L^t-table transform a Multiple-Local table ($(ML$-table), where improves the similarity of local information among agents i and its neighbors, and it serves as the basis for agent i making decisions in the coordination game. Finally, with reinforcement learning iterative updating it and simultaneously send it back to L^t-table. Algorithm 1 describes the overall description of the Multiple-L strategy.

Local Information (L^t-table). The proposed L^t-table for each agent i is iterative updating over time. In particular, it is calculated as follows:

$$L^t = \begin{cases} L^0 & t = 0 \\ ML^{t-1} & t \neq 0 \end{cases} \qquad (1)$$

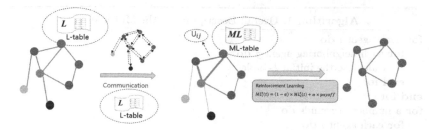

Fig. 1. Overview of the process (ML)-table information strategy

At beginning, each agent i collect the best-response action of each neighbor, and then compute the proportion of these in available action among neighbors, which compose its own State-table, as S-table:$\{P(a_1), P(a_2), ..., P(a_m)\}$, where the proportion as follows:

$$P(a) = \frac{N(a)}{N(nei(i))} \tag{2}$$

where a represent a specific action, $N(a)$ is the sum of the current action in neighbors, and $N(nei(i))$ is the number of neighbors with agent i. Intuitively, in Fig. 2, assuming it has five agents, five available actions with $ABCDE$. The corresponding actions of the agents in time t is A, A, B, C, D. Agent3 have three neighbors, namely agent1, agent2 and agent4. When Agent3 collect the action about neighbors obtain corresponding actions are AAC, Similarly agent4 gets BD. S-table of Agent3 is shown in Table 2 and Agent4 is Table 3. Other agents obtain the S-table in a similar way.

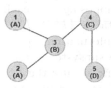

Fig. 2. A simple example of a networks diagram

Table 2. S-table of Agent3

Agent3	A	B	C	D	E
P(a)	$\frac{2}{3}$	0	$\frac{1}{3}$	0	0

Table 3. S-table of Agent4

Agent3	A	B	C	D	E
P(a)	$\frac{2}{3}$	0	$\frac{1}{3}$	0	0

Algorithm 1. Overall Description of the Multiple-L

```
1: for each agent i do
2:      for each neighboring agent j do
3:          Calculate the initial L-table.
4:      end for
5: end for
6: for a number of rounds do
7:      for each agent i do
8:          for each neighboring agent j do
9:              Determine its best-response policy.
10:             Computing its L-table.
11:             Computing its Uij.
12:             Determine its best-action based on ML-table.
13:         end for
14:         for each agent i do
15:             for each neighboring agent j do
16:                 Play game against each neighbor.
17:                 UpdatingStrategy().
18:             end for
19:         end for
20:     end for
```

Then, each agent collects S-table of each neighbor through observation mechanism [6] obtaining L^0-table: $\{S_i, S_{j_1}, S_{j_2} ..., S_{j_n}\}$, which is the probability of agent i and the rest are the probability of its neighbors with the available action.

Multiple-Local Information Table (ML^t-table). To solve the problem of the local observation information in the agent learning process, each agent i compute a weighted ML^t-table replaces the previous L^t-table to evaluate actions by majority vote [13]. Formally, supposing that each agent i can observe the information of their neighbors' L^t-table: $\{L_i^{t-1}, L_{j,1}^{t-1}, L_{j,2}^{t-1}, ..., L_{j,n}^{t-1}\}$. Distributed subsystems rely on neighbor information to determine the dynamic behavior of the system [29], so that the subsystems tend to be similar with a common concern. Specifically in our work, we use the L^t-table as a variable of common concern between agents and propose a way of evaluating L^t-table in adjustment process to compute ML^t-table. ML'-table: $\{L'_{a_1}, L'_{a_2}, ..., L'_{a_m}\}$ is the difference between agent i and its neighbors with the same action a_m in time $t - 1$, and then, we get the weighted difference of agent i and its neighbors, agent$i$ according to following obtain ML^t-table:

$$ML_i^t = L_i^{t-1} + \sum_{j \in nei(i)} U_{ij} \left[L_j^{t-1} - L_i^{t-1} \right] \tag{3}$$

where U_{ij} reflects the relative importance of neighbor j in agent $i's$ neighborhood, namely the correlation degree between i and j. One way [22] to define U_{ij} using the connectivity between agents as follows:

$$U_{ij} = \frac{degree(j)}{totalDegree} \tag{4}$$

where $totalDegree$ is the sum of the degree of the agent i and all the neighbors of the agent i. It is determined by the connection degree of each agent to connect with other agents, so it reflects the degree of influence on other agents. U_{ij} is used as a criterion for considering the neighbor correlation when synthesizing the L^t-table of neighbors under local learning. ML_i^t is used to determine the agent's action a^* with the maximum payoff under state s from its own ML-value function:

$$a_i^* = \arg\max_a ML_i^t(s, a) \tag{5}$$

During each episode, agents use the $\epsilon - greedy$ strategy to choose the action with the highest evaluated value under the current state, which is defined as follows:

$$a_i \leftarrow \begin{cases} a_i^* & \text{with probability } 1 - \epsilon \\ \text{a random action} & \text{with probability } \epsilon \end{cases} \tag{6}$$

If there exist multiple actions with the highest evaluated value, one of them is selected randomly.

ML-Value Function Update. Lastly, based on reinforcement learning, agent updates the $ML^{t'}$ value of its current action a as follows:

$$ML_i^t(a) = ML_i^t(a) \times (1 - \alpha) + \alpha \times payoff \tag{7}$$

Where α is a constant of $[0,1]$, $payoff$ is the payoff of an agent interacting with its neighbor and the updated value of the ML^t table is used as a reference table for L^t-table when each agent communicates at time t.

4 Simulation Studies

This section presents the simulation studies. First, giving the basic settings of the simulation, and then present the results and analysis by evaluating the proposed strategy in a number of different settings.

4.1 Simulation Setup

Our goal in this paper is to address the failure of convention emergence in small-World networks, so the default setting is small-world networks, which is composed of 100 agents and the average node degree is set to 10, the available actions is 10. 90% is the standard for the emergence of the convention. The experiment round is 1000 (part of the experimental results only show part of the number of rounds for convenience of comparison). To eliminate as many random factors as possible to interfere with the experimental results, each experiment was repeated 500 times. It includes two evaluation indicators for the experiment. The first one is the frequency of the convention emergence in 500 experiments, if frequency is less than half, we call the convention progress is failure. Another is the time required to reach convention, which is an average of 500 times.

Three algorithms are comparisons, respectively Q-Learning, Collective Learning (CL), Win-Stay Lose-learn (WSLL), specific parameters as follows (The parameters set in the comparison algorithm follow the optimal parameter settings in previous work):

Q-Learning (Q): The learning rate α is set to 0.3, using the ε exploration method, and the exploration probability is 0.1.

Collective Learning (CL): The learning rate α is set to 0.3, the average voting method is adopted as the integration strategy, the exploration method adopts the ε exploration method, and the exploration rate γ is 0.1.

Win-Stay Lose-Learn (WSLL): The learning rate α in the Bellman equation is 0.3, the exploration rate ε is 0.1, and the state threshold γ is 0.5.

Multiple-L (ML): α value, which is set like the previous study's setting, we set is 0.3.

4.2   Simulation Result

Agents Converged Convention (ACC). Figure 3(a) and Fig. 3(b) respectively represents the process of the percentage with ACC during time step for Q, CL, WSLL and ML in small-world networks and random networks (Because of Q and CL cannot converge in small-world networks, we also shows the dynamics of convention emergence in random networks). We obtain that whether in small-world networks or in random networks, ML converges the fastest, especially improves the convergence time by 85% with WSLL is in small network. Due to the existence of sub-networks in small-world networks, ML not only saves the information of neighbors, but also integrates the L-table of neighbors and then becomes ML-table, which is more effective, thus accelerating the convention emergence (Regular and BA networks are omitted here, because they are similar with random networks).

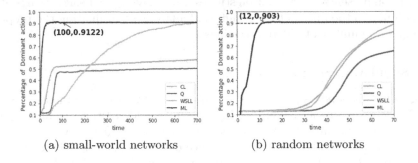

(a) small-world networks   (b) random networks

Fig. 3. Comparison of ACC

Action Space. Figure 4 shows the times to convention by changing the number of available action to 5, 10, 20, and 50 in small-world networks, X-axis and Y-axis respectively represent different algorithms and times. The top line, the middle line and the bottom line correspond to the maximum, average and minimum times for convention needed. Since Q and CL algorithms conventions failed in the small-world network, omitted here. Due to ML taking full advantage of the neighbor table L-table, achieve consistently when synthetic ML- table, which is helpful to choose game action, thus accelerating the formation of the convention, while WSLL merely learns from current action and makes a decision, It is easy to result in randomness, so it has a slow convergence speed.

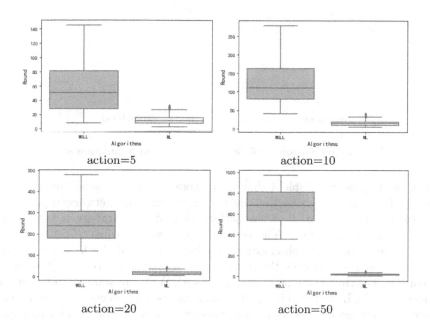

Fig. 4. Comparison of efficiency over action space

Population Size. Figure 5 shows the comparative experiments with different agents' numbers 100, 300, 500, 1000 in a random network (because other algorithms in small world networks cannot converge, there shows the result in random networks). The numerical meaning of the figure is like the Action space. Due to the introduction of the L-table, the speed of ML convergence has improved by at least 80%. L-table contains information about neighbors and merges these, at the same time, increasing the scope of observation information and helpful to chooses the best-response action. It can also be seen that the population size has little effect on each algorithm, but by comparing the maximum time and minimum time of the convention emergence of each algorithm, it can be seen that our algorithm is the most stable, so it is the least sensitive to size. This shows that ML is most suitable for large multiagent networks.

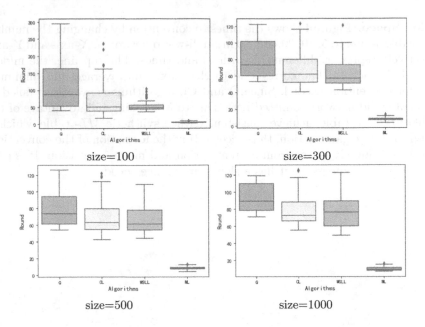

Fig. 5. Comparison of efficiency over different population size

Network Topology. Table 4 shows the average time and standard deviation when 90% of agents choose the convention in different network topology. In all networks, it's obvious that the performance of ML strategy is better than the other three algorithms. The reason is that agent adjust L-table. L-table expands the scope of its observation, making full use of the local observation information and accelerating the convention emergence. Therefore ML improves the convergence time reach 90% comparing with WSLL in small-world networks. Besides, Q and CL are failed in small-world networks, which is owe to easily reach sub-convention, agents may have more time or even be unable reach convention.

Table 4. The performance of Multiple-L in different networks

	Random (p = 0.1)		Small World (k = 10, p = 0.1)		BA ($\gamma = 3$)		Regular (k = 5)	
	t	std	t	std	t	std	t	std
Q	132.8	110.9	✗	✗	215.4	157	283.4	218.2
CL	69.81	56.3	✗	✗	98.9	99.3	154.7	135.7
WSLL	47.27	8.7	146.4	95.4	59.7	16.7	291.4	211.7
ML	5.25	1.8	12.3	4.2	5.3	1.9	9.08	2.6

Communicating Tables. Finally, Fig. 6 shows the necessity of exchanging information with the L^t-table is verified. Because the convergence speed of Q-table is slow which used to make decisions in reinforcement learning algorithm. Giving a comparison about communicating with L^t-table and Q-table (The other experimental steps and settings are the same as this paper algorithm expect communicated table, i.e., Q-table instead of L^t-table.):

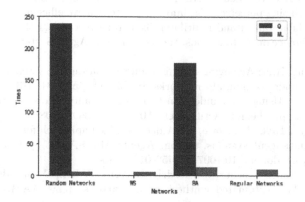

Fig. 6. Comparison of convergence over communicating different table

5 Conclusion and Future Work

In this paper, our goal is to design a framework for making full use of local action information that evolves convention emergence. By increasing the inter-agent communication mechanism, proposes the necessity of communicating the L-table, and proposes to convert the L-table into a ML-table. The Multiple-L has been experimentally proved that, it can be able to accelerate the emergence of a stable convention to a large space, Multiple-L is robust enough, simultaneously with preferable effectiveness and efficiency. In the future, we will consider comparing the Language coordination (larger action space and considering accuracy) with other algorithms.

References

1. Ågotnes, T., Wooldridge, M.: Optimal social laws. In: Proceedings of the 9th International Conference on Autonomous Agents and Multiagent Systems: volume 1, vol. 1, pp. 667–674 (2010)
2. Airiau, S., Sen, S., Villatoro, D.: Emergence of conventions through social learning. Auton. Agents Multi-Agent Syst. **28**(5), 779–804 (2013). https://doi.org/10.1007/s10458-013-9237-x
3. Barabási, A.L., Albert, R.: Emergence of scaling in random networks. Science **286**(5439), 509–512 (1999)

4. Foerster, J., Assael, I.A., De Freitas, N., Whiteson, S.: Learning to communicate with deep multi-agent reinforcement learning. In: Advances in Neural Information Processing Systems, pp. 2137–2145 (2016)
5. Hanneman, R.A., Riddle, M.: Introduction to social network methods (2005)
6. Hao, J., Leung, H.F.: The dynamics of reinforcement social learning in cooperative multiagent systems. In: Twenty-Third International Joint Conference on Artificial Intelligence (2013)
7. Hu, S., Leung, C.W., Leung, H.F., Liu, J.: To be big picture thinker or detail-oriented? Utilizing perceived gist information to achieve efficient convention emergence with bilateralism and multilateralism. In: Proceedings of the 18th International Conference on Autonomous Agents and MultiAgent Systems, pp. 2021–2023 (2019)
8. Hu, S., Leung, H.F.: Achieving coordination in multi-agent systems by stable local conventions under community networks. In: IJCAI, pp. 4731–4737 (2017)
9. Mataric, M.J.: Using communication to reduce locality in distributed multiagent learning. J. Exp. Theoret. Artif. Intell. 10(3), 357–369 (1998)
10. Mihaylov, M., Tuyls, K., Nowé, A.: A decentralized approach for convention emergence in multi-agent systems. Auton. Agents Multi-Agent Syst. 28(5), 749–778 (2013). https://doi.org/10.1007/s10458-013-9240-2
11. Morales, J., Lopez-Sanchez, M., Rodriguez-Aguilar, J.A., Wooldridge, M.J., Vasconcelos, W.W.: Automated synthesis of normative systems. AAMAS 13, 483–490 (2013)
12. Mukherjee, P., Sen, S., Airiau, S.: Norm emergence under constrained interactions in diverse societies. In: AAMAS (2), pp. 779–786 (2008)
13. Polikar, R.: Ensemble based systems in decision making. IEEE Circuits Syst. Mag. 6(3), 21–45 (2006)
14. Savarimuthu, B.T.R., Cranefield, S.: Norm creation, spreading and emergence: a survey of simulation models of norms in multi-agent systems. Multiagent Grid Syst. 7(1), 21–54 (2011)
15. Sen, O., Sen, S.: Effects of social network topology and options on norm emergence. In: Padget, J., Artikis, A., Vasconcelos, W., Stathis, K., da Silva, V.T., Matson, E., Polleres, A. (eds.) COIN-2009. LNCS (LNAI), vol. 6069, pp. 211–222. Springer, Heidelberg (2010). https://doi.org/10.1007/978-3-642-14962-7_14
16. Sen, S., Airiau, S.: Emergence of norms through social learning. In: IJCAI, vol. 1507, p. 1512 (2007)
17. Shoham, Y., Tennenholtz, M.: Robotics laboratory. In: Principles of Knowledge Representation and Reasoning: Proceedings of the Third International Conference, KR 1992, p. 225. Morgan Kaufmann Pub. (1992)
18. Shoham, Y., Tennenholtz, M.: Co-learning and the evolution of social activity. Stanford University California, Department of Computer Science, Technical report (1994)
19. Shoham, Y., Tennenholtz, M.: On the emergence of social conventions: modeling, analysis, and simulations. Artif. Intell. 94(1–2), 139–166 (1997)
20. Villatoro, D., Sabater-Mir, J., Sen, S.: Social instruments for robust convention emergence. In: Twenty-Second International Joint Conference on Artificial Intelligence (2011)
21. Villatoro, D., Sen, S., Sabater-Mir, J.: Topology and memory effect on convention emergence. In: 2009 IEEE/WIC/ACM International Joint Conference on Web Intelligence and Intelligent Agent Technology, vol. 2, pp. 233–240. IEEE (2009)

22. Wang, Y., Lu, W., Hao, J., Wei, J., Leung, H.F.: Efficient convention emergence through decoupled reinforcement social learning with teacher-student mechanism. In: Proceedings of the 17th International Conference on Autonomous Agents and MultiAgent Systems, pp. 795–803 (2018)
23. Watts, D.J., Strogatz, S.H.: Collective dynamics of 'small-world' networks. Nature **393**(6684), 440–442 (1998)
24. Yang, T., Meng, Z., Hao, J., Sen, S., Yu, C.: Accelerating norm emergence through hierarchical heuristic learning. In: Proceedings of the Twenty-Second European Conference on Artificial Intelligence, pp. 1344–1352 (2016)
25. Young, H.P.: The economics of convention. J. Econ. Perspect. **10**(2), 105–122 (1996)
26. Yu, C., Lv, H., Ren, F., Bao, H., Hao, J.: Hierarchical learning for emergence of social norms in networked multiagent systems. In: Pfahringer, B., Renz, J. (eds.) AI 2015. LNCS (LNAI), vol. 9457, pp. 630–643. Springer, Cham (2015). https://doi.org/10.1007/978-3-319-26350-2_56
27. Yu, C., et al.: Modelling adaptive learning behaviours for consensus formation in human societies. Sci. Rep. **6**(1), 1–13 (2016)
28. Yu, C., Zhang, M., Ren, F., Luo, X.: Emergence of social norms through collective learning in networked agent societies. In: Proceedings of the 2013 International Conference on Autonomous Agents and Multi-Agent Systems, pp. 475–482 (2013)
29. Zhang, W., Ma, L., Li, X.: Multi-agent reinforcement learning based on local communication. Cluster Comput. **22**(6), 15357–15366 (2018). https://doi.org/10.1007/s10586-018-2597-x

WikiChain: A Blockchain-Based Decentralized Wiki Framework

Zheng Xu[1,2,3], Chaofan Liu[1,2,3], Peng Zhang[1,2,3(✉)], Tun Lu[1,2,3], and Ning Gu[1,2,3]

[1] School of Computer Science, Fudan University, Shanghai, China
{zxu17,cfliu18,pzhang14,lutun,ninggu}@fudan.edu.cn
[2] Shanghai Key Laboratory of Data Science, Fudan University, Shanghai, China
[3] Shanghai Institute of Intelligent Electronics and Systems, Shanghai, China

Abstract. Wiki-based application, such as Wikipedia, uses collective intelligence to build encyclopedias collaboratively. Most wiki-based applications currently adopt centralized system architectures. Although some work has studied the problem that centralized architectures are unreliable, existing implementations still face some challenges, such as non-disclosure, non-transparency and insecurity. Recently, blockchain has gained extensive attention due to its irreversibility, consensus mechanism and decentralization. We propose a novel blockchain-based wiki framework named WikiChain, which provides decentralized, fair, efficient and collaborative knowledge management. Additionally, we present the workflow of WikiChain by smart contracts and adopt IPFS to store large-scale data. Finally, we implement a prototype on Ethereum and evaluate the efficiency of WikiChain. The experimental results show the utility and scalability of WikiChain.

Keywords: Wiki · Wikipedia · Blockchain · Smart contract

1 Introduction

Wikipedia is a multi-lingual encyclopaedic knowledge base created and maintained collaboratively by the intelligence of crowds based on the Wiki technology. Current wiki engines, including those used by Wikipedia, are based on a centralized architecture that typically consists of a web application containing all services, and a central database where all data resides in [1]. Existing works have addressed some issues on the centralized technical architecture of Wikipedia. For example, lots of requests on centralized servers result in the efficiency decrease of Wikipedia [1].

Some studies have proposed solutions to the problems of the existing centralized architecture. DistriWik [2] and UniWiki [3] were proposed to provide scalable and reliable wiki engines. A Wiki system composed of decentralized nodes was proposed to balance the traffic load in [1]. Although these studies have optimized Wikipedia, Wikipedia still faces many problems. Because Wikipedia

© Springer Nature Singapore Pte Ltd. 2021
Y. Sun et al. (Eds.): ChineseCSCW 2020, CCIS 1330, pp. 46–57, 2021.
https://doi.org/10.1007/978-981-16-2540-4_4

currently relies on a single centralized implementation, this approach is vulnerable to network attacks or failures, resulting in service outages [4]. The rules governing the operation of the Wikipedia system are not transparent. The system often removes a lot of contents, and these articles are deleted and no longer available [5]. Wikipedia's criteria for assessing contents on the web are unclear and unfair, and many of their terms are bureaucratic [6]. Some contents cannot be determined with respect to its source and whether it infringes copyright. It is difficult to confirm the copyright owner of some entries after many people modify them together [7]. Wikipedia relies on the intelligence of crowds to create and edit entries, and currently relies on creators to collaborate spontaneously on editing articles. Therefore, a more effective incentive mechanism is needed to supplement Wikipedia, so that people can actively participate in high-quality article creation [8]. Wikipedia is free to change, resulting in uneven levels of quality [9, 10].

As a novel decentralized technology solution, blockchain has the following advantages to solve above problems:

1. Blockchain is based on a decentralized database. Each node has a complete copy of the data, which can solve problems such as DDoS network attacks faced by centralized technical solutions.
2. Wikipedia's operating rules can be published on the blockchain in the form of smart contracts, which are transparent and automatically executed.
3. Once the article is published to the blockchain, it cannot be deleted. In this way, the complete editing history can be kept and the owner of the article can be identified according to the history, and the infringement can be traced.
4. Blockchain provides native token incentives.
5. Agreements can be reached on the revised articles based on the consensus protocol of blockchain, and only content that meets certain standards will be approved and published on the website.

Thus, we are motivated to design a blockchain-based decentralized wiki framework which is reliable, efficient, fair and autonomous. Based on blockchain technology, this wiki framework can provide stable network services, copyright protection, transparent and autonomous management and native token-based incentive mechanism. We think that our work is the first complete work that combines Wiki and blockchain, which presents a new paradigm for large-scale collaborative applications of knowledge management based on blockchain. Implementation of this decentralized wiki framework is a novel and challenging work. We have to solve two important issues. First, the operation of blockchain-based wiki is based on Turing-complete smart contracts. We need to clarify the operations and types of smart contracts, including how users can participate in the framework, the rules of framework operation and how to contribute contents, etc. Second, blockchain-based wiki needs to store a large amount of data, but the current blockchain system cannot store such magnitude of data, so the interaction scheme between a large-scale data storage and the blockchain system needs to be clarified.

The major contributions of our work are represented as follows:

- Based on the blockchain and IPFS protocol, we propose a decentralized wiki framework named WikiChain. WikiChain is built upon the application module, blockchain module and storage module to provide robust, fair, efficient and collaborative knowledge management.
- We describe important operations of WikiChain in the form of smart contracts. These smart contracts include User Register, Publish Contents, Entry Verify, Vote Proposal, Contents Revert and Query. We implement a prototype of WikiChain on Ethereum test net and evaluate the system overhead to demonstrate the performance of WikiChain.

The rest of this paper is organized as follows. In Sect. 2, we present the related work about blockchain and Wikipedia. In Sect. 3, we present the system model and evaluate the prototype of WikiChain in Sect. 4. Finally, we conclude and analyze the future work of this paper in Sect. 5.

2 Related Work

2.1 Blockchain in the Knowledge Management

The blockchain is considered as the infrastructure of decentralized applications in the new era. Smart contracts allow complex decentralized collaborative applications to be implemented on the blockchain. There have been many studies that combine blockchain with communities and knowledge management. Steemit is a blockchain-based community which is built upon the underlying blockchain called steem. Steemit encourages users to generate high-quality contents through the token incentive mechanism of the blockchain, which is a very good case of knowledge sharing based on the blockchain [11]. Xu et al. implemented a social network Decentralized Application (DApp) based on Ethereum, which utilized blockchain to solve the service availability problem of traditional centralized communities [12]. Antorweep et al. proposed Ushare, a user-centric blockchain community that was democratic [13]. In terms of collaborative knowledge management, with the advent of the big data, large-scale knowledge base is difficult to share and store. Wang et al. proposed a knowledge management model combined with blockchain, which realized a decentralized and transparent knowledge sharing and management framework [14]. The construction of knowledge graph has always been a critical problem. Akhavan et al. used a blockchain-based crowdsourcing system to construct the knowledge graph, which improved the data quality of the knowledge graph [15].

2.2 Decentralized Wikipedia

Wikipedia is a multi-lingual encyclopedia collaborative project based on Wiki technology. There were some studies about the decentralized Wikipedia. Urdanetad et al. designed a decentralized Wiki system in a collaborative approach, which mainly utilized multiple nodes to carry Wiki data and achieved traffic load balancing on the centralized system [1]. Arazy et al. found that increasing the size and diversity of the author group can improve the quality of Wiki content [16]. Forte et al. explored the concept of self-governance in Wikipedia, and gave a detailed description of autonomous organizations in Wikipedia [17]. There are still many flaws in the above mentioned works. We adopt blockchain to design a decentralized wiki framework and provide a novel schema for collaborative knowledge management.

3   System Model

3.1   Overview of WikiChain

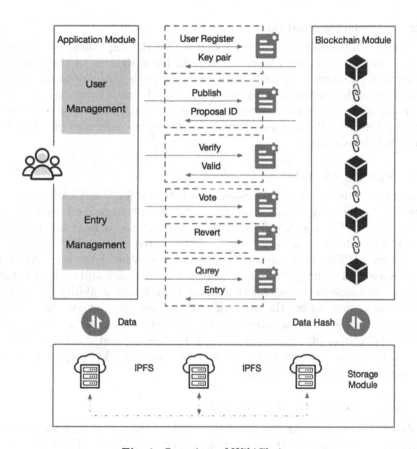

Fig. 1. Overview of WikiChain

In Fig. 1, we present the composition of WikiChain which includes 3 modules, i.e. application module, blockchain module and storage module. Figure 1 only lists the core operations of the system. In fact, any module in WikiChain is pluggable and the architecture is very flexible. Anyone can develop their own DApp and interact with smart contracts and data on the blockchain. WikiChain can be built on public blockchain or permissioned blockchain according to actual scenarios and requirements.

3.1.1 Application Module

The application module is the top layer of this framework that provides services directly to users. Its role is to complete a series of services required for system processing and realize the communication between the application module and the blockchain module. This module provides interface for users to register, publish and modify entries, etc. Application module calls the blockchain module to invoke corresponding smart contracts.

3.1.2 Blockchain Module

The blockchain module is mainly responsible for the implementation of smart contracts and transactions consensus. It links the operation of the application module to data storage. In general, this module deals with transactions in the network to reach consensus and records the global status information that has been confirmed. Meanwhile, smart contracts including User Register, Publish Contents, Entry Verify, Contents Revert and Query are executed on all nodes in the blockchain network.

3.1.3 Storage Module

The storage module provides large-scale data storage for WikiChain. Due to the limited block capacity in the blockchain, We cannot store large-scale data on the blockchain. Thus, a distributed file storage module is built by using IPFS file transfer protocol. Large-scale data can be stored on the IPFS network. As a file system, IPFS can store pictures, videos and other large files, which can greatly make up for the problem of expensive storage space and difficult capacity expansion of blockchain. Beyond that, there is an inherent ability to withstand DDoS attacks that makes the network more secure and the data more secure, because all storage access is spread across different nodes. The security, reliability, irreversibility of blockchain and the efficient distributed storage of IPFS can be integrated into a new ecology. This module mainly works for querying entries and tracing history in WikiChain.

3.2 Smart Contracts of WikiChain

In this section, we present the detailed workflow and implementation of smart contracts. WikiChain consists of six smart contracts which are illustrated below.

3.2.1 User Register Smart Contract

Users can generate a public-private key pair representing the identity of the user. However, numerous identities are generated randomly, which makes the voting mechanism invalid. Users will vote to approve proposals in WikiChain. In order to prevent Sybil attack, each unique identity of the user is associated with some tokens. The user registers through the global user register smart contract (URSC) and stores the ID of users and their corresponding token. Users with fewer tokens have lower voting power, and users often do not register multiple identities to disperse valuable tokens, while those with lower token values have lower voting power. Unlike the traditional centralized system, this system does not store any privacy information of users. Users need a certain amount of deposit to register as authorized account, The system will associate the user's identity with the token information. A proposal list containing its articles will be initialized for each user, so that users can query their own historical operations. Detailed implementation of URSC is presented as follows:

```
Register(user):
    require $deposit(msg.value) >= guaranty
    $lock(guaranty) of user
    initialize the proposal list of user
    initialize the vote list of user
```

3.2.2 Publish Contents Smart Contract

Through the publish contents smart contract (PCSC), users publish contents of the corresponding entry to IPFS in the storage module to get the IPFS hash of the content. The hash value will be used in the subsequent proposal vote process, and the user will publish the hash value to the chain. The user can initiate a proposal for the entry and specify a time limit. For creating entries, the time limit is small, so new entries can be added to wiki as soon as possible. For updating entries, the time limit is long, the purpose is to receive more people's feedback and vote in a longer time, making the editing of the entry more authoritative. Detailed implementation of PCSC is presented as follows:

```
Publish Entry(user, entry, hash, timelock, token):
    require $checkAuthority(msg.sender)
    require $getBalance(msg.sender) >= token
    add proposal to global proposal store
    initialize empty vote list of proposal
    add proposal to proposal list of user
    return proposal.id
```

3.2.3 Entry Verify Smart Contract

The global entry verify smart contract (EVSC) will generate a voting list for the entry and issue an event so that the application layer can receive the proposal, and the application layer can look up all the proposals currently in progress. After the proposal takes effect, other users can vote. After the time limit has passed, the proposer can initiate the finalization through the global contract to determine whether his proposal has passed or not. All this will be automatically executed through EVSC. If the proposal is passed, the IPFS hash associated with the entry will be updated, and the proposer and voters will be rewarded. The rules for checking whether a proposal is valid are configurable and flexibly defined. Detailed implementation of EVSC is presented as follows:

```
Entry Verify(user, proposalId):
    require $checkAuthority(msg.sender)
    require msg.sender = proposal.sender
    get proposal by proposalId
    require now >= proposal.timelock
    if $checkVoteRes(proposal):
        $updateGlobalWiki(proposal.entry, proposal.hash)
        transfer reward to proposal.sender
        transfer reward to voters
    else:
        transfer token of agree voters to global
```

3.2.4 Vote Proposal Smart Contract

After a proposal takes effect, users can vote on the proposal via vote proposal smart contract (VPSC). The system will lock the token when the user applies for the proposal. If the user has enough tokens, the vote will take effect and be added to the voting list of the proposal and the user's voting list. Detailed implementation of VPSC is presented as follows:

```
Vote Proposal(user, proposalId, token, agree):
    require $checkAuthority(msg.sender)
    require $getBalance(msg.sender) >= token
    $lock(token) of user
    add vote(msg.sender, token, agree) to proposal
    add vote to vote list of user
```

3.2.5 Contents Revert Smart Contract

In order to prevent some incorrect and inaccurate information, WikiChain provides the ability to roll back the entry through contents revert smart contract (CRSC). Similarly, users can initiate a proposal. The process is similar to the previous step. If the proposal is approved, WikiChain should automatically roll back the entry to the specific version. All versions of a entry will be stored on the blockchain to form an IPFS hash list, which can be traced back to the correct version approved by most people. Detailed implementation of CRSC is presented as follows:

```
Contents Revert (user, entry, hash, token, timelock):
    require $checkAuthority(msg.sender)
    require $getBalance(msg.sender) >= token.
    add proposal to global proposal store
    initialize empty vote list of proposal
```

```
    add proposal to proposal list of user
    return proposal.id

Revert Verify (user, proposalId):
    require $checkAuthority(msg.sender)
    require msg.sender = proposal.sender
    get proposal by proposalId
    require now >= proposal.timelock
    if $checkVoteRes(proposal):
        get last version hash of the entry
        revertGlobalWiki(proposal.entry, hash)
        transfer reward to proposal.sender
        transfer reward to voters
    else:
        transfer token of agree voters to global
```

3.2.6 Query Wiki Smart Contract

The most important thing in WikiChain is the query function. WikiChain stores the IPFS hash corresponding to all Wiki entries. Similarly, WikiChain provides the Wiki query to users through the query wiki smart contract (QWSC). Users can enter an entry, and DApp will automatically obtain the contents of relevant entries through IPFS and return them to users. At the same time, the user can query all historical versions of the entry and all proposal information. QWSC automatically queries the version, then users can use IPFS hash to query wiki in the off-chain context. Detailed implementation of QWSC is presented as follows:

```
Query Entry(text, version):
    require $checkAuthority(msg.sender)
    require version >= earliest version
    hash = $queryWiki(text, version)
    ipfs cat hash as <ipfs-path>
```

4 Evaluation and Analysis

4.1 System Design

We implement a system prototype of WikiChain and smart contracts of workflow on Ethereum which is the most popular blockchain platform. With the development of consensus algorithms, many blockchain systems can reach thousands TPS nowadays. The implementation of WikiChain can be based on any blockchain system that supports Turing-complete smart contracts. The evolution of the underlying blockchain can greatly improve the performance of WikiChain. Then, we evaluate the time cost and gas consumption of WikiChain. WikiChain provides users with the interface which is developed by Web3 SDK shown in Fig. 2. Users can publish, edit and revert entries through this interface. This

Fig. 2. User interface of WikiChain

interface accepts user input and provides parameters for smart contracts for blockchain modules. Smart contracts required for system operation are implemented by Soliditiy. Solidity is an object-oriented language for smart contracts running on the Ethereum Virtual Machine (EVM) and its syntax is similar to Javascript. Any operation involving Ethereum, whether it's a simple transaction or a smart contract, requires a certain amount of gas. Gas is the consumption unit of calculation that measures the amount of computation required to perform certain operations in Ethereum. The smallest unit of measurement for gas is wei [18].

This DApp monitors all operations on WikiChain. On the leftmost side of the interface, users can find generated proposals on WikiChain, the category of the proposal, and the votes that have been received. The current state of the entry and staking allow users to choose whether to vote on this entry. Users can search for wiki entry that they are interested in by keywords. Then all information and sources of the entry will be displayed on the interface. Based on the traceability function provided by WikiChain, users can easily view the historical versions of this entry.

4.2 System Overhead

We evaluate the time cost and gas consumption of WikiChain by implementing URSC, PCSC, EVSC, CRSC and QSC on Ethereum. We take the core operations including Registry, Vote, Verify and Query as examples to evaluate the overhead of WikiChain. The operation numbers of Registry, Vote, Verify and Query increase by 20. Furthermore, the gas price is set as 2×10^9 wei.

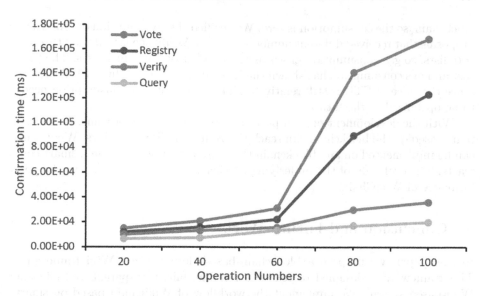

Fig. 3. Time cost of WikiChain

As shown in Fig. 3, when the number of operations is less than 60, the time cost of each operation keeps a slow growth trend with the number of operations. However, when the number of operations is greater than 60, the time cost of Registry and vote increases sharply. We take Vote as an example. When there 60 Vote operations in WikiChain, the time cost is 31237.47ms. When there are 80 Vote operations in WikiChain, the time cost is 141563.39ms. We find the time cost of Verify and Query does not show dramatic increases, because these two operations do not need consensus from WikiChain. Thus, the time cost depends on whether operation needs consensus and how quickly WikiChain processes corresponding requests.

Gas Consumption (gas)		
Registry	Vote	Query
22111	71340	0
Verify (10 Votes)	Verify (50 Votes)	Verify (100 Votes)
56083	186803	350203

Fig. 4. Gas consumption of WikiChain

As shown in Fig. 4, We study the gas consumption of different core operations. As mentioned above, the query operation does not require the consensus process of the blockchain, and the view function does not really change any data on the blockchain, so the consumption is zero. We verified the consumption of processing proposals that received different numbers of votes. Verify needs to iterate through vote lists, so gas consumption grows as the number of Votes increases. The time cost and gas consumption has shown the feasibility and scalability of WikiChain. Since the price of ETH is still relatively high at present, gas consumption needs to be optimized in the future.

With the development of high-performance consensus algorithms, many systems based on the blockchain can reach thousands of TPS nowadays. WikiChain can be implemented on any blockchain that support Turing-complete smart contracts. The evolution of the underlying blockchain can greatly improve the performance of WikiChain.

5   Conclusion and Future Work

In this paper, we propose a blockchain-based decentralized Wiki framework. This framework is designed to realize a reliable, fair, transparent and efficient Wiki-based system. We implement the workflow of WikiChain based on smart contracts. To provide data storage capacity of blockchain-based Wiki framework, we turned to the IPFS protocol for storing a large amount of data. Finally, we implement this framework on Ethereum test net and evaluate the system overhead.

This work is a preliminary attempt to transform wiki system by combing blockchain technology and there are still many problems to be studied in the future. Wiki-based systems aim to form a collaborative editing knowledge base with the help of decentralized and individual wisdom. Malicious behaviors can be avoided by reasonable incentive mechanisms in wiki systems. Thus, how to design an efficient blockchain-based incentive mechanism to motivate people to contribute contents is a significant challenge.

Acknowledgements. This work was supported by the Scientific Research Program of Science and Technology Commission of Shanghai Municipality under Grant No. 19511102203.

References

1. Urdaneta, G., Pierre, G., Van Steen, M.: A decentralized wiki engine for collaborative wikipedia hosting. In: WEBIST, no. 1, pp. 156–163 (2007)
2. Morris, J.C.: DistriWiki: a distributed peer-to-peer wiki network. In: Proceedings of the 2007 International Symposium on Wikis, pp. 69–74 (2007)
3. Oster, G., Molli, P., Dumitriu, S., et al.: UniWiki: a collaborative P2P system for distributed wiki applications. In: 2009 18th IEEE International Workshops on Enabling Technologies: Infrastructures for Collaborative Enterprises, pp. 87–92. IEEE (2009)
4. Newman, S.: DDoS attack on Wikipedia site smacks of hacktivism. https://www.corero.com/blog/ddos-attack-on-wikipedia-site-smacks-of-hacktivism

5. West, A.G., Lee, I.: What Wikipedia deletes: characterizing dangerous collaborative content. In: Proceedings of the 7th International Symposium on Wikis and Open Collaboration, pp. 25–28 (2011)

6. Shen, A., Qi, J., Baldwin, T.: A hybrid model for quality assessment of Wikipedia articles. In: Proceedings of the Australasian Language Technology Association Workshop, pp. 43–52 (2017)

7. Simone, D.: Copyright or copyleft? Wikipedia as a turning point for authorship. King's Law J. **25**(1), 102–124 (2014)

8. Forte, A., Bruckman, A.: Why do people write for Wikipedia? Incentives to contribute to open-content publishing. Proc. of GROUP **5**, 6–9 (2005)

9. Liu, J., Ram, S.: Who does what: collaboration patterns in the Wikipedia and their impact on data quality. In: 19th Workshop on Information Technologies and Systems, pp. 175–180 (2009)

10. Cusinato, A., Della Mea, V., Di Salvatore, F., et al.: QuWi: quality control in Wikipedia. In: Proceedings of the 3rd Workshop on Information Credibility on the web, pp. 27–34 (2009)

11. Steem. https://steem.com/

12. Xu, Q., Song, Z., Goh, R.S.M., et al.: Building an ethereum and IPFS-based decentralized social network system. In: 2018 IEEE 24th International Conference on Parallel and Distributed Systems (ICPADS), pp. 1–6. IEEE (2018)

13. Chakravorty, A., Rong, C.: Ushare: user controlled social media based on blockchain. In: Proceedings of the 11th International Conference on Ubiquitous Information Management and Communication, pp. 1–6 (2017)

14. Wang, S., Huang, C., Li, J., et al.: Decentralized construction of knowledge graphs for deep recommender systems based on blockchain-powered smart contracts. IEEE Access **7**, 136951–136961 (2019)

15. Akhavan, P., Philsoophian, M., Rajabion, L., Namvar, M.: Developing a blockchained knowledge management model (BCKMM): beyond traditional knowledge management. In: The 19th European Conference on Knowledge Management (ECKM 2018), September, Italy (2018)

16. Arazy, O., Morgan, W., Patterson, R.: Wisdom of the crowds: decentralized knowledge construction in Wikipedia. In: 16th Annual Workshop on Information Technologies & Systems (WITS) Paper (2006)

17. Forte, A., Larco, V., Bruckman, A.: Decentralization in Wikipedia governance. J. Manag. Inf. Syst. **26**(1), 49–72 (2009)

18. Tikhomirov, S.: Ethereum: state of knowledge and research perspectives. In: Imine, A., Fernandez, J.M., Marion, J.-Y., Logrippo, L., Garcia-Alfaro, J. (eds.) FPS 2017. LNCS, vol. 10723, pp. 206–221. Springer, Cham (2018). https://doi.org/10.1007/978-3-319-75650-9_14

Dynamic Vehicle Distribution Path Optimization Based on Collaboration of Cloud, Edge and End Devices

Tiancai Li[1,2], Yiping Wen[1,2(✉)], Zheng Tan[1,2], Hong Chen[1,2], and Buqing Cao[1,2]

[1] School of Computer Science and Engineering, Hunan University of Science and Technology, Xiangtan, China
[2] Key Laboratory of Knowledge Processing and Networked Manufacturing, Hunan University of Science and Technology, Xiangtan, China

Abstract. Aiming at the problems of unreasonable distribution routes in the current logistics distribution field, without considering the impact of real-time road conditions, and the inability to reduce the impact on the timeliness of distribution, this paper proposes a dynamic vehicle distribution path optimization method based on the collaboration of cloud, edge and end devices. This method considers the requirements of demand points for the delivery time and considers the changes in road traffic conditions caused by random road traffic incidents. Combining the characteristics of vehicle speed and time penalty cost in the vehicle delivery process establishes a logistics delivery vehicle path optimization model. Solve it and optimize it with the A* algorithm and dynamic schedule. This method collects road condition data in real-time through terminal equipment, evaluates and judges road conditions at the edge, and makes real-time adjustments to the distribution plan made in advance at the cloud data center. Through simulation experiments on application examples, the vehicle path optimization method proposed in this paper that considers real-time road conditions changes and the optimization method that does not consider road conditions are compared and analyzed, verifying the effectiveness of this method. Experimental results show that this method can reduce distribution costs, reduce distribution time, and reduce the impact of changes in road conditions on the distribution results.

Keywords: Dynamic vehicle path optimization · Cloud-side-end collaboration · A* algorithm · Delivery time

1 Introduction

With the rapid development of cloud computing, the Internet of Things, and e-commerce, combining logistics and distribution with new technologies such as cloud computing technology, Internet of Things technology, and intelligent transportation technology, experts and scholars have proposed a cloud distribution model that is different from the traditional logistics distribution model [1]. The cloud distribution model solves the problem that the traditional logistics distribution model is difficult to adapt to the needs of

Y. Sun et al. (Eds.): ChineseCSCW 2020, CCIS 1330, pp. 58–71, 2021.
https://doi.org/10.1007/978-981-16-2540-4_5

modern logistics distribution. However, with the rapid increase in the number of various types of mobile devices and their computing needs, the traditional cloud computing centralized processing model with cloud data centers as the core faces problems such as large network transmission delays, high data transmission costs, and computing security and privacy risks, which cannot be effective. To meet the needs of mobile users, especially users who need an instant response, for computing services. Cloud-side collaboration provides a new way to solve this problem.

As people have higher and higher requirements for the timeliness of logistics and distribution, how to ensure that various logistics and distribution tasks are completed on time and effectively save distribution costs is one of the key issues that the current logistics and transportation industry urgently needs to solve [2]. Through the analysis of the various links of vehicle delivery, the vehicle delivery path is optimized, and at the same time, considering that the road condition changes and other uncertain factors may affect the punctuality of the delivery vehicle, the delivery path is dynamically adjusted after the delivery vehicle starts. Optimization to reduce the impact of road conditions, weather and other factors on the timeliness of the overall distribution is an important method to improve the efficiency of logistics distribution and reduce the overall cost of distribution.

In recent years, domestic and foreign scholars have conducted a lot of research on the optimization of vehicle distribution routes. At present, there are two types of accurate optimization algorithms and heuristic optimization algorithms to solve the optimization problems of vehicle distribution routes. In the literature [3], Ma et al. constructed a well-adaptive adaptive ant colony algorithm to solve the vehicle path optimization problem. This method has low requirements for initial values, but the solution time is longer. Literature [4] uses a genetic algorithm to solve the vehicle routing problem, and the experimental results show that the genetic algorithm has a more obvious improvement effect on the solution. In the literature [5], Wang et al. built a multi-objective model for the logistics vehicle routing problem with time windows and designed two methods of multi-objective local search and multi-objective optimization algorithms such as decomposition and crossover to solve the problem. The modeling of the method is more complicated and the solution time is long. In the process of logistics and distribution, many factors affect the effectiveness of distribution. Under the mutual influence of various factors, it is easy to cause untimely distribution, delay delivery time, affect normal production, etc., according to real-time road conditions, make a reasonable pre-planned distribution path Real-time adjustment is necessary. Reference [6] uses D* algorithm to improve the A* algorithm; Reference [7] proposes a dynamic travel timetable for roads in the road network for the dynamic travel time calculation of roads, and each is recorded in the table. The travel time of each time segment of the road segment, and thus the method for calculating the travel time required by the vehicle to pass the road segment.

Although previous studies on this issue have achieved a lot of results, there are many constraints and uncertainties in actual problems. The comprehensive consideration of dynamic uncertain factors such as traffic, weather, and road conditions require further study. Based on this, this paper proposes a dynamic vehicle delivery route optimization method based on the collaboration of cloud (cloud data center), edge (edge server), and end (terminal equipment) considering the impact of changes in road conditions. In this paper, the goal is to minimize the delivery cost, minimize the delivery time, and minimize the impact of road conditions on delivery, introduce a soft time window penalty function to establish a vehicle path optimization model; use genetic algorithms with elite retention strategies to solve the vehicle distribution path optimization model. And combined with the A* algorithm and dynamic schedule to dynamically adjust and optimize it, reduce the impact of real-time road conditions on the logistics delivery time, and improve the overall satisfaction of the delivery. The experimental results show that this method can achieve better optimization results, while improving the efficiency of vehicle delivery, while reducing the impact of uncertain factors on the delivery results, to achieve the purpose of reducing logistics delivery costs.

2 Model Building

2.1 Problem Description

The vehicle routing optimization problem studied in this paper is the closed vehicle routing problem of a single distribution center corresponding to multiple demand points based on real-time road condition information of simulation. This article first generates an initial optimization plan based on the known information such as the location information of the demand point, the demand of each demand point and the required delivery time window, the type and number of vehicles owned by the distribution center, and when the dynamic information occurs, the initial path planning local adjustments and updates are made on the plan, and fast and efficient solving algorithms are designed to meet the requirements of real-time calculations. The specific description is as follows: the logistics distribution center needs to complete the distribution task to the demand point, deploy several delivery vehicles in the distribution center, make initial optimization for the different demand of each demand point, formulate a suitable distribution plan, and arrange the vehicles at the specified time Complete the distribution task within the internal; then combined with the time series of real-time road condition information generated by the simulation, the unreasonable partial route is dynamically adjusted and optimized until all the logistics distribution tasks are completed. The vehicles are delivered one by one according to the established route and return to the distribution center after the delivery is completed. One vehicle can complete the logistics distribution tasks of multiple demand points, but the total demand of all passing demand points does not exceed the maximum load capacity of a single-vehicle of this type of delivery vehicle, and one demand point can only be delivered by one vehicle. Logistics delivery tasks need to be delivered within the time window specified by the demand point. Early or late arrivals will generate penalty values based on time. The logistics distribution vehicle path optimization problem studied in this paper requires that a reasonable vehicle distribution path be formulated according to the demand at the demand point and real-time road condition information, to reduce the distribution cost and delivery time in the logistics distribution process, and reduce the impact of road conditions on the overall distribution.

2.2 Dynamic Vehicle Distribution Path Optimization

The route optimization problem of logistics distribution vehicles studied in this paper is aimed at minimizing the cost and time of distribution and minimizing the impact of changes in road conditions on distribution. It is a multi-objective optimization problem. This article is based on the vehicle path optimization plan that has been made in the cloud data center after the vehicle departs from the distribution center, according to the road traffic flow, road traffic incidents, weather conditions, and other real-time road condition change information sensed from the edge computing device, Make a judgment on the road traffic conditions, if you continue to complete the delivery according to the original route, it will have a greater impact on the delivery cost, then adjust and optimize the current route. When this paper adopts the method of combining the A* algorithm and dynamic travel schedule to adjust and optimize the original vehicle path planning plan in real-time, on this basis, analyze the delivery time and delivery cost, and establish a dynamic vehicle path optimization model. The network architecture diagram of the vehicle path planning method based on the cloud-side-end collaboration proposed in this paper is shown below.

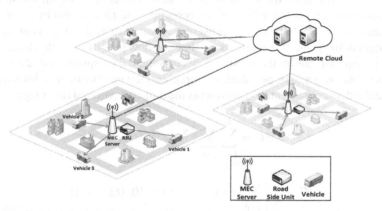

Fig. 1. Vehicle path planning architecture based on cloud - side collaboration

As shown in Fig. 1, the overall logistics distribution area is divided into several areas according to the coverage of edge computing nodes. In the coverage area of each edge computing device, the terminal device perceives the road condition change information in the area and uploads the data to the edge server for processing. The edge server makes a judgment on the road condition change. If an event occurs at a certain moment, Road traffic conditions are affected, and the next delivery point that the delivery vehicle will go to will pass through this road section, the delivery path of the delivery vehicle will be adjusted at the edge according to the improved A* algorithm, and a partial adjustment will be made The new distribution route plan is sent to the terminal vehicle.

2.3 Total Delivery Cost

In the problem of this article, it is necessary to compare the total cost of distribution required to complete the overall distribution task in three cases to test the effectiveness of the method proposed in this article. First, the total cost of the distribution plan obtained by using a genetic algorithm to plan the distribution path of the vehicle is recorded as $Cost_0$; then after the distribution is started, the distribution path in the initial optimization stage is dynamically adjusted according to the changes in the road conditions obtained by the simulation. Adjusted, the total cost of distribution to complete all distribution tasks is recorded as $Cost_1$; finally, the total cost of distribution that does not deal with changes in road conditions and still distributes according to the original plan is recorded as $Cost_2$.

Situation One
First, $Cost_0$ is mainly composed of three parts of cost, calculated as follows.

$$Cost_0 = TC + PC + \sigma_1 \times VN \tag{1}$$

Where, TC represents the transportation cost of vehicles, of which transportation fuel consumption accounts for the largest proportion. Besides, it also includes vehicle maintenance costs. The transportation cost of a certain distribution path is proportional to the length of the path; PC represents the time window penalty cost; the demand point has certain requirements for the logistics delivery time. If the requirements are exceeded, the delivery plan needs to be punished; VN represents the number of vehicles required to complete the delivery task, and σ_1 represents the unit cost required to arrange a vehicle.

$$TC = \sum_{k=1}^{K} \sum_{i=0}^{N} \sum_{j=0, j \neq i}^{N} c_{ij} d_{ij} x_{ijk} \tag{2}$$

$$PC = \sum_{k=1}^{K} \sum_{i=0}^{N} (a * w_{ik} + b * max[0, (t_{ik} - l_i)]) \tag{3}$$

$$VN = \sum_{k=1}^{K} \sum_{j=0}^{N} x_{0jk} \tag{4}$$

Where c_{ij} represents the unit transportation cost between demand point i and demand point j, d_{ij} represents the distance between demand point i and demand point j, the value of x_{ijk} is 0 or 1, and the value 1 means that vehicle k is from demand point i go to demand point j after leaving, otherwise, the value is 0. a, b represents the delivery time window penalty coefficient, w_{ik} represents the waiting time of vehicle k at demand point i, t_{ik} represents the time when delivery vehicle k arrives at demand point i, and li represents the latest service time window of demand point i. The value of x_{0jk} is 0 or 1. When its value is 1, it means that the vehicle k departs from the distribution center 0 to the demand point j.

Situation Two
Compared with $Cost_0$, $Cost_1$ mainly increases the additional cost of adjusting the route, which is composed of the following parts.

$$Cost_1 = TC + PC + \sigma_1 \times VN + EC + \sigma_2 \times g_n(i) \tag{5}$$

Where, EC represents the additional distribution cost due to changes in road conditions; $g_n(i)$ represents the overall time required to complete the real-time adjustment of the distribution path, and $\sigma2$ represents the unit time cost of processing in the edge computing environment.

$$EC = d_e \sum_{k=1}^{K} (t_{back} - t_{dept})_k \tag{6}$$

Where t_{dept} represents the time when the vehicle leaves the logistics distribution center, t_{back} represents the time when the vehicle returns to the distribution center after completing the delivery task, and de represents the unit time cost of delayed delivery.

After the vehicle departs from the distribution center, in the process of completing its established logistics and distribution tasks according to the initial planning plan, if some emergencies occur, the transit time of some roads in the planning plan will change, which will affect the "The path planning made under the assumption that the transit time is proportional to the path length has a great impact, resulting in the delivery task cannot be completed under the time window constraint. Therefore, after starting the delivery, it is necessary to consider the impact of real-time road conditions on the planned delivery plan and make real-time processing and judgments on the road condition change information at the edge, and then dynamically adjust the original plan to ensure the completion of the overall delivery The task process will not be greatly affected due to changes in road conditions, and further ensure the economy and timeliness of logistics distribution.

With the development of edge computing and intelligent transportation technology, various information can be processed in time and the results can be directly transmitted to the destination vehicle [8–10]. Use simulated random events to occur in chronological order for simulation: set up several types of events with different degrees of influence, adjust the transit time of the road, and generate a series of events with time parameters that can trigger changes in the transit time of the road. To simulate real-time changes in road conditions during the logistics distribution process.

In the edge computing environment, from processing real-time road condition changes to adjusting the current distribution plan, refer to the process of sending computing tasks in the computing offloading model proposed in [11]. The time $g_n(i)$ spent in the whole process is as follows Model calculation.

$$\begin{aligned} g_n(i) = \sum_{m=1}^{M} \sum_{n'=1}^{N} F_n^m(i) \cdot Q_n^{n'}(i) \cdot (1 - F_{n'}^m(i)) \cdot \frac{\omega_n}{\lambda_{V2V}} \cdot (\theta_{n,n'} + 1) \\ + \sum_{m=1}^{M} F_n^m(i) \cdot \frac{\omega_n}{\lambda_{V2V}} + \sum_{m=1}^{M} F_n^m(i) \cdot \frac{l_n}{u_n \cdot p} + \frac{\omega_n'}{\lambda_{V2I}} \end{aligned} \tag{7}$$

Where, if v_n is within the range of d_m, $F^m{}_n(i)$ takes the value 0, otherwise it takes the value 1. ω_n represents the data volume of the calculation task being transmitted, and λ_{V2V} and λ_{V2I} respectively represent the data transmission rate based on V2V technology and V2I technology.

Situation Three

If we don't respond to the change of road conditions and continue to complete the distribution according to the original route planning, the final total cost of distribution $Cost_2$ is mainly composed of four parts.

$$Cost_2 = TC + PC + \sigma_1 \times VN + EC \tag{8}$$

In this case, because the road conditions will randomly change after the delivery starts, there is a risk that the vehicle will not be able to reach the delivery point according to the scheduled time during the delivery process, which will affect the delivery schedule and increase the delivery cost.

3   Algorithm Design

The vehicle path planning problem is a typical NP-hard problem. In the dynamic vehicle path planning method based on cloud-side-end collaboration proposed in this paper, the solution to this problem is as follows. First, before the start of the distribution, according to the demand at the demand point and the time window requirements, the genetic algorithm is used in the cloud to plan for the overall vehicle distribution plan. Then, according to the time sequence, a series of road events are generated to simulate the random changes of road traffic conditions in real situations. After the vehicle starts delivery from the distribution center, it senses changes in road conditions through the edge devices installed on the roadside, and runs the improved A* algorithm on the edge server according to the changes in road conditions, and adjusts the affected delivery routes in time. And send the adjusted delivery route to the delivery vehicle terminal.

3.1   Dynamic Passage Schedule

The occurrence of a road event will affect the travel time of roads in the area, thereby forming a dynamically changing road traffic situation. In this article, we use the form of a dynamic travel schedule to express this dynamic change of road conditions. Record the travel time of the road at the current moment and the predicted value of the travel time in the future several times in the table [12].

From the delivery start time t_0, the following period is divided into several paragraphs of length ΔT and these paragraphs are numbered respectively until all vehicles have completed the delivery task. After all the road sections in the delivery area are numbered in sequence, T_{ij} is used to represent the transit time of road section i in time period j.

When the road conditions change, the edge server will determine whether the current delivery route needs to be changed according to the road condition information provided by the edge device, and if necessary, re-plan the route. The optimized path may contain multiple road segments, which are numbered 1, 2, 3, ... k,.... Use $[t_k, t_k']$ to represent the transit time T_k of the vehicle passing through the section k, then $T_k = t_k' - t_k$. It may take multiple time periods for vehicles to pass through the road section k, and these time periods are corresponding to the transit times $T'_{k1}, T'_{k2}, T'_{k3},...$

3.2 Use Improved A* Algorithm for Dynamic Optimization Solution

As a heuristic search algorithm, the A* algorithm is widely used to find the shortest path between two points. In the method proposed in this article, we combine the A* algorithm with the dynamic travel schedule. After obtaining the calculation method for the travel time of the vehicle through section k, the traditional A* algorithm can be adjusted to search for the dynamic optimization path.

Based on the above definition, the main steps of the method proposed in this article can be described by Algorithm 1. In addition, other parts of the traditional A* algorithm do not need to be changed.

Algorithm 1. Improved A*

Input: *OP*: Original path planning scheme
        CE: Set of simulation generated change events
        DPI: Set of demand point location information
Output: Optimized overall path planning scheme
01: while delivery task not completed **do**
02:   {if road event occurs **then**
03:   Determine event type;
04:   Update *DTS*;
05:   if the affected road section on the sub-rout **then**
06:   *FindLeastCostPath* (x_i, x_j, DTS)
07:   Send R_{update} to the vehicle and the cloud;
08:   Caculate total time cost $g_n(i)$;
09:   else
10: R_{update} = NULL;
11:   }
12: Calculate the time consumption of the optimized solution by the cost model;
13: Calculate the cost of the optimized solution by the cost model;

---

4 Experimental Simulation and Result Analysis

In order to verify the effectiveness of the method proposed in this paper, t this paper takes the distribution demand data provided by the enterprise as an example, establishes a logistics distribution vehicle routing model, and adopts the dynamic vehicle routing optimization algorithm based on cloud-side-end collaboration proposed in this paper to solve the problem. The experiment uses the windows 7 operating system; the memory is 8.0 GB; and the IntelliJ IDEA 2020.1.2 x64 software is used for simulation.

4.1 Simulation Settings

During the simulation experiment, we set the experimental parameters reasonably according to the actual situation. The resources possessed by the distribution center are as follows: the distribution center has two different types of vehicles, I and II, with a maximum load of 500 kg for type I and 400 kg for type II vehicles, the number of which is 11 and 2 respectively. A total of 80 distribution points needs to be completed, and the relative position of each distribution point is known. To simplify the problem, suppose that the time window requirements of all delivery points are unified to 11:00–12:00. Correspondingly, we set the soft time windows of the delivery points to 10:30–11:00 and 12:00–12:30. In the soft time window, the demand point can accept the delivery vehicle for delivery, but there will be a time window penalty (ie a and b in the previous article), and the time window coefficients are set to 5.0 and 1.1 respectively. By consulting relevant information, we set the unit distance cost and unit time cost of the delivery vehicle to 6.3 yuan/km and 36 yuan/h, respectively.

We divide the overall distribution area into 1 km*1 km square sub-areas. For each road event that occurs in time series, let it randomly affect the transit time of roads in certain sub-regions. Each sub-area has an RSU, which is responsible for monitoring the road conditions in the area and communicating with the delivery vehicles in the area. The data transmission rates based on V2V technology and V2I technology are respectively set as $\lambda_{V2V} = 1$ Gb/s and $\lambda_{V2V} = 600$ Mb/s.

4.2 Analysis of Results

Based on the known distribution resources of the distribution center and the location coordinates of each demand point, demand and time window constraints, this article first uses a certain multi-objective vehicle path planning method in the cloud to plan and complete the overall logistics before starting the distribution. The vehicle delivery path of the delivery task. Based on the distribution plan planned by the cloud, the distribution path is dynamically adjusted in real time according to the actual situation in the distribution process.

After the delivery vehicle starts to deliver according to the initial optimization plan, the real-time changes of the road conditions in the actual situation are simulated by simulating changes in road conditions. Then through the terminal equipment to perceive changes in road conditions, the road conditions change data is transmitted to the edge server in the area, the data is processed in time at the edge, and the improved A* algorithm is used for the local delivery path that needs to be adjusted. Respond to events that affect road transit time caused by time, and dynamically adjust local distribution routes until the final distribution task is completed.

After starting the delivery, according to the real-time road condition change information of the simulation, the vehicle delivery plan made in the cloud is adjusted in real time, and the path adjustment situation of the vehicle delivery plan is shown in Table 1. In the table, the underlined number of the newly added delivery point indicates the adjustment of the access sequence of the original delivery route.

Figure 2 describes the adjustment of the delivery order of sub-route 1 in Table 1. In the figure, a closed loop represents the driving route of the vehicle that completes the demand point delivery task on the route. The lines of different colors represent the route according to the vehicle in a certain section of the route. The real-time road conditions of the location, and the adjustments made to this segment of the driving path. When the delivery vehicle departs from the distribution center 0, the pre-planned delivery sequence

Table 1. Adjustment of sub-route delivery order

Sub-route	Adjust according to road conditions	Delivery order of demand point
Sub-route 1	Original path	0-78-72-46-28-42-64-34-0
	Adjustment	0-78-72-53-46-28-42-64-34-0
		0-78-72-53-46-28-42-12-14-64-34-0
Sub-route 2	Original path	0-29-53-35-44-55-12-0
	Adjustment	0-29-49-53-35-44-55-12-0
		0-29-49-53-35-44-55-16-12-0
Sub-route 3	Original path	0-57-17-25-10-51-38-0
	Adjustment	0-33-57-17-25-10-51-38-0
		0-33-57-17-25-9-10-51-38-0
		0-33-57-17-25-9-10-51-35-38-0
Sub-route 4	Original path	0-73-33-3-65-58-0
	Adjustment	0-73-45-33-3-65-58-0
		0-73-45-33-43-3-65-58-0
		0-73-45-33-43-60-3-65-58-0
Sub-route 5	Original path	0-41-1-19-54-27-11-0
	Adjustment	0-41-47-1-19-54-27-11-0
		0-41-47-1-19-54-27-5-11-0
Sub-route 6	Original path	0-76-4-31-2-63-0
	Adjustment	0-76-78-4-31-2-63-0
Sub-route 7	Original path	0-26-37-32-77-79-52-0
	Adjustment	0-48-26-37-32-77-79-52-0
		0-48-26-37-32-45-77-79-52-0
		0-48-26-37-32-45-77-79-15-73-52-0
Sub-route 8	Original path	0-22-9-61-69-43-15-49-0
	Adjustment	0-22-12-9-61-69-43-15-49-0
		0-22-12-9-75-59-61-69-43-15-49-0
Sub-route 9	Original path	0-47-80-39-67-60-21-74-0
	Adjustment	0-47-1-80-39-67-60-21-74-0
		0-47-1-80-19-39-67-60-21-74-0
		0-47-1-80-19-39-67-77-60-21-74-0

(continued)

Table 1. (*continued*)

Sub-route	Adjust according to road conditions	Delivery order of demand point
Sub-route 10	Original path	0-36-59-50-40-62-68-0
	Adjustment	0-36-59-73-15-50-40-62-68-0
		0-36-59-73-15-50-40-57-9-62-68-0
Sub-route 11	Original path	0-13-24-30-18-7-56-0
	Adjustment	0-13-24-20-30-18-7-56-0
		0-13-24-20-30-5-18-7-56-0
Sub-route 12	Original path	0-75-71-6-5-23-66-0
	Adjustment	0-75-40-71-6-5-23-66-0
		0-75-40-71-6-30-5-23-66-0
Sub-route 13	Original path	0-48-8-14-20-70-45-16-0
	Adjustment	0-48-26-8-14-20-70-45-16-0
		0-48-26-8-22-14-20-70-45-16-0
		0-48-26-8-22-14-20-64-70-45-16-0

is 78, 72, 46, 28, 42, 64, 34. When the delivery task of the demand point 72 is completed and the demand point 46 is ready, There is an event that affects road traffic in the area where demand point 46 is located. The A* algorithm and the dynamic schedule of nearby paths are used to optimize and adjust. The change of driving route first passes through demand point 53 and then goes to demand point 46. Compared with Going forward according to the original planned route can reduce the waiting time in the middle and reduce the impact on the time to reach the delivery point 46; adopt the same processing method to adjust the driving route from demand point 42 to demand point 64 to route demand point 12, 14 and then reach the demand point 64.

When the vehicle completes its scheduled logistics and distribution tasks in accordance with the initial planning plan, taking into account the road conditions will occur, which will lead to changes in the transit time of some roads in the planning plan, resulting in the delivery tasks not being restricted by the time window carry out. By randomly generating road condition change events with time series, the dynamic traffic schedule of roads in the distribution area is correspondingly generated, thereby simulating the dynamic changes of road conditions. The distribution plan obtained using genetic algorithm (Contrast), the improved A* algorithm (improved A*) is used to optimize and adjust the sub-route driving route according to changes in road conditions, and the impact on the original distribution plan due to changes in road conditions (GA) is added. Consider, in the above three cases, the delivery cost of each sub-route is compared.

In Fig. 3, the total distribution cost of completing the overall distribution task under three conditions is compared. By Fig. 3 according to the results of the experiment it

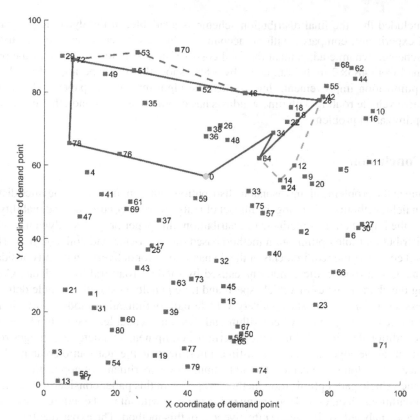

Fig. 2. Adjustment of sub-route 1 delivery order (Color figure online)

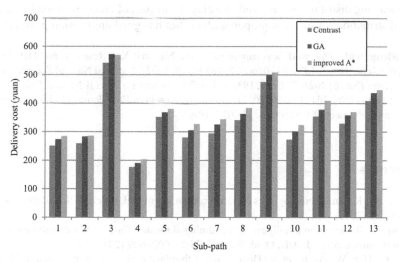

Fig. 3. Comparison of the total cost of distribution in the three situations

is concluded that the final distribution scheme is available, in the dynamic programming experiment, compared with no account of changes in real-time traffic distribution scheme, dynamic adjustment the total costs of the distribution scheme of the path declined from 4838.2 to 4651.6, down by 3.8%, show that the experimental results in the optimization improvement, thus verified the algorithm of this paper is the multi-objective vehicle routing with time windows has certain feasibility and effectiveness of the optimization problem.

5 Conclusion

Aiming at the problems of unreasonable distribution routes in the current logistics distribution field, without considering the impact of real-time road conditions, and inability to reduce the impact on the timeliness of distribution, this paper proposes a dynamic vehicle distribution route optimization method based on cloud-side-end collaboration. This method considers the requirements of the demand point for delivery time, and considers the changes in road traffic conditions caused by random road traffic incidents. Combining the characteristics of vehicle speed and time penalty cost in the vehicle delivery process, establishes a logistics delivery vehicle path optimization model. Solve it and optimize it with improved A* algorithm and dynamic schedule. The method collects road condition data in real time through terminal equipment, evaluates and judges road conditions at the edge, and makes real-time adjustments to the distribution plan made in advance at the cloud data center. Through simulation experiments of application examples, the vehicle path optimization method proposed in this paper considering real-time road conditions changes and the optimization method without road conditions are compared and analyzed, verifying the effectiveness of this method. The experimental results show that the method proposed in this paper has good effects in reducing distribution costs, reducing distribution time, and reducing the impact of changes in road conditions on the distribution results. The proposed algorithm has good application prospects.

Acknowledgment. This work was supported by the National Key Research and Development Project of China (No. 2018YFB1702600, 2018YFB1702602), National Natural Science Foundation of China (No. 61402167, 61772193, 61872139), Hunan Provincial Natural Science Foundation of China (No. 2017JJ4036, 2018JJ2139), and Research Foundation of Hunan Provincial Education Department of China (No. 17K033, 19A174).

References

1. Nowicka, K.: Smart city logistics on cloud computing model. Procedia Soc. Behav. Sci. **151**, 266–281 (2014)
2. Song, L.: Research on intelligent path planning algorithm for logistics distribution vehicles in low-carbon cities. J. Adv. Oxid. Technol. **21**(2), 602–609 (2018)
3. Ma, C., Hao, W., He, R., et al.: Distribution path robust optimization of electric vehicle with multiple distribution centers. PLoS One **13**(3), 189–205 (2018)
4. Zhang, X., Liu, H., Li, D., et al.: Study on VRP in express distribution based on genetic algorithm. Logist. Technol. **32**(05), 263–267 (2013)

5. Wang, J., Ying, Z., Yong, W., et al.: Multiobjective vehicle routing problems with simultaneous delivery and pickup and time windows: formulation, instances and algorithms. IEEE Trans. Cybern. **46**(3), 582–594 (2016)
6. Sui, Y., Chen, X., Liu, B.: D-star Lite algorithm and its experimental study on dynamic path planning. Microcomput. Appl. **34**(7), 16–19 (2015)
7. Su, Y., Yan, K.: Study of the method to search dynamic optimum route for vehicle navigation system. Syst. Eng. **18**(4), 32–37 (2000)
8. Omoniwa, B., Hussain, R., Javed, M.A., et al.: Fog/edge computing-based IoT (FECIoT): architecture, applications, and research issues. IEEE Internet Things J. **6**(3), 4118–4149 (2019)
9. Zhou, H., Xu, W., Chen, J., et al.: Evolutionary V2X technologies toward the internet of vehicles: challenges and opportunities. Proc. IEEE **108**(2), 308–323 (2020)
10. Liao, C., Shou, G., Liu, Y., et al.: Intelligent traffic accident detection system based on mobile edge computing. In: IEEE International Conference on Computer and Communications (ICCC) (2018)
11. Xu, X., Xue, Y., Qi, L., et al.: An edge computing-enabled computation offloading method with privacy preservation for internet of connected vehicles. Future Gener. Comput. Syst. **96**(JUL.), 89–100 (2019)
12. Liu, B., Chen, X., Chen, Z.: A dynamic multi-route plan algorithm based on A* algorithm. Microcomput. Appl. **35**(04), 17–19+26 (2016)

CLENet: A Context and Location Enhanced Transformer Network for Target-Oriented Sentiment Classification

Chao Yang$^{(\boxtimes)}$ ⓘ, Hefeng Zhang ⓘ, Jing Hou ⓘ, and Bin Jiang ⓘ

College of Computer Science and Electronic Engineering, Hunan University,
Lushan Road (S), Yuelu District, Changsha, China
{yangchaoedu,houjing,jiangbin}@hnu.edu.cn, hefengzhang@yeah.net

Abstract. Target-oriented sentiment classification seeks to predict the sentiment polarity of a target in a given text. Previous approaches mainly focus on LSTM-attention structure, but practice shows that LSTM and attention mechanism can not correctly identify and classify partial negation relations and multi-target scenarios on limited datasets. This paper proposes a context and location enhanced transformer network that improves both the modeling of negation relations and its classification accuracy on multi-target scenarios. Experimental results on SemEval 2014 and Twitter datasets confirm the effectiveness of our model.

Keywords: Self-attention · Location enhancement · Multi-target scenarios · Sentiment classification

1 Introduction

Recently, sentiment analysis has become one of the most popular research areas in Natural Language Processing (NLP). Extensive applications are also found in information retrieval, data mining, intelligent recommender system, question answering and so on [4]. Traditional sentiment analysis is usually a sentence-level or document-level task that aims to find the overall sentiment related to one entity in a sentence or document. For the targets of the entity, if we want to determine the sentiment they express, the general sentiment analysis fails to meet this task, target-oriented sentiment classification is raised for this task. Target-oriented sentiment classification seeks to make inferences about the sentiment polarity (e.g. neutral, negative, positive) of each target in the text [8].

One of the challenges of target-oriented sentiment classification is how to effectively model the modification relationship between words. For example, the difficulty in modeling negation relations is that negation modifiers can reverse the sentiment polarities and some negation modifiers have more complicated linguistic rules. Besides, it is also very difficult to identify the negation modifiers, such as "not", "little/few", "barely", "seldom" and so on. In target-oriented

© Springer Nature Singapore Pte Ltd. 2021
Y. Sun et al. (Eds.): ChineseCSCW 2020, CCIS 1330, pp. 72–86, 2021.
https://doi.org/10.1007/978-981-16-2540-4_6

sentiment classification task, most works utilize LSTM as encoder to extract the dependencies between words [1,5,6,10,16], while some models use the word embedding layer as an encoder to reduce time consumption [11]. However, the word embedding layer encodes context on the basis of the bag-of-words assumption, which makes it difficult for the model to identify the negation modifiers, the degree modifiers, and the irony modifiers and so on. Theoretically, LSTM models words according to word order by a recursive structure, which can characterize the modification relationship between words, but LSTM does not achieve the expected effect. Qian et al. [9] argued that the introduction of grammar rules can improve the performance of the model and demonstrated that LSTM is deficient in modeling grammatical relationships. This paper argues that LSTM cannot fully pick up the modification relationship among words on limited datasets due to complex memory and gate mechanisms.

In order to solve the problem that the LSTM-attention structure cannot correctly identify and classify partial negation relations and multi-target scenarios on limited datasets, this paper proposes a context and location enhanced transformer network (CLENet). CLENet introduces the self-attention mechanism and a convolutional layer for modeling the relationship among the current word and local context words and then utilizes the traditional attention mechanism for extracting the interactive features of context and target for classification. Specifically, the encoding process is that CLENet first calculates the semantic correlation between the central word and the surrounding words based on the self-attention mechanism and obtains the context representation of the central word. Then, CLENet models the modification relationship between the central word and the local context based on a convolutional layer. So the encoder is named self_local_attention encoder which can find neighbor words that are more semantically correlated with the central word. Compared to the word embedding layer, the self_local_attention encoder can effectively capture the modification relations between central words and local context words by self-attention mechanisms and convolution operations. Compared to LSTM, the self-local-attention encoder adopts a simple mechanism to consider only the modification relations between local context words without considering long-distance dependencies based on historical information. Experiments show that the self-local-attention encoder has better performance in modeling the modification relations of the local context.

The situation where multiple targets with different sentiment polarities appear in one sentence is the most complex. Attention mechanism is used to extract context related to target, but models solely built on attention mechanism perform poorly on multi-target scenarios. Specifically, attention networks typically utilize alignment functions to retrieve opinion modifiers that are semantically related to the current target from the global context. This paper argues that it is difficult for alignment functions to distinguish different opinion modifiers of multiple targets based on semantic relevance, which causes the model to introduce sentiment features of other targets when inferring the sentiment polarity of the current target, resulting in poor performance of the model on multi-target scenarios.

Usually the opinion modifiers appear near the target, that is, a context word nearer to the current target should be more significant for making inferences about the sentiment polarity of the current target. Relative location information enables the model to focus on opinion modifiers that are closer to the current target, which has become an important method to improve the poor performance of the model on multi-target scenarios. Based on the above ideas, many methods have been derived. Chen et al. [1] weighted relative location information to the context for generating target-specific context representation. Experiments show that location information is beneficial in improving the performance of the model, especially the predictive ability of the model on multi-target scenarios. On this basis, this paper proposes a simple strategy to fuse location information into the attention layer, which aims at improving the performance of the model and the predictive ability of the model on multi-target scenarios. This strategy directly fuses the location information into the final attention alignment score, which enables the model to focus more on context that are closer to target.

2 Related Works

Target-oriented sentiment classification differs from general sentiment classification. Multiple targets could occur in one sentence for this task, and each target has associated words modifying it. In judging the sentiment of current target, additional targets and associated words will be noises. Therefore, it is difficult to establish the connection among context and target for this task. In order to deal with this difficulty, many approaches have been proposed.

The early methods [3,14] for target-oriented sentiment classification task are methods based on machine learning, which mainly relies on manually designed features and utilizes context that are closely associated with target to characterize the relationship among context and target. Features designed according to the context near the target consider only local dependencies and ignore the global context, which limits the effectiveness of features. Usually models leverage syntax parsing tools to account for long-distance dependencies between target and context, but the performance of syntax parsing tools is degraded by transferring, which interferes with the extraction of related features.

Boosted by the development of deep learning techniques, many deep learning-based approaches for target-oriented sentiment classification have been proposed. Tang et al. [10] used two LSTMs to model two clauses. The clause before the last target word was input into the first LSTM forward and the clause after the first target word was input into the other LSTM backward, and finally the coding features of the two LSTMs were used for classification. Zhang et al. [19] utilized GRNN (Gated RNN) to model the context and leveraged the gate mechanism to determine the importance of two clauses for inferring the sentiment of target.

Following the application of attention mechanisms in machine translation successfully, many methods adopt recurrent neural network as encoder and exploit attention mechanism for generating target-specific context features. Wang et al. [16] proposed an attention-based LSTM network. Ma et al. [5] proposed IAN which learned the interactive features of context and target through two parallel attention networks and then connected them to make predictions. Yang et al. [18] proposed a coattention mechanism and alternatively learned the effective features of context and target based on a memory network architecture. Tang et al. [12] proposed an algorithm built on the attention mechanism to automatically mine attention supervision information and retrained the model based on this information to further learn the sentiment connection between words.

Some works obtain better context representation by stacking attention layers. Tang et al. [11] utilized the average of target as target vector and adopted a memory network architecture built on the attention mechanism for acquiring more reasonable features from the context. The model accumulated the historically computed features into a memory vector and continuously modified the attention weight of each context word based on the memory vector to extract more effective context features. On this basis, Chen et al. [1] adopted Bi-GRU to model context and target to obtain nonlinear relationship between words and utilized a GRU-like gate structure to control historical memory.

Ma et al. [6] argued that common sense as an external knowledge can improve the performance of the model, so they proposed Sentic-LSTM based on LSTM which regarded the related concepts of the current word as common sense and input them into the model and then modelled the relationship among target and context by a hierarchical attention mechanism. Wang et al. [15] proposed a CRF-based semantic segmentation attention mechanism to learn the structured feature representation. They utilized CRF to label the importance of each word for classification and finally captured phrase structural features for classification. Xue et al. [17] proposed a model based on convolutional neural network and designed a new gate mechanism for target-oriented sentiment classification, and the model showed very competitive performance.

3   Model Architecture

The overall framework of CLENet as shown in Fig. 1.

3.1   Self_local_attention Encoder

As shown in Fig. 2, the self_local_attention encoder includes three layers: the word embedding layer seeks to map each context word into a word vector; the self local attention layer captures important local context words on the basis of the self-attention mechanism; the transform layer converts the sentiment semantic space of the word into a semantic space modified by the context.

Assuming that the window size in the self_local_attention encoder is set to 3, a sentence s of length n needs to be padded with $<pad>$ so that the final coded sequence is of the same length as the sentence, that is, $s = [w^{<pad>}, w^1, w^2, w^3, \ldots, w^n, w^{<pad>}]$. Coding process can be divided into three steps according to the encoder structure:

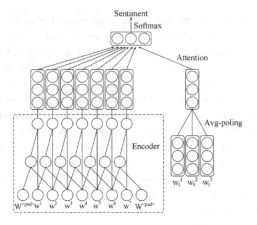

Fig. 1. Overall framework of CLENet

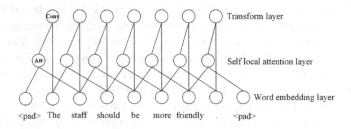

Fig. 2. Self_local_attention encoder with window size set to 3

First, the encoder transforms each word into a vector via the word embedding layer to obtain the sentence representation $E_c = [e^{<pad>}, e^1, e^2, e^3, ..., e^n, e^{<pad>}]$.

Secondly, the encoder calculates the alignment score between the current word and other local words based on the attention mechanism.

We declare an alignment function that computes the semantic correlation between the current word vector e^i and other word vector e^w in the window.

$$f\left(e^i, e^w\right) = \tanh\left(W_t\left[e^i; e^w\right] + b\right) \tag{1}$$

Assuming the size of the window is 3, the alignment scores between the word w_i and the word w_{i-1} or the word w_{i+1} are as follows:

$$a_{i-1} = \frac{\exp\left(f\left(e^{i-1}, e^i\right)\right)}{\exp\left(f\left(e^{i-1}, e^i\right)\right) + \exp\left(f\left(e^i, e^{i+1}\right)\right)} \tag{2}$$

$$a_{i+1} = \frac{\exp\left(f\left(e^i, e^{i+1}\right)\right)}{\exp\left(f\left(e^{i-1}, e^i\right)\right) + \exp\left(f\left(e^i, e^{i+1}\right)\right)} \tag{3}$$

The context representation $C = [c^1, c^2, c^3, \ldots, c^n]$ of all words can be obtained from the alignment score.

$$c^i = a_{i-1}e^{i-1} + a_{i+1}e^{i+1} \tag{4}$$

Finally, E_c and C are fed into a convolutional layer of kernel size 2. The final sequence encoding vector is $H = [h^1, h^2, h^3, \ldots, h^n]$.

$$h^i = \tanh\left(W_h\left[e^i; c^i\right] + b\right) \tag{5}$$

where W_h is a $d \times 2d$ weight matrix, d is the number of the convolution kernel.

3.2 Attention Layer

We initialize the target $t = [w_t^1, w_t^2, w_t^3, \ldots, w_t^m]$ to obtain its matrix representation $E_t = [e_t^1, e_t^2, e_t^3, \ldots, e_t^m]$ and then utilize the average of E_t as the target representation T_r. After that, we calculate the alignment score among the target and the encoding vector of each context word h_i and normalize them and then obtain the context features C_r by weight accumulation.

$$T_r = \sum_{i=1}^{m} e_t^i / m \tag{6}$$

$$f(T_r, h_i) = \tanh(W_a[T_r; h_i] + b) \tag{7}$$

$$\alpha_i = \frac{\exp(f(T_r, h_i))}{\sum_{j=1}^{n} \exp(f(T_r, h_j))} \tag{8}$$

$$C_r = \sum_{i=1}^{n} \alpha_i h_i \tag{9}$$

3.3 Location Enhancement Layer

First, one variable D_i is defined for measuring the relative location information between the word w_i and the target. As an example, for the sentence "The keyboard is too slick." and the target "keyboard". The distance between "keyboard" and "slick" is 3.

Then, We define W_i as the relative position of the word w_i:

$$W_i = 1 - \frac{|len - D_i|}{len} \tag{10}$$

where len means length of the sentence.

The location prior information and the context-level attention network are integrated, which utilizes W_i to directly constrain α_i:

$$R_i = W_i \alpha_i \tag{11}$$

$$\alpha_i = \frac{R_i}{\sum_{i=1}^{n} R_i} \tag{12}$$

3.4 Classifier and Loss Function

In this paper, the softmax layer acts as a sentiment classifier. To begin with, we utilize a linear function to map the feature vector C_r to an l-dimensional classification space. l represents quantity of sentiment categories.

$$x = W_x C_r + b_x \tag{13}$$

Then, we utilize the *softmax* function for calculating the probability that the current feature vector falls into each sentiment category:

$$p_i = \frac{\exp(x_i)}{\sum_{j=1}^{l} \exp(x_j)} \tag{14}$$

Finally, the final loss is obtained by the cross entropy loss function:

$$L = -\sum_{i=1}^{l} y_i \log(p_i) \tag{15}$$

4 Experiments

4.1 Datasets and Parameters

Datasets: We conduct experiments on SemEval 2014 Task4 [8] and Twitter [2] to verify the usefulness of our model. In Table 1, the SemEval 2014 dataset consists of two types of comments, i.e., Restaurant and Laptop. The reviews are marked as having three sentiment polarities: negative, positive and neutral.

Parameters: We use the 300-dimensional Glove vectors for initializing the word embeddings which are trained from web data and the vocabulary size is 1.9M [7]. All out-of-vocabulary words and all weights are randomly generated by the standard distribution $U(0.01, 0.01)$, and all biases are set to zeros. We train the model with a SGD optimizer, and the learning rate is set to 0.01.

4.2 Compared Methods

To justify the effectiveness of our model, we compared it with the following methods.

   Majority: It assigns the largest sentiment polarity in the training set to each sample in the test set.

Table 1. Statistics of SemEval 2014 and Twitter datasets

Dataset	Positive		Neutral		Negative	
	Train	Test	Train	Test	Train	Test
Restaurant	2164	728	637	196	807	196
Laptop	994	341	464	196	870	128
Twitter	1561	173	3127	346	1560	173

LSTM: It models the context using one LSTM network and uses the average of all hidden states as final representation for predicting the category of sentiment [16].

TD-LSTM: It models two clauses through two LSTM networks [10].

AE-LSTM: It is a attention-based LSTM network [16].

ATAE-LSTM: It is an extension of AE-LSTM and adds target embedding to the input of the model [16].

AF-LSTM: It models target and context by a convolutional layer and a attention layer [13].

IAN: It uses two attention networks to learn interactive features between the context and target in parallel [5].

MemNet: It extracts the context features based on the recurrent attention mechanism in the deep memory network [11].

GCN: It extracts context features and target features based on CNN and then uses Gated Units to extract sentiment features for classification [17].

Coattention-MemNet: It learns crucial features alternately from aspects and contexts through an iteration mechanism [18].

All models and their classification accuracy are shown in Table 2. The data in Table 2 mainly comes from Yang et al. [18], Chen et al. [1] and Ma et al. [5].

CLENet achieves 0.2%, 0.6% and 0.6% accuracy improvements on Restaurant, Laptop and Twitter datasets compared with IAN. Compared with AE-LSTM, CLENet achieves 2.8% and 3.8% accuracy improvements on Restaurant and Laptop datasets. Compared with MemNet, CLENet achieves 0.3%, 1.9% and 1.9% accuracy improvements on Restaurant, Laptop and Twitter datasets. Compared with GCN, CLENet achieves 1.4% and 2.2% accuracy improvements on Restaurant, Laptop datasets. The above analysis shows that the encoder proposed in this paper is very effective.

Besides, CLENet achieves approximately 0.3% accuracy improvements on three datasets after considering location information. This proves that the method of fusing location information proposed in this paper is effective. Compared with Coattention-MemNet, CLENet+Location achieves 0.1% and 0.3% accuracy improvements on Laptop and Twitter datasets. Coattention-MemNet outperforms CLENet+Location on Restaurant dataset probably because Coattention-MemNet performs multiple feature extraction operations based on the memory network architecture to extract better sentiment features.

Table 2. The classification accuracy of different methods

Model	Restaurant	Laptop	Twitter
Majority	0.535	0.650	0.500
LSTM	0.743	0.665	0.665
TD-LSTM	0.756	0.681	0.666
AE-LSTM	0.762	0.689	–
ATAE-LSTM	0.772	0.687	–
AF-LSTM	0.754	0.688	–
MemNet	0.787	0.708	0.685
IAN	0.788	0.721	0.698
GCN	0.776	0.695	–
Coattention-MemNet	0.797	0.729	0.705
CLENet (windows = 3)	0.790	0.727	0.704
CLENet+Location(windows = 3)	0.794	0.730	0.708

4.3 Effective Modeling of Modification Relationships

Coding Effect Comparisons. So as to confirm the effectiveness of CLENet, we devise two comparison models, which are versions that replace the encoder in CLENet with the word embedding layer or LSTM. This paper also sets different window sizes for the self_local_attention encoder to explore the importance of the window size. Table 3 provides the classification accuracy of three models.

Table 3. The classification accuracy of models with different encoders

Encoder/dataset	Restaurant	Laptop	Twitter
Word embedding+attention	0.763	0.699	0.681
LSTM+attention	0.775	0.712	0.692
CLENet (windows = 3)	0.790	0.727	0.704
CLENet (windows = 5)	0.793	0.730	0.707

As we can see from Table 3, the model with the LSTM encoder achieves 0.8%, 1.3% and 1.1% accuracy improvements on Restaurant, Laptop and Twitter datasets compared with the model with the word embedding layer, which proves that LSTM could learn some modification relationships between words. CLENet outperforms LSTM over about 1.5% on three datasets, which proves that the self_local_attention encoder could learn more complex modification relationships between words and it is very effective for target-oriented sentiment classification.

CLENet with a window size of 5 outperforms CLENet with a window size of 3 over about 0.3%, which arises from the fact that increasing the window size could increase the field of the self-attention mechanism to extract modification relationships between words over a wider range. Changing the size of the window did not significantly improve accuracy, which may be due to the fact that most valid modification relationships are limited to three words or less.

Case Study. As can be seen from Table 4, the model based on the word embedding layer performs poorly on three sentences, which proves that the model predicts the sentiment polarities only according to the semantics of word vectors and cannot correctly analyze the modification relationship between words. The model based on LSTM cannot make correct predictions on the third sentence, which proves that LSTM cannot effectively model irony modification relationships. The model based on the self_local_attention encoder makes correct predictions on three sentences, which proves that our proposed encoder can effectively model the modification relationship between words compared to LSTM.

Table 4. Some test examples and predictions of the different models

Sample sentence/encoder	Word embedding+attention	LSTM+attention	CLENet
1. Did not enjoy the new Windows 8 and touchscreen functions	False	True	True
2. I trust the people at Go Sushi, it never disappoints	False	True	True
3. The staff should be a bit more friendly	False	False	True

Weight Visualization Analysis. First, to prove the effectiveness of CLENet, we visualize the attention weight distribution of the two sentences. Sentence1 is "I trust the people at Go Sushi, it never disappoints." and the target is "people". Sentence2 is "The staff should be a bit more friendly." and the target is "staff".

From Table 5 and Table 6, it can be seen that the attention layer assigns more weight to opinion phrases, such as "I trust" and "it never disappoints." in sentence1 as well as "staff should be" and "bit more friendly" in sentence2, which proves that the attention layer can pay attention to the opinion phrases and then use them for classification.

Table 5. The attention weight distribution of the sentence1

I	trust	the	people	at	Go	Sushi	,	it	never	disappoints	.
0.04	0.09	0.04	0.03	0.03	0.06	0.06	0.03	0.10	0.22	0.22	0.09

Table 6. The attention weight distribution of the sentence2

The	staff	should	be	a	bit	more	friendly	.
0.02	0.18	0.18	0.05	0.03	0.18	0.22	0.12	0.02

Secondly, in order to verify that self local attention layer is able to model the correlation between the current word and local other words, we take "it never disappoints." and "a bit more friendly." as examples and visualize the attention weights of some context words according to a window of size 3, and the corresponding central words include "never", "disappoints", "bit", "more", and "friendly". The results are shown in Table 7.

From the first row in Table 7, the alignment score between "never" and "it" is 0.20 and the alignment score between "never" and "disappoints" is 0.80, which shows that the encoder believes that "disappoints" has a closer semantic correlation with "never" compared to "it" and also shows that "never" is more suitable for modifying "disappoints". Similarly, the second row of the table shows that "disappoints" has a closer semantic correlation with "never" compared to ".". The above analysis proves that the self local attention layer can capture the local context closely related to the semantic of the current word.

From the third to fifth rows in Table 7, we can see that "bit" has a closer semantic correlation with "more", "more" has the same semantic correlation with "bit" and "friendly", "friendly" has a closer semantic correlation with "more". Therefore, the encoder could correctly model the modification relationship of "bit more friendly" and infer its irony tendency to make correct sentiment prediction.

Table 7. Weight distribution based on the self local attention layer

Central word w_i/context	w_{i-1}	w_{i+1}
never	0.20	0.80
disappoints	0.85	0.15
bit	0.31	0.69
more	0.50	0.50
friendly	0.73	0.27

The above two experimental results show that CLENet can capture important opinion phrases by the attention layer and capture important local context by the self local attention layer, including negative words and sentiment words, as well as extract modification relationship between local context words.

4.4 Analysis of Location Information

Experiment Results. It is evident from Table 8 that the model using the word embedding layer or LSTM utilizes location information to achieve about 2% and

1% accuracy improvements, which proves that it is very effective to use location information as priori information and integrate it into the attention network.

CLENet utilizes location information to achieve about 0.4% accuracy improvement on the three datasets, which proves that the method of fusing location information proposed in this paper can effectively improve the performance of attention mechanism for target-oriented sentiment classification.

Table 8. The classification accuracy of the model with location information

Model/dataset	Restaurant	Laptop	Twitter
Word embedding+attention	0.763	0.699	0.681
Word embedding+attention+Location	0.785	0.720	0.694
LSTM+attention	0.775	0.712	0.692
LSTM+attention+Location	0.780	0.724	0.695
CLENet (windows = 3)	0.790	0.727	0.704
CLENet+Location (windows = 3)	0.794	0.730	0.708
CLENet (windows = 5)	0.793	0.730	0.707
CLENet+Location (windows = 5)	0.795	0.734	0.710

Case Study. For verifying if the location fusion method presented by this paper can capture the sentiment features around the target, we design CLENet with location information. Table 9 presents the prediction results from the model for 6 sample sentences and the context that the model focuses on. Any row in the Table 9 contains a sentence and two triads. The targets are bolded. The triad is composed of target, label, prediction and words with higher weight (>0.1) in the sentence. P, O and N indicate positive, neutral and negative.

In the first sentence, the model would pay attention to the opinion words "great" and "dreadful" for different targets. Different from the traditional attention mechanism, we introduce location information to constrain attention weight. On this basis, for the target "food", the model would assign a higher weight (0.29) to "great" and assign a lower weight (0.13) to "dreadful". For the target "service", the model would assign a higher weight (0.25) to "dreadful" and assign a lower weight (0.16) to "great".

Based on sentences 2, 3, and 4, we find that the model could focus on the more important target words. For example, "selection" in "beer selection" has a closer semantic correlation with "great", "staff" in "staff members" has a closer semantic correlation with "friendly", and "experience" in "dining experience" has a closer semantic correlation with "worst". The analysis proves that CLENet with location information is very effective for extracting key target words.

Sentence 5 involves three targets, and the locations and sentiment tendencies of three targets are different. Among them, "waitress" is negative, and "food" and "wine" are neutral. The model makes correct predictions for the target

Table 9. Some test examples and predictions of CLENet with location information

Sample sentence	Targets, predictions and opinion phrases
1. Great **food** but the **service** was dreadful!	(food, P$\checkmark$, Great dreadful) (service, N$\checkmark$, Great dreadful)
2. Great **beer selection** too, something like 50 **beers**	(beer **selection**, P$\checkmark$, Great beer selection too) (beers, O$\checkmark$, too something beers)
3. The **staff members** are extremely friendly and even replaced my **drink** once when I dropped it outside	(**staff** members, P$\checkmark$, extremely friendly) (drink, O$\checkmark$, even replaced drink)
4. Probably my worst **dining experience** in new york, and I'm a former **waiter** so I know what I'm talking about	(dining **experience**, N$\checkmark$, worst) (waiter, O$\checkmark$, former talking)
5. Our **waitress** had apparently never tried any of the **food**, and there was no one to recommend any **wine**	(waitress, N$\checkmark$, apparently never tried) (food, O$\checkmark$, tried any one) (wine, O$\checkmark$, no one recommend any)
6. Disappointingly, their wonderful **Saketini** has been taken off the **bar menu**	(Saketini, P$\checkmark$, Disappointingly wonderful Saketini) (bar menu, N$\times$, been taken off)

"waitress" because the model could pay attention to the opinion words around "waitress", including "apparently", "never" and "tried". For the target "food", the model could concern the neutral words surrounding the target, including "tried", "any" and "one" and ignores "apparently" and "never", which proves that the location information can guide the attention network to learn the boundaries of relative locations and automatically avoid the interference of distant words. Similarly, the model could also make correct prediction for the target "wine". Sentence 5 proves that CLENet with location information can improve the performance of the model on multi-target scenarios.

We found that introducing location information does not always make correct predictions. For instance, with respect to the target "bar menu" in sentence 6, the model requires an understanding of the semantic relationship of the entire sentence in order to make a correct prediction. However, the model introduces location information to limit the view to "been taken off" and ignores "disappointingly", which causes the model to make incorrect prediction.

5   Conclusion and Future Work

First, this paper designs a new self_local_attention encoder for target-oriented sentiment classification built on the self-attention mechanism. The encoder models the correlation between local context words using the attention mechanism and convolutional layer. It can be seen from the experimental results that the encoder can effectively model the modification relationship between words. Secondly, this paper proposes a simple method to incorporate location information into the attention mechanism. This method uses the relative location information among context words and target as weights to limit the attention network, which makes the attention network to give higher weight to the opinion words surrounding the target. The experimental results prove the usefulness of the location information. The visualization results indicate that the model with the location information can concern the opinion words surrounding the target and achieve the accuracy improvement on the multi-target scenarios.

As the existing attention mechanism and LSTM-based models fail to classify sentences depending on syntactical structures, for future work, we intend to design an unsupervised model based on the unsupervised corpus to encode the potential dependencies in the sentence and enable the model to understand the semantic relationship of the sentence according to the linguistic rules.

References

1. Chen, P., Sun, Z., Bing, L., Yang, W.: Recurrent attention network on memory for aspect sentiment analysis. In: Proceedings of the 2017 Conference on Empirical Methods in Natural Language Processing, pp. 452–461 (2017)
2. Dong, L., Wei, F., Tan, C., Tang, D., Zhou, M., Xu, K.: Adaptive recursive neural network for target-dependent Twitter sentiment classification. In: Proceedings of the 52nd Annual Meeting of the Association for Computational Linguistics (Volume 2: Short Papers), pp. 49–54 (2014)
3. Kiritchenko, S., Zhu, X., Cherry, C., Mohammad, S.: NRC-Canada-2014: detecting aspects and sentiment in customer reviews. In: Proceedings of the 8th International Workshop on Semantic Evaluation (SemEval 2014), pp. 437–442 (2014)
4. Liu, B.: Sentiment analysis and opinion mining. In: Synthesis Lectures on Human Language Technologies, vol. 5, no. 1, pp. 1–167 (2012)
5. Ma, D., Li, S., Zhang, X., Wang, H.: Interactive attention networks for aspect-level sentiment classification. In: Proceedings of the 26th International Joint Conference on Artificial Intelligence, pp. 4068–4074 (2017)
6. Ma, Y., Peng, H., Cambria, E.: Targeted aspect-based sentiment analysis via embedding commonsense knowledge into an attentive LSTM. In: Thirty-Second AAAI Conference on Artificial Intelligence (2018)
7. Pennington, J., Socher, R., Manning, C.D.: GloVe: global vectors for word representation. In: Proceedings of the 2014 Conference on Empirical Methods in Natural Language Processing (EMNLP), pp. 1532–1543 (2014)
8. Pontiki, M., Galanis, D., Pavlopoulos, J., Papageorgiou, H., Androutsopoulos, I., Manandhar, S.: SemEval-2014 Task 4: aspect based sentiment analysis. In: Proceedings of the 8th International Workshop on Semantic Evaluation (SemEval 2014), pp. 27–35 (2014)
9. Qian, Q., Huang, M., Lei, J., Zhu, X.: Linguistically regularized LSTM for sentiment classification. In: Proceedings of the 55th Annual Meeting of the Association for Computational Linguistics (Volume 1: Long Papers), pp. 1679–1689 (2017)
10. Tang, D., Qin, B., Feng, X., Liu, T.: Effective LSTMs for target-dependent sentiment classification. In: Proceedings of COLING 2016, the 26th International Conference on Computational Linguistics: Technical Papers, pp. 3298–3307 (2016)

11. Tang, D., Qin, B., Liu, T.: Aspect level sentiment classification with deep memory network. In: Proceedings of the 2016 Conference on Empirical Methods in Natural Language Processing, pp. 214–224 (2016)
12. Tang, J., et al.: Progressive self-supervised attention learning for aspect-level sentiment analysis. In: Proceedings of the 57th Annual Meeting of the Association for Computational Linguistics, pp. 557–566 (2019)
13. Tay, Y., Tuan, L.A., Hui, S.C.: Learning to attend via word-aspect associative fusion for aspect-based sentiment analysis. In: Thirty-Second AAAI Conference on Artificial Intelligence (2018)
14. Wagner, J., et al.: DCU: aspect-based polarity classification for SemEval Task 4 (2014)
15. Wang, B., Lu, W.: Learning latent opinions for aspect-level sentiment classification. In: Thirty-Second AAAI Conference on Artificial Intelligence (2018)
16. Wang, Y., Huang, M., Zhu, X., Zhao, L.: Attention-based LSTM for aspect-level sentiment classification. In: Proceedings of the 2016 Conference on Empirical Methods in Natural Language Processing, pp. 606–615 (2016)
17. Xue, W., Li, T.: Aspect based sentiment analysis with gated convolutional networks. In: Proceedings of the 56th Annual Meeting of the Association for Computational Linguistics (Volume 1: Long Papers), pp. 2514–2523 (2018)
18. Yang, C., Zhang, H., Jiang, B., Li, K.: Aspect-based sentiment analysis with alternating coattention networks. Inf. Process. Manag. **56**(3), 463–478 (2019)
19. Zhang, M., Zhang, Y., Vo, D.T.: Gated neural networks for targeted sentiment analysis. In: Thirtieth AAAI Conference on Artificial Intelligence (2016)

Long-Tail Items Recommendation Framework Focusing on Low Precision Rate Users

Jing Qin$^{(\boxtimes)}$, Jing Sun, Rong Pu, and Bin Wang$^{(\boxtimes)}$

Northeastern University, Shenyang 110169, China
{1210387,sunjing}@stumail.neu.edu.cn, binwang@mail.neu.edu.cn

Abstract. In the top-k recommendation method, the average precision rate of recommendation results is an important metrics that can be used to measure the precision of a recommendation. However, the average precision rate can only reflect the overall effect of the recommendation results, and not the reflect the recommendation effect of individual users. Therefore, the individual precision of the recommendation results is often low or 0. In addition, coverage and diversity metrics are often ignored in top-k recommendation methods, thus reducing the mining capability of long-tail items. Therefore, users with a precision rate of 0 can get neither accurate recommendation results nor novel results. For users with low levels of accuracy, recommending long-tail items that meet users' preferences is more conducive to mining users' interest preferences than recommending hot items for them. To solve the above-mentioned problems, this paper proposes a long-tail items recommendation framework focusing on low precision rate users and provides an auxiliary recommendation module in the recall module of the framework. The auxiliary recommendation module contains three methods (based on the minimum association value between items, based on the maximum association value between items, and the long-tail items-based diversified methods) to improve the recommendation results of low precision rate users. Experiments on Movielens 1M and Movielens ml-latest-small datasets show that the proposed long-tail items recommendation framework can improve the coverage and diversity as well as the precision of recommendations.

Keywords: Item association · Diversity · Long-tail items · Top-k recommendation

1 Introduction

With the era of the rapid growth of information, people no longer rely on traditional paper media to obtain information, and more often seek to obtain information from network media. Users not only rely on search engines to retrieve information but can also receive possible information of interest through active

© Springer Nature Singapore Pte Ltd. 2021
Y. Sun et al. (Eds.): ChineseCSCW 2020, CCIS 1330, pp. 87–100, 2021.
https://doi.org/10.1007/978-981-16-2540-4_7

recommendations by websites. Goldberg [6] proposed the recommendation system in 1992. Now, it is widely used in various fields, such as e-commerce, tourism, news, and movies. The collaborative filtering method [6] is a landmark method in the field of recommendation systems, and many methods are also developed on this basis. Collaborative filtering methods include nearest neighbor-based methods and model-based methods [1]. Among the nearest neighbor-based methods, the representative ones are UserCF (User-Based Collaborative Filtering) [15] and ItemCF (Item-Based Collaborative Filtering) [16]. The top-k recommendation method mainly uses the neighbor-based recommendation method. Top-k recommendation refers to the recommendation of k items that users may be interested in. The accuracy metric of the top-k recommendation method includes precision, recall, and the F1-Measure metric [7]. However, these metrics all evaluate the average effect of the recommendation results, ignore the recommendation precision of each user in the top-k recommendation method. For example, if the recommended result's average precision rate is 0.3, some users' precision rate is 1 (i.e., All predictions are correct), while some users' precision rate may only be 0.1. In addition, some users' precision rate is 0 (i.e., All predictions are wrong). Taking UserCF and ItemCF methods as examples, 10 items are recommended for all users on the Movielens 1M [9] dataset. The distribution of the correct number of items recommended for each user is shown in Fig. 1.

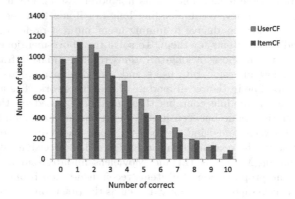

Fig. 1. Quantity distribution of correct items in recommendation results

As can be seen from Fig. 1, in UserCF and ItemCF's recommendation results, the numbers of users with 0 correct items are 566 and 977, while the numbers of users with 1 correct items are 987 and 1143. Due to the uneven distribution of the precision of each user's recommendation result, UserCF and ItemCF's average precision metrics are 0.328 and 0.2937, respectively. In the UserCF and ItemCF methods, some users' recommendation results have all correct items. Thus, the UserCF and ItemCF methods are helpful for users with all correct items, but the recommendation results of this method are still unsatisfactory for users with a low precision rate. Besides, due to the sparsity of user rating data, there is also a phenomenon of low coverage and diversity in the recommendation results for the UserCF and ItemCF methods [17] (i.e., The mining ability of long-tail items is not good). Although the existing recommendation methods combine machine learning, deep learning, and other methods, it is difficult for any recommendation method to make the recommended items for each user correct or the precision rate of the recommendation results evenly distributed. There are two main factors involved in a low individual precision rate in the recommendation results. On one hand, the data are sparse; On the other hand, the objective function of the methods is the design from the perspectives of the average precision rate and recall rate. The precision rate of individual users obtained in each recommendation method is different, so it is hard to improve the precision rate of individuals by using only one recommendation method. To solve the above problems, this paper proposes the addition of an auxiliary recommendation module to the recall module in the recommendation framework. The auxiliary recommendation module contains three methods to improve the recommendation results of users with a low precision rate.

This paper proposes a long-tail items recommendation framework focusing on low precision rate users. The main contributions are as follows:

- This paper analyzes the phenomenon of a low precision rate in the recommendation results.
- This paper proposes a long-tail recommendation framework focusing on low precision rate users.
- This paper adds an auxiliary recommendation module to the proposed framework. Using the auxiliary module, this paper designs three methods to improve the recommendation results of low precision rate users.
- Extensive experimental results show that the auxiliary recommendation module in the recommendation framework is effective.

2   Related Work

The paper proposed the long-tail recommendation framework focusing on the low precision rate users. The phenomenon of low precision rate users can also be regarded as a kind of cold-start problem. Thus, this section introduces the research progress of the cold-start and long-tail recommendation problem.

2.1   Cold-Start Problem

The cold-start problem [4] refers to low precision rate recommendations due to sparse data. In recent years, the main methods used in cold-start research were to supplement sparse data with third-party data [12,14] and to improve the similarity calculation method used between items [2]. Reference [12] proposes the use of social network data and user comment data to make up for the sparse problem in the existing data and implements recommendations based on the GCN (Graph Convolutional Network) model. The cold-start user in [12] refers to users with less interaction data between the user and the item. Reference [2] proposes to improve the cold-start problem by using the similarity calculation method based on an item's genre in the collaborative filtering method. Experimental results show that the proposed method is better than using the similarity calculation methods such as cosine similarity and the Jaccard coefficient in ItemCF. Reference [14] proposes a matrix decomposition method based on the LOD (Linked Open Data) similarity calculation. Movielens 20M and Netflix 20M datasets are respectively retrieved from the corresponding DBpedia knowledge base to retrieve the corresponding information on items in [14]. Experiments show that the LOD similarity calculation method proposed in [14] is superior to the singular value decomposition (SVD) and SVD++ methods. Although the above methods can alleviate the cold-start phenomenon, the problem of low precision rate users in the recommendation results is still not considered in these methods. In this paper, an auxiliary recommendation module is added to the recommendation framework to improve the recommendation results of users with a low precision rate.

2.2 Long-Tail Items Recommendation

The long-tail items recommendation problem [17] is also a method that can be used to solve the issue related to diversity in recommendation results. For example, reference [11] proposed an RNN (recurrent neural network)-based method to improve the diversity of sequence recommendations by using long-tail items. In recent research papers, long-tail recommendation methods are have mainly included graph-based methods [10,13,17] and multi-objective functions [8]. Reference [17] proposed a long-tail items recommendation method based on a bipartite graph. The proposed methods include a Hitting Time-based recommendation method, an Absorting-Time-based recommendation method, an information entropy-based recommendation method, and a latent dirichlet allocation (LDA)-based [3] recommendation method. Experiments showed that the LDA-based recommendation method is better than other methods in terms of the diversity metric. Based on reference [17], reference [10] proposed a tripartite graph and Markov-process based to implement a long-tail items recommendation. The experimental results show that the proposed method is superior to the Hitting Time recommendation method and the Absorbing Time recommendation method in reference [17]. Reference [13], based on references [10,17], proposed the use of an extended tripartite graph method to implement the long-tail items recommendation. The item genres were used to split into single genres of representation methods from composite genres. For example, if the genres of the movie A are *Action* and *Comedy*, then using composite genres, they are represented as {(*Action, Comedy*), A}, and using single genres, they are represented as {*Action, A*},{*Comedy, A*}. From the perspective of graph representation, when using composite genres presentation, there is one edge between the item's genres and the item, and now there are two edges. Reference [8] used the multi-objective function to realize the recommendation of long-tail items, thus increasing the diversity of the recommendation results. Additionally, reference [5] proposed the method of using a knowledge-graph to improve the diversity of the recommended methods at the International Conference on Research and Development in Information Retrieval (SIGIR) in 2020. The long-tail items recommendation problem is a double-edged sword. If the recommendation rate of long-tail items is greatly increased, the precision rate of the recommendation results will decrease. In this paper, a large number of non-repeated recommended items were added for users with low precision rate, which can not only improve the experience of users with a low precision rate but also improve the coverage rate of recommended items.

3   Recommendation Framework and Methods

This section will introduce the long-tail items recommendation framework and the functions of the modules in the framework.

3.1   Long-Tail Items Recommendation Framework

The proposed recommendation framework includes four modules: a data processing module, a recommendation method module, a recall module, and a recommendation list generation module. The structure of the recommendation framework is shown in Fig. 2. As can be seen from Fig. 2, the data processing module contains two tasks. One task is to count the rating times of each item in the dataset, and the other task is to calculate the correlation relationships between items. In the data processing module, all the results obtained from the above two tasks are saved to the file to reduce the time spent on subsequent recommendations. In the recommendation method module, the methods use the classic UserCF and ItemCF methods in the top-k recommendation. Since the efficiency of the auxiliary recommendation module is mainly verified in this paper, only the widely used methods are selected. In addition, the design of the recommendation method module is flexible and adds additional methods or models based on recommended requirements. The recall module contains two sub-modules: one module is used to recall users according to the specified precision rate, and the other module (auxiliary recommendation module) is used to improve the recommendation results of users with a low precision rate. The recommendation list generates module in three steps: 1) the recommendation results obtained in the recommendation method module merge with the recommendation results obtained in the recall module, 2) the results are arranged in descending order according to the rating number of the items (using the rating data file obtained in the data processing module), and 3) the results are returned to the target users. The following section focuses on the functions of the data processing module and the recall module and the methods involved in their use.

Data Processing Module. According to the framework given in Fig. 2, the data processing module obtains two files to save relevant data according to the input dataset.

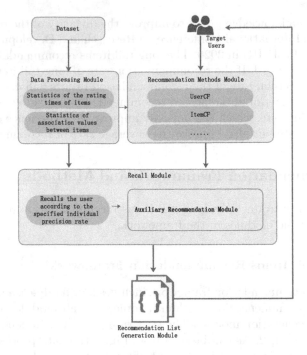

Fig. 2. Long-tail items recommendation framework

(1) Statistics of the rating times of items: the items and the rating times are expressed in the form of a tuple. For example, {*item*, *times*}, where *item* represents the item's ID and *times* represents the rating times of *item*.

(2) Statistics of association values between items: the association between every two items is expressed by triples, for example, {*A*, *B*, 2} means that after selecting item *A*, the number of times item *B* is selected is 2. this is equivalent to representing the associations between items as a weighted directed graph. Assuming there are five items A, B, C, D, E, the relationships among them are as shown in the figure. As can be seen from Fig. 3, the triplet of the relationship between these five items can be expressed as {*A*, *B*, 2}, {*B*, *A*, 4}, {*B*, *C*, 3}, {*C*, *B*, 1}, {*C*, *D*, 3}, {*D*, *C*, 4},{*D*, *E*, 5}, {*E*, *D*, 3}, {*E*, *A*, 1},{*A*, *E*, 4}. If the rating data of users and items are newly added, the associated values in the items can also be directly updated in the file.

Recall Module. The recall module is the core part of the recommendation framework. It consists of a recall method module and an auxiliary recommendation module.

(1) The recall method module: the results for the recommendation method module recall are determined by using the individual precision rate (see Eq. 1). For example, the individual precision is set to 0 and all users with a precision

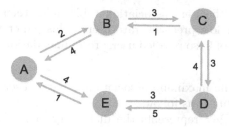

Fig. 3. Relationships among items

rate of 0 are recalled from the recommendation method module. Then, the recalled users pass into the auxiliary module.

$$Individual Precision = \frac{hit}{L * 1.0} \tag{1}$$

where hit denotes the correct number of items recommended in the recommendation list, and L represents the length of the recommendation list.

(2) The auxiliary recommendation module: this proposes three methods: (based on the minimum association value between items, based on the maximum association value between items, and the long-tail items-based diversified methods). The minimum association value between items refers to selecting the item with the smallest edge weight value among the two items, For example, for B and E associated with A in Fig. 3, the selected B is the minimum association value of the A. On the contrary, if E is selected, it is the maximum association value of the A. The maximum association value method contains a large number of items with more rating times, while the minimum association value method contains a large number of items with less rating times (i.e., long-tail items). Taking the method of using the maximum association value between items as an example, the following text introduces how to improve the recommendation results of low precision rate users. The improvement process includes the following three steps:

– Get the top N items with the highest rating among the rating items of users with low precision rate. Here, N is the same length as the recommendation list (L).
– According to the obtained N items, find the items with the largest or smallest association value for each item in the item association table. Get the item with the largest association value, as shown in Eq. 2.

$$x_i = \arg\max_i p(y|x_i) \tag{2}$$

where $p(y|x_i)$ represents the quantity of the item y is associated with x_i, and x_i is the obtained item. If the item with the smallest association value needs to be obtained, this can be done using as Eq. 3.

$$x_i = \arg\min_i p(y|x_i) \tag{3}$$

- According to the item association table, add items to the recommendation list for low precision rate users. This paper recalls users with a precision rate of 0, so recalled users' recommendation list will replace this completely.

Add the item with the maximum association value to users with low precision rate, and the algorithm is as follows.

LowerPrecisionDict represents the dictionary of low precision rate users obtained from the recall method module, and N represents the same number of items as the length of the recommendation list (L). *UserRatingArray* denotes the user's rating of items (i.e., The user's rating table of items in the dataset) and is represented as a triple, i.e., $\{user, item, rating\}$, where *user* represents the user's ID, *item* represents the item's ID, and *rating* represents the rating value. *ItemsAssociateArray* represents the item association value obtained from the data processing module and is represented as a triple. *ListDict* is the recommended list data returned by this method and is represented as a dictionary, i.e., $\{user:\{item_1, item_2,... item_n\}\}$, where *user* represents the user's ID and $item_n$ represents the item's ID.

Algorithm 1. Add the item with the maximum associated value to the users' recommendation list

---

Input: Variables: *LowerPrecisionDict, N, UserRatingArray, ItemsAssociateArray*
Output: *ListDict*
1: **for** *user* in *LowerPrecisionDict* **do**
2:    **if** *user* in *UserRatingArray* **then**
3:       Arrange the items rated by the *user* in descending order of rating value
4:       Add the *user* and their corresponding top N items to *tempdict*
5:    **end if**
6: **end for**
7: **for** *user* in *tempdict* **do**
8:    *items* = *tempdict*[*user*]
9:    **for** *item* in *items* **do**
10:       **if** *item* in *ItemsAssociateArray* **then**
11:          using Equation 2, obtain the *item* with the maximum association value among items.
12:          Add the *item* to *ListDict*[*user*]
13:       **end if**
14:    **end for**
15: **end for**
16: **return** *ListDict*

When adding an item to the recommendation list in line 12 of the Algorithm 1, it is also necessary to determine whether the item exists in the recommendation list, and if so, it is should use Eq. 2 continue to find the next associated item. Similar to Algorithm 1, the method of adding the minimum association value between items to users with low precision rate only replaces Eq. 2 to Eq. 3 in Line 11 of the Algorithm 1.

In the auxiliary recommendation module, the long-tail items-based diversity method is used to enable users with low precision rate to receive more diverse recommendations, rather than just being recommend related items. The long-tail items refer to items with fewer ratings times. The items with the same associated values are arranged in ascending order according to the rating times, and long-tail items are preferentially added to users' recommendation list. In addition, when adding items to the recommendation list, the algorithm judged whether the items are already in the recommendation list of all current low precision rate users. If so, the search continues for the next item, and so on. The long-tail items-based diversified method will recommend most of the items in the dataset. In the following research, we will provide personalized recommendation results for users in fine grains according to the characteristics of the items themselves.

4   Experiments

5   Experiment Environment and Dataset

Experimental environment: the operating system was Windows 10, the CPU was Intel Core i5-4690K@3.5 GHz, and the memory size was 16 GB. The development platform was VSCode and the programming language was Python 3.6.4.

Dataset: MovieLens 1M and the Movielens ml-latest-small dataset [9] were used in the experiment. Since only rating data were used in the experiment and other auxiliary information of items were not needed, only selected datasets in the movie field were used in the experiment. The Movielens 1M dataset contains rating data from 6040 users for 3883 movies, and each user has rating records for at least 20 movies. The Movielens ml-test-small dataset contains rating data from 610 users for 9724 movies, and each user has items rated for at least 20 movies.

5.1   Evaluation Metrics

This experiment used the Precision, Recall, Diversity, and Coverage metrics to evaluate the proposed framework. The equations used to determine Precision and Recall are shown in Eq. 4 and 5.

$$Precision = \frac{\sum_u |R(u) \cap T(u)|}{\sum_u |R(u)|} \tag{4}$$

$$Recall = \frac{\sum_u |R(u) \cap T(u)|}{\sum_u |T(u)|} \tag{5}$$

where $R(u)$ represents the set of items recommended for user u, and $T(u)$ represents the set of items rated by user u on the test set. The equations for the diversity and coverage metrics are shown in Eq. 6 and 7.

$$Diversity = \frac{|\bigcup_{u \in U} R(u)|}{|T|} \tag{6}$$

where T represents the total number of recommended items, and T is the product of the length of the user's recommended list and the number of users (including repeat items).

$$Coverage = \frac{|\bigcup_{u \in U} R(u)|}{|I|} \tag{7}$$

where I represents the total number of items in the dataset.

5.2 Results and Analysis

The comparison methods used in this experiment were as follows.

- UserCF: User-based collaborative filtering method [15]
- ItemCF: Item-based collaborative filtering method [16]
- UserCF+$Association_{max}$: Used to adopt the maximum association value methods to improve the UserCF method's recommendation result
- UserCF+$Association_{min}$: Used to adopt the minimum association value methods to improve the UserCF method's recommendation result
- UserCF+$Association_{max}$+Diversity: The UserCF+$Association_{max}$ combined with long-tail items-based diversity method
- UserCF+$Association_{min}$+Diversity: The UserCF+$Association_{min}$ combined with long-tail items-based diversity method
- ItemCF+$Association_{max}$: Used to adopt the maximum association value methods to improve the ItemCF method's recommendation result
- ItemCF+$Association_{min}$: Used to adopt the minimum association value methods to improve ItemCF method's recommendation result
- ItemCF+$Association_{max}$+Diversity: The ItemCF + $Association_{max}$ combined with long-tail items-based diversity method
- ItemCF+$Association_{min}$+Diversity: The ItemCF + $Association_{min}$ combined with long-tail items-based diversity method

Parameter Settings: The target users were all 6040 users (Movielens 1M) and 610 users (Movielens ml-latest-small) in the dataset; The ratio of the training set to the test set was 7:3. Improved users in the auxiliary recommendation module were users with a precision rate of 0 in the UserCF and ItemCF methods. The recommended list lengths (L) were set to 1 and 3. The number of nearest neighbors for the UserCF was 40; The number of nearest neighbors for the ItemCF was 50. The results shown in bold in the table represent the best values for each evaluation metric.

The experiment results for the Precision, Recall, Coverage, and Diversity metrics in the Movielens 1M dataset are shown in Table 1 and Table 2.

From Table 1 and Table 2, it can be seen that the experimental results for the UserCF and ItemCF methods improve after using the methods in the auxiliary

Table 1. Experimental results on Movielens 1M dataset (the recommend method module uses UserCF)

Method	L	Precision	Recall	Coverage	Diversity
UserCF	1	0.5267	0.0106	0.0966	0.0586
	3	0.4834	0.0292	0.1505	0.0304
UserCF+$Association_{max}$	1	**0.5563**	**0.0112**	0.0776	0.0470
	3	0.4883	0.0295	0.1507	0.0305
UserCF+$Association_{min}$	1	0.5440	0.0109	0.0830	0.0503
	3	**0.4906**	**0.0296**	0.1510	0.0305
UserCF+$Association_{max}$+Diversity	1	0.5392	0.0109	0.7711	0.4674
	3	0.4866	0.0294	0.9335	0.1890
UserCF+$Association_{min}$+Diversity	1	0.5336	0.0108	**0.7826**	**0.4743**
	3	0.4849	0.0293	**0.9476**	**0.1914**

Table 2. Experimental results on Movielens 1M dataset (the recommend method module uses ItemCF)

Method	L	Precision	Recall	Coverage	Diversity
ItemCF	1	0.4939	0.0099	0.0494	0.0299
	3	0.4515	0.0273	0.0803	0.0162
ItemCF+$Association_{max}$	1	**0.5214**	**0.0105**	0.0415	0.0252
	3	0.4574	0.0276	0.0803	0.0162
ItemCF+$Association_{min}$	1	0.5129	0.0103	0.0492	0.0298
	3	**0.4600**	**0.0278**	0.0828	0.0167
ItemCF+$Association_{max}$+Diversity	1	0.5053	0.0102	0.8274	0.5015
	3	0.4547	0.0275	0.9751	0.1970
ItemCF+$Association_{min}$+Diversity	1	0.4993	0.0101	**0.8350**	**0.5061**
	3	0.4530	0.0274	**0.9788**	**0.1961**

recommendation module, which shows that the auxiliary recommendation module is an effective part of in the recommendation framework. For the Precision and Recall metrics, UserCF + $Association_{max}$ and ItemCF + $Association_{max}$ have the best effects when the recommended list length is 1, while UserCF + $Association_{min}$ and ItemCF + $Association_{min}$ have the best effects when the recommended list length is 3. This shows that in this dataset, users with low precision rate (precision rate is 0), the majority of items selected in the test set with fewer rating times. For the Coverage and Diversity metrics, the UserCF + $Association_{min}$+ Diversity method has the best effect when the length of the recommendation list (L) is 1 or 3, and the coverage reached 0.9788.

The experimental results for the Movielens ml-test-small dataset are shown in Table 3 and Table 4.

Table 3. Experimental results for the Movielens ml-test-small dataset (the recommend method module uses UserCF)

Method	L	Precision	Recall	Coverage	Diversity
UserCF	1	0.4540	0.0091	0.0094	0.0132
	3	0.3995	0.0241	0.0148	0.0689
UserCF+$Association_{max}$	1	**0.4803**	**0.0097**	0.0099	0.1377
	3	**0.4093**	**0.0247**	0.0170	0.0790
UserCF+$Association_{min}$	1	0.4721	0.0095	0.0318	0.4426
	3	0.4038	0.0244	0.0519	0.2410
UserCF+$Association_{max}$+Diversity	1	0.4705	0.0095	0.0412	0.5738
	3	0.4093	0.0247	0.0747	0.3464
UserCF+$Association_{min}$+Diversity	1	0.4770	0.0096	**0.0413**	**0.5754**
	3	0.4027	0.0243	**0.0809**	**0.3754**

Table 4. Experimental results for the Movielens ml-test-small dataset (the recommend method module uses ItemCF)

Method	L	Precision	Recall	Coverage	Diversity
ItemCF	1	0.4016	0.0080	0.0143	0.1984
	3	0.3479	0.0211	0.0274	0.1984
ItemCF+$Association_{max}$	1	**0.4328**	**0.0087**	0.0090	0.1246
	3	**0.3579**	**0.0216**	0.0190	0.0880
ItemCF+$Association_{min}$	1	0.4148	0.0083	0.0325	0.4525
	3	0.3536	0.0213	0.0555	0.2574
ItemCF+$Association_{max}$+Diversity	1	0.4262	0.0086	0.0449	0.6246
	3	0.3667	0.0221	0.0835	0.3874
ItemCF+$Association_{min}$+Diversity	1	0.4148	0.0083	**0.0482**	**0.6705**
	3	0.3546	0.0214	**0.0893**	**0.4142**

From the experimental results in Table 3 and Table 4, it can be seen that similar to the test results for the MovieLens 1M dataset, the method using the auxiliary recommendation module is better than UserCF and ItemCF methods, which once again shows that the auxiliary recommendation module is an effective part of in the recommendation framework. For the Coverage and Diversity metrics, the UserCF + $Association_{min}$+ Diversity method is also the best when the recommended list length (L) is 1 or 3. However, for the Precision and Recall tests, UserCF+$Association_{max}$ method has the best effect when the recommended list length (L) is 1 or 3. This shows that in this dataset, for users with low precision rate (precision rate is 0), the majority of items have been rated more times in the test set.

6   Conclusion

To address the phenomenon of low precision rate in recommendation results, this paper proposes a long-tail items recommendation framework focusing on users with low precision rate. The proposed framework includes a data processing module, a recommendation method module, a recall module, and a recommendation list generation module. The recall module in this framework is the core of the whole framework. In the recall module, the corresponding users are recalled according to their individual precision rates, and then the recommendation results are improved for users with low precision rate through the auxiliary recommendation module. Experiments showed an obvious influence of the auxiliary recommendation module used in the framework on the recommendation results. Thus, the auxiliary recommendation module is effective. In follow-up work, we will continue to study the use of a third-party knowledge base to improve the personalized recommendation of users with low precision rate.

References

1. Abdi, M.H., Okeyo, G.O., Mwangi, R.W.: Matrix factorization techniques for context-aware collaborative filtering recommender systems: a survey. Comput. Inf. Sci. **11**(2), 1–10 (2018). https://doi.org/10.5539/cis.v11n2p1
2. Barman, S.D., Hasan, M., Roy, F.: A genre-based item-item collaborative filtering: facing the cold-start problem. In: Proceedings of the 8th International Conference on Software and Computer Applications, ICSCA 2019, Penang, Malaysia, 19–21 February 2019, pp. 258–262 (2019). https://doi.org/10.1145/3316615.3316732
3. Blei, D.M., Ng, A.Y., Jordan, M.I.: Latent dirichlet allocation. J. Mach. Learn. Res. **3**, 993–1022 (2003)
4. Camacho, L.A.G., Souza, S.N.A.: Social network data to alleviate cold-start in recommender system: a systematic review. Inf. Process. Manag. **54**(4), 529–544 (2018). https://doi.org/10.1016/j.ipm.2018.03.004
5. Gan, L., Nurbakova, D., Laporte, L., Calabretto, S.: Enhancing recommendation diversity using determinantal point processes on knowledge graphs. In: Proceedings of the 43rd International ACM SIGIR Conference on Research and Development in Information Retrieval, SIGIR 2020, Virtual Event, China, 25–30 July 2020, pp. 2001–2004 (2020). https://doi.org/10.1145/3397271.3401213
6. Goldberg, D., Nichols, D.A., Oki, B.M., Terry, D.B.: Using collaborative filtering to weave an information tapestry. Commun. ACM **35**(12), 61–70 (1992). https://doi.org/10.1145/138859.138867
7. Gunawardana, A., Shani, G.: A survey of accuracy evaluation metrics of recommendation tasks. J. Mach. Learn. Res. **10**, 2935–2962 (2009)
8. Hamedani, E.M., Kaedi, M.: Recommending the long tail items through personalized diversification. Knowl. Based Syst. **164**, 348–357 (2019). https://doi.org/10.1016/j.knosys.2018.11.004
9. Harper, F.M., Konstan, J.A.: The movielens datasets: history and context. ACM Trans. Interact. Intell. Syst. **5**(4), 19:1–19:19 (2016). https://doi.org/10.1145/2827872

10. Johnson, J., Ng, Y.: Enhancing long tail item recommendations using tripartite graphs and Markov process. In: Proceedings of the International Conference on Web Intelligence, Leipzig, Germany, 23–26 August 2017, pp. 761–768 (2017). https://doi.org/10.1145/3106426.3106439

11. Kim, Y., Kim, K., Park, C., Yu, H.: Sequential and diverse recommendation with long tail. In: Proceedings of the Twenty-Eighth International Joint Conference on Artificial Intelligence, IJCAI 2019, Macao, China, 10–16 August 2019, pp. 2740–2746 (2019). https://doi.org/10.24963/ijcai.2019/380

12. Liu, S., Ounis, I., Macdonald, C., Meng, Z.: A heterogeneous graph neural model for cold-start recommendation. In: Proceedings of the 43rd International ACM SIGIR Conference on Research and Development in Information Retrieval, SIGIR 2020, Virtual Event, China, 25–30 July 2020, pp. 2029–2032 (2020). https://doi.org/10.1145/3397271.3401252

13. Luke, A., Johnson, J., Ng, Y.: Recommending long-tail items using extended tripartite graphs. In: 2018 IEEE International Conference on Big Knowledge, ICBK 2018, Singapore, 17–18 November 2018, pp. 123–130 (2018). https://doi.org/10.1109/ICBK.2018.00024

14. Natarajan, S., Vairavasundaram, S., Natarajan, S., Gandomi, A.H.: Resolving data sparsity and cold start problem in collaborative filtering recommender system using linked open data. Expert Syst. Appl. **149** (2020). https://doi.org/10.1016/j.eswa.2020.113248

15. Resnick, P., Iacovou, N., Suchak, M., Bergstrom, P., Riedl, J.: GroupLens: an open architecture for collaborative filtering of netnews. In: Proceedings of the Conference on Computer Supported Cooperative Work, CSCW 1994, Chapel Hill, NC, USA, 22–26 October 1994, pp. 175–186 (1994). https://doi.org/10.1145/192844.192905

16. Sarwar, B., Karypis, G., Konstan, J., Riedl, J.: Item-based collaborative filtering recommendation algorithms. In: Proceedings of the 10th International Conference on World Wide Web, WWW 2001, pp. 285–295. ACM, New York (2001). https://doi.org/10.1145/371920.372071

17. Yin, H., Cui, B., Li, J., Yao, J., Chen, C.: Challenging the long tail recommendation. Proc. VLDB Endow. **5**(9), 896–907 (2012). https://doi.org/10.14778/2311906.2311916

Learning from Crowd Labeling with Semi-crowdsourced Deep Generative Models

Xuan Wei[1], Mingyue Zhang[2(✉)], and Daniel Dajun Zeng[3,4]

[1] Antai College of Economics and Management, Shanghai Jiao Tong University, Shanghai, China
[2] School of Business and Management, Shanghai International Studies University, Shanghai, China
zhangmy@shisu.edu.cn
[3] State Key Laboratory of Management and Control for Complex Systems, Institute of Automation, Chinese Academy of Sciences, Beijing, China
[4] Shenzhen Artificial Intelligence and Data Science Institute (Longhua), Shenzhen, China

Abstract. Microtask crowdsourcing has become an appealing approach to collecting large-scale high-quality labeled data across a wide range of domains. As the crowd workers may be unreliable, the most fundamental question is how to aggregate the noisy annotations provided by these potentially unreliable workers. Although various factors such as worker reliability and task features are considered in the literature, they are not meaningfully combined in a unified and consistent framework. In this work, we propose a semi-crowdsourced deep generative approach called S-DARFC which combines Bayesian graphical models and deep learning techniques and unifies factors including the worker reliability, task features, task clustering structure, and semi-crowdsourcing. Graphical models are good at finding a structure that is interpretable and generalizes to new tasks easily and deep learning techniques are able to learn a flexible representation of complex high-dimensional unstructured data (e.g., task features). Extensive experiments based on six real-world tasks including text and image classification demonstrate the effectiveness of our proposed approach.

Keywords: Crowdsourcing · Learning from crowds · Deep generative models

1 Introduction

Microtask crowdsourcing has emerged as a cost-effective approach to harnessing the collective efforts of large crowds to complete huge amounts of dividable microtasks in parallel. These microtasks, such as image tagging and sentence translation, are often too repetitive and time-consuming for individuals to complete [9]; they are also challenging for computers because they often rely on

© Springer Nature Singapore Pte Ltd. 2021
Y. Sun et al. (Eds.): ChineseCSCW 2020, CCIS 1330, pp. 101–114, 2021.
https://doi.org/10.1007/978-981-16-2540-4_8

humans' cognitive ability [26]. The unique features of such microtasks make crowdsourcing a particularly suitable and effective solution. In general, large amounts of microtasks are posted by task requesters on some web-based crowd-sourcing platforms (e.g., Amazon Mechanical Turk); crowd workers complete the tasks voluntarily or for a small amount of payment; finally, the results are aggregated and delivered back to the task requesters. The immediate, cheap, and elastic workforce supply in such microtask crowdsourcing systems makes it possible to complete a huge amount of dividable microtasks at low cost and with high throughput [26].

Despite the promise, significant challenges remain when combining crowd efforts. The difficulty mainly stems from the potential low reliability of crowd workers. Actually, they may be unreliably for various reasons. They usually have different levels of competence and dedication [12]. Because the microtasks are often tedious and repetitive, unintentional mistakes are also inevitable [11]. Some malicious workers even disobey the task rules provided by the requesters [8]. Hence, a fundamental challenge is how to aggregate the annotations provided by a set of potentially unreliable crowd workers (i.e., learning from crowds). When simple aggregation strategies such as Majority Voting (MV) are used, the obtained performance may be insufficient to meet the actual business needs.

Previous studies have proposed various models to address the above fundamental challenge. They can be divided into two major categories based on whether the task features are taken into account. In the first category, the proposed models only exploit noisy labels provided by the crowd workers [29]. Most algorithms consider workers' reliability and adopt a probabilistic framework because the relationships among the ground truth labels, noisy labels, and the worker reliability can be described by a probabilistic procedure [13,16]. With the flexibility of probabilistic models, various other factors besides worker reliability are incorporated into the models to further improve the performance, including task difficulty [13,26] and grouping of workers [16,22]. In the second category, the truth inference algorithm also learns a predictive classifier that predicts the ground truth labels based on the task features and in turn helps estimate the ground truth labels. Intuitively, the ground truth labels should be related to the task features, and hence considering the task features will be helpful in the aggregation. In the early work along this research stream, two representative examples for the classifier are logistic regression [1,18] and topic models [19,23]. Inspired by the success of deep neural networks in learning complicated non-linear transformation across various applications [30], recent research has started using deep neural networks to model the relationship between task features and ground truth labels [7,21].

Although various factors such as worker reliability and task features are considered in the literature, they are not meaningfully combined in a unified and consistent framework. First, existing literature often considers only a subset of factors. For example, as previously mentioned, many existing approaches ignore the potential usefulness of task features [13,16]. Second, various factors are often combined in a heuristic manner. Although the overall framework is often probabilistic, most studies lack a probabilistic interpretation when they incorporate task features into the model [7,29]. In this work, we propose a semi-crowdsourced deep generative approach called S-DARFC which unifies factors including the worker reliability, task features, task clustering structure, and semi-crowdsourcing. Our approach is a probabilistic framework which combines Bayesian graphical models and deep learning techniques. Graphical models are good at finding a structure that is interpretable and generalizes to new tasks easily and deep learning techniques are able to learn a flexible representation of complex high-dimensional unstructured data (e.g., task features). Our approach is compared with the state-of-the-art benchmark models on six real-world binary tasks including text and image classification. The results demonstrate the effectiveness of our proposed approach.

2   Method

2.1   Problem Formulation

We first formulate our research scenario. We consider a typical scenario in crowd labeling where task requesters utilize the online crowdsourcing markets (e.g., AMT, CrowdFlower) to collect the ground truth labels of a set of tasks. Like previous literature, we focus on the most frequently used and representative binary labeling task because such tasks are often the building blocks of many other complex tasks [26]. In Sect. 2.3, we show that our approach can be easily extended to the multi-class setting.

Suppose the task requesters want to collect labels for N instances in which the feature of the i-th instance is represented as a vector $\mathbf{o}_i \in \mathbb{R}^d$. Limited by the budget, each task is usually assigned to only several workers (typically three to five) on the crowdsourcing platform. Assume there are J workers in total. Let $y_{i,j} \in \{0, 1, null\}$ represent the annotations of the i-th instance provided by the j-th worker. Here $y_{i,j} = null$ means the i-th instance is not annotated by the j-th worker. Given the above input, the goal of this annotation aggregation problem is to infer the ground truth of each instance, denoted by $\mathbf{z} \in \{0, 1\}^N$.

2.2   Motivating Example

We use an illustrative example to motivate the technical insights behind our proposed approach. Let's assume we want to label six tweets based on whether the authors have the intention to purchase an iPhone, as shown in Fig. 1. The first three tweets are associated with positive labels and the remaining with negative labels. Seven workers with varying reliability annotate the tweets: two reliable workers who provide labels with high accuracy (100.0% and 83.3% respectively), three normal workers whose annotations are mostly correct (accuracy 66.7%), and two random workers with the accuracy being 50.0% and 33.3% respectively. If we use the simple heuristic method Majority Voting (MV), i.e., the label is positive only if at least half of the received annotations are positive, the accuracy

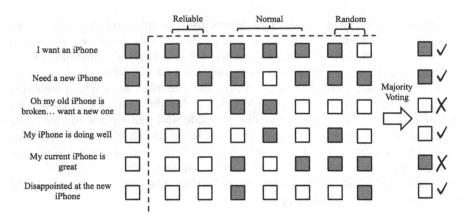

Fig. 1. An illustrative example

of the aggregated results is 66.7%, which will be far from being satisfactory in real-world business applications.

Let's consider a counterfactual scenario where we know the worker's reliability in advance. We can use only the annotations from the reliable workers or simply ignore the random workers; then, applying the majority voting algorithm will generate the correct ground truth. This counterfactual analysis indicates that knowing worker reliability is potentially helpful in ground truth estimation. In the literature, worker reliability has been intensively used to improve the performance of annotation aggregation. Intuitively, it makes sense to estimate each worker's reliability and assign fewer weights to annotations provided by less reliable workers [26]. Hence, we propose to consider worker reliability in this work.

Some recent studies start considering incorporating the task features into the algorithm [7,23,29]. The intuition is that the ground truth labels are associated with the task features and building a predictive classifier can offer valuable information. This idea can be demonstrated by the illustrative example in Fig. 1. Using MV, we are able to recover the ground truth of the first two sentences with high confidence as they both receive 6 positive labels out of 7. Although the estimated label for the third sentence is wrong, we are not confident with this estimation because the number of received negative labels (i.e., 4) is close to the decision boundary (i.e., 3.5). We notice that the first three sentences are all positive and share a similar sentence pattern, i.e., want or need an iPhone. This motivates us to train a predictive classifier which captures this pattern. Given that there are already 3 out of 7 workers providing positive labels, we are likely to correct the results of MV and regard the third sentence as positive. Hence, in this research, we propose to consider task features.

Another way to utilize the task features is clustering instances into groups based on some intrinsic characteristics. Although each task is different, they might be similar to each other if represented in a latent space and form groups.

The instances of the same group are supposed to share the same label [29]. For example, in Fig. 1, although theses sentences are syntactically different, they can be semantically clustered into several clusters. The first three sentences form a group as they all directly express the author's intention to purchase a new iPhone. Assuming such clustering information is available, applying the simple MV algorithm to this cluster will yield a positive prediction. Similarly, the fourth and fifth sentences fall into the same group as they all indicate the authors currently own an iPhone which is doing well, and MV will predict this cluster as negative. As a result, we correctly recover the ground truth labels of all instances even if we apply the simple MV strategy. Hence, we propose to use the task clustering structure in this work.

Both considering task features and considering task clustering heavily rely on learning a good task representation. Considering task features requires us to build a predictive classifier, in which case a good task representation will improve the predictive performance. Task clustering assumes that tasks of the same group are similar in the latent representation space. Motivated by the success of semi-supervised learning [24], we propose to use an extra set of unannotated instances to facilitate representation learning. Following the previous literature on semi-crowdsourced clustering [15,28], we similarly name such a technique as semi-crowdsourced learning and use it in the reported work.

2.3 A Semi-crowdsourced Deep Generative Model: S-DARFC

Combining the aforementioned factors (i.e., worker reliability, task features, task clustering structure, and semi-crowdsourcing) requires our model to handle structured relationships and unstructured data simultaneously. The relationships among the ground truth labels, noisy labels, and the worker reliability are structured; the task features are unstructured data when text and images are involved. To combine these factors in a unified and consistent framework, we propose a deep generative model. Our approach combines Bayesian graphical models and deep learning techniques where graphical models allow us to model structured relationships and the deep leaning techniques allow us to depict the complex non-linear relationships between unstructured task features and the ground truth labels.

We first introduce the model considering all factors except semi-crowdsourcing. The proposed model is called Deep Aggregation model with worker Reliability, task Features, and task Clustering (DARFC). This generative model is composed of two components: the annotation component handles the labeling behaviors of the workers and describes how the observed annotations (i.e., $\mathbf{y}$) are generated; the feature component defines the generative process of the instance features (i.e., $\mathbf{o}$). In the following, we take instance i as an example and introduce the details of each component.

To exploit the potential clustering structure, the DARFC model groups instances into clusters such that instances of the same group share the instance property (i.e., group truth labels and latent representations). The observed annotations depend on the ground truth labels and the observed instance features depend on the latent representations. Hence, in both components, the generation of the observed data (i.e., $\mathbf{o}_i \in \mathbb{R}^D$ or $\{0,1\}^D$ and $\mathbf{y}_i \in \{0,1,null\}^J$ is conditioned on the underlying cluster assignment c_i of instance i. Specifically, let K denote the number of clusters. For each instance, we have a corresponding latent cluster assignment $c_i \in \{1,2,\ldots,K\}$. It indicates which latent cluster the instance i comes from and is drawn from a categorical distribution over $\{1,2,\ldots,K\}$ with mixing coefficient $\boldsymbol{\omega}$. Formally, the conditional distribution of c_i follows

$$c_i \sim \text{Categorical}(\boldsymbol{\omega}), \tag{1}$$

where $\boldsymbol{\omega}$ satisfies $\boldsymbol{\omega} \in [0,1]^K$ and $\sum_{k=1}^{K} \omega_k = 1$.

In the annotation component, conditioned on c_i, we first draw ground truth label z_i from the corresponding Bernoulli distribution:

$$z_i \sim \text{Bernoulli}\left(\lambda_{c_i}\right). \tag{2}$$

As previously mentioned, in a crowdsourcing scenario, each worker may be unreliable for various reasons. We adopt the widely-used two-coin assumption for modeling each worker's labeling behavior [13,18]. Specifically, worker j has a two-by-two confusion matrix $\mathbf{M}_j$ with sensitivity $\alpha_j = p(y_{i,j} = 1 \mid z_i = 1)$ and specificity $\beta_j = p(y_{i,j} = 0 \mid z_i = 0)$. When the ground truth label is one, worker j provides a correct annotation with probability α_j; otherwise, worker j annotates correctly with specificity β_j. Formally,

$$y_{i,j} \sim \text{Bernoulli}\left(\alpha_j\right)^{z_i} \text{Bernoulli}\left(1-\beta_j\right)^{1-z_i}. \tag{3}$$

To explicitly write down the likelihood of the annotations $\mathbf{y}$, we define $I_{i,j} = \mathbb{1}(y_{i,j} \neq null)$. Hence,

$$p(\mathbf{y} \mid \mathbf{z}, \boldsymbol{\alpha}, \boldsymbol{\beta}) = \prod_{i=1}^{N} \prod_{j=1}^{J} \left(\text{Bernoulli}\left(\alpha_j\right)^{z_i} \text{Bernoulli}\left(1-\beta_j\right)^{1-z_i}\right)^{I_{i,j}} \tag{4}$$

The second component (i.e., feature component) is a combination of variational autoencoder (VAE) and Gaussian mixture [2,14]. For each data point $\mathbf{o}_i$, we have a corresponding latent variable $\mathbf{x}_i$, which is sampled from a mixture of multivariate Gaussian distribution. Specifically, depending on the cluster assignment c_i, the latent representation $\mathbf{x}_i$ follows

$$\mathbf{x}_i \sim \mathcal{N}\left(\boldsymbol{\mu}_{c_i}, \boldsymbol{\Sigma}_{c_i}\right). \tag{5}$$

The decoding term $p(\mathbf{o}_i \mid \mathbf{x}_i, \gamma)$ follows a multivariate Gaussian distribution (in case of real-valued data) or multivariate Bernoulli distribution (in case of binary data) whose parameters are calculated from flexible neural networks parameterized by γ. For example, when the data is real-valued (i.e., $\mathbf{o}_i \in \mathbb{R}^D$), $\mathbf{o}_i$ is generated as follows:

$$\mathbf{o}_i \sim \mathcal{N}\left(\boldsymbol{\mu}_\gamma\left(\mathbf{x}_i\right), \text{diag}\left(\sigma_\gamma^2\left(\mathbf{x}_i\right)\right)\right). \tag{6}$$

When the data is binary (i.e., $\mathbf{o}_i \in \mathbb{R}^D$), $\mathbf{o}_i \sim \text{Bernoulli}\left(\boldsymbol{\mu}_\gamma\left(\mathbf{x}_i\right)\right)$.

Finally, following previous probabilistic Bayesian modeling literature [4,27], we place conjugate Beta, Dirichlet, and Normal-Inverse-Wishart (NIW) priors over Bernoulli, Categorical, and multivariate Gaussian distributions respectively. The graphical model representation of DARFC is presented in Fig. 2. The priors are not listed for clarity.

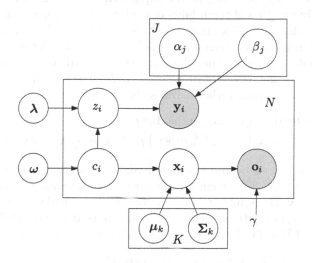

Fig. 2. Graphical representation of the DARFC model

The DARFC model can be extended by introducing an extra set of instances without annotations (denoted as $\mathbf{o}_i^{(s)}, i \in \{1, 2, \ldots, N^{(s)}\}$) to facilitate a better hidden representation learning. The extended model is called Semi-crowdsourced DARFC (i.e., S-DARFC). The generative process of $\mathbf{o}_i^{(s)}$ is similar to the one of $\mathbf{o}_i$:

$$c_i^{(s)} \sim \text{Categorical}(\boldsymbol{\omega}) \tag{7}$$

$$\mathbf{x}_i^{(s)} \sim \mathcal{N}\left(\boldsymbol{\mu}_{c_i^{(s)}}, \boldsymbol{\Sigma}_{c_i^{(s)}}\right) \tag{8}$$

$$\mathbf{o}_i^{(s)} \sim \mathcal{N}\left(\boldsymbol{\mu}_\gamma\left(\mathbf{x}_i^{(s)}\right), \text{diag}\left(\sigma_\gamma^2\left(\mathbf{x}_i^{(s)}\right)\right)\right) \tag{9}$$

Given the flexibility of our Bayesian probabilistic approach, it is straightforward to extend previous models to multi-class settings. In this case, the ground truth labels z_i may take one of multiple potential values. We only need to replace all probabilistic distributions related to z_i with their multivariate variants. Specifically, if involved in the generative process, Bernoulli and Beta distributions should be replaced with Categorical and Dirichlet distributions respectively. Take the S-ARFC model as an example, z_i follows a Categorical distribution; λ_k follows a Dirichlet distribution. The two-by-two confusion matrix is also extended to its multivariate version where the k-th row of the matrix measures the probability of providing each label when the ground truth label is k [13]. Each row follows a Dirichlet distribution and the annotations y is generated based on a Categorical distribution.

2.4 Model Inference

The goal of the model inference is to compute the posteriors of the latent variables, primarily the ground truth label of each instance (i.e.,$\mathbf{z}$), conditioning on the observable data $\mathbf{y}$ and $\mathbf{x}$. However, this conditional density is intractable for exact inference [25]. We take a mean-field variational inference approach for approximate inference [3]. In the following, we use DARFC an example.

Following the same set of notations in Sect. 2.3, the overall joint distribution of all hidden and observed variables takes the form:

$$p(\mathbf{y}, \mathbf{o}, \mathbf{c}, \mathbf{x}, \mathbf{z}, \boldsymbol{\theta}; \gamma) = p(\boldsymbol{\lambda})p(\boldsymbol{\omega})p(\boldsymbol{\alpha})p(\boldsymbol{\beta})p(\boldsymbol{\mu}, \boldsymbol{\Sigma})$$
$$\cdot p(\mathbf{c} \mid \boldsymbol{\omega})p(\mathbf{z} \mid \boldsymbol{\lambda}, \mathbf{c})p(\mathbf{x} \mid \boldsymbol{\mu}, \boldsymbol{\Sigma}, \mathbf{c})p(\mathbf{o} \mid \mathbf{x}; \gamma)p(\mathbf{y} \mid \mathbf{z}, \boldsymbol{\alpha}, \boldsymbol{\beta}). \tag{10}$$

For convenience, we denote the global variables as $\boldsymbol{\Theta} = \{\boldsymbol{\lambda}, \boldsymbol{\omega}, \boldsymbol{\mu}, \boldsymbol{\Sigma}, \boldsymbol{\alpha}, \boldsymbol{\beta}\}$. Our learning objective is to maximize the marginal likelihood of observed data $\mathbf{o}$ and annotations $\mathbf{y}$. By assuming a fully factorized mean-field variational family $\mathcal{Q}$ over the latent variables $\{\mathbf{c}, \mathbf{x}, \mathbf{z}, \boldsymbol{\Theta}\}$, this objective is lower-bounded by evidence lower bound (ELBO) [3]. In our case, the ELBO can be written as:

$$\log p(\mathbf{o}, \mathbf{y}) \geq \mathcal{L}[q(\mathbf{c})q(\mathbf{z})q(\mathbf{x})q(\boldsymbol{\Theta}); \gamma]$$
$$\triangleq \mathbb{E}_{q(\mathbf{c})q(\mathbf{z})q(\mathbf{x})q(\theta)} \left[\log \frac{p(\mathbf{y}, \mathbf{o}, \mathbf{c}, \mathbf{x}, \mathbf{z}, \boldsymbol{\theta}; \gamma)}{q(\mathbf{c})q(\mathbf{z})q(\mathbf{x})q(\boldsymbol{\theta})} \right] \tag{11}$$

We are unable to directly use the variational message passing because of the non-conjugate $p(\mathbf{o} \mid \mathbf{x}; \gamma)$ term. Inspired by the structured variational autoencoders [10], we introduce a recognition network $r(\mathbf{o}_i; \phi)$ to enable fast amortized inference and allow the inference algorithm to leverage conjugacy structure. Specifically, we first assume all mean-field distributions are in the same exponential family and denote their corresponding natural parameters as $\eta_\theta, \theta \in \{\mathbf{c}, \mathbf{z}, \mathbf{x}\} \cup \boldsymbol{\Theta}$. With such an assumption, the ELBO can be written as:

$$\mathcal{L}[q(\mathbf{c})q(\mathbf{z})q(\mathbf{x})q(\boldsymbol{\Theta}); \gamma] \triangleq \mathcal{L}\left[\boldsymbol{\eta}_\mathbf{c}, \boldsymbol{\eta}_\mathbf{z}, \boldsymbol{\eta}_\mathbf{x}, \boldsymbol{\eta}_\theta; \gamma\right]. \tag{12}$$

To enable a conjugate structure, we replace the non-conjugate likelihood $p(\mathbf{o} \mid \mathbf{x}; \gamma)$ in the original ELBO with a conjugate term $\psi(\mathbf{x} \mid \mathbf{o}; \phi)$, which is defined as

$$\psi(\mathbf{x}_i; \mathbf{o}_i, \phi) = \langle r(\mathbf{o}_i; \phi), \mathbf{t}(\mathbf{x}_i) \rangle. \tag{13}$$

In the above formula, the recognition network $r(\mathbf{o}_i; \phi)$ is some parameterized class of functions and $\mathbf{t}(\mathbf{x}_i)$ is the sufficient statistic of $\mathbf{x}_i$. Hence, we obtain the following surrogate objective

$$\tilde{\mathcal{L}}\left[\boldsymbol{\eta}_{\mathbf{c}}, \boldsymbol{\eta}_{\mathbf{z}}, \boldsymbol{\eta}_{\mathbf{x}}, \boldsymbol{\eta}_{\theta}; \phi\right]$$

$$\triangleq \mathbb{E}_{q(\mathbf{c})q(\mathbf{z})q(\mathbf{x})q(\theta)}\left[\log \frac{p(\mathbf{c}, \mathbf{x}, \mathbf{z}, \boldsymbol{\Theta})\exp\{\psi(\mathbf{x}; \mathbf{o}, \phi)\}p(\mathbf{y} \mid \mathbf{z}, \boldsymbol{\alpha}, \boldsymbol{\beta})}{q(\mathbf{c})q(\mathbf{z})q(\mathbf{x})q(\boldsymbol{\Theta})}\right]. \quad (14)$$

Note that this surrogate objective has a conjugate form as the variational inference objective for a conjugate observation model, we can derive the local and global optimizers for each variational parameter (i.e., $\boldsymbol{\eta}_{\mathbf{c}}, \boldsymbol{\eta}_{\mathbf{z}}, \boldsymbol{\eta}_{\mathbf{x}}$, and $\boldsymbol{\eta}_{\Theta}$) in closed form. By using a block coordinate ascent algorithm which updates variational parameters iteratively, we can find a local partial optimizer $\left(\boldsymbol{\eta}_{\mathbf{c}_i}^*(\phi), \boldsymbol{\eta}_{\mathbf{x}_i}^*(\phi), \boldsymbol{\eta}_{\mathbf{z}_i}^*(\phi), \boldsymbol{\eta}_{\Theta}^*(\phi)\right)$ of $\tilde{\mathcal{L}}$ w.r.t. $(\boldsymbol{\eta}_{\mathbf{c}}, \boldsymbol{\eta}_{\mathbf{z}}, \boldsymbol{\eta}_{\mathbf{x}}, \boldsymbol{\eta}_{\Theta})$ given fixed ϕ. By plugging them back into $\mathcal{L}\left[\boldsymbol{\eta}_{\mathbf{c}}, \boldsymbol{\eta}_{\mathbf{z}}, \boldsymbol{\eta}_{\mathbf{x}}, \boldsymbol{\eta}_{\Theta}; \gamma\right]$, the final object $\mathcal{J}(\gamma, \phi)$ is

$$\mathcal{J}(\gamma, \phi) \triangleq \mathcal{L}\left[\boldsymbol{\eta}_{\mathbf{c}_i}^*(\phi), \boldsymbol{\eta}_{\mathbf{x}_i}^*(\phi), \boldsymbol{\eta}_{\mathbf{z}_i}^*(\phi), \boldsymbol{\eta}_{\theta}^*(\phi); \gamma\right], \quad (15)$$

which is a lower bound of the optimized variational inference objective [10], i.e., $\max_{\boldsymbol{\eta}_{\mathbf{c}}, \boldsymbol{\eta}_{\mathbf{z}}, \boldsymbol{\eta}_{\mathbf{x}}, \boldsymbol{\eta}_{\theta}} \mathcal{L}\left[\boldsymbol{\eta}_{\mathbf{c}}, \boldsymbol{\eta}_{\mathbf{z}}, \boldsymbol{\eta}_{\mathbf{x}}, \boldsymbol{\eta}_{\Theta}; \gamma\right]$. Hence, by maximizing $\mathcal{J}(\gamma, \phi)$ using gradient-based optimization, we are maximizing a lower bound on the model log evidence $\log p(\mathbf{o}, \mathbf{y})$. Following previous literature, the gradients of the objective $\mathcal{J}(\gamma, \phi)$ w.r.t. γ and ϕ can be obtained by using the reparameterization trick [14].

3 Empirical Evaluations

3.1 Experimental Design

We compare our proposed model S-DARFC with the state-of-the-art benchmark models on six real-world binary tasks, including text and image classification. The data statistics are shown in Table 1 where Ann., GT, Pos., and Neg. means annotations, ground truth, positive, and negative respectively. Label cardinality refers to the average number of received annotations per instance.

Table 1. Data statistics

Task	Dataset	# Instances (# with Ann.)	# Ann.	Label cardinality	# Workers	# Instances with GT	# Pos./# Neg.
Text	RT	10,662 (4,999)	27,746	5.55	203	10,662	1
	FL	10,801 (3,487)	20,335	5.83	102	903	0.36
	iPhone	7,531 (4,333)	21,889	5.05	133	1,372	0.77
Image	Bill	6,033 (3,016)	15,080	5	356	6,033	0.80
	Shape	6,033 (3,016)	15,080	5	360	6,033	0.80
	Throat	6,033 (3,016)	15,080	5	316	6,033	0.80

Rotten Tomatoes (RT) contains movie reviews from the Rotten Tomatoes website and was labeled for sentiment polarity classification [17]. 27,746 crowd annotations from 203 AMT workers were collected for 4,999 reviews [20]. FL

(FOREO Luna) and iPhone are two datasets collected from Twitter for product purchase intention mining. The products for these two datasets are FOREO Luna facial cleanser and iPhone 7 respectively. Each tweet is labeled based on whether the author of the tweet shows the intention to purchase a specific product [6]. To obtain a set of tweets with ground truth for evaluating the methods, two of the authors dedicatedly label a small set of randomly sampled tweets. The Cohen's Kappa scores for FL and iPhone are 0.941 and 0.926 respectively, suggesting very excellent inter-annotator reliability. The annotations for these two datasets are collected from the Amazon Mechanical Turk. For all sentences in the text classification task, we pre-train its textual feature by applying Google's Bidirectional Encoder Representations from Transformers (BERT) and obtain a vector of size 768 [5].

The image classification datasets are extracted from a larger public dataset Caltech-UCSD Birds 200, which contains several tasks to label binary local characteristics for 6,033 bird images [29]. We use three of them in our experiment: Bill, Shape, and Throat. The labels are whether the bill shape is all-purse or not, whether the shape is perching-like or not, and whether the throat is black or not respectively. To use these datasets for evaluating the semi-crowdsourcing, we keep annotations for only half of the images (i.e., $3,016$) and treat the remaining $3,017$ images as instances without annotations. Last, for each image, we use 287 local attributes as object features [29].

For comparison with our proposed models, we select several state-of-the-art benchmark models from recent literature. Majority Voting (MV) is the most frequently used benchmark, which predicts the label as positive if at least half of the received annotations are positive [26]. As a representative model using the probabilistic approach, iBCC models each worker's reliability with a confusion matrix and estimates the model parameters with the Expectation-Maximization (EM) algorithm [13]. Recent literature starts considering task features to further improve the performance of annotation aggregation. BCCwords models the generation of sentences using a topic model [23]. However, this only works when the task is in the form of sentences. CrowdDeepAE (CDAE) [7] and Deep Clustering Aggregation Model (DCAM) [29] are two deep generative models that utilize deep learning to model the complex non-linear relationship between instance features and task labels. Compared with CDAE, DCAM also considers the clustering structure among tasks. Our method differs from these benchmark models in that it simultaneously considers all factors including worker reliability, task features, task clustering, and semi-crowdsourcing design, whereas the benchmark models only incorporate some of the factors. In addition, S-DARFC is a consistent probabilistic framework where the relationship between variables has a probabilistic interpretation.

As our dataset are slightly skewed (see Table 1), we choose ROC AUC and F1 score as the major metrics. We impose weak priors over all Bayesian priors and set the number of cluster (i.e., K) as 200. The dimension of the latent data representation (i.e., d) is set as 40. The encoder and the decoder are instantiated using a fully connected structure with each layer followed by a batch normalization layer. To examine the robustness of each model to label cardinality and make our approach practically applicable, we randomly sample annotations with the sampling rate varying from 0.6 to 1. Last, all methods were evaluated for 10 runs with a different randomization seed in each run.

3.2 Experimental Results

For all methods, we vary the annotation sample rate from 0.6 to 1 and use line charts to compare the performance across metrics ROC AUC and F1 score. The results are reported in Fig. 3 and Fig. 4 respectively. In text classification tasks, S-DARFC performs comparably to (when evaluated by ROC AUC) or slightly better than (when evaluated by F1 score) DCAM and BCCwords. They are followed by CDAE, iBCC, and MV. In image classification tasks, our proposed model S-DARFC consistently outperforms all other benchmark models with significant margin. Overall, the performance of all methods is in the following order: S-DARFC > DCAM > CDAE > iBCC > MV[1]. The performance gap between S-DARFC and DCAM is mainly due to the consideration of semi-crowdsourced learning. DCAM performs better than CDAE because it also takes into account the task clustering structure. CDAE performs better than iBCC because iBCC completely ignore the task features. The performance gap between iBCC and MV is mainly attributed to modeling of worker reliability. Note that, when comparing these models, the underlying factors considered are one of the major reasons; the performance gap may also stems from the difference of the adopted framework.

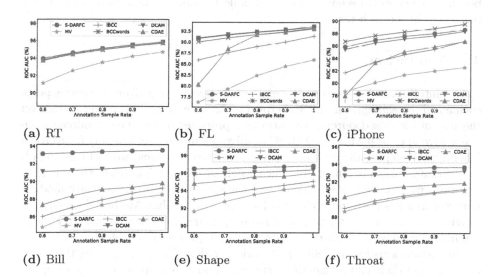

(a) RT   (b) FL   (c) iPhone

(d) Bill   (e) Shape   (f) Throat

Fig. 3. Comparison with benchmark models (ROC AUC as metric)

---

[1] BCCwords is not included because it only works in text classification datasets.

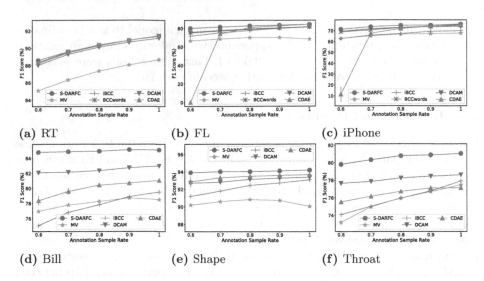

Fig. 4. Comparison with benchmark models (F1 score as metric)

In addition, when less annotations are used, the performance of all methods decreases as expected. However, the performance gap between our model S-DARFC and the benchmark models tends to increase as each instance receives fewer annotations. This indicates our method is less vulnerable to the number of received annotations. This makes our approach more practically applicable because in real-world scenarios, we only collect a limited amount of annotations (e.g., 3 to 5) for each task.

4   Conclusion

Crowd intelligence has been widely explored in recent years from various aspects such as effort aggregation [26] and crowd motivation [31]. This paper focuses on the microtask crowdsourcing scenario and proposes a semi-crowdsourced deep generative approach called S-DARFC to solve the problem of learning from crowd labeling. We empirically evaluate our approach using six real-world tasks including text and image classifications. The experimental results indicate the superiority of our framework in comparison with the state-of-the-art benchmark models. Our research has both methodological and practical value. Methodologically, we provide an example of how to combine deep learning and graphical models in a unified framework. Both deep learning techniques and graphical models hold unique advantages for modeling real-world problems [10]. Integrating these two can combine their complementary strengths and fits a broader range of applications. Practically, our proposed framework serves as a cost-effective approach for aggregating crowd efforts. As our empirical evaluation shows, our framework significantly outperforms the state-of-the-art methods in the literature. Such performance gaps will immediately translate into large differences in cost because businesses usually rely on data collection daily [26].

Acknowledgement. The authors would like to thank Dr. Yong Ge, Dr. Wei Chen, and Dr. Jason Pacheco for their useful comments. Mingyue Zhang is the corresponding author. This work is partially supported by the following grants: the National Key Research and Development Program of China under Grant Nos. 2016QY02D0305 and 2017YFC0820105; the National Natural Science Foundation of China under Grant Nos. 71621002 and 71802024.

References

1. Bi, W., Wang, L., Kwok, J.T., Tu, Z.: Learning to predict from crowdsourced data. In: UAI, pp. 82–91 (2014)
2. Bishop, C.M.: Pattern Recognition and Machine Learning. Springer, Heidelberg (2006)
3. Blei, D.M., Kucukelbir, A., McAuliffe, J.D.: Variational inference: a review for statisticians. J. Am. Stat. Assoc. **112**(518), 859–877 (2017)
4. Blei, D.M., Ng, A.Y., Jordan, M.I.: Latent Dirichlet allocation. J. Mach. Learn. Res. **3**, 993–1022 (2003)
5. Devlin, J., Chang, M.W., Lee, K., Toutanova, K.: Bert: pre-training of deep bidirectional transformers for language understanding. arXiv preprint arXiv:1810.04805 (2018)
6. Ding, X., Liu, T., Duan, J., Nie, J.Y.: Mining user consumption intention from social media using domain adaptive convolutional neural network. In: AAAI, vol. 15, pp. 2389–2395 (2015)
7. Dizaji, K.G., Huang, H.: Sentiment analysis via deep hybrid textual-crowd learning model. In: Thirty-Second AAAI Conference on Artificial Intelligence (AAAI 2018) (2018)
8. Gadiraju, U., Kawase, R., Dietze, S.: A taxonomy of microtasks on the web. In: Proceedings of the 25th ACM Conference on Hypertext and Social Media, pp. 218–223. ACM (2014)
9. Ipeirotis, P.G., Provost, F., Wang, J.: Quality management on Amazon mechanical turk. In: Proceedings of the ACM SIGKDD Workshop on Human Computation, pp. 64–67. ACM (2010)
10. Johnson, M.J., Duvenaud, D.K., Wiltschko, A., Adams, R.P., Datta, S.R.: Composing graphical models with neural networks for structured representations and fast inference. In: Advances in Neural Information Processing Systems, pp. 2946–2954 (2016)
11. Karger, D.R., Oh, S., Shah, D.: Iterative learning for reliable crowdsourcing systems. In: Advances in Neural Information Processing Systems, pp. 1953–1961 (2011)
12. Kazai, G., Kamps, J., Milic-Frayling, N.: Worker types and personality traits in crowdsourcing relevance labels. In: Proceedings of the 20th ACM International Conference on Information and Knowledge Management, pp. 1941–1944. ACM (2011)
13. Kim, H.C., Ghahramani, Z.: Bayesian classifier combination. In: Proceedings of the Fifteenth International Conference on Artificial Intelligence and Statistics, pp. 619–627 (2012)

14. Kingma, D.P., Welling, M.: Auto-encoding variational Bayes. In: International Conference on Learning Representations (ICLR) (2014). arXiv:1312.6114
15. Luo, Y., Tian, T., Shi, J., Zhu, J., Zhang, B.: Semi-crowdsourced clustering with deep generative models. In: Advances in Neural Information Processing Systems, pp. 3212–3222 (2018)
16. Moreno, A., Terwiesch, C.: Doing business with strangers: reputation in online service marketplaces. Inf. Syst. Res. **25**(4), 865–886 (2014)
17. Pang, B., Lee, L.: Seeing stars: exploiting class relationships for sentiment categorization with respect to rating scales. In: Proceedings of the 43rd Annual Meeting on Association for Computational Linguistics, pp. 115–124 (2005)
18. Raykar, V.C., et al.: Learning from crowds. J. Mach. Learn. Res. **11**(Apr), 1297–1322 (2010)
19. Rodrigues, F., Lourenco, M., Ribeiro, B., Pereira, F.C.: Learning supervised topic models for classification and regression from crowds. IEEE Trans. Pattern Anal. Mach. Intell. **39**(12), 2409–2422 (2017)
20. Rodrigues, F., Pereira, F., Ribeiro, B.: Learning from multiple annotators: distinguishing good from random labelers. Pattern Recogn. Lett. **34**(12), 1428–1436 (2013)
21. Rodrigues, F., Pereira, F.C.: Deep learning from crowds. In: The Thirty-Second AAAI Conference on Artificial Intelligence (AAAI), pp. 1611–1618. AAAI Press (2018)
22. Simpson, E., Roberts, S.J., Psorakis, I., Smith, A.: Dynamic Bayesian combination of multiple imperfect classifiers. Decis. Making Imperfection **474**, 1–35 (2013)
23. Simpson, E.D., et al.: Language understanding in the wild: combining crowdsourcing and machine learning. In: Proceedings of the 24th International Conference on World Wide Web, pp. 992–1002 (2015)
24. van Engelen, J.E., Hoos, H.H.: A survey on semi-supervised learning. Mach. Learn. **109**(2), 373–440 (2019). https://doi.org/10.1007/s10994-019-05855-6
25. Wainwright, M.J., Jordan, M.I.: Graphical models, exponential families, and variational inference. Found. Trends Mach. Learn. **1**(1–2), 1–305 (2008)
26. Wang, J., Ipeirotis, P.G., Provost, F.: Cost-effective quality assurance in crowd labeling. Inf. Syst. Res. **28**(1), 137–158 (2017)
27. Wei, X., Zhang, Z., Zhang, M., Zeng, D.D.: Combining crowd and machine intelligence to detect false news in social media. SSRN 3355763 (2019)
28. Yi, J., Jin, R., Jain, S., Yang, T., Jain, A.K.: Semi-crowdsourced clustering: generalizing crowd labeling by robust distance metric learning. In: Advances in Neural Information Processing Systems, pp. 1772–1780 (2012)
29. Yin, L., Liu, Y., Zhang, W., Yu, Y., et al.: Truth inference with a deep clustering-based aggregation model. IEEE Access **8**, 16662–16675 (2020)
30. Zhang, M., Wei, X., Guo, X., Chen, G., Wei, Q.: Identifying complements and substitutes of products: a neural network framework based on product embedding. ACM Trans. Knowl. Discov. Data (TKDD) **13**(3), 1–29 (2019)
31. Zhang, M., Wei, X., Zeng, D.D.: A matter of reevaluation: incentivizing users to contribute reviews in online platforms. Decis. Support Syst. **128**, 113158 (2020)

Multi-view Weighted Kernel Fuzzy Clustering Algorithm Based on the Collaboration of Visible and Hidden Views

Yiming Tang[1,2]([✉]), Bowen Xia[1,2], Fuji Ren[1,2], Xiaocheng Song[1,2], Hongmang Li[1,2], and Wenbin Wu[1,2]

[1] Anhui Province Key Laboratory of Affective Computing and Advanced Intelligent Machine, Hefei University of Technology, Hefei 230601, China
`ren@is.tokushima-u.ac.jp`
[2] School of Computer and Information, Hefei University of Technology, Hefei 230601, China

Abstract. With the development of media technology, data types that cluster analysis needs to face become more and more complicated. One of the more typical problems is the clustering of multi-view data sets. Existing clustering methods are difficult to handle such data well. To remedy this deficiency, a multi-view weighted kernel fuzzy clustering method with collaborative evident and concealed views (MV-Co-KFCM) is put forward. To begin with, the hidden shared information is extracted from several different views of the data set by means of non-negative matrix factorization, then applied to this iterative process of clustering. This not only takes advantage of the difference information in distinct views, but also utilizes the consistency knowledge in distinct views. This pre-processing algorithm of extracting hidden information from multiple views (EHI-MV) is obtained. Furthermore, in order to coordinate different views during the iteration, a weight is distributed. In addition, so as to regulate the weight adaptively, shannon entropy regularization term is also introduced. Entropy can be maximized as far as possible by minimizing the objective function, thus MV-Co-KFCM algorithm is proposed. Facing 5 multi-view databases and comparing with 6 current leading algorithms, it is found that the algorithm which we put forward is more excellent as for 5 clustering validity indexes.

Keywords: Fuzzy clustering · Collaboration · Multi-view data · Clustering validity index

1 Introduction

In the early days, cluster analysis mainly refers to hard clustering algorithms. Hard clustering algorithms rigorously classify data information objects into a certain category, which is an either-or optimization algorithm. The membership degree of the hard clustering algorithm can only have two values: 0 and 1. An object can only belong to a certain class or not belong to a certain class. Recently, Rodriguez and Laio proposed a clustering algorithm RLM [1] that quickly finds the highest relative density, which is a

© Springer Nature Singapore Pte Ltd. 2021
Y. Sun et al. (Eds.): ChineseCSCW 2020, CCIS 1330, pp. 115–129, 2021.
https://doi.org/10.1007/978-981-16-2540-4_9

typical hard clustering algorithm. However, this hard division method cannot fully meet the needs in the actual application process.

In addition to hard clustering, another key clustering method is fuzzy clustering. Among many fuzzy clustering algorithms, the fuzzy c-means (FCM) algorithm [2] is undoubtedly the classical one. According to the iterative update of the minimization of the objective function, FCM can easily measure the membership of the clustering center and each data information target. Subsequently, many scholars proposed improved algorithms for FCM. Krishnapuram and Keller [3] clearly proposes a possible fuzzy clustering method PCM, which eliminates the constraint on membership degree in the FCM algorithm, thereby reducing the noise points and outliers in the data information. However, it is well known that the PCM optimization algorithm can easily cause cluster consistency problems during the entire operation. Pal et al. [4] closely combines the possibility value and the degree of membership, and clearly proposes a new possibility fuzzy clustering algorithm PFCM, which eliminates the clustering consistency problem of the PCM optimization algorithm. In [5], the kernel function was introduced into the whole process of clustering, using kernel distance instead of Euclidean distance for calculation, and clearly proposed the kernel-based fuzzy clustering algorithm KFCM. Zhou and Chen came up with a fuzzy clustering method EWFCM established on weight feature entropy [6], and introduced a feature weight matrix W to represent the weights of different features. Yang and Nataliani put forward a novel fuzzy clustering algorithm FRFCM with feature reduction [7] established on attribute weight entropy. However, although EWFCM and FRFCM realize the selection of features in the process of clustering, they still have trouble in the initialization of cluster centers and being easily affected by noise points.

In the midst of growing development of media technology, the types of information that cluster analysis needs to face have become more complex and diverse. Under such circumstances, the existing cluster analysis methods are difficult to adapt to the complex situation, and many problems inevitably arise. Among them, a more typical problem is the problem of data sets of multi-view. The emergence of multi-view information is an inevitable result of the rapid advancement of human technology and data collection methods.

For example, for a web page, how can we identify a web page? We can identify it by the content information contained in the web page itself, such as the text, images and other information in the web page. Another way is to identify it through the information contained in the hyperlinks connected to this web page. In this way, for the data object of a web page, it has two views composed of different characteristic data: one is a view composed of a feature set describing the content data of a web page, and the other is a feature of hyperlink data. View composed of collections. Generally speaking, the feature data of these two different views differ greatly.

For a multi-view data set, its data can be indicated as $X = \{X^1, X^2, \cdots, X^k\}$. Thereinto the k-th visual view is denoted as a matrix $X^k = \lfloor x_1^k, x_2^k, \cdots x_N^k \rfloor \in R^{m_k \times N}$. N indicates the quantity of features of the k-th view. For traditional single-view-oriented clustering algorithms, they cannot distinguish the differences between different views. Taking video data as an example, its image information is used as a view, and its audio information is used as a view. The characteristic attributes of the two different views are very different. According to the traditional single-view-oriented clustering algorithm, we need an algorithm that can properly process image information and audio information at the same time, which is difficult to achieve. Therefore, using single-view-oriented clustering algorithms to process multi-view data usually cannot achieve good results. Therefore, the multi-view clustering problem has turn into a significant study issue.

Scholars have put forward various methods to deal with multi-view clustering issues. In [8], a multi-view clustering algorithm using non-negative matrix factorization is put forward. And then it uses a clustering algorithm to cluster the clusters immediately. In [9], a novel multi-view clustering algorithm using non-negative matrix factorization (NMF) via resemblance constraint was put forward.

However, although these attribute characteristics of different views in a multi-view data set may be very different, they are after all descriptions of the same data object from different perspectives. Therefore, there is hidden shared information between different views. Therefore, there are both essential topics when learning multi-view data:

1) How can we take full advantage of the difference knowledge among distinct views in the data set?
2) How can we fully find out and use this hidden consistency information among distinct views?

Above algorithms provide a lot of practical solutions to the problem for multi-view clustering, but a number of algorithms only pay attention to the evident knowledge in the evident views, and neglect the concealed point shared among the evident views. At the same time, a lot of other clustering algorithms of multi-view only pays attention to and mine the shared concealed lore of distinct views. As a matter of fact, the shared concealed lore and the specific information in distinct evident views both should be attached importance to. In the research of multi-view clustering algorithm, it's a hard problem to apply shared lore with specific information productively.

To this end, we come up with a multi-view weighted kernel fuzzy clustering algorithm using collaborative visible and hidden views (MV-Co-KFCM). The method fully considers the differences between different visible views in the data, and considers the concealed knowledge shared among distinct views in same data set. By comparing with six related algorithms, the effect of the proposed algorithm is verified.

2   NMF

NMF [10] can be seen as a dimensionality reducing technique employed in pattern recognition, image engineering and other fields. The dimensionality of non-negative data sets can be cut back by NMF. Provided a non-negative data set $X = [x_1, x_2, \cdots x_N]$, the NMF can gain two non-negative matrix factors $W \in R_+^{m \times r}$ and $H \in R_+^{r \times N}$, and $X \approx WH$. Here r represents dimensionality reduction, and W indicates the base matrix and H indicates the coefficient matrix. Consequently, we can regard NMF as the optimization issue:

$$\min_{W,H} \|X - WH\|^2 \quad \text{among them } W > 0, H > 0 \tag{1}$$

Here we assume that the noise obeys Gaussian distribution, then the maximum likelihood function can be obtained:

$$L(W,H)=\prod_{i,j}\frac{1}{\sqrt{2\pi}\sigma_{ij}}\exp\left(-\frac{E_{ij}^2}{2\sigma_{ij}}\right)=\prod_{i,j}\frac{1}{\sqrt{2\pi}\sigma_{ij}}\exp(-\frac{\left(X_{ij}-(WH)_{ij}\right)^2}{2\sigma_{ij}}). \quad (2)$$

In virtue of the largest logarithm likelihood formula, the log likelihood formula can be obtained:

$$\ln L(W,H)=\sum_{i,j}\ln\frac{1}{\sqrt{2\pi}\sigma_{ij}}-\frac{1}{2\sigma_{ij}}\sum_{i,j}\left(X_{ij}-(WH)_{ij}\right)^2 \quad (3)$$

Assuming that the variance of the noise of each data point is the same, then to maximize the value of the likelihood function, it's only necessary to optimize the value of the next function:

$$J(W,H)=-\frac{1}{2}\sum_{i,j}\left(X_{ij}-(WH)_{ij}\right)^2 \quad (4)$$

Because of $(WH)_{ij}=\sum_{k}W_{ik}H_{kj}$,

$$\frac{\partial(WH)_{ij}}{\partial W_{ik}}=H_{kj} \quad (5)$$

Then we can get:

$$\frac{\partial J(W,H)}{\partial W_{ik}}=\sum_{j}\left[-H_{kj}\left(X_{ij}-(WH)_{ij}\right)\right]$$

$$=-\left(\sum_{j}X_{ij}H_{kj}-\sum_{j}H_{kj}(WH)_{ij}\right)=-\left(\left(XH^T\right)_{ik}-\left(WHH^T\right)_{ik}\right) \quad (6)$$

In the same way, we can get:

$$\frac{\partial J(W,H)}{\partial H_{kj}}=-\left(\left(W^TV\right)_{kj}-\left(W^TWH\right)_{kj}\right) \quad (7)$$

Next, we can iterate W and H through the gradient descent algorithm:

$$W_{ik}=W_{ik}-\alpha_1\left[-\left(\left(XH^T\right)_{ik}-\left(WHH^T\right)_{ik}\right)\right] \quad (8)$$

$$H_{kj}=H_{kj}-\alpha_2\left[-\left(\left(W^TX\right)_{kj}-\left(W^TWH\right)_{kj}\right)\right] \quad (9)$$

Here, we select:

$$\alpha_1=\frac{W_{ik}}{\left(WHH^T\right)_{ik}} \quad (10)$$

$$\alpha_2 = \frac{H_{kj}}{\left(W^T WH\right)_{kj}} \tag{11}$$

Substituting formulas (10) and (11) into formulas (8) and (9), we can finally get the iterative formulas of W and H:

$$W_{ik} = W_{ik} \cdot \frac{\left(XH^T\right)_{ik}}{\left(WHH^T\right)_{ik}} \tag{12}$$

$$H_{kj} = H_{kj} \cdot \frac{\left(W^T X\right)_{kj}}{\left(W^T WH\right)_{kj}} \tag{13}$$

Recently, NMF has been widely applied in cluster analysis. For instance, for the term document matrix, a document grouping method based on NMF is proposed [11].

3 Core Algorithm

3.1 Multi-view Clustering of Visible and Hidden View Collaboration

For multi-view data, it can be reasonably assumed that a concealed view data is shared by distinct evident views, and data of distinct evident views can be constructed from the shared concealed view.

Provided a multi-view data set $X = \{X^1, X^2, \cdots X^K\}$, the k-th evident view is expressed as a matrix $X^k = \left[x_1^k, x_2^k, \cdots x_N^k\right] \in R^{m_k \times N}$, where N is the quantity of samples, K indicates the quantity of evident views, and m_k represents the quantity of attributes of the k-th evident view. Because the factors in X^k are possible to be negative, in order to adopt NMF technology, we standardize X^k for the factors in X^k to be non-negative. $H = [h_1, h_2 \cdots, h_N] \in R^{r \times N}$ represents the multi-view data set in the shared concealed space, and $W^k \in R^{m_k \times r}$ denotes the mapping matrix that maps the concealed space data to the k-th evident view. By solving the following optimization problems, we can receive the mapping matrix and concealed view:

$$O = \sum_{k=1}^{K} q_k \left\|X^k - W^k H\right\|^2 + \lambda \sum_{k=1}^{K} q_k \ln q_k \tag{14}$$

The constraints are: $W^k, H \geq 0, \sum_{k=1}^{K} q_k = 1, 0 \leq q_k \leq 1$.

The $q = [q_1, q_2, \ldots q_K]$ is a vector of weights, q_k corresponds to the k-th visual view, $\lambda \geq 0$ indicates the regularization coefficient. The former item represents the experience loss of NMF. The latter item stands for the adaptive weight of the visual view using the maximum entropy mechanism.

It can be resolved by iterative optimization. We divided resolving process divided into three steps:

Firstly, optimizing W^k when fixing H and q.

Secondly, optimizing H when fixing W^k and q.

Finally, optimizing q when fixing H and W^k.

Here are the details. First, we optimize the value of W^k when fixing H and q, then the issue can be transformed into:

$$\min_{W^k} \sum_{k=1}^{K} q_k \left\| X^k - W^k H \right\|^2 + \lambda \sum_{k=1}^{K} q_k \ln q_k \qquad (15)$$

For any W^k, since H and q are fixed, (15) can be changed into:

$$\min_{W^k} \left\| X^k - W^k H \right\|^2 \qquad (16)$$

Through NMF, we can get the formula of W^k:

$$\left(W^k \right)_{i,r} = \left[\frac{\left(X^k H^T \right)_{i,r}}{\left(W^k H H^T \right)_{i,r}} \right] \left(W^k \right)_{i,r} \qquad (17)$$

Then, we optimize the value of H by fixing q and W^k. Under these circumstances, the formula (18) can be obtained:

$$\min_{H} \sum_{k=1}^{K} q_k \left\| X^k - W^k H \right\|^2 + \lambda \sum_{k=1}^{K} q_k \ln q_k \qquad (18)$$

Construct the objective function:

$$J\left(W^k, H \right) = \sum_{k=1}^{K} q_k \cdot \sum_{i,j} \left(X_{ij}^k - \left(W^k H \right)_{ij} \right)^2 + \lambda \sum_{k=1}^{K} q_k \ln q_k \qquad (19)$$

Because of $\left(W^k H \right)_{ij} = \sum_{r} \left(W_{ir}^k H_{rj} \right)$, and $\frac{\partial \left(W^k H \right)_{ij}}{\partial H_{rj}} = W_{ir}^k$, the partial derivative of the objective function can be obtained:

$$\frac{\partial J \left(W^k, H \right)}{\partial H_{rj}} = \sum_{k=1}^{K} q_k \cdot 2 \cdot [-([\left(W^k \right)^T X^k]_{rj} - [\left(W^k \right)^T W^k H]_{rj})] \qquad (20)$$

After this, the value of H is iterated through the gradient descent method:

$$H_{rj} = H_{rj} - \alpha \cdot 2 \cdot \left[-\sum_{k=1}^{K} q_k \cdot [\left(W^k \right)^T X^k]_{rj} + \sum_{k=1}^{K} q_k \cdot [\left(W^k \right)^T W^k H]_{rj} \right] \qquad (21)$$

Here, we set the value of α:

$$\alpha = \frac{1}{2} \cdot \frac{H_{rj}}{\sum_{k=1}^{K} q_k \cdot [(W^k)^T W^k H]_{rj}} \qquad (22)$$

Substituting it into (21) can get the iterative formula:

$$H_{rj} = \frac{\sum\limits_{k=1}^{K} q_k \cdot [(W^k)^T X^k]_{rj}}{\sum\limits_{k=1}^{K} q_k \cdot [(W^k)^T W^k H]_{rj}} \cdot H_{rj} \tag{23}$$

Finally, we optimize the value of q by fixing H and W^k. Lagrangian function is constructed by Lagrangian multiplier method:

$$L = \sum_{k=1}^{K} q_k \left\| X^k - W^k H \right\|^2 + \lambda \sum_{k=1}^{K} q_k \ln q_k - \eta \left(\sum_{k=1}^{K} q_k - 1 \right) \tag{24}$$

Then we can get:

$$\frac{\partial L}{\partial q_k} = \left\| X^k - W^k H \right\|^2 + \lambda (\ln q_k + 1) - \eta = 0 \tag{25}$$

After derivation:

$$q_k = \exp\left(\frac{\eta - \lambda}{\lambda} - \frac{\left\| X^k - W^k H \right\|^2}{\lambda} \right) = \exp\left(\frac{\eta - \lambda}{\lambda} \right) \cdot \exp\left(-\frac{\left\| X^k - W^k H \right\|^2}{\lambda} \right) \tag{26}$$

Because of the constraint $\sum\limits_{k=1}^{K} q_k = 1$ in (14), we can get:

$$\exp\left(\frac{\eta - \lambda}{\lambda} \right) = \frac{1}{\sum\limits_{k=1}^{K} \exp\left(-\frac{\|X^k - W^k H\|^2}{\lambda} \right)} \tag{27}$$

Substituting formula (27) into (26), the iterative formula can be obtained:

$$q_l = \exp\left(-\frac{\left\| X^k - W^k H \right\|^2}{\lambda} \right) \bigg/ \sum_{l=1}^{K} \exp\left(-\frac{\left\| X^l - W^l H \right\|^2}{\lambda} \right) \tag{28}$$

So far, the derivation on extracting hidden information from multiple views has all ended. Next we will give the specific EHI-MV (extracting hidden information from multiple views) algorithm steps:

Algorithm 1 EHI-MV algorithm

Procedure

    Regularize $X^k (1 < k < K)$ for each view;

    Carry on the Initialization;

    Repeat

    For $k = 1 : K$ do

      Change W^k according to (17);

    End for

      Change H according to (23);

    For $k = 1 : K$ do

      Change q^k in the light of (28);

    End for

    Until converge

3.2 MV-Co-kFCM Algorithm

Provided a multi-view data set $X = \{X^1, X^2, \ldots, X^K\}$, we are able to get the concealed view data among the evident view via the EHI-MV method. According to the collaboration of visible and concealed views, we put forward a weighted kernel function clustering algorithm after introducing the kernel function, namely the MV-Co-KFCM algorithm. Its objective function is characterized:

$$
J\left(U, V, \tilde{V}, w\right) = \beta \sum_{i=1}^{C} \sum_{j=1}^{N} u_{ij}^m \left\| \phi(h_j) - \phi(\tilde{v}_i) \right\|^2
$$

$$
+ (1 - \beta) \sum_{k=1}^{K} w_k \sum_{i=1}^{C} \sum_{j=1}^{N} u_{ij}^m \left\| \phi(x_j^k) - \phi(v_i^k) \right\|^2 + \eta \sum_{k=1}^{K} w_k \ln w_k \qquad (29)
$$

The restrictions involved specifically include the following aspects: $H >$ 0, $\sum_{i=1}^{C} u_{ij} = 1$, $\sum_{k=1}^{K} w_k = 1$, $0 \leq w_k \leq 1$, $u_{ij} \in [0, 1]$, $1 < j < N$. U is the membership matrix of size $C \times N$, $V = \{V^1, V^2, \ldots, V^K\}$ represents the cluster center of a set of K evident data, $V^k = [v_1^k, v_2^k, \ldots, v_C^k]$ is the cluster center matrix of the k-th evident data, v_i^k denotes the i-th cluster of the k-th evident data. Class center. $\tilde{V} = [\tilde{v}_1, \tilde{v}_2, \ldots, \tilde{v}_C]$ represent the cluster center matrix of the concealed data, $\tilde{v}_i$ is the i-th cluster center of the evident data. $H = [h_1, h_2, \cdots h_N]$ indicates the shared concealed data of K visible views. w_k is the weight of the k-th evident view, which is similar to the degree of membership, and the view weight also is required to satisfy the constraint that the sum is 1. So as to adaptively regulate the weights, (29) also adopts the Shannon entropy. Assuming that the weights of all evident view data can be regarded as a probability distribution, we have that Shannon's entropy can be indicated as $-\sum_{k=1}^{K} w_k \ln w_k$, and by minimizing the objective function, the entropy can be maximized as much as possible. η is employed to manage the effect of the Shannon entropy during clustering.

The Lagrangian function is generated by

$$L = J\left(U, V, \tilde{V}, w\right) + \lambda\left(1 - \sum_{i=1}^{C} u_{ij}\right) + \alpha\left(1 - \sum_{k=1}^{K} w_k\right) \tag{30}$$

First, by taking the partial derivative of the membership matrix, we can get:

$$\frac{\partial L}{\partial u_{ij}} = \beta m u_{ij}^{m-1} \left\| \phi(h_j) - \phi(\tilde{v}_i) \right\|^2 + (1 - \beta) \sum_{k=1}^{K} w_k m u_{ij}^{m-1} \left\| \phi(x_j^k) - \phi(v_i^k) \right\|^2 - \lambda = 0 \tag{31}$$

Then we have

$$u_{ij} = \frac{\lambda^{1/m-1}}{\left(\beta \cdot m \cdot \left\| \phi(h_j) - \phi(\tilde{v}_i) \right\|^2 + (1 - \beta) \cdot \sum_{k=1}^{K} w_k \cdot m \cdot \left\| \phi(x_j^k) - \phi(v_i^k) \right\|^2 \right)^{1/m-1}} \tag{32}$$

From the constraint conditions, one has

$$\sum_{i=1}^{C} u_{ij} = 1 \tag{33}$$

and then

$$\lambda^{1/m-1} = 1 \Bigg/ \sum_{i=1}^{C} \frac{1}{\left(\beta \cdot m \cdot \left\| \phi(h_j) - \phi(\tilde{v}_i) \right\|^2 + (1 - \beta) \cdot \sum_{k=1}^{K} w_k \cdot m \cdot \left\| \phi(x_j^k) - \phi(v_i^k) \right\|^2 \right)^{1/m-1}} \tag{34}$$

Substituting (34) into (32), we can get

$$u_{ij} = 1 \Big/ \sum_{s=1}^{C} \frac{\left(\beta \cdot m \cdot \|\phi(h_j) - \phi(\bar{v}_i)\|^2 + (1-\beta) \cdot \sum_{k=1}^{K} w_k \cdot m \cdot \|\phi(x_j^k) - \phi(v_i^k)\|^2 \right)^{1/m-1}}{\left(\beta \cdot m \cdot \|\phi(h_j) - \phi(\bar{v}_s)\|^2 + (1-\beta) \cdot \sum_{k=1}^{K} w_k \cdot m \cdot \|\phi(x_j^k) - \phi(v_s^k)\|^2 \right)^{1/m-1}} \tag{35}$$

Then, by taking the partial derivative of the view weight, one has:

$$\frac{\partial L}{\partial w_k} = (1-\beta) \cdot \sum_{i=1}^{C} \sum_{j=1}^{N} u_{ij}^m \left\| \phi(x_j^k) - \phi(v_i^k) \right\|^2 + \eta(\ln w_k + 1) - \alpha = 0 \tag{36}$$

After deduction, we can obtain:

$$\sum_{k=1}^{K} w_k = \exp\left(\frac{\alpha - \eta}{\eta}\right) \cdot \sum_{k=1}^{K} \exp\left(\frac{-(1-\beta) \cdot \sum_{i=1}^{C} \sum_{j=1}^{N} u_{ij}^m \left\| \phi(x_j^k) - \phi(v_i^k) \right\|^2}{\eta} \right) = 1 \tag{37}$$

Combing (37) with (36), we can get the iterative formula of w_k:

$$w_k = \frac{\exp\left(\frac{-(1-\beta)}{\eta} \sum_{i=1}^{C} \sum_{j=1}^{N} u_{ij}^m \left\| \phi(x_j^k) - \phi(v_i^k) \right\|^2 \right)}{\sum_{h=1}^{K} \exp\left(\frac{-(1-\beta)}{\eta} \sum_{i=1}^{C} \sum_{j=1}^{N} u_{ij}^m \left\| \phi(x_j^k) - \phi(v_i^h) \right\|^2 \right)} \tag{38}$$

After that, by obtaining the partial derivative of the cluster center v_i^k of the visual view, here we introduce the Gaussian kernel function for calculation, namely:

$$\left\| \phi(x_j^k) - \phi(v_i^k) \right\|^2 = 2 - 2 \cdot K\left(x_j^k, v_i^k\right) = 2 - 2 \cdot \exp\left(\frac{\left(x_j^k - v_i^k\right)^2}{2 \cdot \sigma^2} \right) \tag{39}$$

Then we can get:

$$\frac{\partial L}{\partial v_i^k} = (1-\beta) \cdot w_k \cdot \sum_{j=1}^{N} u_{ij}^m \cdot (-2) \cdot \exp\left(\frac{\left(x_j^k - v_i^k\right)^2}{2 \cdot \sigma^2} \right) \cdot \frac{\left(x_j^k - v_i^k\right)}{\sigma^2} = 0 \tag{40}$$

Derivation can be:

$$(1 - \beta) \cdot w_k \cdot \sum_{j=1}^{N} u_{ij}^m \cdot (-2) \cdot \exp\left(\frac{\left(x_j^k - v_i^k\right)^2}{2 \cdot \sigma^2}\right) \cdot \frac{x_j^k}{\sigma^2}$$

$$= (1 - \beta) \cdot w_k \cdot \sum_{j=1}^{N} u_{ij}^m \cdot (-2) \cdot \exp\left(\frac{\left(x_j^k - v_i^k\right)^2}{2 \cdot \sigma^2}\right) \cdot \frac{v_i^k}{\sigma^2} \qquad (41)$$

The formula of v_i^k can be obtained as:

$$v_j^k = \frac{\sum_{j=1}^{N} u_{ij}^m \cdot \exp(\frac{\left(x_j^k - v_i^k\right)^2}{2\sigma^2}) \cdot x_j^k}{\sum_{j=1}^{N} u_{ij}^m \cdot \exp(\frac{\left(x_j^k - v_i^k\right)^2}{2\sigma^2})} \qquad (42)$$

Finally, we get for $\tilde{v}_i$ of the hidden view:

$$\frac{\partial L}{\partial \tilde{v}_i} = \beta \sum_{j=1}^{N} u_{ij}^m \cdot (-2) \cdot \exp\left(\frac{\left(h_j - \tilde{v}_i\right)^2}{2 \cdot \sigma^2}\right) \cdot \frac{\left(h_j - \tilde{v}_i\right)}{\sigma^2} = 0 \qquad (43)$$

To sum up, we can obtain:

$$\tilde{v}_i = \frac{\sum_{j=1}^{N} u_{ij}^m \cdot \exp(\frac{(h_j - \tilde{v}_i)^2}{2 \cdot \sigma^2}) \cdot h_j}{\sum_{j=1}^{N} u_{ij}^m \cdot \exp(\frac{(h_j - \tilde{v}_i)^2}{2 \cdot \sigma^2})} \qquad (44)$$

So far, the iterative formulas for each variable of the algorithm have been derived. The operation flow of this algorithm is described below.

Algorithm 2: MV-Co-KFCM algorithm.

Procedure MV-Co-KFCM (X, C, m, σ, r, ε, β, η, *iterMax*)

Obtain the hidden view data H by running the EHI-MV method; Randomly initialize all the parameters. Let $iter = 0$;

repeat

$iter = iter + 1$; Calculate u_{ij} by (35) and update $U^{(iter)} = [\mathrm{u}_{ij}]$;

For $k = 1:K$ do

Calculate v_i^k of the visible view by (42), and update the cluster center matrix of the visible data;

End For

Calculate $\tilde{v}_i$ of the hidden view by (44), and update the cluster center matrix of the hidden data;

For $k = 1:K$ do

Calculate w_k of the visible view by (38), and update W;

End For

Until $\left\| J^{(iter)} - J^{(iter-1)} \right\| < \varepsilon$ or $iter > iterMax$;

Return U, the cluster centers of the visible and hidden view, and W;

end procedure

---

4 Experiment and Analysis

Here we use 5 multi-view data sets to experimentally evaluate the clustering effect of the algorithm. These 5 data sets come from the UCI repository. They are Multiple Features, Image Segmentation, Dermatology, Forest Type and Water Treatment Plant, which are multi-feature data sets, image segmentation data sets, dermatology data sets, forest type data sets, and water treatment plant data sets. Each data set contains two different views.

At the same time, we also compared with the previous 6 algorithms: FCM, KFCM, RLM, VFCM, EWFCM, FRFCM. The platform used in the experiment is Window 10 system and the programming language is Matlab2013b.

Using 5 evaluation indicators ACC, CH, NMI, EARI, XB [12]. The first four are the bigger the better, and the fifth XB index is the smaller the better. The XB(-) in the table here means that the smaller the better. In addition, EARI and XB are indicators specifically for fuzzy algorithms, where RLM is hard clustering, and there is no corresponding value. The results obtained are demonstrated from Table 1, 2, 3, 4 and Table 5.

Table 1. Experimental results of Multiple Features data set

	FCM	KFCM	RLM	V-FCM	EWFCM	FRFCM	MV-Co-KFCM
ACC	0.3370	0.5469	0.5965	0.6267	0.6533	0.6793	0.8972
CH	2.5782	3.1379	3.8730	4.1723	5.6377	7.8784	10.3354
NMI	0.1680	0.2163	0.2389	0.2179	0.3591	0.5398	0.8231
EARI	0.2389	0.3876	–	0.4237	0.4751	0.4838	0.9156
XB(-)	1.8426	0.9257	–	0.8216	0.7910	0.7379	0.1636

Table 2. Experimental results of Image Segmentation data set

	FCM	KFCM	RLM	V-FCM	EWFCM	FRFCM	MV-Co-KFCM
ACC	0.1979	0.2335	0.2754	0.3582	0.3887	0.5183	0.8293
CH	0.0795	1.026	2.3840	2.4283	2.9372	3.7273	9.4739
NMI	0.2669	0.4473	0.5839	0.5993	0.6739	0.7383	0.9976
EARI	0.3969	0.5641	–	0.7342	0.8129	0.8539	0.9585
XB(-)	2.6758	1.8683	–	1.6238	0.9672	0.9373	0.2836

Table 3. Experimental results of Dermatology data set

	FCM	KFCM	RLM	V-FCM	EWFCM	FRFCM	MV-Co-KFCM
ACC	0.3844	0.3942	0.4083	0.4768	0.4910	0.5937	0.8705
CH	1.6304	1.7263	3.9375	5.2364	9.2430	15.9634	23.7451
NMI	0.4652	0.5677	0.7054	0.7344	0.7511	0.7635	0.9233
EARI	0.2326	0.3731	–	0.5308	0.5632	0.7878	0.9953
XB(-)	3.4590	2.8323	–	2.3901	2.3465	1.7692	0.5981

The strong points of the proposed algorithm are mainly reflected:

1) It's glaringly obvious from the above consequents that the MV-Co-KFCM algorithm performs best on ACC, CH, NMI, EARI and XB in 5 multi-view data sets. Experiments illustrate that the effect of MV-Co-KFCM is greater than the other six comparison algorithms.

Table 4. Experimental results of Forest Type data set

	FCM	KFCM	RLM	V-FCM	EWFCM	FRFCM	MV-Co-KFCM
ACC	0.1530	0.1623	0.3256	0.5020	0.5291	0.5802	0.9133
CH	15.5675	32.4543	39.2489	57.9363	59.8535	63.2369	82.3500
NMI	0.1390	0.2277	0.2535	0.3506	0.7487	0.8426	0.9135
EARI	0.3522	0.3809	–	0.4374	0.4399	0.5818	0.9033
XB(-)	5.0374	3.3923	–	3.2856	1.7454	1.3555	0.2929

Table 5. Experimental results of Water Treatment Plant data set

	FCM	KFCM	RLM	V-FCM	EWFCM	FRFCM	MV-Co-KFCM
ACC	0.3290	0.3302	0.4273	0.4701	0.5921	0.5995	0.9312
CH	12.3499	15.2570	18.0887	19.9442	20.0371	26.9373	38.1923
NMI	0.1023	0.1930	0.2288	0.2475	0.2825	0.3972	0.7799
EARI	0.3598	0.4290	–	0.6367	0.6922	0.8800	0.9255
XB(-)	6.3433	4.8947	–	4.3066	2.6805	0.9932	0.0830

2) For 6 clustering algorithms oriented to single view, such as FCM, the test results show that simply directly combining different views to form data for clustering cannot obtain satisfactory results.

3) This article uses NMF to extract hidden shared information from multiple different views of the data set, and then applies it to the iterative process of clustering. In this way, we not only utilize the difference lore among distinct views in the multi-view data set, but also take full advantage of the consistency information in the distinct views. All this is embodied in the previous EHI-MV algorithm.

4) In MV-Co-KFCM, in order to coordinate different views in the iterative process, we assign a weight w_k to each view. Of course, the view weight also needs to meet the constraint condition of 1. So as to adaptively regulate the weight of each evident view, a Shannon entropy regularization term is also introduced. Assuming that the weights of all evident views can be regarded as a probability distribution. By minimizing the objective function, the entropy can be maximized as much as possible.

5 Summary and Outlook

We put forward a multi-view weighted kernel fuzzy clustering algorithm MV-Co-KFCM that synergizes visible and hidden views.

Firstly, the non-negative matrix is decomposed as the kernel, and the hidden shared information is extracted from multiple different views of the data set, and then it is applied to the iterative process of clustering. This not only takes full advantage of the distinct knowledge among distinct views in the multi-view data set, but also employ the consistency information in the different views. It is embodied in the previous EHI-MV preprocessing algorithm.

Secondly, in order to coordinate different views in the iterative process, a weight is allocated to each view, and the view weight needs to meet the constraint condition that the sum is 1.

Thirdly, so as to adaptively regulate the weight, the Shannon entropy is also introduced. By minimizing the objective function, the entropy can be maximized as much as possible, thus establishing the core algorithm of this article. Namely the MV-Co-KFCM algorithm.

Lastly, facing 5 multi-view databases, it's clear to see from the experimental consequent that the MV-Co-KFCM algorithm which we put forward performs best in ACC, CH, NMI, EARI and XB, notably greater than the other 6 comparison algorithms.

In future research, we try to combine fuzzy clustering and heterogeneous implication logical reasoning, and use logical reasoning as the internal drive to establish a new collaborative clustering algorithm.

Acknowledgement. This research was supported by the National Natural Science Foundation of China (Nos. 61673156, 61877016, 61672202, U1613217, 61976078).

References

1. Rodriguez, A., Laio, A.: Clustering by fast search and find of density peaks. Science **344**(6191), 1492–1496 (2014)
2. Tang, Y.M., Ren, F.J., Pedrycz, W.: Fuzzy c-means clustering through SSIM and patch for image segmentation. Appl. Soft Comput. **87**(1), 1–16 (2020)
3. Krishnapuram, R., Keller, J.: A possibilistic approach to clustering. IEEE Trans. Fuzzy Syst. **1**(2), 98–110 (1993)
4. Pal, N.R., Pal, K., Bezdek, J.C.: A possibilistic fuzzy c-means clustering algorithm. IEEE Trans. Fuzzy Syst. **13**(4), 517–530 (2005)
5. Ding, Y., Fu, X.: Kernel-based fuzzy c-means clustering algorithm based on genetic algorithm. Neurocomputing **18**(8), 233–238 (2016)
6. Zhou, J., Chen, L., Chen, C.L.P.: Fuzzy clustering with the entropy of attribute weights. Neurocomputing **19**(8), 125–134 (2016)
7. Yang, M.S., Nataliani, Y.: A feature-reduction fuzzy clustering algorithm based on feature-weighted entropy. IEEE Trans. Fuzzy Syst. **26**(2), 817–835 (2019)
8. Greene, D., Cunningham, P.: A matrix factorization approach for integrating multiple data views. In: Buntine, W., Grobelnik, M., Mladenić, D., Shawe-Taylor, J. (eds.) ECML PKDD 2009. LNCS (LNAI), vol. 5781, pp. 423–438. Springer, Heidelberg (2009). https://doi.org/10.1007/978-3-642-04180-8_45
9. Lee, D.D., Seung, H.S.: Algorithms for nonnegative matrix factorization. Adv. Neural Inf. Process. Syst. **13**(6), 556–562 (2001)
10. Lee, D.D., Seung, H.S.: Learning the parts of objects by non-negative matrix factorization. Nature **401**(6755), 788–791 (1999)
11. Jiang, Y., Liu, J., Li, Z., Li, P., Lu, H.: Co-regularized PLSA for multi-view clustering. In: Lee, K.M., Matsushita, Y., Rehg, J.M., Hu, Z. (eds.) ACCV 2012. LNCS, vol. 7725, pp. 202–213. Springer, Heidelberg (2013). https://doi.org/10.1007/978-3-642-37444-9_16
12. Tang, Y.M., Hu, X.H., Pedrycz, W., Song, X.C.: Possibilistic fuzzy clustering with high-density viewpoint. Neurocomputing **329**(15), 407–423 (2019)

Impacts of Individuals' Trust in Information Diffusion of the Weighted Multiplex Networks

Jianyong Yu[✉], Jie Luo, and Pei Li

School of Computer Science and Engineering, Hunan University of Science and Technology, Xiangtan 411201, China
yujyong@hnust.edu.cn

Abstract. In the recent years, information diffusion and strategy interaction on the multiplex networks have been greatly researched. Some significant progresses came at stochastic spreading models. However, the influences of individuals' trust on strategy interaction and cooperation evolution were generally overlooked. Actually, in real social networks, the different trust levels of individuals are important factors in the information diffusion. Here, we will play some strategy games among individuals, as well as communities or groups in the weighted multiplex networks, to explore the effect of individuals' trust on strategy interaction and cooperation evolution. Each individual uses a cache to storage certain length of previous strategy status, which indicates the individual's trust degree on the strategy. When consecutive identical cooperation strategies occur in the individual's cache, and exceed the certain threshold of strategy memory length, the cooperation strategy will be adopted in next interlayer and intralayer interactions, so as to explore community level diffusion. The results show that the individuals' trust levels highly affect the strategy interaction and cooperation evolution in multiplex networks. The cache length of consecutive identical cooperation strategy is also the key parameter for information diffusion, and the tight degrees between different layers of networks together. This work will be helpful to understand cooperation dynamic and evolution in weighted multiplex networks.

Keywords: Trust · Information diffusion · Strategy interaction

1 Introduction

Nowadays, the wild varieties of information have been spreading on all diversified social networks, such as epidemics, opinions, innovation, gossips and so on. Some researchers focus on the cascade and threshold models [1–3], and try to understand these processes and mechanisms of information diffusion in social networks. Specially, the evolutionary game theories are generally used to model individual's strategy interactions and diffusion with social dilemmas on multiplex networks [4–6], so as to explore the individuals' strategy behaviors and cooperation spread mechanism, as well as communities or groups of sub-networks [7–10].

There are many significant progresses in studying of information spreading and cooperation evolution in multiplex networks. Damon C. [11] focused on the spread of health

© Springer Nature Singapore Pte Ltd. 2021
Y. Sun et al. (Eds.): ChineseCSCW 2020, CCIS 1330, pp. 130–141, 2021.
https://doi.org/10.1007/978-981-16-2540-4_10

behavior and investigated the effects of different network structures. They artificially structured several online groups and communities, and found that the health behavior spread faster and farther in clustered-lattice networks than in random networks. F.C. Santos et al. [12] studied evolution games on different types of graphs and incorporated heterogeneity of the individuals. The graphs covered from single-scale to scale-free graphs. They showed that the cooperative behavior of long-term easily counteracted the defective behavior of short-term for all social dilemmas, by increasing heterogeneity of cooperation. They also showed how cooperative behavior depends on the complicating individuals' relationship in scale-free network. The topological structures of network that may affect the cooperation evolution have been also deeply investigated in [13–18]. Moreover, many recent works studied information diffusion for different contents on real-world social networks [19–26], and finding influential nodes for information diffusion in the independent cascade model [27–29]. It is an important step that identifying the most influential spread leaders in a social network, towards optimizing available use of resources and ensuring the more efficient information diffusion. Maksim Kitsak et al. [30] showed that the best spreaders or the most influential spreaders are not necessarily the most central or highly connected individuals. They found that individuals locating in the core of network are the most influential spread leaders, which could be identified by the k-shell decomposition analysis. They also considered cross-layer spreaders and their distances that determine the spreading extent, simultaneously. However, most of existing works about information diffusion generally overlooked the impact of rational decision of individuals on the outcome of the cooperation evolution process. Actually, in real social networks, individuals may change their interaction strategy and behavior during the information diffusion process according to some factors, such as whether the information is important, whether the neighbor individuals are trustable, individuals' social position and influence force, and so on.

In this work, we will investigate how individuals' trust influence the extent of cooperation evolution in the weighted multiplex networks, and how the different trust degrees of individuals determine the groups' interactions behavior and cooperation evolution in intralayer and interlayer networks, so as to explore community level diffusion. Here, the degrees and levels of trust give expression to the strategy memory span of individuals' cache. The evolutionary games are modeled as the trustable and un-trustable interactions. The strategy interactions are performed between individuals not only in intralayer network, but also in interlayer network. Two classical games are modeled as the strategy interactions, Prisoner's Dilemma (PD) and Snowdrift Game (SG) [31, 32]. The information diffusion is abstracted as a modified version of the threshold model [13]. We try to explore the cooperation evolution of information spread in multiplex networks and considered different strategy cache lengths and different social positions of individuals. The results show that the individuals' trust levels highly affect the strategy interaction and cooperation evolution in multiplex networks. The cache length of consecutive identical cooperation strategy is also important parameter for the information diffusion, and the tight degrees between different layers of networks together.

2 Proposed Method

Here, a two-layered multiplex network is considered, which consists of network $C1$ and $C2$ with weighted values. For each of individual, the sum of weighted values of linked edges is regarded as its social position and its influence force. There are some connected edges between different individuals from different networks, which compose as inter-layer network. The strategy interactions happen on between each pair of individuals in interlayer and intralayer of networks, by playing game PD and SG [32].

In the games, two kinds of strategies would be adopted, cooperators (C) and defectors (D). The two-layered networks are randomly generated, including weighted values of all edges and the individuals' link degrees. According to individuals' social position and influence force, each individual is classified into authority group or non-authority group. After each interaction with linked neighbor from interlayer and intralayer network, an individual will obtain the payoffs. Meanwhile, the ranks of each strategy will be calculated in different groups. Specially, the impact forces and reactive forces of different strategies, whether from several groups or from individuals, are considered in each interaction simultaneously. An individual will accept the strategy from certain intralayer group with probability parameter β, or adopt the strategy from the linked interlayer neighbor with probabilistic parameter a. Each individual has a cache storing certain length of previous strategy status, and the memory length of strategy values indicates different degrees of trust on information spread. The strategies of individuals will be continually updated after each intralayer and interlayer interaction. When consecutive identical cooperation strategies in the individual's cache exceed the setting threshold, the strategy C will be adopted in next interactions. In this way, the evolution process of cooperation diffusion in the weighted multiplex networks is explored.

2.1 The Duplex Network with Weighted Values

Firstly, a two-layered weighted multiplex network is randomly generated. Two layer networks are defined as $C1 = <A, E_A, W_A, K_A>$ and $C2 = <B, E_B, W_B, K_B>$, respectively. $A = \{a_1, a_2, ..., a_m\}$ indicates the set of individuals with number m in $C1$. $\forall(a_i, a_j) \in E_A \Rightarrow \forall E(a_i, a_j) \in W_A$, that means there exist an edge with interaction weight value between node a_i and a_j, where $\left(w_{ij}^a = w_{ji}^a \right) \in W_A$. In addition, $\forall k_i \in K_A$ denotes the link degree of individual a_i. Similarly, $\forall(b_i, b_j) \in E_B \Rightarrow \left(w_{ij}^b = w_{ji}^b \right) \in W_B$. Subsequently, the interlayer network is defined as $DPS = <C1, C2, E_{C1C2}, W_{C1C2}, K_{DPS}>$, where E_{C1C2} is the set of interlayer linked edges, W_{C1C2} is the set of weighted values of interlayer linked edges, and K_{DPS} denotes the average value of link degrees among interlayer linked individuals in multi-layer networks.

2.2 The Groups' Classification

The individual's trust mostly depends on the social position and influence force of the opposite side. Here, the social position of individual i is defined:

$$ SP_i^{C1} = \frac{\sum_{\forall(a_i, a_j) \in E_A} w_{ij}^a}{\sum_{i=1}^{m} \sum_{\forall(a_i, a_j) \in E_A} w_{ij}^a}. \tag{1}$$

If $SP_i^{C1} > SP_j^{C1}$, it indicates that individual i has bigger strategy impact force than individual j. Then, we can calculate the average value of social positions for all individuals in network $C1$:

$$\frac{1}{m}\sum_{i=1}^{m} SP_i^{C1} = \frac{1}{m}. \tag{2}$$

We divide the sub-network $C1$ into two groups, the authority group and the non-authority group. The authority group consists of those individuals with bigger SP than average value. The SPs of individuals in the non-authority group are smaller than average SP.

In network $C1$, G_{C1} and $\overline{G_{C1}}$ represent the authority group with high SP and the non-authority group with low SP respectively. We have $G_{C1} = \{a_i | SP_i^{C1} > \frac{1}{m}\}$, and $\overline{G_{C1}} = A - G_{C1}$. Similarly, we also have G_{C2} and $\overline{G_{C2}}$.

Meanwhile, each individual in G_{C1} would impose strategy impact on target individuals to some extent. Therefore, it is necessary to calculate the impact distance between individual and either of groups.

Let $d(a, u)$ represents the shortest weighted path from individual a to u, $|G|$ means the number of individuals in G. Then, we can define the impact distance between individual a and authority group G:

$$\begin{cases} \text{if } a \in G, \ D_{aG} = \frac{1}{(|G|-1)\times\sum_{u\in G, u\neq a} SP_u} \sum_{u\in\{G-a\}} (SP_u \times d(a,u)) \\ \text{if } a \notin G, \quad D_{aG} = \frac{1}{|G|\times\sum_{u\in G} SP_u} \sum_{u\in G} (SP_u \times d(a,u)) \end{cases} \tag{3}$$

Where, $d(a, u)$ can be calculated by using the graph theory's Dijkstra's algorithm. Similarly, we can also obtain the impact distance $D_{a\overline{G}}$ in sub-network.

2.3 Payoffs of Individual and Ranks of Strategy

The goal of strategy interaction between two individuals is to obtain maximal payoff in games PD and SG. As mentioned above, the payoffs of an individual i will be gotten from intralayer and interlayer interactions with other individual j:

$$\begin{cases} \text{Intralayer: } \sum_{j=1}^{<k_i>} P_{ij} = P_{i1} \\ \text{Interlayer: } \sum_{j=1}^{<k_{C1C2}>} P_{ij} = P_{i2} \end{cases} \tag{4}$$

Here, $\langle k_i \rangle$ is the linked degree parameter of individual i in intralayer network, while $\langle k_{C1C2} \rangle$ is the average link degree of interlayer linked neighbors between sub-network $C1$ and $C2$. P_{ij} is the obtained payoff of individual i by interacting with j. Then, after the intralayer and interlayer interaction, the total payoff of individual i is:

$$P_i = \alpha P_{i1} + (1 - \alpha)P_{i2} \tag{5}$$

Where, α is a crucial parameter reflected the degree of link ties between two sub-networks. Obviously, the close interlayer interaction relation between two sub-networks means that parameter α is smaller. Specially, when $\alpha = 1$, it denotes that two sub-networks have not any interlayer interaction relations, which means two independent sub-networks.

The strategy C and D will be adopted by all of individuals in the two groups. Let G_S denotes the set of individuals with strategy S in the authority group G, and P_u represents the obtained interaction payoffs of individual u. Then, we can calculate the rank of strategy S in G:

$$\forall S, \quad Rank(S) = \sum_{u \in G_S} SP_u \times P_u \tag{6}$$

Where $S = \{C, D\}$. Similarly, we can also obtain the rank of strategy S in $\overline{G}$: $Rank(\overline{S}) = \sum_{u \in \overline{G}_S} SP_u \times P_u$. Therefore, four kinds of ranks in one sub-network can be calculated, according to different strategies and different groups. Each of ranks in each sub-network should be calculated at each step of intralayer and interlayer interaction.

2.4 Strategy Impact Force in Groups

In information diffusion, the strategy of an individual will be highly affected from the different groups with different SP. Here, we define a function to denote the impact force of group G by imposing strategy S to individual a:

$$IF^G_{S \to a} = f\left(\frac{\sigma_1 Rank(S)}{\sigma_2 D_{aG}}\right) = f\left(\frac{\sigma_1 \sum_{u \in G_S}(SP_u \times P_u)}{\sigma_2\left(1/(|G| \times \sum_{u \in G} SP_u)\right) \times \sum_{u \in G}(SP_u \times d(a, u))}\right) \tag{7}$$

Where, function f increases monotonously. σ_1 and σ_2 adjust their relative importance grades. It should be noted that $|G| = |G| - 1$ when individual $a \in G$. Since $S = \{C, D\}$, according to above definition, we have also four kinds of strategy impact forces from different groups and different strategy respectively, e.g., $IF^G_{C \to a}$, $IF^G_{D \to a}$, $IF^{\overline{G}}_{C \to a}$, and $IF^{\overline{G}}_{D \to a}$. Obviously, we also consider the weak strategy impact from the non-authority group, simultaneously.

In addition, although the reactive forces of strategies from an individual to each group are weak, we cannot neglect its influence, because the individual sometimes may obtain greater payoffs in each of interactions. Hence, the reactive force of individual a by imposing strategy S to group G is given:

$$RF^G_{a \to S} = g\left(\frac{SP_a \times P_a}{1/|G_S| \times \sum_{u \in G_S}(SP_u \times P_u)}\right) \tag{8}$$

Here, function g also increases monotonously. Similarly, according to above definition, $RF^G_{a \to C}$, $RF^G_{a \to D}$, $RF^{\overline{G}}_{a \to C}$ and $RF^{\overline{G}}_{a \to D}$, can be calculated respectively.

2.5 Strategy Diffusion in Interlayer and Intralayer Interactions

As for the strategy diffusion in interlayer interactions, the parameter $(1 - \alpha)$ is still used to describe the adopted and updated probability of the strategies from other layer network. It means that an individual would either accept the strategy of different group in intralayer network with the probability α, or accept the strategy of its linked neighbor in interlayer network with probability $(1 - \alpha)$, after once strategy interaction.

For the latter, the linked neighbor j in interlayer network with the Maximum $SP_j \times P_j$, will be choose in neighbors of number $\langle k_{C1C2} \rangle$. Further, if $SP_i \times P_i < SP_j \times P_j$, and $S_i \neq S_j$, then individual i will finally adopts its current strategy with the probability as follows:

$$\prod_{i \rightarrow j}(t) = \frac{SP_j \times P_j(t) - SP_i \times P_i(t)}{Max(k_i, k_j) \times (Max(1, T) - Min(0, S))} \tag{9}$$

Where, t represents the interaction moment, while T and S are key parameters of game PD and SG in interlayer interaction.

For the former, when an individual is ready to adopt the strategy of certain group in intralayer network with the probability α, some factors need to be considered simultaneously in strategy diffusions, such as four kinds of impact forces from different groups in intralayer network ($IF^G_{C \rightarrow a}, IF^G_{D \rightarrow a}, IF^{\overline{G}}_{C \rightarrow a}$, and $IF^{\overline{G}}_{D \rightarrow a}$) and four kinds of reactive forces from individual itself ($RF^G_{a \rightarrow C}, RF^G_{a \rightarrow D}, RF^{\overline{G}}_{a \rightarrow C}$ and $RF^{\overline{G}}_{a \rightarrow D}$). In general, individual a will be inclined to accept the strategy C from group G, when $IF^G_{C \rightarrow a}$ is bigger or $RF^G_{a \rightarrow C}$ is smaller. Here, we use the rate of $IF^G_{C \rightarrow a}$ and $RF^G_{a \rightarrow C}$ to describe the tendency that individual a adopts C of G. Similarly, $IF^{\overline{G}}_{D \rightarrow a}/RF^{\overline{G}}_{a \rightarrow D}$ is used to describe the tendency that individual a adopts D of $\overline{G}$. Meanwhile, we use the probability β to measure the impact degree of two different groups from intralyer to the individual. It means that, individual a would either adopt the strategy of group G with the probability β, or accept the strategy of group $\overline{G}$ with the probability $(1 - \beta)$. Therefore, at next time $t + 1$, the strategy $S_a(t + 1)$ that individual a will adopt can be given as follows respectively.

$$S_a(t + 1) = \begin{cases} S^G_C, & \text{if } IF^G_{C \rightarrow a}/RF^G_{a \rightarrow C} \geq IF^G_{D \rightarrow a}/RF^G_{a \rightarrow D} \\ S^G_D, & \text{otherwise} \end{cases} \tag{10}$$

$$S_a(t + 1) = \begin{cases} S^{\overline{G}}_C, & \text{if } IF^{\overline{G}}_{C \rightarrow a}/RF^{\overline{G}}_{a \rightarrow C} \geq IF^{\overline{G}}_{D \rightarrow a}/RF^{\overline{G}}_{a \rightarrow D} \\ S^{\overline{G}}_D, & \text{otherwise} \end{cases} \tag{11}$$

Where, S^G_C means the strategy C of G, while $S^{\overline{G}}_D$ denotes the strategy D of $\overline{G}$.

2.6 Performance Criterion

In order to measure the strategy diffusion degree in the whole multiplex networks, the density of cooperators at time t is given as follows:

$$c(t) = \frac{1}{N_l} \sum_{l=1}^{N_l} c^l(t) = \frac{1}{2} \times \left(\frac{1}{m} \sum_{i=1}^{m} S^A_i(t) + \frac{1}{n} \sum_{i=1}^{n} S^B_i(t) \right) \tag{12}$$

Here, N_l is the layer number of networks. $S^A_i(t)$ represents the sum of strategy value that individual i in sub-network A obtains at time t. The value of strategy C is supposed for 1, while D is supposed for 0.

As mentioned above, each individual has a cache with certain length, as to storage its own previous strategy values. The length of consecutive identical strategy C in the cache stands for different trust levels of individuals on information diffusion. When the individual's strategy is about to update, the above formula 10 and 11 will be revised. If consecutive identical cooperation strategies in individual's cache exceed the setting threshold, the strategy C will be adopted by the individual in next interactions. Otherwise, the update of strategy will still conform to the formula 10 and 11. Here, the parameter ML is used to indicate the strategy memory span of individual, so as to evaluate the influence of trust. In this way, the evolution process of cooperation diffusion in the weighted multi-layered networks is explored, and the impact of trust in the information diffusion is further reconsidered.

3 Experimental Result and Analysis

In this work, two kinds of classical evolutionary games (PD and SG) are played, and the intralayer and interlayer interactions of strategy diffusion in the weighted multiplex networks are analyzed. Two kinds of strategies would be adopted in each interaction, C or D. The games PD and SG are defined by four parameters. The relative ordering and the range of specified values of four parameters are described and set, as shown in Table 1.

Table 1. The range of four parameters for PD and SG

PD	C	D	SG	C	D
C	$R = 1$	$S \in [-1, 0]$	C	$R = 1$	$S \in [0, 1][0, 1]$
D	$T \in [1, 2]$	$P = 0$	D	$T \in [1, 2]$	$P = 0$

In experiment, parameters T and S increase 0.1step by step in their ranges of values respectively, as to ensure accuracy. For each pair of T and S, 1000 two-layered weighted networks consisting of 1000 individuals are generated randomly, including their interlayer networks. For each of duplex networks, the strategy updates occurring in interlayer and intralayer interactions will be executed 5000 times. Finally, the average densities of C are calculated corresponding to each region pair of T and S. Meanwhile, the individual's different strategy memory lengths (ML) are analyzed.

The cooperation diffusion and evolution in different games owing to the impact of strategy ML is shown in Fig. 1. For the game SG, it is obvious that the densities of C decreases in some regions significantly, e.g. $S \in [0, 0.3]$ and $T \in [1.5, 2]$, when the strategy ML gradually increases. However, there are little impacts of strategy ML in other regions, as Panel 1 shows. When the game PD is played, the influence of individuals' trust mainly focuses on smaller strategy ML values. It is apparent that the coverage proportion of cooperators C is more than 50 percent in some regions, e.g. $S \in [-1, -0.5]$ and $T \in [1, 1.5]$, when strategy ML is 2.

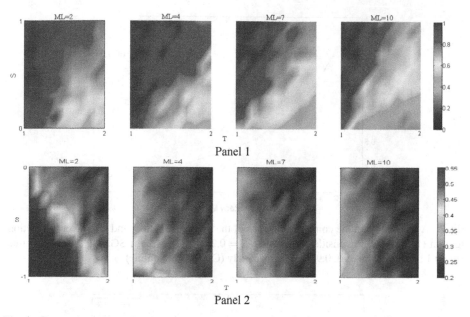

Panel 1

Panel 2

Fig. 1. For each region pair of T and S, 1000 random weighted duplex networks are generated. The average densities of C and corresponding to different strategy ML are shown as cooperation diffusion for game SG (Panel 1) and PD (Panel 2), when the parameter a is 0.9.

As showed in the Fig. 1, we can find that the average density distributions of C are at comparatively low level for game PD. However, the impact of individuals' trust on cooperation diffusion in the game PD is more lager than in game SG, because of its smaller densities of C. In order to obtain the higher densities of C in game PD, the strategy ML should take small value.

In order to investigate the impacts combined interlayer interaction tight degree a and strategy ML of caches on the cooperation densities of C, some experiments are also executed. When the game SG is played, the combined impacts on the cooperation density of C are shown in Fig. 2. We can find that the bigger interlayer tight parameter a and the smaller strategy ML will be obvious to obtain the higher densities of C. When the strategy ML of caches is fixed, the tighter the interlayer interaction between two networks is, the faster the cooperation diffusion is. Meanwhile, under certain interlayer interaction parameter a, the smaller the strategy ML is, the bigger the densities of C are. This illustrates that the trust levels of individuals highly impact the cooperation evolution to some extent, and the individuals' trust degree plays an important role on information diffusion. From the Fig. 2, it is easy to be found when strategy ML is 2, the density of C achieves a maximum value. It is based on one fact that all of next interaction strategies will continuously adopt C, only if the pervious interaction strategy is C. Specially, the impact of individuals' trust will be not considered in information diffusion when strategy ML is 1.

When the game PD is played, the combined impacts on the cooperation density of C are shown in the Fig. 3. From the Fig. 3, we can find that the average densities of C are

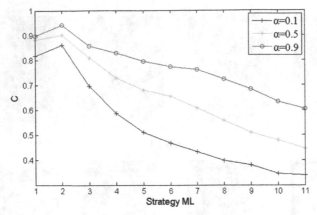

Fig. 2. A case study on the combined impacts of the individuals' trust and interlayer interaction tight on $(1 - \alpha)(1 - \alpha)$ densities of C, when $S = 0.2$ and $T = 1.2$ for SG, and the tight degree a are 0.1 (blue), 0.5 (green), 0.9 (red), respectively (Color figure online).

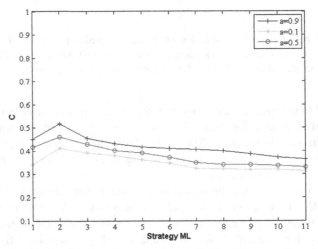

Fig. 3. A case study on the combined impacts of the individuals' trust and interlayer interaction tight on $(1 - \alpha)(1 - \alpha)(1 - \alpha)$ densities of C, when $S = -0.8$ and $T = 1.2$ for PD, and the tight degree a are 0.1 (green), 0.5 (red), 0.9 (blue), respectively (Color figure online).

comparatively lower than those obtained in SG game as a whole. The average densities of C are less than 0.4 in most of regions under certain interlayer interaction parameter a. Therefore, it seems to be more important that a good strategy ML is selected. Just like game SG, the smaller strategy ML and the bigger interaction tight parameter a will obtain the bigger density of C. Similarly, when strategy ML is 2, the density of C achieves a maximum value. In addition, there are not obvious difference on density of C when α is 0.9 and 0.1 respectively. It means the impacts of interlayer interaction on cooperation evolution are relatively less in game PD than in game SG. When the game

PD is played, the strategy ML and intralayer interaction are two important factors for cooperation diffusion.

Finally, since each of two-layered multiplex networks with weighted values and the relations between each individual are all generated randomly, we will also calculate the credible intervals and credible degrees for the cooperation densities in some region pairs, as to verify the confidence degrees of results. On the basis of above method, we firstly obtain 1000 sample sets of cooperation densities for each typical region pairs when the game SG is played, and interlayer and intralayer strategy interactions are performed for thousands of times. Then we can calculate the mean and standard deviation for cooperation densities C, and obtain the credible intervals of cooperation densities C for each selected typical region pair under the setting credible level. Here, we set the strategy ML for 2, the confidence level for 90%. The interlayer interaction parameter a is also set for 0.1 and 0.9. The credible intervals of cooperation densities C are obtained for some selected region pairs.

Our results show that there are the high cooperation densities of C and high incredible intervals of C in the typical region pairs of $S \in [0, 0.3]$ and $T \in [1.5, 2]$ when the game SG is played. In these regions, there are higher cooperation densities of C when $\alpha = 0.1$ than $\alpha = 0.9$. It indicates that the impacts of individuals' trust are more effective in intralayer interaction than in interlayer interaction for these regions. In addition, the densities of cooperation C in most of region pairs are at comparatively low levels for game PD as a whole, despite the high confidence level.

4 Conclusion

In this work, we investigated how individuals' trust impacts on cooperation evolution in the weighted multiplex networks, and how the different trust levels of individuals determine the strategy behavior in intralayer and interlayer interactions, so as to explore the information diffusion in social multiplex networks.

The evolutionary games are modeled as the trustable and un-trustable interactions, and the strategy interactions happen in interlayer and intralayer of networks by playing game SG and PD. Here, each individual uses a cache to storage certain length of previous strategy status, which indicates the individual's trust degree on the strategy. When consecutive identical cooperation strategies occur in the individual's cache, and exceed the setting certain threshold of strategy memory length, the cooperation strategy C will be accepted in next intralayer and interlayer interactions.

We imitated the real-world social networks as much as possible, and considered some factors affecting individual's strategy behavior and interaction, such as the trust degrees among individuals, the social positions of individuals, the impacts of social groups, the interlayer relation tight degrees of sub-networks, and so on. The results show that the individuals' trust levels highly impact the strategy interaction and cooperation evolution in the weighted multiplex networks. The impact of individuals' trust on cooperation diffusion in the game PD is more lager than in game SG, while the impact of interlayer tight degrees on cooperation evolution is relatively less in game PD than in game SG. The cooperation diffusion depends on the memory lengths of individuals with consecutive identical cooperation strategies to a great extent. This work will be helpful to understand cooperation dynamic and evolution process in weighted multiplex networks.

Acknowledgements. This work was supported by The National Social Science Fund of China (No. 20BXW096).

References

1. Buldyrev, S.V., Parshani, R., Paul, G., Stanley, H.E., Havlin, S.: Catastrophic cascade of failures in interdependent networks. Nature **464**, 1025–1028 (2010)
2. Wang, Z., Wang, L., Szolnoki, A., Perc, M.: Evolutionary games on multilayer networks: a colloquium. Eur. Phys. J. B **88**, 124 (2015)
3. Liu, X., Stanley, H.E., Gao, J.: Breakdown of interdependent directed networks. Proc. Natl. Acad. Sci. USA **113**(5), 1138–1143 (2016)
4. Santos, M.D., Dorogovtsev, S.N., Mendes, J.F.F.: Biased imitation in coupled evolutionary games in interdependent networks. Sci. Rep. **4**, 4436 (2014)
5. Marialisa, S., et al.: Combining evolutionary game theory and network theory to analyze human cooperation patterns. Chaos Solitons Fractals **91**, 17–24 (2019)
6. Wang, Z., Szolnoki, A., Perc, M.: Optimal interdependence between networks for the evolution of cooperation. Sci. Rep. **3**, 2470 (2013)
7. Lugo, H., San Miguel, M.: Learning and coordinating in a multilayer network. Sci. Rep. **5**, 7776 (2015)
8. Matamalas, J.T., Poncela-Casasnovas, J., Gómez, S., Arenas, A.: Strategical incoherence regulates cooperation in social dilemmas on multiplex networks. Sci. Rep. **5**, 9519 (2015)
9. Luo, C., Zhang, X., Liu, H., Shao, R.: Cooperation in memory-based prisoner's dilemma game on interdependent networks. Physica A **450**, 560–569 (2016)
10. Huang, K., et al.: Understanding cooperative behavior based on the coevolution of game strategy and link weight. Sci. Rep. **5**, 14783 (2015)
11. Centola, D.: The spread of behavior in an online social network experiment. Science **329**, 1194–1196 (2010)
12. Santos, F.C., Pacheco, J.M.: Tom Lenaerts: evolutionary dynamics of social dilemmas in structured heterogeneous populations. Proc. Natl. Acad. Sci. USA **103**(9), 3490–3494 (2006)
13. Watts, D.J.: A simple model of global cascades on random networks. Proc. Natl. Acad. Sci. USA **99**(9), 5766–5771 (2002)
14. Weng, L., Menczer, F., Ahn, Y.-Y.: Virality prediction and community structure in social networks. Sci. Rep. **3**, 2522 (2013)
15. Li, W., Tang, S., Fang, W., et al.: How multiple social networks affect user awareness: the information diffusion process in multiplex networks. Phys. Rev. E **92**, 042810 (2019)
16. Wang, W., Liu, Q.H., Cai, S.M., et al.: Suppressing disease spreading by using information diffusion on multiplex networks. Sci. Rep. **6**, 29259 (2016)
17. Wang, Z., Wang, L., Perc, M.: Degree mixing in multilayer networks impedes the evolution of cooperation. Phys. Rev. E **89**(5), 052813 (2014)
18. Zhao, D., Wang, L., Li, S., Wang, Z., Wang, L., et al.: Immunization of epidemics in multiplex networks. PLoS ONE **9**(11), e112018 (2018)
19. Liu, T., Li, P., Chen, Y., Zhang, J.: Community size effects on epidemic spreading in multiplex social networks. PLoS ONE **11**(3), e0152021 (2016)
20. Emily, T.: Diffusion of innovations on community based Small Worlds: the role of correlation between social spheres. In: 15th Coalition Theory Network Workshop, Marseille, France (2010)
21. Rasoul, R., Mostafa, S., et al.: Diffusion of innovations over multiplex social networks. In: The International Symposium on Artificial Intelligence and Signal Processing (AISP), pp. 300–304. IEEE (2018)

22. Liu, Q.H., et al.: Impacts of complex behavioral responses on asymmetric interacting spreading dynamics in multiplex networks. Sci. Rep. **6**, 25617 (2016)
23. Johan, U., Lars, B., Cameron, M., Jon, K.: Structural diversity in social contagion. Proc. Natl. Acad. Sci. USA **109**(16), 5962–5966 (2012)
24. Michela, D.V., et al.: The spreading of misinformation online. Proc. Natl. Acad. Sci. USA **113**(3), 554–559 (2016)
25. Azimi-Tafreshi, N.: Cooperative epidemics on multiplex networks. Phys. Rev. E **93**(4), 042303 (2016)
26. Gabriel, E.K., Young, H.P.: Rapid innovation diffusion in social networks. Proc. Natl. Acad. Sci. USA **111**(3), 10881–10888 (2014)
27. Kimura, M., Yamakawa, K., Saito, K., et al.: Community analysis of influential nodes for information diffusion on a social network. In: IEEE International Joint Conference on Neural Networks (IEEE World Congress on Computational Intelligence),pp. 1358–1363. IEEE (2018)
28. Zang, W., Zhang, P., Zhou, C., Guo, L.: Discovering multiple diffusion source nodes in social networks. Procedia Comput. Sci. **29**, 443–452 (2014)
29. Szolnoki, A., Perc, M.: Leaders should not be conformists in evolutionary social dilemmas. Sci. Rep **6**, 23633 (2016)
30. Maksim, K., et al.: Identification of influential spreaders in complex networks. Nat. Phys. **6**, 888–893 (2010)
31. Yu, J., Jiang, J.C., Xiang, L.: Group-based strategy diffusion in multiplex networks with weighted values. Physica A: Stat. Mech. Appl. **469**, 148–156 (2017)
32. Axelrod, R., Hamilton, W.D.: The evolution of cooperation. Science **211**, 1390–1396 (1981)

Intuitionistic Entropy-Induced Cooperative Symmetric Implicational Inference

Yiming Tang[1,2]([✉]), Jiajia Huang[1], Fuji Ren[1], Witold Pedrycz[2], and Guangqing Bao[1]

[1] School of Computer and Information, Hefei University of Technology, Hefei 230601, Anhui, China
[2] Department of Electrical and Computer Engineering, University of Alberta, Edmonton, AB T6R 2V4, Canada

Abstract. In regard to how to find the ideal result of reasoning mechanism, the standard of the largest entropy can offer its proper interpretation. Inspired by this observation, the intuitionistic entropy-induced symmetric implicational (IESI) algorithm is put forward in this study, and then extend it to the corresponding cooperative reasoning version. To begin with, the new IESI standards are given, which are improved versions of the previous symmetric implicational standard. Thenceforth, the inherent characteristics of result of the IESI algorithm are analyzed. In addition, unified results of such algorithm are given, and the fuzzy system via the IESI algorithm is constructed and analyzed. Lastly, its cooperative reasoning version is proposed for multiple inputs and multiple rules, while the corresponding optimal solutions are also gained.

Keywords: Fuzzy inference · Symmetric implicational method · Intuitionistic fuzzy set · Fuzzy system.

1 Introduction

The crucial issues of fuzzy inference [1] are characterized as below:

$$\text{FMP: focusing on } C \longrightarrow D \text{ and } C^*, \text{ achieve } D^*, \tag{1}$$

$$\text{FMT: focusing on } C \longrightarrow D \text{ and } D^*, \text{ achieve } C^*. \tag{2}$$

Hereinto C, C^* belong to $FZY(X)$ ($FZY(X)$ represents the collection of whole fuzzy subsets of X) and $D, D^* \in FZY(Y)$. To find the solutions for FMP and FMT, the CRI (compositional rule of inference) algorithm has been the classical

This research is supported by the National Natural Science Foundation of China (Nos. 61673156, 61877016, 61672202, U1613217, 61976078).

strategy, in which an operator called fuzzy implication $\hookrightarrow$ was utilized. Thereafter, the triple I algorithm was established [2]. Its basic target was to hunt for the appropriate $D^* \in FZY(Y)$ (or $C^* \in FZY(X)$) making

$$(C(c) \hookrightarrow D(d)) \hookrightarrow (C^*(c) \hookrightarrow D^*(d)) \tag{3}$$

acquire its largest value for all $c \in X$, $d \in Y$. Following that, the fuzzy entropy triple I algorithm [3] was proposed. Such idea offered a novel choice for the optimal result of the triple I algorithm.

The triple I algorithm created a huge amount of research enthusiasm [4–6]. It was verified that the triple I algorithm had a large amount of wonderful properties e.g., strict logic basis. Nevertheless it was imperfect from the angle of fuzzy systems (see [7]).

Tang and Yang generalized the core quintessence of the triple I algorithm and adjusted (3) to the next structure [8]:

$$(C(c) \hookrightarrow_a D(d)) \hookrightarrow_b (C^*(c) \hookrightarrow_a D^*(d)). \tag{4}$$

Notice that double fuzzy implications $\hookrightarrow_a, \hookrightarrow_b$ are employed. The strategy induced by (4) was referred to as the symmetric implicational algorithm. In [9], a special case of symmetric implicational algorithm was put forward and researched. The standard of maximum entropy is able to afford an ideal explanation to choose the optimal solution of symmetric implicational algorithm. In [10], the entropy-based symmetric implicational algorithm was put forward.

Following the research of fuzzy sets, the classical fuzzy sets were developed into intuitionistic fuzzy sets, which had a much stronger ability to describe things. As a result, in this study we incorporate such entropy-based symmetric implicational algorithm into the intuitionistic fuzzy environment, and further come up with the intuitionistic entropy-induced symmetric implicational (IESI) algorithm, and then extend it to the intuitionistic entropy-induced cooperative symmetric implicational (IECSI) algorithm.

The innovation of this paper is embodied in the next aspects. First of all, new maximum fuzzy entropy is proposed in the symmetric implicational structure, and then the intuitionistic entropy-induced symmetric implicational algorithm of fuzzy inference is put forward. Furthermore, its cooperative reasoning version is proposed for multiple inputs and multiple rules. Lastly, unified solutions of proposed algorithms are gained revolving around some kinds of implications.

2   Preliminaries

The related definitions of t-norm, t-conorm, fuzzy implicationremaining, R-implication, (S, N)-implication please refer to [10].

Definition 1 ([11]). *An intuitionistic fuzzy set (IFS) on Q is referred to as*

$$C = \{< q, \pi_C(q), \tau_C(q) > \mid q \in Q\}.$$

Thereinto $\pi_C : Q \to [0,1]$, $\tau_C : Q \to [0,1]$ *and*

$$0 \le \pi_C(q) + \tau_C(q) \le 1 \; (q \in Q).$$

$\pi_C(q)$ *and* $\tau_C(q)$ *represent a membership function and a non-membership function of q to C, in turn. Evidently C can be denoted by:*

$$C(q) = (c,d), \quad 0 \le c + d \le 1, \; c, d \in [0,1], q \in Q.$$

IFSs expands the value of normal fuzzy sets from [0, 1] to

$$L^* = \{(c,d) \in [0,1]^2 \mid c + d \le 1\}.$$

When $\pi_C(q) = 1 - \tau_C(q)$ $(q \in Q)$, IFSs degrade into fuzzy sets. We use $IVF(Q)$ to signify the set of all IFSs in Q.

Let $p, q \in L^*$, $p = (p_1, p_2)$, and $q = (q_1, q_2)$. A partial order on L^* is characterized as below:

$$p \le_{PO} q \text{ if and only if } p_1 \le q_1, p_2 \ge q_2.$$

Evidently,

$$p \wedge q = (p_1 \wedge q_1, p_2 \vee q_2), \; p \vee q = (p_1 \vee q_1, p_2 \wedge q_2).$$

$0^* = (0,1)$ and $1^* = (1,0)$ are the least and maximal ones of L^*, in turn.

Here different arithmetic operations $\mathcal{T}_L$, $\mathcal{S}_L$ are provided on L^* as bellow:

$$\mathcal{T}_L(p,q) = (\mathcal{T}(p_1,q_1), \mathcal{S}(p_2,q_2)), \tag{5}$$

$$\mathcal{S}_L(p,q) = (\mathcal{S}(p_1,q_1), \mathcal{T}(p_2,q_2)), \tag{6}$$

Thereinto $\mathcal{S}$ *is the dual t-conorm for* $\mathcal{T}$.

Proposition 1 ([12]). *Suppose that S is a right-continuous t-conorm. There is an operation $\mathcal{M}$ (viz., the S-coresiduum) on $[0,1]$ letting that $(\mathcal{S}, \mathcal{M})$ establish a co-remaining pair, viz.,*

$$x \le \mathcal{S}(y,w) \iff \mathcal{M}(x,w) \le y,$$

and $\mathcal{M}$ is determined by

$$\mathcal{M}(c,d) = \wedge\{w \mid c \le \mathcal{S}(w,d)\}, \quad c, d, w \in [0,1]. \tag{7}$$

Definition 2 ([12]). $\mathcal{I}, \mathcal{S}, \mathcal{M}$ *are referred to as correlative operators of t-norm $\mathcal{T}$ if $(\mathcal{T}, \mathcal{I})$ is an remaining pair, $(\mathcal{S}, \mathcal{M})$ is a co-remaining pair, meanwhile S is the dual t-conorm of the t-norm $\mathcal{T}$.*

Definition 3. *When a mapping $\mathcal{I}_L$ from $L^* \times L^*$ to L^* possesses:*
    (L1'): $\mathcal{I}_L(c,d)$ is decreasing for c,
    (L2'): $\mathcal{I}_L(c,d)$ is increasing for d,
    (L3'): $\mathcal{I}_L(1^*,0^*) = 0^*$, $\mathcal{I}_L(1^*,1^*) = 1^*$ together with $\mathcal{I}_L(0^*,0^*) = 1^*$,
    then $\mathcal{I}_L$ is under the title of an intuitionistic fuzzy implication (IFI).

The definitions of an intuitionistic t-norm induced by a left-continuous $\mathcal{T}$, intuitionistic remaining pair and its properties please refer to [12].

Proposition 2 ([12]). *Suppose that $\mathcal{T}_L$ is an intuitionistic t-norm, while $(\mathcal{T}_L, \mathcal{I}_L)$ is an intuitionistic remaining pair, then one has*

(i) $\mathcal{I}_L(c, d) = 1^* \Longleftrightarrow c \leq_{PO} d$;
(ii) $w \leq_{PO} \mathcal{I}_L(c, d) \Longleftrightarrow c \leq_{PO} \mathcal{I}_L(w, d)$;
(iii) $\mathcal{I}_L(w, \mathcal{I}_L(c, d)) \Longleftrightarrow \mathcal{I}_L(c, \mathcal{I}_L(w, d))$;
(iv) $\mathcal{I}_L(1^*, d) = d$;
(v) $\mathcal{I}_L(d, \wedge_{i \in I} c_i) = \wedge_{i \in I} \mathcal{I}_L(d, c_i)$;
(vi) $\vee_{i \in I} \mathcal{I}_L(c_i, d) = \wedge_{i \in I} \mathcal{I}_L(c_i, d)$;
(vii) $\mathcal{I}_L(c, d)$ *is antitone in x and isotone in d.*

Here $w, c, d, c_i \in L^*$.

Definition 4 ([11]). *$\mathcal{E}$ is an intuitionistic fuzzy entropy if it meets the following conditions:*

i) $\mathcal{E}(C) = 0 \Leftrightarrow C$ *is a crisp set;*
ii) $\mathcal{E}(C) = 1 \Leftrightarrow \pi_C(c) = \tau_C(c)$ *for any $c \in X$;*
iii) $\mathcal{E}(C) = \mathcal{E}(C')$, *in which C' is a complementary set of C.*
iv) $\mathcal{E}(C) \leq \mathcal{E}(D)$ *if C is less fuzzy than D, viz., $\pi_C(c) \leq \pi_D(c)$ and $\tau_C(c) \geq \tau_D(c)$ for $\tau_D(c) \geq \pi_D(c)$; or $\pi_C(c) \geq \pi_D(c)$ and $\tau_C(c) \leq \tau_D(c)$ for $\pi_D(c) \geq \tau_D(c)$ for any $c \in X$.*

3 Intuitionistic Entropy-Induced Symmetric Implicational Algorithm for FMP

In the intuitionistic fuzzy environment, (4) is changed into

$$\mathcal{I}_{L2}(\mathcal{I}_{L1}(C(x), D(y)), \mathcal{I}_{L1}(C^*(x), D^*(y))). \tag{8}$$

Thereinto, $\mathcal{I}_{L1}$, $\mathcal{I}_{L2}$ are two IFIs.

Focusing on FMP (1) in intuitionistic fuzzy environment, from the main idea of intuitionistic entropy-induced symmetric implicational (IESI) algorithm, the standard is given:

IESI Standard for FMP: The perfect result D^* of FMP is the intuitionistic fuzzy set having the maximal intuitionistic fuzzy entropy letting (8) achieve its maximal value.

Because the IESI inference way is an better of the symmetric implicational mechnism, such standard increase the symmetric implicational standard for FMP in [8, 10].

Definition 5. *Let $C, C^* \in IVF(X)$, $D \in IVF(Y)$, if A^* makes (8) to achieve its maximal case for all $x \in X, y \in Y$. For such situation, A^* goes by the name of a foundational P-IESI result.*

Definition 6. *If* $C, C^* \in IVF(X)$, $D \in IVF(Y)$, *and* $\mathbb{IG}$ *is the collection of any foundational P-IESI results, and* A^* *has the maximal intuitionistic fuzzy entropy in* $\mathbb{IG}$, *then* A^* *goes by the name of an official P-IESI result.*

Suppose that the maximal value of (8) for FMP at (x, y) is $\mathbb{IMAX}(x, y)$.

Proposition 3. $\mathbb{IMAX}(x, y) = 1^*$.

Proof. By virtue of the characteristics what IFIs possess, one has

$$\mathcal{I}_{L1}(C^*(x), D^*(y)) \leq_{PO} \mathcal{I}_{L1}(C^*(x), 1^*)$$

and

$$\mathbb{IMAX}(x, y) \leq_{PO} \mathcal{I}_{L2}(\mathcal{I}_{L1}(C(x), D(y)), \mathcal{I}_{L1}(C^*(x), 1^*)).$$

Moreover, (L5') is valid for $\mathcal{I}_{L1}, \mathcal{I}_{L2}$, which implies that $\mathbb{IMAX}(x, y) = 1^*$. ∎

Theorem 1. *If* $\mathcal{I}_{L1}, \mathcal{I}_{L2}$ *are two remaining IFIs generated from left-continuous t-norms* $\mathcal{T}_1, \mathcal{T}_2$, *and* $\mathcal{T}_{L1}, \mathcal{T}_{L2}$ *are the operators remaining to* $\mathcal{I}_{L1}, \mathcal{I}_{L2}$, *then the official P-IESI result are* $(y \in Y)$:

$$D^*(y) = (\frac{1}{2}, \frac{1}{2}) \vee \sup_{x \in X}\{\mathcal{T}_{L1}(C^*(x), \mathcal{T}_{L2}(\mathcal{I}_{L1}(C(x), D(y)), \mathbb{IMAX}(x, y)))\}. \quad (9)$$

Proof. Above all, we mark $(y \in Y)$

$$J^*(y) = \sup_{x \in X}\{\mathcal{T}_{L1}(C^*(x), \mathcal{T}_{L2}(\mathcal{I}_{L1}(C(x), D(y)), \mathbb{IMAX}(x, y)))\}.$$

From J^*, it deduces that

$$\mathcal{T}_{L1}(C^*(x), \mathcal{T}_{L2}(\mathcal{I}_{L1}(C(x), D(y)), \mathbb{IMAX}(x, y))) \leq_{PO} J^*(y).$$

Because $(\mathcal{T}_{L1}, \mathcal{I}_{L1})$ and $(\mathcal{T}_{L2}, \mathcal{I}_{L2})$ are remaining pairs, one has

$$\mathcal{T}_{L2}(\mathcal{I}_{L1}(C(x), D(y)), \mathbb{IMAX}(x, y)) \leq_{PO} \mathcal{I}_{L1}(C^*(x), J^*(y))$$

and

$$\mathbb{IMAX}(x, y) \leq_{PO} \mathcal{I}_{L2}(\mathcal{I}_{L1}(C(x), D(y)), \mathcal{I}_{L1}(C^*(x), J^*(y))).$$

Thus we obtain

$$\mathcal{I}_{L2}(\mathcal{I}_{L1}(C(x), D(y)), \mathcal{I}_{L1}(C^*(x), J^*(y))) = \mathbb{IMAX}(x, y).$$

For this reason, $J^* \in \mathbb{IG}$ and J^* is a foundational P-IESI result.
   In addition, suppose $Q \in IVF(Y)$, and

$$\mathcal{I}_{L2}(\mathcal{I}_{L1}(C(x), D(y)), \mathcal{I}_{L1}(C^*(x), Q(y))) = \mathbb{IMAX}(x, y) \quad (x \in X, y \in Y).$$

Taking into account $(\mathcal{T}_{L1}, \mathcal{I}_{L1})$ and $(\mathcal{T}_{L2}, \mathcal{I}_{L2})$ are remaining pairs, it follows that

$$\mathcal{T}_{L2}(\mathcal{I}_{L1}(C(x), D(y)), \mathbb{IMAX}(x, y)) \leq_{PO} \mathcal{I}_{L1}(C^*(x), Q(y)).$$

It implies

$$\mathcal{T}_{L1}(C^*(x), \mathcal{T}_{L2}(\mathcal{I}_{L1}(C(x), D(y)),\ \mathbb{IMAX}(x, y))) \leq_{PO} Q(y).$$

For this reason, $Q(y)$ is embodied as a higher boundary of

$$\{\mathcal{T}_{L1}(C^*(x), \mathcal{T}_{L2}(\mathcal{I}_{L1}(C(x), D(y)), \mathbb{IMAX}(x, y))) \mid x \in X\}, \quad y \in Y.$$

By the structure of J^*, we obtain that $J^*(y) \leq_{PO} Y(y)$ $(y \in Y)$. For this reason, J^* is the smallest one of $\mathbb{IG}$.

Lastly, from (9), it implies that

$$D^* = (\frac{1}{2}, \frac{1}{2}) \vee J^*.$$

Then $J^*(y) \leq_{PO} D^*(y)$ $(y \in Y)$. It is evident to know that D^* is also a foundational P-IESI result, and hence $D^* \in \mathbb{IG}$.

Let D_0 be any element in $\mathbb{IG}$. Evidently $J^*(y) \leq_{PO} D_0(y)$ $(y \in Y)$. If

$$J^*(y) \leq_{PO} D_0(y) \leq_{PO} (\frac{1}{2}, \frac{1}{2})$$

or

$$J^*(y) \leq_{PO} (\frac{1}{2}, \frac{1}{2}) \leq_{PO} D_0(y),$$

then $D^*(y) = (\frac{1}{2}, \frac{1}{2})$. Then $(y \in Y)$

$$\mathcal{E}(D^*) = 1 \geq \mathcal{E}(D_0)$$

and thus D^* has largest entropy in $\mathbb{IG}$ by virtue of Definition 4. If

$$(\frac{1}{2}, \frac{1}{2}) \leq_{PO} J^*(y) \leq_{PO} D_0(y),$$

then $D^*(y) = J^*(y)$ $(y \in Y)$. So

$$\mathcal{E}(D_0) \leq \mathcal{E}(D^*)$$

in the light of Definition 4.

Together one has

$$\mathcal{E}(D^*) \geq \mathcal{E}(D_0).$$

All in all, D^* as (9) is the official P-IESI result.   ∎

Theorem 2. *If $\mathcal{I}_{L1}$, $\mathcal{I}_{L2}$ are two remaining IFIs generated by left-continuous t-norms $\mathcal{T}_1, \mathcal{T}_2$, and $\mathcal{T}_{L1}, \mathcal{T}_{L2}$ are the functions remaining to $\mathcal{I}_{L1}, \mathcal{I}_{L2}$, then the official P-IESI result are $(y \in Y)$:*

$$D^*(y) = (\frac{1}{2}, \frac{1}{2}) \vee \sup_{x \in X}\{\mathcal{T}_{L1}(C^*(x), \mathcal{T}_{L2}(\mathcal{I}_{L1}(C(x), D(y)), 1^*))\}. \tag{10}$$

Proof. From Proposition 3, (9) can also be written as (10).   ∎

Theorem 2 shows the fuzzy inference process from the intuitionistic fuzzy set C^* to D^* by the intuitionistic entropy-induced symmetric implicational algorithm. In practice, we need fuzzy systems to make fuzzy inference practical. A fuzzy system incorporates fuzzy inference, fuzzier and defuzzier. The most common strategies are the singleton fuzzier and centroid defuzzier.

First of all, we convert the intuitionistic fuzzy number p^* into a singleton intuitionistic fuzzy set

$$C^*(x) = \begin{cases} 1^*, & x = x^* \\ 0^*, & x \neq x^* \end{cases}.$$

In addition, we carry on fuzzy inference with the intuitionistic entropy-induced symmetric implicational algorithm to obtain intuitionistic fuzzy set D^*. Finally, we put to adopt centroid defuzzier to obtain

$$w^* = \int_Y w D^*(w) dw \ / \int_Y D^*(w) dw.$$

Hence, there is

$$w^* = \Phi(x^*)$$

for arbitrary input, and then a single-input single-output (SISO for short) fuzzy system is established.

But, the centroid defuzzier does not make sense if

$$D^*(w) \equiv 0.$$

To overcome this problem, for such case we make use of the defuzzier of mean from the largest, which employs

$$hgt = \{w \in Y \mid D^*(w) = \sup_{w \in Y} D^*(w)\}$$

and then

$$w^* = \int_{hgt} w dw / \int_{hgt} dw.$$

4   Intuitionistic Entropy-Induced Cooperative Symmetric Implicational Algorithm

In this section, we extend the intuitionistic entropy-induced symmetric implicational algorithm extended to the intuitionistic entropy-induced cooperative symmetric implicational (IECSI) algorithm.

To begin with, we show the inference strategy for the situation of multiple-input single-output (MISO).

The case of (1) is only for SISO. In reality, however, the situation of MISO is more common. Now we discuss the situation of double-input single-output

(DISO), which is able to be obviously extended to the MISO case. FMP (1) for DISO is transformed into the next structure:

Aiming at

| Rule: | C_1, C_2 | implies | D |

| and input | C_1^*, C_2^* | | | (11)

Obtain D^*

in which $C_1, C_1^* \in FZY(X_1)$, $C_2, C_2^* \in FZY(X_2)$, and $D, D^* \in FZY(Y)$.

In [13], the hierarchical inference strategy was considered for the CRI algorithm, which possessed a lot of superiorities, especially operation efficiency. Analogously, we introduce hierarchical inference idea to the intuitionistic entropy-induced symmetric implicational algorithm, which incorporates two phases:

(i) Focusing on $C_2 \Rightarrow D$ and C_2^*, by using the intuitionistic entropy-induced symmetric implicational algorithm, we can achieve D_1 from Theorem 2 ($y \in Y$):

$$D_1(y) = (\frac{1}{2}, \frac{1}{2}) \vee \sup_{x_2 \in X_2} \{\mathcal{T}_{L1}(C_2^*(x_2), \mathcal{T}_{L2}(\mathcal{I}_{L1}(C_2(x_2), D(y)), 1^*))\}. \quad (12)$$

(ii) Aiming at $C_1 \Rightarrow D_1$ and C_1^*, by employing the intuitionistic entropy-induced symmetric implicational algorithm, we lastly obtain the result D^* ($y \in Y$):

$$D^*(y) = (\frac{1}{2}, \frac{1}{2}) \vee \sup_{x_1 \in X_1} \{\mathcal{T}_{L1}(C_1^*(x_1), \mathcal{T}_{L2}(\mathcal{I}_{L1}(C_1(x_1), D_1(y)), 1^*))\}.$$

$$= (\frac{1}{2}, \frac{1}{2}) \vee \sup_{x_1 \in X_1} \left\{ \mathcal{T}_{L1}(C_1^*(x_1), \mathcal{T}_{L2}(\mathcal{I}_{L1}(C_1(x_1), \quad (13) \right.$$

$$\left. (\frac{1}{2}, \frac{1}{2}) \vee \sup_{x_2 \in X_2} \{\mathcal{T}_{L1}(C_2^*(x_2), \mathcal{T}_{L2}(\mathcal{I}_{L1}(C_2(x_2), D(y)), 1^*))\}), 1^*)) \right\}.$$

Furthermore, we give the cooperative inference mechanism and we call it the intuitionistic entropy-induced cooperative symmetric implicational algorithm.

When we face multiple rules, (11) is transformed into the following situation (here two rules are empolyed):

Aiming at

| Rule 1 : | C_{11}, C_{12} | implies | D_1 |

| Rule 2 : | C_{21}, C_{22} | implies | D_2 |

| and input | C_1^*, C_2^* | | | (14)

Obtain D^*

in which $C_{11}, C_{21}, C_1^* \in FZY(X_1)$, $C_{12}, C_{22}, C_2^* \in FZY(X_2)$, and $D_1, D_2, D^* \in FZY(Y)$. We can use $C_1 = C_{11} \vee C_{21}$, $C_2 = C_{12} \vee C_{22}$, $D = D_1 \vee D_2$. Then the case (14) is changed into (11).

But this is the simpler case. More generally, many rules are used as follows:

$$
\begin{array}{lll}
\text{Aiming at} & & \\
\text{Rule 1}: & C_{11}, \ C_{12} & \text{implies} \ D_1 \\
\text{Rule 2}: & C_{21}, \ C_{22} & \text{implies} \ D_2 \\
\cdots & \cdots, \ \cdots & \text{implies} \ \cdots \\
\text{Rule } m: & C_{m1}, \ C_{m2} & \text{implies} \ D_m \\
\text{and input} & C_1^*, \ C_2^* & \\
\hline
\text{Obtain} & & D^*
\end{array}
\tag{15}
$$

in which $C_{11}, C_{21}, \cdots, C_{m1}, C_1^* \in FZY(X_1)$, $C_{12}, C_{22}, \cdots, C_{m2}, C_2^* \in FZY(X_2)$, and $D_1, D_2, \cdots, D_m, D^* \in FZY(Y)$.

The basic processing mode of cooperative inference is as follows. First, the general inference tasks are grouped into respective relatively separated sub-tasks, and then these sub-tasks are proceeded separately, and lastly the intermediate results are aggregated into the total results.

The task decomposition here is obtained by describing the attributes of the antecedents of rules, and the process is realized by using fuzzy clustering algorithm, which takes the high similarity within the class and low similarity between the classes as the basic principle of division. The fuzzy C-means clustering algorithm [14] is adopted to form the required clustering results. The obtained clustering results are used as the basis for the division of collaborative tasks. For examples, suppose that there are 12 rules, then after maybe Rules 1–4 belong to the first class, Rules 5–9 belong to the second class, Rules 10–12 belong to the third class.

Then the results of each sub-task are compounded. Suppose that r classes are obtained by the FCM algorithm, and that the reasoning results obtained respectively are $D_1^*, D_2^*, \cdots, D_r^*$, then we employ $D^* = \vee_{i=1,2,\cdots,r} D_i^*$ to get the final results.

5   Intuitionistic Entropy-Induced Symmetric Implicational Algorithm for FMT

In view of FMT (2) in intuitionistic fuzzy environment, from the main structure of the intuitionistic entropy-induced symmetric implicational algorithm, we provide the next standard:

IESI Standard for FMT: The ideal result C^* of FMT is the intuitionistic fuzzy set having the largest intuitionistic fuzzy entropy letting (8) attain its largest value.

Obviously such IESI standard enhances the symmetric implicational standard for FMT in [8,10].

Definition 7. *Let $C \in IVF(X)$, $D, D^* \in IVF(Y)$, if C^* (in $IVF(X)$) makes (8) achieve the maximal cases for $x \in X, y \in Y$, then C^* is referred to as a foundational T-IESI result.*

Definition 8. *If $C \in IVF(X)$, $D, D^* \in IVF(Y)$, and $\mathbb{IH}$ is the series of any foundational T-IESI results, and C^* possesses the biggest intuitionistic fuzzy entropy in $\mathbb{IH}$, then C^* goes by the name of an official T-IESI result.*

Suppose that the maximum of (8) for FMT at (x, y) is $\mathbb{IMAXT}(x, y)$.

Proposition 4. $\mathbb{IMAXT}(x, y) = 1^*$.

Proof. By the properties of IFIs, we have

$$\mathcal{I}_{L1}(C^*(x), D^*(y)) \leq_{PO} \mathcal{I}_{L1}(0^*, D^*(y))$$

and

$$\mathbb{IMAXT}(x, y) \leq_{PO} \mathcal{I}_{L2}(\mathcal{I}_{L1}(C(x), D(y)), \mathcal{I}_{L1}(0^*, D^*(y))).$$

Furthermore, (L4') and (L5') work for $\mathcal{I}_{L1}, \mathcal{I}_{L2}$, which results in that

$$\mathbb{IMAXT}(x, y) = 1^*.$$

∎

Theorem 3. *If $\mathcal{I}_{L1}$, $\mathcal{I}_{L2}$ are remaining IFIs generated from left-continuous t-norms $\mathcal{T}_1, \mathcal{T}_2$, and $\mathcal{T}_{L1}, \mathcal{T}_{L2}$ are the functions remaining to $\mathcal{I}_{L1}$, $\mathcal{I}_{L2}$, then the official T-IESI result are ($x \in X$):*

$$C^*(x) = (\frac{1}{2}, \frac{1}{2}) \wedge \inf_{y \in Y}\{\mathcal{I}_{L1}(\mathcal{T}_{L2}(\mathcal{I}_{L1}(C(x), D(y)), \mathbb{IMAXT}(x, y)), D^*(y))\}(16)$$

Proof. Above all, we sign ($x \in X$)

$$J^*(x) = \inf_{y \in Y}\{\mathcal{I}_{L1}(\mathcal{T}_{L2}(\mathcal{I}_{L1}(C(x), D(y)), \mathbb{IMAXT}(x, y)), D^*(y))\}.$$

From J^*, one has

$$J^*(x) \leq_{PO} \mathcal{I}_{L1}(\mathcal{T}_{L2}(\mathcal{I}_{L1}(C(x), D(y)), \mathbb{IMAXT}(x, y)), D^*(y)).$$

Since $(\mathcal{T}_{L1}, \mathcal{I}_{L1})$ and $(\mathcal{T}_{L2}, \mathcal{I}_{L2})$ are remaining pairs, it follows from Proposition 2(ii) that

$$\mathcal{T}_{L2}(\mathcal{I}_{L1}(C(x), D(y)), \mathbb{IMAXT}(x, y)) \leq_{PO} \mathcal{I}_{L1}(J^*(x), D^*(y))$$

and

$$\mathbb{IMAXT}(x, y) \leq_{PO} \mathcal{I}_{L2}(\mathcal{I}_{L1}(C(x), D(y)), \mathcal{I}_{L1}(J^*(x), D^*(y))).$$

Hence we gain ($x \in X, y \in Y$)

$$\mathcal{I}_{L2}(\mathcal{I}_{L1}(C(x), D(y)), \mathcal{I}_{L1}(J^*(x), D^*(y))) = \mathbb{IMAXT}(x, y).$$

As a result, $J^* \in \mathbb{IH}$ and J^* is a foundational T-IESI result.

Thereafter, suppose $Q^* \in IVF(X)$ and

$$\mathcal{I}_{L2}(\mathcal{I}_{L1}(C(x), D(y)), \mathcal{I}_{L1}(Q^*(x), D^*(y))) = \mathbb{IMAXT}(x, y),$$

$x \in X, y \in Y$. Taking into account that $(\mathcal{T}_{L1}, \mathcal{I}_{L1})$ and $(\mathcal{T}_{L2}, \mathcal{I}_{L2})$ are remaining pairs, from Proposition 2(ii), we obtain that

$$\mathcal{T}_{L2}(\mathcal{I}_{L1}(C(x), D(y)), \mathbb{IMAXT}(x, y)) \leq_{PO} \mathcal{I}_{L1}(Q^*(x), D^*(y))$$

and then

$$Q^*(x) \leq_{PO} \mathcal{I}_{L1}(\mathcal{T}_{L2}(\mathcal{I}_{L1}(C(x), D(y)), \mathbb{IMAXT}(x, y)), D^*(y)).$$

In consequence, $Q^*(x)$ is expressed as a lower margin of

$$\{\mathcal{I}_{L1}(\mathcal{T}_{L2}(\mathcal{I}_{L1}(C(x), D(y)), \mathbb{IMAXT}(x, y)), D^*(y)) \mid y \in Y\}, \quad x \in X.$$

We find from the structure of J^* that $(x \in X)$

$$X^*(x) \leq_{PO} J^*(x).$$

Therefore, J^* is the largest one of $\mathbb{IH}$.

Finally, it follows from (16) that

$$C^* = (\frac{1}{2}, \frac{1}{2}) \wedge J^*.$$

Then $C^*(x) \leq_{PO} J^*(x)$ $(x \in X)$. It is obviously to know that C^* is also a foundational T-IESI result, and hence $C^* \in \mathbb{IH}$.

Let C_0 be any element in $\mathbb{IH}$. Evidently $C_0(x) \leq_{PO} J^*(x)$ $(x \in X)$.

If

$$(\frac{1}{2}, \frac{1}{2}) \leq_{PO} C_0(x) \leq_{PO} J^*(x)$$

or

$$C_0(x) \leq_{PO} (\frac{1}{2}, \frac{1}{2}) \leq_{PO} J^*(x),$$

then $C^*(x) = (\frac{1}{2}, \frac{1}{2})$ $(x \in X)$. Then $\mathcal{E}(C^*) = 1 \geq \mathcal{E}(C_0)$ and thus C^* has largest entropy in $\mathbb{IH}$ according to Definition 4.

If

$$C_0(x) \leq_{PO} J^*(x) \leq_{PO} (\frac{1}{2}, \frac{1}{2}),$$

then $C^*(x) = J^*(x)$ $(x \in X)$. So $\mathcal{E}(C_0) \leq \mathcal{E}(C^*)$ by virtue of Definition 4.

All at once we obtain

$$\mathcal{E}(C^*) \geq \mathcal{E}(C_0).$$

In conclusion, C^* as (16) is the official T-IESI result. ∎

Theorem 4. *If $\mathcal{I}_{L1}, \mathcal{I}_{L2}$ are remaining IFIs generated from left-continuous t-norms $\mathcal{T}_1, \mathcal{T}_2$, and $\mathcal{T}_{L1}, \mathcal{T}_{L2}$ are the operators remaining to $\mathcal{I}_{L1}, \mathcal{I}_{L2}$, then the official T-IESI result are $(x \in X)$:*

$$C^*(x) = (\frac{1}{2}, \frac{1}{2}) \wedge \inf_{y \in Y} \{\mathcal{I}_{L1}(\mathcal{T}_{L2}(\mathcal{I}_{L1}(C(x), D(y)), 1^*), D^*(y))\}. \quad (17)$$

Proof. In line with Proposition 4, (16) can also be depicted as (17). ∎

6 Conclusions

We come up with the intuitionistic entropy-induced symmetric implicational algorithm and extend it to the intuitionistic entropy-induced cooperative symmetric implicational algorithm. Above all, the IESI standard are proposed, which is applied to FMP and FMT, and it is an improvement on the previous symmetric implicational standard. In addition, the basic properties of solutions of the intuitionistic entropy-induced symmetric implicational algorithm are discussed, and then optimal unified solutions of the proposed method are gained for FMP and FMT, which are based upon some characteristics which intuitionistic fuzzy implications possess. Lastly, its cooperative reasoning version is proposed for multiple inputs and multiple rules, while the corresponding optimal solutions are also found.

In the future, we will investigate a more appropriate fuzzy clustering algorithm to better adapt to the cooperative segmentation of fuzzy rules. And, we will put into use the proposed fuzzy inference method in the fields of affective computing together with machine learning.

References

1. Tang, Y.M., Pedrycz, W.: Oscillation-bound estimation of perturbations under Bandler-Kohout subproduct. IEEE Trans. Cybern. (2020). https://doi.org/10.1109/TCYB.2020.3025793
2. Wang, G.J.: On the logic foundation of fuzzy reasoning. Inf. Sci. $117(1)$, 47–88 (1999)
3. Guo, F.F., Chen, T.Y., Xia, Z.Q.: Triple I methods for fuzzy reasoning based on maximum fuzzy entropy principle. Fuzzy Syst. Math. $17(4)$, 55–59 (2003)
4. Luo, M.X., Liu, B.: Robustness of interval-valued fuzzy inference triple I algorithms based on normalized Minkowski distance. J. Logical Algebraic Methods Program. $86(1)$, 298–307 (2017)
5. Tang, Y.M., Liu, X.P.: Differently implicational universal triple I method of (1, 2, 2) type. Comput. Math. Appl. $59(6)$, 1965–1984 (2010)
6. Tang, Y.M., Ren, F.J.: Variable differently implicational algorithm of fuzzy inference. J. Intell. Fuzzy Syst. $28(4)$, 1885–1897 (2015)
7. Li, H.X.: Probability representations of fuzzy systems. Sci. China Ser. F-Inf. Sci. $49(3)$, 339–363 (2006)
8. Tang, Y.M., Yang, X.Z.: Symmetric implicational method of fuzzy reasoning. Int. J. Approximate Reason. $54(8)$, 1034–1048 (2013)
9. Tang, Y.M., Pedrycz, W.: On the $\alpha(u, v)$-symmetric implicational method for R- and (S, N)-implications. Int. J. Approximate Reason. 92, 212–231 (2018)
10. Tang, Y.M., Song, X.C., Ren, F.J., Pedrycz, W.: Entropy-based symmetric implicational method for R- and (S, N)-implications. In: IEEE ISKE 2019, pp. 683–688 (2019)
11. Szmidt, E., Kacprzyk, J.: Entropy for intuitionistic fuzzy sets. Fuzzy Sets Syst. $118(3)$, 467–477 (2001)
12. Zheng, M.C., Shi, Z.K., Liu, Y.: Triple I method of approximate reasoning on Atanassov's intuitionistic fuzzy sets. Int. J. Approximate Reason. $55(6)$, 1369–1382 (2014)

13. Jayaram, B.: On the law of importation $(x \wedge y) \to z \equiv (x \to (y \to z))$ in fuzzy logic. IEEE Trans. Fuzzy Syst. **16**(1), 130-144 (2008)
14. Tang, Y.M., Hu, X.H., Pedrycz, W., Song, X.C.: Possibilistic fuzzy clustering with high-density viewpoint. Neurocomputing **329**(15), 407–423 (2019)

Multi-task Allocation Strategy and Incentive Mechanism Based on Spatial-Temporal Correlation

Zihui Jiang[1]([⊠]) and Wenan Tan[1,2]

[1] Nanjing University of Aeronautics and Astronautics, Nanjing, Jiangsu, China
zihuijiang@nuaa.edu.cn
[2] Shanghai Polytechnic University, Shanghai, China
watan@sspu.edu.cn

Abstract. Task allocation and incentive mechanism are two main research challenges of the mobile crowdsensing system (MCS). Most previous researches commonly focus on sensing users' selection while ignoring the concurrency and spatial-temporal attributes of different tasks performed in the real environment. With the dynamic changes of the opportunity MCS context, the assurance data quality and accuracy tend to be controversial under the heterogeneous characteristics of tasks. To solve this problem, we first elaborate and model the key concepts of the task's spatial-temporal attributes and sensing user mobility attributes, and then propose a novel framework with two fine-grained algorithms incorporating deep reinforcement learning, named DQN-TAIM. By observing the state of the MCS context, DQN-TAIM enriches and accumulates with more decision transitions, which can learn continuously and maximize the cumulative benefits of the platform from the interaction between users and the MCS environment. Finally, we evaluate the proposed method using a real-world data set, then compare and summarize it with a benchmark algorithm. The results confirm that DQN-TAIM has made excellent results and outperforms the benchmark method under various settings.

Keywords: Mobile crowdsensing · Opportunity sensing · Spatial-temporal attributes · Deep reinforcement learning · Cumulative benefits

1 Introduction and Related Work

In recent years, with the continuous development of IoT, MCS has become a universal sensing and computation technology, which has shown great potential in sensing large-scale areas using embedded sensing mobile devices. In the urban environment, surveillance applications that need the sensing data of multi-source tasks to infer useful information are increased progressively. The MCS platform can publish a set of multi-source heterogeneous demands, and recruit workers to execute self-correlation sensing tasks, which has great commercial prospect. For example, Waze [1] and GreenGPS [2] are used to collect road traffic information, DietSensor [3] is used to share and track users' diet and nutrition, and NoiseTube [4] is used to monitor sound pollution.

© Springer Nature Singapore Pte Ltd. 2021
Y. Sun et al. (Eds.): ChineseCSCW 2020, CCIS 1330, pp. 155–166, 2021.
https://doi.org/10.1007/978-981-16-2540-4_12

In several MCS applications, the platform has the requirement for the high quality of the sensing data. To satisfy the requirement, a common approach is to select different users to perform tasks in a certain sensing area during each sensing cycle for multiple data samples. To stimulate and recruit users with mobile devices to participate in perception activities, platforms usually provide users with rewards to compensate for their sensing costs or risks of privacy exposure [5]. Therefore, designing a reasonable incentive strategy that helps maintain the platform's profitable market sustainably is central to state-of-the-art studies. Many outstanding scholars have proposed various incentive mechanisms based on the greedy concept to optimize the utility of the platform [6–8]. However, in real scenarios, the context of the MCS environment is changing dynamically, including the spatial-temporal attribute context of the task and the quality of service (QoS) context. The literature [9] proposed a user allocation algorithm based on a greedy strategy to improve the balance of user coverage with temporal and spatial changes. C. Bassem [10] proposed user mobility prediction and coordination schemes based on the real-time spatial-temporal attributes of tasks. As stated above, data quality and user incentive cost are the two main and opposing issues in MCS. Therefore, how to provide task allocation strategies that can meet the dynamic changes of context for maximizing platform benefits is the central issue that this article focuses on.

In the scenario of opportunity sensing, users register on the MCS platform to obtain sensing qualifications. After the platform assigns tasks to them, users will unconsciously perform perception tasks through constant movement. Considering the probabilistic nature of users passing through different perceptual regions, the probability is included in the platform utility calculation. To guarantee the accuracy of sensing data, multiple users are assigned to independently collect the sensing data of the same area. Turning now to the prediction of individual mobility, we have the following intuitions. When the training data is sparse, the individual's movement trajectory is difficult to predict [11]. However, the mobility pattern of users has certain hidden characteristics. More specifically, individuals' movement trajectories are usually related to daily commuting, and regular users have a high probability to perform tasks in the historical perception area.

More discussion about mobility prediction can be found in this recently published article. Y. Hu pointed out that many optimization algorithms are based on the premise that the platform has a fixed and sufficient history movement trajectory, which is unrealistic in the real world [12]. We do not deny that there is such a possibility, but in actual application scenarios, newly registered users need a certain period of training essential sensing skills and accumulate experience [13]. After their experience value reaches the requirement of the platform, workers can be selected for real sensing tasks. During the training period, the platform can obtain enough historical movement trajectory through GPS. Although there is a risk of exposing privacy by their movement trajectories, the relation between the MCS platform and sensing users is a two-way choice. After all, the platform will reward users based on their sensing performance.

Due to the ability to learn the best strategy from the interaction between the subject and the environment, deep learning (DL) and reinforcement learning (RL) have become popular methods for solving optimization problems. In the literature [14], the author proposed a perceptual vehicle selection algorithm based on deep reinforcement learning, which selects multiple vehicles to perform sensing tasks according to the selection strategy. Literature [15] uses deep reinforcement learning to allocate different resources to the crowd in fog computing. To summarize, reinforcement learning has strong application potential in the domain of MCS [16], and it also has unlimited research value academically.

The main contributions of this paper are summarized as follows:

- We first elaborate on the key issues that need to be solved in the current MCS application and proposed a system framework to solve the spatial-temporal correlation of the sensing task in the real MCS environment.
- Our proposed method DQN-TAIM incorporates the latest technological advances: deep Q-learning, which uses the powerful feature representation capabilities of neural networks as an approximate representation function of the state value. To maximize the cumulative benefits brought to the platform in the interaction between users and the environment, our DQN agent observes the context state of the environment and intelligently distributes tasks to users.
- We implement the proposed algorithm with the real dataset, then compare and summarize it with a benchmark algorithm. The results show that our DQN-TAIM algorithm has brought excellent benefits to the platform in the simulation experiment.

2 Preliminary and System Definition

In this part, the definitions of sensing users and tasks' attribution are given in the first place. Next, we elaborate on the system framework with the spatial-temporal coverage. To better describe and express, the common symbols are listed in Table 1.

Table 1. Symbols for system attributes

Symbols	Explanations
$W = \{w_i\}$	Sensing-user pool
$T = \{t_i\}$	Sensing-task pool
$Region = \{reg_i\}$	Sensing regions
$Cycle = \left\{cy_i^k\right\}$	Sensing cycle where w_k is in
$Pay(w_i)$	w_k's payment
$Cov(t_i)$	$t_i's$ sensing coverage
$CR_{i,j}$	Cycle-region pair (cy_i, reg_i)
$WCR_{u,i,j}$	Chosen worker-cycle-region pair

MCS is a wide-area and spatial-temporal sensing network formed by large-scale users by constant movement. Moreover, the sensing tasks are complex and diverse, which are executed concurrently by sensing workers in different sensing cycles and regions. Therefore, we divide the sensing area into multiple subareas, where users can perform sensing activities during different sensing cycle. We give the definitions of sensing users and tasks down below respectively.

2.1 Sensing Worker Model

Definition 1 (Sensing Workers). Sensing users are the resource users who register on the MCS platform. Each of them can perform different tasks according to the condition their sensing equipment. Additionally, users who carry devices in sensing subregions can execute passively sensing and get corresponding rewards after the centralized distribution and scheduling of the platform.

Since building a reasonable user incentive model helps fair competition among participants, as well as promoting the sustainability and development of the MCS platform. We follow the universal labor incentive model that suggests that workers can get more pay for more work done. The incentive reward of w_u in cy_i is formulated as follows:

$$Pay_i(w_u) = P_0 + P_1 * (N - 1), \tag{1}$$

where N is the total number of tasks performed by w_u in cy_i. P_0 is the basic reward for recruiting users, and P_1 is the ratio reward relative to N, usually $P_1 < P_0$.

The platform follows a centralized design, which collects the historical trajectory of the worker in the target subregions, and predicts the user's mobility model in the same $CR_{i,j}$. Moreover, the platform optimizes the target selection based on the predicted results, thereby maximizing the subregion sensing coverage. When the user connects to the signal tower where reg_i is located, the platform will push the assigned tasks to the workers. The probability that the worker w_u connects to the subarea at least once in $CR_{i,j}$ is formulated as follows [17]:

$$pro_{i,j}(w_u) = 1 - e^{-\lambda_{u,i,j}}, \tag{2}$$

where $\lambda_{u,i,j}$ is the average number of connections to $CR_{i,j}$'s signal towers 7 days ago.

2.2 Sensing Task Model

Definition 2 (Sensing Tasks). We define the task set as $T = \{t_1, t_2, t_3, ..., t_n\}$, each task has spatial-temporal coverage requirements in different $CR_{i,j}$. Furthermore, the feature of spatial-temporal attributes is composed of multiple fine granularity sensing cycles and subregions. Our work aims to maximize the average $Cov_i(T)$ of all the sensing regions in cy_i. The calculation formula for the coverage rate of task t_n is:

$$Cov_i(t_n) = \frac{1}{\gamma} \sum\nolimits_{reg_j \in Reg} ST_{i,j}(t_n) * p_{i,j}(t_n), \tag{3}$$

$$p_{i,j}(t_n) = 1 - \prod\nolimits_{w_u \in WCR} (1 - pro_{u,i,j}), \tag{4}$$

where γ is the number of all subareas, and $ST_{i,j}(t_n)$ is the sensing times of t_n in $CR_{i,j}$. $WCR_{u,i,j}$ is the worker-cycle-region pair which is selected by our DQN agent. $p_{i,j}(t_n)$ is the sensing probability of task t_n in $CR_{i,j}$, which is determined by users in $WCR_{u,i,j}$.

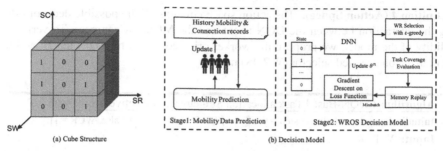

Fig. 1. (a) Cube structure to introduce the solution. (b) Decision model of our proposed algorithm DQN-TAIM.

The utility of the MCS platform in cy_i is calculated as follows:

$$Utility = \sum_{t_n \in T} Cov_i(t_n) - \alpha * \sum_{WCR_{u,i,j}} Pay_{i,j}(w_u), \tag{5}$$

where α is set to 0.1 which is used for the balance between sensing coverage and users' payment.

To better introduce the WCR assignment solution, we abstract the problem into a three-dimensional model, as shown in Fig. 1(a). It is known that tasks can be executed by multiple sensing workers SW in multiple sensing regions SR. We aim to achieve the optimal platform utility during all sensing cycle SC. A selection $WCR_{u,i,j}$ means that the lattice (w_u, cy_i, reg_j) in the cube structure is set to 1. Therefore, the task assignment problem in our scenario is equivalent to optimally fill triple cube lattices with 0 or 1 [18]. It is not feasible to use the traditional brute force search due to time and space complexity is too high to complete. Therefore, an outstanding approximation algorithm is required to calculate within polynomial time.

3 Method Overview

In this part, we further explain how to combine deep reinforcement learning with MCS. The purpose of using DQN agents is to find an optimal state-action strategy to maximize accumulated platform utility.

3.1 Definition of System Components

Definition 3 (System State). The combination of DQN and MCS can be regarded as how to maximize the cumulative long-term utility of the MCS platform. In this way, the platform can obtain relatively the best benefits with the lower sensing payment. According to the spatial-temporal properties of the task, we use the average sensing coverage probability of each task in all subregions to represent the system state. When $Cov_i(t_n)$ meets the spatial-temporal coverage requirement of $Req(t_n)$, $Cov_i(t_n)$ will no longer increase, and the corresponding status item of the task is converted to 1, as shown in the following formula. Through the above method, the DQN agent tends to select tasks that do not meet the perceived requirements in the next iteration to get the total utility increased.

$$S[n] = \begin{cases} 1, & Cov_i(t_n) \geq Req(t_n) \\ 0, & else \end{cases}. \tag{6}$$

Definition 4 (Action Space). The action space contains all the possible decisions during the interaction between the DQN agent and the MCS environment. To verify the validity of the model, we propose the worker-oriented selector (WOS) model and the worker-region oriented selector (WROS) model.

Algorithm 1 DQN-Based Task allocation Strategy and Incentive Mechanism

---
Initialization: step = 0, learn_step = 0, $s^n = [0, 0, 0, ..., 0]$, done = False, WCR = []

Input: W, T

Output: the maximum utility, WCR

1: Initialize DQN with random weights θ

2: **While** True **do**

3:    Get a random number x between 0 and 1

4:    **If** x >= ε

5:       Choose $a = argax_a Q(s, a; \theta)$

6:    **Else**

7:       Choose an action randomly and $action \rightarrow WCR$

8:    **End If**

9:    s', reward, done = WOS/WROS(a)

10:   Store (s, a, reward, s') in replay memory

11:   s = s', step++

12:   **If** step >= LEARN_COUNTER && step % 2 == 0

13:      Randomly sample a minibatch of transition

14:      $Q_{target} = reward + \lambda max_{a'} Q(s', a'; \theta)$

15:      Perform gradient descent on loss function

16:      learn_step ++

17:      **If** learn_step % REPLACE_COUNTER

18:         Update parameters of θ

19:      **End If**

20:   **End If**

21:   **If** done == True

22:      **Return** the maximum utility of this cycle

23:   **End If**

24: **End While**

In the WOS model, the action space is n sensing users. When the worker is selected, all tasks satisfying the condition of the worker's device will be performed in all the perceptive regions he participates in. If the platform has already selected some perceptive users, the chosen probability in the same cy_i is 0. According to the temporal and spatial attributes of tasks, we further propose a more fine-grained decision scheme. In WROS, the behavioral decision space is the Cartesian product for workers and their participating sensing regions denoted as $WR_{u,j}$, which can approach the global optimal through fine-grained allocation and combination.

Definition 5 (Reward). As soon as a decision is made, our DQN agent will receive an immediate reward and the next state by MCS environment as feedback. In each sensing cycle, we make a choice iteratively until the utility of the current cycle meets the platform utility requirement. The reward is initialized to 0 and returns 1 when the platform utility requirement is satisfied.

3.2 Details of the Decision Model

The decision-making process is divided into two stages. The first stage starts before the sensing cycle, our MCS platform calculates the user's sensing probability of all sensing regions by his historical movement trajectory.

In the second stage, before the start of the formal cycle, the DQN agent first conducts exploration and trains the decision model with the return reward. The task will not be assigned during exploration. Then, the latest decision model will make a choice iteratively until the utility of the current cycle meets the platform utility requirement. Due to the combination of system states is various, we corporate neural network functions to approximate the value of states and behaviors. The value function $Q(s, a; \theta)$ takes the current state vector as input, then outputs the value of different actions. Through continuously learning, our DQN agent adjusts the value of the parameter θ to conform to the final state value based on the decision strategy.

Algorithm 2 WOS/WROS Pseudo Code

Initialization: SU = {}, reward = 0

Input: s, action, W, T

Output: s', reward, done

  1: Get w_{action}'s sensing cells and sensing tasks

  2: **Foreach** t_i in sensing tasks

  3:    Add w_{action} in $SU[t_i]$

  4: Calculate payment of w_{action} via (1)

  5: **Foreach** t_i **in** T

  6:    Calculate coverage of t_i via (3)

  7: **If** $Cov(t_n) \geq Req(t_i)$

  8:    $s[i] = 1$

  9: Calculate utility of this cycle via (5)

10: **If** utility >= DEMAND

11:   done = True

12:   reward = 1

13: **Return** s', reward, done

The DQN-Strategy algorithm takes advantage of experience replay (ER) to implement the convergence of the system state value function [19]. The DQN agent selects an action a_t according to the current state matrix s_t by the ϵ-greedy strategy, and receive an immediate reward and the next state s'. Then stores the experience of the state transition $(s, a, reward, s')$ into the memory storage. DQN is a kind of off-policy learning method, which can not only learn the experience in the current sensing cycle, but also the experience in the past cycle. Moreover, when the capacity of the memory storage is large enough, a certain number of state transition experiences are randomly extracted for learning, which disrupts the correlation between experiences and makes the learning of agents more efficient. In addition to ER, fixed Q-Target is also a disruptive mechanism. It constructs two neural networks with the same structure but different parameters. The neural network of Q estimation has the latest parameters, while Q prediction's parameters are outdated.

$$Q_{target} = reward + \lambda max_{a'} Q(s', a'; \theta'),$$ (7)

where s' is the parameter to the value network of the last upgrade. The reward is set to 1 when the platform utility satisfies the target DEMAND. λ is the attenuation coefficient of the long-term reward, usually in the range of [0, 1].

There are two parameters in the input layer, one is the current system state, and the other is the value of Q-Target. The error value is obtained by subtracting Q-Target from the estimated value, which is used for parameter promotion through backpropagation, as shown in the following formula.

$$Q(s, a; \theta) = Q(s, a; \theta) + \alpha (Q_{target} - Q(s, a; \theta)),$$ (8)

where α is the learning ratio for upgrade θ. When the learning count reaches REPLACE_COUNTE, DQN will update the network parameters by the minibatch gradient descent, and replace the Q prediction's parameters with the latest one. Details can be found in Algorithm 1 and 2.

4 Experiment Result

In this section, we evaluate the effectiveness of the DQN-TAIM using real signal tracking data sets. Firstly, we introduced the background of the dataset, the benchmark algorithm, and parameter settings. Then, the evaluation results of WOS and WROS versus the benchmark algorithm are presented and compared in detail.

4.1 Dataset and Simulation Settings

The Called Detail Records (CDR) were collected and maintained by the Semantics and Knowledge Innovation Lab from the City of Milan and the Province of Trentino in Italy in 2014. Users interact through SMS or phone call [20]. When the signal tower collects such information, a CDR record will be generated to record the user's information, interaction time, and the service area. We select the service data within two weeks, including the 20 hot service areas which are divided into 235 m × 235 m. The data in the first week is used as the user's historical trajectory, and the second week's data is used for model training and decision-making. Moreover, we divide the day into 10 sensing cycles of one hour each, starting at 8 a.m.

The benchmark is the EBRA algorithm [21], which selects the user-task pair by the greedy strategy in each iteration. Since the system framework is different from DQN-TAIM, here we only compare the spatial-temporal coverage and the platform utility within the same sensing cycle.

Through trial and error, we select the parameter with the optimal performance. In the exploration stage, the initial value of ϵ is set to 0.8. With the improvement of model performance, the value of ϵ is increased in the continuous iteration which indicates the randomness of selection was gradually reduced. In the decision-making stage, ϵ will be fixed at 0.9 as the model can make correct decisions. In formula (1), we set p_0 to be 5 and p_1 to be 1, and specify the maximum number of task samples to be 5. Besides, $Req(t_n)$ is set to 4. The number of user's ability of tasks follows the Poisson distribution, ranging from 1 to 5. The Learning rate of Q function is 0.1, α is 0.9. The structure of the neural network is set to an input layer, two hidden layers, and an output layer. For initialization, θ is between 0 and 0.3 by random, bias is set to 0.1. At last, the storage of ER is 2000.

4.2 Numerical Results

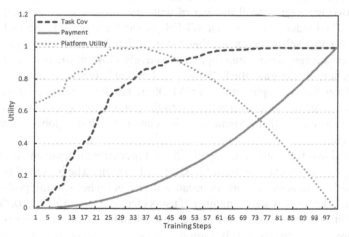

Fig. 2. The trend of task coverage, users' payment and platform utility during the same sensing cycle.

We first set 100 perception users, 20 perception tasks, and selected about 5000 CDR data. The WOS decision model trained in the exploratory stage was used to make decisions iteratively without the stopping conditions. Task Coverage is the average spatial-temporal coverage of the total tasks. As Fig. 2 shows, through the standardization process, these three types of data move forward to a mutually restrictive relationship in the graph. In other words, when the average coverage of the tasks changes from the rising stage to the smoothing stage, the platform utility begins to decline gradually due to users' increasing payment, which shows that our system model accords with the practical circumstances.

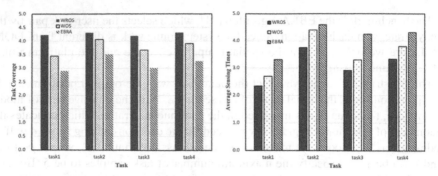

Fig. 3. The performance of three algorithms in task coverage and average sensing time.

In this experiment, we randomly select four tasks to track and record their sensing process. Figure 3(a) shows the average task coverage of all sensing areas in the same sensing cycle for these tasks, and Fig. 3(b) shows the average sensing times of tasks under the condition that the platform utility meets the demand. What makes us confident is that among the three incentive mechanisms including WROS, WOS, and EBRA, WROS has achieved the best performance in both indicators. WROS can achieve the same spatial-temporal coverage with a small number of workers, which means the platform has a fewer cost. The hidden reason is that WROS provides a more fine-grained and more comprehensive task selection scheme than WOS so that the model can further converge to the best parameters during training, and relatively close to the best solution to the optimal solution. By contrast, the EBRA algorithm that is based on the greedy strategy tends to fall into the local optimum. After WROS makes the decision in current episode, it also incorporates the long-term reward into the calculation index according to meet the dynamic change MCS environment, which can approach the global optimum to the greatest extent.

Figure 4 shows the platform utility of different incentive mechanisms when the total task coverage requirement is 80 from 11:00 am to 12:00 am. The benefit of the platform increases with more sensing workers because it brings higher sensing probability and more choices. However, when the number of users increases from 150 to 250, the increase of platform utility is less significant. It is because when the number of users is more than 150, they are already in a relatively saturated state for 20 perception areas. Therefore, platform utility will be relatively stable. When encountering this situation, the platform can add more sensing tasks or expand the perception area. Both method will bring significant improvements to the platform's utility. In general, WROS is 13% higher than WOS in platform utility and about 54% higher than EBRA. WROS has achieved higher overall performance in real data sets.

5 Conclusion

In this paper, we study a novel incentive mechanism DQN-TAIM to solve the spatial-temporal coverage problem of sensing tasks and maximize the utility of the opportunity MCS platform. Specifically, we proposed a task allocation strategy based on deep reinforcement learning, which aims to maximize the cumulative benefits of the platform by

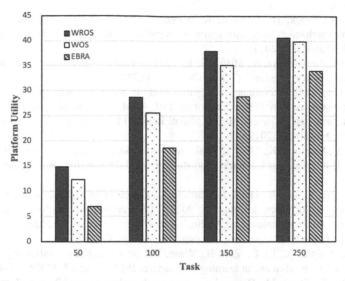

Fig. 4. The platform utility of different incentive mechanisms in the same sensing cycle.

intelligently distributing tasks to users. The simulation that is based on a real dataset confirms the effectiveness and efficiency of our proposed problem-solving approach.

Acknowledgement. This work was supported in part by ***.

References

1. Waze. https://www.waze.com/zh/. Accessed 13 Sept 2020
2. Ganti, R.K., Pham, N., Ahmadi, H., Nangia, S., Abdelzaher, T.F.: GreenGPS: a participatory sensing fuel-efficient maps application. In: Proceedings of the 8th International Conference on Mobile Systems, Applications, and Services, San Francisco, California, USA, pp. 151–164. Association for Computing Machinery (2010)
3. Makhsous, S., Bharadwaj, M., Atkinson, B.E., Novosselov, I.V., Mamishev, A.V.: Dietsensor: automatic dietary intake measurement using mobile 3D scanning sensor for diabetic patients. Sensors **20**(12), 3380 (2020)
4. Maisonneuve, N., Stevens, M., Niessen, M.E., Steels, L.: NoiseTube: measuring and mapping noise pollution with mobile phones. In: Athanasiadis, I.N., Rizzoli, A.E., Mitkas, P.A., Gómez, J.M. (eds.) Information Technologies in Environmental Engineering, pp. 215–228. Springer, Heidelberg (2009). https://doi.org/10.1007/978-3-540-88351-7_16
5. Nie, J., Luo, J., Xiong, Z., Niyato, D., Wang, P.: A stackelberg game approach toward socially-aware incentive mechanisms for mobile crowdsensing. IEEE Trans. Wirel. Commun. **18**(1), 724–738 (2018)
6. Xu, J., Rao, Z., Xu, L., Yang, D., Li, T.: Incentive mechanism for multiple cooperative tasks with compatible users in mobile crowd sensing via online communities. IEEE Trans. Mob. Comput. **19**(7), 1618–1633 (2019)
7. Wang, N., Da, J., Li, J., Liu, Y.: Influence maximization with trust relationship in social networks. In: 14th International Conference on Mobile Ad-Hoc and Sensor Networks (MSN), Shenyang, China, pp. 61–67. IEEE (2018)

8. Suliman, A., Otrok, H., Mizouni, R., Singh, S., Ouali, A.: A greedy-proof incentive-compatible mechanism for group recruitment in mobile crowd sensing. Future Gener. Comput. Syst. **101**, 1158–1167 (2019)
9. Neiat, A.G., Bouguettaya, A., Mistry, S.: Incentive-based crowdsourcing of hotspot services. ACM Trans. Internet Technol. (TOIT) **19**(1), 1–24 (2019)
10. Bassem, C.: Mobility coordination of participants in mobile crowdsensing platforms with spatio-temporal tasks. In: Proceedings of the 17th ACM International Symposium on Mobility Management and Wireless Access, Miami Beach, FL, USA, pp. 33–40. Association for Computing Machinery (2019)
11. Vemula, A., Muelling, K., Oh, J.: Social attention: modeling attention in human crowds. In: 2018 IEEE International Conference on Robotics and Automation (ICRA), pp. 1–7. IEEE (2018)
12. Hu, Y., Wang, J., Wu, B., Helal, S.: Participants selection for from-scratch mobile crowdsensing via reinforcement learning. In: 2020 IEEE International Conference on Pervasive Computing and Communications, Austin, TX, USA, pp. 1–10. IEEE (2020)
13. Baidu Test. https://test.baidu.com/. Accessed 10 Sept 2020
14. Wang, C., Gaimu, X., Li, C., Zou, H., Wang, W.: Smart mobile crowdsensing with urban vehicles: a deep reinforcement learning perspective. IEEE Access **7**, 37334–37341 (2019)
15. Li, H., Ota, K., Dong, M.: Deep reinforcement scheduling for mobile crowdsensing in fog computing. ACM Trans. Internet Technol. **19**(2), 1–18 (2019)
16. Chen, Y., Wang, H.: IntelligentCrowd: Mobile Crowdsensing via Multi-agent Reinforcement Learning. arXiv preprint arXiv:1809.07830 (2018)
17. Wang, J., et al.: Multi-task allocation in mobile crowd sensing with individual task quality assurance. IEEE Trans. Mob. Comput. **17**(9), 2101–2113 (2018)
18. Wang, L., Yu, Z., Zhang, D., Guo, B., Liu, C.H.: Heterogeneous multi-task assignment in mobile crowdsensing using spatiotemporal correlation. IEEE Trans. Mob. Comput. **18**(1), 84–97 (2018)
19. Mnih, V., et al.: Human-level control through deep reinforcement learning. Nature **518**(7540), 529–533 (2015)
20. Barlacchi, G., et al.: A multi-source dataset of urban life in the city of Milan and the Province of Trentino. Sci. Data **2**(1), 1–15 (2015)
21. Tan, W., Jiang, Z.: A novel experience-based incentive mechanism for mobile crowdsensing system. In: Proceedings of the International Conference on Artificial Intelligence, Information Processing and Cloud Computing, Sanya, China, pp. 1–6 (2019)

Domain-Specific Collaborative Applications

Dynamic Key-Value Gated Recurrent Network
for Knowledge Tracing

Bo Xie, Lina Fu$^{(\boxtimes)}$, Bo Jiang, and Long Chen

Zhejiang Gongshang University, Hangzhou 310018, China

Abstract. Knowledge tracing model is one of the important research fields to realize personalized intelligent education. The focus of the model is to trace students' knowledge mastery from the records of students' answering performance. However, the existing knowledge tracing models don't consider the problems of students' knowledge application ability and forgetting rules. Based on the external memory mechanism of Dynamic Key-Value Memory Networks (DKVMN), this paper proposes a knowledge tracing model based on a dynamic key-value gated recurrent network (DKVGRU), which designs a knowledge update network inspired the idea of Gated Recurrent Unit (GRU). DKVGRU calculates the proportion of concepts students apply and measures the degree of forgetting of learned concepts, which traces the knowledge state of each concept well. In this paper, the area under the receiver operating characteristic curve (AUC) of the prediction result is used as an evaluation indicator. The performance of DKVGRU is higher than DKVMN on four public datasets.

Keywords: Knowledge tracing · Deep learning · Data mining · Intelligent education system · Student evaluation

1 Introduction

Nowadays, online learning platform realizes the acquisition of high-quality learning resources without the constraints of time and space. Students can flexibly study on computers and mobile terminals, and can independently arrange study plans and tasks. Because of this, millions of students are learning a variety of courses through online learning platforms. However, there are many obstacles in online learning platform for the supervision of students and the provision of personalized learning guidance due to the large number of learners. In terms of providing personalized guidance, it is very important to evaluate students' knowledge state for online learning platform, which is also an important research topic in the field of intelligent education [1].

Knowledge tracing (KT) is a widely used model for predicting students' knowledge state in intelligent online learning platform [2]. KT can model the interaction process between students and exercises based on the students' past exercise records to trace students' knowledge state dynamically [3]. The goal of KT can be described as: given the interaction sequence of past exercises of a student $X = x_1, x_2, \ldots, x_t$, KT acquires the knowledge state of the student, which is used to predict the probability of the correct

© Springer Nature Singapore Pte Ltd. 2021
Y. Sun et al. (Eds.): ChineseCSCW 2020, CCIS 1330, pp. 169–183, 2021.
https://doi.org/10.1007/978-981-16-2540-4_13

answer to the next exercise. The input $x_t = (q_t, a_t)$ contains the exercise q_t and the actual answer a_t [4].

Using KT model, online learning platforms not only customize learning materials for students based on the knowledge state of students, but also provide to students and teachers with feedback reports. Therefore, students reasonably allocate their study schedules to maximize their learning efficiency, and teachers can timely adjust appropriate teaching plans and schemes.

At present, the traditional KT model and the deep learning-based KT model are two kinds of models provided in the field of knowledge tracing. Among the traditional knowledge tracing models, the most typical one is Bayesian Knowledge Tracing (BKT) [5], which models each concept state separately. Therefore, BKT is limited to capture the correlations between different concepts, which ineffectively simulates the knowledge state transition between complex concepts. Researchers further applied deep learning to KT task and proposed Deep Knowledge Tracing (DKT) [6]. Compared with BKT, DKT uses a hidden state to sum up the knowledge state of all concepts. Considering correlations between multiple concepts, DKT delivers a better simulation in students' knowledge state. But DKT can't pinpoint which concepts a student has mastered like BKT. Consequently, DKT has its weakness in indicating the certain concept that students grasp or not. Combining the advantages of the BKT and DKT, DKVMN uses external memory to store the student's knowledge state [7], and its prediction performance is better than BKT and DKT.

However, existing KT models ignore two aspects in simulating the changes in students' knowledge states. Firstly, in the aspect of knowledge application, students apply different concepts according to their knowledge states for the same exercise. Secondly, according to the Ebbinghaus forgetting curve [8], the process of forgetting is not uniform. Students forget the knowledge they have just learned from the exercises very fast, but the knowledge they have learned before is slow. Existing models have limits in distinguishing the degree of forgetting the learned knowledge.

Based on the external memory mechanism of DKVMN, this paper designed a knowledge update network inspired by the idea of GRU's gating mechanism [9], and proposed a knowledge tracing model based on Dynamic Key-Value Gated Recurrent Network (DKVGRU). In the huge exercise data, DKVGRU uses the Key-Value matrix of DKVMN to explore the relationship between exercises and underlying concepts, while tracing the knowledge state of a certain concept. We provided two knowledge gates for simulating the change of students' knowledge states. The knowledge application gate calculates the proportion of knowledge concepts applied by students in solving exercises, and the knowledge forgetting gate measures the forgetting degree of the learned knowledge.

2   Related Work

There are two main types of KT models. One is the traditional KT model, the other is the KT model based on deep learning. In this chapter, we first introduce BKT, DKT and DKVMN. Besides, DKVGRU is inspired by the gating mechanism, and this chapter also introduces Recurrent Neural Network (RNN) [10] and its variants, which can capture long-term sequence data relations.

2.1 Bayesian Knowledge Tracing

BKT is the most commonly used among traditional KT models, which was introduced in the field of intelligent education by Corbett and Anderson and used to intelligent tutoring systems in 1995 [11]. BKT assumes that each concept is independent of each other and students have only two states for each concept: mastered or not mastered. As shown in Fig. 1, BKT uses Hidden Markov Model (HMM) to model a certain concept separately, and updates the state of a concept with the help of two learning parameters and two performance parameters. The original BKT assumes students do not forget knowledge in learning, which is obviously against students' regular learning pattern [12]. And researchers have proposed several aspects to optimize BKT from forgetting parameters [13], exercise difficulty [14], personalized parameters [15], emotions [16], etc.

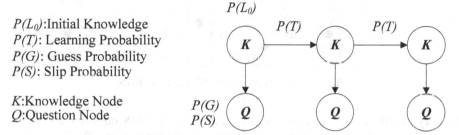

$P(L_0)$:Initial Knowledge
$P(T)$: Learning Probability
$P(G)$: Guess Probability
$P(S)$: Slip Probability

K:Knowledge Node
Q:Question Node

Fig. 1. The architecture of BKT

2.2 Deep Learning-Based Knowledge Tracing

In 2015, Piech et al. firstly applied deep learning to KT tasks and proposed DKT based on RNN and Long Short-Term Memory (LSTM) [17]. As illustrated in Fig. 2, DKT can represent the student's continuous knowledge state using a high-dimensional hidden state. And without the manual annotation, DKT can automatically discover the relationship between concepts from exercises. Using the forgetting gate of LSTM, DKT can simulate the knowledge forgetting that occurs in the learning process. Khajah et al. proved that the advantage of DKT lies in the ability to make good use of some statistical rules in data, which BKT cannot use [18]. Yeung et al. added a regularization term to the loss function to solve two problems of DKT: inaccuracy and instability [19]. Xiong et al. believe that DKT is a potential KT method if more features can be modeled, such as student abilities, learning time, and exercise difficulty [20]. And many variations were raised by adding dynamic student classification [21], side information [22] and other features [23] into DKT.

The Memory Augmented Neural Network (MANN) [24] uses an external memory module to store information, which has a stronger information storage capacity than using a high-dimensional hidden state. And MANN can rewrite local information through the external memory mechanism. Different from the general MANN which uses a simple

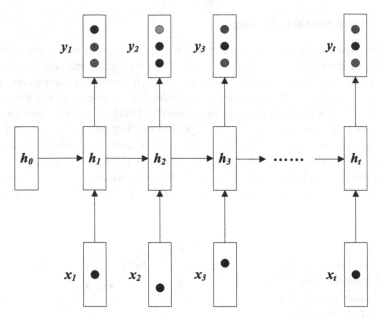

Fig. 2. The architecture of DKT based on RNN

memory matrix or two static memory matrices [25], DKVMN utilizes the key-value matrix to store all concepts and the knowledge state of each concept. The key matrix is used to calculate the correlation between exercises and concepts, and the value matrix is used to read and write the knowledge state of each concept. Ha et al. [26] optimized DKVMN from knowledge growth and regularization.

2.3 Recurrent Neural Network

For sequence data, researchers use RNN to obtain data relationships in general. However, RNN cannot effectively capture long-term sequence data relationships because of its structural defects. And Hochreiter et al. proposed LSTM to solve the problem of long-term in 1997, which used three gates to effectively deal with long-term and short-term dependence. And Cho et al. proposed GRU by optimizing the structure of LSTM in 2014, which not only guarantees model performance but also improves model training efficiency [27]. GRU uses two gates to determine which information needs to be memorized, forgotten, and output respectively, which effectively achieve long-term tracing of information. As shown in Fig. 3, the reset gate generates the weight to decide how much historical information is used according to the input information, and the update gate is used to generate the proportion of historical memory and current memory in new memories.

3 Model

DKVGRU can be divided into three parts: correlation weight, read process and write process, which are represented in Fig. 4. Correlation weight represents the weight of each

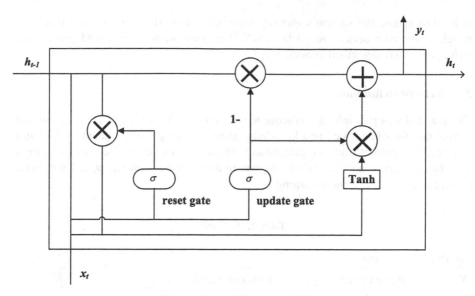

Fig. 3. The architecture of GRU

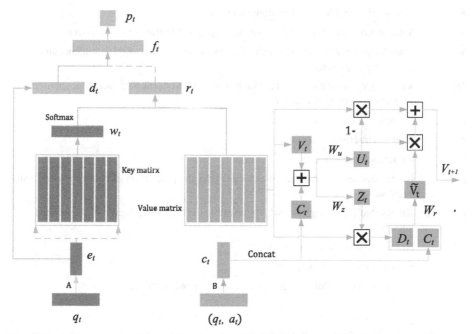

Fig. 4. In the framework of DKVGRU, the green part is write process we designed. The blue and purple parts are correlation weight and read process, which refer to DKVMN.

concept contained in the exercise. Read process can read the student's current memory, which is used to predict students' performance of a new exercise. And write process is used to update the student's memory state after answering a exercise. Correlation weight and read process refer to DKVMN. The correlation weight, read process and write process are described in Sect. 3.2, 3.3, and 3.4.

3.1 Related Definitions

Given a student's past interaction sequence of exercises $X = x_1, x_2, \ldots, x_{t-1}$, our task is to obtain the student's current knowledge state according to the student's interaction sequence and predict students' performance of the next exercise. The interaction tuple $x_t = (q_t, a_t)$ represents the student's answer to the exercise q_t, where a_t is 1 means the answer is correct and 0 means wrong.

Table 1. Symbols

Symbols	Explanation
X	The past exercise sequence of a student: $x_1, x_2, \ldots, x_{t-1}$
E	Number of exercises
N	Number of concepts contained in the exercises
K	Key matrix, which is used to store all concepts
V	Value matrix, which is used to dynamically store the state of each concept
$\tilde{V}$	Knowledge growth matrix, which stores the knowledge growth of students after answering one exercise
D	Knowledge application matrix, which stores the proportion of each concept used by students for a certain exercise
e	Exercise vector
c	Exercise interaction vector
w	Correlation weight
Z	Knowledge application gate, which calculates the proportion of concepts used in an exercise
U	Knowledge forgetting gate, which measures the forgetting degree of the learned knowledge
A	Exercise embedding matrix
B	Interaction embedding matrix
;	The operation of Concatenating two vectors or two matrices

As illustrated in Table 1, the definition of various symbols used in the model is described. The N represents the number of concepts, and the key matrix $K(N \times d_k)$ stores these concepts. Besides, the knowledge state of each concept is stored in the value matrix $V(N \times d_v)$.

3.2 Correlation Weight

Each exercise contains multiple concepts. The exercise q_t is firstly mapped into a vector $e \in R^{d_k}$ by an embedding matrix $A \in R^{d_k}$. The correlation weight $w_t \in R^N$ is computed by taking the softmax activation of the inner product between e_t and each k_i of the key matrix $K = (k_1, k_2, \ldots, k_N)$:

$$w_t = Softmax\left(e_t \cdot K^T\right). \tag{1}$$

k_i is the key memory slot which is used to store the ith concept. And w_t measures the correlation weight between this exercise and concepts.

3.3 Read Process

The probability of answering q_t correctly needs to consider two factors: the student's current knowledge state and exercise difficulty. Above all, w_t is multiplied by the each v_i of the value matrix $V = (v_1, v_2, \ldots, v_N)$, which is to get the read content vector $r_t \in R^{d_v}$:

$$r_t = w_t \cdot V_t. \tag{2}$$

v_i is the value memory slot which is used to store the state of the ith concept. And the read content r_t is regarded as the student's overall mastery of q_t.

Then considering that the difficulty of q_t, the exercise vector e_t passes through the fully connect layer and *Tanh* function to get the difficulty vector $d_t \in R^{d_k}$:

$$d_t = Tanh(e_t \cdot W_1 + b_1), \tag{3}$$

$$Tanh(x) = \frac{1 - e^{-2x}}{1 + e^{-2x}}, \tag{4}$$

and W_i and b_i are the weight and bias of the full connect layer.

The summary vector f_t is obtained after concatenating the read content vector r_t and the difficulty vector d_t:

$$f_t = Tanh([r_t; d_t] \cdot W_2 + b_2). \tag{5}$$

Finally, the probability p_t is computed from the summary vector f_t:

$$p_t = Sigmoid(f_t \cdot W_3 + b_3), \tag{6}$$

$$Sigmoid(x) = \frac{1}{1 + e^{-x}}. \tag{7}$$

And *Sigmoid* function makes the probability p_t between 0 to 1.

3.4 Write Process

The knowledge state of each concept are updated after the student answering the exercise q_t. The interaction tuple $x_t = (q_t, a_t)$ is turned into a number by $y_t = q_t + a_t * E$, which y_t represents the student's interactive information. And y_t is converted into an interaction vector $c_t \in R^{d_v}$ by an embedding matrix $B \in R^{E \times d_v}$. Considering that students apply knowledge to exercise according to their knowledge state, we adds the interaction vector c_t and each value memory slot v_i of the value matrix V_t, and pass the result through the fully connect layer and an activation function to obtain the knowledge application gate $Z_t \in R^{N \times d_v}$:

$$C_t = Concat(c_t, c_t, \ldots, c_t), \tag{8}$$

$$Z_t = Sigmoid([V_t + C_t] \cdot W_z + b_z). \tag{9}$$

Z_t is used to calculate the proportion of concepts used in an exercise. The application knowledge state $D_t \in R^{N \times d_v}$ is obtained by using Z_t to weight the value matrix V_t:

$$D_t = Z_t * V_t. \tag{10}$$

Then, we concatenate the interaction vector c_t and each d_i of the value matrix $D_t = (d_1, d_2, \ldots, d_N)$ to get the knowledge growth matrix $\tilde{V}_t \in R^{N \times d_v}$:

$$\tilde{V}_t = Tanh([D_t; C_t] \cdot W_r + b_r). \tag{11}$$

For the purpose of measuring student's forgetting degrees, We adds the interaction vector c_t and each value memory slot v_i of the value matrix V_t to obtain the knowledge forgetting gate $U_t \in R^{N \times d_v}$:

$$U_t = Sigmoid([V_t + C_t] \cdot W_u + b_u). \tag{12}$$

Each concept state of the value matrix V_t is updated by U_t. $(1 - U_t) * V_t$ represents the unforgettable part of the previous knowledge state, and $U_t * \tilde{V}_t$ represents the unforgettable part of the knowledge gained from this exercise. And V_{t+1} means the new student's knowledge state.

$$V_{t+1} = (1 - U_t) * V_t + U_t * \tilde{V}_t. \tag{13}$$

3.5 Optimization Process

The optimization goal of our model is that the predicted probability p_t is close to the student's answer a_t, that is to minimize the cross entropy loss L.

$$L = -\sum_t a_t \, log(p_t) + (1 - a_t) \, log(1 - p_t). \tag{14}$$

4 Experiments

4.1 Datasets

There are several datasets to test the performance of models in Table 2, including Statics2011, ASSISTments2009, ASSISTments2015 and ASSISTment Challenge. And these datasets come from real online learning systems.

(1) Statics2011: This dataset has 1,223 exercise tags and 189,297 interaction records of 333 students, which comes from an engineering mechanics course of a university.

Table 2. Dataset statistics

Datasets	Students	Exercise tags	Records
Statics2011	333	1,223	189,297
ASSISTments2009	4,151	110	325,637
ASSISTments2015	19,840	100	683,801
ASSISTment Challenge	686	102	942,816

(2) ASSISTments2009: This dataset contains 110 exercise tags and 325,637 interactions records for 4,151 students, which comes from the ASSISTment education platform in 2009.
(3) ASSISTments2015: This dataset is collected from the ASSISTment education platform, which has 100 exercise tags and 683,801 interactions records of 19,840 students.
(4) ASSISTment Challenge: This dataset was used in the ASSISTment competition in 2017, and it contains 102 exercise tags and 942,816 interaction records of 686 students.

4.2 Evaluation Method

In the field of knowledge tracing, we usually use AUC as the evaluation criteria for model classification. The advantage of AUC is that even if the sample is unbalanced, it can still give a more credible evaluation result [28]. This paper also uses AUC as the evaluation of the model. And the higher the value of AUC, the better the classification result. As shown in Fig. 5, ROC curve is drawn according to TPR and FPR, and AUC is obtained from the area under ROC curve.

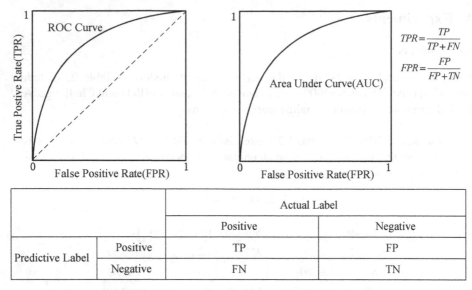

Fig. 5. Description of the AUC calculation

4.3 Implementation Details

In this paper, the training set and test set of each dataset was randomly assigned, 70% of which is the training set and the remaining 30% is the test set. The five-fold cross-validation method was used on the training set, and 20% of the training set was divided into the validation set. We used early stopping and selected hyperparameters of model on the validation set. And the performance of the model was evaluated on the test set.

Gaussian distribution was used to initialize the parameters randomly. Stochastic gradient descent method was adopted as the optimization method for training. And batch size was set to 50 on all datasets. The maximum number of training times of the model was set to 100 epochs. The epoch with the best AUC value on the validation set was selected for testing. And the average value of AUC on the test set was used as the model evaluation result.

Using different initial learning rates, we compared the performance of DKVMN and DKVGRU models when the sequence length was 200. Then, we set sequence lengths of 100, 150, and 200 to compare the performance of DKVMN and DKVGRU.

4.4 Result Analysis

On the four datasets, the experiment used the initial learning rate of 0.02, 0.04, 0.06, 0.08, and 0.1 to measure the AUC scores of DKVMN and DKVGRU. And AUC of 0.5 represents the score that can be obtained by random guessing. The higher the AUC score, the better the prediction effect of the model. As shown in Table 3, there are the test AUC score of DKVMN and DKVGRU of all datasets. It can be clearly seen that DKVGRU performs better than DKVMN on all datasets.

Table 3. The test AUC scores of DKVMN and DKVGRU with different initial learning rates on all datasets

Dataset	The initial learning rate	Test AUC	
		DKVMN	DKVGRU
Statics2011	0.02	0.814900	0.816630
	0.04	0.817625	0.831916
	0.06	0.818041	0.834688
	0.08	0.818591	0.834850
	0.1	0.819070	0.834766
ASSISTments2009	0.02	0.801624	0.806886
	0.04	0.803469	0.808587
	0.06	0.804472	0.808334
	0.08	0.802909	0.806580
	0.1	0.804652	0.808877
ASSISTments2015	0.02	0.726578	0.728503
	0.04	0.725727	0.728791
	0.06	0.724857	0.729177
	0.08	0.724883	0.728526
	0.1	0.724837	0.728371
ASSISTment Challenge	0.02	0.662156	0.676517
	0.04	0.664859	0.684211
	0.06	0.667367	0.687498
	0.08	0.669780	0.689189
	0.1	0.671637	0.689144

For Statics2011 dataset, the average AUC of DKVMN is 81.76%, while the average AUC of DKVGRU is 83.06%, which indicates a 1.29% higher than DKVMN. On the ASSISTments2009 dataset, DKVMN produces the average test AUC value of 80.34%, which shows a 0.44% difference compared with 80.70% for DKVGRU. For ASSISTments2015 dataset, the average AUC of DKVGRU is 72.87% and DKVMN is 72.54%. On the ASSISTment Challenge dataset, DKVGRU achieves the average AUC of 68.53%, which improves 1.82% as DKVMN in 66.72%. Therefore, DKVGRU has a better performance than DKVMN on all four datasets. For both models, the paper observes that a larger initial learning rate might lead to a better AUC score from the aforementioned experiments.

Then, we set the initial learning rate of 0.1 and sequence lengths of 100, 150, and 200 to evaluate these two models. And the experimental results indicate that DKVGRU performs better than DKVMN at different sequence lengths in Table 4.

According to Fig. 6, the AUC results of DKVGRU and DKVMN become better with the increase of sequence length except for Statics2011 dataset. The findings support that the setting of the sequence length of the exercises has a positive correlation with the models performance, which means a longer sequence length results in a better prediction performance for the model. That is, the model can more accurately trace students' knowledge state by utilizing more exercise records.

Table 4. The test AUC scores of DKVMN and DKVGRU with different sequence length on all datasets

Dataset	Sequence length	Test AUC	
		DKVMN	DKVGRU
Statics2011	100	0.833927	0.849142
	150	0.827971	0.842618
	200	0.819070	0.834766
ASSISTments2009	100	0.799399	0.804384
	150	0.800944	0.804276
	200	0.804652	0.808877
ASSISTments2015	100	0.723672	0.727784
	150	0.723034	0.727913
	200	0.724837	0.728371
ASSISTment Challenge	100	0.658292	0.672841
	150	0.664720	0.682050
	200	0.671637	0.689144

On the Statics2011 dataset, the reason why the AUC results have a negative correlation with the sequence length is that exercise tag is the largest among the four datasets, which included 1,223 exercise tags. The more exercise labels in the sequence, the more complex relationships between exercises and concepts need to be considered by the model. Nonetheless, the AUC score of DKVGRU on the Statics2011 dataset is higher, which means DKVGRU can simulate students' knowledge state better than DKVMN.

In summary, DKVGRU performs better than DKVMN with different learning rates and sequence lengths, which shows that the gating mechanism of DKVGRU effectively simulates the changes of students' knowledge state.

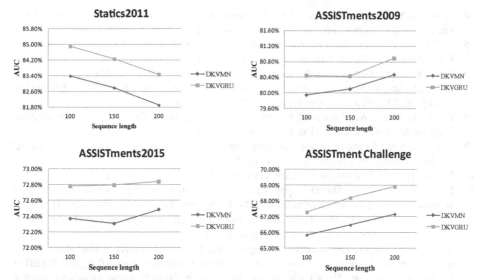

Fig. 6. The test AUC scores of DKVMN and DKVGRU with different sequence length on all datasets

5 Conclusions and Prospects

For the existing shortcomings of knowledge tracing, such as ignoring students apply different concepts to the same exercise and failing to consider the forgetting process of concepts they have learned, we propose a knowledge tracing model DKVGRU, which is based on the dynamic Key-Value matrix and gating mechanism. DKVGRU updates students' knowledge state by the gating mechanism. The experimental data comes from four public datasets. And the experiments demonstrate that DKVGRU performs better than DKVMN.

In addition to the students' exercise records, online learning platforms also record various learning activities of students, such as watching videos, viewing exercise explanations and other learning actions. For future work, we will consider these features in KT tasks. And using these data, we also can classify students according to students' learning attitude and habits, which simulates students' knowledge state reasonably.

References

1. Liu, H., Zhang, T., Wu, P., Yu, G.: A review of knowledge tracking. Comput. Sci. Eng. (5), 1–15 (2019)
2. Pardos, Z., Heffernan, N.: Modeling individualization in a bayesian networks implementation of knowledge tracing. In: De Bra, P., Kobsa, A., Chin, D. (eds.) UMAP 2010. LNCS, vol. 6075, pp. 255–266. Springer, Heidelberg (2010). https://doi.org/10.1007/978-3-642-13470-8_24
3. Xu, M., Wu, W., Zhou, X., Pu, Y.: Research on multi-knowledge point knowledge tracing model and visualization. J. Audio-visual Educ. Res. **39**(10), 55–61 (2018)

4. Yeung, C.K., Lin, Z., Yang, K., et al.: Incorporating features learned by an enhanced deep knowledge tracing model for STEM/Non-STEM job prediction. J. CoRR (2018)
5. Corbett, A.T., Anderson, J.R.: Knowledge tracing: modeling the acquisition of procedural knowledge. User Model. User Adap. Interact. 4(4), 253–278 (1994)
6. Piech, C., et al.: Deep knowledge tracing. Comput. Sci. 3(3), 19–23 (2015)
7. Zhang, J., Shi, X., King, L., Yeung, D.: Dynamic key-value memory networks for knowledge tracing. In: The Web Conference, pp. 765–774. ACM (2017)
8. Deng, Y., et al.: Application of Ebbinghaus forgetting curve theory in teaching medical imaging diagnosis. Chin. J. Med. Educ. 33(4), 555–557 (2013)
9. Cho, K., et al.: Learning phrase representations using RNN encoder-decoder for statistical machine translation. Comput. Sci. 1724–1734 (2014)
10. Elman, J.L.: Finding structure in time. Cogn. Sci. 14(2), 179–211 (1990)
11. Li, F., Ye, Y., Li, X., Shi, D.: Application of knowledge tracking model in education: a summary of related research from 2008 to 2017. Distance Educ. China (007), 86–91 (2019)
12. Wang, Z., Zhang, M.: MOOC student assessment based on Bayesian knowledge tracing model. China Sciencepaper 10(02), 241–246 (2015)
13. Ye, Y., Li, F., Liu, Q.: The influence of forgetting and data volume into the knowledge tracking model on prediction accuracy. China Acad. J. Electron. Publishing House (008), 20–26 (2019)
14. Pardos, Z., Heffernan, N.: KT-IDEM: introducing item difficulty to the knowledge tracing model. In: Konstan, J.A., Conejo, R., Marzo, J.L., Oliver, N. (eds.) UMAP 2011. LNCS, vol. 6787, pp. 243–254. Springer, Heidelberg (2011). https://doi.org/10.1007/978-3-642-22362-4_21
15. Yudelson, M., Koedinger, K., Gordon, G.: Individualized Bayesian knowledge tracing models. In: Chad Lane, H., Yacef, K., Mostow, J., Pavlik, P. (eds.) AIED 2013. LNCS (LNAI), vol. 7926, pp. 171–180. Springer, Heidelberg (2013). https://doi.org/10.1007/978-3-642-39112-5_18
16. Spaulding, S., Breazeal, C.: Affect and inference in Bayesian knowledge tracing with a robot tutor. In: The Tenth Annual ACM/IEEE International Conference, pp. 219–220. ACM (2015)
17. Hochreiter, S., Schmidhuber, J.: Long short-term memory. Neural Comput. 9(8), 1735–1780 (1997)
18. Khajah, M., Lindsey, R.L., Mozer, M.C.: How deep is knowledge tracing? In: Proceedings of the 9th International Conference on Educational Data Mining, pp. 94–101 (2016)
19. Yeung, C.K., Yeung, D.: Addressing two problems in deep knowledge tracing via prediction-consistent regularization. In: 5th Annual ACM Conference on Learning at Scale, pp. 1–10 (2018)
20. Xiong, X., Zhao, S., Van Inwegen, E.G., Beck, J.E.: Going deeper with deep knowledge tracing. In: 9th International Conference on Educational Data Mining, North Carolina, pp. 545–550 (2016)
21. Minn, S., Yu, Y., Desmarais, M.C., Zhu, F., Vie, J.J.: Deep knowledge tracing and dynamic student classification for knowledge tracing. In: 2018 IEEE International Conference on Data Mining, pp. 1182–1187. IEEE (2018)
22. Wang, Z., Feng, X., Tang, J., Huang, G., Liu, Z.: Deep knowledge tracing with side information. In: Isotani, S., Millán, E., Ogan, A., Hastings, P., McLaren, B., Luckin, R. (eds.) AIED 2019. LNCS (LNAI), vol. 11626, pp. 303–308. Springer, Cham (2019). https://doi.org/10.1007/978-3-030-23207-8_56
23. Zhang, L., Xiong, X.L., Zhao, S.Y., Botelho, A.F., Heffernan, N.T.: Incorporating rich features into deep knowledge tracing. In: ACM Conference on Learning (2017)
24. Santoro, A., Bartunov, S., Botvinick, M., Wierstra, D.: Meta-learning with memory-augmented neural networks. In: 33rd International Conference on Machine Learning, pp. 1842–1850. IMLS, New York City (2016)

25. Miller, A., Fisch, A., Dodge, J., Karimi, A.H., Bordes, A., Weston, J.: Key-value memory networks for directly reading documents (2016)
26. Ha, H., Hwang, U., Hong, Y., Yoon, S.: Memory-augmented neural networks for knowledge tracing from the perspective of learning and forgetting. In: 33rd Innovative Applications of Artificial Intelligence Conference Honolulu, Hawaii, USA (2018)
27. Yang, L., Wu, Y., Wang, J., Liu, Y.: Research on recurrent neural network. J. Comput. Appl. **38**(S2), 6–11+31 (2018)
28. Wang, Y., Chen, S.: A survey of evaluation and design for AUC based classifier. Pattern Recogn. Artif. Intell. **24**(001), 64–71 (2011)

Law Article Prediction via a Codex Enhanced Multi-task Learning Framework

Bingjun Liu, Zhiming Luo, Dazhen Lin, and Donglin Cao[✉]

Department of Artificial Intelligent, Xiamen University, Xiamen, China
{zhiming.luo,dzlin,another}@xmu.edu.cn

Abstract. Automatic law article prediction aims to determine appropriate laws for a case by analyzing its corresponding fact description. This research constitutes a relatively new area which has emerged from recommended algorithm. Therefore, the task is still a challenge due to the highly imbalanced long-tail data distribution and lack of discrimination in the feature representation. To deal with these challenges, we proposed a codex enhanced multi-task framework, which consists of two modules. The first one is a codex learning module that estimates the broad codex attributes related to the case fact for alleviating the long-tail issue. The other one is a Bidirectional Text Convolutional Neural Network, which predicts the law articles by considering both local and global information of the facts. These two modules are learning simultaneously through a multi-task loss function. To evaluate the performance of the method proposed in this paper, we construct a new law article prediction data by collecting the judgment documents from the China Judgement online. Experimental results on the dataset demonstrate the effectiveness of our proposed method and can outperform other comparison methods.

Keywords: Law article prediction · Codex attribute learning · Bidirectional Text Convolutional Neural Network

1 Introduction

The main goal of automatic law article prediction is to determine appropriate laws for a case by analyzing its corresponding fact description. Such techniques play vital roles in the legal assistant system, in which legal practitioners can obtain possible judgments via a brief description of the case. Furthermore, it also can be used by a regular civilian to get a legal basis about their cases while without knowing the massive legal materials and the legal jargon.

In the past few years, different algorithms based machine-learning have been proposed. Liu et al. [14] use the k-NN to retrieve the most similar cases. Kim et al. [9] adopt the SVM to classify each case into its corresponding law article. To further combine the context information, Carvalho et al. [3] apply the N-gram for the charge prediction. Li et al. [13] introduce a Markov logical network for enhancing the interpretability of the prediction. Most recently, several

© Springer Nature Singapore Pte Ltd. 2021
Y. Sun et al. (Eds.): ChineseCSCW 2020, CCIS 1330, pp. 184–195, 2021.
https://doi.org/10.1007/978-981-16-2540-4_14

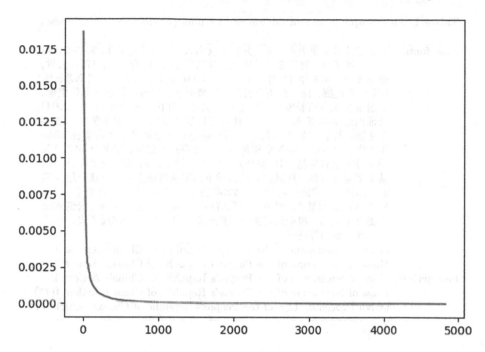

Fig. 1. The sorted case frequency of each law article in our collected dataset. The horizontal axis is the label index of each law article, and the vertical axis is the proportion of cases related to the label.

deep learning-based methods have been introduced into this task [8]. For example, Liu et al. [15] formulate the automatic legal prediction task as a reading comprehension problem and propose an RNN-based deep model.

Despite the progress of previous methods, the law article prediction task is still non-trivial and challenge. The difficulties mainly come from the following two aspects.

- As shown in Fig. 1, the case numbers of the various articles are highly imbalanced according to the real statistic. Notably, the most frequent 2% articles cover more than 76% of cases. This leads to a severely long-tail issue for constructing the model.
- As shown in Table 1. A single case usually involves multiple crimes, which means some articles related to partial fact description, and others may relate to the whole fact description. Then, it needs to have a robust feature representation containing the information from the local and global.

This study explicitly addresses these difficulties with a novel framework consisting of two modules: a codex attribute learning module and a Bidirectional Text Convolutional Neural Network (BiTextCNN) module. The codex attribute learning module alleviates the long-tail law article distribution problem by establishing a middle-level codex attribute estimation. Besides, the BiTextCNN contains the row convolution and column convolution to learn a more robust feature

Table 1. An example judgment document in China (Original text is Chinese)

case facts	XX 与杨 XX 是夫妻，两人共有子女六人，分别是刘某 1、刘某 2、刘某 7、刘某 3、刘某 8、刘某 9。刘 XX 于 2007 年 3 月 17 日去世，杨 XX 于 2013 年 11 月 7 日去世。2004 年 2 月 24 日，刘 XX、杨 XX 订立遗嘱，确定在两人去世以后将诉争三套房屋依法由原告继承，该遗嘱在北京市长安公证处进行了公证，并由公证处出具了（2004）长证内民字第 X 号公证书。现双方因遗产继承事宜发生争议，故诉至法院。被告刘某 2 辩称，不同意原告的诉讼请求。刘某 2 在 1945 年出生，出生后，刘 XX 和杨 XX 就没有抚养过他。刘某 2 和刘 XX 的父母在农村生活，15 岁开始就没再上学了。刘 XX 的父母是由刘某 2 养老送终的，所以我方认为诉争房屋从情理上来说应该分给刘某 2 一部分，要求原告补偿刘某 2200 万元。被告刘某 3、刘某 7、刘某 8、刘某 4、刘某 5、刘某 6 未到庭应诉答辩。当事人围绕诉讼请求依法提交了证据。根据原告陈述及原告提交的证明、干部履历表、公证书、死亡证明等证据
law articles	《Law of Succession of the People's Republic of China》 Article 2 《Law of Succession of the People's Republic of China》 Article 3 《Law of Succession of the People's Republic of China》 Article 5 《Law of Succession of the People's Republic of China》 Article 16(1) 《Civil Procedure Law of the People's Republic of China》 Article 144

representation by considering both the global information and local details in the fact description.

To evaluate the effectiveness of our proposed method, we construct a law dataset by collecting the publicly available judgment documents from the China Judgment Online[1]. This data contains 400,000 cases, and each case consists of the fact description and the corresponding law articles. Experimental results on this dataset show that our method can effectively predict the relevant law articles, and the evaluation metrics demonstrate that our method can achieve better performance than other comparison methods.

2 Related Works

The task of law article prediction aims at finding corresponding law articles based on the fact description. Previous works [1,3,21] address this task mainly by formulating it as a traditional Multi-class Classification problem. However,there is a further problem that these methods usually use the word-level and phrase-level features that are insufficient to distinguish the subtle difference between similar charges. This seems to be a common problem in Natural Language Processing (NLP) research. To deal with these issues, many different improvements have been proposed. Carvalho et al. [3] developed a Legal Question Answering(LQA) system based on the N-gram model, which can cover more word-level features. Aletras et al. [1] introduce a model to predict the law articles only related to

[1] https://wenshu.court.gov.cn/.

the particular field of European human rights. In their model, they adopted a topological structure for improving reasoning.

Along with the development of using neural network methods in NLP, some researches [2,12,16,17,19] attempt to incorporate deep learning into the law article prediction. Luo et al. [16] proposed a two-level RNN scheme in which one for sentence embedding and another for document embedding. They demonstrated the superiority of using deep learning against traditional methods. Besides, the relevant background knowledge plays a significant role in legal affairs. Then, Duan et al. [19] propose the phrase of the relevant charges to enhance the multi-label legal prediction, which can outperform other comparing methods.

On the other hand, the law article correlates with other articles, while previous methods have primarily neglected this correlation. Therefore, Si et al. [4] introduce a Legal Graph Network (LGN) to relieve the defect. Particularly, the LGN fuses all the label information into a unified graph, capturing four types of label relations among legal labels. Experiments show that modeling relationships help improve accuracy. However, the LGN only be applied to a small set of data due to the high memory cost of constructing the legal graph.

Meanwhile, the law article prediction task also suffers the issue of long-tail label distribution. Hu et al. [6] propose a novel multi-task learning to relieve this issue by the assistant of manual attribute annotation. Their method manually annotates ten discriminative attributes for the Chinese criminal law and assigns corresponding attributes for each case. Then they employ an LSTM as an encoder for the fact description and estimate the attributes and final label estimation simultaneously. Nevertheless, their approach heavily relies on the high-cost manual attribute annotation.

To tackle the above issues, in this study, we propose a neural network by establishing an annotation attribute learning module and a novel fact encoder.

3 Our Approach

In this section, we introduce a free-cost annotation attribute learning model by adding legal codex information. In the following parts, we first propose the legal codex attribute which can been discriminative feature of law articles. In the next parts, we propose framework of our approach. Afterward, we give the details of the codex attributes predictor and an novel text encoder named BiTextCNN for fact description embedding.

3.1 Legal Codex Attribute

As we knew, the articles related to the same subject content usually come from the same codex. This means that we can classify the complex law articles according to the subject to which it belongs and establish the connection between the law articles and alleviate the long-tail problem. In order to check on the feasibility of this idea, we counted 100,000 judgment documents, and the results showed that all the articles in the data set belong to the 287 codices. The distribution of the codices are shown in Fig. 2, and the top 10 codices are shown in Table 2.

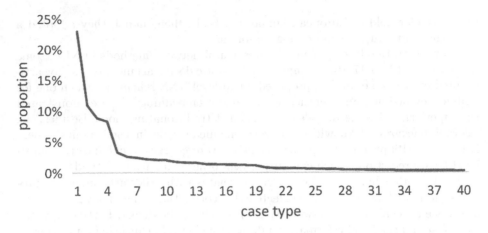

Fig. 2. The sorted case frequency of top 40 codex in our collected dataset.

3.2 The Framework of Our Approach

The model consist of two modules, codex Learning and bi-direction Text Convolution Module (BiTextCNN) module, show as Fig. 4. The fact description, denoted by word embedding , input into these two modules separately for codex attribute learning and text convolution encoding. Then, the output vectors of the two modules are concatenated, and then input into the multi-layer perceptron (MLP) for prediction.

3.3 Legal Codex Learning

In the codex attribute learning module, the fact description text will be encoded by the BiLSTM [22] as follows:

$$\overrightarrow{h}_i = \overrightarrow{LSTM}(T_i), \tag{1}$$

$$\overleftarrow{h}_i = \overleftarrow{LSTM}(T_i), \tag{2}$$

$$h_i = [\overrightarrow{h}_i; \overleftarrow{h}_i]. \tag{3}$$

where T_i denote fact description text, and $\forall i \in [1, n]$, n is the length of the fact.

As show in Fig. 4, the attribute predictor takes the hidden vector $h = [h_1, ..., h_n]$. The attention weights $a = [a_1, ..., a_k]$ calculated by the attentive attribute predictor as follow:

$$a_{i,j} = \frac{\exp(tanh(W^a h_j)^T u_i)}{\sum_t \exp(tanh(W^a h_t)^T u_i)}. \tag{4}$$

where $a_i = [a_{i,1}, ..., a_{i,k}], \forall i \in [1, k], j \in [1, n]$, k is the number of the codex, and u_i is the context vector of the i-th attribute to present the potential information from an element,and W^a is the parameter that all attributes share. Then, each attribute representation predict whether the sample has current attribute by neural network as follow:

Table 2. The top 10 frequent codex

ID	Legal codex
1	《Civil Procedure Law of the People's Republic of China》
2	《Contract Law of the People's Republic of China》
3	《Tort Law of the People's Republic of China》
4	《The Guarantee Law of the People's Republic of China》
5	《General Principles of the Civil Law of the People's Republic of China》
6	《Road Traffic Safety Law of the People's Republic of China》
7	《Interpretation of the Supreme People's Court of Some Issues concerning the Application of Law for the Trial of Cases on Compensation for Personal Injury》
8	《Some Provisions of the Supreme People's Court on Evidence in Civil Procedures》
9	《Interpretation of the Supreme People's Court on Several Issues Concerning the Application of Law in the Trial of Cases on Compensation for Damage in Road Traffic Accidents 》
10	《Provisions of the Supreme People's Court on Several Issues concerning the Application of Law in the Trial of Private Lending Cases》

$$g_i = \sum_t a_{i,t} h_t, \tag{5}$$

$$P_i = f(g_i, \theta_i). \tag{6}$$

Here, P_i is the result of attribute predictor, θ is the parameter of the network.

3.4 BiTextCNN

To address the problem that text encoder cannot balance both whole information and local detail in the legal text, we introduces the encoding method BiTextCNN, and the overall framework is shown in Fig. 3.

The BiTextCNN combined with row convolution module and column convolution module as follow:

$$C_{row}^i = Conv(W_{row} \cdot t_{row}^{i:i+h} + b_{row}), \tag{7}$$

$$L_{row} = [C_{row}^1; ...; C_{row}^{n-h+1}]. \tag{8}$$

Firstly, the text submit to a word embedding layer. You can use pre-trained models such as Word2Vec or GloVe. Then,the fact description text is denoted by the sentence vector T, where $T = [t_{row}^1, ..., t_{row}^n]$, and t_{row}^i is row text vector, n is the length of text, or $T = [t_{col}^1, ..., t_{col}^n]$, and t_{col}^i is column text vector, m is the dimension of word embedding. h is the row-convolution length, W_{row} and b_{row} represent weight and bias respectively. L_{row} is the result of row convolution module as same as classical TextCNN.

Unlike the row convolution, in order to enhance the ability of obtaining whole information, we employ convolution along the vertical axis. The column convolution can cover all the words in the description text at one time, thereby it

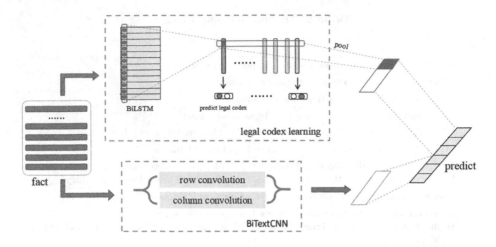

Fig. 3. The overall framework of our method

can alleviate the defect that traditional TextCNN can only consider N-grams. Meanwhile, since the ultra-long convolution will bring huge computational consumption, we employ the deep separable convolution [5,7] to establish column convolution as follow:

$$C_{col}^{i,k} = Conv^k(W_{col}^k \cdot t_{col}^{i:i+i} + b_{col}^k), \tag{9}$$

$$c_{col}^i = [c_{col}^{i,1}; ...; c_{col}^{i,h}], \tag{10}$$

$$C_{col}^i = Conv(W_{col} \cdot t_{col}^{i:i+i} + b_{col}), \tag{11}$$

$$c_{col} = [c_{col}^1; ...; c_{col}^h]. \tag{12}$$

Here, $t_{col}^{i:i+1}$ is the column vector of length one, L_{col} is the output of column module. W_{col}^k and b_{col}^k is weight and bias of separable convolution respectively. After employ the deep separable convolution, we reduce the calculation from $n^2 \cdot h^2$ to $n^2 \cdot h + h^2$, where n is much bigger than h.

3.5 Optimization

The loss function of our network consists of two parts, i.e., one for the law article prediction and another for the codex attribute estimation. The multi-task loss is as follow:

$$Loss = L_{labels} + \alpha \cdot L_{attr}, \tag{13}$$

where L_{attr} formulate from attribute module and L_{labels} formulate from article prediction, α is a parameter to balance the weight in this part, and the details of the two losses formulate are as follows:

$$L_{labels} = -\sum_{i=1}^{C} y_i \log \hat{y}_i, \tag{14}$$

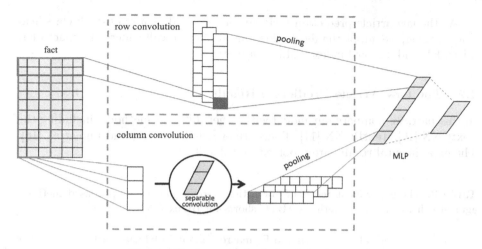

Fig. 4. The framework of the BiTextCNN.

where C is the number of predicted articles, y_i is the truth label, $\hat{y}_i$ is predicted probability distribution;

$$L_{attr} = \sum_{j=0}^{N} -[z_j \log(p_j) + (1 - z_j)log(1 - p_j)]. \tag{15}$$

where N is total number of legal codex, z_j and p_j are the true value and prediction of j-th codex attribute respectively.

4   Experiments

4.1   Dataset and Experimental Setup

Our study constructs a law dataset by collecting the publicly available judgment documents from the China Judgment Online[2]. We collect 400,000 judgment documents, and each document contains the description of the case, law articles with the corresponding codex information. The average length of the fact description is 700 words, and the total distinct number of law articles is 4850 from 287 different codices. We randomly split the data with 90% as the training set and 10% for testing.

For training our model, we adopt the pre-train word embedding with an embedding size of 300, set the hidden state size of BiLSTM to 100, and both row convolution and column convolution filter widths to (3, 4, 5) with each filter size to 100 for consistency. We use Adam for training with an initial learning rate of 10^{-3} and bitch size 32. The weight of α of codex attribute loss is set to 0.5.

---

[2] https://wenshu.court.gov.cn/.

As the law article prediction task can be regarded as a multi-task classification learning, we adopt the most widely used micro-recall (micro-R), macro-F1, micro-F1, and Jaccard index evaluation metrics.

4.2 The Law Article Prediction Result

In this part, we compare our model with other different methods, including the TextCNN [10], TextRCNN [11], Transformer [18], SGM [20] and Luo et al. [16]. The experimental results are reported in Table 3.

Table 3. The experiment result of law articles prediction of our proposed methods against other comparing methods. (BTC shorts for BiTextCNN)

Method	micro-R	macro-F1	micro-F1	Jaccard
TextCNN [10]	0.426	0.530	0.491	0.342
TextRCNN [11]	0.397	0.513	0.488	0.346
SGM [20]	0.293	0.394	0.420	0.292
Luo et al. [16]	0.431	0.533	0.492	0.343
Transformer [18]	0.445	0.547	0.489	0.347
BTC	0.449	0.546	0.498	0.347
BTC+Attr	**0.493**	**0.577**	**0.501**	**0.363**

As can be seen from Table 3, we can get the information that our BiTextCNN model can achieve comparable results with the Transformer [18] while surpassing all other methods. This result indicates that our BiTextCNN can learn a robust feature representation for the fact description while with much fewer parameters. Furthermore, we can observe a significant performance boost after adding the codex attribute learning branch, e.g., 0.044 and 0.031 increments of micro-R and macro-F1, respectively. It demonstrates that the effectiveness of using the codex attribute as an additional supervision signal.

In the handling of actual legal cases, the types of cases usually can be divided according to the situation of fact (e.g., "Disputes over personal obligation", "Motor vehicle traffic accident liability disputes" or "Disputes over the financial loan contracts"). To verify our proposed method's effectiveness in dealing with different types of cases, we select the top 40 types of cases with the highest frequency to conduct experiments and reported the corresponding Jaccard and micro-R results in Fig. 5. From Fig. 5, we can find that BiTextCNN+Attr outperforms the TextCNN and Luo et al. [16] in all different type of cases. This demonstrates the robustness of our proposed method.

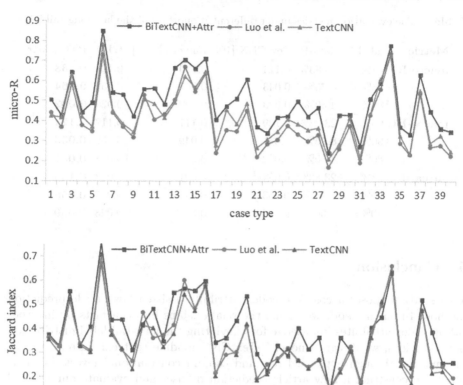

Fig. 5. The comparison our proposed BiTextCNN+Attr against the TextCNN [10] and Luo et al. [16] in different types of cases.

4.3 Experiment Results on the Long-Tail Data

Since our dataset exists a long-tail label distribution, we further compared our methods with others in the long-tail part. We select the labels of different frequency bands for testing, and report the macro-R, macro-F1 and Jaccard metrics in Table 4.

From Table 4, we can find a significant decrease in the three metrics for all different methods, along with the proportion reduction of selected long-tails labels. By comparing with BiTextCNN and BiTextCNN+Attr, we can conclude that using the codex attributes shown a better effect in all ranges of long-tail labels data, and it can indeed alleviate the problem of long-tail labels.

Table 4. The evaluation metrics under different proportion of the last long-tail labels.

Metric	label-P	data-P	TextCNN [10]	Luo et al. [16]	BTC	BTC+Attr
macro-R	98%	23.83%	0.114	0.120	0.121	**0.138**
	90%	8.75%	0.043	0.057	0.059	**0.064**
	80%	4.50%	0.023	0.029	0.029	**0.032**
macro-F1	98%	23.83%	0.109	0.117	0.117	**0.124**
	90%	8.75%	0.041	0.049	0.052	**0.053**
	80%	4.50%	0.022	0.027	0.028	**0.031**
Jaccard	98%	23.83%	0.084	0.091	0.093	**0.148**
	90%	8.75%	0.033	0.041	0.036	**0.058**
	80%	4.50%	0.022	0.028	0.023	**0.039**

5 Conclusion

This study proposes a cost-free codex attribute assisted law article prediction method. In our method, we model the relationships between facts by incorporating codex attributes knowledge for alleviating the long-tail label distribution issue. Besides, we design a novel BiTextCNN module that can fuse all words' information through a parallel row and column convolutional layers. Furthermore, we construct a new article prediction dataset and evaluate our method on it. The experiment results demonstrate the effectiveness of our method and demonstrate that fuses of long-distance information can significantly improve the discrimination of features.

Acknowlegdement. This work is supported by the National Key Research and Development Program of China (No. 2018YFC0831402), and the Nature Science Foundation of China (No. 62076210 & 61806172).

References

1. Aletras, N., Tsarapatsanis, D., Preoţiucpietro, D., Lampos, V.: Predicting judicial decisions of the European court of human rights: a natural language processing perspective. Peer J. **2**, 93 (2016)
2. Bao, Q., Zan, H., Gong, P., Chen, J., Xiao, Y.: Charge prediction with legal attention. In: CCF International Conference on Natural Language Processing and Chinese Computing, pp. 447–458 (2019)
3. Carvalho, D.S., Nguyen, M.T., Tran, C.X., Nguyen, M.L.: Lexical-morphological modeling for legal text analysis. In: JSAI International Symposium on Artificial Intelligence, pp. 295–311. Springer (2015). https://doi.org/10.1007/978-3-319-50953-2_21
4. Chen, S., Wang, P., Fang, W., Deng, X., Zhang, F.: Learning to predict charges for judgment with legal graph. In: International Conference on Artificial Neural Networks, pp. 240–252 (2019)

5. Chollet, F.: Xception: deep learning with depthwise separable convolutions. In: Proceedings of the IEEE Conference on Computer Vision and Pattern Recognition, pp. 1251–1258 (2017)
6. Hu, Z., Li, X., Tu, C., Liu, Z., Sun, M.: Few-shot charge prediction with discriminative legal attributes. In: Proceedings of the 27th International Conference on Computational Linguistics, pp. 487–498 (2018)
7. Kaiser, L., Gomez, A.N., Chollet, F.: Depthwise separable convolutions for neural machine translation. arXiv: ComputationandLanguage (2017)
8. Kanapala, A., Pal, S., Pamula, R.: Text summarization from legal documents: a survey. Artif. Intell. Rev. **51**(3), 371–402 (2017). https://doi.org/10.1007/s10462-017-9566-2
9. Kim, M.Y., Xu, Y., Goebel, R.: Legal question answering using ranking SVM and syntactic/semantic similarity. JSAI International Symposium on Artificial Intelligence, pp. 244–258 (2014)
10. Kim, Y.: Convolutional neural networks for sentence classification. arXiv preprint arXiv:1408.5882 (2014)
11. Lai, S., Xu, L., Liu, K., Zhao, J.: Recurrent convolutional neural networks for text classification. In: Twenty-Ninth AAAI Conference on Artificial Intelligence (2015)
12. Lecun, Y., Bengio, Y., Hinton, G.E.: Deep learning. Nature **521**(7553), 436–444 (2015)
13. Li, J., Zhang, G., Yan, H., Yu, L., Meng, T.: A Markov logic networks based method to predict judicial decisions of divorce cases. In: 2018 IEEE International Conference on Smart Cloud (SmartCloud), pp. 129–132 (2018)
14. Liu, C., Chang, C., Ho, J.: Case instance generation and refinement for case-based criminal summary judgments in Chinese. J. Inf. Sci. Eng. **20**(4), 783–800 (2004)
15. Long, S., Tu, C., Liu, Z., Sun, M.: Automatic judgment prediction via legal reading comprehension. In: China National Conference on Chinese Computational Linguistics, pp. 558–572. Springer (2019). https://doi.org/10.1007/978-3-030-32381-3_45
16. Luo, B., Feng, Y., Xu, J., Zhang, X., Zhao, D.: Learning to predict charges for criminal cases with legal basis. In: Proceedings of the 2017 Conference on Empirical Methods in Natural Language Processing, pp. 2727–2736 (2017)
17. Sulea, O., Zampieri, M., Vela, M., Van Genabith, J.: Predicting the law area and decisions of french supreme court cases. In: Recent Advances in Natural Language Processing Meet Deep Learning, pp. 716–722 (2017)
18. Vaswani, A., Shazeer, N., Parmar, N., Uszkoreit, J., Jones, L., Gomez, A.N., Kaiser, Ł., Polosukhin, I.: Attention is all you need. In: Advances in neural information processing systems, pp. 5998–6008 (2017)
19. Wei, D., Lin, L.: An external knowledge enhanced multi-label charge prediction approach with label number learning. arXiv preprint arXiv:1907.02205 (2019)
20. Yang, P., Sun, X., Li, W., Ma, S., Wu, W., Wang, H.: SGM: sequence generation model for multi-label classification. In: Proceedings of the 27th International Conference on Computational Linguistics, COLING, pp. 3915–3926 (2018)
21. Zhong, H., Zhipeng, G., Tu, C., Xiao, C., Liu, Z., Sun, M.: Legal judgment prediction via topological learning, pp. 3540–3549 . Association for Computational Linguistics (2018)
22. Zhou, P., et al.: Attention-based bidirectional long short-term memory networks for relation classification. Assoc. Comput. Linguist. **2**, 207–212 (2016)

MDA-Network: Mask and Dual Attention Network for Handwritten Mathematical Expression Recognition

Jian Hu[1], Pan Zhou[1], Shirong Liao[1], Shaozi Li[1], Songjian Su[2], and Songzhi Su[1(✉)]

[1] Xiamen University, Xiamen 361005, China
[2] Ropeok Technology Group Co., Ltd., Xiamen, China
ssz@xmu.edu.cn

Abstract. Building a system for automatic handwritten mathematical expressions recognition (HMER) has received considerable attention for its extensive applications. However, HMER remains challenging due to its own many characteristics such as the ambiguity of handwritten symbol, the two-dimensional characteristics of expression structure, and a large amount of context information. Inspired by research on machine translation and image caption, we proposed an Encoder-Decoder structure to recognize the handwritten mathematical expression. Encoder based dual attention is used to extract the features of the expression image and attention-decoder achieves symbol recognition and structural analysis. The mask information is added to the input data allows the model to better focus on the region of interest. In order to verify the effectiveness of our method, we train the model on the CROHME-2016 train set and use the CROHME-2014 test set as the validation set, the CROHME-2016 test set as the test set. The experimental results show that our method is greatly improved compared with other recognition methods, achieved respectively 47.49% and 45.10% ExpRate in the two test sets.

Keywords: Handwritten mathematical expression recognition · Encoder-decoder · Attention

1 Introduction

In many fields like physics, mathematics and other disciplines, mathematical expressions play an essential role in explaining theoretical knowledge and describing scientific problems. With the continuous development of smart devices, it is becoming more and more common to use handwriting devices instead of keyboards to input information in scientific research. Despite the convenience of

Supported by Transformation and industrialization demonstration of marine scientific and technological achievements in Xiamen Marine and Fisheries Bureau (No. 18CZB033HJ11).

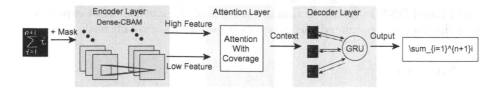

Fig. 1. Pipeline of our proposed recognition method

handwriting input method, it also needs the system to be able to accurately recognize the handwriting mathematical expression.

Different from traditional OCR text recognition, mathematical expressions recognition is more difficult due to the two-dimensional structure of the expression, the scale diversity of mathematical symbols and the existence of implicit symbols. According to the recognition process, the method of mathematical expressions recognition can be divided into two types: single-stage and multi-stage. The multi-stage method sequentially does symbol segmentation, symbol recognition and structure analysis. In contrast, the single-stage method completes all the symbol segmentation, symbol recognition and structure analysis at once. Compared with the single-stage method, the multi-stage method has the problem of error inheritance. The error of symbol segmentation and symbol recognition affect the structural analysis stage lead to error accumulation. This paper proposes the single-stage method to recognize mathematical expressions.

The main contributions of this paper are in the following way: 1)Use convolutional neural networks with a dual attention mechanism to enhance the feature extraction ability of the model. 2)mask channels are added to input data, increased the dimension of input information, improved recognition effect. 3)Use curriculum learning for training to enhance the robustness of the model, enables the model to perform better in the inference process.

2 Related Work

Handwritten Mathematical expression recognition(HMER) can be traced back to 1968 and was first raised by Anderson [29]. Since then, HMER has received intensive attention, many researchers have tried to investigate this problem in different ways. Some approaches use extensive prior knowledge and the corresponding parsing algorithms to build a grammar-based model. Alvaro et al. [19] proposed a model which is based on the stochastic context-free grammar. MacLean et al. [22] use model-based relational grammars and fuzzy sets to recognize mathematical expression. These systems have performed very well in several CROHME competitions. Some approaches treat HMER as the image-to-sequence problem and use the encoder-decoder framework for handwriting expression recognition. Zhang et al. [7] proposed a model named WAP-based encoder-decoder to perform image includes mathematical expressions to the latex representation of the corresponding expression. Compared with the traditional manual design grammar, model-based DNN learns the grammatical rules of the mathematical expression from the data and promoted the recognition performance significantly.

3   Method

The recognition model can be divided into three layers, respectively encoder layer, attention layer, and decoder layer. By inputting the two-dimensional structure expression image into the model, the model outputs the one-dimensional structure Latex sequence result straightly. The complete algorithm model is shown in Fig. 1. In the encoder layer, we use the dense-cbam dual attention module to enhance the feature extraction ability and promote the propagation of the gradient. Considering that there are mathematical symbols of different scales in the expression image, we simultaneously extract the multiple feature sequences to adapt to different scales. In the decoder layer, taking into account over-parse and under-parse, we introduce an attention mechanism that added a coverage vector to store information that has been analyzed in the past. In order to avoid the problem of gradient disappearance and gradient explosion, we used Gated Recurrent Units (GRU) to decode and output the latex sequence.

3.1   Encoder Layer

The encoder layer uses a convolutional neural network to perform feature extraction on the input expression image. The output feature map A has a size of $H \times W \times C$, where W is the width of the image, H is the height of the image, and C is the total number of channels in the image. The output feature A is considered as L C-dimensional vectors, where $L = H \times W$, and each vector corresponds to a region of the image:

$$A = \{a_1, a_2, \ldots, a_L\}, a_i \in \mathbb{R}^C \tag{1}$$

In many tasks of computer vision, the attention mechanism has excellent performance. Xu et al. [6] first proposed the soft attention mechanism and hard attention, the intention mechanism is used to solve the problem of image title; Chen et al. [3] proposed the use of convolutional networks to combine spatial and channel attention named SCA-CNN; Hu et al. [8] proposed channel attention, the force mechanism is used to obtain the importance of each channel and won the championship in the last ImageNet classification competition. Inspired by these works, We added a dual attention mechanism to the encoder. The dual attention mechanism can effectively enhance the convolutional neural network characterization ability to extract features. By combining the dense module and the convolutional attention module (CBAM), the encoder can learn what and where to pay attention to on the channel and spatial axis to improve the expression of the region of interest. To effectively help the information flow in the network, the model learns what information needs to be emphasized or suppressed. Next, we will introduce these two modules separately.

DenseNet. Researches show that if the convolutional neural network includes shorter connections between layers close to the input and close to the output, the network can be significantly deepened, become more accurate , and easier to train. Compared with ResNet, DenseNet uses a more aggressive intensive connection mechanism, connecting all layers to each other. An the network which has L layers contains $(L(L+1))/2$ connections. Moreover, DenseNet directly connects to the channel dimension to achieve feature multiplexing so that it can achieve better performance than ResNet in the case of fewer parameters and computational cost. The calculation formula for each layer of the Dense module is as follows:

$$x_l = H_l \left([x_0, x_1, \ldots, x_{l-1}]\right) \tag{2}$$

where $H(*)$ means a composite function that passes through 4 operations of 3×3 convolutional layer, pooling layer, batch normalization layer, and nonlinear activation function (Relu).

CBAM. Convolutional Block Attention Module (CBAM) is a lightweight, general-purpose module that combines spatial and channel attention mechanisms. Compared to SENet, the attention mechanism that only focuses on channels can achieve better results and be used in any convolutional architecture. The following briefly describes the implementation process of the two attention modules:

1. Spatial attention module: input feature map $F_s (C_s, H_s, W_s)$, after average pooling and maximum pooling, respectively, to obtain two feature maps with the same dimension $F_{\text{savg}} (1, H_{s_1} W_s)$, $F_{\text{smax}} (1, H_{s_1} W_s)$, which are next spliced in the channel dimension. After inputting the spliced one into the convolutional layer of the sigmoid function, we could then obtain a spatial attention matrix, which is finally multiplied with the input feature map to obtain the M_s spatial attention feature map. Since each channel is a feature detector, the channel attention can be effectively used by compressing the spatial dimension, and different pooling operations can be applied simultaneously to achieve better results than a single pooling.
2. Channel attention module: input feature map $F_c (C_o, H_c, W_c)$, respectively, through global average pooling and global maximum pooling to obtain two $1 \times 1 \times c$ channel descriptions, which are inputted to shared MLP to get two kinds of vectors. Then add them, pass the sigmoid function to get a channel attention vector with dimension $c \times 1 \times 1$, finally multiply it with the input feature map to get the M_c space attention feature map.

3.2   Attention Layer

We call the vector calculated by the attention mechanism as the context vector, which is used to capture the context of the regional visual information. We use the formula (3), (4), (5) to simply describe the process of calculating the context vector. The attention mechanism is used to focus on the different image areas at different times of the decoder to generate the corresponding latex sequence. In addition, the attention mechanism also takes into account the possibility that certain areas are over-parsed or under-parsed in the expression recognition, resulting in the phenomenon of missing symbols and repeated symbols in the recognized expression. Inspired by the coverage mechanism of machine translation, we add a coverage vector to the attention mechanism to track the analyzed information. In this way, the future attention can be adjusted, and the unanalyzed areas of the image can also be assigned higher attention probability, thereby reducing repeated analysis and missing analysis.

There are two key quantities in the attention mechanism. One is related to time t, which corresponds to the decoding time. And the other is the area a_i of the input sequence, which corresponds to a region of the image. We calculate the coverage vector by accumulating and summing the weights of past attention, as shown in the formula (6) so that the information analyzed in the past can be better described.

$$\exp_{ti} = f_{att}(a_i, h_{t-1}) \tag{3}$$

$$\alpha_{ti} = \frac{\exp(e_{ti})}{\sum_{k=1}^{L} \exp(e_{tk})} \tag{4}$$

$$z_t = \phi(\{a_i, \alpha_{ti}\}) \tag{5}$$

$$c = \sum_{l=1}^{t-1} \alpha_l \tag{6}$$

where f_{att} represents a multi-layer perceptron, a_i is a vector in the decoder output, corresponding to a region in the image, h_{t-1} is the hidden layer output of the previous unit, and α_{ti} represents the weight of the $i - th$ vector at time t, c represents the coverage vector, and α_l represents the attention weight of the $l - th$ time step. z_t indicates the output of the attention mechanism, and ϕ indicates that the attention weight is applied to the image area.

3.3   Decoder Layer

The decoder layer is used to decode the output of the intermediate vectors by the attention mechanism and output a set of one-hot codes of all mathematical symbols:

$$Y = \{y_1, y_2, \ldots, y_T\}, y_i \in \mathbf{R}^K \tag{7}$$

where k represents the number of mathematical symbols and T represents the number of symbol categories of the latex sequence.

We employ Gated Recurrent Units (GRU) as the decoding unit. GRU as a variant of RNN can effectively alleviate the gradient explosion and gradient disappearance problems in RNN. And there are fewer parameters than the LSTM network, faster convergence speed, and the same effect can be achieved. The calculation of the hidden layer h_t of GRU is as follows:

$$z_t = \sigma(W_z \cdot [h_{t-1}, x_t]) \tag{8}$$

$$z_t = \sigma \left(W_z \cdot [h_{t-1}, x_t] \right) \tag{9}$$

$$\widetilde{h_t} = \tanh \left(W \cdot [r_t * h_{t-1}, x_t] \right) \tag{10}$$

$$h_t = (1 - z_t) * h_{t-1} + z_t * \widetilde{h_t} \tag{11}$$

4   Training and Inference

4.1   Training

Early in the training phase, the prediction ability of RNN is very weak. If a certain neural unit produces a large deviation result, it will seriously affect the learning of the subsequent neural unit. To solve this problem, in the training process, the model can directly use the real label of the training data as the input of the next unit, not to use the output of the previous unit as the input of the next unit. However, there are certain problems in this way. The model has better results in the training process based on the excessive reliance on label data. Whereas the lack of guidance from the real label during the testing process cause that the generated results have a large difference with the training process and the generalization performance of the model is poor. Finally, we decided to use curriculum learning for training, using a probability p to choose whether to use the real label or the output of the previous unit for the input of the current time step.

4.2   Inference

In the inference process, the decoder can only use the output of the previous unit as the input of the current unit. in order to avoid the error output of a unit causing the prediction error of the entire sequence, we decided to use beam search to find the optimal result during decoding. In beam search, there is only one parameter B (beam width), which is used to represent the first B results of each selection. We assume that the beam width B is 3, and the vocabulary size l (The number of mathematical symbol categories) is 110. In the first step, we calculate the distribution probability of $y^{<1>}$ through the model, and select the first three as the candidate results according to the probability, denoted as a1, a2, a3. In the second step, we have selected a1, a2, a3 as the three most likely choices for the first mathematical symbol, considering the probability of the second symbol for the first symbol, and inputting the three symbols into the model respectively, get $B \times l = 3 \times 101 = 303$ probabilities, select the first 3 according to the probability. In the third step, repeat the second step until the end character is reached or the maximum length position is reached, and the sequence with the highest score is finally output.

5 Experiments

5.1 Settings

Table 1. The distribution of the number of expressions in different datasets.

Dataset	Category	Number of expressions
Train set	CROHME2014 train set	8836 expr
Validation set	CROHME2014 test set	986 expr
Test set	CROHME2016 test set	1147 expr

Dataset. Our algorithm uses the CROHME-2016 [10]dataset as the train set. Which was written by 50 volunteers on three different resolution devices, and then the stroke data was sampled by coordinates and saved in InkML format. The data set included a total of 110 mathematical symbols. In order to generate an offline expression image, we render these handwriting coordinates locally to obtain the gray-scale image of the expression image. The size of the generated image is different according to the number of symbols and the structure of the expression. We use the CROHME-2014 test set as the validation set and the CROHME-2016 test set as the test set. The specific distribution of the dataset can be seen in Table 1.

Considering the actual application scenario, meanwhile, if the size of the image is too large and the expression contains too many symbols, which will also increase the difficulty of training the model, reduce model recognition performance. So we did not use all train set images for training, but added restriction for filtering, the maximum size of the input picture is 100,000 pixels, and the expression contains up to 70 mathematical symbols. Based on this restriction, the size of the train set that we use is 8,443 images, the size of the test set is 1,114 images, and the size of the validation set is 935 images.

Experimental Details. The proposed algorithm is based on pytorch, initialized with xavier, the initial learning rate is 10^{-3}, the batch size is set to 2, used SGD optimization algorithm, and L2 regularization and dropout are added to suppress overfitting,90 iterations were performed in the training process.

Evaluation. The CROHME competition is ranked according to the expression recognition rate(ExpRate), Only when the recognized Latex sequence matches the real label's latex exactly is it recognized as correct. However, individual symbol recognition errors may occur in expression recognition, so We also used another evaluation indicator commonly used in speech recognition called word error rate (WER), it measures character-level errors. In order to make the recognized word sequence and standard Keep consistent between word sequences, need to be replaced, deleted, or insert some words. The total number of these

Table 2. Comparison result of different channel Numbers

Channel	CROHOME2014		CROHOME2016	
	ExpRate	WER	ExpRate	WER
Single channel	46.46	**13.80**	43.67	**15.13**
Double channel	**47.49**	17.17	**44.47**	15.42

Table 3. Comparison result of cbam module position

Decoder	CROHME2014		CROMER2016	
	ExpRate	WER	ExpRate	WER
Dense	45.13	**16.75**	43.49	**17.03**
CBAMv1	37.11	20.94	37.47	19.94
CBAMv2	**47.49**	17.17	**45.10**	17.28

inserted, replaced and deleted words is divided by the word in the standard word sequence resulting in WER. The calculation expression is shown in Eq. 12:

$$WER = \frac{N_{ins} + N_{sub} + N_{del}}{N} \tag{12}$$

5.2 Experimental Comparison

The Process of Data. During the data reading process, due to the different size of the expression images, it is necessary to fill the images in a batch. Then add a masked channel to record the filling information in the data reading process to prevent the filling part affects the real loss. In Table 2, we compare the two data processing methods. Although the WER metric of two-channel is higher than that of one-channel on the two data sets, The recognition accuracy of the two-channel with the mask channel is improved by 1.03% compared with the one-channel. This suggests that the increased channel information can reduce some recognition errors in one-channel, effectively improve the recognition accuracy of the model.

Attention Module Position. For the purpose of verifying whether the position of the CBAM module has an effect on the accuracy of recognition, we have done three groups of comparative experiments: Dense means not to join the CBAM attention module; CBAMv1 means that add cbam module to the last convolution layer of each layer of the dense module; CBAMv2 means add CBAM module between each dense module and transform module. The experimental results of these three encoders are shown in Table 3. It can be found that the recognition accuracy of CBAMv2 is improved by 2.36% compared to Dense, indicates that adding the attention information to the encoder part can

Table 4. The test results in CROHME2014 dataset

System	Correct	≤1(%)	≤2(%)	≤3 (%)
I	37.22	44.22	47.26	50.20
II	15.01	22.31	26.57	27.69
IV	18.97	28.19	32.35	33.37
V	18.97	26.37	30.83	32.96
VI	25.66	33.16	35.90	37.32
VII	26.06	33.87	38.54	39.96
WAP$_{vgg}$	46.55	61.20	65.20	66.10
WAP$_{Dense}$	41.49*	54.70*	65.40*	71.30*
Ours	**47.49**	**65.35**	**74.33**	**80.75**

Table 5. The test results in CROHME2016 dataset

System	Correct	≤1(%)	≤2(%)
Wiris	**49.61**	60.42	64.69
Tokyo	43.94	50.91	53.70
São Paolo	33.39	43.50	49.17
Nantes	13.34	21.02	28.33
WAP$_{vgg}$	44.55	57.10	61.60
Ours	45.10	**61.46**	**69.99**

make the extracted features more suitable for decoding the decoder. Moreover, the experimental results of CBAMv1 indicate that too much attention network embedding will reduce the coding effect of the original model, thus reducing the recognition accuracy of the whole model.

Performance Comparison. We verified on the CROHME-2014 test set and CROHME-2016 test set respectively. Table 4 show the comparison results of our system with other systems on CROHME-2014 test set. Considering the fairness of the experimental results, we selected officially provided train set for train to comparison. System I to VII were submitted to CROHME-2014 competition, Details of these systems can refer to [2]. WAP$_{vgg}$ [7] and WAP$_{Dense}$ [1] both adopt the encoding and decoding structure, the * in the table indicates the test results of the official open-source code under a single model, which is lower than the results in the paper. In order to verify the generalization ability of our model and compare it with some new systems. In Table 5, we show the comparison results of our model and other systems on CROHME-2016 test set. Tokyo is based on Stochastic Context Free Grammar, and its model with three modules is extremely complicated. Nantes proposed a system that can consider the invalid symbol segments for preventing errors from one step to another. [10] introduce

Other systems specific implementation details. $\langle=1, \langle=2, \langle=3$ metric indicate that there is at most recognition rates of one error, two errors, and three errors for the prediction result and the real label, which helps to further understand the recognition effect of the model. It can be found that our model achieved the highest recognition rate in CROHME-2014. Although the recognition accuracy of our model is lower than Wiris in CROHME-2016, Wiris uses a formula corpus composed of more than 592000 formulas to train a language model. Considering the train cost, the language model is not used in our proposed method. If the language model is added, our system should be able to achieve better recognition results. And in both test sets, our $\langle=1, \langle=2$ are the highest, indicating that our model has more room for improvement.

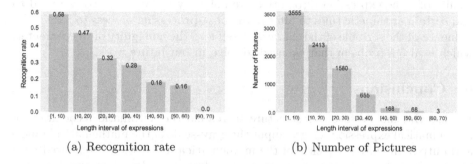

(a) Recognition rate   (b) Number of Pictures

Fig. 2. Recognition results of different length interval of expressions

(a)	4×4 +4+4	gt latex: 4\times4+4+4 pred latex: 4x4+4+4	(d)	γ→∞	gt latex: \gamma\rightarrow\infty pred latex: r\rightrrow\infty
(b)	μ≥0	gt latex: \mu\geq0 pred latex: w\geq0	(e)	βm+1	gt latex: B_{m+1} pred latex: \beta_{m+1}
(c)	√3²+z² =√13	gt latex: \sqrt{3^{2}+2^{2}}=\sqrt{13}} pred latex: \sqrt{3^{2}+z^{2}}=\sqrt{13}	(f)	Logₑx	gt latex: \log_{e}x pred latex: \log_{e}X

Fig. 3. Examples of mathematical expression recognition errors

5.3   Result Analysis

We divided 7 intervals according to the length of the mathematical expression, and calculated their respective Recognition Rate. From the results of (a) in Fig. 2, we can find that the recognition effect of the model in the short sequence is much better than that in the long sequence. First of all, due to the difference in the length of the expression, the size of the corresponding expression image will be different. In this case, the larger the size of the image, the smaller the corresponding image area containing each mathematical symbol, and it also causes a more compact connection between the symbols. Which makes it hard for the model to accurately recognize the expression in the image. Besides, we counted the length distribution of the mathematical expressions in the train set. In Fig. 2, we can see from (b) that the number of train samples for the long sequence in the train set is very letter. The model can't fit the long sequence well, which has an influence on the recognition effect.

There are some examples of wrong recognition in Fig. 3, the symbol is too scribble during the writing process, like (b), (d), (c); the ambiguity of mathematical symbol, like times and x in (a), the difference of symbol size, like upper and lower case X, x in (f). Under the existing model, we found that the wrong results of some expression recognition are not easy to be corrected. we need to add certain arithmetic rules to filter in the post-processing process to reduce the influence of the scribble of the character itself and the ambiguity of the character, which is also a problem that we need to solve in our future work.

6   Conclusions

We use the encoder-decoder structure to realize the automatic recognition of mathematical expression images, input the gray-scale of the mathematical image and output the latex sequence of the mathematical expression. Compared with the traditional expression recognition algorithm, our algorithm simplifies the complex processes of symbol segmentation, symbol recognition, and structural analysis, and directly implements all processes with an end-to-end model. The added dual attention module to the encoder enhanced the representation ability of the convolutional neural network, thereby effectively improving the recognition accuracy of the model. Experimental results show that our algorithm has greatly improved the recognition performance of handwritten mathematical expression compared with other algorithms.

References

1. Zhang, J., Du, J., Dai, L.: Multi-scale attention with dense encoder for handwritten mathematical. In: Conference on Pattern Recognition (ICPR), pp. 2245–2250. IEEE, Beijing (2018)
2. Mouchere, H., Viard-Gaudin, C., Zanibbi, R.: ICFHR 2014 competition on recognition of on-line handwritten mathematical expressions. In: 2014 14th International Conference on Frontiers in Handwriting Recognition, pp. 791–796. IEEE, Hersonissos, Greece (2014)
3. Long, C., Zhang, H., Xiao, J.: SCA-CNN: spatial and channel-wise attention in convolutional networks for image captioning. In: IEEE Conference on Computer Vision and Pattern Recognition, pp. 6298–6306. IEEE, Honolulu (2017)
4. Le, A.D., Nakagawa, M.: Training an end-to-end system for handwritten mathematical expression recognition by generated patterns. In: 14th IAPR International Conference on Document Analysis and Recognition (ICDAR), pp. 1056–1061. IEEE, Kyoto (2017)

5. Vinyals, O., Toshev, A., Bengio, S.: Show and tell: a neural image caption generator. In: IEEE Conference on Computer Vision and Pattern Recognition (CVPR), pp. 3156–3164. IEEE, Boston (2015)
6. Xu, K., Ba, J., Kiros, R.: Show, attend and tell: neural image caption generation with visual attention. In: Proceedings of the 32nd International Conference on Machine Learning, pp. 2048–2057(2015)
7. Zhang, J., Du, J., Zhang, S.: Watch, attend and parse: an end-to-end neural network based approach to handwritten mathematical expression recognition. Pattern Recognit. **71**, 196–206 (2017)
8. Jie, H., Li, S., Gang, S.: Squeeze-and-excitation networks. In: 31st IEEE/CVF Conference on Computer Vision and Pattern Recognition (CVPR), pp. 7132–7141. IEEE, Salt Lake City (2018)
9. Tu, Z., Lu, Z., Liu, Y.: Modeling coverage for neural machine translation. In: 54th Annual Meeting of the Association-for-Computational-Linguistics (ACL), pp. 76–85. Berlin (2016)
10. Mouchère, H., Viard-Gaudin, C., Zanibbi, R.: ICFHR2016 CROHME: competition on recognition of online handwritten mathematical expressions. In: 15th International Conference on Frontiers in Handwriting Recognition (ICFHR), Shenzhen, pp. 607–612 (2016)
11. He, K., Zhang, X., Ren, S.: Deep residual learning for image recognition. In: 2016 IEEE Conference on Computer Vision and Pattern Recognition (CVPR), pp. 770–778. IEEE, Seattle (2016)
12. Huang, G., Liu, Z., Laurens, V.D.M.: Densely connected convolutional networks. In: 30th IEEE/CVF Conference on Computer Vision and Pattern Recognition (CVPR), pp. 4700–4708. IEEE, Honolulu (2017)
13. Bhunia, A.K., Bhowmick, A., Bhunia, A.K.: Handwriting trajectory recovery using end-to-end deep encoder-decoder network. In: 24th International Conference on Pattern Recognition (ICPR), pp. 3639–3644. IEEE, Beijing (2018)
14. Chan, K.F., Yeung, D.Y.: Mathematical expression recognition: a survey. Int. J. Document Anal. Recognit. (ICDAR), 3–15 (2000). https://doi.org/10.1007/PL00013549
15. Chen, X., Ma, L., Jiang, W.: Regularizing RNNs for caption generation by reconstructing the past with the present. In: 31st IEEE/CVF Conference on Computer Vision and Pattern Recognition (CVPR), pp. 7995–8003. IEEE, Salt Lake City (2018)
16. Le, A.D., Indurkhya, B., Nakagawa, M.: Pattern generation strategies for improving recognition of handwritten mathematical expressions. Pattern Recognit. Lett.**128**, 255–262 (2019)
17. Glorot, X., Bengio, Y.: Understanding the difficulty of training deep feedforward neural networks, pp. 249–256 (2010)
18. Woo, S., Park, J., Lee, J.Y.: CBAM: convolutional block attention module. In: Proceedings of the European Conference on Computer Vision (ECCV), pp. 3–19(2018)
19. Alvaro, F., Sánchez, J., Benedí, J.: Recognition of on-line handwritten mathematical expressions using 2D stochastic context free grammars and Hidden Markov models. Pattern Recognit. Lett. **35**, 58–67 (2014)
20. Alvaro, F., Sánchez, J., Benedí, J.: An integrated grammar based approach for mathematical expression recognition. Pattern Recognit. **51**, 135–147 (2016)
21. Zanibbi, R., Blostein, D., Cordy, J.R.: Recognizing mathematical expressions using tree transformation. IEEE Trans. Pattern Anal. Mach. Intell. **24**(11), 1455–1467 (2002)

22. MacLean, S., Labahn, G.: A new approach for recognizing handwritten mathematics using relational grammars and fuzzy sets. Int. J. Document Anal. Recognit. 139–163 (2013). https://doi.org/10.1007/s10032-012-0184-x
23. Mouchere, H., Zanibbi, R., Garain, U.: Advancing the state-of-the-art for handwritten math recognition. Int. J. Document Anal. Recognit. 173–189 (2016)
24. Zanibbi, R., Blostein, D.: Recognition and retrieval of mathematical expressions. Int. J. Document Anal. Recognit. **15**(4), 331–357 (2012). https://doi.org/10.1007/s10032-011-0174-4
25. Chan, K.F., Yeung, D.Y.: Error detection, error correction and performance evaluation in on-line mathematical expression recognition. Pattern Recognit. **34**(8), 1671–1684 (2001)
26. Dauphin, Y.N., Fan, A., Auli, M.: Language modeling with gated convolutional networks. In: Proceedings of the 34th International Conference on Machine Learning, pp. 70:933–941(2017)
27. Mouchere, H., Zanibbi, R., Garain, U.: Advancing the state-of-the-art for handwritten math recognition. Inte. J. Document Anal. Recognit. **19**(2), 173–189 (2016)
28. Zhang, J., Du, J., Dai, L.: A GRU-based encoder-decoder approach with attention for online handwritten mathematical expression recognition. In: 14th IAPR International Conference on Document Analysis and Recognition (ICDAR), pp. 902–907. IEEE, Kyoto (2017)
29. Anderson, R.H.: Syntax-directed recognition of hand-printed two-dimensional mathematics. In: Symposium on Interactive Systems for Experimental Applied Mathematics: Proceedings of the Association for Computing Machinery Inc., Symposium, pp. 436–459 (1967)

A Study Based on P300 Component in Single-Trials for Discriminating Depression from Normal Controls

Wei Zhang[1], Tao Gong[1], Jianxiu Li[1], Xiaowei Li[1], and Bin Hu[1,2,3,4(✉)]

[1] Gansu Provincial Key Laboratory of Wearable Computing, School of Information Science and Engineering, Lanzhou University, Lanzhou, China
bh@lzu.edu.cn
[2] CAS Center for Excellence in Brain Science and Intelligence Technology, Shanghai Institutes for Biological Sciences, Chinese Academy of Sciences, Beijing, China
[3] Joint Research Center for Cognitive Neurosensor Technology of Lanzhou University and Institute of Semiconductors, Chinese Academy of Sciences, Beijing, China
[4] Engineering Research Center of Open Source Software and Real-Time System (Lanzhou University), Ministry of Education, Lanzhou, China

Abstract. The investigation of attentional bias of depression based on P300 component has drawn interest within the last decades. Follow-up of previous research suggested the differential amplitudes between depression and normal controls (NCs) in response to various facial stimuli. In this paper, we used single-trials features in the occurrence of P300 to recognize depression from NCs. EEG activity was recorded from 24 patients and 29 NCs in a dot-probe task. We considered two traditionally used feature-extraction methods: ReliefF and principal component analysis (PCA). Then, the k-nearest neighbor (KNN), BFTree, C4.5, logistic regression and NaiveBayes were adopted in this study to make a performance comparison. The combination of NaiveBayes and PCA was applied to classify the P300 component evoked by sad-neutral pairs, which achieved higher classification accuracy than other classifiers. The classification accuracy was 98%. The classification results support that the P300 component of ERPs may reflect information processing of the specific response of depression to specific stimuli and may be applied as a physiologic index for aided diagnosis of depression in future research.

Keywords: Depression · P300 · Single-trials · Classification

1 Introduction

Depression is a prevalent psychiatric illness with disadvantageous influence on sufferers and society [1]. Studies on cognitive dysfunction and information processing bias in depression have pointed out that patients selectively attend to negative information and have difficulty in disengaging their attention from them [2–4]. According to estimates by the World Health Organization, proximately 340 million people are suffering from varying degrees of depression. By today, the disease has proved to be one of the main

Y. Sun et al. (Eds.): ChineseCSCW 2020, CCIS 1330, pp. 209–221, 2021.
https://doi.org/10.1007/978-981-16-2540-4_16

causes of disability and the second largest factor in the global burden of disease [5]. This has brought huge psychological and economic burdens to individuals, families and even society. However, there are some deficiencies in the clinical diagnosis of depression, which are easily affected by the subjective bias of patients and doctors [6]. Therefore, more objective detection methods are needed. A researchable hypothesis for detecting depression is that brain activity contains rich case information and contributes to the detection of depression. Based on this hypothesis, studies using neurophysiological signals (i.e., functional magnetic resonance imaging (fMRI) and event related potentials (ERP)) for the detection of depressed patients have been frequently reported [4, 7].

The ERPs with a high temporal resolution in the millisecond range, can reflect neuroelectrophysiological changes in the brain during cognitive processes. It has been extensively applied to study information processing of individuals with distinct cognitive modes in neural circuits due to its advantages of real-time and non-intrusive [8]. ERPs are composed of a series of component waveforms with local minimum and maximum values. Kalatzis, et al. [9] designed a computer-based system for classification by extracting seventeen features related to the ERP component from both NCs and patients with depression. The classification accuracy could reach 94%. Specially, as an endogenous ERP, P300 has attracted widespread interest in psychiatry, because the existing evidence available supports the idea that the depressed patients has difficulty in disengaging their attention from negative stimuli. Concretely speaking, P300 is a positive potential in the frontal middle area of the brain, which usually reaches a peak around 300 ms to 600 ms after the stimulation occurs [10]. The information of the time and degree of brain activation hidden behind the cognitive process can be provided by this component [11]. In addition, the recent research has found small P300 amplitudes in the parietal lobe of the brains of depressed patients [12]. In [4], a CFS and KNN combination was reported for the classification of depression by extracting features from 300 component of visual stimuli, and the optimal classification accuracy could reach 94%.

Generally, the dot-probe task is one of the common paradigms for inducing P300 component, which is widely used for investigation abnormal attentional bias in depressed subjects [13, 14]. In the typical version of dot-probe task, a picture or word pair was presented simultaneously on a left and right of computer screen in the visual field. One is emotion-related (e.g., a sad face) and the other neutral. Immediately following the offset of the picture pairs, a target was presented at one of the location which previously presented images. Subjects need to accurately and quickly determine the target's location as accurately and quickly possible. Probe appearing at location of the previous emotional face (happy, sad or fear) was defined as valid trials, whereas probe appearing at the locations of previous neutral was defined as invalid trials.

On the basis of the above, we explored a dot-probe task to investigate which type of stimulus contains more effective information for discriminating normal controls and depressive patients. The features were extracted from the signal of P300 component. We selected 17 scalp electrodes for each subject. These features, together with five classification algorithms (k-nearest neighbor (KNN), BFTree, C4.5, logistic regression and NaiveBayes), were used in the design of the classification. Finally, we compared the accuracies of different classifiers combined with different feature selection techniques, to investigate the ability of the P300 component to discriminate between individuals with depression and NC individuals.

2 Material and Methods

2.1 Participants

In present study, we recruited 24 patients with depression and 29 normal controls (NCs), which age range 18–55 years. Each participant was right handed with normal visual acuity or corrected visual acuity. All of them needed to complete a face-to-face diagnostic interview by the psychiatrist in clinic using *the Mini International Neuropsychiatric Interview (MINI)* [15]. Patients who met the DSM-IV diagnostic criteria for depression, were included in the current research. Furthermore, the current status of depression and anxiety was evaluated by the participants' scores of *the Patient Health Questionnaire-9* (PHQ-9) and *the Generalized Anxiety Disorder* (GAD), and the PHQ-9 score of patients with depression should be greater than or equal to 5.

Normal controls (NCs) were recruited from the community. They were eligible if they had no major medical or psychiatric illness, psychoactive substance abuse and history of depression (including personal and family), and had a PHQ-9 score <5.

All enrolled subjects gave informed consent and received a reward of 100 China Yuan for participation after the experiment. The ethics committee of the Lanzhou University Second Hospital approved the study. Table 1 displays the demographic information of the patients with depression and NC individuals.

Table 1. The demographic information for the patients with depression and NC individuals.

	Depression	Normal controls
Age	30.88 ± 10.15	31.44 ± 8.99
Female/male	11/13	9/20
Education		
Primary school	6	1
Secondary education	6	3
University education	12	25
PHQ-9 score	18.33 ± 3.42	2.66 ± 1.77
GAD score	13.42 ± 4.84	2.10 ± 2.04

2.2 Dot-Probe Task

Emotional Face Stimuli

Facial expressions of one's own country or race can help to elicit emotions. The native Chinese Affective Face Picture System (CAFPS) [16] contains seven emotional types and a total of 870 different Chinese affective facial pictures. In this study, we selected 120 emotional faces from the CAFPS. Among them, there are 20 faces for each of three emotions (happiness, sadness and fear) and 60 neutral faces. The gender and emotional valence of the pictures were balanced across experiment types. Then, we removed non-facial features such as hair and clothing and converted these faces into 8-bit grayscale image.

The task contained 480 trials. Each emotional face was respectively paired with a neutral face, containing three emotions blocks (happy-neutral pairs, sad-neutral pairs and fear-neutral pairs), see Fig. 1A. Depending on the location of dot appeared on the screen (see Fig. 1B), we have six kinds of stimuli: valid happy-neutral pairs, invalid happy-neutral pairs, valid sad-neutral pairs, invalid sad-neutral pairs, valid fear-neutral pairs and invalid fear-neutral pairs. Picture pairs with the same gender were simultaneously presented on both sides of the screen. The number of each emotional condition had 80 trials, and presented in a pseudo-random manner.

The Dot Probe Task

We used DELL 17-I monitor and IBM-PC to present this task, which was run in E-Prime software. Figure 1C shows an example trial of a sequence of events. In each trial, the beginning of the experiment was prompted by a fixation point of 300–600 ms. Then, a CAFPS picture pair emerged in black screen for 500 ms. After blank screen appeared at randomly selected intervals of 100–300 ms, a white target appeared at the central position of previously presented pictures, disappeared after 150 ms. Participants were told that the pictures were not related to the task and needed to quickly and accurately determine the location of the target within 2000 ms. Then, a blank screen occurred on the interval of 600 ms. In order to prevent the pollution of EEG recording, participants were asked to keep eye fixed at center of the screen throughout the whole trial.

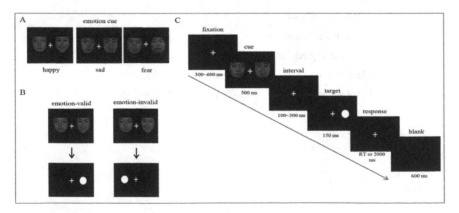

Fig. 1. Example trial of a sequence of events for the dot probe task.

The test procedure was carried out in a sound insulation and obscure room. Participants sat before a 17-inch LCD screen at a distance of about 60 cm a way. They were asked to hit the "1" key (marked "left") of the keyboard with their left middle finger and the "4" key (marked "right") of the keyboard with their right middle finger. Before starting the experiment, participants completed 10 practice trials and then completed all 480 trials in this task.

2.3 EEG Data Record

A HydroCel Geodesic Sensor Net sites of 128-channel was applied to record EEG data at 250 Hz sampling rate, with electrode impedance $<70\,\mathrm{k\Omega}$. All of the electrodes were re-referenced by subtracting the average activity recorded at all channels in the off-line analysis. Then, the continuous EEG was conducted using the EEGLAB (https://sccn.ucsd.edu/wiki/EEGLAB_Wiki) toolbox, including FIR band-pass filter which set to 0.1–30 Hz. Decomposition of independent component analysis (ICA) [17] was used to remove eye blinking activities for each participant. Finally, cleared EEG data were segmented into 600 ms epochs (100 ms before the target appeared to 500 ms after the target appeared). Baseline correction was performed using a 100 ms before the onset of target.

2.4 Classification Methodology

Feature Generation

Negative stimuli to the occurrence of depression symptoms and maintain plays an important role. Previous research has shown that depression selectively focus on negative information, especially in late stage (such as P300 achieved the largest peak in centro-parietal sites) [18]. So we selected 17 electrodes from these areas, including CP1, CP2, CP3, CP4, CP5, CP6, CPz, P1, P2, P3, P4, P5, P6, Pz, PO3, PO4 and POz. Then, the 14 time-domain features and 2 frequency-domain features at 300–400 ms time interval were calculated from single trials in each subject.

Time-Domain Characteristics. 14 time-domain characteristics are calculated and defined as follows:

1) PPmean:

$$pp_{mean} = \frac{s_{max} - s_{min}}{2} \tag{1}$$

Among them, s_{max} and s_{min} represent the extreme value or minimal value of the EEG data in the interval 300–400 ms after stimulus, respectively.

2) Mean square:

$$meanSquare = \frac{1}{N}\sum_{t=1}^{N} s_t^2 \tag{2}$$

3) Variance:

$$Variance = \frac{1}{N-1} \sum_{t=1}^{N} (S_t - \overline{S_t})^2 \tag{3}$$

4) The Hjorth parameter [19] characterizes EEG data by defining activity, mobility and complexity variables. The three parameters correspond to amplitude, gradient and change rate. The value of m moments are obtained by the following formula:

$$m_0 = \int_{-\infty}^{+\infty} s_{\varpi} d\varpi = \frac{1}{T} \int_{-\infty}^{+\infty} s_t^2 dt \tag{4}$$

$$m_2 = \int_{-\infty}^{+\infty} \varpi^2 s_{\varpi} d\varpi = \frac{1}{T} \int_{t-T}^{t} \left(\frac{ds}{dt}\right)^2 \tag{5}$$

$$m_4 = \int_{-\infty}^{+\infty} \varpi^4 s_{\varpi} d\varpi = \frac{1}{T} \int_{t-T}^{t} \left(\frac{d^2 s}{dt^2}\right)^2 dt \tag{6}$$

where $s(\varpi)$ is the energy density spectrum function and m represents the moment of ith. Then, the hjorth parameters can be obtained by the following formula:

$$Activity = m_0 \tag{7}$$

$$Mobility = \sqrt{m_2/m_0} \tag{8}$$

$$Complexity = \sqrt{m_4/m_2 - m_2/m_0} \tag{9}$$

5) Latency (LAT), which is defined as the time when the maximum value of the signal occurs over a period of time.

$$LAT = \{t | s_t = s_{max}\} \tag{10}$$

6) Maximum wave amplitude (AMP), which is defined as:

$$AMP = \max\{s_t\} \tag{11}$$

7) Ratio value of Latency and amplitude (LAR, LAT/AMP).
8) Absolute value of amplitude (AVOA, $|AMP|$).
9) Absolute value of LAR (ALAR, $|LAT/AMP|$).
10) Positive area (PAR), which is defined as the sum of all positive values at 300–400 ms time interval:

$$PAR = \sum_{t=300ms}^{400ms} \frac{(s_t + |s_t|)}{2} \tag{12}$$

11) Negative area (NAR), which is defined as the sum of negative values at 300–400 ms time interval:

$$NAR = \sum_{t=300ms}^{400ms} \frac{(s_t - |s_t|)}{2} \tag{13}$$

12) Total area (TAR):

$$TAR = PAR + NAR \tag{14}$$

Frequency-Domain Characteristics. The power spectrum density (PSD) on each signal s_t was obtained by using the adaptive AR (Auto Regressive) model. Suppose that $P(e^{jw})$ is signal PSD, which was defined as:

$$P\left(e^{jw}\right) = \alpha^2 w \left| \frac{1}{1 + \sum_{i=1}^{M} \alpha_i e^{-jwi}} \right|^2 \tag{15}$$

Among them, the white noise variance is expressed as M, and the coefficient of AR model is expressed as α. Afterwards, two frequency domain characteristics are computed as follows:

$$Max\,power\,spectrum = max\{P(e^{jw}), w \in [0, \pi]\} \tag{16}$$

$$Sumpower = \sum P(e^{jw}) \tag{17}$$

Following the feature extraction, we obtained 4240 data samples (53 subject * 80 trials) with 272 dimensional feature vectors (17 electrodes * 16 features) for each stimuli condition. Afterwards, these feature samples of each stimuli condition were divided into depression group and normal controls, with category labels of 1 and 2 respectively.

Feature Selection Technique
The high feature dimension leads to the increase of system running time and the reduce of generalization performance. Therefore, when constructing classifiers, the selection of the optimum discriminant features is crucial. Thus, two feature selection algorithms, ReliefF [20] and principal component analysis (PCA) [21, 22], were applied to select the optimum feature subset for further classification.

Due to high operational efficiency, ReliefF has been widely used in feature extraction and classification. The gist of ReliefF is to sort the features by distance by searching the nearest neighbor of each category of samples, and then weight the features according to the extent to which they distinguish the samples in different categories. The specific operation is to randomly take one sample S from the training sample set at a time. And then, it find k approximate Hits (Misses) of S from sample collection with same type (different type) as S. Finally, the weights of each feature are updated according to the following formula:

$$W_A = W_A - \frac{\sum_{j=1}^{k} diff(A, S, H_j)}{mk}$$

$$+ \frac{\sum_{C \neq class(R)} \left[\frac{p_c}{1 - p(class_S)}\right] \sum_{j=1}^{k} diff(A, S, M_j(C))]}{mk} \tag{18}$$

Where the distinction between the sample S1 and the sample S2 on the feature A is expressed as $diff(A, S1, S2)$, and its calculation formula, $M_j(C)$ represents the bottom j nearest neighbor samples in the class C. The proportion of categories is expressed as $p(C)$. The proportion of randomly selected sample categories is expressed as $p(class(S))$. It can be seen that the meaning of weight is to subtract the difference of the feature of the same classification and add the difference of the feature of different classifications. Finally, the appropriate features can be sorted according to the weight.

PCA converted multiple original features into several comprehensive features by calculating the transformation matrix $W(Y = Wx)$. Firstly, find out the dimension with the greatest variance in the matrix and orient it, and determine it as the first component. Secondly, orient along the vector that is the maximum of the remaining variance and determine it as the second component, and it was orthogonal with the first one, and so on. In order to maintain the selected percentage of data variance after conversion, the experiment retained 95% of the number of components.

Classifiers

After feature selection of six stimuli conditions by two feature selection methods, five classifiers were applied to classify the selected features and the classification accuracy was obtained respectively. The five classifiers were the k-nearest neighbor (KNN), BFTree, C4.5, Logistic Regression and NaiveBayes [23–25]. This classification program is implemented in the Waikato environment for knowledge analysis software. It can be obtained online (https://www.cs.waikato.ac.nz/ml/weka). In the KNN method, k was set as 3. We used default parameter settings for other classifiers. Then, three performance parameters were calculated: sensitivity, specificity and accuracy. Sensitivity is the proportion of positive instances recognized by the classifier to all positive instances. Specificity is the proportion of false instances identified by the classifier to all false instances. We used the Leave-One-Out-Cross-Validation (LOOCV) to assess classification performance of the training set. In LOOCV program, the classifier is trained on all subjects except for the one to be classified.

3   Results and Discussions

As shown in Fig. 2, we analyzed the P300 component of depressive patients and NC individuals in six different stimuli conditions, and compared the brain topology map of P300 activation between the two groups. The results showed that the latency of P300 in NC individuals was significantly later than that in depressive patients, and the prefrontal lobe activation of healthy controls was stronger than that of depressive patients when emotional faces appeared, especially in the stimuli conditions of fear faces. This is consistent with previous study [26].

Average classification accuracy, for five classifiers used (KNN, BFTree, C4.5, logistic regression, and NaiveBayes), for each feature selection method and for six different stimuli conditions are shown in Table 2. For happy-neutral invalid, sad-neutral valid and sad-neutral invalid, PCA was the optimum method for screening features, being able to classify correctly 77%. Using the ReliefF, happy-neutral valid reaches the highest value (77.88%).

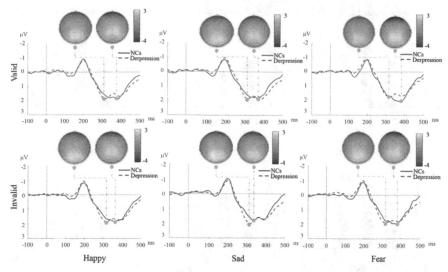

Fig. 2. The waveforms were averaged across CP1, CP2, CP3, CP4, CP5, CP6, CPz, P1, P2, P3, P4, P5, P6, Pz, PO3, PO4 and POz electrodes to reflect the P300 component in depression group and NCs. The corresponding scalp topographies evoked by the emotion valid stimuli and emotion invalid were obtained in the latencies of the P300 peaks.

Table 2. Comparison of accuracy (mean accuracy ± standard deviation) of five classifiers (KNN, BFTree, C4.5, logistic regression and NaiveBayes) for two different feature selection algorithms.

	Happy-Neutral valid	Happy-Neutral invalid	Sad-Neutral valid	Sad-Neutral invalid	Fear-Neutral valid	Fear-Neutral invalid
PCA	$73.79 \pm 11\%$	**$77.28 \pm 14\%$**	**$77.12 \pm 15\%$**	$77.42 \pm 15\%$	$68.43 \pm 9\%$	$67.19 \pm 9\%$
ReliefF	**$77.88 \pm 10\%$**	$68.99 \pm 9\%$	$75.22 \pm 11\%$	**$76.27 \pm 7\%$**	**$71.68 \pm 4\%$**	**$70.47 \pm 3\%$**

Figure 3 shows the classification accuracies of five classification models under the six different stimuli using PCA feature selection. As shown in Fig. 3, NaiveBayes gave the optimum overall results for the sad-neutral valid and sad-neutral invalid, which obtained an average accuracy of 98%. The lowest performances were obtained by IBK ($k = 3$), when attempting to classify depression from normal controls with mean rates of approximately 57% for six stimuli. The mean classification rates combined with ReliefF obtained for each classifier configuration are showed in the Fig. 4. The classifier of NaiveBayes yielded an accuracy of over 90% for the sad-neutral invalid and happy-neutral valid in discriminating between depressive patients and normal controls.

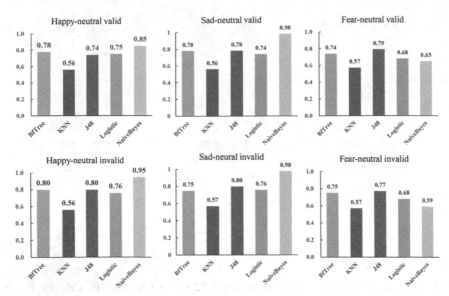

Fig. 3. Classification accuracies (%) of the five classifiers (KNN, BFTree, C4.5, logistic regression and NaiveBayes) corresponding to the P300 components evoked by the six stimulus conditions with PCA as the feature selection method.

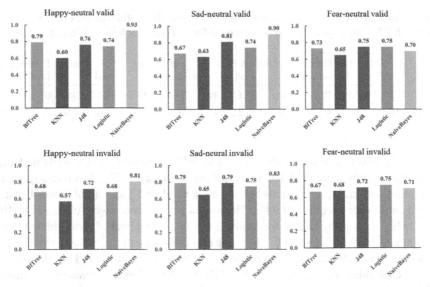

Fig. 4. Classification accuracy (%) of the five classifiers (KNN, BFTree, C4.5, logistic regression and NaiveBayes) corresponding to the P300 components evoked by the six stimulus conditions with ReliefF as the feature selection method.

Additionally, for PCA and ReliefF, the sensitivity and specificity values correspond-ing to the accuracy of classification in depression and NC subjects respectively, are

Table 3. Sensitivity (%) and specificity (%) are given for per classifier, when PCA was selected as feature selection method.

PCA	Happy-Neutral valid		Happy-Neutral invalid		Sad-Neutral valid		Sad-Neutral invalid		Fear-Neutral valid		Fear-Neutral invalid	
BFTree	98%	72%	100%	73%	100%	71%	95%	70%	97%	69%	97%	69%
IBK	52%	60%	52%	60%	52%	60%	53%	60%	52%	60%	52%	60%
J48	100%	68%	100%	73%	99%	72%	100%	74%	98%	70%	98%	70%
Logistic	98%	69%	98%	70%	96%	69%	96%	70%	77%	66%	77%	66%
NaiveBayes	76%	97%	98%	93%	100%	97%	100%	96%	69%	58%	69%	58%

displayed in Table 3 and Table 4. For the same classifier (KNN, BFTree, C4.5, logistic regression and NaiveBayes) and the feature selection method, there are great differences between sensitivity and specificity. For example, it can be seen in Table 3 and Table 4 that for sad-neutral invalid, for PCA, and for C4.5, sensitivity was 100%, while specificity was 73%. For happy-neutral valid, for ReliefF and for logistic, sensitivity was 97%, while specificity was 69%. However, the use of NaiveBayes, for happy-neutral pairs and sad-neutral pairs, provided the best results of classification, and the sensitivity and specificity value were stable across the two feature selection methods (>80%).

Table 4. Sensitivity (%) and specificity (%) are given for per classifier, when ReliefF was selected as feature selection method.

ReliefF	Happy-Neutral valid		Happy-Neutral invalid		Sad-Neutral valid		Sad-Neutral invalid		Fear-Neutral valid		Fear-Neutral invalid	
BFTree	87%	75%	68%	68%	82%	64%	94%	73%	90%	68%	71%	66%
IBK	67%	67%	52%	61%	61%	64%	64%	66%	64%	66%	68%	68%
J48	90%	71%	72%	71%	100%	74%	98%	73%	85%	71%	77%	70%
Logistic	97%	69%	100%	69%	97%	68%	100%	69%	100%	69%	100%	68%
NaiveBayes	97%	91%	81%	81%	100%	85%	100%	76%	100%	65%	100%	65%

4 Conclusions

In present study, the depressive patients and normal controls were used as the research object to determine the optimal way of the feature selection and classifier combination for depression detection. Data analysis pointed out the classifier which can effectively identify patients with depression was NaiveBayes. The accuracy of NaiveBayes and PCA combination achieved 98% for sad-neutral pairs. We have achieved better classification performance than previous studies [9], indicating that ERP component P300 may contribute to the auxiliary diagnosis of depression. However, single-trials classification systems make performance of classifiers unstable, showing the robustness of the trip. It is a limitation of the current research. We will further increase the amount of data in future research to enhance the stability of the classifier and obtain more accurate results. In addition, more stable and valid classification algorithms should be developed for single-trials classification systems, it is challenging problems in future research.

Acknowledgement. This work was supported in part by the National Key Research and Development Program of China (No. 2019YFA0706200), in part by the National Natural Science Foundation of China (Nos. 61632014, 61627808, 61210010), in part by the National Basic Research Program of China (973 Program, No. 2014CB744600), in part by the Program of Beijing Municipal Science & Technology Commission (No. Z171100000117005), in part by the Typical Application Demonstration Project of Shandong Academy of Intelligent Computing Technology (No. SDAICT2081020), and in part by the Fundamental Research Funds for the Central Universities (Nos. lzujbky-2017-it74, lzujbky-2017-it75, lzujbky-2019-26).

References

1. Santini, Z.I., Koyanagi, A., Tyrovolas, S., Mason, C., Haro, J.M.: The association between social relationships and depression: a systematic review. J. Affect. Disord. **175**, 53–65 (2015). https://doi.org/10.1016/j.jad.2014.12.049
2. Delaveau, P., et al.: Brain effects of antidepressants in major depression: a meta-analysis of emotional processing studies. J. Affect. Disord. **130**, 66–74 (2011). https://doi.org/10.1016/j.jad.2010.09.032
3. Etkin, A., Schatzberg, A.F.: Common abnormalities and disorder-specific compensation during implicit regulation of emotional processing in generalized anxiety and major depressive disorders. Am. J. Psychiatry. **168**, 968 (2011). https://doi.org/10.1176/appi.ajp.2011.10091290
4. Li, X., et al.: Attentional bias in MDD: ERP components analysis and classification using a dot-probe task. Comput. Methods Programs Biomed. **164**, 169–179 (2018). https://doi.org/10.1016/j.cmpb.2018.07.003
5. Spielberger, C.D., Reheiser, E.C.: Assessment of emotions: anxiety, anger, depression, and curiosity. Appl. Psychol.: Health Well-Being **1**(3), 271–302 (2009). https://doi.org/10.1111/j.1758-0854.2009.01017.x
6. Sung, M., Carl, M., Alex, P.: Objective physiological and behavioral measures for identifying and tracking depression state in clinically depressed patients (2010)
7. Li, X., et al.: A resting-state brain functional network study in MDD based on minimum spanning tree analysis and the hierarchical clustering. Complexity **2017**, 1–11 (2017). https://doi.org/10.1155/2017/9514369
8. Kim, E.Y., et al.: Gender difference in event related potentials to masked emotional stimuli in the oddball task. Psychiatry Investig. **10**, 164–172 (2013). https://doi.org/10.4306/pi.2013.10.2.164
9. Kalatzis, I., et al.: Design and implementation of an SVM-based computer classification system for discriminating depressive patients from healthy controls using the P600 component of ERP signals. Comput. Methods Programs Biomed. **75**, 11–22 (2004). https://doi.org/10.1016/j.cmpb.2003.09.003
10. Kaiser, S., Unger, J., Kiefer, M., Markela, J., Mundt, C., Weisbrod, M.: Executive control deficit in depression: event-related potentials in a Go/Nogo task. Psychiatry Res. Neuroimaging **122**(3), 169–184 (2003). https://doi.org/10.1016/s0925-4927(03)00004-0
11. Luck, S.J., Woodman, G.F., Vogel, E.K.: Event-related potential studies of attention. Trends Cogn. Sci. **4**, 432–440 (2000)
12. Dai, Q., Feng, Z.: More excited for negative facial expressions in depression: Evidence from an event-related potential study. Clin. Neurophysiol. **123**(11), 2172–2179 (2012). https://doi.org/10.1016/j.clinph.2012.04.018

13. Delle-Vigne, D., Wang, W., Kornreich, C., Verbanck, P., Campanella, S.: Emotional facial expression processing in depression: Data from behavioral and event-related potential studies. Neurophysiologie Clinique-clinical Neurophysiol. **44**, 169–187 (2014)
14. Ham, K., Chin, S., Suh, Y.J., Rhee, M., Chung, K.-M.: Preliminary results from a randomized controlled study for an app-based cognitive behavioral therapy program for depression and anxiety in cancer patients. Front. Psychol. **10** (2019)
15. Lecrubier, Y., et al.: The mini international neuropsychiatric interview (MINI). A short diagnostic structured interview: reliability and validity according to the CIDI. Eur. Psychiatry **12**, 224–231 (1997). (in English). https://doi.org/10.1016/S0924-9338(97)83296-8
16. Lu, B., Hui, M.A., Huang, Y.X.: The development of native Chinese affective picture system-a pretest in 46 college students. Chin. Mental Health J. **19**(11), 719–722 (2005)
17. Jung, T.P., et al.: Removing electroencephalographic artifacts by blind source separation. Psychophysiology **37**, 163–178 (2000)
18. Leutgeb, V., Sarlo, M., Schöngassner, F., Schienle, A.: Out of sight, but still in mind: electrocortical correlates of attentional capture in spider phobia as revealed by a 'dot probe' paradigm. Brain Cogn. **93**, 26–34 (2015)
19. Hjorth, B.: EEG analysis based on time domain properties. Electroencephalogr. Clin. Neurophysiol. **29**, 306–310 (1970)
20. Robnik-Šikonja, M., Kononenko, I.: Theoretical and empirical analysis of ReliefF and RReliefF. Mach. Learn. **53**, 23–69 (2003)
21. Kota, S., Gupta, L., Molfese, D.L., Vaidyanathan, R.: A dynamic channel selection strategy for dense-array ERP classification. IEEE Trans. Bio-med. Eng. **56**, 1040 (2009). https://doi.org/10.1109/TBME.2008.2006985
22. Kuncheva, L.I., Rodríguez, J.J.: Interval feature extraction for classification of event-related potentials (ERP) in EEG data analysis. Progress Artif. Intell. **2**, 65–72 (2013)
23. Pechenizkiy, M.: The impact of feature extraction on the performance of a classifier: kNN, Naïve Bayes and C4.5. In: Kégl, B., Lapalme, G. (eds.) AI 2005. LNCS (LNAI), vol. 3501, pp. 268–279. Springer, Heidelberg (2005). https://doi.org/10.1007/11424918_28
24. Khatun, S., Morshed, B.I., Bidelman, G.M.: A single-channel EEG-based approach to detect mild cognitive impairment via speech-evoked brain responses. IEEE Trans. Neural Syst. Rehabil. Eng. **27**, 1063–1070 (2019). https://doi.org/10.1109/TNSRE.2019.2911970
25. Mao, X., Hou, J.: Object-based forest gaps classification using airborne LiDAR data. J. Forestry Res. **30**, 241–251 (2019). CNKI:SUN:LYYJ.0.2019-02-023
26. Liu, X., et al.: Relationship between the prefrontal function and the severity of the emotional symptoms during a verbal fluency task in patients with major depressive disorder: a multichannel NIRS study. Progress Neuropsychopharmacol. Biol. Psychiatry **54**, 114–121 (2014). https://doi.org/10.1016/j.pnpbp.2014.05.005

Research on Ship Classification Method Based on AIS Data

Pu Luo[1,2(✉)], Jing Gao[1,2], Guiling Wang[1,2], and Yanbo Han[1,2]

[1] Beijing Key Laboratory on Integration and Analysis of Large-Scale
Stream Data, Beijing, China
[2] North China University of Technology, Beijing, China

Abstract. It is important for maritime authorities to effectively identify and classify unknown types of ships in historical trajectory data. A method of using trajectory image and training based on deep residual network (ResNet) to obtain ship type classifier is proposed. First, a method of integrating speed information into the trajectory image is proposed, then the trajectory image is input into the ResNet model for training to obtain the classifier. Finally, the real AIS data of 5 ship types are used for experiments. The experimental results show that the method can meet the requirements of type identification and classification of ships based on AIS data, and provides technical support for further research on the identification of camouflaged ships, mining of ship behavior patterns and anomaly detection.

Keywords: Ship classification · Residual network · AIS data · Data pre-processing

1 Introduction

Ship Automatic Identification System (AIS) is one kind of aids the navigation equipment. The AIS can exchange, detect and identify information such as speed, longitude, latitude and Maritime Mobile Service Identify (MMSI) between ships and between ships and shore. It greatly enhanced the ship navigation safety [1]. However, some of the ships with AIS equipment close the AIS equipment, falsify the data or transmit other data to cheat the detection of illegal operations, or even illegal exploration [2]. And there are quality issues with the data itself, such as some errors in latitude, longitude and speeds. For the above reasons, it's difficult to identify some ship types, which brings a threat to marine surveillance and maritime security defense. Identifying the unknown types of ships has therefore become an important task.

With the upgrade of hardware, computer computing power has been greatly improved. Based on large-scale trajectory data, using data mining technology to classify and predict moving objects has become a common phenomenon [3]. For example, the research of using AIS trajectory data to identify ship types includes: literature [4] based on AIS data of five hotspots, calculating the features of various types of ships through attributes such as latitude, longitude and speed, using Decision Tree, Fuzzy

© Springer Nature Singapore Pte Ltd. 2021
Y. Sun et al. (Eds.): ChineseCSCW 2020, CCIS 1330, pp. 222–236, 2021.
https://doi.org/10.1007/978-981-16-2540-4_17

Rule, K Nearest Neighbor, Neural Network and Naive Bayes five classifiers for experiments, three sets of classification labels, two evaluation criteria Accuracy and Cohen's Kappa are adopted to evaluate the results. The experimental results show that Decision Tree, Fuzzy Rules, and K-Nearest Neighbors have better accuracy, and the classification results are analyzed and summarized in depth. The literature [5] proposed three basic movement patterns for ships: Anchored-off pattern, Turning pattern and Straight-sailing pattern. Then extract 17 types of features based on the basic movement patterns and the sub-trajectory segments. Finally, the logistic regression model was used to train the classifier. The paper uses only two types of fishing vessels and cargo ships, and its generalization ability needs to be verified.

Some scholars have used image data such as aperture radar images of ships, satellite remote sensing images, and visible light images [6] to identify and classify ships. For example, in literature [7], a feature fusion-based approach for ship detection and classification is proposed by mining two types of features: the Completed Local Binary Pattern (CLBP) and the Histogram of Oriented Gradient (HOG). And the use of Image Pyramid Slip Detection to achieve multi-scale detection of ship targets, but it have an impact on detection speed. As deep learning makes great progress in image classification, some scholars also use convolutional neural networks (CNN) to classify images of ships. In the literature [8], the images of transport ships are trained by an 8-layer convolutional neural network structure based on AlexNet, which can provide better local features, then extract the features of the sixth layer of this network for classification and identification by SVM classifier. However, many neural networks now have much more than 8 layers, so there is room to improve the method. A complete workflow for SAR marine target detection and classification based on TerraSAR-X high-resolution imagery is detailed in paper [9]. However, additional experiments are needed to understand how image variations affect the CNN internals and there is room for improvement in the pre-processing steps.

As deep learning continues to evolve, there are more and more layers of models and more complex network structures. Theoretically, assuming that the newly added layers are constant mappings, the deep model structure will be as good as the original model structure as long as the original layer learns the same parameters as the original model. In other words, the solution of the original model is just a subspace of the solution of the new model, and better results should be found in the space of the new model solution than in the corresponding subspace of the original model solution. However, practice shows that the training error tends to rise rather than fall after increasing the number of layers of the network. Kaiming He et al. proposed the residual network (ResNet) [10] to solve the above problem.

According to the above analysis, the existing ship type identification research is mainly divided into two types according to the adopted data types: One is based on information such as speed and course from AIS data and the other is based on the image data, such as photos of ships taken in the port, while the research combining AIS data and image classification is relatively rare. And ResNet has an advantage in the field of image classification. Therefore, a method for using image classification models ResNet-50 to identify unknown types of vessels is proposed: firstly, denoising, sorting and other preprocessing operations are carried out on the original AIS data; then the combination of latitude, longitude and speed is visualized to generate image data; finally, the ResNet-50 neural network model is established, and input the image data into the ResNet-50 model for training. The ResNet model is trained based on image data to obtain the classifier. The ResNet-50 ship classification model is built with real AIS data and compared with the K-Nearest Neighbor and Decision Tree classification models. The experimental results show that the ResNet-50 model obtained by this method has the best performance and satisfies the need of identifying ship types from history ship trajectory data.

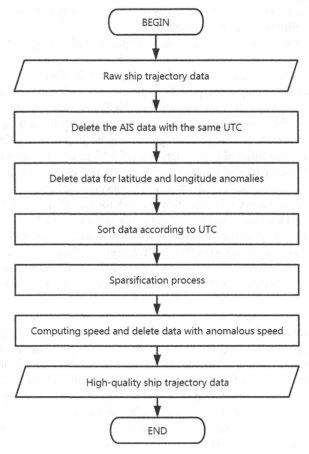

Fig. 1. Pre-processing flow chart

2   Method for Identifying Ship Types Based on ResNet Model

2.1   AIS Data Pre-processing

There are several columns in the original AIS data, this method only uses three columns: time stamp (UTC), longitude (LON) and latitude (LAT). Speed over ground (SOG) and course over ground (COG) are calculated from these three columns.

Real AIS data has quality issues, such as data duplication, time disorder, etc. Therefore, pre-processing operations are needed to improve data quality. The pre-processing flow chart is shown in Fig. 1:

1) Delete the AIS data with the same *UTC*. There are a large number of duplicates in the raw trajectory data, which can be quickly identified by *UTC*. Deleting this data can speed up the process later.
2) Delete data for latitude and longitude anomalies. Normally, the range of longitude is [−180, 180] and the range of latitude is [−90, 90], however, there are some data in the raw trajectory data that are not in the normal range of longitude and latitude. And these data need to be removed to improve the quality of the data.
3) Sort data according to *UTC*. Most of the data is ordered by *UTC* in the raw trajectory data. But there is still a small number of disordered data that affect the calculation of the speed attribute, so the data needs to be sorted by *UTC*.

```
ALGORITHM:Sparsification
INPUT:trajectory,threshold_time=300
OUTPUT:sparsified_trajectory
PROCESS:
1. cur_p = trajectory[0]
2. sparsified_trajectory=[cur_p]
3. for i, p in enumerate(trajectory):
4.    dt = p[utc] - cur_p[utc]
5.    if dt > threshold_time or i == (len(trajectory)-1):
6.        cur_p = p
7.        sparsified_trajectory.append(cur_p)
8.    end if
9. end for
Return sparsified_trajectory
```

Fig. 2. Sparsification pseudo-code

4) Sparsification process. Set the sparse time interval threshold (*THRE_SPARSE_TIME*) to 300 s. If the time interval *DIS_UTC* between two points P_i and P_{i+1} is less than *THRE_SPARSE_TIME*, then discard P_{i+1}. The pseudo-code of algorithm of sparsification process is shown in Fig. 2. Sparsing the data can reduce the speed error, because there is certain error between the ship's true position and the reported position. But the error is within a certain range, that is, the distance error *DIS_ERR* between the real distance and the calculated distance is within a certain range, the time interval for *T*, the speed error *SPEED_ERR* can be obtained from Formula (1):

$$SPEED\_ERR = DIS\_ERR \,/\, DIS\_UTC \tag{1}$$

so when the appropriate increase in *DIS_UTC* can reduce the speed error;

5) Computing speed. The distance between two adjacent points P_i and P_{i+1} is S, the time interval is T, get the speed V by the Formula (2):

$$V = S / T \qquad (2)$$

V is recorded as the speed of the track point P_i, the speed of the last track point is recorded as 0. And setting the speed threshold *THRE_SPEED* to 20 m/s, treat data with a speed greater than *THRE_SPEED* as anomalous data and delete it.

2.2 Conversion of AIS Data to Image Data

All experimental AIS data are from the Bohai Sea. The latitude and longitude ranges of the Bohai region are obtained by observation, where the longitude range is [117°E, 123°E] and the latitude range is [36°N, 41°N]. Then set the minimum longitude value *MIN_LON* is 117 and the minimum latitude value *MIN_LAT* is 36. Also get the latitude and longitude coordinates of the outline of the Bohai Sea and record it as *BH_COORDINATES*.

Convert the pre-processed AIS data to image data by following 2 steps:

1) Subtract *MIN_LON* from the longitude and *MIN_LAT* from the latitude in *BH_COORDINATES* to get *BH_COORDINATES_2*. For each ship's AIS trajectory data, subtract *MIN_LON* from each value in the *LON* to get *LON_2*, and subtract *MIN_LAT* from each value in the *LAT* to get *LAT_2*. See Formula (2, 3, 4, 5) for details of this step.

$$\text{BH\_COORDINATES\_2[LON]} = \text{BH\_COORDINATES[LON]} - \text{MIN\_LON}$$
$$(3)$$

$$\text{BH\_COORDINATES\_2[LAT]} = \text{BH\_COORDINATES[LAT]} - \text{MIN\_LAT} \qquad (4)$$

$$\text{LON\_2} = \text{LON} - \text{MIN\_LON} \qquad (5)$$

$$\text{LAT\_2} = \text{LAT} - \text{MIN\_LAT} \qquad (6)$$

2) The Bohai contour *BH_COORDINATES_2* is visualized as a line graph, and the ship trajectories *LON_2* and *LAT_2* are visualized as a scatter graph, while the speed is mapped to the color of each track point in the scatter plot. The mapping rule is shown in Fig. 3.

The horizontal axis indicates the speed, the colorbar indicates the corresponding color of the speed. The final image is shown in Fig. 4. Note that since track points are dense and track point color has important speed information, set the track point size slightly smaller so that P_{i+1} does not obscure the P_i.

color of trajectory point

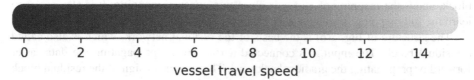

0   2   4   6   8   10   12   14

vessel travel speed

Fig. 3. Mapping speed to color

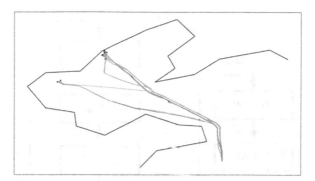

Fig. 4. Trajectory image example diagram

2.3 Residual Learning Framework

The residual network solves the problem that the training burden increases and the accuracy even decreases as the network deepens. Figure 5 [10] show its basic idea.

Figure 5(a): represents that when increasing the number of network layers, map x to $y = F(x)$ and then output. Figure 5(b): an improvement on Fig. 5(a) with output $y = F(x) + x$, instead of learning the representation of output feature y directly, $y - x$ is learned. If you want to learn the representation of the original model, just set all the parameters

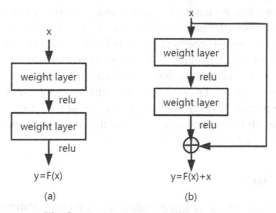

Fig. 5. The basic idea of residual block

of $F(x)$ to zero, then $y = x$ is a constant mapping. $F(x) = y - x$ is also called the residual block, and if the mapping of $x \rightarrow y$ is close to a constant mapping. It is also easier to learn the full mapping form by learning the residual term in Fig. 5 (b) than in Fig. 5(a).

The structure of Fig. 5(b) is the basis of the residual network, which is also called a residual block. The input x is connected across layers, propagating the data faster forward or propagating the gradient backward. The specific design of the residual block is shown in Fig. 6, which is also called a *BottleNeck*.

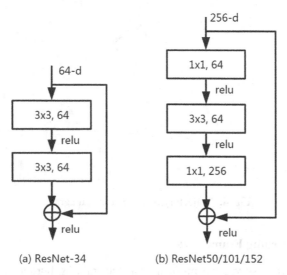

(a) ResNet-34   (b) ResNet50/101/152

Fig. 6. Diagram of the residual block structure

Figure 7 shows the structure of ResNet-50, which contains a total of 49 layers of convolution and 1 layer of full connectivity, hence the name ResNet-50. In the Fig. 7, K stands for *Kernel*, which is the size of the convolution kernel or the size of the pooling kernel of the pooling layers; F stands for *Filter*, which is the number of feature channels output from the convolution operation; and S stands for *Stride*, which is the step size of the convolution kernel (the number of pixels skipped between two convolutions). The first is a two-dimensional convolution kernel, which reduces the image size and goes through the batch regularization, ReLU activation, and maximum pooling layers in sequence. Then the spatial information is mapped to the feature information through 3, 4, 6 and 4 Bottleneck Block layers. And a 2048-dimensional feature space is created through an average pooling layer. Since there are 5types of trajectory image data, set the parameter F for the full connection layer to 5. Finally, the feature space is mapped to a classification target by a fully connected layer.

2.4 Training ResNet-50

General image classification tasks will do data enhancement operations such as rotating and grayscale on the data before training. In the task of classifying ship trajectories, however, the location and color of the track points contain important geographic and speed information. Information-destroying operations such as cropping and grayscale should not be performed. So we will train the Resnet-50 model is trained according to the following steps:

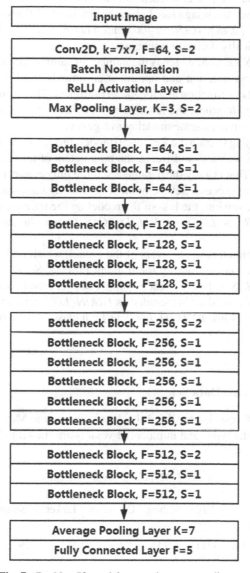

Fig. 7. ResNet-50 model network structure diagram

1) Resize all images to *(256, 256)* and normalize the image data to obtain the new image data. The normalization process is shown in Formula (7).

$$NEW\_IMAGE\_DATA = (IMAGE\_DATA - MEAN) / STD \qquad (7)$$

MEAN is the mean parameter of each channel and *STD* is the variance parameter of each channel. Since the image has 3 channels RGB, *MEAN* and *STD* have 3 values, which *MEAN* = *[0.485, 0.456, 0.406]* and *STD* = *[0.229, 0.224, 0.225]*. These values are empirically derived by scholars.

2) Set the *BATCH_SIZE* to 8, which is the number of samples selected for one training session. Using loss function and optimizer commonly used in classification task, they are the *NLLLoss* function and the *Adam* optimizer. The learning rate adjustment strategy uses *StepLR*, where the learning rate decreases as the number of iterations increases in order to accelerate model convergence.

3) Use transfer learning to train the Resnet-50 model. First load trained resnet-50 model from the Pytorch website. and freeze the parameters of all layers except the fully connected layer. That is, only the parameters of the fully connected layer will change during the training process. Set the number of training sessions *EPOCHES* to 100. For each training session, the loss of the model on the training set and the valid set is recorded as *TRAIN_LOSS* and *VALID_LOSS*. Output the model with the lowest *VALID_LOSS*.

4) Unfreeze the parameters in all layers of the model, which makes it possible for all parameters to be changed during the training process. Set the number of training sessions *EPOCHES* to 100. For each training session, the loss of the model on the training set and the valid set is recorded as *TRAIN_LOSS* and *VALID_LOSS*. Output the model with the lowest *VALID_LOSS* as final classification model.

3 Case Study

3.1 Introduction to the Data Set

The AIS data of 7828 vessels near the Bohai Sea in January to March 2018 are used for the experiment, and the types and numbers of vessels are shown in Table 1.

Table 1. Type and number of ships

Ship type	Cargo	Fishing	Container	Tanker	Passenger
Amount	3311	3211	866	311	129

The AIS data schema is shown in Table 2. The data columns needed in the experiment include: Maritime Mobile Service Identify *MMSI*, time stamp *UTC*, longitude *LON* and latitude *LAT*.

Table 2. AIS data schema

Field	Meaning
MMSI	Maritime Mobile Service Identify
UTC	Time of reporting location
LON	Longitude at reporting position
LAT	Latitude at reporting position
SEG	Subsection number
BEARING	Course over ground

Part of the AIS trajectory data is shown in Fig. 8.

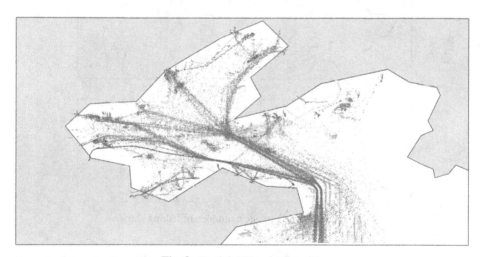

Fig. 8. Partial AIS trajectory data

3.2 Pre-processing and Convert AIS Data into Image Data

Using the programming language Python (Version = 3.6.5) to implement the data pre-processing process proposed in Sect. 2.1 and the conversion of AIS data into image data in Sect. 2.2. Some examples of cargo ship and container ship are shown in Fig. 9. Although the trajectory shapes of cargo ships and container ships are somewhat similar, the colors of the trajectory points are not quite the same.

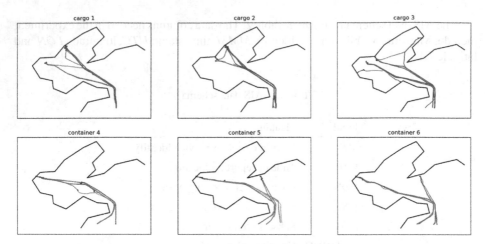

Fig. 9. Trajectory image examples of cargo ships and container ships

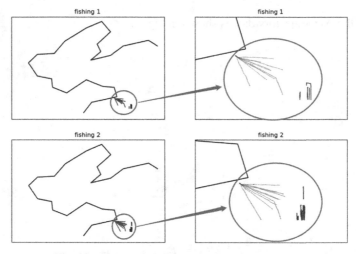

Fig. 10. Trajectory image examples of fishing ships

Trajectory image examples of fishing ships are shown in Fig. 10, the trajectory of fishing ships is relatively complex and has a small range of activities.

Trajectory image examples of passenger ships and tanker ships are shown in Fig. 11. Passenger ships tend to be frequent trips between two coastal cities or around the coast; and the tanker trajectory has its own characteristics.

3.3   Image Classification Experiments Based on ResNet-50

A stratified sampling was used to divide all trajectory images of the five classes of ships into training and valid sets. Setting the parameter *pretrained* of ResNet-50 provided by the deep learning library to *True,* means that the pre-trained model will be loaded. Then use transfer learning to train this model to get the final classification model.

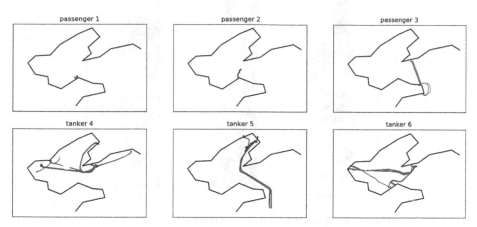

Fig. 11. Trajectory image examples of fishing ships

3.4 Comparative Analysis of Experimental Results

The Decision Tree is one of the most widely used inference algorithms. It is a method of approximating discrete value functions, usually based on decision trees for classifier models and prediction models. The decision tree is good at reasoning out the classification rules formed by the decision tree representation from a set of unordered and irregular data [11]. The K-Nearest Neighbor method (KNN) is one of the most popular machine learning classification algorithms. Its basic idea is: in a data set, if most of the K samples that are most similar to a sample belong to a certain category, then the sample is judged to belong to that category as well [12]. Also, Decision Tree and KNN are the best classifiers in the literature [4]. Therefore, the decision tree and the K-nearest neighbor method will be chosen as comparison models to train on the features such as average speed of sailing mentioned in the literature [5] for ship classification.

The overall evaluation of the three models is shown in Fig. 12, where RN represents ResNet-50, KNN represents K-Nearest Neighbor, and DT represents Decision Tree. And it can be seen that the ResNet-50 classification model has better performance, with 92% accuracy, 83% precision, 75% recall, and 78% f1-score.

The F1-Score of the five types of ships in the three classifiers is shown in Fig. 13, where RN represents ResNet-50, KNN table K-Nearest Neighbor and DT represents Decision Tree. It can be seen that ResNet-50 has better classification performance on cargo ships, container ships, fishing vessels and especially tankers. The F1-Score values of ResNet-50 model for cargo ships, container ships, fishing vessels, passenger ship and tankers are 0.977, 0.960, 0.970, 0.714, 0.631, respectively.

In summary, the trajectory image-based Resnet-50 model has better classification performance than traditional KNN and Decision Tree classifiers based on speed statistics.

234   P. Luo et al.

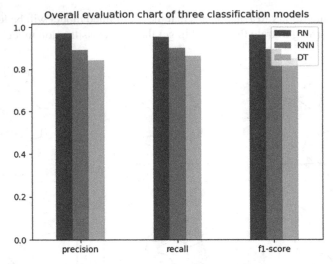

Fig. 12. Overall evaluation chart of three classification models

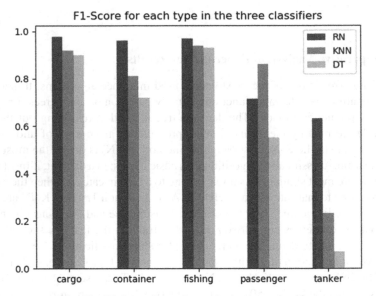

Fig. 13. F1-Score for each type in the three classifiers

Confusion matrix for ResNet-50 classification results is shown in Table 3. As can be seen from the table, it is relatively easy for the classifier to identify tankers as cargo vessels and passenger vessels as fishing vessels, after all, both sometimes have a relatively small range of activities. This provides a guide for the direction of our subsequent research, where we will explore the differences between tankers and cargo ships, passenger ships and fishing vessels, and experiment with other classifier models.

Table 3. Confusion matrix for ResNet-50 classification results

Real	Predict				
	Cargo	Container	Fishing	Passenger	Tanker
Cargo	449	0	3	0	0
Container	0	73	3	0	0
Fishing	0	0	291	3	0
Passenger	0	0	9	15	0
Tanker	18	3	0	0	18

4 Conclusion

In this paper, a novel method of vessel type identification using AIS data based on the image classification model ResNet-50 is proposed. Firstly, the data quality is improved by removing anomalous and duplicate data, sorting according to timestamps, calculating speed and other pre-processing operations. Then visualize the longitude and latitude as a scatter plot, and map the speed of each track point to the color of the track point, add the line graph of Bohai Sea outline to generate the trajectory image data. Finally, use transfer learning to train models on trajectory image data analyzed in comparison to KNN and Decision Tree. The experimental results show that ResNet-50 is better than KNN, Decision Tree classification model based on speed statistics. However, there are still some areas that can be improved, such as the time attribute is not used. And the classifier still has a lot of room for improvement in the classification of passenger ship and tanker injuries. Further work will be done to add more data, to look for differences between passenger ships and tankers and other ships, and to further explore the integration of image features and speed-related statistics for classification.

References

1. Yong, Z.: Application of automatic identification system AIS for ships. Technol. Enterprise 000.012, 115 (2013)
2. Mccauley, D.J., et al.: Ending hide and seek at sea. Science **351**(6278), 1148–1150 (2016)
3. Feng, Z., Zhu, Y.: A survey on trajectory data mining: techniques and applications (2017)
4. Krüger, M.: Experimental comparison of ad hoc methods for classification of maritime vessels based on real-life AIS data. In: International Conference on Information Fusion (FUSION) (2018)
5. Sheng, K., et al.: Research on ship classification based on trajectory features. J. Navig. 1–17 (2017)
6. Wu, K.: Deep learning based target detection for surface ships. Diss.
7. Liu, S.S.: Research on feature fusion-based ship detection and classification method. Dalian Maritime University (2019)
8. Chi, C., Wang, X.: Convolutional neural network based method for classification and identification of transport vessels. Micro Mach. Appl. **17** (2017)

9. Bentes, C., Velotto, D., Tings, B.: Ship classification in TerraSAR-X images with convolutional neural networks. IEEE J. Oceanic Eng. **PP**(99), 1–9 (2017)
10. He, K., et al.: Deep residual learning for image recognition (2015)
11. Yu, X.: Parallel algorithm design and performance analysis of decision tree in data mining grid. China New Commun. **20**(04), 231 (2018)
12. Guo, G., Wang, H., Bell, D., Bi, Y., Greer, K.: KNN model-based approach in classification. In: Meersman, R., Tari, Z., Schmidt, D.C. (eds.) OTM 2003. LNCS, vol. 2888, pp. 986–996. Springer, Heidelberg (2003). https://doi.org/10.1007/978-3-540-39964-3_62

A Low-Code Development Framework for Constructing Industrial Apps

Jingyue Wang[1,2], Binhang Qi[1,2], Wentao Zhang[1,2], and Hailong Sun[1,2(✉)]

[1] SKLSDE Lab, School of Computer Science and Engineering, Beihang University,
Beijing 100191, China
{wangjingyue,sunhl}@buaa.edu.cn,
{qibh,zhangwt}@act.buaa.edu.cn
[2] Beijing Advanced Innovation Center for Big Data and Brain Computing,
Beijing 100191, China

Abstract. With the advent of the Industry 4.0, intelligent manufacturing has become a technological highland to conquer in the process of enterprise digitalization. As the core competitiveness of intelligent manufacturing, industrial apps, with new features such as customization and lightweight, has emerged as a new type of industrial software. Traditional development methods and tools can hardly meet the large demand of industrial software on account of its long development cycles while low-code development can greatly improve the productivity of industrial software, lower the barriers and reduce costs for development. Therefore, the research and application of low-code development for industrial apps has received much attention. Industrial Internet platforms such as Siemens, OutSystems have successively launched low-code tools. However, there is still a lack of an open, unified low-code development framework in industry. In response to the above problems, we propose a low-code framework to develop industrial apps quickly and easily, which paves the way for leveraging the crowd intelligence of worldwide developers to improve the productivity of developing industrial apps. Based on BPMN2.0 and Apache Activiti engine, this framework provides drag-and-drop process design, one-click process deployment and operation, data monitoring and other functions. In this paper, we present a prototype system of a low-code development framework and demonstrate its functions through a use case of developing a predictive maintenance application. Finally, the aircraft turbine life data is used to verify the effectiveness of the system.

Keywords: Industrial app · Low code · Crowd intelligence · Predictive maintenance · BPMN

1 Introduction

With the development of the real economy and the advancement of artificial intelligence technology, the digitalization process of the manufacturing industry is gradually accelerating. Intelligent manufacturing, which is mainly characterized by data-driven, platform support, and intelligent guidance, has gradually

Y. Sun et al. (Eds.): ChineseCSCW 2020, CCIS 1330, pp. 237–250, 2021.
https://doi.org/10.1007/978-981-16-2540-4_18

become the technological highland of the manufacturing industry [2]. As mobile devices are highly popularized, conventional industrial software, which is the core competitiveness of intelligent manufacturing, appears to be too heavy and bulky to handle. Based on this status quo, the concept of industrial app came into being gradually. Compared with traditional style, industrial app highlight customization and lightweight. On the other hand, different from the mobile app development which benefits from open ecosystems and the contribution of a large number of global developers, the development of industrial apps is usually conducted by a limited group of developers due to the lack of productive development tools. This leads to the fact that the number of industrial apps is far less than that of mobile apps. Hence, constructing an open and productive development platform is pivotal for enriching the kinds and quantities of industrial applications through the crowd intelligence of a plethora of developers. However, industrial workers with industrial knowledge do not have sufficient programming capabilities, which limits the efficiency and quality of development. These problems have brought widespread attention to low-code development in recent years. Low-code style can develop industrial app rapidly without or with little amount of coding. Due to the advantages of high efficiency, low threshold, and easy integration, low-code development allows business personnel who possess less software development skills to develop industrial applications quickly and easily [3]. Covering business process design management, workflow and BPMN2.0, Activiti serves as an executable standard process language framework based on Apache license [4]. The Activiti project is open source and widely used, and its drag-and-drop, process-based development format conforms to the low-code development mode. Activiti, however, only applies to ordinary business processes, which does not meet needs of industries. Recently, low-code development platforms have emerged one after another, such as OutSystems, which is mature, widely used, safe and reliable, and Siemens Mendix, which is closely integrated with the Internet of Things. Compared with Activiti, these low-code platforms achieve higher industrial integration, while most of them are not open source and expensive to use. In general issues in low-code development platforms impose restrictions on their practicality for instance, low integration of business processes and industrial knowledge, lack of components and functions.

In response to the above problems, this paper proposes to establish a generic industrial app low-code development framework through investigation and analysis of typical industrial fields. The specific work to realize this low-code development framework includes the packaging of industrial knowledge in common industrial fields, the establishment of basic components for industrial app development, and the description of logic to connect those components. The main contributions of this research are divided into the following two parts: 1. After analyzing and summarizing the commonalities of typical industrial tasks, a series of modules and components with a certain generality are designed; 2. In Predictive Maintenance, a typical industrial field, the use case of aircraft engine life prediction with NASA's public dataset is used to verify the effectiveness of the framework.

The organization of the rest of this article is as follows: Sect. 2 introduces the existing low-code industrial software development platform and Apache Activiti engine; Sect. 3 details the overall structure and four functional modules of the development framework; Sect. 4 verifies the effectiveness of the development framework through the development and testing of the aircraft turbine predictive maintenance app [5]. Finally, summarize and prospect.

2   Related Work

Low-code development usually requires no code or a small amount of code to develop software efficiently. Furthermore, Low-code development platforms adopt the drag-and-drop approach to develop instead of the conventional way of editing code and realize the development logic by means of connecting graphical elements in a flowchart. At present, loads of low-code development platforms have come out on the market, and applied to CAE, predictive maintenance preliminarily. On the other hand, this paper carries out research combining the mature workflow technology in recent years on account of the established pattern of industrial business processes. Among numerous open source workflow engines, Activiti is wildly used with the features of full functions and robust operation.

2.1   Low-Code Development Platform

In recent years, low-code development platforms emerged one after another such as OutSystems, Mendix, Microsoft PowerApps and JogetCloud. OutSystems is mature enough to accomplish the development of commercial software in major industries which is fully functional, user friendly, safe, and reliable. However, it is too expensive to promote in small and medium scale enterprises. Siemens Mendix provides two development modes: website development which is simple and fast, client development which is complete and complex. The reusability of components in this platform is high, and developers can interact with the Siemens IoT (Internet of Things) platform conveniently. Mendix, nevertheless, is fit for business developers who master programming knowledge such as branch loops, which has a certain threshold to manipulate. Microsoft PowerApps integrates a large number of scripting tools and plugins. Meanwhile this platform could be used with Visual Studio, on which users are allowed to develop web applications compatible with IOS, Android and Windows devices at the same time. However, the business process components of PowerApp are not rich enough and most applications needs to develop based on preset templates, which may lead to poor customization of its products. At present the market share of low-code platforms is not high, but their development trend is flourishing. Some platforms with aPaaS have gradually transformed into low-code development platforms, such as APICloud, DaDayun, H3Yun, JiandaoYun, iVx [12]. APICloud's low-code platform came out earlier correspondingly. It provides a one-stop solution for enterprise customize applications through its embedded development tool,

Plus Mode. DadaYun is designed to quickly develop specialized industrial software for small and medium scale enterprises. It has realized the synchronous development of the mobile terminal, which greatly lower the development time cost. In short, these platforms gradually have complete functions, but they still are insufficient in aspect of reliability and stability. As advanced and efficient productivity tools, a few excellent low-code development platforms have emerged in the industry. However, the research and application of low-code development mode stay at the exploratory stage. Large-scale application, in other words, is still a distant goal to achieve.

2.2   Activiti Workflow Technology

BPMN (Business Process Model Annotation) is a standardized graphical for business process models proposed and maintained by OMG (Object Management Organization). The current business model based on BPMN2.0 can be executed on any process engine that follows its standard and guidelines, which improves the portability and flexibility of business processes [15]. Figure 1 shows a simple process example that complies with BPMN2.0, which covers common components such as Activiti Task, Start Events and End Events, Exclusive Gateways, and Text Annotation.

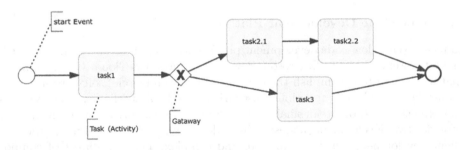

Fig. 1. A BPMN2.0 diagram

Workflow refers to the abstraction, generalization and description of business rules and business process in enterprises [13]. It assists business personnel in standardizing the work process under the support of Internet technology, sharing information in process of the business flow execution and management. In addition to solving simple tasks such as leave approval, workflow technology is also applied in industrial production. Activiti is a lightweight workflow platform that complies with the BPMN2.0 standard. As a leader in workflow engines, Activiti has the following advantages: 1. Separation of runtime data and historical data; 2. Support for SpringBoot framework; 3. Data persistence [15]. Activiti has four important components: Engine, Modeler, Designer, and Explorer. Activiti Engine is the most critical part of the workflow platform, which is used to execute a designed workflow. Activiti Explorer is a web application for users to deploy and start process in a visual form [14]. Activiti Modeler is used to manage the Activiti process [15]. Activiti Designer is an Eclipse plugin, used to draw flowcharts on the IDE [16].

3 Low-Code Development Framework for Industrial Apps

The low-code industrial app development framework presents a novel visualization method for industrial software development and real-time operation, it reduces the use threshold and maintenance cost of industrial apps, which is based on BPMN process designer and Activiti engine. Figure 2 shows the hierarchical structure of the overall architecture of the development framework and the relationship between the various modules. This article builds a series of related components through the User-customized Module, combined with Activiti's native components, users design equipment maintenance processes on the web, then the Data Flow Process Module receives the source data uploaded by users. The Process Management Module is responsible for Activti engine to invoke the process instance designed by the user. The Data Visualization Module can visualize the key result selected by the user which is generating in the process. As a typical industrial task, predictive maintenance is used to demonstrate the development framework. This section will introduce some preliminary knowledge of predictive maintenance, and then introduce each functional module section by section.

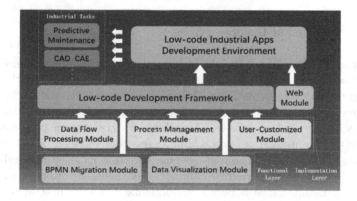

Fig. 2. The system architecture of our low-code development framework

3.1 Low-Code Development Requirements for Predictive Maintenance Industrial Apps

In industrial production, for realizing a return on investment, most companies hope that critical equipment will run in high efficiency and utilization. In order to ensure the efficient use of equipment, companies must consider the risk of equipment failure and time lost due to the failure. According to a study by the American Electric Company Emerson, the annual cost of unplanned downtime to manufacturing companies is estimated to be US$50 billion [6]; at the same time, a survey by PTC also shows that poor maintenance methods will cause 5% to 20% loss of factory production [11]. How to efficiently maintain equipment has always been a concern in the industry. Under normal circumstances, most companies adopt corrective maintenance and preventive maintenance, that is, only replace parts when they have failed, or perform regular maintenance before failures. The emergence of predictive maintenance (PdM) optimizes and balances corrective maintenance and preventive maintenance. Its purpose is to replace parts when they are close to failure, thereby extending equipment life and reducing maintenance costs [1]. However, as a maintenance method with a certain technical threshold, the usage of PdM often faces many problems such as imbalance of source data and model adjustment. Therefore, it is necessary to provide enterprises with a simple and easy-to-use platform to process data, select models, perform the corresponding work of predictive maintenance, and provide corresponding technical support. Low-code development methods meet all requirements. Drag-and-drop development method and flow-based logic enable industrial business personnel to develop predictive maintenance apps quickly and effectively, even if they have few machine learning knowledge. This development framework encapsulates the complex algorithm technology into blocks and provides them to developers in the form of components to meet the low-code development needs of enterprises for such industrial apps.

3.2   Implementation Layer

Data Flow Processing Model. Industrial app development and application are inseparable from a large amount of industrial data, but the Activiti architecture pays more attention to business processes, the data-related parts only involve process ID, user information, process variables and other simple parameters of the control process. Therefore, a data stream processing module is designed for uploading, embedding, exporting datasets and transferring data between components. The data storage adopts a Key-Value database. The upload of the dataset is realized by calling the Attachment interface in Activiti. For data flow maintenance between components, the overall strategy is static table creation and dynamic read and write. The specific method is to statically establish a data table for each type of task component according to the intermediate result data specification, the key value is the task component id, and the value is the data that this type of component may generate during the running process. When the process is running, each type of component dynamically fills in its own generated data in the corresponding data table for other components to read. The data types of the intermediate results are roughly divided into three categories: simple data, file type, and json type. See Table 1 for details.

Process Management Module. The Process management module is responsible for invoking the process, by calling the service interface provided by Activ-

Table 1. Data types related to components.

Type	Storage method	Remark
Sample	Direct storage	None
File	Store url	url: dir/task_id/file_name
json	Store json object	json object is stored as blob type in mysql

iti to realize the creation of the Activiti engine, the creat of the process, and the operation of the process instance. This module generates the ActivitiServiceImpl tool class by implementing the Activiti Service interface, and finally uses the methods provided in the tool class to construct the Activiti Controller class. The process control page is mainly divided into four parts: process list and related information, process operation, and process model import. First, there are two ways to create a process model: click "New" at the top left of the page to jump to the process design page; select a file in the import box at the top right of the page to upload and submit. The created process will appear in the main part of the page and display the process ID, model name and other information. Click the model name of each process to jump to the process design interface for secondary design. The designed process can be deployed and run by clicking "Deploy" and "Start Process". The corresponding page is shown in Fig. 3.

Fig. 3. A screenshot of process control

User-Customized Module. This module is used for component development with specific functions and belongs to the implementation layer of the development framework. In general, the user-customized component module implements support for the deployment of customized components, component customization attributes. Activiti components usually have some attributes to describe themselves or adapt to the configuration in different scenarios. The attributes of native components include component id, name, type, and whether it is asynchronous or not. To implement user-customized attributes, it is necessary to deploy modified codes on the front-end page and back-end respectively. Activiti components are mostly general-purpose components and do not have practical meaning. Therefore, this framework needs to develop some user-customized components to achieve specific functions, such as data cleaning, feature extraction, model training, result prediction. This paper adopts the technical solution of Task Listener mounting + calling python script to realize the logic function of the new component. Task Listener is a native attribute of Activiti, which is used to execute a piece of java logic code during task execution. On the other hand, considering that many logic parts need to call machine learning and deep learning frameworks, when writing logic code, we chose python with dependent packages such as numpy and pytorch. After completing the definition description and functional design of the user-customized components, they need to be registered during deployment to avoid exceptions caused by the Activiti engine not identifying new components. The designed custom component appears on the process design page, as shown in Fig. 4. The process design page is divided into four parts: component menu, canvas, toolbar, and property bar. On the left side of the page is the component menu. Developers can drag the components to the canvas on the right for process design. Below the canvas is the property bar of the selected component, you can modify the related properties of the component. After the process design is completed, click the save button at the top left of the page.

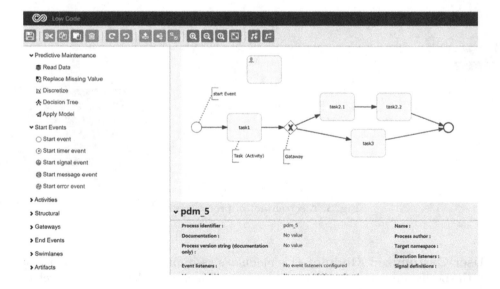

Fig. 4. A screenshot of process design

3.3   Functional Layer

BPMN Migration Module. Activiti follows the BMPN2.0 standard, so the process model that conforms to this specification can be executed on the Activiti engine, and the portability and reusability of the process model are greatly improved. According to this characteristic, the BPMN model import and export module is designed. The main function of this module is to export the designed process model and import the existing process model into the development framework.

Visualization Module. Taking into account that after designing the business process, the developer may need to monitor some key parameters and data during the process of running the process, and design a data visualization module. The general idea of data visualization is to standardize all the intermediate results that this component can provide during the implementation of user-customized components, and to open the data monitoring attributes of the component when users design the process for them to choose the parameters to be monitored. This module integrates the data display panel of the open-source Internet of Things platform ThingsBoard, including various statistical charts, dashboards, etc. Furthermore, a static table is added to the database to map the intermediate result data of various components and the visualization panel. When the process is running, the Activiti engine will dynamically read the intermediate data of the component from the database, and then check the table to select the corresponding visualization panel to display the data. Figure 5 shows the intermediate data generated by the data analysis component during operation.

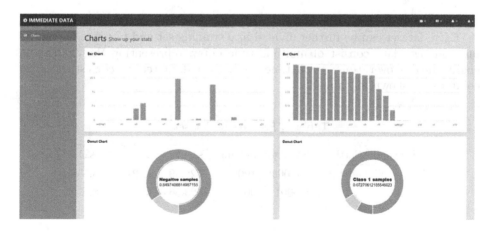

Fig. 5. A screenshot of data visualization

4   Use Case

This paper proposes a low-code development platform for industrial apps based on Activiti engine and tomcat server. Aiming at the industrial task of predictive maintenance of aircraft turbines, we have developed a corresponding application on this platform, which is regarded as a use case. The use case is carried out on the Windows 10 OS equipped with four 2.70 GHz CPUs and one 4G GPU. Web application container is Apache Tomcat 9.0.33, browser is Google Chrome 84.0.4147.105 and database is MySQL Ver 8.0.19 for Win64 on x86_64.

4.1   Dataset

The aircraft turbine health dataset CMAPSSData[1] comes from NASA's Prognostics database. It records the degradation simulation and failure evolution process of aircraft turbine engines. CMAPSSData contains four data subsets, which record the simulation data of four types of engines, as shown in Table 2:

Table 2. CMAPSData Dataset

Dataset	# of training data	# of testing data	Type of fault
FD001	100	100	HPC
FD002	260	259	HPC
FD003	100	100	HPC, flabellum
FD004	248	249	HPC, flabellum

Each dataset can be further divided into training set and test set. The training and test data consist of multiple time entry, representing the changes in engine health over time. Each time entry is a data record consisting of 26 columns, as shown in Table 3:

Table 3. Features of dataset

Id	Cycles	Setting1	Setting2	Setting3	s1	s2	...	s21
1	1	−0.0007	−0.0008	100	528.99	643.21		8.556
1	2	0	0.0001	90	529.56	643.88		8.579

Among these columns, id represents the engine number, cycles represents the time period, setting1 to setting3 represent the operating settings that have a great impact on the engine performance [17], and s1 to s21 are the sensors measurements.

---

[1] https://ti.arc.nasa.gov/tech/dash/groups/pcoe/prognostic-data-repository/.

4.2   Evaluation Metrics

Time Cost of Model Training and Testing. Record the time spent for this development framework to run the designed process to evaluate the performance of the framework.

Predictive Maintenance Regression Model Metrics. This paper uses the following three indicators to evaluate the performance of the predictive maintenance application developed by this framework.

- RMSE (Root Mean Square Error). It is the root of the mean square error, which well reflects the deviation of the forecast.

$$RMSE = \sqrt{\frac{\sum_{i=1}^{n}(p_i - a_i)^2}{n}} \tag{1}$$

- MAE (Mean Absolute Error). It is the average value of the absolute value error, used to measure the deviation between the predicted value and the true value.

$$MAE = \frac{\sum_{i-1}^{n}|p_i - a_i|}{n} \tag{2}$$

- R^2 (Coefficient of Determination). It is the combination of the mean square error and variance. The closer R^2 is to 1, the higher accuracy model presents.

$$R^2 = 1 - \frac{\sum_{i=1}^{n}(p_i - a_i)^2}{\sum_{i=1}^{n}(a_i - \overline{a})^2} \tag{3}$$

In Eqs. (1) to (3), p_i represents the predicted value and a_i represents the ground truth, and $\overline{a}$ represents the mean value of the ground truth.

4.3   Functional and Performance Test

Development Framework Function Test. To verify the functional usability of the development framework, this paper used the low-code development framework to complete the development of the aircraft turbine maintenance software. Figure 6 shows the workflow chart of the app. After data wrangling and data analysis, three models are used in the workflow: binary classification model, multi-classification model [7] and regression model. The Binary Classification Model predicts whether the engine will break down which includes algorithms such as LR, Decision Tree, KNN; Multi-Class Classification Model predicts the cause of a breakdown which includes algorithms such as SVM, ANN; Regression Model predicts the remaining life of the engine which includes algorithms such as LASSO, Polynomial Regression, Random Forest [8–10]. Since these model components are designed to solve different types of predictive maintenance problems, they are functionally parallel. Regression Model is selected here in the use case to predict the remaining life of the engine.

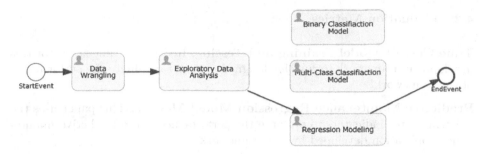

Fig. 6. Predictive maintenance workflow for aircraft turbines

In this use case, the Activiti engine is started to deploy and run the aircraft turbine prediction workflow. The results of each component are as follows: In Data Wrangling, the source dataset was tagged with TTF (Time-To-Failure) tag meanwhile the new dataset extracted the average and standard deviation as features; in Exploratory Data Analysis, some features showed high correlation with TTF labels which will be input in the model, and Fig. 7 shows the correlation heat map; in Regression Model six algorithms were used: LR, LASSO, Ridge, Decision Tree, Polynomial Regression, and Random Forest. Table 4 shows the RMSE, MAE and R^2 of the six algorithms. The results claim that the performance of nonlinear models such as Random Forest and Polynomial Regression is significantly better than other models. In the way of editing code, Random forest and Polynomial Regression models present higher accuracy. To sum up, the results of the optimal model between this use case and the coding way are basically the same, which means that the framework is functionally effective.

Table 4. Regression model results

Metric	Linear regression	LASSO	Ridge	Decision tree	Random forest	Polynomial regression
RMSE	33.90	31.25	32.03	32.22	28.63	29.01
MAE	25.59	25.53	25.07	24.43	23.17	22.38
R^2	0.41	0.41	0.41	0.39	0.51	0.20

Development Framework Performance Test. To evaluate the performance of this development framework, the use case compared the time spent on completing aircraft turbine life prediction by writing code and the time spent on that by using this framework. The results are shown in Table 5. The time cost of this framework is about 5 s longer than that of coding. Over-consuming time accounts for less than 2% of total time in the last two models. The results explain that this framework can design and complete industrial tasks quickly and effectively while the extra time overhead brought by it is negligible.

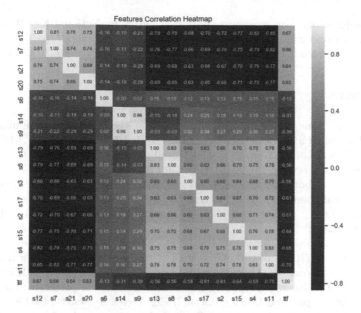

Fig. 7. Correlation between features and tags [17]

Table 5. Time overhead

Model	Low code method(s)	Traditional code method(s)
Regression	44.77	49.39
Binary classification	374.19	378.57
Multiclass classification	397.61	402.35

5 Conclusion

Through the research and analysis of typical industrial tasks processing process and implementation mechanism, based on the Activiti engine and the BPMN2.0 process design specification, this paper establishes a low-code industrial app development framework with a certain versatility, including a series of generic modules such as draggable process design, process deployment and operation, data monitoring, and model management. An app for predictive maintenance, a typical industrial task, has been developed and evaluated on CMAPSSData, which proves the effectiveness of the framework.

In the future, we intend to improve the framework with streaming data processing architecture to process the massive real-time industrial data. Besides, we plan to improve the framework by designing and developing various components for various industrial tasks.

Acknowledgment. This work was supported partly by National Key Research and Development Program of China under Grant No. 2019YFB1705902, partly by National Natural Science Foundation under Grant No. (61972013, 61932007, 61421003).

References

1. Zhang, W., Yang, D., Wang, H.: Data-driven methods for predictive maintenance of industrial equipment: a survey. IEEE Syst. J. **13**(3), 2213–2227 (2019)
2. Zhong, R.Y., Xu, X., Klotz, E., et al.: Intelligent manufacturing in the context of industry 4.0: a review. Eng. **3**(5), 616–630 (2017)
3. Daniel, G., Cabot, J., Deruelle, L., et al.: Xatkit: a multimodal low-code chatbot development framework. IEEE Access **8**, 15332–15346 (2020)
4. Rademakers, T.: Activiti in Action: executable business processes in BPMN 2.0. Manning (2012)
5. Mathew, V., Toby, T., Singh, V., Rao, B.M., Kumar, M.G.: Prediction of remaining useful lifetime (RUL) of turbofan engine using machine learning. In: IEEE International Conference on Circuits and Systems (ICCS), pp. 306–311. IEEE (2017)
6. Proto, S., Di Corso, E., Apiletti, D., et al.: REDTag: a predictive maintenance framework for parcel delivery services. IEEE Access **8**, 14953–14964 (2020)
7. Susto, G.A., Schirru, A., Pampuri, S., et al.: Machine learning for predictive maintenance: a multiple classifier approach. IEEE Trans. Ind. Inform. **11**(3), 812–820 (2014)
8. Carvalho, T.P., Soares, F.A., Vita, R., et al.: A systematic literature review of machine learning methods applied to predictive maintenance. Comput. Ind. Eng. **137**, p.106024 (2019)
9. Kanawaday, A, Sane, A.: Machine learning for predictive maintenance of industrial machines using IoT sensor data. In: 2017 8th IEEE International Conference on Software Engineering and Service Science (ICSESS), 87–90. IEEE (2017)
10. Cline, B., Niculescu, R.S., Huffman, D., et al.: Predictive maintenance applications for machine learning. In: Annual Reliability and Maintainability Symposium (RAMS) 2017, pp. 1–7. IEEE (2017)
11. Sipos, R., Fradkin, D., Moerchen, F., et al. Log-based predictive maintenance. In: Proceedings of the 20th ACM SIGKDD International Conference on Knowledge Discovery and Data Mining, pp. 1867–1876 (2014)
12. Rymer, J.R., Koplowitz, R., Leaders, S.A., et al.: The Forrester wave^{TM} : low-code development platforms For AD&D professionals, Q1 (2019)
13. Yang, S.L., Hu, J.P.: Design of task workflow based on activiti technology. In: Applied Mechanics and Materials, vol. 740, pp. 802–805. Trans Tech Publications Ltd. (2015)
14. Yu, Y., Liu, Z., Tang, J.: Design of MES based on open source Activiti5 Workflow. Ind. Control Comput. **27**(09), 121–122 (2014)
15. Sun, W., Zheng, C., Deng, C., Jiang, T.: Research and implementation of CPPCC public opinion system based on activiti. Inf. Commun. **8**, 175–177 (2016)
16. Yi, Z.: Activiti: The Definitive Guide. Tsinghua University Press (2017)
17. Saxena, A., Goebel, K., Simon, D., et al.: Damage propagation modeling for aircraft engine run-to-failure simulation. In: International Conference on Prognostics and Health Management, 1–9. IEEE (2008)

Research on Machine Learning Method for Equipment Health Management in Industrial Internet of Things

Zheng Tan[1,2] and Yiping Wen[1,2(✉)]

[1] School of Computer Science and Engineering, Hunan University of Science and Technology, Xiangtan, China
[2] Key Laboratory of Knowledge Processing and Networked Manufacturing, Hunan University of Science and Technology, Xiangtan, China

Abstract. Much industrial equipment integrates multiple types of sensors for data collection and real-time connection with the Industrial Internet of Things (IIoT) at now. With the popularity and application of miniaturized and low-cost sensors, to manage the health of equipment in the IIoT, such as evaluating the current health indicator (HI) of the equipment, and predicting its remaining useful life (RUL), the machine learning-based method reflects a broader application prospect due to its good at data mining and analysis. A RUL prediction method of equipment in this paper is proposed. First, a specific data processing method is proposed according to the characteristics of the data, the outputs of the hidden layer of stacked denoising autoencoder (SDAE) as the features of the data set which are extracted to complete the HI construction. Then, the trajectory pointwise difference and similarity method are proposed to generate predicted equipment RUL. This paper uses the C-MAPSS engine data set for experimental verification. The experimental results show that the method proposed in this paper can effectively use the data of multi-dimensional sensor data to get the degradation progress of the equipment, thereby assessing the health of the equipment and more accurately estimating the RUL.

Keywords: Industrial Internet of Things · Health management · Health indicator · Remaining useful life · Stacked denoising autoencoder · Similarity · C-MAPSS

1 Introduction

The Industrial Internet of Things (IIoT) is an important technical foundation that drives the upgrade of industrial intelligence. With the popularization and application of miniaturized and low-cost sensor nodes, the IIoT can provide generalized sensing services for various target equipment [1].

At present, much industrial equipment integrates multiple types of sensors for data collection. With the development of the industrial Internet, how to effectively analyze the large amount of monitoring data generated by mechanical equipment and provide

Y. Sun et al. (Eds.): ChineseCSCW 2020, CCIS 1330, pp. 251–262, 2021.
https://doi.org/10.1007/978-981-16-2540-4_19

beneficial decisions for the independent health management and maintenance of the equipment, independent guarantees which have become a challenging problem, among them, remaining useful life (RUL) prediction of equipment has great significance and practical value for improving the safety and reliability of system operation. However, for the possible types and causes of downtime for different equipment and industrial processes which cannot be described by precise mathematical models, traditional mechanism models and empirical models will lose their ability to explain the unknown equipment degradation process [2]. Artificial intelligence uses machine learning algorithms due to its powerful feature extraction capabilities and nonlinear function mapping capabilities to discover the laws of the degradation process. Therefore, to achieve effective health management of equipment, machine learning-based models are used to predict the RUL of equipment from industrial data.

The RUL prediction program based on machine learning usually consists of three processes, including data collection, health indicator (HI) construction, and prediction. The construction of HI plays an important role in the prediction of RUL for mechanical equipment. A suitable HI can simplify the prediction model and produce accurate prediction results. According to different HI construction methods, there existing mechanical equipment HI can be divided into two types: HI with physical meaning and virtual HI without physical meaning [3]. The construction of HI with physical meaning is often obtained by processing equipment monitoring data with the traditional mathematical statistics or signal processing methods, such as root mean square value, energy entropy. Frequency domain characteristic parameters include such as harmonic frequency [4], harmonic amplitude [5], a method of logarithmic transformation of the equipment spectrum to obtain the health factor was proposed in [6]; Virtual without physical meaning HI is to further integrate it on the basis of the former, with variety of signal processing methods and machine learning methods, to obtain the HI trajectory that does not have actual physical meaning, which is only used as a representation of the degradation state for mechanical equipment. Widodo [7] first extract multiple features from the equipment monitoring signal, and then use the PCA method to reduce the dimensionality of the feature set, and construct HI by calculating the deviation between the degradation state and the equipment health state feature vector, a multilayer perceptron was proposed [8] to fuse multiple time/frequency domain features as input into hidden Markov model for constructing HI.

The construction of HI is finally applied to the prediction of RUL. For example, Miao [9] combined wavelet analysis and SVM to predict the remaining life of the gyroscope, Wang [10] used a combination of time series analysis and BP neural network to predict the remaining life of the cooling fan. Time series analysis can predict the trend of data changes, and the BP neural network adjusts the prediction error in real-time to ensure the accuracy of the RUL prediction. An adaptive RNN to predict the RUL of the Li battery was proposed [11] and optimized the weight of the network structure online by the cyclic Levenberg-Marquardt method.

In addition, Wang [12] proposed a prediction method based on similarity theory that successfully achieved an effective improvement in prediction accuracy on engine simulation data. A method based on similarity theory and back-propagation (BP) neural network was used in [13] for engine simulation data. This experimental result shows that the prediction accuracy of the improved method based on similarity is better than the original method. Therefore, the effective combination of similarity theory with machine learning will improve prediction accuracy which can use the fitting ability of neural network models and similarity rules.

In response to the above problems, an RUL prediction method based on SDAE with similarity theory is proposed in this paper, the proposed method uses similarity theory with machine learning to increase the prediction accuracy. First, the feature of the data set was extracted based on the SDAE, virtual HI of the equipment is generated. Then, the trajectory pointwise difference and similarity method are proposed to generate predicted equipment RUL. To evaluate the performance of the method, the C-MAPSS engine data set is selected for experimental verification and makes a comparison with this method and four traditional methods for mean square error (MSE), and cumulative accuracy (CRA).

The main contributions of this article are as follows:

1. A specific data processing method is proposed according to the characteristics of the data.
2. Propose an unsupervised learning-based method for HI construction. No specific expert experience and a large number of personnel are required to manually input features.

The rest of the paper is organized as follows: Sect. 2 introduces the basic concepts of SDAE. Section 3 introduces the HI construction method based on SDAE, the method for RUL prediction. Section 4 verifies the effectiveness of our method with an example of the C-MAPSS engine data set and conducts a comparative experiment. Section 5 summarizes this work and proposes future work.

2 Preliminaries

2.1 Stacked Denoising Autoencoder

A single autoencoder (AE) is a three-layer unsupervised neural network that can be divided into input layer-hidden layer-output layer, the purpose of the network is to reproduce the input signal as much as possible and let the output is equal to the input. The network is divided into two parts: encoding network and decoding network, when adding noise interference to the input data will become a denoising autoencoder (DAE). As shown in Fig. 1, during the training process, DAE will encode the noise-added data through the encoding network, and then use the decoding network to decode the encoding result to generate reconstructed data, the difference between the original input and reconstructed data is used as the reconstruction error, and the gradient descent algorithm is used for training.

For DAE, when given an equipment data sample set $X = \{x^{(j)}\}_{j=1}^{M}$, among them M is the number of data samples, $x^{(j)}$ is the j-th input data. The input data of the DAE is $x^{(j)}$ which is destroyed to $\tilde{x}^{(j)}$ by add random noise according to q_D distribution, $\tilde{x}^{(j)}$ through the encoding network function f_θ to generate the output $h^{(j)}$ of the hidden layer.

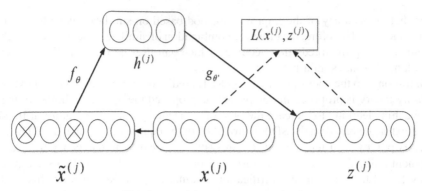

Fig. 1. Denoising Autoencoder model

The output $h^{(j)}$ through the decoded network function g'_θ to generate reconstructed data $z^{(j)}$ and the difference between the input data $x^{(j)}$ and the reconstructed data $z^{(j)}$ is used as the reconstruction error for training. The reconstruction error is given by

$$L\left(x^{(j)}, z^{(j)}\right) = \left\| x^{(j)} - z^{(j)} \right\| \tag{1}$$

DAE uses the minimum *MSE* as the cost function and uses the gradient descent method to minimize the reconstruction error function as

$$J = \frac{1}{M} \sum_{j=1}^{M} L(x^{(j)}, z^{(j)}) \tag{2}$$

Multiple DAE stacks can form SDAE with a certain depth, and train the SDAE based on the layer by layer greedy algorithm.

3   Proposed Method

The flowchart of our proposed method is shown in Fig. 2, including data pre-processing, features extraction, and RUL prediction.

3.1   Data Pre-processing

1. **Data division.** The data set has been divided into train set and test set, which can be used directly in this experiment.
2. **Handling data redundancy.** Delete the redundant data at the same point in time, and only keep one piece of available data.
3. **Handling missing data.** When a quarter or more of the sensor data is missing, the data is deemed to be seriously missing, and the data has been scrapped and has no reference. When there is only a small amount of missing sensor data, consider deleting the missing data directly or filling in the average of the upper and lower time points of the missing data.
4. **Normalization.** Normalize the original data to the interval [0, 1] to eliminate the dimensional difference between the sensors.

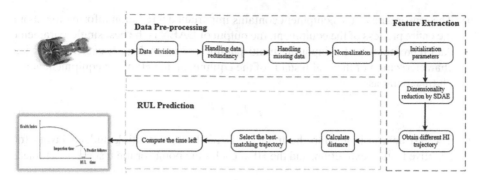

Fig. 2. Flowchart of our proposed method

3.2 Feature Extraction

This paper uses the deep neural network's strong nonlinear expression ability and good discriminative ability and uses SDAE to fuse the data characteristics of the equipment's multiple sensor data to evaluate the health of the equipment. The specific steps:

1. Initialize the SDAE network structure, such as setting the number of SDAE layers, the number of nodes in each layer, etc.
2. Select the m-dimensional sample $x^{(j)} = [x^{(j,1)}, x^{(j,2)}, \ldots .x^{(j,m)}]$ at the jth time point of the normalized training set data, where $x^{(j)}$ is the m-dimensional sensor data at each time point extracted from the training set data, $j = 1, 2, \ldots, T$, T is the number of time points, and all samples at T time points are input to SDAE, the number of nodes in the input layer of the network is consistent with the number of sensors.
3. Use the layer-by-layer training algorithm to train the hidden layer and extract the deep features. Noise is added to the data samples $x^{(j)}$ to obtain $\tilde{x}^{(j)}$ to train DAE1 and $\tilde{x}^{(j)}$ is coded into $h_1^{(j)}$ as

$$h_1^{(j)} = f\left(w_1 x^{(j)} + b_1\right), \quad \theta_1 = \{w_1, b_1\} \tag{3}$$

4. θ_1 is the parameter of DAE1, and because $h_1^{(j)}$ can be reconstructed as input data, the main information of $x^{(j)}$ is obtained. Then use $h_1^{(j)}$ as the input data to train DAE2, and $h_1^{(j)}$ is coded into $h_2^{(j)}$ as

$$h_3^{(j)} = f\left(w_2 h_1^{(j)} + b_2\right), \theta_2 = \{w_2, b_2\} \tag{4}$$

5. Repeat this process until the DAEN pre-training is completed. The layer-by-layer training connects multiple DAEs to form the SDAE. The output of the hidden layer N is $h_N^{(j)}$ that completes the dimensionality reduction and denoising of the input data by the network is given by

$$h_N^{(j)} = f\left(w_N h_{N-1}^{(j)} + b_N\right), \quad \theta_N = \{w_N, b_N\} \tag{5}$$

6. The sensor data of the equipment contains important degradation information about the entire process of the equipment, the output of DAEN is the best single-parameter data representation for inputting multi-parameter sensor data. It can be used as a characterization of the health status of the equipment, the HI of the equipment at the jth time point as

$$HI^j = h_N^{(j)} \tag{6}$$

7. The test data is input into the trained SDAE through multiple hidden layers for adaptive feature extraction, and the HI at each time point for the test set is obtained.

3.3 RUL Prediction

Similarity-based methods belong to data-driven RUL prediction. The training unit refers to all unit s extracted from the train set, and test unit refers to a unit extracted from the test set. The main idea can be expressed as: if the test unit and the reference unit have similar degradation processes, they may have similar RUL.

The distance relationship S^i between HI trajectory of equipment i in the train unit and the HI trajectory of the test unit is defined by an array

$$S^i = (t^i, s), \quad i = 1, 2, \ldots..m \tag{7}$$

$t^i = \left[t_1^i, t_2^i, \ldots t_{l_i}^i \right]$ represents the HI trajectory of the equipment i with the running time of l_i in the train unit, m means the number of train unit, $s = \left[s_1, s_2, \ldots s_{l_s} \right]$ means the HI trajectory of the test unit with the running time of l_s.

Sub-trajectory Extraction. According to the trajectory pointwise difference method, extract HI sub-trajectories of length from each train unit, then all sub-trajectories of train unit t^i can be expressed as

$$H^i = \begin{bmatrix} t_1^i & \cdots & t_{l_s}^i \\ \cdots & \cdots \cdots \\ t_{l_i-l_s+1}^i & \cdots & t_{l_i}^i \end{bmatrix} \tag{8}$$

The matrix H^i represents the set of sub-trajectories of the train unit t^i, and each row vector $r_j^i = \left[t_j^i, t_{j+1}^i, \ldots t_{j+l_s-1}^i \right], j = 1, 2, \ldots l_i - l_s + 1$ in the matrix represents the jth HI sub-trajectory in the train unit i with the same length as the test unit time length l_s.

Distance Measure. Based on the sub-trajectory set matrix H^i representing the train unit t^i, calculate the Euclidean distance between the HI sub-trajectories of all train unit i and the HI trajectory of the test unit .

$$D^i = \begin{bmatrix} e_1^i \\ e_2^i \\ \cdots \cdots \\ e_{l_i-l_s+1}^i \end{bmatrix} \tag{9}$$

Euclidean distance between the HI trajectory of the test unit and the jth HI sub-trajectory of the train unit i as

$$e_j^i = \left\| s - r_j^i \right\|_2 \tag{10}$$

Prediction. The best matching sub-trajectory can be expressed as

$$d = \arg \underset{j}{find} \left(e_j^i = \min \left(D^i \right) \right) \tag{11}$$

d indicates the value of j when the distance between test unit and each HI sub-trajectory in the train unit i are the smallest.

The HI sub-trajectory $r_d^i = \left[t_d^i, t_{d+1}^i, \ldots t_{d+l_s-1}^i \right]$ of the train unit i is the most similar to the HI trajectory of the test unit. Since HI reflects the health status of the equipment, similar HI trajectories mean similar RUL. Therefore, the RUL of the train unit i when the running time is d can be regarded as the RUL of the test unit when the running time is l_s.

The RUL of the train unit i when the running time is d can be expressed as

$$RUL^i = l_i - (d + l_s) \tag{12}$$

Equipment i in the train unit set can generate an estimated RUL for the test unit, and the final RUL of the test unit is calculated by the average of RUL obtained based on the m train unit

$$RUL = \frac{1}{m} \sum_i^m RUL^i \tag{13}$$

4   Experiments

4.1   Dataset Description

The data set in this paper comes from the C-MAPSS aerospace engine simulation of NASA [14].

This data set is one of the most used data sets for studying the RUL prediction. The data set records the 24-dimensional performance parameters of aero-engines in each flight cycle, including 3 condition variables and 21 sensor measurement variables. For each cycle of the operation simulation, the engine undergoes a complete transition from the normal state (with varying degrees of initial wear and manufacturing changes) to the fault state and records 3 operating condition values and 21 sensors measurement for each flight cycle during the process. The data set has been divided into train set and test set.

Figure 3 shows the raw data of multiple sensors of the equipment in the data set that the raw data is highly dispersed, and some sensor readings have many differences, so it cannot be directly used to characterize the degradation process of the equipment.

Based on the above reasons, it is difficult to directly construct the HI from the raw sensor data to predict the RUL, so the data needs to be pre-processed.

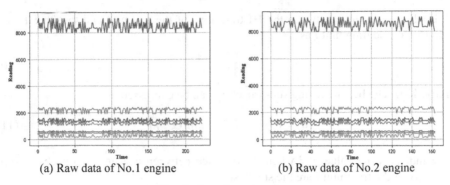

(a) Raw data of No.1 engine   (b) Raw data of No.2 engine

Fig. 3. Raw data of sensors

4.2 Data Processing

After the redundant processing and missing processing of the data, because the equipment is in different working conditions, this is also the reason why the original data fluctuates up and down, and the sensor readings are affected by the working conditions. Therefore, it needs to identify different working conditions before normalization.

We use the K-means clustering method to identify the equipment operating conditions in this paper because the operating conditions are affected by three operating settings (the altitude, the number of flight speeds, and the ambient temperature), among them, the setting of ambient temperature is 6 constants. This paper uses the different settings of this kind of ambient temperature to identify different working conditions and considers that there are 6 working conditions, and clustering according to 6 clusters.

The clustering results are shown in Fig. 4.

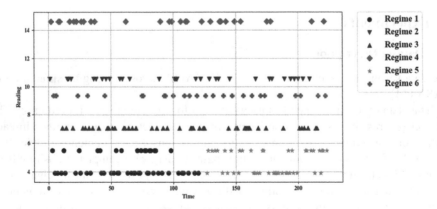

Fig. 4. Clustering result

Based on the clustering results of working conditions, Table 1 shows the sensor readings of each sensor under different working conditions.

As shown in Figure 4 and Table 1, all sensor data are successfully clustered into different working conditions. The raw data can be normalized according to the working

Table 1. Sensor data based on clustering results

Regime		1	2	3	4	5	6
Sensor1	Max	491.1	489.1	445.0	449.4	462.5	518.6
	Min	491.1	489.1	445.0	449.4	462.5	518.6
Sensor2	Max	609.5	606.7	551.5	557.5	538.3	644.4
	Min	605.9	603.4	548.2	554.4	535.6	641.3
Sensor3	Max	1508.9	1523.2	1378.6	1389.6	1284.9	1614.6
	Min	1467.1	1484.5	1335.7	1348.3	1245.4	1571.5
·············							
Sensor21	Max	14.9	17.4	6.5	9.1	8.7	23.9
	Min	14.3	16.7	6.1	8.6	8.2	22.8

conditions, and this paper uses linear function normalization, and the normalization method as

$$N(x^{(t)}) = \frac{x^{(t)} - x_{\min}}{x_{\max} - x_{\min}} \tag{14}$$

This method normalizes the data of the sensor at t time point to [0, 1], and converts the raw data $x^{(t)}$ into normalized data $N(x^{(t)})$, which $x_{\max}$ represents the maximum value of the data belonging to the operating condition at time t, and $x_{\min}$ represents the data the minimum value of the data belonging to the working condition at time t. Figure 5 shows the normalized dimensions after normalization.

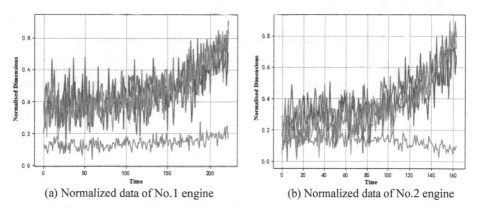

(a) Normalized data of No.1 engine   (b) Normalized data of No.2 engine

Fig. 5. Normalized data of sensors

4.3 Experimental Results

In order to quantitatively evaluate the predictive accuracy of our method, we used two widely used evaluation metrics such as MSE, and CRA are employed in this paper.

MSE is the average of squared errors

$$MSE = \frac{1}{n} \sum_{i=1}^{n} (y_i - \hat{y}_i)^2 \tag{15}$$

Cumulative relative accuracy (CRA) as

$$CRA = \sum_{k=1}^{K} w_k RA(t_k) \tag{16}$$

$$w_k = k / \sum_{k=1}^{K} k \tag{17}$$

$$RA(t_k) = 1 - \frac{|ActRUL(t_k) - RUL(t_k)|}{ActRUL(t_k)} \tag{18}$$

The closer the CRA value is to 1, the higher the accuracy of the prediction method is.

To illustrate the superiority of the proposed hybrid prediction method over the traditional data-driven method, we introduced two commonly used data-driven methods for constructing HI, the PCA-based method [15] and the BP network-based method; and two commonly used models for fitting RUL, the LSTM network-based method, and the BP network-based method. The results of various predictive evaluation indicators are shown in Table 2.

Table 2. Experimental results for five prediction approaches

Engine	Proposed method		BP+Similarity		PCA+Similarity		BP		LSTM	
	MSE	CRA	MSE	CRA	MSE	CRA	MSE	CRA	MSE	CRA
1	129.9	0.7	975.2	0.3	326	0.4	3214	0.1	1476	0.2
2	81.2	0.8	782.3	0.4	297	0.5	2917	0.1	1387	0.2
3	78.2	0.7	527.1	0.5	287	0.5	2633	0.1	1676	0.2
4	48.9	0.8	386.6	0.6	223	0.5	2432	0.1	1493	0.2
....										

Table 2 shows the experimental results of five RUL prediction approaches. We can see that the proposed approach has a lower MSE value and higher CRA value than the other traditional machine learning methods for tested equipment clearly from this table, this shows that the proposed method in this paper can provide more accurate RUL prediction results. Therefore, the proposed approach is better than the other prediction approaches for RUL prediction.

5 Conclusion

In this paper, a prediction approach based on machine learning was proposed for prediction RUL in IIoT, similarity is fused with machine learning to improve prediction accuracy. First, the feature of the data set was extracted based on the SDAE, virtual HI of the equipment is generated. Then, the trajectory pointwise difference and similarity method are proposed to generate predicted equipment RUL.

The equipment HI feature extraction based on SDAE solves the problems of traditional mechanical equipment degradation status modeling methods relying on many signal processing technologies and expert experience, difficulty in label selection in supervised training, and insufficient versatility, and have more accurate extraction of equipment health status. Then the deep learning and similarity theory is validly combined and achieved better results than the traditional approach for RUL prediction.

The selection of hyperparameters in the deep learning model will directly affect the prediction accuracy. For future works, a more general hyperparameter selection strategy with higher prediction accuracy will be developed to improve the feature extraction capability of the machine learning model.

Acknowledgment. This work was supported by the National Key Research and Development Projectof China (No. 2018YFB1702600, 2018YFB1702602), National Natural Science Foundation of China (No. 61402167, 61772193, 61872139), Hunan Provincial Natural Science Foundation of China (No. 2017JJ4036, 2018JJ2139), and Research Foundation of Hunan Provincial Education Department of China (No. 17K033, 19A174).

References

1. Huang, Z., Lin, K.J., Tsai, B.L., et al.: Building edge intelligence for online activity recognition in service-oriented IoT systems. Future Gener. Comput. Syst. **87**, 557–567 (2018)
2. Wang, S., Wan, J., Zhang, D., et al.: Towards smart factory for industry 4.0: a self-organized multi-agent system with big data based feedback and coordination. Comput. Netw. **101**, 158–168 (2016)
3. Hu, C., Youn, B.D., Wang, P., Yoon, J.T.: Ensemble of data-driven prognostic algorithms for robust prediction of remaining useful life. Reliab. Eng. Syst. Saf. **103**, 120–135 (2012)
4. Gebraeel, N., Lawley, M., Liu, R., et al.: Residual life predictions from vibration-based degradation signals: a neural network approach. IEEE Trans. Ind. Electron. **51**(3), 1694–700 (2004)
5. Mark, W.D., Hines, J.A.: Frequency-domain assessment of gear-tooth bending-fatigue damage-progression using the average-log-ratio, ALR, algorithm. Mech. Syst. Signal Process. **45**(2), 479–487 (2014)
6. Wang, D., Tse, P.W.: Prognostics of slurry pumps based on a moving-average wear degradation index and a general sequential Monte Carlo method. Mech. Syst. Signal Process. **56**, 213–229 (2015)
7. Widodo, A., Yang, B.S.: Application of relevance vector machine and survival probability to machine degradation assessment. Expert Syst. Appl. **38**(3), 2592–2599 (2011)
8. Giantomassi, A., Ferracuti, F., Benini, A., et al.: Hidden Markov model for health estimation and prognosis of turbofan engines. In: ASME 2011 International Design Engineering Technical Conferences & Computers and Information in Engineering Conference, pp. 1–6 (2011)

9. Miao, J.Q., Li, X.G., Ye, J.H.: Predicting research of mechanical gyroscope life based on wavelet support vector. In: First International Conference on Reliability Systems Engineering, pp. 1–5 (2016)
10. Wang, L.X., Wu, Z.H., Fu, Y.D., et al.: Remaining life predictions of fan based on time series analysis and BP neural networks. In: Information Technology, Networking, Electronic and Automation Control Conference, pp. 607–611 (2016)
11. Liu, J., Saxena, A., Kai, G., et al.: An adaptive recurrent neural network for remaining useful life prediction of lithium-ion batteries. In: Annual Conference of the Prognostics and Health Management Society, pp. 1–9 (2010)
12. Wang, T., Yu, J., Siegel, D., Lee, J.: A similarity-based prognostics approach for engineered systems. In: International conference on prognostics and health management, pp. 4–9 (2008)
13. Bektas, O., Jones, J.A., Sankararaman, S., et al.: A neural network filtering approach for similarity-based remaining useful life estimation. Int. J. Adv. Manuf. Technol. 101, 87–103 (2019)
14. Saxena, A., Goebel, K., Simon, D., Eklund, N.: Damage propagation modeling for aircraft engine run-to-failure simulation. In: International Conference on Prognostics and Health Management, pp. 1–9 (2008)
15. Lasheras, F.S., et al.: A hybrid PCA-CART-MARS-based prognostic approach of the remaining useful life for aircraft engines. Sensors (Basel, Switzerland) 15(3), 7062–7083 (2015)

A Transfer Learning Method for Handwritten Chinese Character Recognition

Meixian Jiang[1], Baohua Zhang[2], Yan Sun[1], Feng Qiang[2], and Hongming Cai[1(✉)]

[1] School of Software, Shanghai Jiao Tong University, Shanghai, China
{jiangmeixian,sun_yan}@sjtu.edu.cn, cai-hm@cs.sjtu.edu.cn
[2] Industrial and Commercial Bank of China, Shanghai, China
{zhangbh,qiangfeng}@sdc.icbc.com.cn

Abstract. The deep convolutional neural network has achieved high accuracy in handwritten Chinese character recognition (HCCR). Large scale handwritten data collection as well as labor-intensive labeling work are required to train an effective model suitable for various writing styles. Typically synthetic data are generated with data augmentation to alleviate the scarcity of labeled training data in real applications. However, the domain shift between synthetic data and real data results in unsatisfying recognition accuracy, bringing a significant challenge. A transfer learning method is proposed for synthetic-to-real handwriting recognition to alleviate the issue. In the framework, a proposed convolutional neural network is pre-trained with the synthetic data as the source model. Then, the source model is optimized with real samples of a specific handwriting style in an unsupervised manner. The effectiveness of the proposed transfer learning method is validated through systematic experiments on the public dataset as the target domain data.

Keywords: Handwritten Chinese character recognition ·
Convolutional neural network · Transfer learning · Writer adaptation.

1 Introduction

Optical character recognition (OCR) has become a widely studied domain, since digitizing the information in paper documents can facilitate the storage, editing, and management of information [8]. Using machine learning and computer vision to achieve the automatic recognition of Chinese handwritten characters is a challenging research topic of OCR. In contrast to phonological languages such as English, Chinese has a much larger vocabulary. According to the GB2312-80 standard, there are 6763 Chinese characters, including 3755 first-level Chinese characters and 3008 s-level Chinese characters [15]. Compared with printed Chinese characters, handwritten characters are more random and lack standardization. There will be adhesions between adjacent Chinese characters, which

Y. Sun et al. (Eds.): ChineseCSCW 2020, CCIS 1330, pp. 263–274, 2021.
https://doi.org/10.1007/978-981-16-2540-4_20

increases the difficulty of recognition. A series of quite variable glyphs of the same Chinese character can be produced by the different writers, therefore one sample apears completely different from another. Even the same person can write the same character quite differently relying on the context in which it was written.

Recently, convolutional neural networks achieve high accuracy on HCCR with a larger number of labeled samples from different people provided in the training phase. However, collecting such a large number of labeled training dataset is too labor-intensive and time-costing. Employing a fully synthetic training dataset is a common and cheap solution to handle the scarcity of labeled training data [1]. However, the visual features extracted from synthetic samples may be different from that of real handwritten ones, meaning the existence of the domain shift problem between the two. Thus, the model solely trained with synthetic data may not perform high accuracy on real ones, and cannot be put into use directly.

To address the aforementioned issue, a transfer learning method is proposed to mitigate the gap between the synthetic data and the real data in HCCR. Firstly, the recognition model is pre-trained using synthetic images as the source model. Based on the source model, the parameters of layers are optimized with unlabeled real handwritten images. The unsupervised transfer learning method adapts the model to real handwriting data, which largely improves the initial recognition accuracy by the model trained with synthetic data. The main contributions of this paper can be summerized as follows:

- A novel model combining the classic VGG convolutional neural network and the adaptation layer suitable for transfer is proposed to perform handwritten character recognition.
- An unsupervised transfer learning strategy is proposed to adapt the pre-trained source model to various handwriting styles.

The remainder of this paper is organized as follows. Section 2 reviews related work in handwritten recognition and transfer learning methods. Section 3 describes the complete framework of the proposed method. Section 4 introduces the algorithms in the two phases of the method. In Sect. 5, the performance of the unsupervised transfer learning method is validated. Section 6 concludes the paper and illustrates the future direction of this work.

2   Related Work

The challenges of handwritten Chinese character recognition (HCCR) problem lies in its large scale vocabulary, great diversity in handwriting styles, too many similar and confusable glyphs [12]. With the blooming of convolutional neural networks (CNN), HCCR with high performance is achieved. Among them, MCDNN proposed by Cireşsan et al. [5] is the early reported the successful application of CNN for HCCR. Recent benchmarks in HCCR are even beyond human performance, reached over 96%, including HCCR-Ensemble-GoogLeNet-10 [15], CNN-Voting-5 [3], Ensemble-DCNN-Similarity ranking [4] and so on.

In order to alleviate the cost and difficulty of collecting and labeling large amounts of handwriting images for training CNN, the use of data augmentation techniques and synthetically generated characters as source data has lately been taken into consideration [2]. However, recognition models that are fully trained on source data inevitably suffer from the domain shift problem when encountering new coming handwriting in different application scenarios. Therefore, unsupervised domain adaptation techniques, one of the branches of transfer learning have been applied in image recognition problems [10]. The adversarial learning strategy [9] is a well-studied and cutting-edge approach of unsupervised domain adaptation techniques. In this approach, the source domain data are correctly classified by the recognizer, while the difference between tow domains is distinguished by the discriminator. Ganin et al. [6] trained a digit recognizer with the MNIST dataset and adapted it towards other datasets like MNIST-M and SVHN. The integrateion of discrimination steps in a sequence encode-decoder network with attention mechanism was prposed by Zhang et al. [14] for robust text image recognition. Lei et al. [7] incorporated a discriminator in handwritten text recognition model fully trained with synthetic samples to adapt towards different real writing styles in an unsupervised manner. Even if the adversarial learning strategy shows effectiveness in the literature, it still possesses several disadvantages. The adversarial-based method requires an extra structure called discriminator as well as an additional loss in the training phase, which increases the complication of the original network. The adversarial training strategy requires both source and target data to be input and trained from scratch, which is computationally intensive. It also faces the problem of slow convergence in the training process, resulting in the high training cost.

Additionally, writer adaptation as a specific topic of domain adaptation has been researched in the document analysis community to address the differences in individual handwriting styles [13]. Yang et al. proposed a writer adaptation scheme for Chinese character recognition called style transfer mapping (STM), and further extended it to the deep network structure by adding an adaptive layer [11]. [12] even utilized STM to reduce the domain gap between training and testing data to achieve a new benchmark in HCCR. However, algorithms extended from STM are narrowed within a specific scope, targeting at mitigating the domain shift between various real handwriting styles, which is smaller than that between synthetic and real data. Therefore, STM may not be an appropriate algorithm to be directly applied to solve the problem brought out by removing the expensive manual labeling step in real world applications.

To sum up, researches still need to be filled in the gap field of transfer learning for synthetic-to-real HCCR. A transfer learning method is proposed for handwritten Chinese character recognition. Compared with the above mentioned approaches, the proposed transfer method is straightforward without modifying the original network structure and performs effective results when the domain shift is large, which cannot be achieved by the related work.

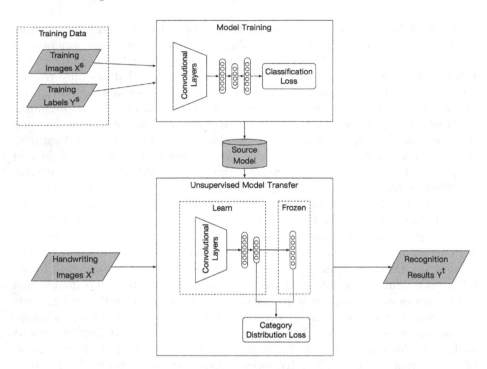

Fig. 1. The proposed framework of unsupervised transfer learning.

3   Framework

The main goal of this paper is to provide a transfer learning method for handwriting recognition. As shown in Fig. 1, the framework of the proposed method consists of two phases, including model training and unsupervised model transfer.

a) Phase I: Model training: A proposed convolutional neural network pretrained on the synthetic images with its corresponding character labels is saved as the source model.

b) Phase II: Unsupervised model transfer: This phase utilizes unlabeled handwritten character images to transfer the source model to the target model in an unsupervised manner.

The above shows the framework of the entire system, and the algorithms and techniques used in the two phases will be introduced.

4   Method

In this section, algorithms in each phase of the method is elaborated. Firstly, a proposed neural network model is trained with synthetic data as the source model. Secondly, the source model is transferred using unlabeled real handwriting images.

Table 1. The structure of the proposed model.

Type	Configurations
Convolution	64, k:3*3, s:1, p:0
Convolution	128, k:3*3, s:1, p:0
Max-pool	2*2
Convolution	256, k:3*3, s:1, p:0
Convolution	256, k:3*3, s:1, p:0
Max-pool	2*2
Convolution	512, k:3*3, s:1, p:0
Convolution	512, k:3*3, s:1, p:0, dropblock:7*7, 0.1
Max-pool	2*2
Convolution	512, k:3*3, s:1, p:0
Convolution	512, k:3*3, s:1, p:0, dropblock:3*3,0.1
Max-pool	2*2
Full connected	900, dropout:0.5
Full connected	200
SoftMax	3755

4.1 Model Training

The state-of-the-art recognition models are mainly voting ensemble models, composed of several sub-models to boost accuracy with sacrifice for prediction time. The main focus of our proposed model is to perform transfer learning effectively, thus the accuracy on the target data instead of source data is the first consideration. The classic VGG model is proven to be an effective and light-weighted backbone network for transfer learning. Therefore, the proposed convolutional layers are based on it and a few modifications are added to make it more suitable for our recognition and transfer task. Since the size of the input images is small, the last two convolutional layers are no more needed and removed. The reduction on model parameters contributes to high real-time. To avoid overfitting, drop blocks and dropout are respectively applied in some convolutional layers as well as fully connected layers. A fully connected layer with small number of parameters as bottleneck layer is especially provided before the softMax layer for the application of the model transfer. The configurations of each layer of the proposed neural network for recognition is shown in Table 1.

During the training phase, the synthetic handwriting images x_i^s are the input of the source model. They are normalized to the size of 32*32, and then goes through multi convolutional layers, max-pooling layers, and fully-connected layers, and finally, its character category is computed.

$$L_{classification} = \sum_{i=1}^{n} L(y_i^s, convNet(x_i^s)) \tag{1}$$

$L()$ denotes the standard cross-entropy loss function. The classification loss between the label y_i^s and the output $convNet(x_i^s)$ is minimized by the mini-batch gradient descent with momentum. When the model converges, network parameters learned under an enormoud amount of synthetic training data are saved as the source model.

4.2   Unsupervised Model Transfer

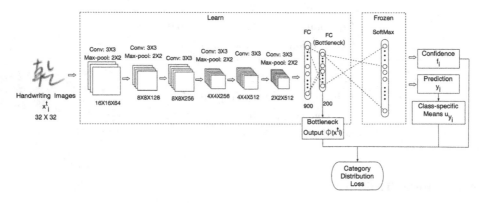

Fig. 2. The proposed unsupervised model transfer algorithm.

The synthetic images used for training differ from the real handwriting images, that is, the data from two domains do not obey a uniform distribution. In this case, directly using the source model for handwriting recognition cannot achieve satisfactory results. This paper proposes the unsupervised model transfer algorithm for synthetic-to-real handwritten Chinese character recognition, inspired by style transfer mapping in writer adaptation. Writer adaptation aims to mitigate the domain gap between handwriting samples by different writers. For instance, the HCCR model proposed by [12] considers it between the public training and testing dataset to guarantee a higher recognition accuracy. Since synthetic data are more standard than real handwriting samples, the proposed unsupervised model transfer needs to be more effective to mitigate a larger domain shift than writer adptation. The general framework of combining STM with the deep network is to view it as an additional linear layer, which adaptively learns to minimize the domain gap of the category distribution. However, the unsupervised model transfer proposed by this paper is achieved directly on the original layers of the source model without modifying its structure. The unsupervised model transfer combining several layers is non-linear and can fit the target data more properly. Figure 2 illustrates the mechanism of our unsupervised model transfer method on the proposed recognition model.

The first step of unsupervised model transfer is to calculate the category distribution on the bottleneck layer from training data $\{(x_i^s, y_i^s)\} \in D_s$, where $y_i^s \in \{1, 2, ..., K\}$ and K denotes the number of classes. Suppose $\phi(x_i^s)$ is the output of the bottleneck layer of 200 dimensions. Class-specific means u_k, where $k \in \{1, 2, ..., K\}$ is defined to represent the category distribution:

$$u_k = \frac{1}{\sum_{i=1}^{N} l(y_i^s = k)} \sum_{i=1}^{N} \phi(x_i^s) l(y_i^s = k) \tag{2}$$

where $l() = 1$ when the condition is true and otherwise 0. The $\{u_1...u_K\}$ will be used to perform adaptation towards the handwriting data.

Suppose some unlabeled handwritten data $\{(x_i^t)\} \in D_t$ are used for adaptation. Since the last layer of the model is softMax, the output is the probability vector of x_i^t belonging to each classes. Therefore, the class of the maximum one in the probability vector is chosen as the prediction, which can be denoted as $y_i^t = convNet(x_i^t) \in \{1, 2, ..., K\}$. Additionally, the confidence about the prediction is the maximum number in the probability vector and can be denoted as $f_i = convNet_{softmax}(x_i^t) \in [0, 1]$. f_i is a float number between 0 and 1, indicating the probability of the sample x_i^t belonging to the predicted classification y_i^t. The purpose of adaptation is to reduce the mismatch between the category distribution of the training data and the handwriting data. Therefore, the adaptation problem can be expressed as minimizing the category distribution loss function:

$$L_{category\ distribution} = \frac{1}{n} \sum_{i=1}^{n} f_i |\phi(x_i^t) - u_{y_i^t}|^2 \tag{3}$$

The Euclidean distance between the bottleneck output $\phi(x_i^t)$ and the class-specific means $u_{y_i^t}$ is denoted by $|\phi(x_i^t) - u_{y_i^t}|$, and (3) can be viewed as the weighted mean square entropy loss. The objective of minimizing the loss function is to adapt each $\phi(x)$ towards the class-specific mean u_{y_i} on the bottleneck layer. Each loss is weighted by the confidence f_i given by the softMax of the network, since the reliability of each prediction y_i varies.

The model transfer can be carried out with the last softMax layer frozen, and the other layers updated by minimizing the weighted mean square entropy loss using full-batch gradient descent with momentum iteratively. Since y_i^t and f_i computed by the gradually updating model are part of the loss calculation, it is necessary to take control of the right updating orientation. Full-batch gradient descent considers the gradient descent orientation of the whole dataset, which can reduce the fluctuation in the model transfer process. After each iteration, the network prediction y_i^t as well as the softMax confidence f_i of it will be more accurate. Therefore, after transfer learning for several iterations, the accuracy can reach a higher level. Algorithm 1 illustrates the proposed unsupervised model transfer algorithm.

Algorithm 1: Unsupervised model transfer

Input: $\{x_i^t\}_{i=1}^n$, $convNet$, $iterNum$
Output: $optimized \;\; convNet$
1 estimate class-specific means $\{u_i...u_K\}$ by (2);
2 freeze the softMax layer after the bottleneck layer;
3 **for** $j=1{:}iterNum$ **do**
4   **for** $i=1{:}n$ **do**
5   compute $\phi(x_i^t)$;
6          $y_i = convNet(x_i^t) \in \{1,...,K\}$;
7          $f_i = convNet_{softmax}(x_i^t) \in [0,1]$;
8          $i++$;
9   **end**
10   update $convNet$ by (3) using full-batch gradient descent;
11     $j++$;
12 **end**

By means of our proposed unsupervised model transfer, the model needs not to learn from scratch with labels, and each writing style can be adapted based on the source model. The loss function is computed after the bottleneck layer, which has the smallest number of parameter, can largely reduce the computation cost.

5 Experiment

In this section, the performance of the unsupervised transfer learning is validated by taking a comparision between the source model, target model transferred by related work of STM [12] and target model transferred by the proposed method on the ICDAR-2013 offline competition dataset as the target domain data.

5.1 Dataset

The labeled source dataset is applied to pre-train a source model, while the target dataset is employed to transfer the source model without labeling.

- Source dataset: The synthetic character images are automatically generated using traditional image processing techniques, with 102 electronic fonts and the 3755 first-level Chinese character set. Data augmentation is applied to ensure the variability and quantity of the source dataset.
- Target dataset: The ICDAR-2013 offline competition dataset contains real handwritten character images from 60 writers, each of which possesses about 3755 samples. Due to the difference between writing styles, handwriting images of a specific writer from the ICDAR-2013 dataset are selected respectively as the target domain data for model transfer. Since the number of images of each writing style remains small, they are all used as target domain data for unsupervised model transfer.

5.2   Implementation Details

The source model is trained with the batch size of 1024 and SGD optimizer where the learning rate is 0.01, and the momentum is 0.9.

To control variables, STM and our prposed unsupervised model transfer are both carried out on the same source model after the bottleneck layer. In STM, the dimensionality of the adaptative layer is 200, the same as the bottleneck layer. The hyper-parameters $\alpha, \beta, iterNum$ of STM are set as suggested by [12]. Our proposed model transfer is carried out using the same optimizer in the model training phase, but with the batch size of the total number. The unsupervised model transfer stops at 35 iterations when the error rate reaches its lowest level, the time cost of which is acceptable.

5.3   Evaluation Metric

Error reduction rate (ERR) is adopted to measure the performance of unsupervised model transfer:

$$Error \quad reduction \quad rate = \frac{Error_{initial} - Error_{transferred}}{Error_{initial}} \quad (4)$$

$Error_{initial}$ refers to the error rate of the source recognition model trained with synthetic data solely, while $Error_{transferred}$ denotes to the error rate of the target recognition model after transferring with unlabeled real data.

5.4   Results

Experiments are conducted respectively with real handwritten images from 60 writers in the ICDAR-2013 dataset.

Table 2. Results of initial error rate, error rate after transfer and ERR of 10 representative writers.

Writer no.	Initial Error rate(%)	STM		Proposed unsupervised method	
		Error rate after transfer(%)	ERR(%)	Error rate after transfer(%)	ERR(%)
C009	39.54	39.03	1.29	30.26	23.47
C015	20.3	16.7	17.73	7.3	64.04
C016	24.05	23.87	0.75	18.09	24.78
C019	15.29	13.66	10.66	10.55	31.0
C022	9.71	8.94	7.93	6.23	35.84
C034	55.23	56.07	−1.52	50.89	7.86
C038	14.14	12.68	10.33	8.36	40.88
C044	26.66	24.74	7.2	19.32	27.53
C051	16.81	15.81	5.95	9.35	44.38
C052	11.27	10.45	7.28	6.18	45.16
Average	23.12	21.89	5.33	15.58	32.63

Table 2 shows the result of 10 representative writers respectively as well as the average results of all the 60 writers. The initial error rate computed by the source model as the baseline is presented in the second column, the results of adaptation by STM are shown in the third and forth columns, while the results of the unsupervised model transfer by our proposed method are provided in the fifth and sixth columns. We observe that for all the 10 writers, the error rate is enhanced obviously after our proposed unsupervised transfer method. However, the error rate after STM fails to show significant promotion. Particularly, the error rate of the wrtier C034 after transfer by STM even increases, meaning that STM is not applicable for some writing styles. On the whole, the average error reduction rate over all 60 writers by STM is 5.33%, while our proposed unsupervised method achieves 32.63%, which outperforms STM.

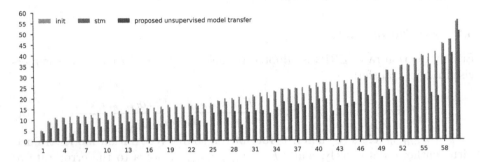

Fig. 3. Error rates (percent) over 60 writers arranged in ascending order. Initial denotes the error rate of the source model, while STM and proposed unsupervised model transfer denote that of the target model by the two method.

The graphical comparisons between the error rate with STM and our proposed unsupervised method over all 60 writers arranged in ascending order of the initial error rates presents a comprehensive results in Fig. 3. The ranges of boost by STM and our proposed unsupervised method are different across the writers have no relation with the initial error rate computed by the source model. It can be seen from the figure that the performance of STM is similar to the initial error rate. Moreover, it performs even worse on four writing styles than the initial error rate of the source model without any transfer, indicating that STM is prone to negative transfer. However, our proposed method promotes positive transfer for all writing styles, and the error reduction rate of each individual writer is significant compared with STM. In conclusion, the proposed unsupervised transfer method is more effective and stable across different writing styles.

6   Conclusion

In this paper, a transfer learning method is proposed to deal with the challenge in application scenarios of HCCR, where labeled data is not easily acquired, synthetic training dataset is used, and the existence of the domain gap between the training data and real handwriting data can not be ignored. The framework incorporates two phases, including model training and unsupervised model transfer. The comparison with STM validates the performance of the proposed unsupervised model transfer. The advantage of our proposed approach over the existing transfer method is shown in error reduction rate.

In the future, it is expected that our method can be hopefully extended as a general transfer learning method and applied to other application scenarios, like domain adaptation in image classification, speaker adaptation in speech recognition, and so on.

References

1. Ahmad, I., Fink, G.A.: Training an Arabic handwriting recognizer without a handwritten training data set. In: 2015 13th International Conference on Document Analysis and Recognition (ICDAR), pp. 476–480. IEEE (2015)
2. Chang, B., Zhang, Q., Pan, S., Meng, L.: Generating handwritten Chinese characters using cyclegan. In: 2018 IEEE Winter Conference on Applications of Computer Vision (WACV), pp. 199–207. IEEE (2018)
3. Chen, L., Wang, S., Fan, W., Sun, J., Naoi, S.: Beyond human recognition: a CNN-based framework for handwritten character recognition. In: 2015 3rd IAPR Asian Conference on Pattern Recognition (ACPR), pp. 695–699. IEEE (2015)
4. Cheng, C., Zhang, X.Y., Shao, X.H., Zhou, X.D.: Handwritten Chinese character recognition by joint classification and similarity ranking. In: 2016 15th International Conference on Frontiers in Handwriting Recognition (ICFHR), pp. 507–511. IEEE (2016)
5. Cireşan, D., Meier, U.: Multi-column deep neural networks for offline handwritten Chinese character classification. In: 2015 International Joint Conference on Neural Networks (IJCNN), pp. 1–6. IEEE (2015)
6. Ganin, Y., Ustinova, E., Ajakan, H., Germain, P., Larochelle, H., Laviolette, F., Marchand, M., Lempitsky, V.: Domain-adversarial training of neural networks. J. Mach. Learn. Res. **17**(1), 2030–2096 (2016)
7. Kang, L., Rusiñol, M., Fornés, A., Riba, P., Villegas, M.: Unsupervised adaptation for synthetic-to-real handwritten word recognition. In: 2020 IEEE Winter Conference on Applications of Computer Vision (WACV), pp. 3491–3500. IEEE (2020)
8. Memon, J., Sami, M., Khan, R.A., Uddin, M.: Handwritten optical character recognition (OCR): a comprehensive systematic literature review (SLR). IEEE Access **8**, 142642–142668 (2020)
9. Sankaranarayanan, S., Balaji, Y., Castillo, C.D., Chellappa, R.: Generate to adapt: aligning domains using generative adversarial networks. In: Proceedings of the IEEE Conference on Computer Vision and Pattern Recognition, pp. 8503–8512 (2018)
10. Tan, C., Sun, F., Kong, T., Zhang, W., Yang, C., Liu, C.: A survey on deep transfer learning. In: Kůrková, V., Manolopoulos, Y., Hammer, B., Iliadis, L., Maglogiannis, I. (eds.) ICANN 2018. LNCS, vol. 11141, pp. 270–279. Springer, Cham (2018). https://doi.org/10.1007/978-3-030-01424-7_27
11. Yang, H.M., Zhang, X.Y., Yin, F., Sun, J., Liu, C.L.: Deep transfer mapping for unsupervised writer adaptation. In: 2018 16th International Conference on Frontiers in Handwriting Recognition (ICFHR), pp. 151–156. IEEE (2018)

12. Zhang, X.Y., Bengio, Y., Liu, C.L.: Online and offline handwritten Chinese character recognition: a comprehensive study and new benchmark. Pattern Recognit. **61**, 348–360 (2017)
13. Zhang, X.Y., Liu, C.L.: Writer adaptation with style transfer mapping. IEEE Trans. Pattern Anal. Mach. Intell. **35**(7), 1773–1787 (2012)
14. Zhang, Y., Nie, S., Liu, W., Xu, X., Zhang, D., Shen, H.T.: Sequence-to-sequence domain adaptation network for robust text image recognition. In: Proceedings of the IEEE Conference on Computer Vision and Pattern Recognition, pp. 2740–2749 (2019)
15. Zhong, Z., Jin, L., Xie, Z.: High performance offline handwritten Chinese character recognition using googlenet and directional feature maps. In: 2015 13th International Conference on Document Analysis and Recognition (ICDAR), pp. 846–850. IEEE (2015)

An Aided Vehicle Diagnosis Method Based on the Fault Description

Xiaofeng Cai, Wenjia Wu, Wen Bo, and Changyou Zhang[(⊠)]

Laboratory of Parallel Software and Computational Science, Institute of Software, Chinese Academy of Sciences, Beijing, People's Republic of China
changyou@iscas.ac.cn

Abstract. To promote the efficiency of fault diagnosis and improve user experience, we proposed an aided vehicle fault diagnosis method based on user's fault description. According to the description of faults phenomenon, we combined both natural language processing technologies and vehicle structure knowledge to diagnose to locate the fault. First, we collected descriptions of failure from users and made a standardized processing of vehicle fault to build the vehicle fault database and summarized a vehicle proper noun dictionary. Second, according to the design of vehicle, we create a vehicle structure tree and combined both distribution statistics and topic semantic which to build a semantic vector model of vehicle fault description. Finally, we use cosine distance to evaluate the semantic similarity of fault description, based on the result, we can conclude the reason of fault and location. The experiment shows that the precision of fault diagnosis has reached 86.7% and the precision of fault locating has also reached 81.8%. The vehicle diagnosis method will help providers of vehicle maintenance to build a self-service fault inquiry system based on the database from internet. It can collect user's maintenance requirements ahead, optimizing the service progress and improving the service efficiency.

Keywords: User description · Vehicle fault · Semantic vector · Structure tree · Aided fault diagnosis

1 Introduction

With the development of relative technologies, the vehicle's internal structure has become more and more complex. This brings lots of challenges to the traditional maintenance mode. The traditional vehicle fault diagnosis mode usually takes a lot of time in the fault determination, which is lack of efficiency. In this paper, we designed an aided vehicle diagnosis method based on the fault description of users. This method will simplify the service process of vehicle diagnosis and improve the service efficiency and user experience. Depend on the aided vehicle diagnosis method, we can transfer the traditional diagnosis process to the network. In this way, it can realize the self-service fault query of users, saving their time, and also collect the user's fault descriptions for service provider in advance. By collecting the past fault descriptions for model

© Springer Nature Singapore Pte Ltd. 2021
Y. Sun et al. (Eds.): ChineseCSCW 2020, CCIS 1330, pp. 275–289, 2021.
https://doi.org/10.1007/978-981-16-2540-4_21

training, the aided diagnosis method can predict the possible cause and location of the fault. It can be helpful for maintainer to narrow the fault range, and have a more targeted diagnosis and maintenance. Aided vehicle fault diagnosis method can be a supplement to the traditional mode, focus on the analysis of user's fault description. At present, the system has accumulated a certain number of user fault descriptions, with high availability and reliability. The fault cases we collected can combined with knowledge map and other advanced technologies to achieve a more intelligent diagnosis mode. The construction of aided vehicle diagnosis system can optimize the process of traditional vehicle maintenance service, also provide an interactive service platform between users and service providers.

2 Related Works

There are mainly three kinds of methods of fault diagnosis which includes artificial experience diagnosis, equipment diagnosis and intelligent diagnosis [1]. Currently, artificial experience and equipment diagnosis are the mainstream methods in this field. Intelligent diagnosis method is often combined with artificial intelligence, big data and other advanced computer technology, which improves many disadvantages of traditional diagnosis methods and has a wide application prospect. In recent years, it appears a lot of research on vehicle fault diagnosis, such as the remote detection system based on wireless communication, vehicle fault diagnosis method based on fuzzy fault tree. The remote detection system based on wireless communication is mainly combined with the IoT technologies to implement an active data push service, which can actively remind the abnormal information during the running of vehicle [2]. The method of vehicle fault diagnosis based on fuzzy fault tree establishes a fuzzy fault tree model, defines the fault diagnosis strategy, and achieve the fault early warning function [3].

The processing of user fault description relies on the related technologies of natural language processing. Word segmentation is a basic technology in NLP, it will directly influence the result of the following process steps. The word segmentation of Chinese includes three categories which are the segmentation based on dictionary, the segmentation based on statistics, and the segmentation based on understanding [4]. In application, statistical segmentation is usually combined with the dictionary, which give full play to their own advantages to achieve a better segmentation result. At present, the mainstream word segmentation tools are as follows: Jieba, THULAC and SnowNLP [5, 6].

The vectorization of text is the process of transforming the result of word segmentation into a series of word vectors which can represent the semantic meaning of text in vector space. Text vectorization can be divided into discrete representation and distributed representation [7, 8]. The BOW model is a classical discrete representation method which regard words as the basic processing unit. The Bow model will establish a dictionary according to the recurrence frequency of word in the corpus. Each nonrepetitive word can be seen as a dimension to represent the text in a vector. Word2vec is a distributed representation method, and its under layer adopts neural network model based on CBOW and Skip-Gram algorithm [9].

TF-IDF is a statistical calculation method, which is often used to evaluate the importance of a word in a text [10]. TF-IDF algorithm relies on the statistical distribution characteristics of words in text set to finish keyword extraction, word importance evaluation and other functions [11]. Latent semantic analysis (LSA) is a topic model algorithm based on semantic information. LSA model connects words and documents with a new topic dimension, which solves the deficiency of traditional space vector model in using semantic information [12, 13].

At present, the vehicle parts numbering rules implemented in China are QC/T 265-2004 standard proposed and issued by China Automobile Industry Association in 2004. After that, each automobile enterprise has also proposed its own parts numbering rules based on this standard. In the construction of vehicle structure tree, this paper mainly refers to the vehicle parts numbering rules of JB-SJGF-0003-2011 standard issued by Fujian Motor Industry Group in 2011. This standard covers the vast majority of vehicle parts, but there still remains differences between the standard and the realities.

3 Aided Vehicle Fault Diagnosis Method

3.1 The Working Process of Vehicle Fault Diagnosis System

The design of vehicle fault diagnosis system in this research can be described as Fig. 1. The working process can be divided into four functional parts which includes the pre-processing of user's description, semantic vectorization model training, optimal case matching, and fault locating based on vehicle structure tree.

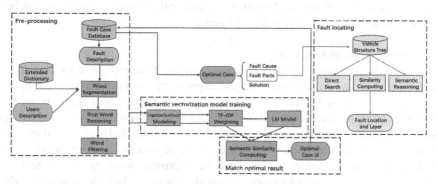

Fig. 1. The design of vehicle fault diagnosis system.

3.2 Pre-processing of User's Description

The Database of Vehicle Fault Cases

The database of vehicle fault cases is the knowledge foundation of the entire diagnosis process. We trained the semantic vectorization model based on the cases in this database. The data resource is mainly from the actual fault cases on the Internet. In this research, we collect the cases and process them into a uniform data format. The vehicle fault cases in this database are formed with the fault description, fault reason, solution and

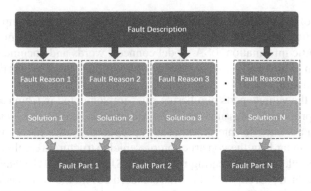

Fig. 2. The relationships of four dimensions in vehicle fault case.

fault location. The relationships of four dimensions in each cases can be summarized as Fig. 2.

For now, the database has included 439 vehicle fault cases and it's still growing. To solve the lack of fault cases data, we design a learning mechanism for this system, it allows the system to collect diagnosis result and transform into the fault cases. The learning ability can help fault diagnosis system to achieve a better performance. When the vehicle fault diagnosis system put into use, it can record the cases after diagnosis. The system will automatically transform the diagnosis result into the standard case format. As the system runs, the learning mechanism will increase the amount of fault case database, and it will be helpful for semantic model training.

Vehicle Proper Nouns Dictionary
Fault description is the colloquial expression of users. It requires us to have a series of standardized process to transform colloquial expression into a standard corpus input. In this research, we choose Jieba as the Chinese word segmentation tool. Since the research covers a lot of professional knowledge in vehicle field, it is necessary to build a proper nouns dictionary to improve the word segmentation effect. Jieba is a combines both statistical method and dictionaries to complete word segmentation. When we import the vehicle proper noun dictionary, it can reduce the incorrectly word segmented frequency.

The components of vehicle proper nouns dictionary include three sources which are vehicle dictionaries form Sogou, vehicle structure tree and vehicle nouns on the Internet. After we construct the vehicle structure tree, we extract the proper nouns to build the dictionary. We summarized all vehicle proper nouns as a uniform format as Jieba requires, and remove the duplicate items. The vehicle proper nouns dictionary contains 7085 items, 4786 items are from Sogou, 1092 items are from vehicle structure tree and 1207 items are from the Internet. The vehicle dictionary has prominently promoted the word segmentation result, and it also brought a significant help to the model training and other following works.

Filter of Word Segmentation Result
Due to the fault description is the user's colloquial expression, the word segmentation result always contains lots of redundant information. The redundant information can be

divided into two different types, one is the stop words in Chinese, another is the unrelated words of semantic analysis.

The stop word in Chinese mainly refers those which appear frequently in the text but have no actual meaning. Typical stop words include the modal particles, adverbs, prepositions, conjunctions and so on. The filter of segmentation result mainly depends on the list we build of stop words. We collect and summarize lots of stop words in Chinese to build a stop word list. For word segmentation results, if the word belongs to the stop word list we can directly remove it. The stop word list we build includes 1599 words, covers vast majority of the Chinese stop words. When we finished the stop words filter, we also have a selectively words filter. In this research, nouns are regarded as a more important element in semantics extraction. In the research, we have a tendentious filter on Non-noun words to improve the accuracy of semantics extraction.

3.3 The Statistical and Semantic Vectorization Model

Vectorization of Word Segmentation Result
Vectorization modeling is a fundamental skill in NLP, which can transform word segmentation results into word vectors. In our research, we use the BOW method to represent the word segmentation result. The vectorization model accept the word segmentation results as input from the pre-processing step. The BOW model collects the occurrence frequency and distribute word id to each word. With the occurrence frequency information, BOW model would remove repetitive words and build a dictionary which forms like [Word id, frequency]. Each word can be seen as a dimension of the vectorization model, the fault description is transformed as a unique vector in the model. A vector set containing all fault cases is used as a corpus input for model training. Vector representation of word segmentation results is the necessary basis for statistical information weighting, semantic extraction and similarity calculation.

Weighting with Statistical Information
TF-IDF algorithm is a statistical information based weight computing method. TF-IDF algorithm always being used to estimate the importance of the word, and extract the key word of a text. TF-IDF algorithm is formed with TF and IDF algorithm. TF algorithm compute the occurrence frequency of a word in the current text. IDF algorithm calculate the amount of descriptions which a certain word has existed in the entire fault description set. When a word only appears in few fault descriptions, it can better distinguish the fault descriptions.

The calculating formula of TF is shown as formula 3.1.

$$\text{tf}_{i,j} = \frac{n_{i,j}}{\sum_k n_{k,j}} \tag{3.1}$$

In formula 3.1, $n_{i,j}$ represents the occurrence frequency of word i in text j, $\sum_k n_{k,j}$ represents the total words number of text j. The meaning of denominator is to normalize the word frequency, and avoid the effect caused by the difference of each fault description.

The calculating formula of IDF is shown as formula 3.2.

$$\text{idf}_i = log \frac{|D|}{1 + |D_i|} \tag{3.2}$$

In the formula 3.2, $|D|$ is the amount of fault description set, $|D_i|$ is the amount of text which contains word i. The target of denominator is using Laplacian smoothing to ensure the algorithm can run well when the $|D_i|$ equals zero. This method can effectively improve the robustness of the algorithm.

The TF-IDF algorithm combines the method of TF and IDF, and the calculating formula is shown as formula 3.3.

$$\text{tf} * \text{idf}(i, j) = \text{tf}_{i,j} * \text{idf}_i = \frac{n_{i,j}}{\sum_k n_{k,j}} * log \frac{|D|}{1 + |D_i|} \tag{3.3}$$

By using this algorithm, we can compute the tf-idf value of every words in the data set. According to the tf-idf value, we can further obtain the word's weight based on the statistical information. According to the result of TF-IDF, we weight the word vectorization model. Besides that, we also considered the effect of word type. With the help of part-of-speech tagging, we increase the weight of nouns to make the key word extraction result more accurate.

Weighting with Semantic Information
In fault data set, the semantic information may hide in the text. In this research, we using LSI topic model to discover the latent semantic information. The structure of LSI topic model can be summarized as Fig. 3.

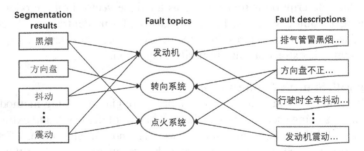

Fig. 3. The structure of LSI topic model.

In the LSI topic model, each fault description is corresponded with certain topics, and the fault topic has its own set of words. Depending on the topic, we can compute the word distribution of each fault description, the calculating formula is shown as formula 3.4.

$$p(w_i|d_j) = \sum_{k=1}^{K} p(w_i|t_k) * p(t_k|d_j) \tag{3.4}$$

Before compute the word distribution of each fault description $p(w_i|d_j)$, we need to get the word distribution of topic $p(w_i|t_k)$ and the topic distribution of fault description $p(t_k|d_j)$.

After training the semantic model, each fault description can be mapped as a vector in a K dimension topic space. Different from the common vectorization space model, LSI topic model can map the fault descriptions to a low-dimensional semantic space. Then, exploring the latent semantic information to have a more essential expression of the description [14, 15].

3.4 Optimal Case Matching Based on the Similarity

After weighting the vectorization model with two algorithms we mentioned above, the model contains both statistical and semantic information. When we get the fault description input from user, it will be mapped as a vector in this vector space. To find the most similar case in fault database, we use the cosine distance as a target to measure the similarity. The angle between two vectors and the cosine value is a positive correlation, which means the smaller the cosine value is, the more similar two vectors are [16, 17].

The dimension of the semantic vectorization model is determined by the amount of items in the dictionary. Each fault description in cases can be mapped as a vector in the semantic vectorization model. The weight of each dimension is trained by both statistical and semantic information. The system will choose the optimal case based on the similarity result. According to the case id, the related information such as fault reason, fault location and solution can be easily found from the database. The optimal case can be helpful for users to estimate the circumstance of the fault. It can also provide the extra information support to the fault locating step.

4 Fault Locating Based on the Vehicle Structure Tree

The innovation of the fault locating method is that we completed a visualized tree representation of the vehicle structure. We construct a vehicle structure tree based on the vehicle divided rules. We also distribute the value weight of each nodes according to their repair values. The fault locating algorithm based on vehicle structure tree, combined with fault description to infer the fault's location and level. For the evaluation standard, we not only consider the accuracy but also the repair value of parts to meet the actual requirements.

4.1 The Structure Tree of Vehicle

The structure tree of vehicle mainly refers to the JB-SJGF-0003-2011 vehicle parts numbering standard. In this standard, the part group number represents the classification of functional system layer which under the root "vehicle". The functional system layer includes all the functions of a vehicle. Next level is assembly layer, which is an integrated structure include few parts, sub-assembly layer or their random combination. The last level is sub-assembly, commonly it is made of two or more parts through assembly or combined processing. The vehicle structure tree has four layer start from the root node.

If we further subdivide the layer, we will encounter the following problems:

(1) The information provided by user's input is not sufficient to support us for further deduction;

(2) The system is not aimed at a certain vehicle brand. Different vehicle has different structure which will cause inelasticity for the system;
(3) We still lack of an effective way to get more detailed parts information.

In the second layer, we numbered functional systems by a two-digit group number. For example, "10, Engine" is an independent functional system, group numbers prefixed with 10 are all belong to this function system. According to vehicle classification standards, there are 86 functional systems in the second layer. The third layer are the assembly layers contains in a function system. Each assembly has a 0 or 9 in the last digit. 0 represent the left part assembly, 9 represent right part assembly. In the fourth layer, it includes sub-assemblies and a few parts, each of them has a 1–8 in the last digit. The odd number represents it belongs to the left parts, even numbers represents it belongs to the right parts. Such numbering rules make the distinction between various functional systems much easier. When we locating the child nodes in the structure tree, we can find related information from upper layers according to the structure division.

4.2 Weight Design Based on Maintenance Value

After the construction of the vehicle structure tree, we design the weight distribution method based on maintenance value. In this article, we select a repair project list of a vehicle repair manufacturer as the reference for the maintenance value. We first correspond the nodes between maintenance project and vehicle structure tree, recording the maintenance price of all parts. For those nodes which has no suitable correspondence in the maintenance list, we use average price of this layer to represent its maintenance value. After matching the maintenance price for each node in the sub-assembly layer, we sum all the maintenance prices of all sub-assemblies as the maintenance price of assembly based on the vehicle structure tree. The process can be expressed as 4.1

$$R_A = \sum_i^n R_s \tag{4.1}$$

where R_A represent the maintenance price of assembly, n is the amount of sub-assembly included in assembly, R_s is the maintenance price of sub-assembly. Then calculate the maintenance price of the functional system in the same way as shown in 4.2.

$$R_F = \sum_i^n R_A \tag{4.2}$$

Where R_F is the maintenance price of the functional system, n is the nodes amounts in the functional system, R_A represent the maintenance price of assembly. In the same layer of the vehicle structure tree, we compute the ratio of each maintenance price according to formula 4.3.

$$C_{ij} = 1 * \frac{R_{ij}}{MaxR_i} \tag{4.3}$$

This calculation method can ensure the price ratio of each node is between 0–1. The nodes of the functional system layer are denoted as $C1$, the nodes of the assembly layer are denoted as $C2$, and the nodes of the sub-assembly layer are denoted as $C3$. After that, count the average number of lower nodes under all functional system nodes and assembly nodes $\overline{M_A}$ and $\overline{M_F}$. The structure tree constructed in this paper is divided into 86 functional system nodes, 207 assembly nodes and 821 sub-assembly nodes, so $M_A = 2.4$, $\overline{M_F} = 4.0$. The division of the vehicle structure tree is such that within the same branch, the maintenance value of the nodes is gathered from bottom to top, the maintenance value of the functional system level is the largest, and the assembly level in the middle the sub-assembly layer is the lowest. The calculation formula of the repair value weight V_i of each node is shown as formula 4.4.

$$V_i = \frac{\omega 1 * c1 + \omega 2 * c2 + \omega 3 * c3}{\omega 1 + \omega 2 + \omega 3} \tag{4.4}$$

In actual calculations, functional system layer nodes only bring into C_1, assembly layer nodes bring into C_1, C_2, and sub-assembly layer nodes bring into C_1, C_2, C_3 for common calculation. The parameter ω in the formula is set as $\omega 1 = \overline{M_A} * \overline{M_F} = 9.6$, $\omega 2 = \overline{M_F} = 4.0$, $\omega 3 = 1$. The setting method of the parameter ω reflects the process that the maintenance value of the nodes in the vehicle structure tree continuously converges upward from the bottom nodes. According to this formula, all nodes in the vehicle structure tree are traversed, and the maintenance value weight V_i of each node is obtained, which not only preserves the relationship of the maintenance price, but also combines reasonably with the hierarchy of the vehicle structure tree.

At this point, the data format of each node of the vehicle structure tree has become a triple representation of [node number-parts name-maintenance value weight], as shown in Fig. 4, which not only contains the hierarchical relationship of the vehicle structure, but also it integrates the value relationship of actual maintenance.

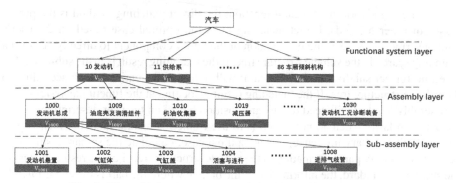

Fig. 4. The vehicle structure tree weighted by maintenance value.

4.3 Fault Locating Algorithm Based on Vehicle Structure Tree

In order to help vehicle repair service providers to more quickly determine the fault location and reduce the range of troubleshooting, it is necessary to determine the location and level of the fault based on the optimal case and the vehicle structure tree. The locating process contains three main parts: direct matching method, similarity computing method and semantic reasoning method. The pseudo code of the locating process is summarized in Fig. 5.

Input: Fault parts from fault diagnosis result: **P**;

Vehicle structure tree: $T = \{(id_1, name_1, value_1), (id_2, name_2, value_2), \ldots, (id_m, name_m, value_m)\}$;

Similarity method: **similarity()**;

Similarity threshold: **C1**;

Semantic method: **semantic()**;

Confidence: **C2**;

Process:

1: **for** t **in** T:
2:      **if** P == t.name:
3:          **if** t.id in thirdlayer:
4:              findResult = t **then break**;
5:          **else: do** semantic(P,t);
6:              **if** senmantic(P,t) > C2:
7:                  findResult = senmantic(P,t) **then break**;
8:              **else:**
9:                  findResult = t **then break**;
10:     **if** P **not in** T:
11:         **do** similarity(P,t);
12:         **if** similarity(P,t) > C1:
13:             findResult = t **then break**;
14:         **else: do** semantic(P,t) ;
15:             **if** senmantic(P,t) > C2:
16:                 findResult = senmantic(P,t) **then break**;

Output: Fault locating result: **findResult**

Fig. 5. The pseudo code of locating algorithm.

Among the algorithms for fault location, the direct matching method is the predecessor of other methods. Direct matching finds the optimal case based on the user's fault description and uses the information of the faulty component to directly perform a matching search in the vehicle structure tree. If the matching result is in the sub-assembly layer, no further subdivision is allowed, it will return to the entire branch according to the vehicle structure tree. If the matching result is located at the functional system level or assembly level, it will need downward position based on the user description.

The similarity calculation method is to use trained semantic model, which is process of fault auxiliary inference, for the optimal case, calculate the semantic similarity between the faulty part information in the vehicle structure tree. For the results that meet the similarity standard, the information of the branch where the part is located is directly taken out and returned the entire branch to the service provider's technicians.

The semantic inferential method is a supplement to the above two methods. In addition to relying on the fault component information in the case, it also refers to the user's fault description to make further semantic reasoning. The user's description of the fault contains a lot of useful information, such as fault location information, phenomenon information, etc. The semantic reasoning method is to classify and extract the useful information and combine the information of the faulty component to jointly judge the location of the fault. The extraction of location information also needs to have a word filter of the fault description. When the fault location information is filtered out, the intermediate location obtained by direct matching or similarity matching is used to infer the lower level parts that meet the location information. A confidence coefficient is necessary to measure the inference result's reliability. If the confidence condition is not met, the representative is not enough to support downward semantic reasoning. In this case, keep the original judgment result and no longer reason down, and leave it to the technician to determine the specific fault location.

4.4 Evaluation Standard for Fault Locating

In order to better serve customers and repair manufacturers, we have considered both the positioning accuracy and the repair value of parts in the design of the locating evaluation standard. The locating accuracy is to examine whether the algorithm can locate the corresponding position and level in the fault tree. The repair value of parts is based on the repair price of each part of the repair manufacturer and weights each part.

The first dimension is locating accuracy. The results of fault positioning can be roughly divided into correct and wrong. The correct positioning results include three situations: Correct locating to the sub-assembly and sub-assembly layer; Correctly locating to the assembly layer; Correctly locating to the functional system layer. The wrong positioning results include: Locating to other functional systems and their branches; Not locating to the position in the structure tree. Sort these 5 situations according to the accuracy of the positioning results and assign scores. The scores for correct positioning are 100, 80, 60, and incorrect locating result is not scored.

The basis of the second dimension is the maintenance value. According to the manufacturer's repair price list, we corresponded to the nodes of the vehicle structure tree and assigned the repair value weight. The repair value weight of each component is affected by the repair price, and the calculation process of the weight has been clarified before. The final comprehensive locating score is determined by the accuracy score and the repair value weight. In the calculation of the final comprehensive score, the accuracy of locating is still prioritized, and the maintenance value is used as an auxiliary reference factor. Therefore, in the calculation of the comprehensive score, the locating accuracy accounts for 70% and the maintenance value accounts for 30%. The calculation of comprehensive score is shown as formula 4.5.

$$P = 0.7 * C + 0.3 * (100R) \tag{4.5}$$

In the calculation formula of the locating comprehensive score, P is the comprehensive score, C is the locating accuracy score, and R is the maintenance value weight corresponding to the positioning accessories.

5   Experiments Verification

5.1   Aided Vehicle Fault Diagnosis Verification

To ensure the comprehensive and representativeness of the data gathered in the verification process, we covered different kinds of faults as much as possible in the verification set, which contains 60 new fault cases. This comparison result by comparing the result of aided diagnosis system with the real fault cause is shown in Table 1.

Table 1. Experiment results of fault diagnosis.

Experiment condition	Correct	Error	Accuracy
Only default dictionary	46	14	76.7%
Loaded customized dictionary	52	8	86.7%

When the default dictionary was used alone for word segmentation, the accuracy of the aided fault diagnosis system was 76.7% in this validation set. After loaded the vehicle dictionary, the accuracy was improved by 10% to 86.7%.

The result verifies that the vehicle dictionary has a significant improvement for vehicle fault diagnosis. And it helps the accuracy of the aided fault diagnosis system reach a high level of availability, which proves the system has the potential of practical application in the future.

5.2 Vehicle Fault Locating Verification

The work flow of our system is that after completing the aided vehicle fault diagnosis, the system will use a fault locating algorithm to estimate the fault locating according to the existing diagnosis results. The experimental result is shown in Table 2. The comprehensive score shown in the table, which calculation formula is defined in the Sect. 3.4, is determined by the locating accuracy and maintenance value weight.

Table 2. Experiment results of fault locating.

Experiment condition	Right locating result			Wrong	Accuracy	Comprehensive score
Default dictionary only	30	7	7	16	73.3%	67.73
Loaded customized dictionary	32	8	9	11	81.7%	73.16

When the default dictionary was used alone, the accuracy of the fault location was 76.7% and the comprehensive score was 67.73 in this validation set. After loaded the vehicle dictionary, the accuracy was improved by 8.4% to 81.7% and the comprehensive score was 73.16.

The experimental result of fault location verifies that the vehicle dictionary effectively improves the accuracy of fault locating. The main reason of the location error is that the information in the bottom "sub-assembly layer" of the structure tree we built in Sect. 4 is less than that of the fault parts in the case information. When there is an error in the locating of the system, the maintainer need to confirm the actual location of the fault manually according to the existing information. In addition to the locating accuracy, the comprehensive score of locating results can also reflect the maintenance value of the fault. In practical application, the system can increase the diagnosis priority of more important faults. The result of the vehicle fault locating experiment shows that the locating method can effectively help the maintainer reduce the examine range and promote the efficiency and quality of the service provider's vehicle maintenance.

6   Reconstruction of Vehicle Maintenance Service Value Chain

In vehicle maintenance market, the multi-value-chain collaboration mainly refers to the interaction process of data and service among multiple value chains which can create additional revenue and value [18–20]. In the traditional vehicle maintenance service process, no matter whether there is mechanical fault or not and whether the fault is serious enough, customers need to go to the store to describe the fault. However, the process of fault confirmation consumes a lot of time of maintainer. The aided vehicle fault diagnosis system in our paper can not only provide more effective information before manual detection, but also help maintainers preliminarily confirm the fault type and cause. And then maintainers can accurately detect the fault location through professional instruments. Besides, the system provides an information platform which can make the existing process of vehicle maintenance service more simple, and also establishes a new collaborative relationship between service chain and parts chain in vehicle market.

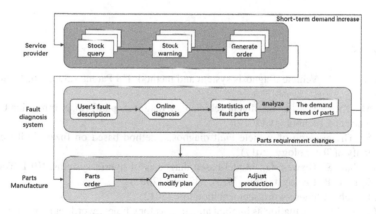

Fig. 6. The ideal collaborative process.

The aided vehicle fault diagnosis system in our paper can help to reconstruct the value chain of vehicle maintenance service. As shown in Fig. 6, an ideal collaborative process is that when the aided fault diagnosis process is finished, the system will record the information of the corresponding fault case and the fault parts, etc. After that, the system will summarize the consumption of fault parts for a certain period of time and connect to the inventory database of the service provider to view the real-time parts inventory. When the demand for a certain type of accessories is on the rise, the system will go to check the inventory status in time and give an early warning on the possible shortage of inventory. According to the real-time consumption of fault parts, the system can realize automatic inventory management for service providers, so that it can timely apply for purchase of accessories with large demand in the near future. From the perspective of the vehicle-parts-supply-chain, the system can also help parts suppliers adjust production demand timely by feeding back the consumption of parts, so as to reduce the waste caused by the production of spare parts.

7 Conclusion

At present, there are still some drawbacks in this aided vehicle fault diagnosis system. First of all, the amount of vehicle fault case in the database is still not enough, which affects the final effect of the model. Secondly, the representation of semantic vectorization model still needs to be studied. In the follow-up work, we will expand the scale of the vehicle fault case database and optimize the structure and representation of the semantic vector model so that the system can be really applied to the actual fault maintenance business.

Acknowledgement. The authors would like to express their thanks to the editors and experts who participated in the review of the paper for their valuable suggestions and comments. This research was supported by National Key Research and Development Project--Multi core value chain collaborative business technology resources and service integration technology (2017YFB1400902).

References

1. Shumate, D.M.: Vehicle diagnosis system and method. US Patent 7,590,476, 15 September 2009
2. Xue, M.: The research on intelligent vehicle fault diagnosis platform. North China University of Science and Technology (2019)
3. Xia, S.: The research of vehicle fault diagnosis method based on fuzzy fault tree. Hubei University of Technology (2017)
4. Narasimhan, S., Biswas, G.: Model-based diagnosis of hybrid systems. IEEE Trans. Syst. Man Cybern. Part A: Syst. Hum. **37**(3), 348–361 (2007)
5. Sun, J.: Jieba' Chinese word segmentation tool (2012)
6. Li, Z., Sun, M.: Punctuation as implicit annotations for Chinese word segmentation. Comput. Linguist. **35**(4), 505–512 (2009)
7. Mikolov, T., Chen, K., Corrado, G., Dean, J.: Efficient estimation of word representations in vector space. arXiv preprint arXiv:1301.3781 (2013)
8. Maas, A.L., Daly, R.E., Pham, P.T., Huang, D., Ng, A.Y., Potts, C.: Learning word vectors for sentiment analysis. In: Proceedings of the 49th Annual Meeting of the Association for Computational Linguistics: Human Language Technologies, vol. 1, pp. 142–150. Association for Computational Linguistics (2011)
9. Erk, K.: Vector space models of word meaning and phrase meaning: a survey. Lang. Linguist. Compass **6**(10), 635–653 (2012)

10. Ramos, J., et al.: Using TF-IDF to determine word relevance in document queries. In: Proceedings of the First Instructional Conference on Machine Learning, Piscataway, New Jersey, vol. 242, pp. 133–142 (2003)
11. Zhang, W., Yoshida, T., Tang, X.: A comparative study of TF* IDF, lSI and multiwords for text classification. Expert Syst. Appl. **38**(3), 2758–2765 (2011)
12. Dumais, S.T., et al.: Latent semantic indexing (lSI) and trec-2, p. 105. Nist Special Publication Sp (1994)
13. Hofmann, T.: Probabilistic latent semantic indexing. ACM SIGIR Forum **51**(2), 211–218 (2017)
14. Landauer, T.K., Foltz, P.W., Laham, D.: An introduction to latent semantic analysis. Discourse Process. **25**(2–3), 259–284 (1998)
15. Hofmann, T.: Unsupervised learning by probabilistic latent semantic analysis. Mach. Learn. **42**(1–2), 177–196 (2001)
16. Tata, S., Patel, J.M.: Estimating the selectivity of TF-IDF based cosine similarity predicates. ACM Sigmod Rec. **36**(2), 7–12 (2007)
17. Dehak, N., Dehak, R., Glass, J.R., Reynolds, D.A., Kenny, P.: Cosine similarity scoring without score normalization techniques. In: Odyssey, p. 15 (2010)
18. Roy, R., Whelan, R.: Successful recycling through value-chain collaboration. Long Range Plan. **25**(4), 62–71 (1992)
19. Normann, R., Ramirez, R.: From value chain to value constellation: designing interactive strategy. Harvar. Bus. Rev. **71**(4), 65–77 (1993)
20. Dania, W.A.P., Xing, K., Amer, Y.: Collaboration behavioural factors for sustainable agri-food supply chains: a systematic review. J. Clean. Prod. **186**, 851–864 (2018)

Blockchain Medical Asset Model Supporting Analysis of Transfer Path Across Medical Institutions

Qingqing Yin[ID], Lanju Kong[(✉)][ID], Xinping Min[ID], and Siqi Feng[ID]

Shandong University, Jinan 25000, Shandong, China
klj@dareway.com.cn

Abstract. The complete transfer path pays attention to the synergy among medical institutions, the result of treatment and the temporal sequence. However, the patient's visiting behavior usually spans many medical institutions. It is not only difficult for inter-agency medical records, examination results and treatment process to be transmitted comprehensively and efficiently, but also difficult to trace the origin of the complete transfer path. This paper proposes a blockchain medical asset model to support the analysis of transfer paths across medical institutions. Firstly, this method establishes a sharing mechanism based on blockchain across medical institutions, and proposes a mapping algorithm between visiting data and blockchain assets. To solve the problem of lack of traceability and reduce the cost of using medical assets, the blockchain is used to transfer the status and inspect structure of diagnosis and treatment process among institutions. Then, aiming at the problem of lack of referral for patient transfer paths, a blockchain based full-chain transfer path analysis method is designed to find the optimal transfer paths for local medical institutions and overcome the bottleneck of the lack of referral for medical institutions transfer paths. Experiments show that the blockchain medical asset model proposed in this paper can cover the whole chain data of transfer, and can meet the needs of cross-medical institutions tracing the complete transfer path. The prediction algorithm used in this model has better performance than other prediction algorithms in mining the optimal path.

Keywords: Blockchain · Medical assets · Transfer path · Integrity

1 Introduction

With the development of social medical service system, the imbalance of resources and information in medical institutions is widespread. When medical institutions cannot afford the patients' condition, it is an unavoidable link under the current system to transfer medical treatment. At present, there are two main sources for the choice of patient transfer paths. On the one hand, doctors in medical institutions give advice to patients according to their professional abilities and clinical experience. Under the weak trust relationship between patients and doctors, blind belief in doctors makes the recommendation limited and risky. On the other hand, patients refer to the transfer path of similar

© Springer Nature Singapore Pte Ltd. 2021
Y. Sun et al. (Eds.): ChineseCSCW 2020, CCIS 1330, pp. 290–306, 2021.
https://doi.org/10.1007/978-981-16-2540-4_22

patients with the same disease. Taking heart disease patients as an example, patients can refer to the provincial or national transfer medical institutions for heart disease patients, and select the local optimal treatment transfer path according to their own background (geographical location, family affordability). However, under the current medical background, it is difficult for patients' medical records and transfer data to be shared and interacted safely among different institutions, and there is a serious lack of transfer paths. Both patients and doctors lack sufficient data to support their recommendation or selection of local optimal results, which is of poor reference and increases the risk of transfer. Therefore, we need to collaborate and share among various medical institutions, integrate patient data, improve the patient transfer paths, solve the problem of path missing, to provide patients with complete spatial data support. So, data integration is the key to solve the problem of path missing.

With the wide application of digital medical information system, there are many research programs for medical data integration. In developed countries, such as the United States, the Netherlands, the United Kingdom, large-scale medical institution integration projects research have been carried out. For example, Integrating Health Care the Enterprise (IHE) [1] is a common data integration framework constructed by American IT departments, clinical practitioners and consultants to provide technical specifications for medical data integration. In [2], a medical integration information model developed by Stanford University is introduced to help integrate and manage heterogeneous clinical data and transformed medical data. In [3], personalized biomedical data integration method is proposed by using domain ontology integrated entity connection model. The related research work in China started late. The existing research results, such as Zhou et al. [4], put forward the TCM clinical reference information model for large-scale structured electronic medical record data, constructed clinical data warehouse, realized the task of large-scale clinical data integration and pretreatment. In addition, Zhejiang University has also proposed a standardized, open and extensible platform for medical data integration, which gives access right of information to real data demanders to ensure the construction of medical big data.

We require data integration to describe facts safely, reliably, consistently and accurately. However, there are many problems in current research on medical data integration, such as high fragmentation of data information, low standardization, and disconnection from clinical behavior. On the one hand, due to the dispersibility and privacy of patient's medical data, it is difficult to achieve collaborative services and universal sharing of data across medical institutions, and different medical institutions have different censorship systems and designated standards. On the other hand, data tampering, malicious access, human error and hardware failure are all threats to data security. However, the birth of blockchain technology has changed the storage structure of the medical data set. Through the de-centralization of blockchain, it ensures that the medical information cannot be tampered with, and solves the problems of opaque medical or transfer records and untrustworthy medical data sources in the actual process of diagnosis and treatment. In addition, due to the traceability of blockchain, access control method can be used to realize the safe sharing and integration of diagnostic and therapeutic data across medical institutions in the chain.

Therefore, we propose a blockchain medical asset model to support the analysis of hospital transfer path, which maps the patient's visiting data to the block-chain assets, realizes the collaborative sharing of trans-medical institutions. Based on the blockchain transfer path analysis method, we find the optimal transfer path and solve the problem of path missing under the current medical background. The following is the elaboration of each chapter: the second part of this paper introduces the related work; The third part introduces the relevant definition and introduction of the model; The fourth part introduces the cross-medical collaboration mechanism based on blockchain and its path traceability; The fifth part introduces the transfer path analysis method based on the whole chain data of the fourth part to recommend the optimal transfer path for patients; The sixth part is the experiment and result analysis of the prediction algorithm in this paper; The seventh part is the summary of the characteristics of the model and the follow-up work.

2 Related Work

Blockchain research in the medical field has been receiving attention. The application of blockchain in the medical field mainly includes integrated health care system, medical decentralization, electronic healthcare record, traceability management of medical device and hospital information integration platforms. The most current research is the use of blockchain to store electronic medical records [5, 6]. Due to the traceability and temper-proof of blockchains, the blockchain creates a secure environment for storing and analyzing EHR. Currently, the problem faced by patients is that the inability to share data sets between medical institutions leads to a waste of medical resources. Sharing patient electronic medical data can promote more effective care. Storing medical data in the blockchain not only provides a secure storage environment, but also enables patient medical data sharing well based on blockchain technology. For example, in reference [7], in order to ensure the integrity of electronic health data and support fine-grained access control, a medical data sharing architecture was proposed, it use the smart contract to solve data access control problems. Based on this research, Zonyin Shae [8] proposed a method for building clinical trials and precision medical applications on blockchain platforms. Healthcare Data Gateway [9] was a successful data sharing architecture that uses blockchain technology to make healthcare systems smarter. All of the above studies have facilitated the effective sharing of medical data. However, in the face of complex unstructured medical data, the large capacity and small granularity of medical diagnostic data pose new challenges for blockchain storage. There is still room for improvement in storage performance.

Encryption algorithms can be used in the blockchain to store and verify content to ensure the security and privacy of medical data. Reference [10] proposed a digital signature based on blockchain-based encryption algorithm, which provides a good help for patient data access and privacy protection. Based on the intangible modification and traceability of blockchains, we can use blockchains to implement traceable data sources. Reference [11] established a drug traceability model based on Ethereum, but no specific design details. Based on previous research, the Ref. [12] established a drug traceability blockchain system to track the source and destination of drugs in the drug supply chain. There is also a combination of blockchain and medical internet of things (IoMT). Reference [13] proposed a blockchain-based authorization framework to manage IoMT devices and medical files by creating a distributed regulatory and health data privacy solution chain. Kristen N. Griggs [14] proposed a healthcare blockchain system that uses smart contracts to safely and automatically remotely monitor patients, bringing the blockchain into the healthcare industry. The above two models can only establish trust relationships for small-scale personnel and medical devices to achieve interaction and fine-grained access, but the scalability of cross-region data distribution remains to be explored.

In summary, the use of medical and blockchain is still in its infancy, and domestic research in the medical field is still in the exploratory stage. Based on the above research, we have innovatively proposed a blockchain medical asset model supporting analysis of transfer path across medical institution.

3   Blockchain Medical Asset Model Supporting Analysis of Transfer Path Across Medical Institutions

3.1   Related Definitions

Medical Patient (MDP). Medical patient is used to describe the applicant for the medical services and the owner of medical diagnostic records in the medical treatment process.

MDP = <MDP_ID, MDP_prk, MDP_pk> where MDP_ID is the unique identification on the blockchain; <MDP_prk, MDP_pk> are the public key and private key of MDP to identify MDP, indicating the ownership of records of records of diagnosis and treatment by signing with the private key. On the blockchain, MDP accredits the jurisdiction of operating medical records to the hospital through transactions which are called accredit transactions and then the hospitals operate the medical records through accredit transactions so as to read the medical records.

Medical Institution (MDI). Medical institutions are institutions that produce and process information. Each medical institution possesses its own workflow engine to acquire diagnostic records from blockchain, completing the specific diagnostic process and upload the diagnoses to the blockchain in the form of transactions.

MDI = <MDI_ID, MDI_pk, MDI_prk>, where MDI is the unique identification on the blockchain; <MDI_pk, MDI_prk> represents the public key and private key of a hospital institution used to identify itself in a medical diagnosis.

Medical History (MDH). MDN is the original records of patients during the whole process of medical diagnosis and treatment.

MDH = <MDH_ID, MDI_ID, MDH_Content, MDH_Status, LTxHash>, where MDH_ID is the unique identification of medical records on the Blockchain; MDH_Content represents the specific content of medical records; MDH_Status represents a change in status of a medical treatment, such as authorized, used, etc. Departments of the hospital trigger changes in diagnostic records status through transactions in which hospital institutions set the originator, receiver, medical records and corresponding operation of the transaction. LTxHash indicates the latest transaction of the assets, which can realize the quick retrieval of assets.

Transaction (Tx). Tx refers to the agreement between the two parties on the medical assets. Blockchain is mainly made up of individual transactions which are the real data stored on the Blockchain. The blocks are the records that when and in what order certain transactions become part of the Blockchain. The medical assets trading model is:

Tx = <TxID, TxType, From, To, MDP_ID, MDH, PreHash, MDHStatus, Operation, Input, Output> where TxID is the unique identification of a transaction, that is, the address of a transaction; From indicates the supply-side of medical assets, that is, the address of the originator of medical assets; To indicates the demand-side, that is, the address of the receiver medical assets; MDH refers to the medical records of diagnosis and treatment of patients; MDHStatus: represents the status change of medical before and after transaction Tx. Operation represents the operations on assets; PreHash records the account address of the latest transaction of the medical assets; Input, output indicates the incidence relation between assets.

Where input = <ObjectID, MDHLatestTxID> which indicates the other assets that the asset MDH needs to associate with. ObjectID indicates the ID of the asset being associated. MDHLatestTxID indicates the latest transaction address of associated assets.

Output = <ObjectID, MDH_ID, Operation, MDHStatus>, indicates the operation of the transaction on assets. MDHStatus refers to current status of associated assets. The medical records MDH of patients in different medical institutions is correlated to realize the sharing across medical institutions.

3.2 Model Introduction

For more secure and complete preservation of this data, a growing number of patients are willing to store it on the blockchain, considering the decentration and traceable characteristics. Storing the medical records on the Blockchain realizes information sharing among organizations, promoting the efficiency of medical treatment and avoid extra consumption and other waste of medical.

Resources caused by repeated inspection across medical institutions. In this model, during the diagnosis and treatment of patients across institutions, patients or hospitals store the records of each hospital transfer on the blockchain, realizing the preservation and traceability of medical records.

This model firstly realizes cross-medical institution collaboration sharing based on blockchain, mapping medical records to blockchain assets, creating blockchain accounts for each medical asset. The incidence relation between medical assets realizes the information sharing across medical institutions, which not only provides the patients with complete transfer path, but also reduces the use of medical resources and enhances the efficiency of diagnosis and treatment. According to the complete transfer path of patients with the same disease provided by blockchain, a transfer route analysis method is proposed. It effectively integrates the blockchain data with the third-party data, completing the transfer path data set, recommending the optimal transfer path to patients based on the prediction algorithm and solves the bottleneck of the lack of reference of transfer.

As shown in Fig. 1, the blockchain provides a new data sharing and storage mechanism. This model transfers patients' medical records MDH into digital assets. On the blockchain, patients MDP and medical institutions MDI are principal accounts and each principal account is able to possess multiple asset addresses. Each medical institution is required to deploy blockchain nodes so as to jointly form a blockchain network. The blockchain node of each institution keeps a complete distributed ledger, that is, all medical records and their use records. Patients or medical institutions operate medical records in the form of transactions. The consensus algorithm is used to verify whether the transaction is correct and to ensure that the medical institutions have consistent data.

Medical institutions are responsible for diagnosing and treating patients and generating their medical records, which are then uploaded to the blockchain for storage after being validated by the consensus algorithm encapsulated in the smart contract. For each patient, the system authorization center will generate the key pair of the patient at random automatically and all medical records about the patient will be encrypted by the public key of the patient, digitally signed by the hospital and then broadcasted to the blockchain so as to protect patients' privacy. Access and control of patients' medical records is realized through blockchain technology. Patients can delegate data to medical institutions and revoke access at any time. After a patient initiates an authorized transaction with a medical institution, the institution obtains the patient's private key, acquiring the patient's medical treatment data in all medical institutions. At the same time, all transaction records will be stored on the blockchain, making the medical treatment process more transparent and more convenient for users or the country to supervise.

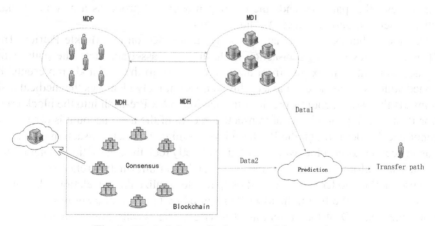

Fig. 1. Blockchain medical asset model supporting.

4 Sharing Mechanism Based on Blockchain Across Medical Institutions

4.1 Model Design

Model Principle. The application of blockchain technology in digital assets can reduce the value of digital assets and the cost of trust building, and also guarantee the safety of digital assets. Therefore, this paper converts the medical records of patients into digital assets and traces the complete transfer records of patients, which in essence is to explore the correlation relationship between assets. Existing bitcoin and Ethereum [17, 18] have their own transaction models. Bitcoin adopts UTXO (Unspent Transaction Output) transaction mode, in which each user can have multiple UTXO transaction addresses. Since it is difficult to link the users' UTXO transactions, it is not convenient to conduct digital asset management. Nowadays, the forms of data assets are diverse, and not all of them are presented in the form of coins, obviously UTXO transaction mode is not applicable. Ethereum uses the Account model. Although this model can refer to the data associated with the account, it mainly provides a quick retrieval of the account balance and a consistency check of the final transaction. With the diversification of medical assets and the diversity of medical asset states, transactions are not limited to the final consistency of balance. Medical asset transactions with diverse contents and forms have higher requirements on blockchain accounts from consistency of state to consistency of state, content and timing. In addition, we need to traverse the block and parse the transaction data in order to trace the patient's complete transfer information. Therefore, problems such as low efficiency and high index space overhead will occur. To meet the complex requirements of data storage, query and privacy protection in the medical field, this paper extends the existing model and proposes a new blockchain medical asset transaction model. As shown in Fig. 2.

This model builds the asset account status tree based on the Merkle Patricia Tree [19], which stores the last transaction LT-hash of the asset account. The path of the asset account status index tree from the root directory to the leaf node represents the encoded address of the medical asset. This address uniquely identifies the medical asset and inserts the asset account's previous transaction index Pre-hash into the block transaction data. A complete historical transaction record of the asset account is obtained in chronological order through the Pre-hash link, forming an asset transaction chain with the asset account as the clue, which is used to directly lock the transaction content information, status information, transaction history and other information related to the asset. In the transaction model, the Input and Output fields realize the correlation relationship between assets. In the Input, the asset ID required by the current asset is saved; in the Output, the asset ID of the current asset is saved. The synchronous transfer of medical records among institutions is realized through the association relationship between assets to realize the essence of sharing across medical institutions.

As shown in Fig. 3, TX01, TX02 are two consecutive transactions of digital asset MDH01 associated with the transaction chain of MDH01, which is provided to the patient in the medical procedure record of the medical facility. MDH01 is associated with MDH02 through the Input attribute in the transaction model, even for medical records belonging to two medical institutions. Medical institutions with MDH02 digital

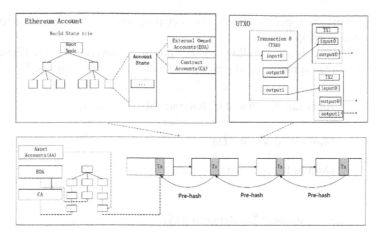

Fig. 2. Medical assets transaction structure.

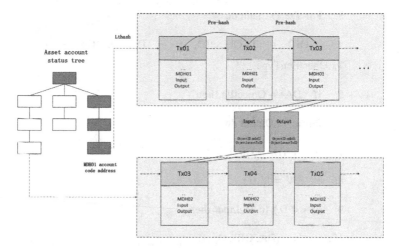

Fig. 3. Asset correlation diagram.

assets can operate digital asset MDH01. This reduces the cost of using medical assets, thereby enabling a sharing model across medical institutions.

4.2 Time Sequence Diagram of Transfer Relations

The traceability of a complete time sequence diagram of transfer relations is basically obtained by acquiring the relationship between digital assets. The time sequence diagram of transfer is defined as G = (V, L, T, O), where V represents a set of digital assets, L represents a set of edges in the graph, O represents a set of operational types in the system, and T represents a set of time information. Time sequence diagram of transfer relations is essentially a directed network, it records the associated assets, the types of asset relationships, asset operations and associated time. Therefore, we can get a complete digital asset relationship in space.

The algorithm is shown as follows: The input to the algorithm is the digital asset address, or DA. The output of the algorithm is the directed edge set of the DA.

Algorithm 1.   Procedure **Relation_ Retrieval** (*DA*)

Input: DAO account: *DA*

Output: R'={(a₁,a₂,op,t) | a₁, a₂ ∈ ExistedMDH, a₁=DA or a₂=DA,op ∈ O}

1:   TxChain = Tx_ Retrieval (**DA**)

2:   R′ = {}

3:   **For** Tx: TxChain **do**

4:      OID = Tx.ObjectID, t = Tx.time, Op = Tx.OperationType

5:   **If**   DA ═ ObjectID **then**

6:   **For** a: Tx.Output **do**

7:       r=(DA,a,Op,t)

8:       R′ = R′ ∪ r

9:   **Else if** DA **in** Tx.Input **then**

10:      r = (DA, OID, Op, t)

11:      R′ = R′ ∪ r

12:   **Else if** DA **in** Tx.output **then**

13:      r = (OID, DA, Op, t)

14:      R′ = R′ ∪ r

15:   **Return** R′

The specific steps for generating the time sequence diagram of transfer relations are as follows:

Step 1: Initialize the directed edge set R' to null and then extract the transaction Tx from the transaction chain.

Step 2: Parse out the specific content of Tx, OID represents the digital asset address of transaction Tx, Op represents the type of operation performed by the asset, and t represents transaction time. The location of the DA has the following three conditions: If the DA appears in the input of the transaction, go to step 3. If the DA appears in the output of the transaction, go to step 4. If the DA is a transaction OID, go to step 5.

Step 3: If the DA appears in the input of the transaction, it indicates that the asset DA is associated with the asset OID, and then constructs the corresponding relationship edge (DA, OID, Op, t).

Step 4: If the DA appears in the OID input, it means that the asset OID is associated with the asset DA and then constructs the corresponding relationship edge (DA, OID, Op, t).

Step 5: If the DA is a transaction OID, construct a corresponding relationship edge (DA, a, Op, t) for each asset a in the output.

Step 6: Add the relationship edge constructed in steps 3, 4, 5 to the set R', and obtain the next transaction from the transaction chain, and perform step 2.

Step 7: After traversing the transaction in the transaction chain, R' already contains the relationship edge set of all Das.

Step 8: Based on the time series, you can record asset relationships, asset status, and asset operations. According to the directed edge set R', the complete asset relationship timing diagram in the asset life cycle can be restored, that is, time sequence diagram of transfer relations (Fig. 5).

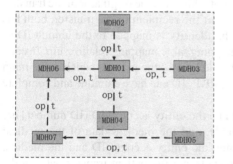

Fig. 4. Time sequence diagram of transfer relation.

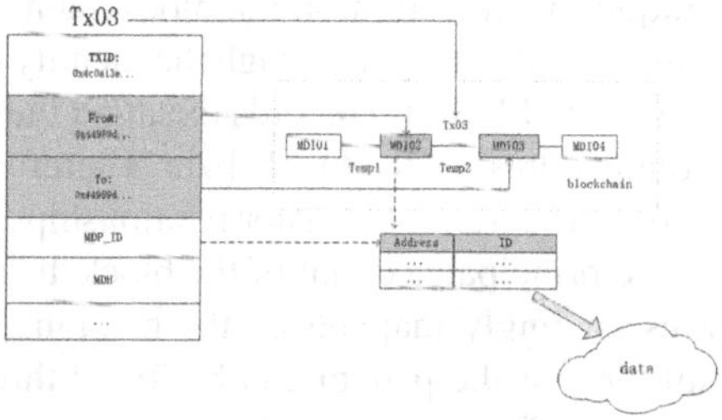

Fig. 5. Data analysis method.

Figure 4 depicts the complete transfer log of the patient. Each link represents a transaction. The time on the connection line indicates when the transaction occurred and records the type of operation between the two assets. Depending on the relationship between the assets, the patient can be provided with a spatially complete transfer record of the hospital.

5 Analysis of Transfer Routes Across Medical Institution

To provide the best transfer path for patients, we propose a transfer route analysis method based on sharing mechanism based on blockchain across medical institutions. This paper innovates on the data set to break the phenomenon of dispersion and path missing of traditional centralized data, the complete hospital transfer path provided on the blockchain is integrated with the third-party data to form a complete hospital transfer data set, based on the LSTM [20] (Long Term Memory) prediction algorithm to train the data set and then more accurate and reliable recommendation programs can be provided for patients according to their medical treatment results in current medical institutions.

5.1 Data Analysis Method

In order to predict the optimal transfer path, accurate and complete data sets are the basis of improving the reliability of the prediction model. At present, most available medical data exist in hospitals or third-party data centers (such as commercial insurance companies or data companies). The dispersion of data leads to bottlenecks in the information integration of hospital transfer path, which affects the accuracy of prediction algorithm to some extent. Therefore, this paper innovatively integrates third-party data with information on the chain based on sharing mechanism based on blockchain across medical institutions provided which solved the phenomenon of missing path of traditional data integration and improved the efficiency of integration.

Patients can store each discharge records or operation by medical institutions on the blockchain, a complete time sequence diagram of transfer relations can be obtained based on sharing mechanism based on blockchain across medical institutions. As shown in the figure, each line between the two associated medical assets records the transaction of the assets. Tx03 is the transaction of the medical asset MDP02 and the medical asset MDP3, which is the transaction of the transfer. According to the data structure of the transaction, from records the blockchain account address of the originator of the transfer, and to records the blockchain account address of the recipient of the transfer, both has a unique identification on the blockchain. The identity is mapped to the unique ID of the medical institution through the identity mapping table mapping relationship. In each transaction, MDP_ID records the id on the blockchain of the patient, which is the main account address of the blockchain. Therefore, MDP_ID can have a unique and complete sequence diagram of transfer relationships.

The principal account of the blockchain and the entity account ID (ID number) are correspondingly mapped by the mapping table, and the medical data of the medical institution of the patient can be found through the entity account ID and the medical institution ID, or through the entity account ID to get the patient transfer path on the blockchain. This mapping table is stored on the blockchain by the Smart Contract and the unique blockchain address can be obtained by logging in the medical institution system according to the entity ID, and this mapping table can be viewed by authorization of the medical institution. According to the complete transfer path provided by blockchain, the third-party company integrates the complete path provided by blockchain with the local data of medical institutions, so as to form a local data set with accurate path and complete data.

In addition to providing the address of the medical institution transferred to the hospital, the blockchain can also provide the time for medical treatment in this medical institution. Each transaction of the blockchain will record a timestamp(temp), the time of hospitalization of the medical institution can be estimated according to the timestamp difference of each transaction in the time sequence diagram of the transfer route, (for example, the time of hospitalization in MDP02 is temp2-temp1). This time is only for reference. Because the discharge records are in the form of transaction on the chain and it takes a certain amount of time to package the transaction to generate blocks, and there may be a certain time interval between the patient's information and discharge time (the patient did not immediately put the discharge records on the chain, he did it when returned to the hospital with the doctor's order). Therefore, the length of hospital stays provided on the blockchain is only for reference, it can be taken as a feature if and only if the third party provides incomplete (lost length of hospital stays) data.

5.2 Transfer Path Prediction

Structure and Frame. Local information: Local information is the discharge forms information provided by the hospital, including: type of disease, medication, length of hospital stays, amount spent, this information provided by the hospital when the patient is discharged from the medical institution. Throughout the medical treatment process, the patient's medical records in different hospitals can be expressed as $\{x_1^{loacl}, x_2^{loacl}, \cdots x_t^{loacl}, \cdots\}$. Here, we count the patient's history in the local hospital and compress it into a vector x_t^{locol}, x_t^{locol} represents the statistical summary of the patient's historical medical records provided by the patient's hospital at time t.

Blockchain information: Blockchain can provide a complete hospital transfer path for patients, which represents a medical institution that spans the patient's diagnosis and treatment process, in addition to this, the blockchain can also roughly calculate the patient's hospital stay. We use x_t^{chain} to represent the information provided by the blockchain at time t. $\{x_1^{chain}, x_2^{chain}, \cdots x_t^{chain}, \cdots\}$ indicates the information that the blockchain system can provide at different times.

Transfer information: The transfer information records the patient's transfer to the medical institution at time t, we use the one-hot vector y_t to represent this information. When $y_{j,t} = 1$, it means that the patient transfers the hospital j at time t.

We use the $f(\cdot, \cdot, \cdot)$ function to predict the medical institution that the patient is about to transfer to, and then predict the optimal transfer path based on the current diagnostic record of the patient: $y_t = f(x^{local}, x^{chain}, h_{t-1})$. Where h_{t-1} represents the historical information left at the previous moment, that is, the external state of the previous moment, which is what we call short-term memory in the LSTM.

Prediction Model. RNN [21] (Recurrent Neural Network, RNN) is a kind of neural network with short-term memory ability. In RNN, neurons can not only receive information from other neurons, but also accept their own information to form a network structure with loops. When the input sequence is relatively long, there will be gradient explosion and disappearance problem, which is well solved by LSTM. Therefore, this paper predicts the transfer path based on LSTM algorithm. We use time series $\{x_1^{loacl}, x_2^{loacl}, \cdots x_t^{loacl}, \cdots\}$ and $\{x_1^{chain}, x_2^{chain}, \cdots x_t^{chain}, \cdots\}$ as input. At time t, by input the information x_t^{chain} obtained through the blockchain, the information x_t^{loacl} provided by the local hospital and the hidden state h_{t-1} update the state h_t, the calculation formula of each internal gate is shown below:

$$f_t = \sigma\left(W_f\left[x_t^{locol}, x_t^{chain}, h_{t-1}\right] + b_f\right)$$
$$i_t = \sigma\left(W_i\left[x_t^{locol}, x_t^{chain}, h_{t-1}\right] + b_i\right)$$
$$o_t = \sigma\left(W_o\left[x_t^{locol}, x_t^{chain}, h_{t-1}\right] + b_o\right)$$
$$\tilde{C}_t = tanh(W_c\left[x_t^{locol}, x_t^{chain}, h_{t-1}\right] + b_c)$$
$$C_t = f_t * C_{t-1} + i_t * \tilde{C}_t$$
$$h_t = o_t * tanh(C_t)$$

The calculation process is as follows:

Firstly, by using external state h_{t-1} at the previous moment and the current input x_t^{loacl}, x_t^{chain} to calculate three gate: input gate i_t, output gate o_t, forgetting gate f_t and candidate state $\tilde{c}_t$; then update the memory unit c_t by combining the forgetting gate f_t and the input gate i_t; Finally Combined with the output gate o_t, the information of the internal state is transmitted to the external state h_t.

The structure of LSTM is shown in the Fig. 6, which shows the cyclic unit structure of the network. Its prediction is to add the new predicted value into the input, and then forget some information in some samples in turn to form a new sample for the next prediction, that is, to realize the prediction through continuous iteration. According to the basic information of patients in the current medical institutions, the model can be used to predict the optimal medical institutions and solve the problem of the lack of reference of the inter-institutional transfer route.

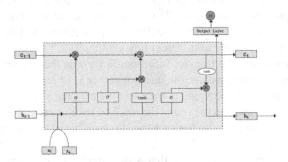

Fig. 6. Timing diagram of transfer relationship.

6   Experiment

6.1   Experimental Data Sources and Analysis

First, we trace the patient's complete transfer path based on the patient's records order on the blockchain. Then look for local data sets based on the unique identifiers of the patient and the medical institution on the blockchain and integrate based on the complete transfer path on the blockchain to form a complete and accurate data set. A total of 260 transfer path information for patients with heart disease were collected in this experiment, the data is divided into two parts: the training set and the test set in a ratio of 5:1. The data needs to be normalized before the neural network, in order to avoid the unit selection dependence of the data, and is more conducive to model training. Commonly used normalization processes have min-max normalization and z-score normalization. Because of z-score normalization is less affected by noise in the sample, so this article uses z-score standardization, also called standard scores, this is the process of dividing the difference between a number and an average by the standard deviation (Table 1).

Table 1. Optimal learning rate for training different neurons.

	100 cells	200 cells	300 cells	400 cells
0.01	13.34%	24.6%	25.1%	21.1%
0.001	15.23%	44.8%	50.01%	49.61%
0.0001	11.72%	15.62%	48.37%	50.21%

6.2 Experimental Performance Evaluation and Comparison of Results

To test the prediction results, this paper introduces the macro F1 for evaluation, which is the average value of each category F1. Each of F1 is calculated as

$$F1 = \frac{2 \times P \times R}{P + R}$$

where P is the precision of each category and R is the recall rate for each category. The macro F1 can be sensitive to reaction data and the prediction can be better evaluated.

$$Macro\_F = \frac{1}{n} \sum_{i=1}^{n} F_i$$

In addition, we hope to reduce the difference between the prediction and the actual data, which is measured by the loss function and called cost, the loss function selected in this paper is cross entropy, which can be regarded as the difficulty of probability distribution p(x) through probability distribution q(x). The cross entropy characterizes the distance between two probability distributions, that is, the smaller the cross entropy, the closer the two probability distributions are. The expression is as follows:

$$H(p, q) = \sum_{i=1}^{n} p(x_i) \log \frac{1}{q(x_i)}$$

The selection of learning rate η has an important impact on the performance of the model and is also the most difficult parameter to debug. In order to achieve better performance of the model, this paper will experiment the changes of F1 of different neurons at different learning rate, and then select a learning rate that is most suitable for neuron training. The table shows the changes of different neurons (Cells = 100, 200, 300) in training LSTM model F1 at different learning rates (= 0.01, 0.01, 0.001). It can be seen that: the F1 has the best effect when Cells = 100, 200, 300 = 0.001 and when Cells = 400, the selected learning rate = 0.0001 s can achieve the best effect of the model. The table can be used to obtain the optimal learning rate for training different neurons.

Figure 7 and 8 respectively describe the changes of F1 and Cost in different neuron Cells, it can be seen from the figure that when Cells = 300 and Cells = 400, the average F1 value can reach 0.45. However, when Cells = 300, the cost value is slightly less than Cells = 400, so this paper will select Cells = 300 to train LSTM.

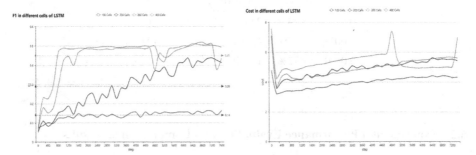

Fig. 7. F1 in different cells of LSTM.   **Fig. 8.** Cost in different cells of LSTM.

Common medical recommend neural network models include RNN model, LSTM model and GRU model. The RNN model can capture the time dependence between features and deal with some time-related problems. On the medical side, you can learn to process some longitudinal time data, such as patient arrival time and follow-up time [22]. But if the interval between visits is large, this can lead to "long-term dependence", the LSTM model solves this problem well and LSTM is a variant of the RNN model, which can obtain information dependence for a long time. There are many applications for predicting the probability of illness in health informatics [23, 24]. The GRU is also a kind of circulating neural network, which is more concise than LSTM model. The experimental effect is similar to LSTM, but the parameters are 1/3 less. In comparison, LSTM has more parameters, which is more suitable for large-scale samples and can train better models. In this paper, LSTM is compared with RNN and GRU models under the same network parameters, as shown in Figs. 9 and 10, the loss values of LSTM, GRU and RNN are all stable in a certain range, and the loss of LSTM is slightly less than that of RNN. However, the average F1 value of LSTM is stable at 0.45, the average value of F1 for GRU is 0.16, while the performance of RNN is only 0.07. This shows that LSTM has higher performance than other models.

7   Conclusion

We propose a blockchain medical asset model supporting analysis of transfer path across medical institutions, including sharing mechanism based on blockchain across medical

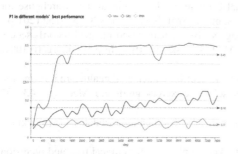

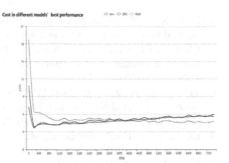

Fig. 9. F1 in different models' best performance.

Fig. 10. Cost in different models' best performance.

institutions and analysis method of transfer path across medical institutions, experiment shows:the mapping between medical treatment data and blockchain digital assets can cover all the data of medical treatment to provide patients with a complete cross-medical treatment path and realize cross-medical data sharing, improve the efficiency of medical treatment. Through the association relationship of medical assets of blockchain, a spatially complete transfer timing path diagram can be obtained through the traceability algorithm. To solve the lack of reference of the transition path, this paper innovates the data integration, integrates the data of the third-party data center according to the time sequence diagram of transfer relations by the blockchain, solves the problems of traditional data dispersion and path loss, and further improves the accuracy of the prediction model. The LSTM prediction algorithm used in this paper has higher performance than RNN and GRU in the same network environment.

This model solves the problem of lack of reference to the transfer path between medical institutions to find the optimal transfer path for local medical institutions, overcomes the bottleneck of the implementation of the transfer path of medical institutions. Based on the current work, we can continue to study the lack of clinical pathways, improve the path of clinical pathways, and further extend the blockchain technology to the medical field, and contribute to the medical industry with the greatest benefit.

Acknowledgement. This research is supported, in part, by the National Natural Science Foundation, China (No. 61772316); the major Science and Technology Innovation of Shandong Province (No. 2019JZZY010109); the Industrial Experts Program of Spring City; Joint SDU-NTU Centre for Artificial Intelligence Research (C-FAIR) (NSC-2019-011).

References

1. United States: Healthcare Information and Management Systems Society, Chicago (1997)
2. Lowe, H.J., Ferris, T.A., Hernandez, P.M., Weber, S.C.: STRIDE – an integrated standards-based translational research informatics platform. In: AMIA Annual Symposium Proceedings, vol. 2009, pp. 391–395 (2009)
3. Wang, X., et al.: Translational integrity and continuity: personalized biomedical data integration. J. Biomed. Inform. **42**, 100–112 (2009)

4. Zhou, X., et al.: Development of traditional Chinese medicine clinical data warehouse for medical knowledge discovery and decision support. Artif. Intell. Med. **48**, 139–152 (2010)
5. Chen, Y., Ding, S., Xu, Z., Zheng, H., Yang, S.: Blockchain-based medical records secure storage and medical service framework. J. Med. Syst. **43**, 5 (2018). https://doi.org/10.1007/s10916-018-1121-4
6. Zhou, T., Li, X., Zhao, H.: Med-PPPHIS: blockchain-based personal healthcare information system for national physique monitoring and scientific exercise guiding. J. Med. Syst. **43**, 305 (2019). https://doi.org/10.1007/s10916-019-1430-2
7. Li, C., Cao, Y., Hu, Z., Yoshikawa, M.: Blockchain-based bidirectional updates on fine-grained medical data (2019)
8. Shae, Z., Tsai, J.J.P.: On the design of a blockchain platform for clinical trial and precision medicine. In: IEEE International Conference on Distributed Computing Systems (2017)
9. Yue, X., Wang, H., Jin, D., Li, M., Jiang, W.: Healthcare data gateways: found healthcare intelligence on blockchain with novel privacy risk control. J. Med. Syst. **40**, 218 (2016). https://doi.org/10.1007/s10916-016-0574-6
10. Wang, H., Song, Y.: Secure cloud-based ehr system using attribute-based cryptosystem and blockchain. J. Med. Syst. **42**, 152 (2018). https://doi.org/10.1007/s10916-018-0994-6
11. Bocek, T., Rodrigues, B.B., Strasser, T., Stiller, B.: Blockchains everywhere - a use-case of blockchains in the pharma supply-chain. In: Integrated Network and Service Management (2017)
12. Huang, Y., Wu, J., Long, C.: Drugledger: a practical blockchain system for drug traceability and regulation, In: 2018 IEEE International Conference on Internet of Things (iThings) and IEEE Green Computing and Communications (GreenCom) and IEEE Cyber, Physical and Social Computing (CPSCom) and IEEE Smart Data (SmartData), pp. 1137–1144 (2018)
13. Malamas, V., Dasaklis, T., Kotzanikolaou, P., Burmester, M., Katsikas, S.: A forensics-by-design management framework for medical devices based on blockchain. In: 2019 IEEE World Congress on Services (SERVICES), pp. 35–40 (2019)
14. Griggs, K.N., Ossipova, O., Kohlios, C.P., Baccarini, A.N., Hayajneh, T.: Healthcare blockchain system using smart contracts for secure automated remote patient monitoring. J. Med. Syst. **42**, 130 (2018). https://doi.org/10.1007/s10916-018-0982-x
15. Spooner, S.H., Yockey, P.S.: Assessing clinical path effectiveness: a model for evaluation. Nurs. Case Manage. Managing Process Patient Care **1**, 188–198 (1996)
16. Panella, M., Marchisio, S., Stanislao, F.D.: Reducing clinical variations with clinical pathways: do pathways work? Int. J. Qual. Health Care **15**, 509–521 (2003)
17. Nakamoto, S.: Bitcoin: a peer-to-peer electronic cash system. https://bitcoin.org/bitcoin.pdf
18. Wood, G.: Ethereum: a secure decentralised generalised transaction ledger. Ethereum Proj. Yellow Pap. **151**, 1–32 (2014)
19. Morrison, D.R.: PATRICIA – practical algorithm to retrieve information coded in alphanumeric. J. ACM **15**, 514–534 (1968)
20. Graves, A.: Long short-term memory. In: Graves, A. (ed.) Supervised Sequence Labelling with Recurrent Neural Networks, vol. 385, pp. 37–45. Springer, Heidelberg (2012). https://doi.org/10.1007/978-3-642-24797-2_4
21. Informatik, F.F.J., Bengio, Y., Frasconi, P., Schmidhuber, J.: Gradient flow in recurrent nets: the difficulty of learning long-term dependencies (2001)
22. Chao, C., Cao, X., Jian, L., Bo, J., Zho, J., Fei, W.: An RNN architecture with dynamic temporal matching for personalized predictions of Parkinson's disease (2017)
23. Che, Z., Purushotham, S., Cho, K., Sontag, D., Yan, L.: Recurrent neural networks for multivariate time series with missing values. Sci. Rep. **8**, 6085 (2016)
24. Maragatham, G., Devi, S.: LSTM model for prediction of heart failure in big data. J. Med. Syst. **43**, 1–13 (2019). https://doi.org/10.1007/s10916-019-1243-3

Cooperative Anomaly Detection Model and Real-Time Update Strategy for Industrial Stream Data

Tengjiang Wang[1(✉)], Pengyu Yuan[1], Cun Ji[2], and Shijun Liu[3]

[1] Inspur General Software Co. Ltd., Ji'nan, China
[2] School of Information Science and Engineering, Shandong Normal University, Ji'nan, China
[3] School of Software, Shandong University, Ji'nan, China
lsj@sdu.edu.cn

Abstract. With the development of industrial Internet technology, more and more devices are brought into the industrial big data platform. To improve the efficiency of device maintenance, the industrial big data platform needs to monitor the abnormal data of the device. However, most of the current anomaly detection algorithms are offline and they can't be updated in real-time. To solve this problem, this paper proposes an anomaly detection model for the industrial stream. The model realizes anomaly detection by cooperatively calling 3σ and DBSCAN algorithm. The model has the advantages of low cost, fast speed, and easy to use. On this basis, this paper presents a real-time update strategy for this model, which further improves the accuracy of the model. The experimental results of water pump equipment monitoring data show the effectiveness of this method.

Keywords: Industrial big data · Streaming data · Anomaly detection · Collaborative verification

1 Introduction

Along with the global digital economy development and the implementation of industrial strategies, the manufacturing industry is moving towards a new situation of rapid development with a series of new technologies such as cloud computing, big data, and the Internet of Things. In an industrial Internet platform, integrates equipment, production lines, factories, suppliers, products, and customers are connects. So, the Industrial Internet platform can efficiently share various element resources in the industrial economy. Thereby, an Industrial Internet platform can reduce costs through automation and intelligence. Also, it is helpful to extend the industrial chain and to promote the transformation of the manufacturing industry.

In an industrial Internet platform, the automation equipment and programs generated data in chronological order. So, stream data is formed naturally. Besides, industrial big data also has the characteristics of a high degree of structure and high correlation among various parameters. They are suitable for analysis [1]. Because of industrial noise, sensor failure, poor I/O interface contact, malicious network fake data injection attacks, and

Y. Sun et al. (Eds.): ChineseCSCW 2020, CCIS 1330, pp. 307–319, 2021.
https://doi.org/10.1007/978-981-16-2540-4_23

other factors, there are some abnormal data when collecting data. So, the control system is difficult to achieve stable operation in a closed-loop environment [2]. For this, we should detect the abnormal stream data in real-time and recovery them with high precision [3].

In the industrial big data analysis platform, the industrial stream data are characterized by a large amount, fast frequency, low attention paid by factories, and imperfect edge calculation. So, it is very important to detect outliers [4]. In recent years, a large number of anomaly detection algorithms have been proposed. However, most of the current anomaly detection algorithms are offline. These anomaly detection algorithms trained models offline and then deployed the models for anomaly detection. When the model needs to be updated, these anomaly detection algorithms will retrain and redeploy a new model offline according to the collected data. With this strategy, the old model cannot be updated in real-time and the model may not adapt to the new data, which affects the accuracy. At the same time, the cost of retraining the model based on a large amount of data is very high.

To solve this problem, an anomaly detection model for industrial stream data is proposed in this paper. This model monitors the anomaly of industrial stream data by using the 3σ algorithm and DBSCAN algorithm [5] in a cooperatively way. This model can avoid the impact of outliers on data analysis and ensure the rationality of data. On this basis, aiming at solving the problems of slow updating iteration and a high error rate of old model monitoring new data, a real-time updating strategy is put forward. This strategy can update the anomaly detection model in real-time. The accuracy of anomaly monitoring can be improved through this strategy.

The rest of this paper is structured as follows: In Sect. 2, we briefly introduce the related works on anomaly detection. Our anomaly detection model is given in Sect. 3, and the real-time update strategy of the model is shown in Sect. 4. Section 5 presents the results from experiments to show the effectiveness of the anomaly detection model and real-time update strategy. Finally, we draw our conclusions from this study in Sect. 6.

2 Related Work

In recent years, researchers have extensively studied the problem of anomaly detection [6]. The goal of anomaly detection is to find unusual values or patterns that do not satisfy the normal, constraints, rules, or given models [7, 8]. The characteristics of abnormal mode are described in detail in Ref. [7]. The difficulties of anomaly detection tasks are also described in [7]. Anomaly detection methods can be divided into categories based on classification, clustering, statistics, and information theory. The principles and time complexities of these categories are summarized in [7].

In the industrial Internet platform, anomaly detection is mainly used in industrial damage detection, equipment anomaly detection, sensor fault detection, etc. For example, Yintao et al. performed a machine learning algorithm for water pump data to predict whether the water pump would fail within a specified time. Industrial data analysis has a strong mechanism, which needs some techniques of feature engineering, such as interpolation, handling outliers, boxing, log conversion, one hot coding, grouping operation, feature segmentation, scaling, extraction date, correlation analysis, and so on [10].

With the attention to the temporal and timeliness of data, researchers have carried out anomaly detection on temporal data. In [11], Gupta et al. introduced the anomaly detection problems, research methods, and applications of all kinds of temporal data, such as time-series data, stream data, spatiotemporal data, and temporal network data. In the researches of time series anomaly detection, abnormal detection objects can be divided into three types: point anomaly, subsequence anomaly, and pattern anomaly. For abnormal detection methods, there are mainly four types of methods: statistical model-based methods [12] (such as Arima, GARCH, etc.), clustering-based methods [13, 14] (such as k-means, EM, SVM model, etc.), similarity measurement-based methods [15], constraint rules-based methods [16].

In [17], Hadley Wickham put forward the requirements for clean data, namely: 1) each feature forms a column; 2) each observation forms a row, and 3) each type of observational unit forms a table. The clean data can improve the performance of the data analysis model. It can significantly improve the accuracy of anomaly detection by using the structured nature and correlation nature of industrial stream data. For example, Ji et al. detected outliers based on relative outlier distance (ROD) in [18], and they identified abnormal points and change points based on the correlation among time series. By mining the correlation knowledge of high-dimensional time series data, it not only saves the calculation cost but also realizes the accurate identification of abnormal data of complex patterns [19].

At present, in most of the industrial big data analysis platforms, anomaly detection algorithm uses off-line training model for data filtering, which lacks the steps of automatic updating model. These methods also do not implement offline training to update and re-deploy anomaly detection models. For example, the Spark batch data calculation tool, which is widely used in many industrial big data analysis platforms, integrates a machine learning tool Spark ML. This tool helps the platform save and load the exception monitoring model. It realizes the process of offline training and online detecting. However, this method has the following disadvantages: 1) it is slow to update and has a high error rate to detect new data using the old model; 2) each deployment process is cumbersome and easy to make mistakes, and 3) the cost of retraining every iteration is too high and it is not suitable for real-time streaming computing.

Using the real-time algorithms which can update the anomaly detection model online, can improve model accuracy. And these methods can be long-term used after a successful deployment. So FTRL (follow the regularized leader) algorithm [20] is favored by many industries. For each training sample, FTRL first calculated the category and loss of the sample. And then, FTRL uses the loss generated by the sample to calculate the gradient. Finally, FTRL updates the anomaly detection model backpropagation once. For example, the Flink stream data processing tool, which integrated A link algorithm tool, is an algorithm that can update the anomaly detection model online. This tool has good real-time performance. FTRL algorithm works well for the model based on logistic regression. But this algorithm is difficult to understand, and this algorithm requires logistic regression model. Moreover, it is more complicated to introduce the norm as a supplementary loss function, which increases the calculation cost.

3 Anomaly Detection Model for Industrial Stream Data

Based on the idea based on FTRL algorithm, an anomaly detection model, that uses the 3σ algorithm and DBSCAN algorithm in a cooperatively way, for industrial stream data is proposed in this paper. The proposed model has the characteristics of low cost, fast speed, and easy to use, and it supports dynamic updating.

The workflow of the proposed anomaly detection model is shown in Fig. 1. As shown in Fig. 1, this anomaly detection model firstly uses the 3σ algorithm and DBSCAN algorithm to detect the anomaly of industrial stream data, respectively. After that, this anomaly detection model collaborative verifies the suspicious data found by the 3σ algorithm or DBSCAN algorithm based on mechanism. Finally, this anomaly detection model outputs the abnormal data.

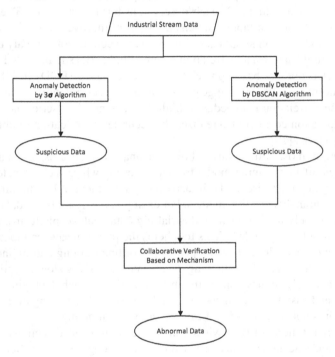

Fig. 1. The workflow of the proposed anomaly detection model.

Next, we will introduce the 3σ algorithm, DBSCAN algorithm, and collaborative verification in the following three subsections, respectively.

3.1 3σ Algorithm

3σ algorithm is a simple algorithm that is easy to understand and use. This algorithm is based on one statistical distribution theory. This statistical distribution theory is shown in Fig. 2. Suppose the data is approximately normal distribution, about 68% of the data will be within a standard deviation of the mean in statistics. About 95% of the data will be in the range of two standard deviations and about 99.7% of the data will be in the range of three standard deviations.

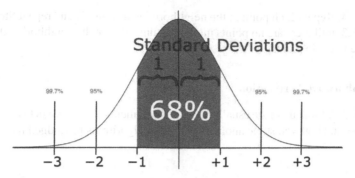

Fig. 2. Normal distribution and the range of 3σ.

In the 3σ algorithm, the standard deviation of data can be calculated as Eq. (1). In Eq. (1), X_i is the value of data, N is the number of data items, and $\overline{X}$ is the mean value of one data item. $\overline{X}$ can be calculated as Eq. (2).

$$S = \sqrt{\frac{1}{N-1} \sum_{i=1}^{N} \left(X_i - \overline{X}\right)^2} \tag{1}$$

$$\overline{X} = \frac{1}{N} \sum_{i=1}^{N} X_i \tag{2}$$

Therefore, according to the data distribution, it can be determined that several times of standard deviation is required as the threshold value, to monitor the abnormal value.

3σ algorithm determined the times of standard deviation according to the data distribution. And the 3σ algorithm uses it as the threshold value to detect abnormal data. This rule can be formally described by Eq. (3).

$$X_i \in \textit{Outliers if } X_i - \overline{X} < n \cdot S \textit{ or } X_i + \overline{X} > n \cdot S, \quad n = 1, 2, 3 \tag{3}$$

3.2 DBSCAN Algorithm

DBSCAN algorithm is a density-based clustering algorithm, which is widely used in anomaly detection. The model itself is a marked cluster of data objects. The process of DBSCAN algorithm steps are shown as follows:

– Step 1: Input D (the dataset), ε (the parameter of neighborhood radius), and *MinPts* (the density value of neighborhood)
– Step 2: Mark all data points in D as unvisited
– Step 3: Repeat the following steps, until all data point has been visited.

- Step 3.1: Randomly select an unvisited point P and mark it as visited.
- Step 3.2: Calculate the distance from point P to other points.
- Step 3.3: If there are at least *MinPts* points in the neighborhood of P and P do not belong to any clusters, a new cluster is created and P is added to the new cluster. Otherwise, P is marked as noise.

- Step 3.4: Repeat each point in the neighborhood as point P, and repeat Step 3.2 and Step 3.3 until there are no points marked as unvisited in the neighborhood of all the marked visible points.

3.3 Collaborative Verification

In anomaly detection, there are usually some misjudgments. As shown in Fig. 3, Point 3 is a change point. However, some anomaly detection algorithms misjudged it as abnormal points.

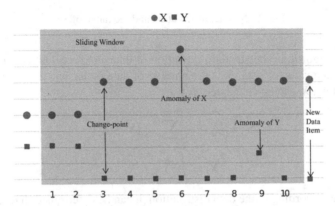

Fig. 3. Abnormal points and change points [18].

In our anomaly detection model, the 3σ algorithm assumes that the group of detection data contains a random error, and then determine abnormal data according to probability. DBSCAN algorithm divides the region into clusters according to the density of data. It finds clusters of arbitrary shape in the spatial database of noise, thus isolating abnormal data. However, both methods may be misjudged to abnormal data in principle.

Our anomaly detection model collaborative validates the results of the two methods. Only the outliers judged by two algorithms are the real abnormal data. Thus, we can distinguish outliers which may be "false".

4 Real-Time Update Strategy

The standard deviation and mean value of the whole model need to be recalculated whenever the 3σ algorithm is updated offline. It needs a lot of calculation when updating. Similarly, the DBSCAN algorithm has many iterations and a large amount of computation. For this, we introduce a real-time update strategy for the anomaly detection model as shown in Fig. 4.

Next, we will describe how to update the 3σ algorithm and DBSCAN algorithm, respectively.

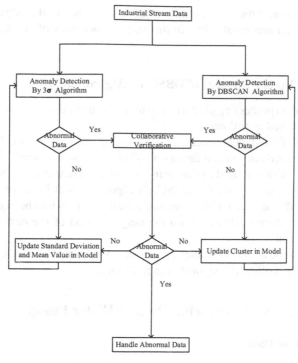

Fig. 4. Real-time update strategy.

4.1 Real-Time Update Strategy of 3σ Algorithm

In the streaming data processing platform, the real-time update strategy for the 3σ algorithm is described as following:

Firstly, a certain amount of data is collected for initialization training. The mean value and standard deviation of the data are calculated and saved. Whenever industrial data is generated, 3σ algorithm uses the standard deviation, mean value, and coefficient, which are calculated previously, to determine whether the new data is abnormal.

If the new data is abnormal, the data will pass to the collaborative verification process. Otherwise, the mean value and standard deviation are recalculated as Eq. (4) and Eq. (5), respectively. And the new standard deviation and mean value are used in 3σ Algorithm.

$$\overline{X}_{new} = \frac{\overline{X} \cdot N + X_I}{N + 1} \qquad (4)$$

$$S_{new} = \sqrt{\frac{S^2 \cdot (N - 1) + \left(X_i - \overline{X}_{new}\right)^2}{N + 1 - 1}} \qquad (5)$$

Firstly, a certain amount of data is collected for initialization training. The mean value and standard deviation of the data are calculated and saved.

As shown in Fig. 2 in Sect. 3.1, if there is more data on both sides, the curve in 3σ algorithm will change in real-time according to the above real-time update strategy.

The curve will become flat. So, the new value that is close to the edge of $n \cdot S$ is no longer marked as an abnormal value. In this way, the accuracy of anomaly detection is improved.

4.2 Real-Time Update Strategy of DBSCAN Algorithm

In the streaming data processing platform, the real-time update strategy for the DBSCAN algorithm is described as following:

Firstly, a part of the data is collected in real-time as the training set. Then the model initialization is carried out on the training set. Whenever new industrial data is generated, the DBSCAN algorithm is used to determine whether the new data is abnormal.

If the new data is normal, the DBSCAN algorithm will look for outliers in the neighborhood of the new data. If there are outliers, the outliers will be marked as normal and belong to the normal cluster. Then the neighborhood of the new normal data is checked by iteration.

In this way, some abnormal values are corrected to normal values, which improves the accuracy of the DBSCAN Algorithm in real-time.

5 Experiments on Monitoring Data of Water Pump

5.1 Experimental Data

The data in the experiment come from the real sensor data of some water pumps. In this paper, we use voltage to judge the state of the water pump. In the experimental process, we first collected 1000 data for training. After training, we used the offline model and real-time model to monitor the conditions of the water pump. Finally, the results of this to models were compared and analyzed.

Table 1. Experimental data

Training data	1–1000
Testing data	1001–2829
Abnormal data	data[667, 1] = 161.4 data[941, 1] = 292.3 data[1134, 1] = 256.2 data[1534, 1] = 372.5 data[1799, 1] = 301.1 data[2180, 1] = 200.2 data[2587, 1] = 188.7

In addition to the abnormal data listed in Table 1, abnormal data also includes data that value is 0. Since these data do not affect the subsequent calculation, they are not listed.

In this experiment, the parameter of the 3σ algorithm is set as shown in Table 2.

Table 2. The parameter of real-time 3σ algorithm.

Threshold coefficient	2
Mean value	avg_train = 236.23
	avg_update1 = 232.98
	avg_update2 = 226.20
Standard deviation	std_train = 20.20
	std_updata1 = 34.07
	std_updata2 = 45.47

5.2 Analysis of Experimental Results

After training on 1000 data items, the following 1001–2829 data items were detected. And the detection results were obtained. So that, we can calculate the detection accuracy of 1001–2000, 1001–2829 twice. DBSCAN algorithm and the real-time 3σ algorithm have the same results on this set of data. And the original 3σ algorithm is inferior to DBSCAN in anomaly detection accuracy. The results of the training data are shown in Fig. 5.

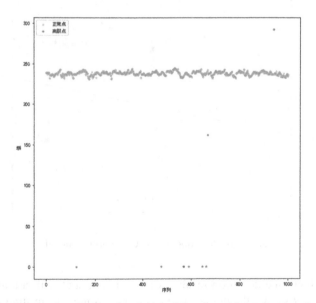

Fig. 5. The detection results of the training data. (Color figure online)

Figure 5 shows the 1000 data item for model training. The mean value of the training data was 236.23. In Fig. 5, the blue points mark normal data items, and the red points mark abnormal data items. There is 6 data item which value is 0. 0 means the device sensor transmission is abnormal. Also, there are two points with a large deviation from the mean value in Fig. 5, which indicated the abnormal data of the water pump. The details of the abnormal data are shown in Table 3.

Table 3. Abnormal data.

Indexes of abnormal data	Values of abnormal data
667	161.4
941	292.3

The detection results of the offline model are shown in Fig. 6, and the abnormal data are shown in Table 4. As Fig. 6 and Table 4 shows, only 1134 and 2180 serial numbers with less data deviation were included in the normal value by the offline training model, and the rest points were judged as abnormal points.

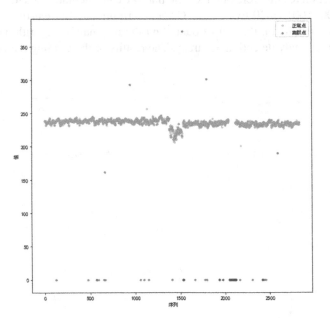

Fig. 6. The detection results of the offline model.

Figure 7 shows the detection results of the online model with a real-time update strategy. As Fig. 7 shows, the online model with a real-time update strategy judged the points with indexes 1134, 1799, 2180, and 2587 as normal points. The range of judgment is expanded, which is conducive to the further processing of these data. This strategy makes the model updated in real-time. So the model can adapt to the new data in time, and the detection accuracy of the online model is improved.

The precision of the two models is shown in Table 5. As Table 5 shows, the precision of the offline model and online model with real-time update strategy is improved with

Table 4. The abnormal data of offline models.

Indexes of abnormal data	Values of abnormal data
1134	256.2
1534	372.5
1799	301.1
2180	200.2
2587	188.7

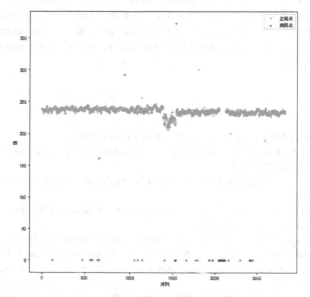

Fig. 7. The detection results of the online model with real-time update strategy.

the increase of training data. The precision of the online model with a real-time update strategy increases faster.

Table 5. The precision of two model

Data index	Precision	
	Offline model	Online model with real-time update strategy
1–1000	99.798%	99.798%
1001–2000	99.796%	99.898%
1001–2829	99.826%	99.942%

6 Conclusion

In this paper, an anomaly detection model for industrial stream data is proposed. This model detects the anomaly of industrial stream data by using the 3σ algorithm and DBSCAN algorithm in a cooperatively way. Besides, a real-time updating strategy is put forward to further improve the accuracy of our model. When updating the anomaly detection model, the data involved in this strategy is greatly reduced, thus this strategy saves a lot of computing costs. The real-time update strategy is based on the principle of the original algorithm. There is no need to add an algorithm library in the update strategy. The strategy also has the following advantage: the code to implement this strategy is simple and the strategy updates the anomaly detection model very fast.

In the future, we will use the characteristics of the Flink window function to screen the points in the DBSCAN model to improve the accuracy of the anomaly detection model. Combined with the two algorithms, we will propose a faster and more accurate real-time update algorithm.

References

1. Lee, J.: Industrial Big Data. Mechanical Industry Press (2015)
2. Hu, X., Tang, X., Hu, R.: Modern Detection Technology and System. China Machine Press, Beijing (2015). (in Chinese)
3. Zhou, D., Ding, X.: Theory and applications of fault tolerant control. Acta Autom. Sin. **26**(6), 788–797 (2000)
4. Ji, C., Shao, Q., Sun, J., et al.: Device data ingestion for industrial big data platforms with a case study. Sensors **16**(3), 279 (2016)
5. Ester, M., Kriegel, H.P., Sander, J., et al.: A density-based algorithm for discovering clusters in large spatial databases with noise. In: KDD, vol. 96, no. 34, pp. 226–231 (1996)
6. Chandola, V., Banerjee, A., Kumar, V.: Anomaly detection: a survey. ACM Comput. Surv. **41**(3), 1–58 (2009)
7. Wu, J.F., Jin, W., Tang, P.: Survey on monitoring techniques for data abnormalities. Comput. Sci. **44**(S2), 24–28 (2017)
8. Toledano, M., Cohen, I., Ben-Simhon, Y., et al.: Real-time anomaly detection system for time series at scale. In: KDD 2017 Workshop on Anomaly Detection in Finance, pp. 56–65 (2018)
9. Liu, Y., Yao, K.T., Liu, S., et al.: System and method for failure prediction for rod pump artificial lift systems. U.S. Patent 8,988,236, 24 March 2015
10. Rençberoğlu, E.: Fundamental techniques of feature engineering for machine learning. Towards Data Sci. (2019)
11. Gupta, M., Gao, J., Aggarwal, C., et al.: Outlier detection for temporal data. Synth. Lect. Data Min. Knowl. Discov. **5**(1), 1–129 (2014)
12. Gao, C.: Network traffic anomaly detection based on industrial control network. Beijing University of Technology (2014)
13. Xu, C., Lin, T.: Improved K-means algorithm based on anomaly detection. Softw. Guide **18**(4), 74–78 (2019)
14. Fujun, J., Min, C., Lei, W., et al.: SVM based energy consumption abnormality detection in ARM system. Electr. Measur. Instrum. **51**(6), 64–69 (2014)
15. Chen, Y.: Research on time series anomaly detection based on similarity analysis. Southwest Jiaotong University (2011). (in Chinese)

16. Ding, G., Sun, S.: Research on outlier detection algorithm based on dynamic rule constraint. Technol. Inf. (008), 41–42 (2019). (in Chinese)
17. Wickham, H.: Tidy data. J. Stat. Softw. **59**(10), 1–23 (2014)
18. Ji, C., Zou, X., Liu, S., et al.: ADARC: an anomaly detection algorithm based on relative outlier distance and biseries correlation. Softw. Pract. Experience **50**(11), 2065–2081 (2019). https://doi.org/10.1002/spe.2756
19. Ding, X.O., Yu, S.J., Wang, M.X., Wang, H.Z., Gao, H., Yang, D.H.: Anomaly detection on industrial time series based on correlation analysis. J. Softw. **31**(3), 726–747 (2020). (in Chinese)
20. McMahan, B.: Follow-the-regularized-leader and mirror descent: equivalence theorems and L1 regularization. In: Proceedings of the Fourteenth International Conference on Artificial Intelligence and Statistics, pp. 525–533 (2011)

An Analytical Model for Cooperative Control of Vehicle Platoon

Nianfu Jin, Fangyi Hu, and Yanjun Shi[✉]

Mechanical Engineering Department, Dalian University of Technology, Dalian, China
syj@ieee.org

Abstract. The cooperative control of the vehicle platoon is an important research field in smart transportation. In the existing research, the vehicle platoon cooperative control model's establishment and analysis are always tricky. This paper tries to make an analysis model to reduce the vehicle platoon's fuel consumption by controlling the vehicle's acceleration input for the vehicle platoon's cooperative control on the highway. The vehicle platoon's cooperative control is modeled as the optimization problem with the relationship between acceleration input and fuel consumption under certain constraints. And a bacterial foraging algorithm is employed to solve this problem. The preliminary results showed the feasibility of our model.

Keywords: Cooperative control · Platoon · Analysis model

1 Introduction

With the increasingly serious traffic problems, to improve the safety and intelligence of transportation system, the system concept of intelligent transportation is gradually rising. Intelligent transportation can make use of the new generation of communication networks and data processing capacity, improve the overall efficiency of the transportation system, reduce energy consumption, and increase the safety and convenience of transportation. In recent years, the development of an intelligent transportation system mainly focuses on the field of intelligent highway transportation system, also known as the Internet of vehicles. Among them, V2V (vehicle to vehicle) and V2I (vehicle to infrastructure), or V2X for short, refer to the communication systems between vehicles, vehicles and roadsides. This kind of system is called vehicle ad hoc networks (VANETs), which belongs to a kind of mobile ad hoc networks (MANETs) [1, 2]. In this network, each vehicle can become a network node with mobility. Each node cooperates with other nodes to reduce blind areas and avoid accidents. Developed countries such as Europe and the United States are actively promoting the research of related technologies. At present, based on the "integration of automobile and road infrastructure" (VII) program, the United States has developed and integrated various on-board roadside equipment and communication technologies to enable drivers to make better and safer decisions in driving [3]. More than 70 eSafety R & D projects of ERTICO, an organization of European ITS (Intelligent Traffic System), focus on vehicle road communication and collaborative control [4]. Japan's "SmartWay" plan focuses on the development and integration of Japan's ITS functions and the establishment of a common platform for vehicle units so that roads and vehicles can become Smartway and Smart cars by the two-way transmission of ITS consultation, to reduce traffic accidents and alleviate traffic congestion. At present, it has entered the stage of technology popularization [5]. The implementation of vehicle road coordination in China started late and is still in the preliminary stage. Some universities and research institutions have researched intelligent vehicle road collaborative control technology. For example National Science and technology research project "intelligent highway technology tracking", National 863 project "intelligent road system information structure and environmental perception and reconstruction technology research". At present, the primary research topic is to establish the framework of vehicle road coordination technology system in China, combine with the development status of China's road traffic infrastructure and the application status of intelligent vehicle information terminal, and develop the vehicle road integrated transportation management system suitable for China's national conditions and meeting the market demand.

Platoon control is a special form of formation control. In the environment of Internet of Vehicles, vehicles can obtain relevant information (such as speed and location) of surrounding vehicles within a certain range through V2V communication, which can promote efficient cooperation between vehicles and facilitate the realization of vehicle queue control [6, 7]. Vehicle queue control is an important driving behavior in vehicle coordinated control under the environment of Internet of vehicles. It is one of the important research directions of vehicle road cooperative system based on vehicle workshop cooperation and cooperation. At present, many scientific research institutions and units have carried out on-site real vehicle tests, and have carried out relevant tests and inspections on the vehicle queue control system.

In [8], the Chandler model of multi-leader vehicles was established, and a speed control algorithm was designed for the model so that the control of the platoon only needs to be based on the speed information of one or two vehicles. The control algorithm was verified on four intelligent cars. Reference [9] adopts the controlling idea of two-way following, which makes the current vehicle only need to know the relevant information of the front and rear vehicles. Compared with the leader tracking method, the two-way follow-up method is closer to practical application. In [10], a nonlinear speed function is proposed to control the formation of a motorcade. Compared with the constant time interval method, it can improve the traffic capacity on the premise of ensuring the stability of the platoon, while the traffic capacity of the constant time distance method is lower. In [11], the constant distance method without considering interference is adopted for formation control of vehicle platoon, and an improved lead-following method is established. This method is a heterogeneous platoon control method, that is, the control weight of each vehicle is different. In [12], a model prediction method is used to control the platoon, and the simulation experiments are carried out for three situations of vehicle following, platoon stability and constraint compensation, which verifies that the model prediction method can ensure the stability of the platoon. Reference [13] adopts an optimal distributed control method, which makes each vehicle in the platoon only need to obtain the locally available information, and does not need to master the global information of the platoon.

2 Problem Formulation

Considering that n automatic driving vehicles on the same lane of the highway all have the ability of V2V and V2I, after entering a certain area, the n automatic driving vehicles are identified as the vehicles that need to be formed during driving through the information exchange with RSU. At this time, the area where the n automatic driving vehicles are located is named as the confirmation area. The confirmation area is a one-way single lane road section with a length of C, which is confirmed at this section For the end of the formation vehicle, the vehicle can be identified as the vehicle whose body has just fully entered the confirmation area from the ideal state, and the length of the head car under the road section from the next area (merging area) is enough to ensure that all vehicles are confirmed as formation vehicles. The information flow between the vehicles and RSU in the confirmation area is shown in Fig. 1.

After the i_{th} automatic driving vehicle enters the merging area of length M, the control input is executed immediately, which ensures that the i_{th} automatic driving vehicle can adjust to the specified formation speed within the M-length one-way single lane road section. Each vehicle only pays attention to the distance and speed information between itself and the vehicle in front. After walking out of the merging area, each automatic driving vehicle obtains a final constant speed and a specified distance range from the front vehicle (the distance between the first vehicle and the former vehicle is not considered). When the body of the last confirmed formation vehicle of all n automatic driving vehicles has completely walked out of the merging area, the whole vehicle formation with a length of n is realized on the way. The information flow between the vehicle and RSU in the merging area is shown in Fig. 2.

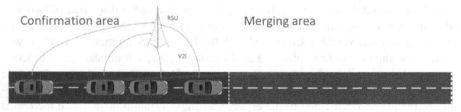

Confirmation area   RSU   Merging area

V2I

Fig. 1. Confirmation area traffic flow information.

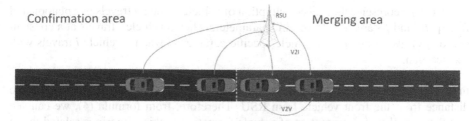

Fig. 2. Information flow when some vehicles entering the merging area.

2.1 Vehicle Dynamics Model

In the scenario proposed in this paper, it is assumed that there is only one one-way single lane road section, so the establishment of vehicle lateral dynamic model is not considered. According to Newton's second law, the longitudinal dynamic model of vehicle is as follows:

$$\begin{cases} p_i(t) = \int_{t_i^b}^{t} v_i(t)dt \\ v_i(t) = v_i^b + \int_{t_i^b}^{t} u_i(t)dt \end{cases} \tag{1}$$

Among them, t_i^b represents the time point when the i_{th} automatic driving vehicle begins to enter the merging area, p_i represents the distance between the i_{th} automatic driving vehicle at a time t and the starting position of the merging area, $v_i(t)$ represents the speed of the i_{th} automatic driving vehicle at a time t, and $u_i(t)$ represents the acceleration value of the i_{th} automatic driving vehicle at a time t.

The distance between the automatic driving vehicle and the starting position of the merging area in (1) has the following constraints:

$$0 \le p_i \le M, \forall p_i\{p_1 \dots p_n\} \tag{2}$$

The restriction of formation area length is as follows:

$$M = v_i(t_i^b) + \frac{1}{2} \int_{t_i^b}^{t_i^e} t u_i(t)d(t) \tag{3}$$

2.2 Fuel Consumption Model

The purpose of this paper is to achieve the optimal fuel consumption and time consumption of n autonomous vehicles after formation by control strategy. This paper cites the polynomial fuel consumption model in reference [14, 15], in which acceleration u is the input variable and velocity v is the output variable:

$$\begin{aligned} f_i = \beta_0 &+ \beta_1 v_i(t) + \beta_2 v_i^2(t) + \beta_3 v_i^3(t) \\ &+ a_i(t)(\partial_0 + \partial_1 v_i(t) + \partial_2 v_i^2(t)) \end{aligned} \tag{4}$$

In (4), f_i represents the fuel consumption of vehicle i when it travels in a platoon and β_0, β_1, β_2 and β_3 are vehicle-specific parameters when a vehicle i travels at a constant speed $v_i(t)$, ∂_0, ∂_1 and ∂_2 are vehicle-specific parameters when a vehicle i travels with acceleration.

In this paper, the final speed of formation vehicles is consistent, that is, in the confirmation area, n autonomous vehicles have obtained the final speed constant and the distance from the front vehicle from RSU. Therefore, from formula (4), we can see that the overall fuel consumption of vehicle formation in this scenario is related to the acceleration input of each automatic driving vehicle. After formation completion, the speed constraint is as follows:

$$v_i^e = v_{constant}, \forall v_i^e \in \{v_1^e...v_n^e\} \tag{5}$$

2.3 Optimization Model of Platoon Control Strategy

According to the principle of internal combustion engine fuel consumption, when the vehicle braking action, it can be considered that the vehicle no longer consumes fuel. Therefore, the optimization model in this paper does not consider the fuel consumption of vehicle deceleration, but the automatic driving vehicle allows the vehicle to slow down to achieve the required final formation speed when the control input is made in the merging area. Besides, the time consumption of vehicle formation behavior is considered Therefore, the time consumption function is increased by weight. The optimization model of this paper is as follows:

$$\min\{f_{all} = f_{fuel} + f_{time}\} \tag{6}$$

The model is expanded as follows:

$$\min_{u_{min} \leq u_i \leq u_{max}} \{\lambda_1 \frac{1}{2} \sum_{i=1}^{n} \int_{t_i^b}^{t_i^e} u_i^2(t)dt + \lambda_2 t_n\} \tag{7}$$

In (7), t_i^e is the time point that vehicle i leaves the merging area, λ_1 represents the weight of fuel consumption, λ_2 is the weight of time consumption, and t_n represents the overall time that platoon merging spend.

In this model, the velocity and acceleration have the following constraints:

$$v_{min} \leq v_i \leq v_{max}, \forall v_i \in \{v_1...v_n\}$$
$$u_{min} \leq u_i \leq u_{max}, \forall u_i \in \{u_1...u_n\} \tag{8}$$

As mentioned above, acceleration can be negative as a control input, but when acceleration is negative or zero, the fuel consumption can be regarded as zero. The constraint can be expressed as follows:

$$s.t. \{f_{fuel}(u_i(t)) = 0 | u_i(t) \leq 0\} \tag{9}$$

In this model, the time consumption is the time difference between the first vehicle entering the merging area and the last vehicle leaving the merging area completely. The completion time of the j_{th} vehicle formation can be expressed by the following formula:

$$t_j = \sum_{i=1}^{j} t_i + (t_j^e - t_j^b) \tag{10}$$

The range of safe distance after vehicle formation is shown as follows:

$$d_{\min} \leq d_i^e \leq d_{\max}, \forall d_i^e \in \{d_1^e...d_n^e\} \tag{11}$$

Each vehicle only cares about the distance from the previous vehicle and its own speed. If the distance between the front vehicle and the rear vehicle is less than the safe distance, the latter vehicle will slow down first and then deal with it (chain reaction may occur).

3 Analytical Solution

In the scenario proposed in this paper, vehicles follow the FIFO (First In First Out) rule, that is, the vehicles that enter the merging area first start to receive the acceleration input control. Each formation vehicle automatically obtains its ID when entering the confirmation area, and has its ID number i, $i \in N = \{1, 2...n\}$ before entering the merging area.

The automatic driving vehicle in the scenario in this paper belongs to the same vehicle type, that is, ignoring the constraints of different vehicle types in the formation process. The time when the i_{th} automatic driving vehicle enters the merging area is t_i^b, which is the initial time when the automatic driving vehicle begins to receive the control signal until the automatic driving vehicle drives out of the merging area. The vehicle has obtained a final constant speed $v_{constant}$ and a constant distance $d_{end} \in (d_{\min}, d_{\max})$ from the previous autonomous vehicle at the time t_i^e.

Since the initial distance and speed difference between autonomous vehicles are uncontrollable, only the time of the first automatic driving vehicle is considered In the definition of vehicle platoon, the vehicle spacing in the row is much smaller than that of every two vehicles in the normal traffic flow, which is the characteristic of vehicle formation to improve traffic efficiency. Therefore, in the scenario of this paper, it is assumed that all the acceleration inputs are to reduce the car spacing and achieve constant speed, and there is no due vehicle spacing Too small to affect the input.

For each automatic driving vehicle, the weight of acceleration control variable is far greater than the weight of time consumption. Therefore, the optimization model of the i_{th} automatic driving vehicle can be expressed by the following functions:

$$\min_{u_i} \{\frac{1}{2} \int_{t_i^b}^{t_i^e} u_i^2(t)dt\} \tag{12}$$

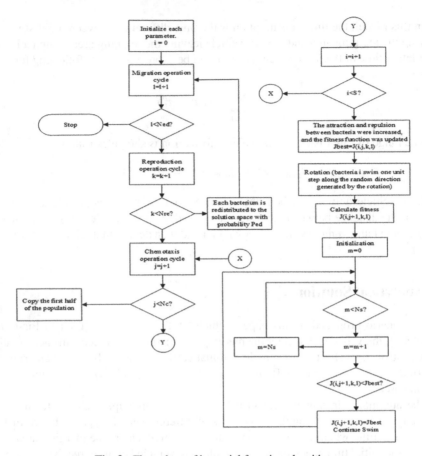

Fig. 3. Flow chart of bacterial foraging algorithm.

Since the control input of each vehicle depends on the speed of the vehicle itself and the distance from the previous vehicle, the solution can be transformed into an acceleration input string of n vehicles when the number of formation vehicles is known. The acceleration input string is used as the variable of evaluation function, and the fuel consumption index of the whole formation of n vehicles is evaluated by calculating the total square sum of accelerations. The string can be represented as follows:

$$[[u_1^{t_1^b}, \ldots, u_1^{t_1^e}], [u_2^{t_2^b}, \ldots, u_2^{t_2^e}], \ldots, [u_n^{t_n^b}, \ldots, u_n^{t_n^e}]] \tag{13}$$

The initial speed of each vehicle is not necessarily the same, and the residence time of each vehicle in the merging area is not the same. In this paper, the acceleration input of each vehicle is divided into the acceleration input within the equal time segment (ensure that the unit time segment is long enough, and the number and length of time segments of each vehicle are consistent, and the acceleration input in the time segment beyond the control of the merging area is zero) In other words, under the assumption that the merging area is long enough, the starting time of the vehicle input control is fixed.

Therefore, the constraints of the acceleration string of the i_{th} vehicle are as follows:

$$\int_{t_i^b}^{t_i^e} u_i^t dt = (v_{constant} - v_i^b) \tag{14}$$

Under the condition that the merging region is long enough and the acceleration in the constraint is sufficient to fill all time segments N_f, the acceleration string of each vehicle has $N_f!$ permutation and combination modes. If there is n a formation vehicle, the solution space has at most $n \cdot N_f!$ solution vector.

In this paper, a bacterial foraging algorithm [16] is used as the optimization algorithm of vehicle formation on an expressway, and the acceleration input string of each vehicle is taken as the solution vector, the flow chart is in Fig. 3.

4 Parameter Settings and Preliminary Results

4.1 Parameter Settings

In this paper, Matlab is used to simulate the vehicle formation scenario given in the previous chapter. In this scenario, the length of the road section in the confirmation area is C and the length of the road section in the merging area is M. Assuming that the total number of automatic driving vehicles that need to be formed in the confirmation area is $n = 4$, the speed of each automatic driving vehicle is the same (Theoretically, the speed difference and vehicle spacing are not calculated in the confirmation area, so it can be assumed that the vehicles in the confirmation area have the same initial speed), and different vehicle distances are used for simulation analysis. The initial data is shown in Table 1.

Table 1. Initial data of formation vehicles.

Vehicle number	Distance from the front vehicle	Initial speed
1	No considering	60 km/h
2	60 m	60 km/h
3	75 m	60 km/h
4	80 m	60 km/h

The constraint parameters are shown in Table 2.

Bacterial foraging algorithm is a bionic random search algorithm based on the process of bacterial foraging behavior. The algorithm simulates the behavior of bacterial population, including chemotaxis, reproduction and dispersal.

Table 2. Constraints and value.

Constraints	Value
n	4
$v_{min}, v_{max}, v_{constant}$	$v_{min} = 58$ km/h, $v_{max} = 82$ km/h, $v_{constant} \approx 80$ km/h
u_{min}, u_{max}	$u_{min} = -6$ m/s^2, $u_{max} = 6$ m/s^2
d_{min}, d_{max}	$d_{min} = 45$ m, $d_{max} = 50$ m
C, M	$C = 500$ m, $M = 500$ m

Bacterial foraging algorithm mainly includes a three-layer cycle, the outer layer is dispersal operation, the middle layer is propagation operation, and the inner layer is chemotactic operation. The parameters of bacterial foraging algorithm are selected in Table 3.

Table 3. BFO parameters and value.

Parameters	Value
Population size	50
Problem size	4
d_attr, d_repel	0.1, 0.1
w_attr, w_repel	0.2, 10
Search space	$[[-6, 6] [-6, 6] [-6, 6] [-6, 6]]$ for n
Step size	0.2
Dispersal step	2
Reproduction step	4
Chemotaxis step	25
Swim length	4

4.2 Preliminary Results

In this paper, according to the vehicle formation model established above, by selecting a group of basic data, the simulation environment is established. In this group of data, a total of four vehicles are determined. The initial data of the distance between each vehicle and the front vehicle and the speed are shown in Table 1. The maximum speed, minimum speed, final speed, minimum acceleration, maximum acceleration and minimum distance of each vehicle in the formation process, the maximum distance of each vehicle, the length constraints of confirmation area and the merging area are shown in Table 2. Finally, the optimal solution or suboptimal solution of acceleration control input is obtained by a bacterial foraging algorithm within the constraints. The parameter settings of bacterial foraging algorithm are shown in Table 3.

The results of the optimization are shown in Table 4.

Table 4. Results of BFO.

Number of iterations	The result of solution
1	[[4, 5, 1, 2], [2, 4, 3, 3], [6, 1, 3, 2], [1, 5, 2, 4]]
2	[[4, 5, 1, 2], [2, 4, 3, 3], [6, 1, 3, 2], [1, 5, 2, 4]]
3	[[3, 3, 0, 6], [2, 6, 0, 4], [0, 3, 6, 3], [4, 5, 1, 2]]
4	[[3, 6, 2, 1], [1, 5, 1, 5], [4, 5, 0, 3], [6, 5, 0, 1]]
5	[[0, 5, 4, 3], [3, 2, 3, 4], [0, 6, 0, 6], [6, 5, 0, 1]]
6	[[0, 5, 4, 3], [3, 2, 3, 4], [0, 6, 0, 6], [6, 5, 0, 1]]
7	[[0, 5, 4, 3], [3, 2, 3, 4], [0, 6, 0, 6], [6, 5, 0, 1]]
8	[[5, 2, 3, 2], [4, 4, 3, 1], [2, 3, 4, 3], [3, 1, 6, 2]]
9	[[5, 2, 3, 2], [4, 4, 3, 1], [2, 3, 4, 3], [3, 1, 6, 2]]
10	[[5, 2, 3, 2], [4, 4, 3, 1], [2, 3, 4, 3], [3, 1, 6, 2]]
11	[[5, 2, 3, 2], [4, 4, 3, 1], [2, 3, 4, 3], [3, 1, 6, 2]]
12	[[5, 2, 3, 2], [4, 4, 3, 1], [2, 3, 4, 3], [3, 1, 6, 2]]
13	[[5, 2, 3, 2], [4, 4, 3, 1], [2, 3, 4, 3], [3, 1, 6, 2]]
14	[[5, 2, 3, 2], [4, 4, 3, 1], [2, 3, 4, 3], [3, 1, 6, 2]]
15	[[5, 2, 3, 2], [4, 4, 3, 1], [2, 3, 4, 3], [3, 1, 6, 2]]
16	[[5, 6, 0, 1], [5, 6, 0, 1], [2, 6, 1, 3], [5, 2, 4, 1]]
17	[[5, 6, 0, 1], [5, 6, 0, 1], [2, 6, 1, 3], [5, 2, 4, 1]]
18	[[5, 6, 0, 1], [5, 6, 0, 1], [2, 6, 1, 3], [5, 2, 4, 1]]
19	[[5, 6, 0, 1], [5, 6, 0, 1], [2, 6, 1, 3], [5, 2, 4, 1]]
20	[[5, 4, 1, 2], [6, 0, 2, 4], [0, 5, 1, 6], [1, 4, 1, 6]]
21	[[5, 4, 1, 2], [6, 0, 2, 4], [0, 5, 1, 6], [1, 4, 1, 6]]
22	[[5, 6, 0, 1], [6, 2, 0, 4], [0, 0, 6, 6], [5, 0, 4, 3]]
23	[[5, 6, 0, 1], [6, 2, 0, 4], [0, 0, 6, 6], [5, 0, 4, 3]]
24	[[5, 6, 0, 1], [6, 2, 0, 4], [0, 0, 6, 6], [5, 0, 4, 3]]
25	[[5, 6, 0, 1], [6, 2, 0, 4], [0, 0, 6, 6], [5, 0, 4, 3]]

The time-distance graph of the suboptimal solution is shown in Fig. 4:

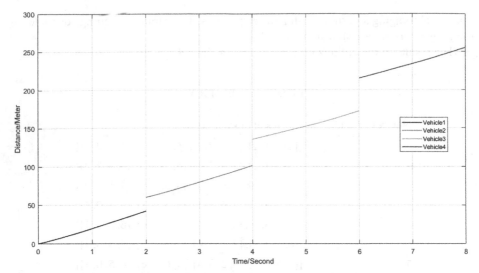

Fig. 4. Time-distance image of formation vehicles.

5 Conclusion

In this paper, a model for cooperative control of vehicle platoon on a freeway is proposed and analyzed. Taking fuel consumption and time cost of vehicle formation as evaluation functions, the model is analyzed as the functional relationship between acceleration input and fuel consumption output. Finally, bacteria foraging optimization algorithm is used to optimize the model, and the analytical solution of the model is obtained through preliminary calculation, which proves the practical significance of the proposed model.

Future research can be extended to multiple heterogeneous vehicles and more complex traffic environment simulation.

Funding. This work was funded by China National Key Research and Development Program (No. 2018YFE0197700).

References

1. Qiu, H.J.F., Ho, W.H., Chi, K.T., et al.: A methodology for studying 802.11p VANET broadcasting performance with practical vehicle distribution. IEEE Trans. Veh. Technol. **64**(10), 4756–4769 (2014)
2. Jia, D., Lu, K., Wang, J., et al.: A survey on platoon-based vehicular cyber-physical systems. IEEE Commun. Surv. Tutor. **18**(1), 263–284 (2016)
3. Ma, Y., Chowdhury, M., Sadek, A., et al.: Real-time highway traffic condition assessment framework using vehicle–infrastructure integration (VII) with artificial intelligence (AI). IEEE Trans. Intell. Transp. Syst. **10**(4), 615–627 (2009)
4. https://ec.europa.eu/information_society/activities/esafety/index_en.htm
5. Makino, H.: The Smartway Project. National Institute for Land and Infrastructure Management in Japan, San Francisco (2010)

6. Jia, D.Y., Dong, N.: Enhanced cooperative car-following traffic model with the combination of V2V and V2I communication. Transp. Res. Part B **90**, 172–191 (2016)
7. Shao, C.X., Leng, S., Zhang, Y., et al.: Performance analysis of connectivity probability and connectivity-aware MAC protocol design for platoon-based VANETs. IEEE Trans. Veh. Technol. **64**(12), 5596–5609 (2015)
8. Kita, E., Sakamoto, H.: Robot vehicle platoon experiment based on multi-leader vehicle following model. In: IEEE 2014 Second International Symposium on Computing and Networking, pp. 491–494 (2014)
9. Kwon, J.W., Chwa, D.: Adaptive bidirectional platoon control using a coupled sliding mode control method. IEEE Trans. Intell. Transp. Syst. **15**(5), 2040–2048 (2014)
10. Santhanakrishnan, K., Rajamani, R.: On spacing policies for highway vehicle automation. IEEE Trans. Intell. Transp. Syst. **4**(4), 198–204 (2003)
11. Peters, A.A., Mason, O.: Leader following with non-homogeneous weights for control of vehicle formations. In: IEEE Conference on Control Applications (CCA), pp. 109–113 (2016)
12. Kianfar, R., Falcone, P.: A control matching model predictive control approach to string stable vehicle platooning. Elsevier Control Eng. Pract. **45**, 163–173 (2015)
13. Sabău, S.: Optimal distributed control for platooning via sparse coprime factorizations. IEEE Trans. Autom. Control **62**(1), 305–320 (2017)
14. Kamal, M., Mukai, M., Murata, J., Kawabe, T.: Model predictive control of vehicles on urban roads for improved fuel economy. IEEE Trans. Control Syst. Technol. **21**(3), 831–841 (2013)
15. Rios-Torres, J., Malikopoulos, A.A.: Automated and cooperative vehicle merging at highway on-ramps. IEEE Trans. Intell. Transp. Syst. **18**, 780–789 (2016)
16. Passino, K.M.: Bacteria foraging optimization. In: Gazi, V., Passino, K.M. (eds.) Swarm Stability and Optimization, pp. 233–249. Springer, Heidelberg (2011). https://doi.org/10. 1007/978-3-642-18041-5_11

Collaborative Mechanisms, Models, Approaches, Algorithms, and Systems

Research on Scientific Workflow Scheduling Based on Fuzzy Theory Under Edge Environment

Chaowei Lin[1,2], Bing Lin[2,3]([✉]), and Xing Chen[1,2]

[1] College of Mathematics and Computer Science, Fuzhou University, Fuzhou 350108, China
[2] Fujian Provincial Key Laboratory of Networking Computing and Intelligent Information Processing, Fuzhou 350108, China
[3] College of Physics and Energy, Fujian Normal University, Fuzhou 350117, China

Abstract. With the rapid development of Internet of Everything (IoE) and the popularization of 4G/5G wireless networks, edge computing has gradually become the mainstream. In actual edge environment, some factors have an uncertain impact on scientific workflow scheduling, such as CPU load and bandwidth fluctuations of servers. Aiming at scientific workflow scheduling under uncertain edge environment, based on fuzzy theory, triangular fuzzy numbers (TFNs) were used to represent task computation time and data transmission time. In addition, an Adaptive Discrete Fuzzy GA-based Particle Swarm Optimization (ADFGA-PSO) is proposed to reduce the fuzzy execution cost while satisfying the scientific workflow's deadline. The uncertainty of workflow scheduling was introduced, which caused by server execution performance fluctuations during task computation and bandwidth fluctuations during data transmission. Meanwhile, two-dimensional discrete particle is adopted to encode the fuzzy scheduling strategy of workflow, and the two-point crossover operator, neighborhood mutation and adaptive multi-point mutation operator of genetic algorithm (GA) were introduced to improve the diversity of population and avoid the local optimum. Experimental results show that compared with other algorithms, ADFGA-PSO can obtain better fuzzy execution cost for deadline-constrained workflow scheduling under uncertain edge environment.

Keywords: Edge computing · Workflow scheduling · Uncertainty · TFN

1 Introduction

With the rapid development of IoE and the popularization of 4G/5G wireless networks, mobile devices and applications at the network edge have grown rapidly [1]. Faced with massive terminal device access and explosive growth of data traffic, the traditional cloud

This work was supported by the National Key R&D Program of China (2018YFB1004800), the National Natural Science Foundation of China (62072108), the Guiding Project of Fujian Province under Grant (2018H0017), and the Natural Science Foundation of Fujian Province under Grant (2019J01286).

computing has some problems such as high response delay, high bandwidth pressure. Based on the traditional cloud computing architecture, network service providers deploy computing and storage capabilities in the wireless access network so that some network services can be performed without the core network, which is defined as Multi-Access Edge Computing (MEC). This mode realizes ultra-low latency data transmission, saves bandwidth resources.

Although MEC provides the possibility to achieve the goal of low latency under edge computing, it still needs to solve how to perform reasonable task scheduling between cloud and edge server. Scientific workflow scheduling under edge environment is an NP-hard problem [2], because there are complex time and data dependencies among a series of tasks in workflow. Therefore, a reasonable and efficient workflow scheduling strategy is needed to minimize the execution cost of the workflow while meeting the deadline as much as possible.

In practical applications, the computation time of tasks and the data transmission time between tasks are uncertain. Some factors will have an uncertain impact on workflow scheduling, such as CPU load and bandwidth fluctuation. Inspired by the literature [3], task computation time and data transmission time are represented as TFNs to establish the workflow fuzzy scheduling model, and an Adaptive Discrete Fuzzy GA-based Particle Swarm Optimization (ADFGA-PSO) is proposed to minimize the execution cost of workflow under edge environment while meeting deadline constraint. The main contributions of this paper include the following parts:

1) The uncertainty of scientific workflow scheduling is introduced, which caused by server execution performance fluctuations during task computation and bandwidth fluctuations during data transmission.
2) Two-dimensional discrete particle is adopted to encode the fuzzy scheduling strategy of workflow, and the two-point crossover operator, neighborhood mutation and adaptive multipoint mutation operator of genetic algorithm (GA) were introduced to improve the diversity of population and avoid the local optimum.
3) With full consideration of task computation cost and data transmission cost, a cost-driven fuzzy scheduling strategy is designed for deadline-based workflow.

2 Related Work

In recent years, workflow scheduling has gradually shifted from cloud computing to edge-cloud collaborative computing, and edge environment has gradually become the mainstream. Scholars at home and abroad have also conducted a lot of research on workflow scheduling under edge environment, whose goal were mostly to minimize the execution cost or completion time of the workflow. Xie et al. [4] designed a novel directional and non-local-convergent PSO to optimize the completion time and execution cost of workflow simultaneously. Huang et al. [5] proposed a safe and energy-efficient computing offloading strategy, whose goal is to optimize the energy consumption of the workflow under the constraints of risk probability and deadline. Experiments show that this strategy can achieve security and high energy efficiency for mobile applications. Peng et al. [6] developed an algorithm based on Krill-Herd to evaluate the reliability of resources in mobile edge computing (MEC) environments. The results show that this method is significantly better than traditional methods in terms of success rate and completion time. However, the above work mainly focuses on workflow scheduling under deterministic environment.

In practical applications, uncertainty is bound to exist. Before workflow execution is completed, task computation time and data transmission time are only estimated. There is no clear boundary or the boundary is uncertain, so that fuzzy theory [7] is an effective tool to deal with this problem. Recently, scholars have also done a lot of research. Lei [8] uses triangular fuzzy numbers to represent fuzzy completion time and trapezoidal fuzzy numbers to represent fuzzy due-date, while introducing an improved fuzzy max operation to study the fuzzy job shop scheduling problem (FJSSP) with availability constraints. Sun et al. [3] used triangular fuzzy numbers to represent processing time, and then a fuzzification method was proposed to fuzzify the processing time in classic data sets into triangular fuzzy numbers to study FJSSP. Fortemps [9] expressed the uncertain duration as a six-point fuzzy number, and established a fuzzy job shop scheduling model with the goal of minimizing the fuzzy completion time. However, existing research is mainly focused on FJSSP, but little on fuzzy workflow scheduling. Therefore, this paper will establish a novel approach to cost-driven scheduling for deadline-based workflows under fuzzy uncertain edge environment.

3   Problem Definition

The scientific workflow scheduling model based on fuzzy theory under edge environment mainly includes three parts: edge environment, the deadline-based workflow and the uncertainty cost-driven scheduler.

Edge environment S consists of cloud S_{cloud} and edge S_{edge}, as shown in Fig. 1. The cloud $S_{cloud} = \{s_1, s_2, \ldots, s_n\}$ contains n cloud servers, and the edge $S_{edge} = \{s_{n+1}, s_{n+2}, \ldots, s_{n+m}\}$ contains m edge servers. Also, a server s_i can be expressed as:

$$s_i = \left(T_i^{boot}, p_i, c_i^{com}, f_i \right), \tag{1}$$

where T_i^{boot} represents the booting time of s_i; p_i represents the specific time unit of s_i; c_i^{com} represents the computing cost of s_i per unit time p_i; $f_i = \{0, 1\}$ represents which environment s_i belongs to: when $f_i = 0$, it belongs to the cloud; when $f_i = 1$, it belongs to the edge.

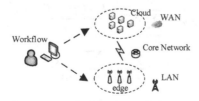

Fig. 1. Edge computing environment

The bandwidth $b_{i,j}$ between servers s_i and s_j is denoted by Eq. (2):

$$b_{i,j} = \left(\beta_{i,j}, c_{i,j}^{tran} \right),\tag{2}$$

where $\beta_{i,j}$ represents the value of bandwidth $b_{i,j}$, and $c_{i,j}^{tran}$ represents the cost of transmitting 1 GB of dataset from s_i to s_j.

Scientific workflow can be represented by a directed acyclic graph $G = (T, D)$, where T represents a set of l tasks $\{t_1, t_2, \ldots, t_l\}$, and D represents a set of m data dependencies $\{d_{12}, d_{13}, \ldots, d_{ij}\}$, $\forall i \neq j$ between tasks. For $d_{ij} = \langle t_i, t_j \rangle$, task t_j is the successor node of task t_i, task t_i is the predecessor node of task t_j, and the value of d_{ij} is the amount of data transferred from t_i to t_j. Each workflow has a corresponding deadline $D(G)$. In a scheduling strategy, if the workflow can be executed before the corresponding deadline, the scheduling strategy is called a feasible solution.

Given that $T_{com}(t_i, s_i)$ is the computation time of task t_i executed on server s_i. For the data-dependent edge d_{ij}, the data transmission time $T_{tran}(d_{ij}, s_i, s_j)$ of transmitting dataset d_{ij} from server s_i to server s_j is

$$T_{tran}(d_{ij}, s_i, s_j) = \frac{d_{ij}}{\beta_{ij}}.\tag{3}$$

The scheduling strategy of workflow can be defined as:

$$\Psi = (G, S, M, T_{total}, C_{total}),\tag{4}$$

where $M = \{(t_k, s_k) \cup (d_{ij}, s_i, s_j) | t_i \in T, d_{ij} \in D, s_i, s_j, s_k \in S\}$ represents the mapping from $G = (T, D)$ to S, T_{total} represents the completion time of the workflow, and C_{total} represents the execution cost of the workflow.

In the scheduling strategy Ψ, once the mapping M is determined, the server performing each task is also determined. Therefore, each task t_i has its corresponding start time $T_{start}(t_i)$ and completion time $T_{end}(t_i)$, so the completion time T_{total} of the workflow is calculated by Eq. (5).

$$T_{total} = \max_{t_i \in T} \{T_{end}(t_i)\},\tag{5}$$

Similarly, each server s_i also has a corresponding opening time $T_{on}(s_i)$ and closing time $T_{off}(s_i)$. Therefore, the task computation cost C_{com} and data transmission cost C_{tran} of the workflow are calculated as follows:

$$C_{com} = \sum_{i=1}^{|S|} c_i^{com} \left\lceil \frac{T_{off}(s_i) - T_{on}(s_i)}{p_i} \right\rceil,\tag{6}$$

$$C_{tran} = \sum_{t_j \in T} \sum_{t_k \in T, \langle t_j, t_k \rangle \in D} c_{jk}^{tran} d_{jk}.\tag{7}$$

Therefore, the execution cost of the workflow can be expressed as:

$$C_{total} = C_{com} + C_{tran}.\tag{8}$$

In summary, workflow scheduling under edge environment is to minimize the execution cost C_{total} of the workflow while the completion time T_{total} meets its deadline constraint. Therefore, the problem can be formally expressed as:

$$\begin{cases} \min \ C_{total} \\ s.t. \ T_{total} \le D(G) \end{cases}. \tag{9}$$

Due to server performance fluctuations during task computation and bandwidth fluctuations during data transmission, workflow scheduling is uncertain. Based on fuzzy theory, TFNs are used to represent the uncertainty of task computation time T_{com} and data transmission time T_{tran}.

The membership function of TFN $\tilde{t} = \left(t^l, t^m, t^u\right)$ is shown in Fig. 2, where the t^m indicates the most possible time, while t^l and t^u indicate the possible range of fuzzy time $\tilde{t}$. The real number t is treated as a special TFN $\tilde{t} = (t, t, t)$ for fuzzy operation.

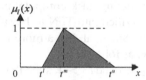

Fig. 2. Membership function of TFN

In this paper, the fuzzy variable $\tilde{\tau}$ corresponds to the variable τ for certainty scheduling. Based on model (9), this problem can be formally expressed as:

$$\begin{cases} \min \ \tilde{C}_{total} \\ s.t. \ \tilde{T}_{total} \le D(G) \end{cases}. \tag{10}$$

For the objective function $\tilde{C}_{total}$, its value is a TFN, and we adopt the method of minimizing its mean value C_μ and standard deviation C_σ to optimize it. Therefore, the objective function can be transformed into:

$$\min \ \tilde{C}_{total} = \left(c^l, c^m, c^u\right) \Rightarrow \min \ C_\mu + C_\sigma. \tag{11}$$

Based on fuzzy event probability measurement, Lee and Li defined the mean and standard deviation of fuzzy sets in the uniform distribution and proportional distribution, respectively [10]. In this paper, fuzzy execution cost $\tilde{C}_{total}$ is based on proportional distribution, so that C_μ and C_σ are given by Eqs. (12) and (13), respectively.

$$C_\mu = \frac{\int x \tilde{C}_{total}^2(x) dx}{\int \tilde{C}_{total}^2(x) dx} = \frac{c^l + 2c^m + c^u}{4}, \tag{12}$$

$$C_\sigma = \left[\frac{\int x^2 \tilde{C}_{total}^2(x) dx}{\int \tilde{C}_{total}^2(x) dx} - C_\mu^2 \right]^{1/2} = \left[\frac{2\left(c^l - c^m\right)^2 + \left(c^l - c^u\right)^2 + 2\left(c^m - c^u\right)^2}{80} \right]^{1/2}, \tag{13}$$

For the constraint $\tilde{T}_{total} \leq D(G)$, suppose that scheduling strategy should meet the deadline constraint even in the worst case. Therefore, it can be transformed into:

$$s.t. \ \tilde{T}_{total} = \left(t^l, t^m, t^u\right) \leq D(G) \Rightarrow s.t. \ t^u \leq D(G). \tag{14}$$

In summary, based on fuzzy theory, workflow scheduling under uncertain edge environment can be formalized as:

$$\begin{cases} \min \ C_\mu + C_\sigma \\ s.t. \ t^u \leq D(G) \end{cases}. \tag{15}$$

4 Fuzzy Theory in Workflow Scheduling

4.1 Fuzzification of Dataset

During workflow uncertain scheduling, task computation time and data transmission time in the dataset need to be fuzzified into TFN to be suitable for the fuzzy situation. Based on the method adopted by Sun et al. [3], a more realistic method is proposed to describe the uncertainty of time, as follows:

For two given parameters $\delta_1 < 1$ and $\delta_2 > 1$, such that $\delta_2 - 1 > 1 - \delta_1$. For a certain time t, t^l is selected randomly from $[\delta_1 t, t]$, t^m is taken as t, t^u is selected randomly from $\left[2t - t^l, \delta_2 t\right]$, and then obtain the corresponding fuzzy time $\tilde{t} = \left(t^l, t^m, t^u\right)$.

4.2 Fuzzy Operation in Workflow Scheduling

In uncertain scheduling, when constructing the fuzzy scheduling strategy of workflow, some operations of fuzzy numbers need to be redefined as follows.

The addition operation is used to calculate the fuzzy completion time of the task. For two TFNs $\tilde{s} = \left(s^l, s^m, s^u\right), \tilde{t} = \left(t^l, t^m, t^u\right)$, the addition operation is given by (16).

$$\tilde{s} + \tilde{t} = \left(s^l + t^l, s^m + t^m, s^u + t^u\right). \tag{16}$$

The ranking operation is used to compare the maximum fuzzy completion time of the precursor task. The ranking criteria proposed by Sakawa et al. [11] is used to compare $\tilde{s} = \left(s^l, s^m, s^u\right)$ and $\tilde{t} = \left(t^l, t^m, t^u\right)$ as follows:

1) if $c_1(\tilde{s}) = \left(s^l + 2s^m + s^u\right)/4 > c_1(\tilde{t}) = \left(t^l + 2t^m + t^u\right)/4$, then $\tilde{s} > \tilde{t}$;
2) if $c_1(\tilde{s}) = c_1(\tilde{t})$ and $c_2(\tilde{s}) = s^m > c_2(\tilde{t}) = t^m$, then $\tilde{s} > \tilde{t}$;
3) if $c_1(\tilde{s}) = c_1(\tilde{t})$, $c_2(\tilde{s}) = c_2(\tilde{t})$ and $c_3(\tilde{s}) = s^u - s^l > c_3(\tilde{t}) = t^u - t^l$, then $\tilde{s} > \tilde{t}$.

The max operation is used to calculate the maximum value of the fuzzy completion time of the precursor task and the fuzzy completion time of the current server to determine the fuzzy start time of the task. For two TFNs $\tilde{s} = \left(s^l, s^m, s^u\right)$ and $\tilde{t} = \left(t^l, t^m, t^u\right)$, the membership function $\mu_{\tilde{s} \vee \tilde{t}}(z)$ of $\tilde{s} \vee \tilde{t}$ is defined as follows:

$$\mu_{\tilde{s} \vee \tilde{t}}(z) = \sup_{z = x \vee y} \min\left(\mu_{\tilde{s}}(x), \mu_{\tilde{t}}(y)\right) \triangleq \bigvee_{z = x \vee y} \left(\mu_{\tilde{s}}(x) \wedge \mu_{\tilde{t}}(y)\right). \tag{17}$$

In this paper, Lei criterion [8] is used to approximate the maximum value as follows:

$$\tilde{s} \vee \tilde{t} \cong \begin{cases} \tilde{s}, \tilde{s} \geq \tilde{t} \\ \tilde{t}, \tilde{s} < \tilde{t} \end{cases}. \tag{18}$$

The multiplication operation is used to calculate the fuzzy computation cost and fuzzy transmission cost of the server. For TFN $\tilde{t} = (t^l, t^m, t^u)$, its multiplication operation is given by Eq. (19).

$$\lambda \cdot \tilde{t} = \left(\lambda t^l, \lambda t^m, \lambda t^u\right), \forall \lambda \in \mathbb{R}. \tag{19}$$

In addition, the division operation between the TFN and the real number can be equivalent to the multiplication operation, which is given by Eq. (20).

$$\tilde{t} \div \mu \triangleq \lambda \cdot \tilde{t}, \lambda = 1/\mu, \forall \mu \in \mathbb{R}. \tag{20}$$

5 Scheduling Strategy Based on ADFGA-PSO

5.1 PSO

The traditional PSO is a swarm intelligence optimization technology based on the population social behavior. It was first proposed by Kennedy and Eberhart [12] in 1995. Each particle is used to simulate an individual of bird groups, corresponds to a candidate solution of optimization problem. Its velocity can be adjusted according to its current situation, the optimal position of the individual particle and the global optimal position of the population. The particle motion process is the search process of the problem, and fitness is introduced to evaluate the quality of each particle. Each particle updates its own velocity and position by learning from the experience of itself and other particles, which are denoted by Eqs. (21) and (22), respectively.

$$V_i^{t+1} = w \cdot V_i^t + c_1 r_1 \left(pBest_i^t - X_i^t\right) + c_2 r_2 \left(gBest^t - X_i^t\right), \tag{21}$$

$$X_i^{t+1} = X_i^t + V_i^{t+1}, \tag{22}$$

where t represents the current number of iterations, V_i^t and X_i^t represents the velocity and position of the ith particle after t iterations. $pBest_i^t$ and $gBest^t$ respectively represent the individual optimal position of the ith particle and the global optimal position of the population. w is the inertia factor, which determines the search ability of the algorithm, thereby affecting its convergence. c_1 and c_2 are the learning factors, which respectively reflect the learning ability of the particle on the individual cognitive part and the social cognitive part. r_1 and r_2 is a random number in the interval [0, 1] to strengthen the algorithm's random search ability in the iterative process.

5.2 ADFGA-PSO

Problem Encoding. In order to make the traditional PSO solve the discrete optimization problem better, inspired by the literature [13], this paper proposes a new encoding method, namely priority-server nested discrete particles to encode the workflow scheduling problem. Each particle in the particle swarm corresponds to a potential fuzzy scheduling strategy of the workflow under edge computing environment, and P_i^t is represented by the ith particle of the particle swarm in the tth iteration, as shown in Eq. (23).

$$P_i^t = \left((\mu_{i1}, s_{i1})^t, (\mu_{i2}, s_{i2})^t, \ldots, (\mu_{i|T|}, s_{i|T|})^t \right) \tag{23}$$

For a two-tuple $(\mu_{ij}, s_{ij})^t$, μ_{ij} represents the priority of the jth task. The code is a real number. The larger the value, the higher the priority of the task in the queue to be executed. If there is a task with the same code value, it will be entered first The task priority of the queue to be executed is higher; s_{ij} represents the execution position of the jth task, the code is an integer, and its value represents a different server number. Figure 3 shows the coded particles of a workflow scheduling containing 8 tasks.

Task	0	1	2	3	4	5	6	7
Particale	0.2 ¦ 0	1.3 ¦ 3	0.5 ¦ 2	1.2 ¦ 1	2.4 ¦ 1	1.6 ¦ 4	0.7 ¦ 2	2.1 ¦ 2

Fig. 3. A coded particle corresponding to workflow scheduling

Fuzzy Fitness Function. This paper aims to meet the deadline constraint of workflow while minimizing the fuzzy execution cost $\tilde{C}_{total}$ of workflow. It can be seen that the encoding strategy proposed in this paper may have infeasible solutions that do not meet the deadline constraint. Therefore, the fuzzy fitness function used to compare two candidate solutions must be differentiated and defined as the following situations.

1) Both particles are feasible. Choose particles with less fuzzy execution cost $\tilde{C}_{total}$, and the fuzzy fitness function is defined as

$$F(P_i^t) = \tilde{C}_{total}(P_i^t) = C_\mu(P_i^t) + C_\sigma(P_i^t). \tag{24}$$

2) One particle is feasible, and the other is infeasible. Obviously, the feasible solution must be selected, and its fuzzy fitness function is defined as

$$F(P_i^t) = \begin{cases} \tilde{C}_{total}(P_i^t), & \tilde{T}_{total}(P_i^t) \leq D(G) \\ \infty, & \tilde{T}_{total}(P_i^t) \leq D(G) \end{cases} = \begin{cases} \tilde{C}_{total}(P_i^t), & t^u(P_i^t) \leq D(G) \\ \infty, & t^u(P_i^t) \leq D(G) \end{cases}. \tag{25}$$

3) Both particles are infeasible. Choose a particle with a less fuzzy completion time $\tilde{T}_{total}$, because this particle is more likely to become feasible after evolution, and its fuzzy fitness function is defined as

$$F\left(P_i^t\right) = \tilde{T}_{total}\left(P_i^t\right) = t^u\left(P_i^t\right). \tag{26}$$

Update Strategy. The update of each particle is affected by its current situation, optimal position of the individual and global optimal position of the population [14]. In the $(t+1)^{\text{th}}$ iteration, the update method of the ith particle is shown in (27).

$$P_i^{t+1} = c_2 \otimes \left(c_1 \otimes \left(w \odot P_i^t, pBest_i^t\right), gBest^t\right). \tag{27}$$

For the inertia part, mutation operator of GA is introduced to update the corresponding part of Eq. (21), and the update method is shown in Eq. (28).

$$A_i^{t+1} = w \odot P_i^t, r < w. \tag{28}$$

where r is a random number in $[0, 1]$. Only when $r < w$, the dual mutation operator $\odot$ is executed on the particle P_i^t, which is composed of the following two parts.

For task priority, the neighborhood mutation operator[15] randomly selects 3 particles, generates all sort combinations for the task priority of these 3 locations, obtains the neighborhoods of these particles, and then randomly selects one as the result, as shown in Fig. 4. For server number, the adaptive multi-point mutation operator randomly selects k locations of the particles, and randomly mutates the server number of each locations in the interval $[0, |S|)$ to generate new particles, as shown in Fig. 5.

Particle: | 0.2; 0 | 1.3; 3 | 0.5; 2 | 1.2; 1 | 2.4; 1 | 1.6; 4 | 0.7; 2 | 2.1; 2 |

Neighborhood Particle:
0.2; 0	1.3; 3	0.5; 2	0.7; 1	2.4; 1	1.6; 4	1.2; 2	2.1; 2
0.2; 0	1.2; 3	0.5; 2	1.3; 1	2.4; 1	1.6; 4	0.7; 2	2.1; 2
0.2; 0	1.2; 3	0.5; 2	0.7; 1	2.4; 1	1.6; 4	1.3; 2	2.1; 2
0.2; 0	0.7; 3	0.5; 2	1.3; 1	2.4; 1	1.6; 4	1.2; 2	2.1; 2
0.2; 0	0.7; 3	0.5; 2	1.2; 1	2.4; 1	1.6; 4	1.3; 2	2.1; 2

Fig. 4. Neighborhood mutation operator on task priority in inertia part

For individual cognition part and social cognition part, crossover operator of GA is introduced to update the corresponding part of Eq. (21), and the update method is shown in Eq. (29) and Eq. (30), respectively.

$$B_i^{t+1} = c_1 \otimes \left(A_i^{t+1}, pBest_i^t \right), r_1 < c_1. \tag{29}$$

$$P_i^{t+1} = c_2 \otimes \left(B_i^{t+1}, gBest^t \right), r_2 < c_2. \tag{30}$$

where r_1 and r_2 is a random number in [0, 1]. Only when $r_1 < c_1$ (or $r_2 < c_2$), execute the two-point crossover operator $\otimes$ for $A_i^{t+1}(B_i^{t+1})$, which randomly select the 2 locations of the particles for updating, and then replace the values between two locations with corresponding locations in *pBest*(or *gBest*), as shown in Fig. 6.

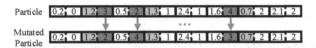

Fig. 5. Adaptive multi-point mutation operator for server number in the inertia part

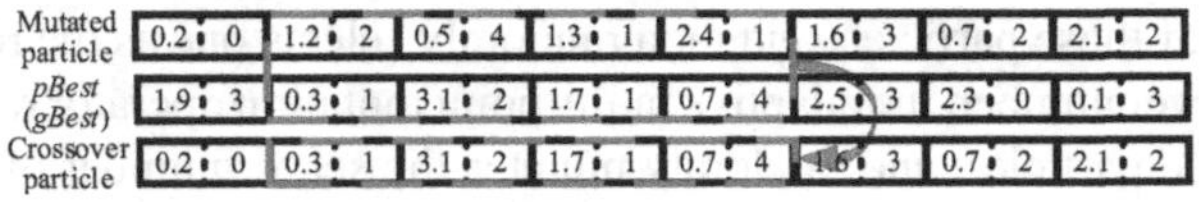

Fig. 6. The two-point crossover operator of the individual (or social) cognitive part

With the iterations running, the inertia factor w and the learning factor c_1, c_2 are dynamically adjusted by linear increase or decrease [16], with not to repeat.

In addition, for the adaptive multi-point mutation operator, its mutation number k is adaptively adjusted with the change of the inertia factor w, and its adjustment strategy is shown in Eq. (31).

$$k = k_{max} + (k_{max} - k_{min}) \cdot \frac{w - w_{min}}{w_{max} - w_{min}} \tag{31}$$

where k_{max} and k_{min} respectively represent the maximum and minimum of k.

Mapping from Encoding Particle to Workflow Scheduling Strategy. The mapping from encoded particle to workflow scheduling strategy under edge environment is shown in Algorithm 1.

Algorithm 1: Mapping from encoding particle to workflow scheduling strategy

Input: G, S, P

Output: $\Psi = \varnothing$ or $\Psi = \left(G, S, M, T_{total}, C_{total} \right)$

1: **Initialization:** $M \leftarrow null, T_{total} \leftarrow 0, C_{total} \leftarrow 0, C_{tran} \leftarrow 0$
2: Based on 4.1, Calculate $T_{com} \left[|T| \times |S| \right]$ and $T_{tran} \left[|D|, |S| \times |S| \right]$
3: **for** $i = 0$ **to** $i = |T| - 1$
4:     $M = M \cup \left(t_i, s_i \right)$ // Based on the encoded particle P.
5:     **if** t_i *is an entry task* **then** // t_i has no precursor task.
6:         **if** s_i is off **then**
7:             Turn on s_i, $T_{off}\left(s_i \right) = T_{s_i}^{boot}$, $T_{on}\left(s_i \right) = T_{off}\left(s_i \right) - T_{s_i}^{boot}$
8:         **end if**
9:         $T_{start}\left(t_i \right) = T_{off}\left(s_i \right)$
10:     **else**
11:         $maxT = 0$
12:         **for each** *parent* t_p *of* t_c **do**
13:             $maxT = maxT \vee \left(T_{end}\left(t_R \right) + T_{tran}\left[d_{pc} \right]\left[s_p \right]\left[s_c \right] \right)$
14:             $C_{tran} += fuzzy\left(c_{pc}^{tran} \cdot d_{pc} \right)$ // $fuzzy(*)$ is the fuzzification function.
15:         **end for**
16:         **if** s_i is *off* **then**
17:             Turn on s_i, $T_{off}\left(s_i \right) = maxT \vee T_{s_i}^{boot}$, $T_{on}\left(s_i \right) = T_{off}\left(s_i \right) - T_{s_i}^{boot}$
18:         **end if**
19:         $T_{start}\left(t_i \right) = maxT \vee T_{off}\left(s_i \right)$
20:     **end if**
21:     $T_{end}\left(t_i \right) = T_{start}\left(t_i \right) + T_{com}\left[t_i \right]\left[s_i \right]$
22: **end for**
23: Calculate T_{total} and C_{total} based on equation (5) and equation (8)
24: **if** $T_{total} > D\left(G \right)$ **then**
25:     set P as an infeasible particle, and **return** $\Psi = \varnothing$
26: **end if**
27: **return** $\Psi = \left(G, S, M, T_{total}, C_{total} \right)$

Algorithm Flow. The algorithm flow of ADFGA-PSO includes the following steps:

1) Initialize the parameters of the ADFGA-PSO, such as population size S_{pop}, maximum iteration number $iter_{max}$, inertia factor, learning factor, and mutation number, and then randomly generate the initial population.
2) According to Algorithm 1, calculate the fitness of each particle based on Eq. (24)–(26). The initial particle is set as its individual optimal particle, and the particle with the smallest fitness is set as the current global optimal particle.
3) According to Eq. (27)–(30), update the encode of the particle, and calculate the fitness of the updated particle.

4) If the fitness of the updated particle is less than its individual optimal particle, set the current particle as its individual optimal particle. Otherwise, go to step 6.
5) If the fitness of the updated particle is smaller than the global optimal particle of the population, set the current particle as the global optimal particle of the population.
6) Check whether the maximum number of iterations is reached. If so, the algorithm is terminated and output the global best particle; otherwise, return to step 3.

6 Experimental Simulation and Results

To validate the effectiveness of ADFGA-PSO for minimizing fuzzy execution cost with deadline constraint, experimental simulation will be carried out in this section. Both ADFGA-PSO and comparison algorithm are implemented in Python 3.7 and run on the win10 system with 8 GB RAM 2.70 GHz and Intel i5-7200U CPU. Based on the literature [16], the parameters of ADFGA-PSO are set as: $S_{pop} = 100$, $iter_{max} = 1000$, $w_{max} = 0.9$, $w_{min} = 0.4$, $c_1^{start} = 0.9$, $c_1^{end} = 0.2$, $c_2^{start} = 0.4$, $c_2^{end} = 0.9$, $k_{max} = |V|/10$, $k_{min} = 1$.

6.1 Experimental Setup

The scientific workflows for simulation come from five different scientific fields researched by Bharathi et al. [17], including: CyberShake, Epigenomics, LIGO, Montage, and SIPHT. Each workflow has a different structure, with more detail information stored in xml files. For each scientific field, three scales of workflows are selected: tiny (about 30 tasks), small (about 50 tasks) and medium (about 100 tasks).

Assume that there are 3 cloud servers (s_1, s_2, s_3) and 2 edge servers (s_4, s_5) under edge environment. The cloud server s_3 has the strongest computing power, the computing time of each task executed on s_3 is obtained from the corresponding xml file. Besides, the computing power of the cloud server $s_1(s_2)$ is about 1/2 (1/4) of s_3, and the computing power of the edge server $s_4(s_5)$ is about 1/8 (1/10) of s_3; set the computation cost per hour of the cloud server s_3 is 15.5$, and the computing cost per hour of the remaining servers is approximately proportional to their computing power. In addition, assumed that the initialization time T_i^{boot} of the server is 97 s.

According to the scale of workflows, tiny and small workflows set 60 s as a time unit p_i, while medium-sized workflow set 1 h as a time unit p_i.

The relevant parameters for the bandwidths between different types of servers are set as shown in Table 1.

In addition, Sect. 4.1 has introduced the fuzzification method, which fuzzify task computation and data transmission time into TFNs, where δ_1 is 0.85, and δ_2 is 1.2.

Finally, each workflow has a corresponding deadline $D(G)$, which is used to test the performance of algorithms, defined as Eq. (32).

$$D(G) = 1.5 * Minw \tag{32}$$

where $Minw$ represents the execution time of workflow using HEFT algorithm [18].

In order to test the performance of the ADFGA-PSO, the following comparison algorithms are used for workflow scheduling under edge environment.

Table 1. Parameters for bandwidth between different types of servers

$f_i \leftrightarrow f_j$		$\beta_{i,j}$(MB/s)	$c_{i,j}^{tran}$($/GB)
0	$\leftrightarrow$ 0	2.5	0.4
0	$\leftrightarrow$ 1	1.0	0.16
1	$\leftrightarrow$ 1	12.5	0.8

- FPSO: This method uses the same task priority encoding value as ADFGA-PSO, while the server encoding value uses the traditional continuous encoding method of PSO [12], with its rounded value used as server number. The update strategy of particles and the setting of parameters refer to literature [16].

- FGA: This method uses the same encoding strategy as ADFGA-PSO. According to the traditional GA [19] update strategy, elitist preservation strategy, binary tournament selection, two-point crossover and exchange mutation operator are used to update each individual, with the final elite individual as the optimal solution.

- FRAND: This method also uses the same encoding strategy as ADFGA-PSO. A random search strategy [20] is used to search the solution space. In addition, each iteration does not influence each other, and the optimal solution in the population is obtained until the algorithm is terminated.

6.2 Experimental Results

In order to compare ADFGA-PSO and comparison algorithms for scientific work-flow scheduling performance under edge environment, 10 sets of independent repeated experiments were carried out on different types and different scales of workflows.

In order to compare the quality of results, the fuzzy execution cost and fuzzy completion time need to be defuzzified, whose method is introduced in detail in Sect. 3.

Next, we will analyze the fuzzy scheduling results of each algorithm for scientific workflows of different scales, including the optimal cost and the average cost. In order to show the performance of the proposed algorithm more intuitively, Tables 2, 3 and 4 respectively record the fuzzy execution cost and its fuzzy fitness of scientific workflow scheduling of different scales. At each table, the best solutions among all algorithms are bolded, and all infeasible solutions are marked with "*".

Table 2 shows the fuzzy scheduling results of the tiny-size workflows for 10 repeated experiments. For tiny workflows, ADFGA-PSO obtains all the best solutions, except for the optimal cost of SIPHT; FPSO and FGA is second; FRAND is the worst, with some infeasible solutions. This is because ADFGA-PSO introduces mutation operator and crossover operator of GA to improve the traditional PSO, so as to obtain better fuzzy scheduling strategy. In addition, the size of solution space for fuzzy workflow scheduling problem is generally exponential, so that the random strategy adopted by FRAND has low search efficiency.

Table 3 shows the fuzzy scheduling results of 10 repeated experiments for small-size workflows. ADFGA-PSO obtains the best solution of optimal cost and average cost in all workflows. In addition, the optimal cost of ADFGA-PSO is better than FPSO

348   C. Lin et al.

Table 2. Fuzzy scheduling result of tiny workflows

Workflows	Algorithms	Optimal cost	Average cost
CyberShake	ADFGA-PSO	**(8.94,8.98,9.08),9.02**	**(11.25,11.48,12.14),11.74**
	FPSO	(11.08,11.19,11.60),11.36	(13.42,13.73,14.66),14.09
	FGA	(9.86,9.93,10.22),10.05	(12.18,12.49,13.31),12.81
	FRAND	(17.96,18.66,20.65),19.43	(21.07,21.96,24.42),22.91
Epigenomics	ADFGA-PSO	**(146.64,154.25,174.53),162.05**	**(151.50,159.47,181.36),167.92**
	FPSO	(150.71,157.72,173.19),163.51	(157.01,164.49,177.16),169.03
	FGA	(162.41,166.14,177.28),170.49	(165.85,171.71,181.42),175.17
	FRAND	(168.93,174.57,186.42),178.98	(173.47,180.76,198.27),187.40
LIGO	ADFGA-PSO	**(63.14,64.08,66.81),65.14**	**(63.77,65.81,70.11),67.41**
	FPSO	(64.80,67.05,72.19),68.98	(67.28,69.06,73.61),70.80
	FGA	(64.59,66.27,70.48),67.88	(66.90,68.27,72.09),69.75
	FRAND	(79.77,81.74,86.06),83.36*	(87.39,89.05,93.78),90.88
Montage	ADFGA-PSO	**(3.89,3.99,4.17),4.05**	**(4.11,4.24,4.48),4.32**
	FPSO	(4.79,4.92,5.33),5.08	(5.44,5.64,6.16),5.84
	FGA	(4.79,5.00,5.32),5.11	(5.52,5.76,6.30),5.96
	FRAND	(12.45,13.07,14.53),13.62	(13.76,14.49,16.28),15.17
SIPHT	ADFGA-PSO	(56.26,58.05,64.33),60.54	**(56.58,58.38,64.79),60.93**
	FPSO	**(58.51,59.79,61.88),60.53**	(59.41,60.90,63.78),61.95
	FGA	(60.69,60.84,61.31),61.02	(62.21,63.41,65.96),64.36
	FRAND	(68.62,69.60,72.03),70.52	(70.62,71.72,74.62),72.84

by up to 26.4%, while FGA and FRAND respectively is 15.9% and 125.5%, all of which are for CyberShake. Besides, the average cost is better than FPSO, FGA and FRAND up to 19.9% (Montage), 16.3% (CyberShake) and 117.6% (Montage). It can be considered that scheduling performance of ADFGA-PSO on the cases of CyberShake and Montage is better than comparison algorithms. It is worth noting that these two types of workflows have similar computationally intensive structures, including a large number of data aggregation and data partitioning tasks [17], that is to say, ADFGA-PSO has better performance for scheduling computationally intensive workflows.

Table 3. Fuzzy scheduling results of small workflows

Workflows	Algorithms	Optimal cost	Average cost
CyberShake	ADFGA-PSO	**(16.50,16.71,17.32),16.94**	**(19.77,20.31,21.67),20.83**
	FPSO	(20.20,20.73,22.48),21.42	(22.61,23.36,25.37),24.13
	FGA	(18.72,19.25,20.29),19.63	(22.54,23.40,25.57),24.23
	FRAND	(35.50,36.83,40.38),38.20	(37.83,39.41,43.37),40.92
Epigenomics	ADFGA-PSO	**(345.83,363.89,411.18),382.04**	**(376.84,386.03,410.86),395.60**
	FPSO	(391.26,398.33,418.60),406.20	(407.82,416.06,435.23),423.30
	FGA	(363.74,379.11,428.36),398.48	(388.50,397.94,425.33),408.59
	FRAND	(496.64,505.48,523.12),511.98*	(564.06,568.60,579.76),572.85
LIGO	ADFGA-PSO	**(107.82,111.24,118.70),114.03**	**(113.38,116.20,122.40),118.52**
	FPSO	(120.41,123.28,131.47),126.46	(124.62,127.64,134.89),130.39
	FGA	(114.56,116.30,120.01),117.68	(123.37,126.19,132.77),128.68
	FRAND	(152.70,154.54,160.22),156.76*	(164.98,168.11,175.94),171.09
Montage	ADFGA-PSO	**(14.04,14.74,16.51),15.41**	**(14.81,15.63,17.55),16.35**
	FPSO	(15.19,15.97,17.73),16.63	(17.79,18.70,21.08),19.61
	FGA	(15.18,15.98,17.69),16.62	(16.21,17.10,19.18),17.89
	FRAND	(27.20,28.72,32.28),30.06*	(32.01,33.87,38.39),35.58
SIPHT	ADFGA-PSO	**(121.53,125.81,131.61),127.79**	**(127.25,130.62,137.58),133.20**
	FPSO	(126.75,129.94,138.62),133.29	(131.67,135.11,143.00),138.08
	FGA	(116.28,121.37,138.58),128.18	(130.06,131.97,139.32),134.92
	FRAND	(158.74,164.08,172.71),167.14	(169.13,172.17,178.76),174.63

Table 4 shows the fuzzy scheduling results of 10 repeated experiments for medium-sized workflows. Similar to the small-size, ADFGA-PSO obtains the best solution of the optimal cost and average cost of all workflows. However, as task scale increases, FRAND's performance becomes worse, and it is almost impossible to obtain a feasible scheduling strategy. In addition, ADFGA-PSO obtains better results for workflows with larger task scales, with better robustness for large-scale workflows.

Consequently, compared with comparison algorithms, ADFGA-PSO can effectively jump out of local optimum by combining the characteristics of PSO and GA, so as to obtain better fuzzy execution cost and show better performance.

Table 4. Fuzzy scheduling results of medium workflows

Workflows	Algorithms	Optimal cost	Average cost
CyberShake	ADFGA-PSO	**(44.17,45.38,48.34),46.51**	**(47.44,48.76,52.23),50.09**
	FPSO	(45.97,47.64,51.34),49.03	(49.40,51.19,55.32),52.75
	FGA	(50.24,51.94,56.05),53.50	(52.92,54.64,58.70),56.18
	FRAND	(93.39,96.68,105.27),99.97*	(98.46,101.74,110.10),104.94
Epigenomics	ADFGA-PSO	**(3386.73,3466.08,3589.69),3509.62**	**(3504.47,3596.05,3774.37),3661.50**
	FPSO	(3626.87,3666.34,3815.99),3726.23	(3977.96,4007.34,4085.84),4037.54
	FGA	(3554.42,3610.30,3761.11),3668.39	(3898.24,3962.29,4105.97),4016.23
	FRAND	(5627.38,5650.88,5724.76),5679.87*	(7282.51,7315.38,7398.61),7347.17
LIGO	ADFGA-PSO	**(211.42,212.46,218.45),214.94**	**(217.55,219.01,224.18),221.06**
	FPSO	(241.59,243.66,263.88),252.26	(255.12,257.34,263.90),259.90
	FGA	(217.03,218.52,227.99),222.46	(242.46,243.76,250.33),246.45
	FRAND	(370.15,371.98,376.40),373.66*	(407.01,410.40,416.49),412.60
Montage	ADFGA-PSO	**(33.12,34.10,36.36),34.95**	**(35.63,36.56,38.82),37.42**
	FPSO	(49.53,51.48,56.65),53.47	(56.11,58.42,63.99),60.53
	FGA	(35.43,36.28,38.23),37.01	(44.48,45.86,49.01),47.05
	FRAND	(76.85,80.90,90.94),84.72*	(80.16,84.64,95.54),88.78
SIPHT	ADFGA-PSO	**(179.68,179.88,183.85),181.61**	**(196.88,198.53,207.24),202.11**
	FPSO	(202.81,211.04,222.32),214.91	(217.00,221.85,231.43),225.37
	FGA	(183.41,185.18,187.36),185.91	(197.15,199.97,208.19),203.17
	FRAND	(286.66,289.82,298.31),293.09	(310.68,315.97,324.83),319.14

7 Conclusion

Aiming at the scientific workflow scheduling under uncertain edge environment, based on fuzzy theory, task computation time and data transmission time are expressed as TFNs. Besides, ADFGA-PSO is proposed, which introduces genetic operators into PSO to avoid falling into local optimum so as to improve the search ability. Through the uncertainty scheduling experiment for five scientific workflows of different scales, results show that as for the deadline-based workflow scheduling under uncertain edge environment, the optimal fuzzy execution cost of ADFGA-PSO is better than that of FPSO, FGA, and FRAND up to 26.4%, 15.9% and 125.5% respectively, showing better performance.

In the future work, we will further study the characteristics of scientific workflow and preprocess the structure of workflow. In addition, the performance fluctuation and bandwidth fluctuation of the server will be directly modeled to enrich the uncertain scheduling model of the workflow.

References

1. Shi, W., Zhang, X.: Edge computing: state-of-the-art and future directions. J. Comput. Res. Dev. **56**(1), 69–89 (2019)

2. Hosseini, S.: A hybrid meta-heuristic algorithm for scientific workflow scheduling in heterogeneous distributed computing systems. Eng. Appl. Artif. Intell. **90**, 103501 (2020)
3. Sun, L., Lin, L., Gen, M., et al.: A hybrid cooperative coevolution algorithm for fuzzy flexible job shop scheduling. IEEE Trans. Fuzzy Syst. **27**(5), 1008–1022 (2019)
4. Xie, Y., Zhu, Y., Wang, Y., et al.: A novel directional and non-local-convergent particle swarm optimization based workflow scheduling in cloud-edge environment. Future Gen. Comput. Syst. **97**(AUG.), 361–378 (2019)
5. Huang, B., Li, Z., Tang, P., et al.: Security modeling and efficient computation offloading for service workflow in mobile edge computing. Future Gen. Comput. Syst. **97**(AUG.), 755–774 (2019)
6. Peng, Q., Jiang, H., Chen, M., et al.: Reliability-aware and deadline-constrained workflow scheduling in mobile edge computing. In: 2019 IEEE 16th International Conference on Networking, Sensing and Control (ICNSC), pp. 236–241 (2019)
7. Zadeh, L.A.: Fuzzy sets. Inf. Control **8**(3), 338–353 (1965)
8. Lei, D.: Fuzzy job shop scheduling problem with availability constraints. Comput. Ind. Eng. **58**(4), 610–617 (2010)
9. Fortemps, P.: Jobshop scheduling with imprecise durations: a fuzzy approach. IEEE Trans. Fuzzy Syst. **5**(4), 557–569 (1997)
10. Lee, E.S., Li, R.J.: Comparison of fuzzy numbers based on the probability measure of fuzzy events. Comput. Math. Appl. **15**(10), 887–896 (1988)
11. Sakawa, M., Kubota, R.: Fuzzy programming for multiobjective job shop scheduling with fuzzy processing time and fuzzy duedate through genetic algorithms. Eur. J. Oper. Res. **120**(2), 393–407 (2000)
12. Kennedy, J., Eberhart, R.: Particle swarm optimization. In: Icnn95-International Conference on Neural Networks (1995)
13. Rodriguez, M.A., Buyya, R.: Deadline based resource provisioning and scheduling algorithm for scientific workflows on clouds. IEEE Trans. Cloud Comput. **2**(2), 222–235 (2014)
14. Li, H., Yang, D., Su, W., et al.: An overall distribution particle swarm optimization MPPT algorithm for photovoltaic system under partial shading. IEEE Trans. Ind. Electron. **66**(1), 265–275 (2019)
15. Li, X., Gao, L.: An effective hybrid genetic algorithm and tabu search for flexible job shop scheduling problem. Int. J. Prod. Econ. **174**(Apr.), 93–110 (2016)
16. Shi, Y.: A modified particle swarm optimizer. In: Proceedings of IEEE Icec Conference (1998)
17. Bharathi, S., Chervenak, A., Deelman, E., et al.: Characterization of scientific workflows. In: Third Workshop on Workflows in Support of Large-Scale Science, WORKS 2008 (2008)
18. Topcuoglu, H., Hariri, S., Min-You, W.: Performance-effective and low-complexity task scheduling for heterogeneous computing. IEEE Trans. Parallel Distrib. Syst. **13**(3), 260–274 (2002)
19. Cui, L., Zhang, J., Yue, L., et al.: A genetic algorithm based data replica placement strategy for scientific applications in clouds. IEEE Trans. Serv. Comput. **11**(4), 727–739 (2018)
20. Zhou, B., Xie, S.S., Wang, F., et al.: Multi-step predictive compensated intelligent control for aero-engine wireless networked system with random scheduling. J. Franklin Inst.-Eng. Appl. Math. **357**(10), 6154–6174 (2020)

Fast Shapelet Discovery with Trend Feature Symbolization

Shichao Zhang, Xiangwei Zheng$^{(\boxtimes)}$, and Cun Ji

School of Information Science and Engineering, Shandong Normal University, Jinan, China
jicun@sdnu.edu.cn

Abstract. Time series classification (TSC) is a hot topic in data mining field in the past decade. Among them, classifier based on shapelet has the advantage of interpretability, high accuracy and high speed. Shapelet is a discriminative sub-sequence of time series, which can maximally represent a class. Traditional fast shapelet algorithm uses SAX to represent time series. However, SAX usually loses the trend information of the series. In order to solve the problem, a trend-based fast shapelet discovery algorithm has been proposed. Firstly, the method of trend feature symbolization is used to represent time series. Then, a random mask is applied to select the candidate shapelets. Finally, the best shapelet is selected. The experimental results show that our algorithm is very competitive.

Keywords: Shapelet · Trend feature · Symbolization · Time series classification

1 Introduction

Time series classification (TSC) is one of the classical and hot issues in time series data mining. Time series come from a wide range of sources, including weather prediction, malware detection, voltage stability assessment, medical monitoring, and network anomaly detection [1]. In general, time series $T = \{t_1, t_2, \ldots t_m\}$ is a series of the values of the same statistical index arranged according to the time sequence of their occurrence [2]. The main goal of TSC is to divide an unlabeled time series into a known class.

In the existing time series classification algorithms, shapelet based algorithms are promising. Shapelet is a sub-sequence in time series which can represent a class maximally. These sub-sequences may appear anywhere in the time series and are generally shorter in length. Compared with other TSC algorithms, shapelet based classification method has the advantages of high classification accuracy, fast classification speed and strong interpretability [3]. In order to improve the speed of shapelet dicovery, a fast shapelet algorithm (FS) which uses Symbolic Aggregate Approximation (SAX) representation is proposed by Rakthanmanon and Keogh. However, SAX only uses the mean of sequence to represent time series, and it may cause the loss of trend information of time series. To solve this problem, a fast shapelet discovery algorithm based on Trend SAX (FS-TSAX) is proposed in this work. The main contributions of this paper are as follows:

© Springer Nature Singapore Pte Ltd. 2021
Y. Sun et al. (Eds.): ChineseCSCW 2020, CCIS 1330, pp. 352–363, 2021.
https://doi.org/10.1007/978-981-16-2540-4_26

(1) A new shapelet discovery algorithm is proposed, combining FS and TSAX. It solves the shortcoming that SAX is easy to lose the trend information of time series and improves the accuracy of shapelet classification.

(2) Experiments are conducted on different data sets to evaluate the performance of the proposed algorithm. Experimental results show that the accuracy of our algorithm is at a leading level.

The remainder of this paper is structured as follows. Section 2 gives some related works on shapelet based algorithms and TSAX. Section 3 gives some definitions about FS-TSAX algorithm. Section 4 introduces our proposed FS-TSAX algorithm. Experimental results are presented in Sect. 5 and our conclusions are given in Sect. 6.

2 Related Works

2.1 Shapelet Based Algorithms

Since the concept of the shapelet was first proposed in 2009 [4], algorithms based on Shapelet have been proposed in large numbers.

However, shapelet-based algorithms are complex and take a long time to train [5]. For this, Rakthanmanon and Keogh proposed fast shapelet algorithm (FS) [6]. It uses SAX to reduce the dimension of the original data and uses the mean of sequence to represent time series. Then random masking the SAX string and construct Hash table statistics scores. Finally select the best shapelet according to the scores.

In addition, Wei et al. [7] combined existing acceleration techniques with sliding window boundaries and used the maximum correlation and minimum redundancy feature selection strategy to select appropriate shapelets. To dramatically speed up the discovery of shapelet and reduce the computational complexity, a random shapelet algorithm is proposed by Renard et al. [8]. In order to avoid using online clustering/pruning techniques to measure the accuracy of similar candidate predictors in Euclidean distance space, Grabocka et al. proposed a new method denoted as SD [9], which includes a supervised shapelet discover that filters out only similar candidates to improve classification accuracy. Ji et al. proposed a fast shapelet discovery algorithm based on important data point [12] and a fast shapelet selection algorithm [15]. The former accelerated the discovery of shapelet through important data points. The latter was based on shapelet transformation and LFDPs identification of the sampling time series, and then select the sub-sequences between two non-adjacent LFDPs as candidate sub-sequences of shapelet.

2.2   Trend-Based Symbolic Aggregate Approximation (TSAX) Representation

The symbolic representation of time series is an important step in data preprocessing, which may directly leads to the low accuracy of data mining. SAX is one of the most influential symbolic representation methods at present. SAX is a discrete method based on PAA, which can carry out dimensionality reduction processing simply and mine time series information efficiently. The main step of SAX is dividing the original time series into equal length sub-sequences, and then calculate the mean value of each subsequences and use the mean value to represent the subsequences, that is PAA. Then the breakpoint is found in the breakpoint table with the selected alphabet size, and the mean value of the PAA computed subsequences is mapped to the corresponding letter interval, finally the time series is discretized into strings [10].

However, SAX uses the letters after the mapping of PAA to represent each sub-sequence after segmentation, which may lose important features or patterns in the time series and lead to poor results in subsequent studies. As shown in Fig. 1, the result of SAX string is the same between two time series with completely different trend information.

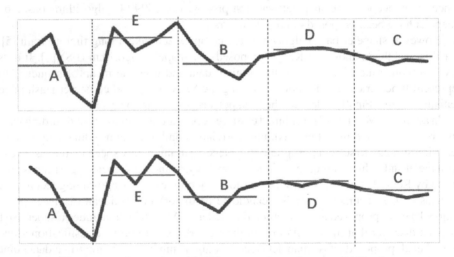

Fig. 1. Two time series with different trends get the same SAX string.

To solve the problem that SAX is easy to lose trend information of time series, Zhang et al. proposed a symbolic representation method based on series trend [11]. Specifically, after PAA is used on the time series, the sub-sequence with equal length are evenly divided into three segments, and the mean value of each segment is calculated respectively. Then a smaller threshold ε is defined and the size of the subsequence is calculated according to the formula (1). The letter "u" represents the upward trend, "d" represents the downward trend and "s" represents for horizontal trend. For example: if the mean of the first sub-sequence is less than the second one, represented as "u", and the mean of the second sub-sequence is larger than the third one, represented as "d", then this piecewise trend information is represented as "ud", so the trend information of the time series is represented. TSAX is a combination of the SAX letters which represent trends.

$$Trend_{indicator} = \begin{cases} d, \overline{C_{i1}} - \overline{C_{i2}} > \varepsilon \\ s, \left| \overline{C_{i1}} - \overline{C_{i2}} \right| \leq \varepsilon \\ u, \overline{C_{i1}} - \overline{C_{i2}} < \varepsilon \end{cases} \tag{1}$$

3 Definition

Definition 1: $T = \{t_1, t_2, \ldots t_m\}$ is a time series which contains an ordered list of numbers. Each value t_i can be any finite number and assume that m is the length of T.

Definition 2: S is a continuous sequence on time series, which can be expressed by formula (2). Where l is the length of S, i is the start position of S.

$$S = T_i^l = \{t_i, t_{i+1}, \ldots, t_{i+l-1}\} \tag{2}$$

Definition 3: Time series dataset D is a set of N time series, each of which is m in length and belongs to a specific class. The class number in D is C.

Definition 4: $(dist\,(T, R))$ is a distance function, whose input is two time series $T = \{t_1, t_2, \ldots t_m\}$ and $R = \{r_1, r_2, \ldots r_m\}$. It returns a non-negative value. This paper uses Euclidean distance, and its calculation method is shown in formula (3).

$$dist(T, R) = \sqrt[2]{\sum_{i=1}^{m} (t_i - r_i)^2} \tag{3}$$

Definition 5: The distance between the subsequence S and the time series $T(subdist(T, S))$ is defined as the minimum distance between subsequence S and any subsequence of T of the same length as subsequence S. It is a distance function, which inputs time series T and sub-sequence S, returns a non-negative value. Intuitively, this distance is the distance between S and the best matching point at a certain position in T, as shown in Fig. 2, and its calculation method is given by formula (4).

$$SubDist(T, S) = min(dist(T_1^l, S), dist(T_2^l, S), \ldots, dist(T_{m-l+1}^l, S)) \tag{4}$$

Definition 6: Entropy is used to indicate the level of clutter in a dataset. The entropy of dataset D shown in formula (5). Where D are datasets, C are different classes, and p_i is the proportion of time series in class i.

$$e(D) = -\sum_{i=1}^{c} (p_i \log p_i) \tag{5}$$

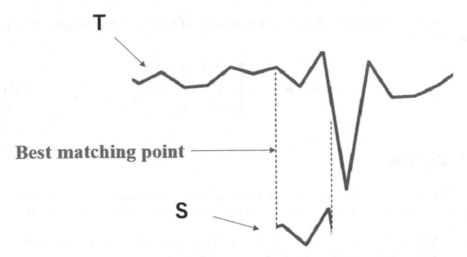

Fig. 2. Best matching point.

Definition 7: Information gain represents the degree of uncertainty reduction in a dataset under a partition condition. For a spilt strategy, the information gain calculation method is shown in Formula (6).

$$gain(TSAX) = e(D) - \frac{n_C}{n}e(D_C) - \frac{n_N}{n}e(D_N) \tag{6}$$

Definition 8: Optimal Split Point (*OSP*). When the information gain obtained by splitting at a threshold is larger than any other point, this threshold (distance value) is the *optimal split point*.

4   Fast Shapelet Discovery Algorithm Based on TSAX(FS-TSAX)

4.1   Overview of the Algorithm

The TS-TSAX algorithm is shown in Algorithm 1 and Fig. 3. Figure 3(a) shows the four processes of the FS-TSAX algorithm: (1) Generating TSAX words (Line 1–Line 3 in Algorithm 1); (2) Random masking of TSAX Words (Line 5–Line 7 in Algorithm 1); (3) Choose the top-k TSAX words with highest scores (Line 9 in Algorithm 1); (4) Find the best shapelet (Line 17–Line 21 in Algorithm 1).

Algorithm 1 *FastShapeletBasedOnTSAX*

Input: Time series dataset D
            The number of iterations r
            TSAX candidate number k
Output: *shapelet*
1: [TS,Label]← ReadData(D) // Read the time series and
                                        their labels
2: **for** (length = 1, length<*m, length++*)
3:      TSAXList ← creatTSAXList (TS,length) //creat
                                        TSAXList
4:      S ← {}
5:      **for** (i=1, i<r, i++)
6:          Calculate ← RandomMask (TSAXList,TS) // random
                                        masking TSAX words
                                        and calculating
                                        score
7:              S ← New S (S,Calculate)   //update scores
8:      **end for**
9:      TSAXCand ← FindTopKTSAX (SList,Calculate,k,r)
                                        //choose the best
                                        top-k shapelets
10:     Candshapelets ← Remap (TSAXCand,TS) // Select can-
                                        didate shapelets
11:
12:     Gain ← Inf, Gap ← 0
13:     **for** (i=1, i<Candshapelets, i++)
14:         candidate ← Candshapelets [i]
15:         Dist ← Euclidean distance (TS, candidate) //
                                        Calculate the distance be-
                                        tween subsequence and time
                                        series
16:         [gain,gap] ← CalInfoGain(Dist) // Calculated in
                                        formation gain
17:         **while** (gain>Gain) ||
18:            ((gain==Gain)&&(gain>Gap))
19:            Gain ← gain
20:            Gap ← gap
21:            shapelet ← candidate //find shapelet
22:         **end while**
23:     **end for**
24: **end for**

Figure 3(b) shows the visual description of these four steps. Next, these four steps are described in detail.

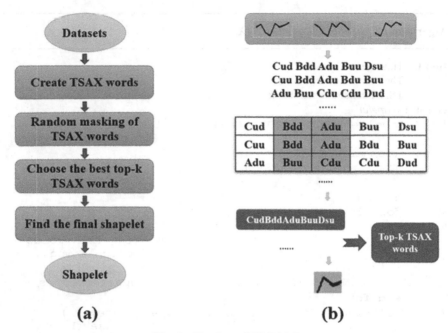

Fig. 3. The flow of FS-TSAX.

4.2 Create TSAX Words

For an original time series, after normalization, it is divided into equal length subsequences, then calculate the mean value of each segment and use the mean to find the corresponding letter in the breakpoint table, and a string is used to represent the time series, this process is SAX. After the time series is segmented with equal length by PAA, the sub-sequence segments are evenly divided into three segments. Calculate the mean of each segment and use formula (1) to get the trend letters. Then the trend letters of each segment with the SAX letter are combined, and the time series is represented in TSAX. Figure 4 shows an example of this process.

4.3 Random Masking TSAX Words

Two time series with similar real values may produce two different TSAX words just because of a minimal difference. Therefore, the best shapelet in the original time series may map to different TSAX words. The solution to this problem is to use random masking, which is the idea of projecting all the higher-dimensional TSAX words into smaller dimensions. The process of random masking TSAX words is as follows:

(1) Randomly select a character in a TSAX word.
(2) Select another character to generate the new TSAX word.

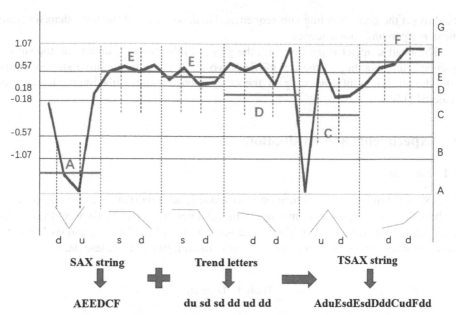

Fig. 4. Combine the SAX string "AEEDCF" and the trend letters "du sd sd dd ud dd" and get the TSAX string "AduEsdEsdDddCudFdd".

Typically, this process requires 10 iterations, which means masking 10 times TSAX words. After the TSAX words are randomly masked, the TSAX words and the words after masking are applied together for subsequent processing.

4.4 Choose the Best k TSAX Words

For a TSAX word, note 1 when it appears in a time series and 0 when it does not. By statistics of this information, training data set D can be divided into two sub-datasets: dataset D_C composed of time series containing the TSAX words, and dataset D_N composed of time series not containing the TSAX words. The information gain for this TSAX word can be calculated according to the following formula.

$$gain(TSAX) = e(D) - \frac{n_C}{n}e(D_C) - \frac{n_N}{n}e(D_N) \tag{7}$$

Where D is the dataset, D_C is the data set formed by the time series containing the TSAX words, D_N is the dataset formed by the time series without the TSAX words, n, n_C, and n_N are the number of time series contained in each of the three datasets. After calculating the information gain of TSAX, we find the best k TSAX words which has the best information gain and obtain the final shapelet.

4.5 Discover the Best Shapelet

Each TSAX word represents a corresponding time series, so after getting the top-k TSAX words, the corresponding relationship between TSAX words and time sequence can be used to get the corresponding sub-sequence. Then, we can find the final shapelet from the corresponding sub-sequence.

The finall shapelet is the one with the greatest information gain among the subsequences corresponding to the top k TSAX words. If there are multiple subsequences with the maximum information gain, the sub-sequence with the maximum clearance is selected [11].

5 Experiments and Evaluation

5.1 Datasets

UEA&UCR time series classification warehouse is an important open source data set in the field of time series data mining. In this chapter, we select 12 datasets from it for comparative experiments [13]. These data sets are set in "arff" format and each dataset sample carries a category label. Table 1 shows the information of these datasets.

Table 1. Dataset.

Data sets	Number of classes	Size of training dataset	Size of test dataset	Length of time series
Beef	5	30	30	470
ChlorineConcentration	3	467	3840	166
Coffee	2	28	28	286
ECG200	2	100	100	96
FaceFour	4	24	88	350
Haptice	5	155	308	1092
ItalyPowerDemand	5	67	1029	24
OSULeaf	6	200	242	427
Trace	2	810	3636	500
TwoLeadECG	2	23	1139	82
WordSynonyms	25	267	1139	82
Yoga	2	60	61	637

5.2 Effect of the Number of TSAX Segments

To verify the effect of the number of segments, we compared the classification accuracy on different data sets when the number of TSAX segments is 2 and 3. The experimental results are shown in Fig. 5. One thing to explain, theoretically, the number of segments is artificially selected. But our code uses binary to symbolize the time series, and the int in Java is 32 bits, and representing a letter needs two bits. A TSAX word is 15 characters, consisting of 5 SAX words and 10 trend letters, so it needs to be represented with 30 bits, and the maximum value of segment number is 3. Therefore, we compared the influence on classification accuracy when the number of segments is 2 and 3. As shown in Fig. 5, the accuracy of FS-TSAX (three-segments) on the 10 data sets was higher than that of FS-TSAX (two-segments), the accuracy of FS-TSAX (three-segments) on the 1 data sets was lower than that of FS-TSAX (two-segments), and they are equally accurate on 1 data set. In general, FS-TSAX (three-segments) is more competitive than FS-TSAX (two-segments).

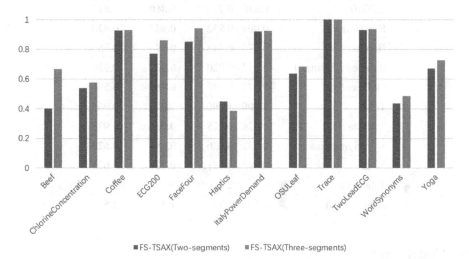

Fig. 5. Comparison of accuracy between FS-TSAX (three-segments) and FS-TSAX (two-segments).

5.3 Accuracy Comparison

In this section, we calculate the accuracy of our method and other three shapelet discovery algorithms. Table 2 shows the accuracy of these algorithms. The algorithm with the best result in each data set is shown in bold. The algorithms we selected are FS [6], SD [9] and the TSAX (Two-segments) algorithm mentioned in the previous section. As Table 2 shows, our approach gets the highest accuracy on 9 data sets, and the average rank is 1.25.

Table 2. Comparison of accuracy of different algorithms on different data sets.

	FS	FS-TSAX (Two-segments)	FS-TSAX (Three-segments)	SD
Beef	0.567	0.400	**0.667**	0.507
ChlorineConcentration	0.546	0.537	**0.577**	0.553
Coffee	0.929	0.927	0.929	**0.961**
ECG200	0.810	0.770	**0.860**	0.818
FaceFour	0.909	0.852	**0.943**	0.820
Haptice	0.376	**0.448**	0.386	0.356
ItalyPowerDemand	0.917	0.920	**0.922**	0.920
OSULeaf	0.678	0.636	**0.681**	0.566
Trace	**1.000**	**1.000**	**1.000**	0.965
TwoLeadECG	0.924	0.928	**0.936**	0.928
WordSynonyms	0.431	0.436	0.484	**0.625**
Yoga	0.695	0.671	**0.727**	0.625
Total wins	1	2	9	2
Average rank	2.5	2.75	1.25	2.58

6 Conclusion

The algorithms based on shapelet have attracted great attention in recent years. Shapelet discovery algorithm is the basis of shapelet transformation algorithm and shapelet learning algorithm, and it has various of acceleration strategies. We propose a shapelet discovery algorithm of FS-TSAX, which uses TSAX to represent time series, and can retain the trend information of time series well, and then carry out the process of shapelet discovery. Experiments on different data sets show that compared with other shapelet discovery algorithms, the accuracy of the proposed algorithm is at a leading level, especially in some time series with obvious trends.

References

1. Yan, W., Li, G.: Research on time series classification based on shapelet. Comput. Sci. **046**(001), 29–35 (2019)
2. Zhao, C., Wang, T., Liu, S., et al.: A fast time series shapelet discovery algorithm combining selective ex-traction and subclass clustering. J. Softw. **000**(003), 763–777 (2020)
3. Zhang, Z., Zhang, H., Wen, Y., Yuan, X.: Accelerating time series shapelets discovery with key points. In: Li, F., Shim, K., Zheng, K., Liu, G. (eds.) APWeb 2016. LNCS, vol. 9932, pp. 330–342. Springer, Cham (2016). https://doi.org/10.1007/978-3-319-45817-5_26
4. Ye, L., Keogh, E.: Time series shapelets: a novel technique that allows accurate, interpretable and fast classification. Data Mining Knowl. Discovery **22**(1–2), 149–182 (2011)

5. Bagnall, A., Lines, J., Bostrom, A., et al.: The great time series classification bake off: a review and experimental evaluation of recent algorithmic advances. Data Mining Knowl. Discov. **31**(3) (2017)

6. Rakthanmanon, T., Keogh, E.: Fast shapelets: a scalable algorithm for discovering time series shapelets. In: Proceedings of the 2013 SIAM International Conference on Data Mining. Philadelphia: Society for Industrial and Applied Mathematics, pp. 668–676 (2013)

7. Wei,Y., Jiao, L., Wang, S., et al.: Time series classification with max-correlation and min-redundancy shapelets transformation. In: International Conference on Identification, Information, and Knowledge in the Internet of Things (IIKI), Beijing, China, pp. 7–12 (2015)

8. Renard, X., Rifqi, M., Detyniecki, M.: Random-shapelet: an algorithm for fast shapelet discovery. In: IEEE International Conference on Data Science and Advanced Analytics, pp. 1–10. IEEE (2015)

9. Grabocka, J., Wistuba, M., Schmidt-Thieme, L.: Fast classification of univariate and multivariate time series through shapelet discovery. Knowl. Inf. Syst. **49**(2), 1–26 (2015)

10. Lin, J., Keogh, E., Li, W., et al.: Experiencing SAX: a novel symbolic representation of time series. Data Mining Knowl. Discov. **15**(2), 107–144 (2007)

11. Zhang, K., Li, Y., Chai, Y., et al.: Trend-based symbolic aggregate approximation for time series representation. In: Chinese Control and Decision Conference (CCDC). Shenyang, pp. 2234–2240 (2018)

12. Ji, C., Zhao, C., Lei, P., et al.: A fast shapelet discovery algorithm based on important data points. Int. J. Web Serv. Res. **14**(2), 67–80 (2017)

13. Bagnall, A., Lines, J., Keogh, E., et al.: The UEA and UCR time series classification repository (2016). www.timeseriesclassification.com

14. Ji, C., Zhao, C., Liu, S., et al.: A fast shapelet selection algorithm for time series classification. Comput. Netw. **148**, 231–240 (2019)

A Distributed Storage Middleware Based on HBase and Redis

Lingling Xu[1,2], Yuzhong Chen[1,2,3], and Kun Guo[1,2,3](✉)

[1] College of Mathematics and Computer Science, Fuzhou University, Fuzhou 350108, China
gukn@fzu.edu.cn
[2] Fujian Provincial Key Laboratory of Network Computing and Intelligent Information Processing, Fuzhou, China
[3] Key Laboratory of Spatial Data Mining and Information Sharing, Ministry of Education, Fuzhou 350108, China

Abstract. In the era of big data, traditional relational databases face challenges in massive data storage. HBase is a non-relational database based on column storage which is used widely for big data storage. The writing performance of HBase is high, but the unbalanced load caused by its uneven data storage strategy is the bottleneck of its reading performance. HBase needs to access disks to get query results. Therefore, the efficiency of disk access has a great impact on the query performance of HBase. In view of the above problems, this paper proposes a storage middleware which utilizes HBase for load balancing and Redis for memory caching. Specifically, we improve HBase's original load balancing algorithm for Regions and RegionServers and customize a Redis cache eviction algorithm according to the data's query and update frequency. Furthermore, the coprocessor of HBase is used to synchronize data between HBase and Redis. Experiments on synthetic datasets show that the proposed storage middleware achieves better writing and reading performance than HBase. The load balancing algorithm employed in the middleware is better than HBase's original algorithm. The hit rate of the customized cache eviction algorithm is also higher than that of the LRU algorithm.

Keywords: Distributed storage · Load balancing · Memory cache · HBase · Redis

1 Introduction

With the advent of the era of big data, traditional relational databases can no longer meet people's requirements. In some circumstances, data storage does not require strict compliance with the ACID model, therefore new distributed NoSQL databases [1] are gaining more and more attention, such as HBase. HBase is derived from Bigtable [2] proposed by Google. Its bottom layer uses HDFS which is part of the Hadoop ecosystem to store data. With the increase of data volume, load balancing of the distributed database has become one of the bottlenecks of performance. In HBase, data is written to RegionServers in an unbalanced way, making some RegionServers overloaded and

© Springer Nature Singapore Pte Ltd. 2021
Y. Sun et al. (Eds.): ChineseCSCW 2020, CCIS 1330, pp. 364–380, 2021.
https://doi.org/10.1007/978-981-16-2540-4_27

others underloaded. Hence, RegionServers have become a bottleneck of HBase. HBase continuously read and write requests to the disks, which makes its query performance limited by hardware. Load balancing in HBase aims to balance the load of its Region-Servers [3]. However, the size of Regions in different RegionServers varies greatly, which may lead to the phenomenon that only a small part of the RegionServers are busy when reading data. Using memory caching technology to cache HBase's hot data has become a common method to improve its query performance. The memory cache technology for HBase employs third-party tools (such as Redis, Memcache, etc.) and a set of auxiliary tools [4]. The way of writing auxiliary tools for caching according to the LSM-Tree model will decrease HBase's writing performance. The third-party tools provide few cache eviction strategies, and they cannot support cache updating.

To solve the above problems, this paper proposes a high efficient distributed storage Middleware (HEDSM) which employs HBase to persist data and Redis to cache hot data to speed up data query. The main contributions of this paper are as follows:

(1) In order to solve HBase's problem of unbalanced loading, a global load balancing strategy based on Region and RegionServer levels are proposed to improve the reading and writing performance of HBase.
(2) In order to solve Redis's problem of not supporting update caching, a heat value cache eviction strategy based on the time smoothing method of query frequency and update frequency is proposed to improve the cache hit rate of Redis.
(3) In order to improve the performance of data synchronization between HBase and Redis, a strategy of rewriting the preput function of HBase coprocessor is proposed to avoid the generation of dirty data.
(4) A middleware that contains a storage module to implements the above improvements and a query module to verify the improvements is developed. The experimental results on synthetic datasets demonstrate the effectiveness of our strategies.

2 Related Work

2.1 Load Balancing Strategies of HBase

HBase inherently contains two load balancing strategies: SimpleLoadBalancer and StochasticLoadBalancer [5]. After version 0.98 was published, HBase uses the StochasticLoadBalancer strategy by default, which balances the load of RegionServers in HBase by reducing the cost of the evaluation function. The scheme's goal is to keep the load of the Region managed by RegionServer as equal as possible. However, the strategy of Region migration it adopts will increase the number of iterations.

Several improvements of the HBase-based load balancing algorithms have been proposed. Huang et al. [6] proposed an improved method of load balancing based on subtable restrictions. By distributing Regions of the same table to different RegionServers, the request of a certain RegionServer is reduced. To perform HBase load balancing, the method considers HBase load balancing locally without considering the integrity of HBase. Guo et al. [7] proposed a task scheduling algorithm based on load balancing, which maps fixed frequency data to the server. The algorithm only considers the load balancing at the level of RegionServer resulting in a small number of regions and large region capacity. Shao et al. [8] used the heat of the predicted data as the load of RegionServers and reduces the evaluation function by migrating Region. However, the migration is random and stops only when the traversal of the Region in the specified RegionServer is completed, which takes a long time.

2.2 Cache Strategies of HBase

Currently, HBase-oriented data caching technology is considered from the hardware and software aspects. TBF [9] proposed a cache eviction strategy based on solid-state drives. The algorithm can reduce the space overhead of metadata by combining the CLOCK and the Bloom filter. However, data query information outside a CLOCK cycle cannot be saved. Microsoft [10] uses the exponential smoothing method to distinguish the frequency of the recorded data query and the difference between the cold and hot data to dynamically adjust the boundaries of the cold and hot data. However, the method is used mainly for the analysis of the log data query. Zhang et al. [11] proposed a method that uses Redis's exponential smoothing to predict the frequency of record query. The method has a higher hit rate than the LRU algorithm, but the method directly uses Redis as a persistent database, which is not suitable for big data processing. HiBase [12] uses Redis to index HBase's hotspot data and employs a cache eviction strategy based on heat accumulation. However, it does not consider the impact of data update frequency on the cache. Zhai et al. [13] introduced TwemProxy into Redis clusters, which is a Redis middleware proxy service that can improve the performance of Redis. However, the solution has to maintain an additional set of TwemProxy clusters, which incurs additional overhead. Li et al. [14] proposed a hot cumulative Redis cache eviction algorithm based on query and update frequency. The algorithm considers the frequency of data updates, but it neglects the impact of historical data.

2.3 Data Synchronization Strategies of HBase

Currently, methods for data synchronization between HBase and other frameworks are mainly divided into two categories: the methods based on HBase and ElasticSearch data synchronization and the methods based on HBase and custom memory caching tools. In the first category of methods, ElasticSearch is used to store HBase's secondary indexes [15–17] and a coprocessor for index management is employed to achieve data synchronization. In the second category of methods, a coprocessor is used to synchronously cache the data from HBase to the memory cache tool [4].

3   The Proposed Distributed Storage Middleware

As shown in Fig. 1, HEDSM uses HBase and Redis as the underlying storage framework, and stores the data in the HBase database for persistent storage. HBase stores the data in disks and the query to the data requires multiple times of disk access, which incurs inefficient data reading. This paper uses Redis as a memory database to store the frequently queried data to speeds up data reading.

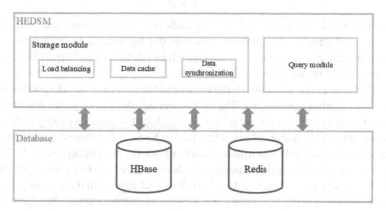

Fig. 1. The HEDSM architecture

3.1   The Storage Module

The storage module contains three components: load balancing, data cache, and data synchronization. The load balancing component realizes the load balancing of Region and RegionServer to improve the reading and writing performance. The data cache component implements a cache elimination strategy based on the heat value calculated by the time smoothing method to improve the cache hit rate. The data synchronization component realizes the data synchronization problem of HBase and Redis based on the coprocessor to avoid the generation of dirty data.

3.1.1   The Load Balancing Component

HBase always writes data to a Region with the largest Startkey. When the data storage reaches the upper limit of a Region's space which is restricted by a preset threshold., the Region is split and the data is transferred to the new Region with the largest Startkey. In this way, the writing operations will concentrate on the same node, leading to the problem of data writing hotspots. This paper develops in load balancing component two load balancing strategies for Region and RegionServer respectively.

(1) Load balancing for Region.

First, Region-level load balancing performs pre-partitioning operations so that data can be written to different regions when data is written. The prepartition generates the Startkey and Endkey of each partition.

The number of HBase partitions is determined by the following Eq. 18 :

$$N = \frac{M \times F}{S \times A} \qquad (1)$$

M represents the memory size of RegionServer. F represents the proportion of RegionServer allocated to MemStore, which is 0.4 by default in HBase. S indicates the size of MemStore, the default value of which in HBase is 128 MB. A is the number of column families in the table.

Even if HBase is pre-partitioned, the unreasonable design of Rowkey will still make the size of the Region unbalanced. Therefore, second, this paper generates Rowkey by a consistent hash function with virtual nodes which can write data evenly to Regions.

Consistent hash algorithm 19 is widely used in NoSQL databases. it can be regarded as a load balancing method to distribute data to different computation nodes. A consistent hash algorithm without virtual computation nodes uses a certain hash function to map the data into a large space, as shown in Fig. 2(a). K1 to K6 represent data items (or records) and A to D represents the computation nodes. When storing data, the algorithm first gets a hash value for each record that reflects its position in the ring. Second, it finds the nearest computation node in a clockwise direction to store the record. A consistent hash algorithm with virtual nodes first adds "virtual computation nodes" which are mapped to the corresponding real computation nodes to the ring. Then, the nearest computation node (which may be virtual) is searched in a clockwise direction of the ring, as shown in Fig. 2(b).

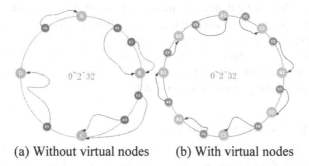

(a) Without virtual nodes   (b) With virtual nodes

Fig. 2. Consistent hash algorithm

Rowkey is generated by a consistent hash algorithm with virtual nodes to map the original unique identifier of the data. In this way, a region is used as a computation node, which means the number of regions is equals to the number of real computation nodes. This paper chooses the FNV1_32_HASH algorithm [20] to generate Rowkey. The basic process is as follows: first, this paper chooses the number of virtual computation nodes and map them to the real computation nodes; second, the primary key of each record needs to be hashed., The hash values are used as the Startkey of the Region when storing the records; finally, the Startkeys are concatenated with a universal unique identifiers to generate the Rowkeys. The details of the Rowkey generation is described in Algorithm 1.

Algorithm 1 Generate Rowkey

Input: *input* //the primary key of data

Output: *Rowkey* //String

1. $N \leftarrow \frac{M \times F}{S \times A}$ // Pre-split HBase and get the number of regions
2. $N_v \leftarrow N_t \times N \times 3$ // Determine the number of virtual nodes
3.   *Startkey* $\leftarrow$ FNV1_32_HASH (*input*) //Get Startkey
4.   *Rowkey*←*Startkey*||UUID   //Generate Rowkey
5.   Return *Rowkey*

---

(2) Load balancing on RegionServer.

Only using region-level load balancing may cause regions with frequent reads and writes to be stored on the same RegionServer, resulting in the phenomenon that only some of the RegionServers in the HBase cluster are busy and the others are idle. In this case, highly concurrent queries will cause congestion, which greatly reduces the performance of HBase.

The load balancing strategy of HEDSM on RegionServer periodically monitors the reading and writing requests of each Region and RegionServer and records the over-loaded RegionServer and low-load RegionServer. A greedy algorithm is used to exchange the maximum Region in the maximum overloaded RegionServer and the minimum Region in the minimum underloaded RegionServer to make the reading and writing requests of each RegionServer as close as possible. Therefore, the load of HBase is more balanced.

The realization of the load balancing strategy on RegionServer is shown in Algorithm 2:

Algorithm 2 Load balancing on RegionServer

Input: $loadRS(loadN(loadR))$, the initial set of region servers with load; t, the number of iterations; n, the number of nodes

Output: $S(maxN, minN, maxR, minR)$, the switching strategy

$maxN$ represents region server with the maximum load

$minN$ represents region server with the minimum load

$maxR$ represents region with the maximum load in $maxN$

$minR$ represents region with the minimum load in $minN$

1. Set the visited *flag* of each region to false

2. $load_{avg} \leftarrow \frac{\sum_{i=1}^{n} loadN_i}{n}$ // the average load of all region servers

3. $load_{max} \leftarrow 1.02 \times load_{avg}$ // the ideal upper bound of load

4. $load_{min} \leftarrow 0.98 \times load_{avg}$ // the ideal lower bound of load

5. **FOR** each $loadN_i$ in $loadRS$ **DO**

6.   **IF** $loadN_i > load_{max}$ **THEN**

7. $overLoadQueue \leftarrow loadN_i$ // Add region server with load greater than the upper bound to the $overLoadQueue$

8.   **ELSE IF** $loadN < load_{min}$ **THEN**

9. $underLoadQueue \leftarrow loadN$ // Add region server with load less than the lower bound to the $underLoadQueue$

10.   **ELSE**

11. $normalLoadQueue \leftarrow loadN_i$

12.   **END IF**

13. **END FOR**

14. **WHILE** $t>0$ **DO**

15. $(loadN_{max}, loadN_{min}) \leftarrow Sort(loadRS)$ // Search region servers with the maximum load and the minimum load in $loadRS$

16. $T_{current} \leftarrow \frac{loadN_{max}}{loadN_{min}}$ // $T_{current}$ represents current value of evaluation indicator

17. $maxN \leftarrow$ sort($overLoadQueue$)   // $maxN$ represents region server with the maximum load in the sorted $overLoadQueue$

18. $minN \leftarrow$ sort($underLoadQueue$) // $minN$ represents region server with the minimum load in the sorted $underLoadQueue$

19. $maxR \leftarrow$ sort($maxN$) // $maxR$ represents non-visited region with the maximum load in the sorted set of regions in $maxN$

20. $minR \leftarrow$ sort($minN$)// $minR$ represents region with the minimum load in the sorted set of regions in $minN$

21.   **IF** $maxR$ is null **THEN**

22.   remove $maxN$ from $overLoadQueue$

23.   **CONTINUE**

24.   **END IF**

25.   Set the visited *flag* of *maxR* to true
26. $h \leftarrow load\_maxR - load\_minR$ // *load_maxR* represents load value of *maxR*, *load_maxR* represents load value of *minR*
27.   **IF** $h>0$ **AND** $load\_maxN - h > load_{min}$ **AND** $load\_minN + h < load_{max}$ **THEN**
28. $S \leftarrow (maxN, minN, maxR, minR)$
29.   exchange the position of *maxR* and *minR*
30.   update(*loadRS*) // update load value of related region server
31. $T_{new} \leftarrow \frac{load_{max}}{load_{min}}$ // T_{new} represents current new value of evaluation indicator
32.   **IF** $T_{new} > T_{current}$ **THEN** // undo exchange
33.   exchange the position of *maxR* and *minR*
34.   update(*loadRS*) // update load value of related region server
35.   update(*S*)
36.   **END IF**
37.   **END IF**
38. $t \leftarrow t - 1$
39. **END WHILE**
40. Return *S*

3.1.2 The Data Cache Component

Each reading and writing request to HBase involves disk access, which limits the performance of data reading and writing. Therefore, Redis is introduced to cache data. Redis has 6 kinds of cache elimination strategies, of which LRU is the most frequently used. LRU is based on the idea that recently more queried data is more likely to be queried. However, it does not consider the frequency of data query, which is the long suit of the LFU strategy.

This paper designs a cache eviction algorithm based on the heat value calculated by the time smoothing method according to query and update frequency. This paper proposes the more frequently queried data will be more likely to be queried in the future, and the more frequently updated data will be less likely to be queried. In view of this, this paper designs the following equations to calculate the heat value of a record:

$$
\begin{aligned}
heatvalue_n &= \alpha \times \frac{queryf}{updatef} + (1 - \alpha) \times heatvalue_{n-1} \\
&= \alpha \times \frac{\frac{queryCount}{T}}{\frac{updateCount}{T}} + (1 - \alpha) \times heatvalue_{n-1} \\
&= \alpha \times \frac{queryCount}{updateCount} + (1 - \alpha) \times heatvalue_{n-1}
\end{aligned}
\tag{2}
$$

queryCount is the number of visits in a cycle, *updateCount* is the number of updates in a cycle, *T* is a heat calculation cycle, *queryf* is the query frequency, *updatef* is the update frequency, $0 < \alpha < 1$, and the historical value $heatvalue_{n-1}$ reflects the historical heat. The parameter α is the attenuation coefficient which is used to determine the weights of the accumulated heat and the historical heat in $heatvalue_{n-1}$. The larger the weight of recent visit is, the smaller the impact of the historical query records on data heat is, and vice versa. The historical heat of the data in the cache decayed at the rate of $1 - \alpha$ in each cycle. The accumulated heat of the earlier cycles will experience more decay. Therefore, the impact of the early accumulated heat on the data heat is reduced gradually.

The idea of the data cache component is to calculate the number of times the cached data being queried or updated and calculate the heat values of each record according to Eq. (2) and store them in an ordered hot value set. At the beginning of data caching, the cache is empty, the data can be written to HBase and Redis directly. When the amount of cached data reaches the preset threshold, the cache eviction algorithm is invoked which will delete K records with the lowest heat values. The basic steps of cache eviction are shown in Algorithm 3:

Algorithm 3   Cache eviction

Input: *Rn(key,value)* //the string data;
way //operation way: insert : 1, query : 0;
T //period time;
α // attenuation coefficient
Output: *result* //operation result*(success/failure)*
1. **WHIILE** *Time* == *T* **DO**
2.   *heatvalue = α×queryCount/updateCount+(1-α) ×heavalue*
3.   *Zset(score,key,hotvalue)* //Ordered set Zset
4. **END WHILE**
5. **IF** way = 0 **THEN**
6.   **IF** Zset.size()>N AND not *key*.exists **THEN**
7.   Zset.delete(*n*) // Eliminate n data with low hot value
8.   *visitCount ← visitCount + 1*
9.   flag← insert*(key, value, visitCount, udateCount)*
10. **END IF**
11. **ELSE**   //insert data
12.   **IF** Zset.size()>N AND !*key*.exists **THEN**
13.   Zset.delete(*n*) // Eliminate n data with low heat value
14.   *updateCount ← updateCount + 1*
15.   *flag ← insert(key, value, visitCount, udateCount)*
16. **END IF**
17. **END IF**
18. **IF** flag **THEN**
19.   return *success*
20. **ELSE**
21.   return f*ailure*
22. **END IF**

3.1.3   The Data Synchronization Component

A data synchronization usually needs to consider the consistency of data between the cache and the database. Data inconsistency will cause "dirty" data. HBase's Observer

Coprocessor is similar to the trigger in RDMS. The hook function can be customized and will be called automatically by the server. RegionServer provides hook functions for client-side data manipulation, such as Get, Put, Delete, Scan, etc. The data synchronization component uses the Put function to write data to Redis by rewriting the preput method. When the client sends a put request, the request is distributed to the appropriate RegionServer and Region. Then, the Observer Coprocessor intercepts the request and calls the preput function. After the preput function is executed, HBase executes the Put function to perform persistent storage to realize data synchronization. The details of data synchronization are given in Algorithm 4.

Algorithm 4   Data synchronization

Input: *input(Rowkey, column family, attributes)* // the put data input
Rowkey // generated by load balancing on Region algorithm
column family // the column family of HBase
attributes // data attribute list
Output: *result* // put result(*success/failure*)
1.   **IF** Put *input* **THEN**   // Intercept put request
2.   *flag* ← preput(*input*) //Execute function preput
3.   **IF** *flag* **THEN**
4.   *result* ← Put *input*
5.   **END IF**
6.   **END IF**
7.   **Function** preput(*input*)
8.   *Rn* ← (*Rowkey* , *attributes*) //Generate the Redis input string data
9.   Cache eviction(*Rn, 1, T, α*) //Call algorithm 3
10. **END Function**

3.2   The Query Module

The query module is designed based on the above-mentioned improvements of HB-ase and Redis to speed up data query. When the data is hit in Redis, the query task is ended without accessing the disk, which greatly reduces the disk overhead and query time. HEDSM supports batch data query. The basic steps are as follows:

(1)   Read the query conditions in batches and use Algorithm 1 to generate Rowkeys.
(2)   Traverse Rowkeys and initiate a query request to Redis. If Rowkey exists in Redis, the record matching the Rowkey will be returned from Redis's cache and the query is ended.
(3)   Otherwise, initiates a request to the specified region of HBase. If the querying record exists in HBase, write it to Redis by Algorithm 3 and return the record. The query is ended. Otherwise, return a flag indicating that the querying record does not exist.

4   Experiments

4.1   Experimental Environment

An HBase cluster with 5 nodes was deployed for the experiments. The hardware of each machine in the cluster is dual-core CPU 2.60 GHz, 16G memory, 300G hard disk. The operating system is Ubuntu 16.04.3. Other software installed on the cluster is given in Table 1.

Table 1. Cluster softwares

Software name	Version
Hadoop	2.6.5
Zookeeper	3.4.6
HBase	1.1.5
Redis	4.0.7

4.2   Datasets

In order to evaluate the effectiveness of the modules in HEDSM, this paper uses a synthetic raw dataset from a province's annual power consumption, as shown in Table 2. This paper has built several specialized datasets from the raw dataset for the experiments of load balancing, cache hit ratio, and batch query, as shown in Table 3.

Table 2. Power consumption data

USER_NO	USER_NAME	AREA_NO	TRADE_NO	USER_TYPE
2346467	ZhangSan	1	01	01
2346458	LiSi	2	02	01
131686	WangQian	1	04	01
307945	LiJin	4	06	02

4.3   Experimental Scheme

In the experiments verifying the load balancing component, the basic steps are as follows:

Step 1: Determine the optimal number of regions through HBase cluster configuration and calculate the number of regions for each RegionServer according to Eq. (1). In this experiment, the number of regions for each RegionServer is 13.

Table 3. Experimental datasets

Experiment	Size of dataset						
Load balancing	1 million		4 million		7 million		10 million
Cache hit ratio	5,000	3,333	2,500	2,000	1,667	1,429	1,250
Batch Query	5,000		10,000		15,000		20,000

Step 2: Generate Rowkey according to Algorithm 1.
Step 3: Set two periods (T = 5, T = 10) and periodically execute Algorithm 2 on the RegionServers.

In order to verify the performance of the data cache component, this paper employs 7 datasets, whose details are shown in Table 4. Each dataset will perform 10,000 read operations, and the number of cacheable records cached in memory is 1000.

Table 4. Datasets for cache experiments

Data set	Reading times	Number of cacheable records	Percentage of cacheable data (%)
5,000	10000	1000	20
3,333	10000	1000	30
2,500	10000	1000	40
2,000	10000	1000	50
1,667	10000	1000	60
1,429	10000	1000	70
1,250	10000	1000	80

In order to verify the performance of the query module, the number of records stored is 1 million, four batch query datasets were set, as shown in Table 3. This paper compares the query time of HBase with HEDSM.

4.4 Experimental Results and Analysis

4.4.1 HBase Load Balancing

(1) Number of Regions

From Fig. 3, this paper can find that HEDSM can make Regions more uniformly distributed in RegionServers than HBase. Moreover, by comparing the results of HEDSM-5 and HEDSM-10, this paper can conclude that the larger the period of HEDSM is, the more uniform the number of Region is. This is because a large period of load balancing will reduce the number of load balancing calls.

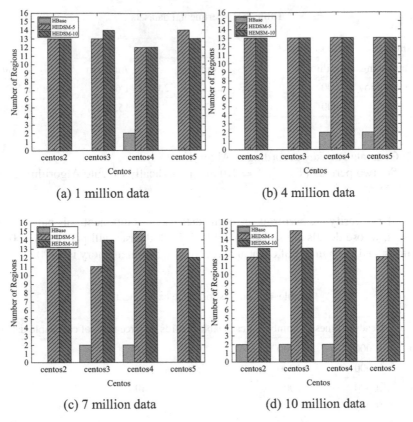

(a) 1 million data

(b) 4 million data

(c) 7 million data

(d) 10 million data

Fig. 3. The number of regions written to RegionServer with different amounts of data

(2) Load of RegionServers

As shown in Fig. 4, the reading and writing requests after load balancing by HBase are concentrated on certain RegionServers. On the contrary, the reading and writing requests after load balancing by HEDSM are distributed evenly to all RegionServers. Also, the load of RegionServers with the 5-min-cycle HEDSM is more balanced than the 10-min-cycle HEDSM. This is because the load balancing strategy of HEDSM is dynamically adjusted according to the reading and writing requests of the RegionServers during each period. The shorter the cycle is, the faster the dynamic load balancing, and the more balanced of the load of each RegionServer is.

(3) Data writing

From Fig. 5, this paper can observe that as the amount of data increases, the data writing time increases. The writing time of HEDSM-10 is similar to that of HEDSM-5, which indicates that the load balancing strategy can reduce the data writing time and the load balancing operation has only little impact on the data writing performance. The

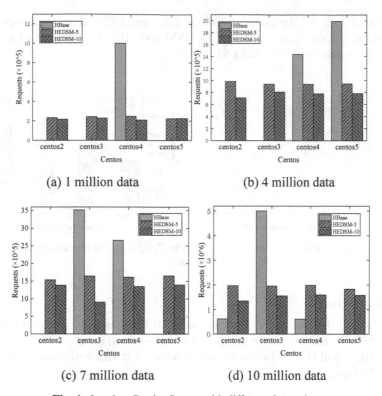

(a) 1 million data   (b) 4 million data

(c) 7 million data   (d) 10 million data

Fig. 4. Load on RegionServer with different data volume

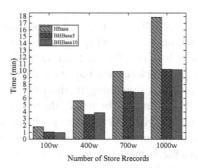

Fig. 5. Write time for different amounts of data

larger the amount of data, the better the effect of HEDSM, which is about 50% higher than HBase's default implementation.

4.4.2 Cache Hit Ratio

As shown in Fig. 6, the cache hit ratio of HEDSM is higher than that of the LRU strategy with varying percentages of the storage capacity. With the increase of the percentage of the storage capacity, the cache hit radio of LRU and HEDSM also increases, which is consistent with the characteristics of HEDSM and LRU. And the cache hit radio of HEDSM is higher about 10% than LRU.

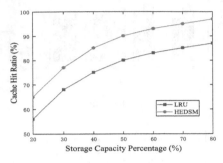

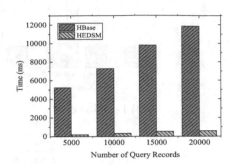

Fig. 6. Cache hit ratio

Fig. 7. Query time for different amounts of data

4.4.3 Batch Query

Figure 7 shows that the query performance of HEDSM is much greater than that of HBase. With the increase of the number of query records, the gap between the query time of HBase and HEDSM enlarges, which means that HEDSM is more suitable for massive data query.

5 Conclusions

To improve the performance of HBase, this paper proposes a distributed storage middleware named HEDSM whose writing and querying performance far exceeds that of HBase. The middleware employs an efficient load balancing strategy for Region and RegionServer which has a great performance improvement over HBase. The cache eviction algorithm based on the time smoothing method of query frequency and update frequency improves the cache hit ratio. The newly designed preput function according to the hook function provided by HBase's coprocessor helps achieve data synchronization. All the functions are packed into the storage module whose effectiveness is proved in the experiments through a batch query interface. In the future, this paper will continue to study more complex query functions such as non-primary key query and range query.

Acknowledgments. This work is partly supported by the National Natural Science Foundation of China under Grant No. 61672159, No. 61672158 and No. 61300104, the Fujian Collaborative Innovation Center for Big Data Applications in Governments, the Fujian Industry-Academy Cooperation Project under Grant No. 2017H6008 and No. 2018H6010, the Natural Science Foundation of Fujian Province under Grant No. 2019J01835 and Haixi Government Big Data Application Cooperative Innovation Center.

References

1. Konstantinou, I., Tsoumakos, D., Mytilinis, I.: DBalancer: distributed load balancing for NoSQL data-stores. In: The 2013 ACM SIGMOD International Conference on Management of Data, pp. 1037–1040. ACM (2013)
2. Chang, F., et al.: Bigtable: a distributed storage system for structured data. ACM Trans. Comput. Syst. (TOCS) **26**(2), 1–26 (2008)
3. Xia, L., Chen, H., Sun, H.: An optimized load balance based on data popularity on HBASE. In: The 2nd International Conference on Information Technology and Electronic Commerce, pp. 234–238. IEEE, Dalian (2014)
4. Dang, P.: Design and Implementation of HBase Hierarchical Auxiliary Index System. Xidian University (2019)
5. HBase Load Balancing Algorithm. https://hbase.apache.org/. Accessed 05 July 2020
6. Huang, W.J., Song, Y.Y.: HBase load balancing algorithm analysis and improvement. Microelectron. Comput. **33**(4), 125–128 (2016)
7. Guo, H., Fang, J., Li, D.: A multi-source streaming data real-time storage system based on load balance. Comput. Eng. Sci. **39**(4), 641–647 (2017)
8. Levandoski, J.J., Larson, P., Stoica, R.: Identifying hot and cold data in main-memory databases. In: 2013 IEEE 29th International Conference on Data Engineering (ICDE), pp. 26–37. IEEE, Brisbane (2013)
9. Shao, F.: The analysis and optimization of load balancing algorithm for big data platform based on HBase. Beijing University of Technology (2018)
10. Levandoski, J.J., Larson, P.Å., Stoica, R.: Identifying hot and cold data in main-memory databases. In: 2013 IEEE 29th International Conference on Data Engineering (ICDE), pp. 26–37. IEEE, Tianjin (2013)
11. Zhang, C., Li, F., Jestes, J.: Efficient parallel kNN joins for large data in MapReduce. In: The 15th International Conference on Extending Database Technology, pp. 38–49. IEEE (2013)
12. Wei, G., et al.: HiBase: a hierarchical indexing mechanism and system for efficient HBase query. Chinese J. Comput. **39**(01), 140–153 (2016)
13. Qu, L., Li, X.: A HBase index buffer solution based on TwemProxy. Inf. Technol. **10**, 103–107+117 (2017)
14. Li, K., Guo, K., Guo, H.: Financial big data hot and cold separation scheme based on hbase and redis. In: 2019 IEEE International Conference on Parallel & Distributed Processing with Applications, Big Data & Cloud Computing, Sustainable Computing & Communications. In: Social Computing & Networking (ISPA/BDCloud/SocialCom/SustainCom), pp. 1612–1617. IEEE, Fuzhou (2019)
15. Ye, F., Zhu, S., Lou, Y., Liu, Z., Chen, Y., Huang, Q.: Research on index mechanism of hbase based on coprocessor for sensor data. In: 2019 IEEE 43rd Annual Computer Software and Applications Conference (COMPSAC), pp. 598–603. IEEE, Milwaukee (2019)
16. Ma, T., Xu, X., Tang, M., Jin, Y., Shen, W.: MHBase: a distributed real-time query scheme for meteorological data based on HBase. Future Internet **8**(1), 6 (2016)
17. Cao, C., Wang, W., Zhang, Y., Lu, J.: Embedding index maintenance in store routines to accelerate secondary index building in hbase. In: 2018 IEEE 11th International Conference on Cloud Computing (CLOUD), pp. 500–507. IEEE, San Francisco (2018)

18. Zhang, Y., Ma, Y.Z., Meng, X.F.: Efficient processing of spatial keyword queries on HBase. J. Chinese Comput. Syst. **33**(10), 2141–2146 (2012)
19. Karger, D., Lehman, E., Leighton, T., Panigrahy, R., Levine, M., Lewin, D.: Consistent hashing and random trees: distributed caching protocols for relieving hot spots on the world wide web. In: The Twenty-Ninth Annual ACM Symposium on Theory of Computing, pp. 654–663. ACM (1997)
20. Zhang, S.: Design and security analysis on several cryptography hash functions. Beijing University of Posts and Telecommunications, Beijing (2011)

Cooperative Full-Duplex Relay Selection Strategy Based on Power Splitting

Taoshen Li[1,2(✉)], Anni Shi[2], Jichang Chen[1], and Zhihui Ge[2]

[1] School of Information Engineering, Nanning University, Nanning 530299, China
tshli@gxu.edu.cn
[2] School of Computer, Electronics and Information, Guangxi University,
Nanning 530004, China

Abstract. To optimize the full-duplex relay cooperation system throughput, a cooperative full-duplex relay selection strategy based on power splitting is proposed. This strategy introduces self-energy recycling technology to eliminate self-interference problem in system, which makes the self-interference signal become beneficial to the system. The strategy also solves the relay selection problem under the optimal system throughput by optimizing the parameters such as relay transmission power and power allocation factor. It transforms the original non-linear mixed integer programming problem into two coupled optimization problems by using mathematical transformation, and then obtains the optimal solution by KKT condition and Lagrange function. Simulated experiments and analysis results show that the proposed strategy can retrieve the performance loss caused by self-interference and the spectrum efficiency loss caused by half duplex relay. Compared with the full duplex random selection strategy, it can select the best channel and achieve system gain.

Keywords: Simultaneous Wireless Information and Power Transfer (SWITP) · Full duplex · Relay selection · Power splitting · Self-interference energy recovery

1 Introduction

The simultaneous wireless information and power transfer (SWIPT) is a new type of wireless communication technology, which can realize cooperative transmission of information and energy in wireless networks, and effectively provide energy for various wireless devices by extracting the energy in the received signal [1].

In the research of wireless energy capture cooperative relay network, some works had been done for various optimal transmission strategies and power splitting (PS) strategies [2–6]. In the full-duplex (FD) mode, the receiving and forwarding of signals are performed at the same time and frequency band, which leads to strong self-interference at the relay end and greatly reduces the efficiency of information transmission [7]. With the progress of self-interference elimination technology, the technology of self-energy recycling has gained more and more attention in FD relay networks [8]. [9] put the energy harvesting into the second slot to eliminate the self-interference and make the self-interference become beneficial to the system. [10] proposed a FD relay system based on

© Springer Nature Singapore Pte Ltd. 2021
Y. Sun et al. (Eds.): ChineseCSCW 2020, CCIS 1330, pp. 381–394, 2021.
https://doi.org/10.1007/978-981-16-2540-4_28

PS protocol, in which the receiving and forwarding of relay information were completed in one time slot. [11] used SWITP scheme and the self-energy recycling technology in FD system, and minimized the weighted sum transmit power. [12] studied the application of self-energy recycling protocol to FD cooperative non-orthogonal multiple access system. [13] proposed a joint time and power allocation scheme suitable for FD wireless relaying network. [14–16] studied the security beamforming technology of FD wireless relay system.

However, most of the existing researches focused on a single relay system, and did not consider the operation planning of the SWITPS system with multiple relays. In addition, the existing wireless relay systems are based on the traditional three-stage, two-stage time-sharing transmission protocol, In addition, the existing wireless relay systems were based on the traditional three-stage and two-stage time-sharing transmission protocol, information and energy are transmitted separately, and the SWITP technology has not been widely used.

This paper will establish the FD relay network model by using PS-SWITP technology with self-energy recycling, propose a new optimal FD relay selection strategy based on PS cooperation. On the basis of satisfying the conditions of communication quality of service (QoS) and transmission power of source node, the optimal relay selection problem is solved by jointly optimizing the parameters such as relay transmission power and power allocation factor. Finally, the feasibility and effectiveness of the proposed model and strategy are illustrated by simulation experiments and performance analysis.

2 System Model

We set up a full-duplex SWITP relay cooperation system based on PS protocol as shown in Fig. 1. Where, S is the source node, $R_i(i = 1, 2,..., K)$ is the relay nodes, D is the destination node, and all devices are equipped with a single antenna. Due to the limitation of network coverage, it is assumed that the S node is not directly connected to the D node, and the communication between them is only conducted through the relay node. Suppose that the application scenario of the model is: the base station (S node) sends information to the remote user (D node) through multiple small base stations (relay nodes).

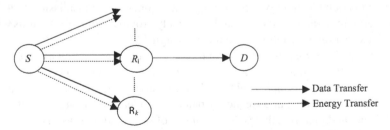

Fig. 1. The full-duplex SWITP relay cooperation system

The relay receiver structure of the system is as shown in Fig. 2. First, the S node transmits the signal to the relay node. Then the relay node forwards the amplified signal to the D node. The relay node operates in amplify-and-forward(AF) mode [17], and can eliminate information self-interference and realize energy self recycling. According to the different power ratio of information receiving and energy harvesting, the relay node processes the received signal. The harvested energy is converted into electrical energy that can be used by relay nodes, and stored in a storable battery.

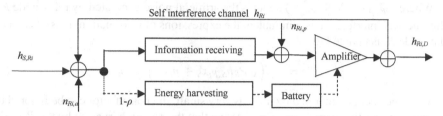

Fig. 2. Structure of the relay receiver in full-duplex SWITP relay cooperation system

The time allocation of PS protocol of full-duplex SWITP relay cooperation system is shown in Fig. 3. Different from other relay systems, the information receiving, energy acquisition and information forwarding are all carried out in the same time slot T, which can maintain uninterrupted information flow.

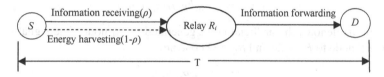

Fig. 3. The transmission time allocation

In Fig. 3, all channels in the system are set to be static fading channels and their states are basically unchanged in the same time slot, so the status information of all channels is known. Let h_{S,R_i} denotes the channel factor between the S node and the $R_i(i = 1, 2,..., K)$ node, $h_{R_i,D}$ denotes the channel factor between the R_i node and the D node, h_{R_i} denotes the self-interference channel factor of the R_i node, $\sigma^2_{n_{R_i,a}}$ and P_{R_i} denote the transmission power of the S node and the R_i node respectively. The signal received by the R_i node is

$$y_{R_i} = \sqrt{P_S}h_{S,R_i}x_S + \sqrt{P_{R_i}}h_{R_i}x_{R_i} + n_{R_i,\,a} \tag{1}$$

Where, $x_S \in C$ is the signal from the S node which satisfies $E\{|x_S|^2\} = 1$, $x_{Ri} \in C$ is the signal from the R_i node. $n_{R_i,a} \sim CN (0, \sigma^2_{n_{R_i,a}})$ is the additive white Gaussian noise produced by the R_i node, and its variance is $\sigma^2_{n_{R_i,a}}$ and the mean value is 0. $\sqrt{P_{R_i}}h_{R_i}x_{R_i}$ is the loop-back self-interference signal of relay node. According to the PS method and factor $\rho \in (0, 1)$, the relay node R_i divides the y_{R_i} into $\rho \cdot y_{R_i}$ and $(1 - \rho) \cdot y_{R_i}$, in which the $\rho \cdot y_{R_i}$ is used for information transmission and the $(1 - \rho) \cdot y_R$ is used for energy transmission. The corresponding formulas are as follows:

$$y_{R_i}^{ID} = \sqrt{\rho}\left(\sqrt{P_S}h_{S,R_i}x_S + \sqrt{P_{R_i}}h_{R_i}x_{R_i} + n_{R_i,a}\right) + n_{R_i,p} \tag{2}$$

$$y_{R_i}^{EH} = \sqrt{1-\rho}\left(\sqrt{P_S}h_{S,R_i}x_S + \sqrt{P_{R_i}}h_{R_i}x_{R_i} + n_{R_i,a}\right) \tag{3}$$

Where, $n_{R_i,p} \sim CN(0, \sigma_{n_{R_i,p}}^2)$ denotes the artificial noise generated by relay node R_i. After the self-interference is eliminated, the expressions of information transmission of relay node R_i is updated to

$$y_{R_i}^{ID-SIC} = \sqrt{\rho}\left(\sqrt{P_S}h_{S,R_i}x_S + n_{R_i,a}\right) + n_{R_i,p} \tag{4}$$

Since the energy generated by noise is very small, the noise factor can be ignored in the energy harvesting expression. Considering that the energy harvested by the R_i node is composed of the energy of signal from S node and the energy recovered from the self-interference signal, the total harvested energy is

$$E_{R_i} = \eta(1-\rho)\left(P_S\left|h_{S,R_i}\right|^2 + P_{R_i}\left|h_{S,R_i}\right|^2\right) \tag{5}$$

Where, η is the energy transformation efficiency.

Therefore, the signal that the destination node D can receive is

$$y_D = \sqrt{P_{R_i}}h_{R_i,D}x_{R_i} + n_d \tag{6}$$

Where, n_d denotes the artificial noise generated by R_i node. The signal noise ratio (SNR) of S node to R_i node and R_i to D node are

$$\gamma_{S,R_i} = \frac{\rho P_S\left|h_{S,R_i}\right|^2}{\rho\sigma_{n_{R_i},a}^2 + \sigma_{n_{R_i},p}^2} \tag{7}$$

$$\gamma_{R_i,D} = \frac{P_{R_i}\left|h_{R_i,D}\right|^2}{\sigma_{n_d}^2} \tag{8}$$

Since the relay node adopts the AF protocol, the total SNR of the system is expressed as

$$\gamma = \frac{\gamma_{S,R_i}\gamma_{R_i,D}}{1 + \gamma_{S,R_i} + \gamma_{R_i,D}} = \frac{\rho P_S\left|h_{S,R_i}\right|^2}{\rho\sigma_{n_{R_i},a}^2 + \sigma_{n_{R_i},p}^2 + \frac{\left(\rho A + \sigma_{n_{R_i},p}^2\right)\sigma_{n_d}^2}{P_{R_i}\left|h_{R_i,D}\right|^2}} \tag{9}$$

Where, $A = \left(P_S\left|h_{S,R_i}\right|^2 + \sigma_{n_{R_i},d}^2\right)$, the system throughput expression is

$$R = \log_2(1+\gamma) = \log_2\left(1 + \frac{\rho P_S\left|h_{S,R_i}\right|^2}{\rho\sigma_{n_{R_i},a}^2 + \sigma_{n_{R_i},p}^2 + \frac{\left(\rho A + \sigma_{n_{R_i},p}^2\right)\sigma_{n_d}^2}{P_{R_i}\left|h_{R_i,D}\right|^2}}\right) \tag{10}$$

3 Our Relay Selection Strategy

Under the condition that the communication QoS is guaranteed and the source transmit power does not exceed the limit, our strategy can select the best relay to maximize the throughput of energy-limited full duplex relay system by jointly optimizing the parameters such as relay transmission power and power allocation factor.

3.1 Optimization Problem

The optimization goal of our strategy is to maximize the system throughput. The global optimization problem can expressed as

$$P0: \quad \max_{i \in \mathcal{K}, P_{R_i}, \rho} R = \log_2(1 + \gamma) \tag{11}$$

$$\text{s.t} \quad \gamma \leq \gamma_0 \tag{11a}$$

$$P_{R_i} \leq E_{R_i} + Q \tag{11b}$$

$$0 < P_S \leq P^* \tag{11c}$$

$$0 \leq \rho \leq 1 \tag{11d}$$

Where, $\mathcal{K} = \{1, \dots, K\}$ is the code of K relay nodes; γ is the total SNR of the system(as shown in (9)); γ_0 denotes the SNR threshold; E_{Ri} denotes the total harvested energy (as shown in (5)); Q represents the external energy required for relay to transmit the signal further; P^* is the maximum transmitting power constraint of S node.

Because the expression equation of throughput optimization problem is complex and not easy to calculate, the optimization problem can be transformed to a easily solved minimum objective function. The transformed expression is as follows

$$\max R = \log_2(1 + \frac{P_S|h_{S,R_i}|^2}{\sigma_{n_{R_i},a}^2 + \gamma^*})$$
$$\Rightarrow \min \gamma^* = \frac{\sigma_{n_{R_i},p}^2}{\rho} + \frac{A\sigma_{n_d}^2}{P_{R_i}|h_{R_i,D}|^2} + \frac{\sigma_{n_{R_i},p}^2 \sigma_{n_d}^2}{\rho P_{R_i}|h_{R_i,D}|^2} \tag{12}$$

Because γ is greater than 1, the equation for throughput R is a function that increases as γ increases. The equation of maximizing throughput can be transformed into maximizing γ. After simplifying γ, it can be transformed into the objective function of minimizing γ^*. The optimization problem of P0 becomes

$$P1: \quad \min_{i \in \mathcal{K}, P_{R_i}, \rho} \gamma^* = \frac{\sigma_{n_{R_i},p}^2}{\rho} + \frac{A\sigma_{n_d}^2}{P_{R_i}|h_{R_i,D}|^2} + \frac{\sigma_{n_{R_i},p}^2 \sigma_{n_d}^2}{\rho P_{R_i}|h_{R_i,D}|^2}$$
$$\text{s.t} \quad \gamma \leq \gamma_0$$
$$P_{R_i} \leq E_{R_i} + Q$$

$$0 < P_S \le P^*$$
$$0 \le \rho \le 1 \tag{13}$$

After simplification, the whole optimization problem becomes a non-linear mixed integer planning problem, which means relay selection and joint optimization of P_{Ri} and ρ are still difficult to solve. Therefore, problem P1 needs to be redefined as a pair of coupling optimization problems, that is the external optimization problem of selecting the optimal relay and the internal optimization problem of joint calculation of P_{Ri} and ρ. The optimal solutions for internal optimization and external optimization are discussed below.

3.2 Optimization of Dynamic P_S Factor and Relay Transmission Power

This section discusses the internal optimization of P1, which mainly involves the joint optimization of P_{Ri} and ρ parameters. Assuming that the relay node R_i ($i = 1, 2,..., K$) is active, the corresponding sub-problem is

$$\mathbf{P2}: \min_{P_{R_i},\rho} \quad \gamma^* = \frac{\sigma_{n_{R_i,p}}^2}{\rho} + \frac{A\sigma_{n_d}^2}{P_{R_i}\left|h_{R_i,D}\right|^2} + \frac{\sigma_{n_{R_i,p}}^2 \sigma_{n_d}^2}{\rho P_{R_i}\left|h_{R_i,D}\right|^2}$$

$$\text{s.t} \quad \gamma \ge \gamma_0$$
$$P_{R_i} \le E_{R_i} + Q$$
$$0 < P_S \le P^*$$
$$0 \le \rho \le 1 \tag{14}$$

Obviously, this is a nonlinear planing problem, and its optimal solution involves the joint optimization of P_{R_i} and ρ. In our strategy, the Karush Kuhn Tucker (KKT) condition is used to find the solution of this problem. Firstly, the Lagrange function corresponding to problem P2 can be expressed as

$$\begin{aligned}
&\mathcal{L}\left(P_{R_i}, \rho; \lambda_1, \lambda_2, \lambda_3\right) \\
&= F\left(P_{R_i}, \rho\right) + \lambda_1 G\left(P_{R_i}, \rho\right) + \lambda_2 H\left(P_{R_i}, \rho\right) + \lambda_3 I\left(P_{R_i}, \rho\right)
\end{aligned} \tag{15}$$

Where,

$$F\left(P_{R_i}, \rho\right) = \frac{\sigma_{n_{R_i,p}}^2}{\rho} + \frac{A\sigma_{n_d}^2}{P_{R_i}\left|h_{R_i,D}\right|^2} + \frac{\sigma_{n_{R_i,p}}^2 \sigma_{n_d}^2}{\rho P_{R_i}\left|h_{R_i,D}\right|^2} \tag{16}$$

$$G\left(P_{R_i}, \rho\right) = \gamma_0 - \gamma = \gamma_0 - \frac{\rho P_S\left|h_{S,R_i}\right|^2}{\rho\sigma_{n_{R_i,a}}^2 + \sigma_{n_{R_i,p}}^2 + \frac{\left(\rho A + \sigma_{n_{R_i,p}}^2\right)\sigma_{n_d}^2}{P_{R_i}\left|h_{R_i,D}\right|^2}} \le 0 \tag{17}$$

$$\begin{aligned}
H\left(P_{R_i}, \rho\right) &= P_{R_i} - Q - E_{R_i} \\
&= P_{R_i} - Q - \eta(1-\rho)\left(P_S\left|h_{S,R_i}\right|^2 + P_{R_i}\left|h_{R_i}\right|^2\right) \le 0
\end{aligned} \tag{18}$$

$$I\left(P_{R_i}, \rho\right) = \rho - 1 \le 0 \tag{19}$$

In order to obtain local optimum, $\nabla \mathcal{L}(P_{R_i}, \rho; \lambda_1, \lambda_2, \lambda_3) = 0$ must be guaranteed. Therefore, the equation satisfying the optimal condition is

$$\frac{\partial \mathcal{L}(P_{R_i}, \rho; \lambda_1, \lambda_2, \lambda_3)}{\partial P_{R_i}}$$

$$\Rightarrow -\frac{A\sigma_{n_d}^2}{P_{R_i}^2 |h_{R_i,D}|^2} - \frac{\sigma_{n_{R_i,p}}^2 \sigma_{n_d}^2}{P_{R_i}^2 \rho |h_{R_i,D}|^2}$$

$$-\lambda_1 \rho P_S |h_{S,R_i}|^2 \frac{\dfrac{\left(\rho A + \sigma_{n_{i,p}}^2\right)\sigma_{n_d}^2}{P_{R_i}|h_{R_i,D}|^2}}{\left(\rho \sigma_{n_{R_i,a}}^2 + \sigma_{n_{R_i,p}}^2 + \dfrac{\left(\rho A + \sigma_{n_{R_i,p}}^2\right)\sigma_{n_d}^2}{P_{R_i}|h_{R_i,D}|^2}\right)^2}$$

$$+\lambda_2\left(1 - \eta(1-\rho)|h_{R_i}|^2\right) \tag{20}$$

$$\frac{\partial \mathcal{L}(P_{R_i}, \rho; \lambda_1, \lambda_2, \lambda_3)}{\partial \rho}$$

$$\Rightarrow -\frac{\sigma_{n_{R_i,p}}^2}{\rho^2} - \frac{\sigma_{n_{R_i,p}}^2 \sigma_{n_d}^2}{\rho^2 P_{R_i}|h_{R_i,D}|^2}$$

$$-\lambda_1 P_S |h_{S,R_i}|^2 \frac{\dfrac{\sigma_{n_{R_i,p}}^2}{\rho^2} + \dfrac{\sigma_{n_{R_i,p}}^2 \sigma_{n_d}^2}{\rho^2 P_{R_i}|h_{R_i,D}|^2}}{\left(\sigma_{n_{R_i,a}}^2 + \dfrac{\sigma_{n_{R_i,p}}^2}{\rho} + \dfrac{A\sigma_{n_d}^2}{P_{R_i}|h_{R_i,D}|^2} + \dfrac{\sigma_{n_{R_i,p}}^2 \sigma_{n_d}^2}{\rho P_{R_i}|h_{R_i,D}|^2}\right)^2}$$

$$+\lambda_2 \eta\left(P_S |h_{S,R_i}|^2 + P_{R_i}|h_{R_i}|^2\right) + \lambda_3 \tag{21}$$

The feasible condition expressions of the objective function are shown as (16), (17) and (18). The complementary relaxation expression is as follows:

$$\lambda_1 G(P_{R_i}, \rho) = 0, \tag{22}$$

$$\lambda_2 H(P_{R_i}, \rho) = 0 \tag{23}$$

$$\lambda_3 I(P_{R_i}, \rho) = 0, \tag{24}$$

Where, The nonnegative parameters are P_{R_i}, ρ, and $\lambda_1, \lambda_2, \lambda_3 \geq 0$. Obviously, if $\lambda_3 \neq 0$, then $I(P_{R_i}, \rho) = 0$ which means $\rho = 1$ and this is not a viable solution. So there is $\lambda_3 = 0$. If $\lambda_2 = 0$, then (21) and (22) do not hold. Therefore, two possible solutions are derived, and are described in the following two propositions.

Proposition 1. If $\lambda_1 \neq 0$, then $G(P_{R_i}, \rho) = 0$. If $\lambda_2 \neq 0$, there is $H(P_{R_i}, \rho) = 0$. If $\lambda_3 = 0$ then $I(P_{R_i}, \rho) \neq 0$. Thus, we can get the following expression of the optimal solution:

$$\frac{\rho' P_S |h_{S,R_i}|^2}{\rho' \sigma_{n_{R_i,a}}^2 + \sigma_{n_{R_i,p}}^2 + \dfrac{\left(\rho' A + \sigma_{n_{R_i,p}}^2\right)\sigma_{n_d}^2}{P_{R_i}'|h_{R_i,D}|^2}} = \gamma_0, \tag{25}$$

$$P'_{R_i} = Q + \eta(1 - \rho')\left(P_S|h_{S,R_i}|^2 + P'_{R_i}|h_{R_i}|^2\right), \tag{26}$$

Proposition 2. If $\lambda_1 = 0$, then $G(P_{R_i}, \rho) \neq 0$. If $\lambda_2 \neq 0$, then $H(P_{R_i}, \rho) = 0$. If $\lambda_3 = 0$ then $I(P_{R_i}, \rho) \neq 0$. Thus, the expression of the optimal solution is as follows.

$$-\frac{A\sigma_{n_d}^2}{\left(P''_{R_i}\right)^2|h_{R_i,D}|^2} - \frac{\sigma_{n_{R_i,p}}^2\sigma_{n_d}^2}{\left(P''_{R_i}\right)^2\rho''|h_{R_i,D}|^2} + \lambda_2(1 - \eta(1 - \rho'')|h_{R_i}|^2) = 0, \tag{27}$$

$$-\frac{\sigma_{n_{R_i,p}}^2}{(\rho'')^2} - \frac{\sigma_{n_{R_i,p}}^2\sigma_{n_d}^2}{(\rho'')^2 P''_{R_i}|h_{R_i,D}|^2} + \lambda_2\eta\left(P_S|h_{S,R_i}|^2 + P''_{R_i}h_{R_i}|^2\right) = 0, \tag{28}$$

$$P''_{R_i} = Q + \eta(1 - \rho'')\left(P_S|h_{S,R_i}|^2 + P''_{R_i}|h_{R_i}|^2\right), \tag{29}$$

According to the above solutions, we design the internal optimization algorithm. The algorithm's goal is to solve the value of the minimum objective function γ^*, and put it into the formula of maximizing the system throughput R, and then get the maximum throughput value R_{max}. The algorithm is described as follows:

Input: Parameters h_{S,R_i}, $h_{R_i,D}$, h_{R_i}, Q;
Output: Maximum system throughput R_{max} under the minimum objective function (γ^*) value after parameter optimization.

1) Initialization:$\eta \in (0, 1]$, $P_s \in (0, \mu P_{max}]$; $0.5 < \mu < 1$; $\sigma_{n_{R_i,a}}^2 = 10^{-4}$; $\sigma_{n_{R_i,p}}^2 = 10^{-10}$; $\sigma_{n_d}^2 = 10^{-4}$
2) Calculating the parameters P'_{R_i}, ρ' by Eqs. (25) and (26);
3) Defining: $\gamma' = \frac{\sigma_{n_{R_i,p}}^2}{\rho'} + \frac{A\sigma_{n_d}^2}{P'_{R_i}|h_{R_i,D}|^2} + \frac{\sigma_{n_{R_i,p}}^2\sigma_{n_d}^2}{\rho' P'_{R_i}|h_{R_i,D}|^2}$, $R' = \log_2(1 + \frac{P_S|h_{S,R_i}|^2}{\sigma_{n_{R_i,a}}^2 + \gamma'})$
4) Calculating the parameters P''_{R_i}, ρ'' by Eqs. (27), (28) and (29);
5) Defining: $\gamma'' = \frac{\sigma_{n_{R_i,p}}^2}{\rho''} + \frac{A\sigma_{n_d}^2}{P''_{R_i}|h_{R_i,D}|^2} + \frac{\sigma_{n_{R_i,p}}^2\sigma_{n_d}^2}{\rho'' P''_{R_i}|h_{R_i,D}|^2}$, $R'' = \log_2(1 + \frac{P_S|h_{S,R_i}|^2}{\sigma_{n_{R_i,a}}^2 + \gamma''})$
6) $R_{max} = \max\left(R', R''\right)$;
7) Return: R_{max}.

3.3 Selection of Relay Nodes

This section discusses optimal selection of relay nodes to find the external optimization problem solution of P1. Its idea is that the optimal relay node is searched in one dimension by using the maximum throughput of the relay nodes obtained in the internal optimization problem solving algorithm. The index of the best relay node can be expressed as

$$j^* = \arg\min_{j\in\{1,\ldots,k\}}\gamma^* = \arg\max_{j\in\{1,\ldots,k\}} R \tag{30}$$

Where, γ^* denotes the optimal solution of P2, and the exchange expression of R and γ^* is shown in (12). It is worth noting that this relay selection is based on exhaustion search, which has high performance cost and high complexity. Looking for compromise relay selection is the focus of future research.

4 Experiments and Analysis

The experiments use Monte Carlo and other experimental methods to compare our strategy with with the full-duplex random relay selection strategy (FDRRSS), half-duplex optimal relay selection strategy(HDORSS) and half-duplex random relay selection strategy (HDRRSS) given in [18], and then analyze the performance of these strategies. The experimental parameters are set as follows: network relay nodes $K = 8$, energy conversion rate $\eta = 0.5$; channel coefficients are independent and identically distributed, and obey Rayleigh distribution; $R_0 = $ 3bps, $Q = -30$ dbm, $\sigma^2_{n_{R_i,a}} = \sigma^2_{n_d} - 10$ dbm, $\sigma^2_{n_{R_i,p}} = -70$ dbm.

In the experiment, the FDRRSS strategy can randomly select relay nodes in the feasible solution set to perform information amplification and forwarding, and other configurations and condition assumptions are the same as our strategy. In the HDORSS strategy and HDRRSS strategy, the whole information transmission phase T is divided into two stages. The first T/2 by the relay node to receive information and harvest energy, and the second T/2 is used by the relay node to amplify the information and then forward to the D node. Since there is no interference signal, there is no self-energy recycling.

Firstly, we compare our strategy with the FDRRSS strategy, HDORSS strategy and HDRRSS strategy. Figure 4 shows the experimental comparison results when $P_S = 35$ dbm. The cumulative distribution function (CDF) graph is used to describe the probability distribution of the maximum throughput of the system under different strategies. In the experimental data curve of proposed strategy, the y value corresponding to $R > 11.54$ bps on the horizontal axis is about 0.4, which indicates that the proportion of the data points with $R > 11.54$ bps is about 60%. In the experimental data curve of the full-duplex random relay selection strategy, the y value corresponding to $R > 11.54$ bps on the horizontal axis is about 0.78, and the proportion of the data points with $R > 11.54$ bps is about 22%. In the HDORSS strategy and HDRRSS strategy, the probability of $R < 7$ bps is 1. The experimental results show that in the relay selection cooperative communication system based on SWITP, the performance of the optimal relay selection strategy is superior to the random relay selection strategy, and the our strategy is obviously better than the traditional HDORSS strategy and HDRRSS strategy.

Secondly, we observe and analyze the influence of antenna noise produced by relay nodes on throughput. Figure 5 shows the experimental results of $P_S = 35$ dbm and $P_S = 40$ dbm.

From the experimental results, we can see that when the antenna noise is greater than -15 dbm, the throughput of the proposed strategy begins to decrease, and the rate of descend is faster. When the noise is large, the throughput under different P_S values becomes smaller and smaller. In the full-duplex random relay selection strategy (FDRRSS), because the selected relay nodes are stochasticly extracted from the feasible set, the experimental curve is different, but its maximum throughput does not exceed the

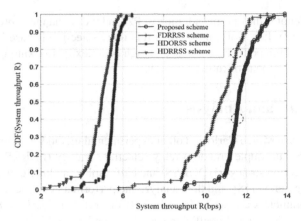

Fig. 4. Comparison of system throughput

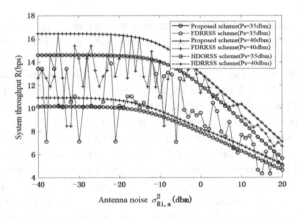

Fig. 5. System throughput comparison of different strategies when relay antenna noise changes

our strategy. The results of simulated experiment show the uncertainty of the random relay selection strategy. When the antenna noise of relay node is larger than −20 dbm, The throughput of the HDORSS strategy begins to decrease, and the decline speed gradually increased. When the noise is very high, the difference between its throughput and that of the full-duplex relay selection strategy becomes smaller. Experimental results also show that the maximum throughput of the half-duplex relay selection strategy is smaller than our strategy under different antenna noise conditions.

Thirdly, we analyze the change of system throughput R with relay transmit power P_{Ri}. The experimental results of $P_S = 35$ dbm and $P_S = 40$ dbm are given in Fig. 6. The simulation results show that when relay transmit power continues to grow, the system throughput of the HDORSS strategy, FDRRSS strategy and our strategy also increases. The number of P_{Ri} is determined by the energy E_{Ri} harvested by relay node. The larger the P_{Ri}, the more energy is harvested. This shows that the harvested energy has a significant influence on the system throughput. When $P_{Ri} > 40$ dbm, the speed growth curve of system throughput growth gradually flattens. The main reason is that when the value of P_{Ri} exceeds a certain limit, the part allocated to the relay information receiving according to the power splitting factor becomes very small, which seriously affects the system throughput. Therefore, it is very important to get an equilibrium point of energy rate.

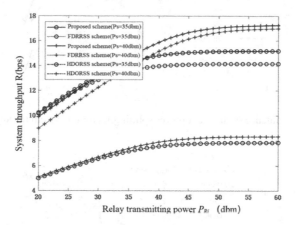

Fig. 6. Influence of relay transmit power on system throughput under different strategies

The influence of the change of power splitting factor ρ on the system throughput R is shown in Fig. 7. This experiment is very important to find a suitable balance between energy and velocity. In the experiment, P_S is set to 34, 36, 38, 40 and 42 dbm. As shown in Fig. 7, when ρ is small, the system throughput is not seriously affected and the change is very small. However, when ρ is close to 0.8, the system throughput begins to decrease obviously, and the decline speed is faster. When ρ approaches 1, the throughput reaches the minimum. The main reason is that when ρ is large, the proportion of information receiving part of relay node in power allocation becomes larger, which results in the imbalance of energy rate and impacts on system throughput seriously.

Finally, we observe and analyze the change of system throughput R with relay transmit power P_{Ri} under different relay antenna noises. Figure 8 shows the experimental results of $\sigma^2_{n_{R_i,a}} = -20$ dbm and $\sigma^2_{n_{R_i,a}} = -10$ dbm. Experimental results show that when P_S increases, the system throughput R also increases. The larger the P_S, the smaller the impact of antenna noise on system throughput. When $\sigma^2_{n_{R_i,a}} = -20$ dbm and $P_S = 30$ dbm, the throughput of proposed strategy is 4.6 bps more than that of the FDRRSS strategy, and is 8.051 bps more than that of the HDORSS strategy. With the increase of the source transmitting power, the throughput difference between half-duplex strategy and full-duplex strategy is larger and larger, and the influence of different relay antenna noise to the system throughput tends to be the same. Experimental results also show that the higher the source transmission power, the greater the performance advantage of our strategy.

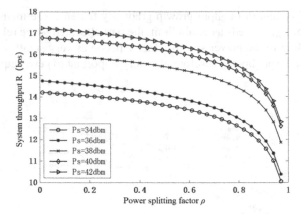

Fig. 7. Influence of different power splitting factor on system throughput

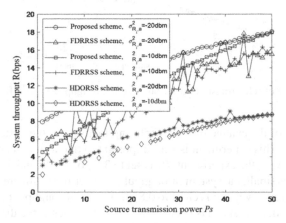

Fig. 8. The influence of different relay antenna noise to the system throughput

5 Conclusions

Aiming at the optimal relay selection problem of full-duplex SWITP relay cooperation system, a cooperative full-duplex relay selection strategy based on power splitting is proposed to optimize system throughput. By using the self-energy recycling technology, the relay node can recover the energy of self-interference signal, which makes the self-interference signal beneficial to the system. The problem model is established based on the constraints of communication service quality and transmitting power of source node. The original nonlinear mixed integer programming problem is transformed by mathematical transformation into two coupled optimization problems, namely, the external optimal relay selection problem and the internal transmission power control problem. This strategy has the following advantages: (1) it can recover the performance loss caused by self-interference and recover the relay's self-interference energy effectively; (2) it can overcome the spectrum efficiency loss caused by half duplex relay. Compared with the random relay selection strategy, the proposed relay selection strategy is more efficient The optimal relay selection strategy can select the optimal channel and achieve the goal of optimal system throughput. Compared with the random relay selection strategy, the proposed strategy can select the optimal channel and obtain the better system throughput.

Acknowledgments. These works are supported by the NNSF of China (No. 61762010).

References

1. Clerckx, B., Zhang, R., Schober, R., et al.: Fundamentals of wireless information and power transfer: from RF energy harvester models to signal and system designs. IEEE J. Sel. Areas Commun. **37**(1), 4–33 (2018)
2. Gurakan, B., Ozel, O., Ulukus, S.: Optimal energy and data routing in networks with energy cooperation. IEEE Trans. Wirel. Commun. **15**(2), 857–870 (2016)
3. Abuseta, A.A.M., Daryl, R.: Energy harvesting networks: energy versus data cooperation. IEEE Commun. Lett. **22**(10), 2128–2131 (2018)
4. Baidas, M.W., Alsusa, E.A.: Power allocation, relay selection and energy cooperation strategies in energy harvesting cooperative wireless networks. Wirel. Commun. Mob. Comput. **16**(14), 2065–2082 (2016)
5. Li, T.S., Ning, Q.L., Wang, Z.: Optimization scheme for the SWIPT-NOMA opportunity cooperative system. J. Commun. **41**(8), 141–154 (2020). (in Chinese)
6. Zhou, Y.N., Li, T.S., Wang, Z., et al.: Non-time-switching full-duplex relay system with SWIPT and self-energy recycling. Comput. Res. Dev. **57**(9), 1888–1897 (2020). (in Chinese)
7. Nasir, A.A., Zhou, X., Durrani, S., et al.: Relaying protocols for wireless energy harvesting and information processing. IEEE Trans. Wirel. Commun. **12**(7), 3622–3636 (2013)
8. Shende, N.V., Gurbuz, O., Erkip, E.: Half-duplex or full-duplex relaying: A capacity analysis under self-interference. In: 2013 47th Annual Conference on Information Sciences and Systems, pp. 1–6. IEEE Press, Baltimore (2013)
9. Mohammadi, M., Suraweera, H.A., Zheng, G., et al.: Full-duplex MIMO relaying powered by wireless energy transfer. In: 16th IEEE International Workshop on Signal Processing Advances in Wireless Communications, pp. 296–300. IEEE Press, Stockholm (2015)
10. Zeng, Y., Zhang, R.: Full-duplex wireless-powered relay with self-energy recycling. IEEE Wirel. Commun. Lett. **4**(2), 201–204 (2015)
11. Hu, Z., Yuan, C., Zhu, F., Gao, F.: Weighted sum transmit power minimization for full-duplex system with SWIPT and self-energy recycling. IEEE Access **4**, 4874–4881 (2016)
12. Wang, Z., Yue, X., Peng, Z., et al.: Full-duplex user relaying for NOMA system with self-energy recycling. IEEE Access **6**, 67057–67069 (2018)
13. Yang, S., Ren, Y., Lu, G., et al.: Optimal resource allocation for full-duplex wireless-powered relaying with self-energy recycling. In: 2019 11th International Conference on Wireless Communications and Signal Processing, pp. 1–6, IEEE Press (2019)
14. Dong, Y., Shafie, A.E., Hossain, M.J., et al.: Secure beamforming in full-duplex SWIPT systems with loopback self-interference cancellation. In: 2018 IEEE International Conference on Communications, pp.1–6, IEEE Press, Kansas City (2018)
15. Chen, P.P., Li, T.S., Fang, X.: Secure beamforming and artificial noise design in full-duplex wireless-powered relay networks. In: 2019 IEEE 11th International Conference on Communication Software and Networks, pp. 789–794. IEEE Press, Chongqing (2019)

16. He, L., Li, T.S., Baber, B., et al.: Secure beamforming design of SWIPT relay network with multiple users and eavesdroppers. In: 2020 International Symposium on Electronic Information Technology and Communication Engineering, pp. 1–10. IOP Publishing, Jinan (2020)
17. Torabi, M., Haccoun, D., Frigon, J., et al.: Relay selection in AF cooperative systems: an overview. IEEE Veh. Technol. Mag. 7(4), 104–113 (2012)
18. Ikhlef, A., Bocus, M.Z.: Outage performance analysis of relay selection in SWIPT systems. In: 2016 IEEE Wireless Communications and Networking Conference, pp. 1–5. IEEE Press, Doha (2016)

A Two-Stage Graph Computation Model with Communication Equilibrium

Yanmei Dong[1,2], Rongwang Chen[4,5], and Kun Guo[1,2,3(✉)]

[1] College of Mathematics and Computer Science, Fuzhou University, Fuzhou 350108, China
gukn@fzu.edu.cn
[2] Fujian Provincial Key Laboratory of Network Computing and Intelligent Information Processing, Fuzhou, China
[3] Key Laboratory of Spatial Data Mining and Information Sharing, Ministry of Education, Fuzhou 350108, China
[4] College of Mathematics and Computer Science, Wuyi University, Wuyishan 354300, Fujian, People's Republic of China
[5] Digital Fujian Tourism Big Data Institute, Wuyishan 354300, Fujian, People's Republic of China

Abstract. Distributed graph computing aims at performing in-depth analysis on large networks in a parallel manner. Iterative communication computation is an important model to perform graph analysis. Moreover, high-efficiency iterative communication computation is necessary for assuring the quality of graph partitioning. Many strategies to improve graph computing are usually based on the hypothesis of single communication iteration which focuses on the optimization of load balance in a single graph partitioning and the improvement of parallel granularity or communication manners. However, a graph in the real-world usually requires complex communication iterations to achieve good analysis results, which often have the problems of data and communication tilting and low analysis efficiency. In this paper, we propose a two-stage graph computing model with communication equilibrium (TSMCE). The model employs a communication-equilibrated graph partitioning strategy (CEGP) for load balancing and a two-stage graph computing mode (TS) to reduce the crossing-partition communication. With the change of graph density, various factors such as load balancing, communication balancing, cross-partition transmission traffic, communication delay, and convergence speed can always maintain an efficient balance. The experiments conducted on the real-world datasets show that the communication-equilibrated graph partitioning strategy can divide a graph with high quality compared with Hash and Metis. Besides, the overall performance of our two-stage graph computing model with communication equilibrium is higher than that of the BSP model.

Keywords: Graph computing · Graph partitioning · Communication iteration

1 Introduction

Large-scale graph computing research is a massive network analysis based on a distributed graph computing model. There are two mainstream graph computing models:

Y. Sun et al. (Eds.): ChineseCSCW 2020, CCIS 1330, pp. 395–409, 2021.
https://doi.org/10.1007/978-981-16-2540-4_29

one is the overall synchronous parallel computing model BSP [1–3], and the other is the typical asynchronous parallel computing model GAS [4–6]. The BSP defines the graph computing process as a series of super-steps. Each super-step is completed through communication, aggregation, and computing. Its biggest feature is the global vertex parallel processing, and the super-steps are separated by synchronous barriers. When all tasks are completed, the next iteration of super-step begins. Therefore, there is a shortage of long waiting. The GAS divides the computing process into three stages: Gather, Apply, and Scatter. There is no need to wait for each other between the tasks, so its running is faster than the synchronous parallel process, but it is necessary to design approaches to maintain data consistency. Because the BSP has no use for considering too many programming details in practical applications, the graph computing system/framework based on BSP is widely used. Typical representatives include Pregel [2], Graphx [7], Giraph [8], Pregelix [9], etc.

Aiming at the shortcomings of the long waiting of the BSP when the running time of the tasks is different [10], existing work has improved the following three aspects. Designing efficient algorithms to speed up the partitioning of graphs [11–13]. Reducing the communication delay by coarsening the minimum execution unit [14, 15]. Improving the parallel mechanism [10, 16], such as asynchronous parallel, to solve the long waiting problem.

Graphs like social and biological networks have the characteristics of uneven data distribution [17], and the density of different graphs also varies. Graph density is used to describe the distribution of vertices and edges on the graph. There have been more improvements in the algorithm efficiency of BSP, and usually, a single application algorithm for communication is used for model testing, which is not enough for the comprehensive performance of the graph computing model. This paper integrates factors such as load balancing, communication balancing, traffic of cross-network, communication efficiency, and convergence speed. A two-stage graph computing model with communication equilibrium is proposed to ensure that as the graph density changes, various factors can still maintain an efficient balance, further improve the operational efficiency of graph computing. The main contributions of this article are as follows.

(1) We propose a two-stage graph computing model with communication equilibrium. The graph partitioning method is optimized based on the vertex degree, and the communication iteration method of the BSP is improved to improve the graph computing performance.

(2) We propose a communication-equilibrated graph partitioning strategy. The cluster vertex dichotomy strategy, partition balance division, and partition cycle distribution principle are introduced to balance the density of each partition and solving the problem of data tilt and communication tilt caused by the increase of data volume and the change of graph density.

(3) We propose a two-stage graph computing model. The model is made up of message communication, message aggregation, local computing in the first stage and local computing in the second stage. It brings in data separation cache structure, data mapping, and cache discard function to reduce communication delay, accelerate algorithm convergence, and reduce message traffic.

2   Related Work

Existing work mainly improved the graph computing model in terms of graph partitioning and communication models.

In terms of graph partitioning. The PowerGraph proposed by Gonzalez et al. [4] divides the multi-degree vertices, reducing the traffic of cross-network, and solving the problem of significant imbalance in partitions. Zhao et al. [11] introduced a splitter-merger manager to improve message communication efficiency. Zhang et al. [12] proposed the LSH graph partitioning algorithm to reduce the problem of irregular memory access. Liu et al. [13] proposed a distributed re-encoding partitioning method to make the vertex serial number, to ensure the aggregation of adjacent vertices. Xiao et al. [14] cut the multi-degree vertices globally or locally according to the characteristics of the combination of graph computing and machine learning to reduce redundant partitioning. These improvements can effectively improve the efficiency of graph segmentation, but it does not combine the analysis of the communication model and complex communication algorithms on the segmentation effect, communication performance, and stability.

In terms of communication models. There are some types of point-to-point, subgraph-to-subgraph, and block-to-block communication [17]. In contrast, the point-to-point communication model represented by BSP can support most graph computing algorithms. However, because BSP has the characteristics of overall synchronization and parallelism, for large graph, it has the problems of slow communication speed and low parallel efficiency. Existing work has been carried out to improve these problems. Liu et al. [10] proposed the CSA-BSP model of asynchronous BSP to solve the problem of mutual task waiting and optimize the data throughput rate. Redekopp et al. [18] accelerated graph computing by controlling the use of memory based on the BSP. Lai et al. [19] proposed an improved GPregel model based on BSP, which coarsened the minimum execution unit to reduce the memory access rate and parallel data. Ji et al. [16] used the local approximation method to map the local to the global graph according to the similarity of the vertices of different partitions, thereby reducing the running time of graph computing. Heintz et al. [20] proposed an improved MESH framework based on BSP, which can flexibly switch the minimum execution unit according to algorithm requirements. These improvements can speed up the efficiency of the algorithm, but the communication process of BSP highly depends on the quality of graph partitioning, which ignoring the impact of graph partitioning on communication iteration efficiency.

Existing graph partitioning methods focus on improving the partition load problem, and most on balancing the number of communication edges or vertices of the partition. For graph algorithms with relatively simple communication, good load balancing can be achieved, but for complex, the density balance of each partition can ensure the stability of load balancing. The methods of asynchronization, optimized memory access, and parallel granularity in the existing BSP improvement schemes are highly targeted and are not suitable for complex communication graph algorithms. This paper proposes a two-stage graph computing model with communication equilibrium combining multiple factors such as traffic of cross-partitions, load balancing, communication balance, iterative methods, and memory utilization to improve.

3 Preliminary

3.1 Graph Partitioning

Common partitioning methods include Metis and Hash, both of which focus on the division of local graph data.

Metis [21] is based on k-way partitioning, which can effectively decrease the communication edge across partitions and ensure the load balance of partitions to a certain extent. There are three steps in the partitioning process, including coarsening, k-way division and refinement. As the density of the graph increasing, this method cannot guarantee the balance and stability of the partition density.

There are four common Hash [7] partition methods. EdgePartition2D employs matrix two-dimensional division to partition the edges of the source and destination vertex. EdgePatition1D uses one dimension to partition the edge to which the source vertex belongs. RandomVertexCut partitions according to the HashCode of the source and destination vertex. CanonicalRandomVertexCut first sorts the source and destination vertex of the edge, and then uses RandomVertexCut. Hash utilize vertex cutting, which is easy to generate duplicate vertices. When the number of duplicate vertices is plenty, the overhead will increase accordingly.

3.2 BSP Model

BSP [1] is a vertex-centered overall synchronous parallel graph computing model. It consists of a series of iterative super-steps, where each of it needs to go through three stages of message communication, message aggregation, and local computing. Each iteration super-step is separated by a synchronization barrier. Because BSP has the characteristics of overall synchronization and parallelism, it is necessary to wait for all tasks to complete before entering the next iteration super-step. When the task time difference is large, there is a lack of long waiting.

When using the BSP for graph computing, a Shuffle operation is required. Shuffle is a process of serializing parallel data using a buffer, and the amount of data in the buffer will affect the operating efficiency.

4 A Two-Stage Graph Computing Model with Communication Equilibrium

A two-stage graph computing model with communication equilibrium (TSMCE) includes the communication-equilibrated graph partitioning strategy (CEGP) and the two-stage graph computing mode (TS). On the basis of the edge cutting and vertex degree, CEGP employs the cluster vertex dichotomy strategy, the balanced partition division principle, and the cyclic partition allocation principle to partition graphs. TS brings in two-stage local computing, at the same time designs data separation cache structure, data mapping function, and cache discard function to assist it.

4.1   Communication-Equilibrated Graph Partitioning Strategy

CEGP is different from the traditional partition method. It considers two aspects.

Considering globally how to ensure the balance of partition density. On the basis of the degree, vertex with the same degree are grouped into a cluster. According to the cluster vertex dichotomy strategy, the cluster vertices that can be evenly distributed to each partition are putting into the sharable area, otherwise the extra area. Among them, the vertices in the sharable area have continuity, and most vertices with the same degree have an adjacent relationship.

Considering locally how to divide adjacent vertices into the same partition. The key is to evenly divide the vertices of the same degree in the sharable area into each partition in the form of neighbor aggregation. For the division of vertices in extra areas, it is necessary to ensure the balance of the density of each partition. The principle of balanced partition division and cyclic partition allocation can be used for sharable area and extra area.

The design of CEGP is mainly divided into the following methods.

(1)   Cluster vertex dichotomy. Obtaining the vertex with the smallest degree and merging into a subgraph. The vertex on the subgraph is the cluster vertex. Taking multiple sets of data in the subgraph, and the elements of each set including degree and vertex. Calculating the number of vertices of subgraph and that cannot be divided equally into each partition. Storing the data that can be evenly divided into each partition into the sharable area, otherwise into the extra area. The already processed subgraph is deleted from the remaining graph. Then repeat the above process until the algorithm stops when the remaining graph is empty.

(2)   Balanced partition division. The data in the sharable area are evenly distributed to each partition according to the same degree. It needs to obtain the data of the smallest degree. Dividing the data into the number of partitions and putting them into each partition respectively. Deleting the processed data from the sharable area. Then repeat the above process until the algorithm stops when the sharable area is empty.

(3)   Cyclic partition allocation. According to the number of partitions, the data in the extra area is cyclically allocated to each partition. To stop the algorithm until the data in the extra area is processed.

The specific division process of CEGP is shown in Fig. 1.
The specific implementation of CEGP is shown in Algorithm 1.

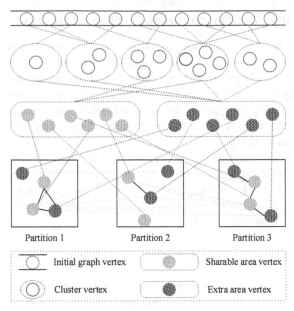

Fig. 1. CEGP specific division process

Algorithm 1. CEGP

Input: The initial graph $G(V,E)$, the number of partitions p.

Output: The updated graph G_{new}.

1.   Initialize flag←TRUE, G_ψ←G;
2.   **WHILE** (flag) **DO**
3. $X=(x_1(d_{min},v_1)\cup ... \cup x_n(d_{min},v_n))$, $X\in G_\psi$; //d_{min} is the smallest degree.
4. G_φ←X; // Forming subgraph G_φ.
5.   m←n%p
6.   **IF** (m == 0) **THEN**
7.   averData.add(X);
8.   **ELSE**
9.   **IF** (n > p) **THEN**
10.   otherData.add($x_1\cup ... \cup x_m$);
11.   averData.add(($x_{(m+1)}\cup ... \cup x_n$);
12.   **ELSE**
13.   otherData.add($x_1\cup ... \cup x_m$);
14.   **END IF**
15.   **END IF**
16. G_ψ←(G_ψ - G_φ);
17.   **IF** (G_ψ is empty) **THEN**
18.   flag←FALSE;
19.   **END IF**
20.   **END WHILE**
21.   averData←*equalAllocation*(averData, p); //The principle of balanced division.
22.   otherData←*roundAllocation*(otherData, p); //The principle of circular allocation.
23.   partData←(averData∪otherData);
24. G_{new}← G.add(partData);
25.   return G_{new};

Function 1. *equalAllocation*
Input: The sharable area averData, the number of partitions p.
Output: newData.
1.   Initialize flag←TRUE, n←1;
2.   **WHILE** (flag) **DO**
3. X←$(x_1(d_{min}, v_1) \cup \ldots \cup x_t(d_{min}, v_t))$, X∈averData; // d_{min} is the smallest degree.
4.   tag←1;
5.   **WHILE** (tag ≤ p) **DO**
6.   m←(t/p)× tag;
7.   newData.add $(x_n(d_{min}, v_n,$ tag$) \cup \ldots \cup x_m(d_{min}, v_m,$ tag$))$;
8.   tag←(tag+1);
9.   n←m+1;
10.   **END WHILE**
11.   averData←(averData−X);
12.   **IF** (averData is empty) **THEN**
13.   flag←FALSE;
14.   **END IF**
15.   **END WHILE**
16.   return newData;

Function 2. *roundAllocation*
Input. The extra area otherData, the number of partitions p.
Output: newData.
17.   Initialize flag←TRUE, tag←1, part←1;
18. X←$(x_1(d_1, v_1) \cup \ldots \cup x_t(d_t, v_t))$, X=otherData;
19.   **WHILE** (flag) **DO**
20.   **IF** (part > p) **THEN**
21.   part←1;
22.   **END IF**
23.   newData.add($x(d_{tag}, v_{tag},$ part$)$);
24.   **IF** (tag > t) **THEN**
25.   flag←FALSE;
26.   **END IF**
27.   part←part + 1;
28.   tag←tag + 1;
29.   **END WHILE**
30.   return newData;

4.2 Two-Stage Graph Computing Mode

On the basis of the BSP parallel graph computing model, the two-stage graph computing mode improves communication iteration mode by adding two-stage local computing. It forms a two-stage computing mode (TS) of message communication, message aggregation, first stage local computing, second stage local computing. TS merges two adjacent iterations of BSP super-step, using multi-step local computing to reduce communication time and delay.

Shuffle operations are required for parallel computing. Shuffle is a process of serializing parallel data using a buffer, and the amount of data in the buffer will affect the operating efficiency. To reduce the amount of data buffer generated by the Shuffle process during iteration, a data separation cache structure is introduced, which including graph data and partition shared memory. The former is mainly used for data interaction during communication, while the latter is mainly used to assist local computing. Graph data memory is further subdivided into vertex and edge data memory, and separating the management of vertex and edge data of each partition. Partitioned shared memory is distributed on each partition and is used to store serialized messages/data. It mainly stores the latest news of the vertices adjacent to the same partition after the local computing in the first stage, to avoid the need to communicate to obtain the latest news of the neighboring vertices of the same partition when entering the second stage of the local computing.

The data separation cache structure is shown in Fig. 2.

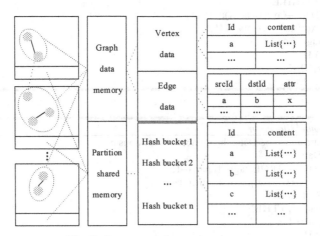

Fig. 2. Data separation cache structure

Besides, data mapping and cache discard functions are also introduced to assist local computing. The data mapping function is mainly used to obtain the message to be transmitted from the vertex data memory during communication, to obtain data from the vertex data memory/partition shared memory during local computing, and to store the latest message into the vertex data memory/partition shared memory. The cache discard function is applied to clear the time-sensitive data in the partition shared memory in time after the local computing is completed, and reserving more memory for the next communication.

The comparison between the parallel mechanism of BSP and TS is shown in Fig. 3. (a) is the parallel mechanism of BSP and (b) is the parallel mechanism of TS.

The running process of TS single iteration super-steps is as follows:

(1) Message initialization. Initializing vertex data only needs to be initialized before the start of the first iteration super-step.
(2) Message communication. Each vertex in the graph sends a message to its neighbor along the edge, and then each vertex will receive the corresponding message.
(3) Message aggregation. The vertices on the graph aggregate the received messages to facilitate subsequent local computing.

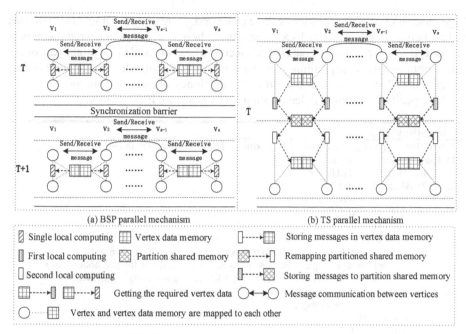

Fig. 3. Comparison of parallel mechanisms

(4) The first stage of local computing. The application algorithm needs to process the aggregated messages. To facilitate the second stage of local computing, the time-sensitive data of the vertices adjacent to the same partition is stored in the partition shared memory to realize the separate cache of the graph data and the time-sensitive data, and the time-sensitive data is updated to the vertex data memory.

(5) The second stage of local computing. At this time, it is necessary to use the data remapping function to obtain the adjacent data of the same partition from the partition shared memory. Using the data remapping function to obtain the adjacent data from the vertex data memory, and then executing the application algorithm. When completed, each vertex will generate a new message. Updating the new message to the vertex data memory to provide data for the next iteration super-step.

(6) Super-step ends. Before the end of a super-step, you need to use the cache discard function to clear the time-sensitive data in the partition shared memory and reserve memory for the next communication.

From the analysis of the TS parallel mechanism, it can be seen that the two-stage local computing converts part of the traffic to the second-stage local computing by controlling the cache utilization rate, reducing the traffic within the partition. Also, the data separation cache structure introduced separately designed graph data and partition shared memory for parallel and serialization process. For parallel processes, all vertices access the graph data memory without interfering with each other and without waiting. Most of the input in the local computing stage is serialized data, and the advantage of graph data memory is only for parallel processes. If it interacts with local computing, a large number of Shuffle operations will be generated, increasing the amount of buffering and reducing the efficiency of local computing. Therefore, partition shared memory is introduced to store serialized data. In the second stage of local computing, part of the data is obtained from the partition shared memory, which can reduce the amount of data buffer in the Shuffle process and improve the efficiency of local computing.

The specific algorithm implementation of TS is shown in Algorithm 2.

Algorithm 2. TS

Input: The graph $G_{new}(V,E)$ after CEGP partition, The number of iterations *maxIter*.
Output: The iteration result C.
1.   Initialize flag←TRUE, iter←1;
2.   **WHILE** (flag) **DO**
3.   **IF** (iter == *maxIter*) **THEN**
4.   flag←FALSE;
5.   **END IF**
6.   **FOR EACH** $v \in V$ **DO**
7.   msg←getMsg(msg); // Getting message from vertex data memory.
8.   sendMessage(msg); // Sending message to $N(v)$. $N(v)$ is the neighbor set of v.
9.   mergeData←receiveMessage(msg); // Receiving and merging messages.
10.   msg←*localCompute*(v, mergeData); // Local computing.
11. G_{new}←G_{new}.add(msg);
12.   **END FOR**
13.   iter←iter + 1;
14.   **END WHILE**
15. $C←((v_1,c_1) \cup ... \cup (v_n,c_n))$; // Extracting results from G_{new}.
16.   return C;

Function 1. *localCompute*

Input: The serialize messages *mergeData*, The number of iterations *maxIter*.
Output: The updated *vertexMem*.
Explanation: *shareMem* is the partition shared memory, *vertexMem* is the vertex data memory, *dealMsg* is the message processing function, *mergeMsg* is the message merge function.
1.   *shareMem*←*dealMsg*(v, *mergeData*); // First stage local computing
2.   pos←getPosition(v); // Getting the location of v from shareMem.
3.   *vertexMem*←(v,pos); //Mapping v and pos to vertex data memory.
4.   **FOR EACH** v **DO**
5.   msg←mergeMsg($N(v)$); // Getting and integrating the latest messages of $N(v)$.
6.   **END FOR**
7.   *vertexMem*←*dealMsg*(v, msg); // Second stage local computing
8.   *shareMem*.clear();
9.   return *vertexMem*;

---

5 Experiments

5.1 Experimental Environment

The experiment was carried out in a blade cluster with 1 master node and 4 worker nodes. The configuration of each node is: CPU*2.6 GHz, Intel Xeon CPU E5-2660v3, 256G memory, 300G hard disk.

5.2 Experimental Datasets

The experiment uses 5 real datasets from Stanford University [22], as shown in Table 1.

Table 1. Experimental datasets

Name	Nodes	Edges
Amazon	334863	925872
DBLP	317080	1049866
Google	875713	5105039
Youtube	1134890	2987624
Patents	3774768	16518948

5.3 Evaluation Metrics

The evaluation metrics used in the experiment are the number of cross-partition communication edges, the variance of partition density, the number of algorithm iterations, the number of super-step communication edges, and the algorithm running time.

The graph is represented as $G(V,E)$, where V and E are the set of vertices and edges, respectively.

Ahn et al. [23] proposed the concept of community link density, which measures the density of the internal edges of the community, and it can well express the quality of the partition. Therefore, this metric is introduced in this paper and defined as partition density. The partition density of each partition is defined as:

$$D_p = \frac{|E_{in}|_p - (n_p - 1)}{n_p * (n_p - 1)/2 - (n_p - 1)} \tag{1}$$

Where p is the number of partitions, $|E_{in}|_p$ is the number of communication edges in partition p, and n_p is the number of vertices in partition p.

When the density of each partition is equal, it is the ideal partition density. The density of each partition is recorded as $\overline{D}$.

The variance of partition density is used to measure the deviation of a group of actual partition density, which is relative to ideal. It is defined as:

$$S^2 = \frac{\sum_1^q (D_p - \overline{D})^2}{p} \tag{2}$$

Where p is the number of partitions, $1 \le n \le p$.

5.4   Experimental Results and Analysis

CEGP Performance Analysis. Experimental research on the influence of Hash, Metis, and CEGP partition methods on multi-label propagation algorithm [24]. The metrics include the number of cross-partition communication edges, the variance of partition density, and the running time of the algorithm.

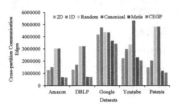

Fig. 4. Number of cross-partition edges.   **Fig. 5.** Variance of partition density.

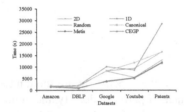

Fig. 6. Running time of the algorithm.

As shown in Fig. 4, the number of cross-partition communication edges of Metis and CEGP is always at a low level, and other partition methods are unstable. The effect of CEGP is more significant on large datasets. CEGP is divided based on edge cutting and degree averaging. Compared with Hash based on vertex cutting, CEGP doesn't need to maintain huge duplicate vertices. Compared with Metis based on Breadth-First Search, CEGP can better guarantee the gathering of adjacent vertices.

As shown in Fig. 5, the variance of using CEGP is always the smallest, indicating that the density fluctuation of each partition is not obvious, and the density of each partition is relatively similar and closer to the ideal. From the perspective of the fluctuation state of the difference between the partition methods on different datasets, the variance of CEGP is more balanced, while the others have obvious peaks or valleys. The consistency division of CEGP ensures the balance and stability of the partition density under different datasets.

As shown in Fig. 6, the running time of the algorithm using CEGP is the shortest, followed by Metis. The changes in 2D and Canonical are relatively stable, while Random is not. The increase in the dataset has a greater impact on 1D.

For Hash and Metis, the number of communication edges of each partition is easily affected by the change in graph density, resulting in skewed partition data. CEGP ensures that the vertices of high degrees are distributed centrally, allowing communication to occur within the partition as much as possible to reduce cross-partition transmission. Besides, it can Guarantee the balanced distribution of data, achieve a better communication balance, and reduce the waiting time for synchronization.

TSMCE Performance Analysis. Using a multi-label propagation algorithm [24] to analyze BSP and TSMCE, both of which are partitioned by CEGP. The metrics include the number of algorithm iterations, the number of super-step communication edges, and the algorithm running time.

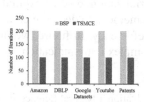

Fig. 7. Number of iterations.   **Fig. 8.** Number of communication edges.

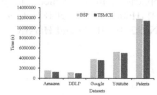

Fig. 9. Running time of the algorithm.

As shown in Fig. 7, the algorithm implemented using TSMCE has half the number of iterations compared to BSP. The reason is that TS using two-stage local computing, which merging two adjacent super-steps.

As shown in Fig. 8, The decrease in the number of communication edges of TSMCE raise with the increase of the graph density. This is because CEGP ensures the consistency of the division, and also allows the adjacent vertices to be more concentrated in the same partition, reducing cross-partition transmission. Furthermore, the number of communication edges of the merged super-step only including the BSP single super-step and CEGP division to obtain the communication edge in the partition.

As shown in Fig. 9, The running time of the algorithm implemented by TSMCE is less than BSP. TS elastically uses memory utilization to exchange adjacent communications in the partition, thereby reducing the communication delay.

6 Conclusions

This paper proposes a two-stage graph computing model with communication equilibrium (TSMCE). Consistency partitioning of the communication-equilibrated graph partitioning strategy (CEGP) can divide vertices of the same degree into each partition in a balanced manner, and ensure the aggregation of adjacent vertices. When the load of each partition is balanced, the communication between higher-degree vertices is concentrated on the same partition. For the complex communication algorithm, the communication density will be more balanced. Two-stage local computing of the two-stage graph computing mode (TS) merges two adjacent super-steps to convert part of the communication delay into the local computing. Auxiliary measures such as data separation cache structure, data mapping, and cache discarding are adopted to manage data according to the characteristics of parallel or serialization, discard expired data in time to reduce memory pressure, and elastically use memory to decrease communication traffic and delay, thereby improving operating efficiency. In future, we intend to further improve TSMCE in combination with the complex communication method of local iteration.

Acknowledgement. This work is partly supported by the National Natural Science Foundation of China under Grant No. 61672159, No. 61672158 and No. 61300104, the Fujian Collaborative Innovation Center for Big Data Applications in Governments, the Fujian Industry-Academy Cooperation Project under Grant No. 2017H6008 and No. 2018H6010, the Natural Science Foundation of Fujian Province under Grant No. 2019J01835 and Haixi Government Big Data Application Cooperative Innovation Center.

References

1. Valiant, L.G.: A bridging model for parallel computation. Commun. ACM **33**(8), 103–111 (1990)
2. Malewicz, G., et al.: Pregel: a system for large-scale graph processing. In: Proceedings of the ACM SIGMOD International Conference on Management of Data (SIGMOD 2010), pp. 135–146. Indianapolis (2010)
3. Zhang, W.D., Cui, C.: Delta-stepping synchronous parallel model. J. Softw. **30**(12), 3622–3636 (2019)
4. Khorasani, F., Vora, K., Gupta, R., Bhuyan, L.N.: PowerGraph: distributed graph-parallel computation on natural graphs. Oper. Syst. Des. Implementation 17–30 (2012)
5. Low, Y., Gonzalez, J.E., Kyrola, A., Bickson, D., Guestrin, C.E., Hellerstein, J.: Graphlab: a new framework for parallel machine learning. Comput. Sci. (2014)
6. Khorasani, F., Vora, K., Gupta, R., Bhuyan, L.N.: CuSha: vertex-centric graph processing on GPUs. High Performance Distrib. Comput. 239–252 (2014)
7. Reynold, S., Xin, J.E., Gonzalez, M.J., Franklin, I.S.: GraphX: a resilient distributed graph system on spark. In: Proceedings of the 1st International Workshop on Graph Data Management Experiences and Systems (GRADES 2013), pp. 1–6. ACM, New York (2013)
8. Salihoglu, S., Shin, J., Khanna, V., Truong, B., Widom, J.: Graft: a debugging tool for Apache Giraph. In: Proceedings of the ACM SIGMOD International Conference on Management of Data (SIGMOD 2015), pp. 1403–1408. ACM, Melbourne (2015)

9. Bu, Y., Borkar, V., Jia, J., Carey, M.J., Condie, T.: Pregelix: Big(ger) graph analytics on a dataflow engine. Very Large Data Bases **8**(2), 161–172 (2014)

10. Liu, F.A., Liu, Z.Y., Qiao, X.Z.: An asynchronous BSP model and optimization techniques. Chinese J. Comput. **25**(004), 373–380 (2002)

11. Zhao, X., Li, B., Shang, H.C., Xiao, W.D.: A revised BSP-based massive graph computation model. Chinese J. Comput. **40**(01), 223–235 (2017)

12. Zhang, W., Zhang, M.: Graph partitioning algorithm with LSH: poster extended abstract. In: 2018 IEEE International Conference on Cluster Computing (CLUSTER), pp. 166–167. International Conference on Cluster Computing, Belfast (2018)

13. Liu, Q., Dong, X., Chen, H., Zhang, X.: H2Pregel: a partition-based hybrid hierarchical graph computation approach. Futur. Gener. Comput. Syst. **104**, 15–31 (2020)

14. Xiao, W., et al.: Distributed graph computation meets machine learning. IEEE Trans. Parallel Distrib. Syst. **31**(7), 1588–1604 (2020)

15. Heintz, B., Hong, R., Singh, S., Khandelwal, G., Tesdahl, C., Chandra, A. MESH: a flexible distributed hypergraph processing system. In: 7th IEEE International Conference on Cloud Engineering (IEEE IC2E), pp. 12–22, International Conference on Cloud Engineering, Prague (2019)

16. Ji, S., Zhao, Y.A.: Local approximation approach for processing time-evolving graphs. Symmetry **10**(7), 2073–8994 (2018)

17. Heidari, S., Simmhan, Y., Calheiros, R.N., Buyya, R.: Scalable graph processing frameworks: a taxonomy and open challenges. ACM Comput. Surv. **51**(3), 1–53 (2018)

18. Redekopp, M., Simmhan, Y., Prasanna, V.K.: Optimizations and analysis of BSP graph processing models on public clouds. In: 27th IEEE International Parallel and Distributed Processing Symposium (IPDPS), pp. 203–214. International Parallel and Distributed Processing Symposium, Boston (2013)

19. Lai, S., Lai, G., Lu, F., Shen, G., Jin, J., Lin, X.: A BSP model graph processing system on many cores. Clust. Comput. **20**(2), 1359–1377 (2017). https://doi.org/10.1007/s10586-017-0829-0

20. Heintz, B., Hong, R.., Singh, S., Khandelwal, G., Tesdahl, C., Chandra, A.: MESH: a flexible distributed hypergraph processing system. In: 7th IEEE International Conference on Cloud Engineering (IEEE IC2E), pp. 12–22. International Conference on Cloud Engineering, Prague (2019)

21. Karypis, G., Kumar, V.: METIS, a Software Package for Partitioning Unstructured Graphs, Partitioning Meshes, and Computing Fill-in Reducing Ordering of Sparse Matrices Version 4.0. Department of Computer Science, pp. 44–78, University of Minnesota, Minneapolis (1998)

22. Stanford University. https://snap.stanford.edu/data/index.html. Accessed 03 June 2020

23. Ahn, Y., Bagrow, J.P., Lehmann, S.: Link communities reveal multiscale complexity in networks. Nature **466**(7307), 761–764 (2010)

24. Gregory, S.: Finding overlapping communities in networks by label propagation. New J. Phys. **12**(10) (2010)

Video Pedestrian Re-identification Based on Human-Machine Collaboration

Yanfei Wang, Zhiwen Yu$^{(\boxtimes)}$, Fan Yang, Liang Wang, and Bin Guo

Northwestern Polytechnical University, Xi'an 710072, China
zhiwenyu@nwpu.edu.cn

Abstract. The purpose of pedestrian re-identification is to find pedestrian targets from many real surveillance videos. It has multiple realistic application scenarios such as criminal search, intelligent security, and cross-camera tracking. However, the current stage of pedestrian re-identification faces many challenges, including blurring of surveillance video, changes in surveillance angles, changes in pedestrian posture, and pedestrian occlusion. Although the existing research methods have a good performance on some datasets, there is still a big gap from the practical application. Taking into account the advanced intelligence, flexibility, and adaptability of human beings in the face of complex problems and environmental changes, this paper attempts to establish a model based on human-machine collaboration, in order to improve the accuracy and enhance the practicability. Along this line, we first construct a new human-machine collaboration video pedestrian re-identification model (HMRM) and design the task allocation strategies and result fusion strategies accordingly. Then, based on the YOLOv3 object detection algorithm and the ResNet-50 network structure, we perform the pedestrian detection part and the pedestrian re-identification part respectively. Moreover, since HMRM allocates human-machine tasks based on overlapping thresholds, we do exploratory experiments on human workload, overlapping thresholds, and model accuracy. Finally, based on pyqt5 and multi-threading technology, a human-machine collaboration video pedestrian re-identification interface is developed. Experiments on the public datasets show that HMRM is superior to other pedestrian re-identification approaches.

Keywords: Human-machine collaboration · Deep learning · Pedestrian detection · Pedestrian re-identification

1 Introduction

The purpose of video pedestrian re-identification is to find pedestrian targets from many real surveillance videos, and output the video frame and the area where it appears. This technology is suitable for multiple application scenarios such as criminal search, cross-region tracking and personnel activity analysis, therefore has become an important research issue in academia and industry. Of course, the research on pedestrian re-identification faces many complex practical problems: most of the pedestrian images are recorded at different time, but the clothing and posture of pedestrians will change

Y. Sun et al. (Eds.): ChineseCSCW 2020, CCIS 1330, pp. 410–423, 2021.
https://doi.org/10.1007/978-981-16-2540-4_30

with different degrees over time; the video obtained by the surveillance camera is generally not clear enough, so the external characteristics of different pedestrians are likely to be difficult to distinguish due to similar body shapes and similar clothes; in crowded scenes, huge traffic would cause a lot of occlusions. Although more public datasets have provided, and many methods have achieved high accuracy on them, but most of the datasets and methods are based on manually cropped pedestrian images. Most existing pedestrian re-identification methods assume that pedestrian detection is ideal [1]. However, in real applications, due to the complex pedestrian background and interference, it is not possible to directly use the manually cropped pedestrian images. In addition, the existing pedestrian detection methods still have some shortcomings. In the detection process, pedestrian omissions, detection errors, and inaccurate detection frames will inevitably occur, which will cause the accuracy to fail to meet the actual application requirements. Although pedestrian re-identification technologies continue to innovate and improve, none of them can completely replace humans. As a results of the openness of the problems and the high degree of uncertainty in living environment, humans have great advantages in dealing with changes and complex issues.

In this paper, we take into account the advanced intelligence of humans, and accomplish the task of video pedestrian re-identification utilizing human-machine collaboration. Human-machine collaboration means that in a constantly changing environment, machines and humans can perform tasks interactively so that the machine can achieve the best performance under a small amount of manual guidance. we construct a new human-machine collaboration video pedestrian re-identification model(HMRM) and formulate task allocation strategies and result fusion strategies accordingly. Then, based on the YOLOv3 object detection algorithm and the ResNet-50 network structure, we perform the pedestrian detection part and the pedestrian re-identification part respectively. Since HMRM allocates human-machine tasks based on overlapping thresholds, we do extensive exploratory experiments on human workload, overlapping thresholds, and model accuracy. It is concluded that higher accuracy can be achieved with less human incorporation under the threshold of 0.3. Finally, based on pyqt5 and multi-threading technology, a human-machine collaboration video pedestrian re-identification interface is developed.

The main contributions of this paper are:

- A new human-machine collaboration video pedestrian re-identification model (HMRM) is proposed, and task allocation strategies and result fusion strategies are formulated accordingly. Experiments on the public datasets show that HMRM is superior to other pedestrian re-identification approaches.
- Exploratory experiments on human workload, overlapping thresholds and model accuracy are Completed. It is concluded that higher accuracy can be achieved with less human participation under the threshold of 0.3.
- An interactive pedestrian re-identification operation interface is developed, based on pyqt5 and multi-threading technology. Furthermore, we propose a concept named virtual file queue.

2 Related Work

Pedestrian re-identification task includes two parts: pedestrian detection part and pedestrian re-identification part. For the pedestrian detection part, based on CifarNet, Hosang [2] established a CNN pedestrian detection model. The input is changed from traditional artificial design features to RGB three-channel images, which means that there is no need to manually design complex features. In addition, he conducted experiments on various factors, including proposal extraction strategy, CNN model structure, parameter settings, data processing and various training tricks. Tian [3] uses scene attributes to improve the accuracy, inputs pedestrians and scene attribute tags into the CNN, and trains the model in a multi-task mode. The model fully considers the relationship between pedestrian attributes and scene attributes, and uses the relevance of semantic information to remove part of the interference information contained around pedestrians. Cai [4] proposed a perceptual enhancement algorithm for the training of CNN cascade detectors to achieve the best trade-off between accuracy and speed.

For pedestrian re-identification models, Li [5] applied deep learning for the first time and proposed a neural network called FPNN, which automatically learns the best features of pedestrians and features of photometric and geometric transformations. Ahmed [6] designed a specific CNN model for pedestrians. The input of the model is two cropped photos, and the binary verification is used as the loss function. The parameter training is completed through the loss function, finally, the similar value of two pedestrians is output. The network joins the layer that calculates the neighborhood difference of the cross input, extracting local relationship of the input pedestrians. Ding [7] and Cheng [8] used triplet samples to train convolutional neural networks.

At the same time, many researchers have proposed end-to-end pedestrian re-identification methods. In 2016, Xiao Tong [1] proposed an end-to-end pedestrian re-identification model and improved it in 2017, using a fifty-layer residual neural network as the backbone for feature extraction. The original pixels are extracted into a convolutional feature map through a special Stem CNN structure, and then the boundary box of the candidate pedestrian is generated through the pedestrian candidate frame generation network. In 2019, Han [9] proposed a position correction model of detection results based on pedestrian re-identification and corrected the detected pedestrians through the ROI transform layer. The model learns the laws of affine transformation for oblique pedestrian photos. In the training phase, a new proxy triplet loss is proposed, and the softmax loss function is combined with it.

Due to the complexity of the pedestrian re-identification problem, some researchers have also proposed a human-machine collaboration model. Abir [10] added human to the model to expand its multi-class based model. Liu [11] proposed a POP model, which improves the speed and accuracy of re-identification. In 2016, Wang [12] added human guidance in pedestrian re-identification. Given a pedestrian to be queried, use the model to re-identify the pedestrian pictures in the gallery, and get the rank list results sorted by confidence. People find a true match from the rank list and feed it back to the model to update the model. After three iterations, the final model is obtained.

Researches mentioned above focus on innovation and improvement of target detection algorithms and pedestrian re-identification algorithms. However, the algorithm itself still has certain limitations, and human-machine collaboration can make up for this deficiency to a certain extent. Even though some researchers propose to apply human guidance to pedestrian re-identification, the object detection algorithm and a pedestrian re-identification algorithm at that time were not mature enough. With the introduction of a variety of deep learning networks, pedestrian detection algorithms and pedestrian re-identification models have also been greatly improved and developed [13–15]. Therefore, it is necessary to explore new human-machine collaboration methods to improve the accuracy and practicability.

3 Human-Machine Collaboration Video Pedestrian Re-identification Model

3.1 Human-Machine Collaboration Framework

The traditional video pedestrian re-identification framework is shown as Fig. 1. First, the video is split into video frames, and the pedestrian detection model detects all pedestrians in the video frame, crops the detected pedestrians, and inputs them into the gallery. Input the query pedestrian step, including a single or multiple pictures of a pedestrian, the pedestrian re-identification module extracts the characteristics of the pedestrian and processes the characteristics of the pedestrian in the gallery, and compares the characteristics to find the same pedestrian in the gallery.

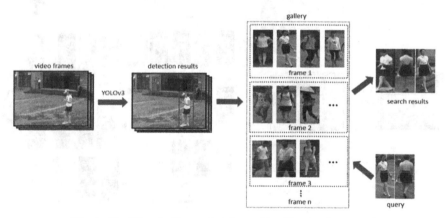

Fig. 1. Traditional video pedestrian re-identification framework

However, the traditional pedestrian re-identification framework has great limitations. In practical applications, there will inevitably be very crowded scenes, and pedestrians may have a large number of occlusions, as shown in Fig. 2. In places with high pedestrian traffic such as subway entrances and shopping malls, after object detection, a large number of detected pedestrians will be stacked in certain period, resulting in incomplete or redundant characteristics of the pedestrians' frames. It is certain to affect the pedestrian features extracted by the re-identification model and reduces the accuracy of re-identification.

414   Y. Wang et al.

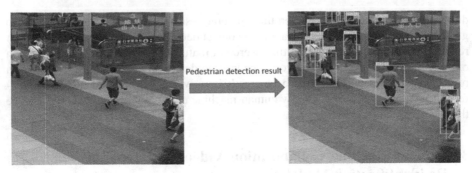

Fig. 2. Pedestrian detection result in crowded scenes

Based on this thinking, we propose a human-machine collaboration pedestrian re-identification framework, making full use of human intelligence. With the limited guidance of humans, the occlusion phenomenon that is difficult for the machine to handle is recognized in some extent for our humans, thereby effectively improving the accuracy and practicability. The human-machine collaboration video pedestrian re-identification framework is shown in Fig. 3.

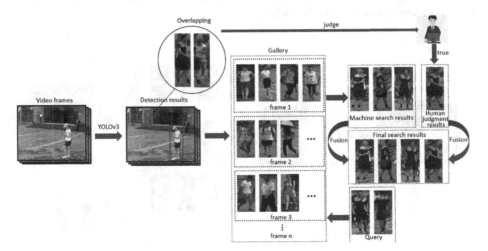

Fig. 3. Human-machine collaboration video pedestrian re-identification framework

The beginning part of the human-machine collaboration framework is the same as the traditional framework, that is, the video is divided into video frames. Then the pedestrian detection module detects all pedestrians in the video frames. The difference is that when the pedestrian detection module detects pedestrians, the model will calculate the degree of overlap of all pedestrians detected in each frame, using Intersection over Union to measure metric. IoU is calculated as follows:

$$IoU_{ij} = \frac{Area(det_i) \cap Area(det_j)}{Area(det_i) \cup Area(det_j)} \tag{1}$$

When IoU is greater than the predefined threshold, it can be considered that the detected pedestrians have overlapped. For the overlapping cropped pedestrian pictures, the model will be conducted by humans, and the humans will give their own judgments and get the feedback results of the same pedestrian or not the same pedestrian. Finally, the model will fusion the human judgment result and the machine re-identification result to give the final pedestrian re-identification result.

3.2 Task Allocation Strategy and Result Fusion Strategy

In the human-machine collaboration video pedestrian re-identification framework, the task allocation strategy is the core part. In our proposed framework, the machine refers to the computing unit, which is composed of a detection module and a pedestrian re-identification module. People refer to experts involved in the video pedestrian re-identification task who need to complete certain subtasks to re-identify pedestrians. Task analysis and allocation are what the task allocation strategy needs to accomplish. For a video file, analyze its frame by frame, and decompose the video pedestrian re-identification task into three different sub-tasks: pedestrian detection sub-task, pedestrian re-identification sub-task and overlap processing sub-task. The pedestrian detection subtask is in charge of the pedestrian detection module computing unit. For video frames, the pedestrian detection module locates all pedestrians and outputs the cropped pedestrian photos. At the same time, the pedestrian detection module calculates the degree of overlap of pedestrian frames in the same frame of video. When IoU exceeds the threshold set in the task allocation strategy, we believe that the pedestrian has overlapped, and use it as the input of the overlap processing sub-task to the expert for artificial judgment and processing. In parallel execution, when IoU does not exceed the threshold set in the task allocation strategy, the detection frame is used as the input of the re-identification subtask to the pedestrian re-identification module for judgment.

There are two interactive modes for human-machine task allocation, explicit and implicit. Explicit interaction mode means that people participate in the task execution process and assigned to certain tasks. For example, people analyze the reliability of the results in the result evaluation stage, re-modifying the model to output acceptable results through feedback. The implicit interaction mode means that people do not participate in the execution of actual tasks, for example, experts mark data in advance. In this framework, the human-machine task allocation strategy is explicit, that is, humans participate in the execution of tasks.

In the framework, result fusion is the key part. Result fusion means that the intermediate results of the sub-tasks executed by humans and machines are combined in a certain way and the final results are output. In this framework, the result fusion strategy is formulated as follows:

- If the model receives feedback from people that they are the same pedestrians, then the model marks this pedestrian photo as the highest confidence level and set this picture first in the rank list.
- For pedestrians in different frames, if the model gets multiple feedbacks that are all true, the newly judged picture is placed in the first position of the rank list, and the existed picture will be moved back one position in turn.

3.3 Pedestrian Detection Module Based on YOLOv3

The network structure is shown in Fig. 4.

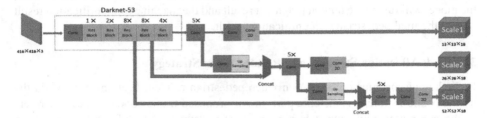

Fig. 4. Pedestrian detection module network structure

The input of the pedestrian detection module is a three-channel video frame with a size of 416 × 416. Darknet-53 is the backbone network for feature extraction. After a series of convolution operations, up sampling and concatenating, it is extracted to 13 × 13 × 18, 26 × 26 × 18, 52 × 52 × 18 three different scale feature maps. Each scale sets a priori boxes of three sizes, that is, a total of nine different anchor boxes are generated, and then multiple independent logistic classifiers are used to predict the class probability. Conv is the basic component of the detection module, including a convolution layer, normalization layer, and activation function layer. UP Sampling is an up-sampling operation. In the model, the features are first down-sampled, and then up-sampling is performed to obtain deeper features. Concatenating is spliced in the channel dimension to integrate the features of the feature map, and merge the features extracted from Darknet middle layer with the features extracted later. Conv2D is a layer of convolution operation, the convolution kernel size is 1 × 1 × 18, at last, feature vectors of different scales are obtained. The loss of pedestrian detection model is shown as follows:

$$Loss = \lambda coord * LossXY + \lambda coord * LossWH - LossConf - LossClass \qquad (2)$$

Among them, LossXY represents the loss of the center coordinate of the prediction boxes; LossWH represents the size loss of the prediction boxes; LossConf represents the IoU loss, which can also be called the confidence loss, and LossClass represents the category loss of the prediction boxes.

LossXY, LossWH, LossConf and LossClass are shown as follows expressions:

$$LossXY = \sum_{i=0}^{S \times S} \sum_{j=0}^{M} l_{ij}^{obj} \beta_i \left[(px_i - gx_i)^2 + (py_i - gx_i)^2 \right] \qquad (3)$$

$$LossWH = \sum_{i=0}^{S \times S} \sum_{j=0}^{M} l_{ij}^{obj} \beta_i \left[(pw_i - gw_i)^2 + (ph_i - gh_i)^2 \right] \qquad (4)$$

$$LossConf = \sum_{i=0}^{S \times S} \sum_{j=0}^{M} l_{ij}^{obj} \left[\hat{C}_i \log(C_i) + \left(1 - \hat{C}_i\right) \log(1 - C_i) \right] +$$

$$\lambda_{noobj} \sum_{i=0}^{S \times S} \sum_{j=0}^{M} l_{ij}^{noobj} \left[\hat{C}_i \log(C_i) + \left(1 - \hat{C}_i\right) \log(1 - C_i) \right] \quad (5)$$

$$LossClass = \sum_{i=0}^{S \times S} l_i^{obj} \sum_{c \in classes} [\hat{G}_i(\alpha) \log(G_i(\alpha)) + \left(1 - \hat{G}_i(\alpha)\right) \log(1 - G_i(\alpha))] \quad (6)$$

S represents the number of grids to be detected; M represents the number of boxes predicted by each grid, and β is a scale factor. l_{ij}^{obj} is used to judge whether the j-th prediction box in the i-th grid is responsible for this pedestrian, if responsible, set it to 1, otherwise it is 0. $G(\alpha)$ is the probability that the prediction box is a pedestrian. If α is a pedestrian, the value is 1, and if it is not a pedestrian, it is 0.

3.4   Pedestrian Re-identification Module Based on ResNet50

The module is shown in Fig. 5.

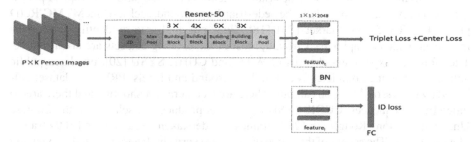

Fig. 5. Pedestrian re-identification module network structure

The input of the module is P × K cropped photos of pedestrians. P is the entered pedestrian identities, and K is the photos entered for every identity. The feature extraction part of the module uses a residual neural network, which will output a pedestrian feature vector of 2048 dimensions, denoted as $feature_t$. Previously, most training of pedestrian re-identification models combined pedestrian ID loss and triple loss. However, they will constrain the same features, but their optimization goals are not the same in space [16]. In the embedded space, the former is mainly to optimize the cosine distance, but the latter is computed by Euclidean distance.

In order to solve this problems, a batch normalization is added between the obtained $feature_t$ and the fully connected layer of the classifier. The output features from the ResNet are represented by f_t, then f_t is normalized to f_i. During backpropagation, f_t and f_i calculate these two loss functions respectively. Finally, the center loss is used to reduce the feature difference within a class (for example, the same pedestrian).

The loss of the pedestrian re-identification module is shown as follows:

$$Loss = L_{ID} + L_{Triplet} + \beta L_{Center} \quad (7)$$

L_{ID}, $L_{Triplet}$ and L_{Center} are shown as follows:

$$L_{ID} = - \sum_{i=0}^{B} [G_i(a) \log(G_i(a)) + \left(1 - \hat{G}_i(a)\right) \log(1 - \hat{G}_i(a))] \quad (8)$$

$$L_{Triplet} = [d_p - d_n + \alpha]_+ \tag{9}$$

$$L_{Center} = \frac{1}{2}\sum_{j=1}^{B} \left\| f_{t_j} - c_{y_j} \right\|_2^2 \tag{10}$$

L_{ID} represents the pedestrian ID loss, which is Cross-Entropy loss function. B represents the number of pedestrian ID. $\hat{G}_i(a)$ is the probability of predicting pedestrian ID, The purpose of the L_{Center} calculation is the quadratic sum of the distance between the pedestrian feature and the feature center.

4 Experiment

4.1 Datasets

At present, most researches on pedestrian re-identification are based on cropped pedestrian photos, which do not include the pedestrian background, such as DukeMTMC-ReID [17], CUHK03 [5] and Market-1501 [18]. Since our goal is to re-identify pedestrians in a scene with a complex background, most of the existing datasets are not available. Therefore, the experiment chooses PRW [1] and CUHK-SYSU [20], two large public datasets containing complete pedestrian background and labels. PRW is a dataset collected by Professor Zheng on campus. There are 6 cameras for shooting and there are 16 video frame sequences; CUHK-SYSU is a dataset produced by scholars of the Chinese University of Hong Kong, with 12490 frames of video taken on campus and 5694 frames of a movie clip. The scenes of these two datasets are very challenging, including viewing angle changes, relatively low pixels, and severe occlusion.

The experimental dataset is shown as Table 1.

Table 1. PRW and CUHK-SYSU datasets

Dataset	All frames	All boxes	All identities
PRW	11,816	43,100	932
CUHK-SYSU	18,184	96,143	8,432

The PRW dataset has 11,816 frames of video, including 43,100 pedestrian boxes and 932 pedestrians with different identities. The training set includes 5,704 frames, 18,048 pedestrian boxes, and 482 pedestrians with different identities. The CUHK-SYSU dataset has 18,184 frames of video, including 96,143 pedestrian boxes and 8,432 pedestrians with different identities. The training set includes 11,206 frames of video, 55,272 pedestrian boxes, and 5,532 pedestrians with different identities.

4.2 Exploratory Experiments

Obviously, the greater the human workload feedback, the higher the accuracy of the model. So, what threshold will be set to maximize the role of humans? In response to this problem, we conducted some exploration experiments to explore the impact on workload and accuracy by setting different IoU thresholds. Figure 6 shows the curve of the number of overlapping pedestrian boxes with different IOU threshold settings. It can be seen that the changing trend of the number of overlapping pedestrian frames in PRW dataset and CUHK-SYSU dataset is similar.

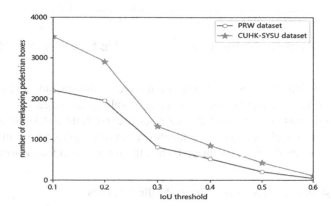

Fig. 6. Curve of overlapping pedestrian boxes with IoU threshold

Both of them showed a decreasing trend with the increase of the threshold, and the decreasing speed was slightly slow, then fast and then slow. When the IoU threshold is less than 0.2, the number of overlapping pedestrian boxes is at a relatively high level, and the number of overlapping pedestrian boxes decreases sharply from 0.2 to 0.3. After 0.3, the number of overlapping pedestrian boxes tends to be steady, and it is nearly 0 above 0.6. It can be seen that most of the overlapped pedestrian boxes have IoUs above 0.3.

Figure 7 and Fig. 8 are the curves of the mAP of the model on PRW and CUHK-SYSU datasets with different IoU thresholds respectively.

With the increase of the IoU threshold, mAP shows a downward trend, and the decline rate shows a trend of a slowly decreasing at beginning, then become fastly and gradually slow in the end. The analysis shows that when the IoU threshold is set small, there are more cropped photos of pedestrians that overlap. Through human processing, the model's mAP can reach up to 48%. As the IoU threshold increases, the average detection accuracy will decrease to a certain extent. If the threshold is 0.2 and 0.3, the average detection accuracy is still at a high level. If the threshold is 0.4, the mAP has been significantly reduced to a large extent, because when the IoU threshold exceeds 0.4, the output overlapped pedestrian cropped photos are in a low number. If the threshold is 0.3, there is already a large amount of overlap and occlusion in the cropped photos of people, and a large number of overlapped pedestrian cropped photos are not output for

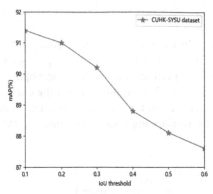

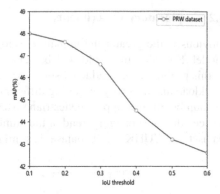

Fig. 7. mAP curve on CUHK-SYSU   **Fig. 8.** mAP curve on PRW

human judgment. However, the re-identification effect of the machine on the pedestrian cropped photos with a larger overlap is not good, resulting in a large drop in mAP.

From experiments, we can conclude that when the threshold is 0.3, human judgment can make up for the lack of machine re-identification to the greatest extent, so that the higher mAP of the model can be achieved when the human workload is low.

4.3   The Results of the Experiments on Public Datasets

Table 2 and Table 3 show the results on PRW and CUHK-SYSU dataset. Our human-machine collaboration model's Rank-1 accuracy reached 74.0% and 94.6% on PRW and CUHK-SYSU dataset respectively. Moreover, our human-machine collaboration model's mAP accuracy reached 42.2% and 92.8% on PRW and CUHK-SYSU dataset respectively.

Table 2. Rank-1 and mAP on PRW

Model	Rank-1(%)	mAP(%)
ACF-Alex + IDEdet + CWS [19]	45.2	17.8
CLSA [20]	65.0	38.7
MGTS [21]	72.1	32.6
Localization Refinement [9]	70.2	**42.9**
HMRM	**74.0**	42.2

Our human-machine collaboration model has a better performance on the PRW dataset, because the PRW dataset has more crowded scenes than the CUHK-SYSU dataset. Traditional methods or deep learning methods can not handle the occlusion situation well, but people can easily judge by other features of pedestrians, so the performance of our approach is testified effectively.

Table 3. Rank-1 and mAP on CUHK-SYSU

Model	Rank-1(%)	mAP(%)
ACF + LOMO + XQDA	63.1	55.5
CLSA [20]	88.5	87.2
MGTS [21]	83.7	83.0
Localization Refinement [9]	94.2	**93.0**
HMRM	**94.6**	92.8

5 Human-Machine Collaboration Interactive Interface

5.1 Human-Machine Collaboration Interactive Interface

Based on HMRM, a human-machine collaboration interactive interface is developed. The interactive interface is shown in Fig. 9. It provides a demonstration for the input and output of tasks, and is the interface of the human-machine collaboration system to the user. Users can upload videos through this interface, handle overlaps and view the results. The interface is developed based on pyqt5 technology, including the function of selecting pedestrians to be queried, the function of re-identifying pedestrians, the function of playing the output video, and the function of human-machine cooperation processing overlap. We use multi-threading technology to prevent problems such as interface stuck.

There are many ways of human-machine collaboration, including serial collaboration, parallel collaboration, and serial-parallel alternate collaboration. This interface applies multi-threading technology to achieve parallel collaboration between human and machine. It uses two threads to execute machine tasks and human tasks to improve processing speed. On the one hand, the machine processes video segmentation tasks, detection tasks, and re-identification tasks in thread one. On the other hand, the user handles the overlap and submits feedback in thread two.

5.2 Virtual File Queue

We propose a concept named virtual file queue. the virtual file queue still follows the concept of queue mentioned in computer science, that is, first in first out, or it can be called first in first processing. For the overlapping pedestrian pictures output by the model, we temporarily store them in the virtual file queue. During the process of human processing, the overlapping pictures are transferred to the display interface for human judgment according to the principle of first-in-first-processing. The virtual file queue is essentially the storage of files, but when storing, we get the timestamp of the current system and use the timestamp as the mark of the file, and the timestamp keeps increasing over time. For each judge operation, the system will refresh the file mark list in the folder, sort by timestamp mark, and take the first one. In this way, a simple virtual file queue is realized through simple coding. It can ensure that the pictures judged by people are strictly following the input order so that when the final result is fused, all the video frames only need to be traversed once.

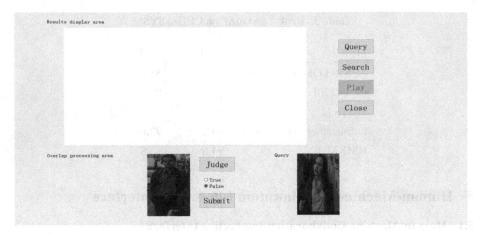

Fig. 9. Human-machine collaboration interactive interface

6 Conclusion and Future Work

In this paper, we propose a new human-machine collaboration video pedestrian re-identification model(HMRM) and formulate a task allocation strategy as well as a result fusion strategy accordingly. Furthermore, we conduct exploratory experiments on human workload, overlapping thresholds, and model accuracy. It is concluded that higher accuracy can be achieved with less human incorporation. Finally, we develop a human-machine collaboration video pedestrian re-identification interface and propose a concept named virtual file queue. Our future work will consider exploring a more highly collaborative pattern to improve efficiency. Humans can be used as modifiers of model results, or as instructors of the model itself, and ultimately achieve the task of re-identification of pedestrians quickly and well.

References

1. Xiao, T., Li, S., Wang, B.: Joint detection and identification feature learning for person search. In: IEEE Conference on Computer Vision and Pattern Recognition, pp. 3415–3424. IEEE (2017)
2. Hosang, J., Omran, M., Benenson, R., Schiele, B.: Taking a deeper look at pedestrians. In: IEEE Conference on Computer Vision and Pattern Recognition, pp. 4073–4082. IEEE (2015)
3. Tian, Y., Luo, P., Wang, X., Tang, X.: Pedestrian detection aided by deep learning semantic tasks. In: IEEE Conference on Computer Vision and Pattern Recognition, pp. 5079–5087. IEEE (2015)
4. Cai, Z., Saberian, M., Vasconcelos, N.: Learning complexity-aware cascades for deep pedestrian detection. In: IEEE International Conference on Computer Vision, pp. 3361–3369. IEEE (2015)
5. Li, W., Zhao, R., Xiao, T., Wang, X.: DeepReid: deep filter pairing neural network for person re-identification. In: IEEE Conference on Computer Vision and Pattern Recognition, pp. 152–159. IEEE (2014)

6. Ahmed, E., Jones, M., Marks, T.K.: An improved deep learning architecture for person re-identification. In: IEEE International Conference on Computer Vision, pp. 3908–3916. IEEE (2015)
7. Ding, S., Lin, L., Wang, G., Chao, H.: Deep feature learning with relative distance comparison for person reidentification. In: Pattern Recognition, pp. 2993–3003. Elsevier Ltd (2015)
8. Cheng, D., Gong, Y., Zhou, S., Wang, J., Zheng, N.: person re-identification by multi-channel parts-based CNN with improved triplet loss function. In: IEEE Conference on Computer Vision and Pattern Recognition, pp. 1335–1344. IEEE (2016)
9. Han, C., Ye, J., Zhong, Y.: Re-ID driven localization refinement for person search. In: IEEE/CVF International Conference on Computer Vision, pp. 9814–9823. IEEE (2019)
10. Das, A., Panda, R., Roy-Chowdhury, A.: Active image pair selection for continuous person re-identification. In: 2015 IEEE International Conference on Image Processing. IEEE (2015)
11. Liu, C., Loy, C.C., Gong, S., Wang, G.: Pop: person re-identification post-rank optimisation. In: IEEE International Conference on Computer Vision, pp. 441–448. IEEE (2013)
12. Wang, H., Gong, S., Zhu, X., Xiang, T.: Human-in-the-loop person re-identification. In: Leibe, B., Matas, J., Sebe, N., Welling, M. (eds.) ECCV 2016. LNCS, vol. 9908, pp. 405–422. Springer, Cham (2016). https://doi.org/10.1007/978-3-319-46493-0_25
13. Zheng, L., Yang, Y., Hauptmann, A.G.: Person re-identification: Past, Present and Future. arXiv preprint arXiv:1610.02984 (2016)
14. Chen, W., Chen, X., Zhang, J., Huang, K.: Beyond triplet loss: a deep quadruplet network for person re-identification. In: IEEE Conference on Computer Vision and Pattern Recognition, pp. 403–412. IEEE (2017)
15. Wang, C., Zhang, Q., Huang, C., Liu, W., Wang, X.: Mancs: a multi-task attentional network with curriculum sampling for person re-identification. In: Ferrari, V., Hebert, M., Sminchisescu, C., Weiss, Y. (eds.) ECCV 2018. LNCS, vol. 11208, pp. 384–400. Springer, Cham (2018). https://doi.org/10.1007/978-3-030-01225-0_23
16. Luo,H., Gu,Y., Liao, X., Lai, S., Wei, J.: Bag of tricks and a strong baseline for deep person re-identification. arXiv preprint arXiv:1903.07071 (2019)
17. Zheng, Z., Zheng, L., Yang, Y.: Unlabeled samples generated by GAN improve the person re-identification baseline in vitro. In: IEEE International Conference on Computer Vision, pp. 3754–3762. IEEE (2017)
18. Zheng, L., Shen, L., Tian, L., Wang, S., Wang, J., Tian, Q.: Scalable person re-identification: a benchmark. In: IEEE International Conference on Computer Vision, pp. 1116–1124. IEEE (2015)
19. Zheng, L., Zhang, H., Sun, S., Chandraker, M., Yang, Y., Tian, Q.: Person re-identification in the wild. In: IEEE Conference on Computer Vision and Pattern Recognition, pp. 1367–1376. IEEE (2015)
20. Lan, X., Zhu, X., Gong, S.: Person search by multi-scale matching. In: Ferrari, V., Hebert, M., Sminchisescu, C., Weiss, Y. (eds.) ECCV 2018. LNCS, vol. 11205, pp. 553–569. Springer, Cham (2018). https://doi.org/10.1007/978-3-030-01246-5_33
21. Chen, D., Zhang, S., Ouyang, W., Yang, J., Tai, Y.: Person search via a mask-guided two-stream CNN model. In: Ferrari, V., Hebert, M., Sminchisescu, C., Weiss, Y. (eds.) ECCV 2018. LNCS, vol. 11211, pp. 764–781. Springer, Cham (2018). https://doi.org/10.1007/978-3-030-01234-2_45

Cooperative Scheduling Strategy of Container Resources Based on Improved Adaptive Differential Evolution Algorithm

Chengyao Hua, Ningjiang Chen[✉], Yongsheng Xie, and Linming Lian

School of Computer, Electronics and Information, Guangxi University, Nanning 530004, China
chnj@gxu.edu.cn

Abstract. The resource scheduling of the container cloud system can be handled as a path planning problem. In response to the need for container resource scheduling that comprehensively considers the interests of users and service providers, this paper combines user quality of service (QoS) models and resource load balancing to study multi-objective container resource scheduling solutions, and proposes an improved Dynamic Adaptive Differential Evolution Algorithm (DADE), which adds adaptive changes to the mutation factor and crossover factor, and optimizes the mutation strategy and selection strategy, so that the algorithm has a broad solution space in the early stage; and a small-scale local search is carried out in the later stage, the resource scheduling strategy based on this algorithm is realized. Perform simulation experiments on the proposed algorithm and scheduling strategy. Experimental results show that the DADE algorithm is superior to mainstream heuristic algorithms in the evaluation of average function evaluation times, solution accuracy, convergence speed and other indicators. The resource scheduling effect has obvious advantages in task completion time, completion cost and resource load balancing.

Keywords: Container · Resource scheduling · Differential evolution algorithm

1 Introduction

Container technology, a new generation of cloud computing virtualization technology has received wide attention since it was proposed, compared with traditional virtual machine technology, containers have the advantages of high efficiency, flexibility, light weight, easy deployment and fast response [1]. Container technologies, represented by Docker [2], are widely used by cloud computing service providers, and container clouds are a replacement for traditional virtual machines. The construction of container clouds requires consideration of multiple physical resources, user needs, system costs, robustness, and load balancing, which makes resource allocation and management a challenge. For dynamic and complex cloud computing runtime environment, how to manage and schedule container cloud resources efficiently is a problem worth studying.

The resource scheduling problem of container cloud system belongs to a typical path planning problem, which can be regarded as an NP-hard problem, while the traditional

© Springer Nature Singapore Pte Ltd. 2021
Y. Sun et al. (Eds.): ChineseCSCW 2020, CCIS 1330, pp. 424–438, 2021.
https://doi.org/10.1007/978-981-16-2540-4_31

scheduling algorithm, such as first come first service algorithm and MinMin algorithm, is difficult to solve such problems effectively. At present, one of the main directions of relevant research is the introduction of intelligent planning optimization algorithms, such as genetic algorithms, particle group optimization algorithms, ant colony algorithms, etc. There have been researches on planning and scheduling container cloud computing resources. In the literature [3], a resource scheduling method for container virtualized cloud environment is proposed to reduce the response time of customer jobs and improve the resource utilization of the provider, but the cost of user concern is not involved. The literature [4] improves the ant colony algorithm, constructs the multi-target scheduling model for the utilization of resources, and the experimental results show that it can maintain the reliability of the system service and reduce the network transmission overhead, but the user-oriented QoS demand is still insufficient. The literature [5] describes a dynamic load management method for assigning tasks to virtual machines, using parameter processing time and response time to evaluate performance on the cloud analyzer, which can achieve load balancing at the user level or virtual machine level, without taking into account the time and cost issues that the user is most concerned about. The literature [6] integrates the global search algorithm and firefly algorithm to guarantee the user's QoS demand, but does not consider the load of resources. In the literature [7], the ant colony optimization algorithm and genetic algorithm are used to obtain better scheduling results with short total task completion time and average task completion time, but there is no systematic load balancing.

Although the above work from different angles to achieve the container cloud resource scheduling allocation, but because the user is concerned about whether the task can be completed, pay more attention to the cost and time to complete the scheduling, and for service providers, it pays more attention to the cost of services, system stability and system robustness. In this regard, this paper takes into account the interests of users and service providers, and constructs a multi-target container cloud computing resource scheduling model based on task completion time, completion cost and load balancing of the system. Differential Evolution is a vector meta-heuristic random search algorithm based on group differences [8], which performs well in solving NP-hard problems. For the traditional intelligent planning algorithm, there are problems such as poor robustness, small spatial search range and easy to get caught up in local optimal solution to different degrees. This paper improves the traditional differential evolution algorithm, designs a dynamic adaptive differential evolution algorithm (DADE), increases adaptive changes to the variation and cross factors of differential evolution algorithm, and optimizes the variation strategy and selection strategy, so that the algorithm in the early stage has a broad solution space, and in the later stage of a small range of local search. The results of this paper show that the improved algorithm efficiency is obviously improved, which is obviously improved in handling the problem of resource allocation of container computing, and improves the load balancing of the system while reducing cost and saving task time.

2 Description of the Problem

Pay-as-you-go is an important feature of the container cloud service model. Users pay a certain fee to obtain the corresponding services. Users consider their spending power at the same time to the cloud service providers to put forward the corresponding needs, and the corresponding cloud service providers for their own benefit also want to minimize costs, which requires a model that can take into account the interests of both. This section combines the user service quality QoS, in order to reduce the task completion time and cost while ensuring the load balancing of resources, put forward a multi-target container cloud resource scheduling model.

2.1 Problem Modeling

Task scheduling algorithm is to divide a task into several relatively independent sub-tasks, and then assign each sub-task to the corresponding container nodes at the same time to calculate, and finally the results of each calculation will be summarized processing, to obtain the final results. This paper only analyzes the parallel calculations of relatively independent sub-tasks in the above cases. T is a task set that contains n elements, represented as: $T = \{t_1, t_2, \cdots, t_i, \cdots, t_n\}$, T represents the task that the user needs to handle.

When a task is assigned to a container node, the container node collection is $Ct = \{ct_1, ct_2, \cdots, ct_j, \cdots, ct_m\}$, which contains m container nodes, where task t_i corresponds to ct_i, represented by the relationship matrix TC, that is:

$$TC = \begin{bmatrix} tc_{11} & tc_{12} & \cdots & tc_{1m} \\ tc_{21} & tc_{22} & \cdots & tc_{2m} \\ \vdots & \vdots & \ddots & \vdots \\ tc_{n1} & tc_{n2} & \cdots & tc_{nm} \end{bmatrix}$$

TC is a 0,1 matrix, tc_{ij} means that the i-th task is running on the j-th container node, because a task can only run in one container node, there can only be one 1 per row for the relationship matrix TC.

(1) The time the task was completed

$$C\_ET(ct_j) = \frac{\sum\limits_{i=1}^{n} M(t_i)}{Abality(ct_j)} \tag{1}$$

$$C\_TT(ct_j) = \frac{\sum\limits_{i=1}^{n} Data(t_i)}{Band(ct_j)} \tag{2}$$

Among above, $C\_ET(ct_j)$, $C\_TT(ct_j)$, $Abality(ct_j)$ and $Band(ct_j)$ are the execution time, transmission time, processing power and bandwidth of container node ct_j. $M(t_i)$ and $Data(t_i)$ represent the task amount of t_i and the amount of data it transmits, respectively. The total time it takes for each container node to complete the task is calculated as follows:

$$totalTime(ct_j) = C\_ET(ct_j) + C\_TT(ct_j) \tag{3}$$

It can be known that for a container node ct_j, its total completion time T consists of execution time and transfer time. When U is recorded as a task schedule, its task completion time can be represented by the following:

$$Time(U) = \max(totalTime(ct_j)), \tag{4}$$

where $Time(U)$ represents the task completion time, $\max(totalTime(ct_j))$ represents the maximum time the task completes, and the smaller $Time(U)$ indicates that the task has the shortest total completion time, while more meeting the user's time requirements.

(2) Task completion cost

For each container node, the execution and transfer costs per unit of time are different, and we calculate the total task completion cost by suming the costs of the container nodes used, as shown in the following equation:

$$Cost(U) = \sum_{j=1}^{m} C\_ET(ct_j) * pet_j + C\_TT(ct_j) * ptt_j, \tag{5}$$

where $Cost(U)$ represents the cost of the task, you can see that the smaller the $Cost(U)$ indicates that the task is less expensive to complete, while more to meet the user's cost requirements, pet_j and ptt_j represent the container node ct_j's unit execution costs and unit transfer costs, respectively.

(3) Load balancing

The load on container resources can be indirectly reflected by the usage of container nodes. Standard deviation can reflect the volatility and stability of a set of data, we use the container node execution time standard deviation BL to evaluate the load situation of the system, the smaller the BL, indicating that the virtual node execution time fluctuations are smaller, the load is balanced, there is no underload or heavy load, resources are effectively utilized. It can evaluate the load of the system to some extent, and its standard deviation is calculated as follows:

$$BL(U) = \sqrt{\frac{\sum_{j=1}^{m} (C\_ET(ct_j) - C\_ET_{avg})^2}{m}}, \tag{6}$$

where $BL(U)$ represents load balancing of container cloud resources, C_ET_{avg} represents the average execution time of the current container node.

2.2 Problem Normalization and Model Construction

Based on the completion time of the task, the cost of completion and the load balancing of resources, this paper constructs a multi-objective resource scheduling model, and the linear weighting method is the method to solve the multi-objective problem, which can be converted into a single target for processing. The linear normalization processing is explained as follows.

First, time normalization is:

$$NormTime = \frac{Time(U) - TotalTime_{min}}{TotalTime_{max} - TotalTime_{min}}, \tag{7}$$

where $TotalTime_{min}$ and $TotalTime_{max}$ represent the execution time of task set T on the optimal and worst container nodes, respectively.

Second, cost normalization is:

$$NormCost = \frac{Cost(U) - TotalCost_{min}}{TotalCost_{max} - TotalCost_{min}}, \tag{8}$$

where $TotalCost_{min}$ and $TotalCost_{max}$ represent the minimum and maximum costs that Task Collection T performs on the container node, respectively.

Then, load balancing is normalized as:

$$NormBL = \frac{BL(U)}{BL_{max}}, \tag{9}$$

where $BL_{max} = \max(|C\_ET(ct_j) - C\_ET_{avg}|)$ is the maximum difference between the execution time and the average execution time of the task.

Finally, after normalization, the target function can be expressed as:

$$F = c_1 * NormTime + c_2 * NormCost + c_3 * NormBL, \tag{10}$$

where $c1 + c2 + c3 = 1, c1, c2, c3$ represent weight relationships for time, cost, and load balancing, respectively. $c1$, $c2$, $c3$ can be automatically adapted to the needs of users and service providers to meet the diverse needs of cloud users and the complexity of cloud platforms.

3 DADE: Improved Differential Evolution Algorithms

The process of Differential Evolution Algorithm is similar to genetic algorithm, which is a greedy genetic algorithm, and compared to other intelligent algorithms, DE algorithm keeps the global search mode based on the population, using the evolutionary strategy of variation, crossover, and selection to reduce the complexity of the evolutionary operation. However, the DE algorithm is susceptible to early maturity convergence and search stagnation in the later stages of evolution due to the reduced level of differentiated information between individuals in the population. The DADE algorithm proposed in this paper mainly optimizes the traditional DE algorithm in terms of the value of the variation factor F, the value of the cross-factor CR, the improvement of the variation strategy, and the adjustment of the selection operation.

3.1 Introduction of Adaptive Variation Factor

It can be seen from the variation operation of differential evolution algorithm that the variation factor F determines the variation of the assignment value of the difference vector obtained in the variation operation, and it is closely related to the convergence and convergence speed of DE algorithm. When F is large, the algorithm has a large solution range and a high diversity of populations, but due to relatively slow convergence, the global optimal solution may still not be found later in the algorithm iterations. When F is small, the result is opposite to the above description, it is easy to reciprocate around the current local optimal solution, and it is difficult to jump out. Accordingly, this paper introduces an adaptive factor F that changes according to the current number of iterations CI. The current number of iterations CI can well represent the evolution stage of the population, which can be used to determine the change of the current adaptive factor F. In the early stage of execution, the value of F is large, that is, the convergence speed is reduced, extensive global search can be conducted, and diversity can be increased; while in the later stage of execution, the value of F is small, that is, the convergence speed is increased, and a local search phase with a small range is started near the optimal individual found before, so as to obtain better results. The adaptive variation factor is calculated by the following equation.

$$F = F_{max} - (F_{max} - F_{min}) * (CI/NI) \tag{11}$$

Among them, NI indicates the total number of iterations, F_{min} and F_{max} represent the minimum and maximum values of F respectively A large number of numerical simulation experiments have been carried out in the literature [9] on relevant Standard test function sets, which can be used as a reference for this work. In general, the optimal value range of F can be set as [0.2, 0.9]. When the constructed population size exceeds more than 10 individuals, the optimal value range of F is [0.2, 0.6]. Therefore, the value of F_{max} is determined according to the dimension N of the relevant solution space.

3.2 Improvement of Variation Strategies

The variation strategy is the most important step in the DE algorithm, and it is the key indicator for the global and local search of the algorithm. The DADE algorithm selects a novel variation strategy based on the adaptive cross-factor F constructed above. Its expression is:

$$v_i^{g+1} = (1 - F) * x_{r_1}^g + F * (x_{r_2}^g - x_{r_3}^g) \, i \neq r_1 \neq r_2 \neq r_3 \tag{12}$$

Since the variation factor F is a linear decreasing function, $1-F$ is a linear increasing function. According to the search properties of the DE algorithm, a good differential evolutionary algorithm should do a global search of the extensive solution space in the early stages and a small local search strategy near the current optimal solution in the later stages. The variation strategy chosen in this paper satisfies the above content. The population is scattered in the solution space in the early stage for large-scale detection for global search [9], that is, the weight coefficient of $x_{r_1}^g$ is small, and the weight of $x_{r_2}^g - x_{r_3}^g$ is large, which helps the algorithm to detect and search the current solution space. And as the search process proceeds, the population gradually converges to perform local search near some local optimal solution, i.e., $x_{r_1}^g$ has a larger weighting factor and $x_{r_2}^g - x_{r_3}^g$ has a smaller weighting, which helps the algorithm to be in the vicinity of the optimal individual currently being sought, a local search phase with a small span begins.

3.3 Dynamic Change Cross-Factor

In traditional algorithms, the value of the cross factor CR is often taken as a fixed value, which will make the algorithm fall into a local optimum to a large extent and affect the performance of optimization. According to the study of the differential evolution algorithm in the literature [10], the differential evolution algorithm performs a global search at an early stage and a local search at a later stage, among them, the setting should be small in the global search and large in the local search. In this paper, a strategy for dynamically varying cross-factors is proposed to enhance the global and local search capabilities of the algorithm, as shown in the following equation.

$$CR = CR_{\min} + (CR_{\max} - CR_{\min}) * (CI/NI) \qquad (13)$$

Among them, $CR_{\min}$ and $CR_{\max}$ are the minimum and maximum values of the cross-factor respectively. Thus, the global search capability of the algorithm is relatively strong at the beginning of the iteration and the local search capability of the algorithm increases as the CI increases. Using the settings of the interval of change for the cross-factor CR in the literature [11], among them, $CR_{\min} = 0.5$, $CR_{\max} = 0.9$.

3.4 Improvements in Selection Strategies

The traditional selection strategy of the DE algorithm is to compare the adaptation values of trial individuals u_i^{g+1} and x_i^g generated by mutations and crossover manipulations, and select better adapted individuals as the next generation of new members. Although this crossover process contributes to the evolution of populations, it can destroy good genes and cause good individuals to be lost. In order to prevent this kind of damage, this article improves the selection strategy, comparing the fitness value of the variant individual v_i^{g+1}, the test individual u_i^{g+1} after the crossover operation and the target individual x_i^g. Selection of better-adapted individuals as new members of the next generation, as shown by the following equation:

$$x_i^{g+1} = fitness\_best(x_i^g, u_i^{g+1}, v_i^{g+1}). \qquad (14)$$

Among them, the operation means to select the optimal value of fitness in the current set. Based on the above improvement methods, the innovation of the DADE algorithm proposed in this paper is that the traditional DE algorithm has a small global search space during the iteration process and is easy to fall into a local optimal solution at the later stage of the iteration, and so on. Including the linear decreasing of the variation factor and the linear increasing of the cross-factor to ensure that the algorithm has a relatively large solution space at the early stage of the iteration process; at the same time, the optimal selection of selection strategies can enhance the diversity of the solution space and optimize the individual populations, which can effectively avoid the situation of falling into the local optimal solution. In the latter stage of the iteration, it can quickly converge and perform a local search to find an excellent solution result. The implementation steps of the DADE algorithm are shown below.

Algorithm : Dynamic adaptive differential evolution algorithm (DADE)

Input : N : Individual dimension, NP : Population size, *Fitness* : Objective function, NI : The maximum number of iterations, $[x_{i,j}^{min}, x_{i,j}^{max}](i=1,2,K,NP, j=1,2K,N)$: Solution space size

Output: BestIndividual: Optimal individual, BestFitness: Optimal individual objective function value

1	$CI \leftarrow 1, CR_{min} = 0.5, CR_{max} = 0.9, F_{min} = 0.2, F_{max} = 0.9$
2	*if* $N \geq 10$:
3	$F_{max} = 0.6$;
4	Initial generation population
5	while CI ≤ NI:
6	Update the variation factor F according to equation 11;
7	Select the optimized mutation strategy equation 12 for the mutation operation and save the mutated individuals v_i^{g+1};
8	Update cross-factor CR according to equation 13;
9	Cross manipulation to preserve experimental individual u_i^{g+1};
10	Use equation 14 to selects the best individual, and performs a selection operation to generate a new generation population;
11	$CI = CI + 1$;
12	Output the best-adapted individual in the last generation population BestIndividual and its objective function value BestFitness;

An analysis of the steps of the algorithm shows that the DADE algorithm does not add any cyclic conditions to the traditional DE algorithm, the time complexity of the DADE and DE algorithms are $O(N * NP * NI)$. The specific flowchart of the container resource scheduling policy based on the DADE algorithm is shown in Fig. 1.

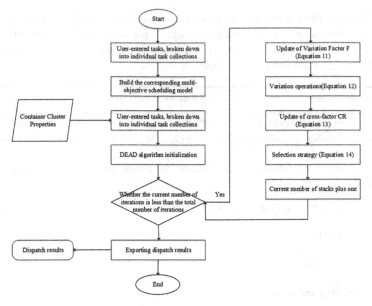

Fig. 1. Container resource scheduling policy process based on DADE algorithm

4 Experiments

This section experimentally verifies the work of this paper from the level of algorithm and resource scheduling strategy respectively.

4.1 Algorithm Performance Experiment

Experiments are mainly based on different algorithms in the standard test function set of test results and optimization curves to compare and analyze the algorithm's solvency. The closer the test results are to the optimal solution, the steeper the optimization curve and the more it can illustrate the performance of the algorithm. The basic parameters of the algorithm selected for the comparison experiment are shown in Table 1, and the population size is $NP = 5N$, which is replaced according to each test function. All algorithms run 100 times, with the maximum number of iterations set to 2000.

In order to verify the validity of the DADE algorithm, five representative functions were selected from the standard test function collection [12]. Three of them are peak functions (f_1, f_2, f_3) and two are multimodal functions (f_4, f_5). They are used to conduct comparative experiments on the above algorithms and compare them with the comparison algorithms selected above. The description of the test benchmark function and the global optimal solution are shown in Table 2.

Table 3 shows the test results of the five algorithms of DE, JDE, GA, PSO, and DADE. The execution times are 20 times, and the average and variance are calculated. Table 4 shows the optimal analytical solutions of these five algorithms. As can be seen from Table 3, in all test cases, DADE has the lowest average error and standard deviation, which indicates that the DADE algorithm has higher stability and availability. It can be

Table 1. The algorithm selected by the experiments and its basic parameters.

The selected algorithm	The relevant parameters of algorithm
Genetic algorithms (GA)	$prob_{mut} = 0.001 \quad prob_{cor} = 0.5$
Particle swarm optimization algorithm (PSO)	$c_1 = c_2 = 0.5 \quad w = 0.8$
Differential evolutionary algorithm (DE)	$F = 0.5 \quad CR = 0.3$
Adaptive differential evolutionary algorithm (JDE)	$F = 0.5 \quad CR = 0.3 \quad F_{min} = 0.1, \ F_{max} = 0.9$

Table 2. The standard test function selected.

Function name	Expression	Number of dimensions	Ranges	Optimal solution
f_1	$\sum_{i=1}^{N} x_i^2$	$N = 30$	$x_i \in [-100, 100]$	0
f_2	$\sum_{i=1}^{N} \lvert x_i \rvert + \prod_{i}^{N} \lvert x_i \rvert$	$N = 30$	$x_i \in [-10, 10]$	0
f_3	$\sum_{i=1}^{N} [\frac{1}{N-1} \sqrt{s_i} * (\sin(50.0 s_i^{\frac{1}{5}}) + 1)]$ $s_i = \sqrt{x_i^2 + x_{i+1}^2}$	$N = 30$	$x_i \in [-100, 100]$	0
f_4	$\sum_{i=1}^{N} [x_i^2 - 10 \cos(2\pi x_i) + 10]$	$N = 30$	$x_i \in [-5.12, 5.12]$	0
f_5	$-20 \exp\left(-0.2 * \sqrt{\frac{1}{N} \sum_{i=1}^{N} x_i^2}\right)$ $- \exp\left(\frac{1}{N} \sum_{i=1}^{N} \cos(2\pi x_i)\right) + 20 + e$	$N = 30$	$x_i \in [-32, 32]$	0

seen from Table 4 that the optimal solution given by the DADE algorithm is better than the other four algorithms, closer to the actual optimal solution. These results benefit from the algorithm's improved strategy, variation factor and strategy, and the adjustment of selection strategy, which makes the algorithm have a wider solution space in the early stage. Therefore, to a certain extent to avoid getting caught up in the problem of local optimal solution. The change of cross factor makes the algorithm be able to search locally in a small range of solving space later in execution, and find the optimal solution as far as possible.

The results of the execution of the five algorithms are shown in Fig. 2, Fig. 3, Fig. 4, Fig. 5 and Fig. 6. From these diagrams, it can be seen that the DADE algorithm can

Table 3. Comparison of test results of DE, JDE, GA, PSO and DADE

Test function	Algorithm type	Average error	Standard error
f_1	DE	2.48e−28	8.55e−29
	JDE	4.49e−24	2.20e-24
	GA	1.04e−12	5.22e−13
	PSO	1.43e−1	1.82e−1
	DADE	0.0	0.0
f_2	DE	4.22e−17	6.33e−18
	JDE	2.04e−14	4.96e−15
	GA	1.11e−06	0.0
	PSO	28.21	11.37
	DADE	2.15e−183	0.0
f_3	DE	1.01e−1	1.67e−2
	JDE	2.80e−1	7.23e−3
	GA	3.40e−4	8.41e−4
	PSO	1.91e−2	3.32e−2
	DADE	1.14e−06	1.97e−06
f_4	DE	27.14	2.56
	JDE	6.33	2.74
	GA	31.74	5.62
	PSO	111.97	33.95
	DADE	0.0	0.0
f_5	DE	8.61e−15	1.62e−15
	JDE	5.34e−13	1.46e−13
	GA	2.08e−07	1.05e−07
	PSO	16.58	6.74
	DADE	4.44e−16	0.0

achieve better results for the selected test functions, and also shows the validity of the algorithm. Compared with DE algorithm and JDE algorithm, DADE algorithm has good results in the early and late stages, which is due to the DADE algorithm has a wide range of solving space in the early stage of running, and has a good ability to jump out of local optimal solution in the later stages of algorithm execution. Compared with GA algorithm and PSO algorithm, it can be found that DADE algorithm may be second to GA algorithm in the early stage of execution, but the results of DADE algorithm in the later stage of execution are obviously better than them. DADE algorithm has a good ability to jump out of local optimal solution, especially compared with the traditional adaptive

Table 4. Comparison of optimal solutions of DE, JDE, GA, PSO and DADE

Test function	f_1	f_2	f_3	f_4	f_5
Theoretical solution	0.0	0.0	0.0	0.0	0.0
DE	1.27e−28	3.24e−17	7.30e−2	21.86	7.54e−15
JDE	1.22e−24	1.46e−14	1.46e−2	0.65	3.52e−13
GA	3.07e−13	1.11e−06	2.41e−07	19.89	1.19e−07
PSO	2.37e−4	1.79	1.00e−4	57.70	2.58
DADE	0.0	7.47e−184	1.13e−11	0.0	4.44e−16

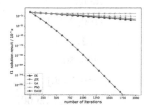 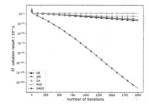 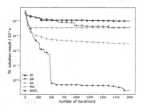

Fig. 2. Convergence curve of f_1 **Fig. 3.** Convergence curve of f2   **Fig. 4.** Convergence curve of f3

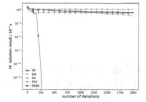 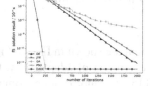 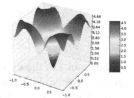

Fig. 5. Convergence curve of f4   **Fig. 6.** Convergence curve of f5   **Fig. 7.** Image of f5

differential evolution algorithm (JDE), DADE algorithm shows superior performance. This is related to the characteristics built by the DADE algorithm. It can get better solution results, which shows that the algorithm is improved in the right direction. The linear decrease and linear increase operations of the mutation factor and the cross factor can effectively ensure the search range of the solution space in the early stage of the algorithm. At the same time, the optimization of the selection strategy increases the diversity of the population solution. The variation strategy which changes according to the variation factor can make the algorithm more in line with the ability of large-scale global search and later small-scale local search, and avoid the algorithm getting caught in the local optimal solution. Figure 7 shows the *Fine grain* image of the test function f_5 *(Ackley's Function)*.

4.2 Experiment of Resource Scheduling

This article uses CloudSim [13] 5.0 as a simulation framework. This comparison experiment selects two traditional scheduling strategies MinMin, FCFS and two intelligently optimized scheduling strategies GA, DE and the improved DADE scheduling strategy. Simulating the real cloud computing platform, randomly generating different virtual nodes and task properties, selecting the unit pricing method of one of the three pricing methods of container calculation [14], the specific properties of which are shown in Table 5, where the number of container nodes is 30, the number of tasks range is [50,400], increments of 50.

Table 5. Properties of task and container node.

Name	Property	Range
Ct	Computing power (Mips)	[100, 2000]
	Number of CPUs	[1, 2]
	Bandwidth (MB/s)	[300, 500]
	Running memory (MB)	[256, 2560]
	Storage memory (MB)	[102400, 512000]
	Computing power unit pricing (MIPS/yuan/s)	0.03
	Bandwidth unit pricing (MB/yuan/s)	0.01
Cloudlet	The length of the task (Millions of instruction lengths)	[500, 10000]
	The size of the task transfer data (MB)	[0.5, 20]

The results of the experiment are shown in Fig. 8, Fig. 9, Fig. 10 and Fig. 11. After many full experiments on the solution of adaptability, the DADE algorithm obtained 71.32%, 53.73%, 42.37% and 15.51% improvement compared with the above-mentioned algorithm for comparison. Figure 8, Fig. 9, Fig. 10 and Fig. 11 respectively show the result comparison of the total target function, the task completion time, the task completion cost, the load balancing. It can be seen that the solution result of DADE resource scheduling strategy is optimal, and the target result solved by DADE is the smallest, which shows that changing the scheduling strategy can take good account of the multi-target container cloud resource scheduling model, which is closely related to the improvement of DADE algorithm.

Based on the above experimental results, it is shown that the improvement direction of DADE algorithm is successful. In view of the shortcomings of the DE algorithm, the following improvements have been obtained: In the early stage of a wide range of global search, increase the solution space, while increasing the diversity of populations; In the later stage of the algorithm to quickly converge, since the population solution space to be explored is already large, it is difficult to fall into the local optimal solution, so a small range of local search is required. For the DADE algorithm, its unique dynamic adaptive variation and cross factors, as well as the improved genetic strategy and selection strategy, can make the algorithm have a larger solution space and diverse population individuals, which can effectively avoid the late into the problem of local optimal solution. At the same time, the test results in CloudSim are also superior to other classical algorithms, which shows that the DADE algorithm can be used for container cloud resource scheduling, and can obtain a better allocation strategy.

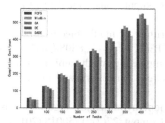

Fig. 8. The results of the target function

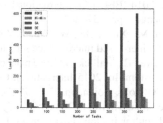

Fig. 9. Task completion time

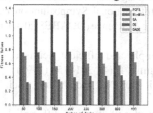

Fig. 10. Task completion cost chart

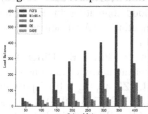

Fig. 11. Load balancing

5 Conclusion

In this paper, in order to protect the common interests of users and services providers, we constructed a multi-objective container cloud resource scheduling model according to user's task completion time and cost in QoS combined with the load balance of container cloud resources, and based on the traditional differential evolution algorithm, adaptive selection of mutation and cross factors and optimization of mutation and selection strategies are added, named as DADE. The simulation results show that the container cloud resource scheduling model based on DADE algorithm can significantly reduce the task completion time and cost, and has a good effect in load balancing test. In the following work, we will further consider the more complex resource scheduling situation, such as the correlation between tasks, the usage of computing storage resources, etc., and further optimize the algorithm.

Acknowledgments. The work is supported by the Natural Science Foundation of China (61762008), the National Key Research and Development Project of China (2018YFB1404404), the Guangxi Natural Science Foundation Project (2017GXNSFAA198141), and the Major special project of science and technology of Guangxi (No. AA18118047-7).

References

1. Lu, W., Li, B., Wu, B.: Overhead aware task scheduling for cloud container services. In: 2019 IEEE 23rd International Conference on Computer Supported Cooperative Work in Design (CSCWD), Porto, Portugal, pp. 380–385 (2019)

2. Merkel, D.: Docker: lightweight Linux containers for consistent development and deployment. Linux J **2014**(239), 2 (2014)
3. Xu, X., Yu, H., Pei, X.: A novel resource scheduling approach in container based clouds. In: 2014 IEEE 17th International Conference on Computational Science and Engineering, Chengdu, pp. 257–264 (2014)
4. Lin, M., Xi, J., Bai, W., Wu, J.: Ant colony algorithm for multi-objective optimization of container-based microservice scheduling in cloud. IEEE Access **7**, 83088–83100 (2019)
5. Panwar, R., Mallick, B.: Load balancing in cloud computing using dynamic load management algorithm. In: 2015 International Conference on Green Computing and Internet of Things (ICGCIoT), Noida, pp. 773–778 (2015)
6. Neelima, P., Rama Mohan Reddy, A.: An efficient hybridization algorithm based task scheduling in cloud environment. J. Circuits Syst. Comput. **27**(2), 1850018 (2017)
7. Liu, C.Y.: A task scheduling algorithm based on genetic algorithm and ant colony optimization in cloud computing. In: Proceedings of the 13th International Symposium on Distributed Computing and Applications to Business, Engineering & Science (DCABES 2014), pp. 81–85 (2014)
8. Storn, R., Price, K.: Differential evolution–a simple and efficient heuristic for global optimization over continuous spaces. J. Global Optimiz. **11**(4), 341–359 (1997)
9. Jia, L., Zhang, C.: Self-adaptive differential evolution. J. Cent. South Univ. (Sci. Technol.) **44**(9), 3759–3765 (2013)
10. Li, Z.W., Wang, L.J.: Population distribution-based self-adaptive differential evolution algorithm. Comput. Sci. **47**(2), 180–185 (2020)
11. Gao, C., Ma, J., Shen, Y., Li, T., Li, F., Gao, Y.: Cloud computing task scheduling based on improved differential evolution algorithm. In: International Conference on Networking and Network Applications, pp. 458–463 (2019)
12. Brest, J., Greiner, S., Boskovic, B.: Self-adapting control parameters in differential evolution: a comparative study on numerical benchmark problems. IEEE Trans. Evol. Comput. **10**(6), 646–657 (2006)
13. CloudSim: a toolkit for modeling and simulation of cloud computing environments and evaluation of resource provisioning algorithms. Softw.: Pract. Exp. **41**(1), 23–50 (2011)
14. Youseff, L., Butrico, M., Silva Da, D.: Toward a unified ontology of cloud computing. In: Grid Computing Environments Workshop 2008, pp. 1–10 (2008)

Social Media and Online Communities

Personalized Recommendation Based on Scholars' Similarity and Trust Degree

Lunjie Qiu, Chengzhe Yuan[✉], Jianguo Li, Shanchun Lian, and Yong Tang

School of Computer Science of South China Normal University, Guangzhou 510631, Guangdong, China

{qiulunjie,jianguoli,acsamson,ytang}@m.scnu.edu.cn

Abstract. With the rapid development of online social networks, academic social networks (ASNs) platforms are increasingly favored by researchers. For the scholar services built on the ASNs, recommending personalized researchers have become more important, as it could promote academic communication and scientific research for scholars. We propose a personalized recommendation method combining similarity and trust degree in an academic social network. First, the text-similarity hybrid model of LDA and TF-IDF is used to calculate the similarity of scholars' interests, moreover, the social similarity between scholars is combined as the final similarity. Second, the trust degree is calculated according to the multi-dimensional interactive behavior among scholars. Finally, the combined similarity and trust degree between scholars are used as a ranking metric. We demonstrate and evaluate our approach with a real dataset from an academic social site SCHOLAT. The experiment results show that our method is valid in recommending personalized researches.

Keywords: Personalized recommendation · Academic social networks · Trust degree · Similarity

1 Introduction

With the beginning of online social networking, the mode of academic communication and interaction among scholars has gradually changed from offline mode to academic social network, which is a fine classification social network platform for researchers. The emergence of academic social networks has played an important role in making scientific discoveries and seeking opportunities for cooperation among researchers and promoting academic exchanges [1]. The main friend recommendation algorithms in general social networks have many methods such as graph-based recommendation algorithms [2], Recommendation Mechanism based on user influence [3], Make recommendations based on users' content interests [4, 5], mixed recommendation [6] and similarity calculation [7].

Most traditional college website portals are static display pages, which leads to many problems, such as difficulty in collecting and modifying teachers' information, uneven data quality, untimely updating of teachers' information. It has been well solved on the

© Springer Nature Singapore Pte Ltd. 2021
Y. Sun et al. (Eds.): ChineseCSCW 2020, CCIS 1330, pp. 441–455, 2021.
https://doi.org/10.1007/978-981-16-2540-4_32

teacher information management platform. However, the exchange of teachers between colleges is often limited to some related cooperation within the college, which leads to a small communication circle among scholars, and it is difficult to find and effectively establish friend relationships related to their fields.

In this paper, we take the real ASNs platform SCHOLAT[1] as the source of massive data, and use the hybrid model of integrating LDA [8] and TF-IDF to calculate the text-similarity for the scholar information data, and integrates the social similarity among scholars as to the final similarity. We also prose calculation method of trust degree in an academic social network based on the literature [9], and further refine the trust degree model among users. Add more realistic factors in line with the current trust model under the current social network from the two aspects of cognitive trust and interactive trust. Under the academic social network, scholars' cognitive trust is divided into three relationships: mutual attention, attention, and fans combined with the influence factor of the common team of scholars to improve the trust calculation model. Finally, the personalized recommendation results of scholars are obtained by combining similarity and trust degree, which effectively solves the problem of a too-small circle of teachers' friends and single recommendation of friends in the faculty information management platform[2]. Summarized as follows:

1) Through the hybrid model of LDA and TF-IDF, the interest similarity between scholars is calculated; combining the social similarity among scholars as to the final similarity, it is helpful for scholars to find friendships in similar research fields.
2) Improve the calculation model of mutual trust among scholars, and integrate team trust factor as the trust degree among scholars. Ability to obtain more realistic social relationships and alleviate data sparseness.
3) We use SCHOLAT's real data set for experiments. The results show that our recommendation effect is significantly improved compared to the past.

2 Related Work

In modern society, social networking sites are an indispensable part of people's work and life. Due to the increasing amount of information available on the website and the variety, personalized recommendations have become a major difficulty for researchers. It can be subdivided into five aspects, The recommendation based on collaborative filtering is mainly based on the information matrix between the user and the product. The degree to which the content-based recommendation matches the system recommendation projects through the user's interest. Based on the graph structure, the similarity between nodes is usually used to recommend for users. The hybrid recommendation algorithm is mainly aimed at the fusion recommendation of the first three algorithms. The similarity calculation method is based on the Pearson correlation and cosine similarity calculated by users for common scoring items [10–13].

Hoi et al. [14] presented an estimate of the similarity between Twitter users' entries and behaviors on social networks, and the experimental results have a high accuracy

---

[1] https://www.scholat.com/.

[2] https://faculty.scholat.com/.

rate. Zeng et al. [15] put forward a general recommendation algorithm, which integrates user's personalized preferences, geographic factors, and social influence, and achieves a good recommendation result. Du et al. [16] take advantage of the structural relevance of shoppers and products by using a neural co-autoregressive model for CF tasks, The model combines analysis by distinguishing different attributes under different tasks. Zhang et al. [17] proposed a two-layer neighbor recommendation model that integrates user influence and credibility, and improves the accuracy of the model by selecting the most influential and credible neighbors. Zheng et al. [18] by verifying the authenticity of user identities in the community and the distribution of topics in the community. They use the HIOC method to recommend users' personalized models Liu et al. [19] presented an adaptive method that considers trust and score, which to measure the implicit similarity between users and friends when they are trustworthy. They considered the biases of users and considered implicit feedback from users to improve the accuracy of the algorithm. Yuan et al. [20, 21] introduce the semantics of the sentences between the two, combined with the contextual summary of scientific articles to measure the similarity to obtain a better semantic similarity relationship. Zhang et al. [22] designed a new type of Top-N friend recommendation model based on quantifying users' implicit trust strength and social influence. Wang et al. [23] proposed a text similarity hybrid model (L-THM) integrating LDA and TF-IDF for calculating text similarity. The model uses the semantic information contained in the text itself and the keyword information reflecting the text to comprehensively analyses and calculates the similarity between the texts.

Combined with the above-mentioned user friend recommendation methods, most of them are carried out in the field of general social networks, which are limited to a certain aspect of the user's information, and often have many problems, such as sparse user information, cold start and so on. On the other hand, they fail to fully consider the social network structure and the characteristics of users' multi-dimensional information interaction to recommend friends with users. Therefore, aiming at the more interactive behavior among users in specific fields and scenarios, this paper synthesizes the real situation under the academic social and analyzes the specific interactive behavior among scholars from many angles. We propose a calculation method that integrates the similarity and trust degree with scholars under the academic social network, and the final result of TOP-N recommendation is to recommend friends of teachers under the faculty information management platform, which proves the feasibility and effectiveness of this method.

3 Methodology

In the specific field of the academic social network, in addition to the similarity extracted from the LDA topic model and the TF-IDF model according to the scholars' own profile information and academic research achievements. More intimacy among scholars is reflected in communication, cooperation, and interaction within the community, and the more frequent interaction between them, the higher the degree of trust between them. Therefore, combined with the similarity between scholars, taking into account the close relationship between the interaction between scholars from the cognitive trust and interactive trust to measure, and finally the integration of similarity and trust for the personalized recommendation of scholar friends.

3.1 Calculation of Similarity in Scholars

The interesting similarity among scholars is calculated by the weighted hybrid LDA and TF-IDF model, which improves the effect of similarity calculation under a single factor, effectively displays the semantics of the text to be expressed and alleviates the problem of data sparsity when the text is too short.

LDA Similarity Calculation. As shown in Fig. 1 LDA model, the LDA model is a typical directed probability graph model, which is mainly used to train a set of potential topics containing a certain probability from the existing text set. The text is generated as follows:

(1) Drawing generation a topic distribution θ_m of text d from a Dirichlet prior α;
(2) Drawing generation the vocabulary distribution φ_k of the topic k from the Dirichlet prior β;
(3) Drawing generation the topic $z_{m,n}$, of the n-th word of the text d from the multinomial θ_m of the topic;
(4) Drawing generation the word $w_{m,n}$, from the multinomial φ_k of the word;

Where is α prior distribution parameters of topic distribution; β is prior distribution parameters of word distribution; θ_m is the topic distribution of the m-th text; φ_k is the word distribution of the topic k; $z_{m,n}$ is the subject corresponding to the n-th word of text m; $w_{m,n}$ is the n-th word of the text m.

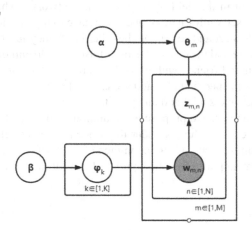

Fig. 1. LDA model

Text similarity based on the LDA can map high-dimensional text content is transformed into a low-dimensional semantic space theme, effectively reduce the dimension, and make the most of the semantic information contained in the text itself. Therefore, the scholars' papers and profile data sets are uniformly merged into a long text for processing, and then after data cleaning and word segmentation; the text topic probability distribution is generated by the LDA model, and the calculation formula is as follows

(1).

$$S_{LDA}(u_i, v_i) = \frac{\theta_u \cdot \theta_v^{\mathsf{T}}}{\|\theta_u\| \cdot \|\theta_v\|} = \frac{\sum_{k=1}^{K} \theta_{u,k} \cdot \theta_{v,k}}{\sqrt{\sum_{k=1}^{K} \theta_{u,k}} \cdot \sqrt{\sum_{k=1}^{K} \theta_{v,k}}} \tag{1}$$

Where θ_u and θ_v are topic probability vector of the u-th text and the v-th text; $\|\theta_u\|$ and $\|\theta_v\|$ are the norms of the topic vector of the u-th text and the v-th text; $\theta_{u,k}$, and $\theta_{v,k}$ is the probability value of the k-th topic of the u-th text and the v-th text; $S_{LDA}(u_i, v_i)$ is Similarity between scholar u_i and scholar v_i based on LDA model.

TF-IDF Similarity Calculation. In the above LDA model similarity calculation, the influence of words on the text is ignored, a lot of valuable knowledge contained in the text is lost, and the calculation accuracy of text similarity is affected, while the TF-IDF models similarity calculation can solve this problem very well. First of all, traverse all the scholars, and get the corresponding text data according to the scholar ID; Then the TF-IDF algorithm is used to design the scholar's interest vector; Finally, the interest similarity between any two scholars is calculated according to the ID list of scholars' concerns $S_{TF-IDF}(u_i, v_i)$, and the design algorithm is shown in Table 1 below.

Finally, the weighted hybrid calculation is carried out by combining the above LDA and TF-IDF text similarity calculation method, which not only considers the potential semantic relationship between texts, but also considers the impact of words on the text. The final interest similarity among scholars is obtained, and the calculation formula is as follows (2).

$$Msim(u_i, v_i) = \alpha S_{LDA}(u_i, v_i) + \beta S_{TF-IDF}(u_i, v_i) \tag{2}$$

Where $S_{LDA}(u_i, v_i)$ is interest similarity under LDA Model between Scholar u_i and Scholar v_i; $S_{TF-IDF}(u_i, v_i)$ is interest similarity under TF-IDF Model between Scholar u_i and Scholar v; α and β are set as linear fusion parameters with equal weight. $Msim(u_i, v_i)$ is the ultimate interest similarity between the two scholars.

The academic social network topology is a directed graph. The more similar the distribution of friends among scholars, the more likely they are to become friends with each other. If the scholar u_i pays attention to the scholar v_i, then the scholar u_i is called the fan of the scholar v_i, and the scholar v_i is called the follower of the scholar u_i. Fans of scholar u_i form a set, and followers from another set. The calculation formula based on the follower is as follows (3)

$$Rsim(u_i, v_i) = \frac{|\ Follower(u_i) \cap Follower(v_i)\ |}{\sqrt{|\ Follower(u_i| \times |Follower(v_i)\ |}} \tag{3}$$

Where $Follower(u_i)$ is a collection of followers of scholar u_i; $Follower(v_i)$ is a collection of followers of scholar v_i; $|Follower(u_i)|$ and $|Follower(v_i)|$ are the number of scholars in the collection; $Rsim(u_i, v_i)$ is the ultimate social similarity among scholars.

3.2 Calculation of Trust Degree in Scholars

When calculating the similarity of scholars, they often face the problem of data sparseness and cold start. Increasing the trust metric among scholars can alleviate the problem of

Table 1. TF-IDF Scholars' interest similarity algorithm.

TF-IDF Scholars' interest similarity algorithm
Input: In academic social networks Scholar data S, Scholar-Scholar Relational data R, Scholar text data T
Output: User interest similarity $S_{\text{TF-IDF}}$

1.   For every $S_i \in S$ do
      // Traverse every scholar
2.   For every $T_i \in T$ do
      // Traverse the text data of each scholar
3. $TF = \frac{m}{M}$
      // M: the total number of words in the text
      //m: the number of times the current word appears in the text
4. $IDF = \log(\frac{D}{D_w})$
      //D: total number of text, D_w:the number of times the word w appears in D
5. $TFIDF = TF * IDF$
6.   Set $vector_i = \{(p_1, w_1), (p_2, w_2), (p_3, w_3) \cdots \}$
      // p_i:Attribute, w_i: Weight of p_i
7.   End for
8.   End for
9.   For every $S_u, S_v \in S$ do
      // Traverse every scholar
10.   For every $R_{u,v} \in R$ do
      // Traverse every scholar's friend relationship
11. $S_{TF-IDF}(u_i, v_i) = \frac{\sum_{i=1}^{n}(p_{ui} \times p_{vi})}{\sqrt{\sum_{i=1}^{n}(p_{ui})^2} \times \sqrt{\sum_{i=1}^{n}(p_{vi})^2}}$
      //p_{ui}, p_{vi}: The i-th value of the eigenvector of scholars u_i and v_i
12.   End for
13.   End for

dataset sparseness and cold start. The trust between friends is a measure of the closeness of the relationship between two people, which is obtained by gathering and analyzing the interaction between users [24]. By analyzing the relationship of friends and interactions among scholars in social networks to calculate the trust between scholars, a more realistic social relationship can be obtained.

Aiming at the problem that the trust between users in reference [9] includes insufficient consideration of cognitive trust, and interactive trust, refine and improve the model of cognitive trust, and interactive trust among scholars. Cognitive trust is to subdivide the cognitive trust among users into different levels of relationships. In the current field of academic social design network, the main cognitive relationships can be divided into three types: mutual attention, attention, and fans, the results of cognitive trust formed by this should be different. Interactive trust is the main factor influencing the integration of more interactive trust in the current academic social network. For example, whether scholars join a common team or not represents a stronger interaction between the two.

In the past, the trust of scholars in academic social networks was calculated mainly based on dynamic comments, retweets, and likes, as well as online chat and on-site e-mail communications. The trust degree of the number of collections is increased, and the trust calculation formula considering the directionality of behavior interaction is as follows (4)–(9):

$$Cr(u_i, v_i) = \frac{Ic(u_i, v_i) + Ic(v_i, u_i)}{2 \times \sqrt{Max(Ic(u_i, i)) \times Max(Ic(v_i, i))}} \tag{4}$$

$$Tr(u_i, v_i) = \frac{It(u_i, v_i) + It(v_i, u_i)}{2 \times \sqrt{Max(It(u_i, i)) \times Max(It(v_i, i))}} \tag{5}$$

$$Zr(u_i, v_i) = \frac{Iz(u_i, v_i) + Iz(v_i, u_i)}{2 \times \sqrt{Max(Iz(u_i, i)) \times Max(Iz(v_i, i))}} \tag{6}$$

$$Ch(u_i, v_i) = \frac{Ic(u_i, v_i) + Ic(v_i, u_i)}{2 \times \sqrt{Max(Ic(u_i, i)) \times Max(Ic(v_i, i))}} \tag{7}$$

$$Er(u_i, v_i) = \frac{Ie(u_i, v_i) + Ie(v_i, u_i)}{2 \times \sqrt{Max(Ie(u_i, i)) \times Max(Ie(v_i, i))}} \tag{8}$$

$$Sr(u_i, v_i) = \frac{Is(u_i, v_i) + Is(v_i, u_i)}{2 \times \sqrt{Max(Is(u_i, i)) \times Max(Is(v_i, i))}} \tag{9}$$

$Cr(u_i, v_i)$, $Tr(u_i, v_i)$, $Zr(u_i, v_i)$, $Chr(u_i, v_i)$, $Er(u_i, v_i)$, $Sr(u_i, v_i)$ respectively represent the comment interactive trust, retweet interactive trust, like interactive trust, online chat interactive trust, email communication interactive trust and favorite interactive trust of scholar u_i and trust of scholar v_i. $Ic(u_i, v_i)$, $It(u_i, v_i)$, $Iz(u_i, v_i)$, $Ich(u_i, v_i)$, $Is(u_i, v_i)$ respectively represent the number of comments, likes, retweets, online chats, email exchanges and favorites of scholar u_i and scholar v_i, On the contrary, it is the interactive behavior of scholar v_i to scholar u_i. $Max(Ic(u_i, i))$, $Max(It(u_i, i))$... represents the maximum interaction value between scholars u_i, v_i and all scholars I, and the formula for calculating the trust of six items is as follows (10).

$$Ir(u_i, v_i) = \alpha Cr(u_i, v_i) + \beta Tr(u_i, v_i) + \gamma Zr(u_i, v_i) + \\ \lambda Chr(u_i, v_i) + \mu Er(u_i, v_i) + \delta Sr(u_i, v_i) \tag{10}$$

Where $Ir(u_i, v_i)$ is fusion interactive trust, α, β, γ, λ, μ and δ are the fusion parameters of each interaction, and the sum of the parameters is 1. Due to the difference of influence factors among different items, to solve the lack of objective rationality of the simple fusion of parameters, the normalization method min-max is used for standardization in the later stage, and its formula is as follows (11).

$$Ir(u_i, v_i) = \frac{\frac{Ir(u_i, v_i) - Min(Ir(u_i, i))}{Max(Ir(u_i, i)) - Min(Ir(u_i, i))},}{0,} \quad \begin{matrix} Max(Ir(u_i, i)) - Min(Ir(u_i, i)) \neq 0 \\ Max(Ir(u_i, i)) - Min(Ir(u_i, i)) = 0 \end{matrix} \tag{11}$$

Where $Ir(u_i, v_i)$ is the ultimate trust between the author u_i and the scholar v. $Max(Ir(u_i, i))$ is the maximum degree of trust between scholar u_i and all concerned scholars; $Min(Ir(u_i, i))$ is the minimum degree of trust.

To make the degree of trust among scholars more specific and in line with the current description of the relationship between scholars under the network, the trust factor of the common team between scholars u_i and v_i are added, and the calculation formulas are shown below (12).

$$Tr(u_i, v_i) = \frac{|\ Team(u_i) \cap Team(v_i)\ |}{|\ Team(u_i) \cup Team(v_i)\ |} \tag{12}$$

Where $Team(u_i)$ and $Team(v_i)$ represents the set of teams that exist for scholar u_i and scholar v_i, respectively. $Tr(u_i, v_i)$ is the calculated trust factor of the common team among scholars.

3.3 Personalized Scholar Recommendation Algorithm

Combined with the above-mentioned four parts of interest and social similarity, interactive trust, and team trust factors, the weighted fusion calculation is carried out as the final recommendation score of scholars in the academic social network is calculated as follows (13).

$$Fr(u_i, v_i) = \eta Rsim(u_i, v_i) + \theta Msim(u_i, v_i) + uIr(u_i, v_i) + \kappa Tr(u_i, v_i) \tag{13}$$

Where $Fr(u_i, v_i)$ is Scholars' personalized final recommendation score. $\eta, \theta, \mu, \kappa$ are for each fusion parameter, the algorithm flow is shown in Fig. 2 below.

4 Experiments and Results

4.1 Datasets

We make a personalized recommendation for scholars based on the data from the real academic social networking site-SCHOLAT, and finally, apply it to the teacher service platform to recommend for teachers' friends. After preprocessing the data set, 3,134 scholars, 4,4896 one-way friend relationships among scholars where each user has an average of 16 real friends, and 2,7478 papers were selected. The data set mainly includes basic information of scholars' profiles, basic information of academic papers, and behavioral interactive information such as scholars' comments, likes, retweets, online chats, email exchanges, and collections, etc.

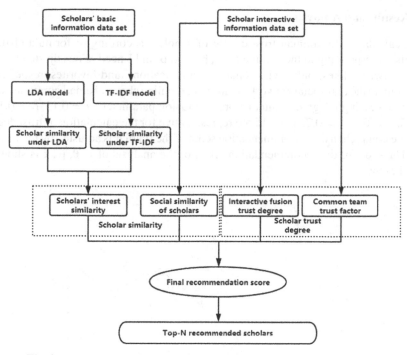

Fig. 2. A recommendation model combining scholars' similarity and trust

4.2 Evaluation Metrics

To verify the accuracy of the recommendation, the commonly used evaluation criteria recommended by Top-N are adopted. The recommendation results are verified by comparing the *Precision*, the *Recall*, and the harmonic average *F*1 − *measure* of the comprehensive Precision and Recall. The calculation formulas are as follows (14–16).

$$Precision = \frac{|R(u) \cap T(u)|}{|R(u)|} \tag{14}$$

$$Recall = \frac{|R(u) \cap T(u)|}{|T(u)|} \tag{15}$$

$$F1 - measure = \frac{2 \times Precision \times Recall}{Precision + Recall} \tag{16}$$

Where $R(u)$ is A set of scholars recommended in the experiment, $T(u)$ is a Set of scholars in the verification set, $R(u) \cap T(u)$ is a collection of recommended scholars who have become friends. Finally, the comprehensive index of *F*1 − *measure* is used to reflect the measurement of recommendation effect. According to the index of F1, the personalized recommendation algorithm in this paper is compared with the recommendation algorithm based on matrix decomposition, and the traditional collaborative filtering algorithm.

4.3 Results and Analysis

When calculating the mutual trust degree of scholars according to formula (10), the weights corresponding to the parameters α, β, γ, λ, μ and δ need to be adjusted as comments, retweets, likes, online chat, email communications, and favorites, respectively. According to the real social network environment, online chat and email communication usually have a high degree of interaction. The fusion parameters $\alpha = 0.15$, $\beta = 0.15$, $\gamma = 0.15$, $\lambda = 0.2$, $\mu = 0.2$, $\delta = 0.15$ are representative for the calculation of trust degree. Then the comprehensive recommendation score of the scholar is adjusted by the formula (13). The score of the recommendation effect of the analysis of η, θ, μ, κ is shown in Fig. 3 below.

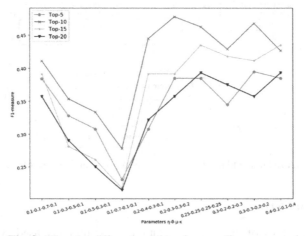

Fig. 3. F1 under different weights of personally recommend

It can be seen from Fig. 3 that when the proportion of information similarity among scholars decreases and the proportion of interactive trust increases, the score of F1 will increase, indicating that the social relationship between scholars has a great impact on the effect of recommendation. When the proportion of trust among scholars is constant, the similarity has little effect on the recommendation effect of scholars. When η-θ-u-κ is 0.2, 0.3, 0.3, and 0.2 respectively, it has better-personalized recommendation results for scholars.

Figure 4 shows the influence of the number of recommendations on the score of F1-measure when only scholar similarity, scholar trust degree, fusion similarity and trust degree are considered. It can be seen that the fusion similarity and trust degree are better than the other two cases, and the F1-measure is also greatly improved, and the effect is the best when the number of recommendations is 10.

Figure 5 shows the comparison of experiments under different recommendation algorithms. Especially when the number of scholars' recommendations is 10, the F1 value of this recommendation algorithm is 16.49% and 29.52% higher than that of the recommendation algorithm based on matrix decomposition and collaborative filtering, respectively.

Figure 6 shows the evaluation and analysis of personalized recommendation in Top-N recommendation results with scholars' similarity and trust degree when the parameters η-θ-u-κ are 0.2, 0.3, 0.3 and 0.2. The N values are 5, 10, 15, and 20 pairs of accuracy, recall, and F1-measure results, respectively.

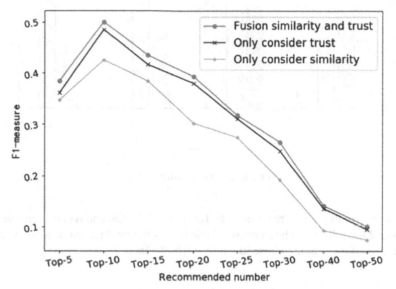

Fig. 4. F1 under different methods of own experiment

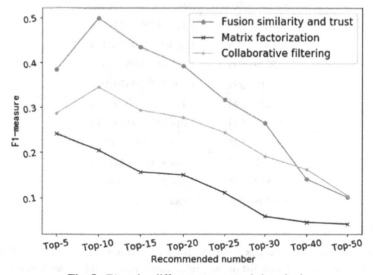

Fig. 5. F1 under different recommended methods

Finally, the recommendation score calculated by the personalized scholar recommendation under the academic social network is imported into the res_score field in the

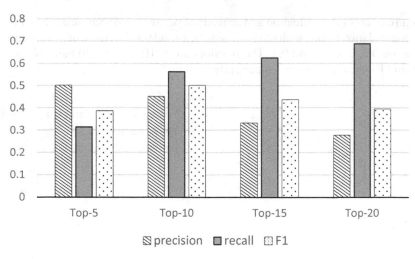

Fig. 6. Each evaluation index

database, where the sender represents the ID value of the fan and receive represents the ID value of the follower. The following Table shows the final recommendation score of any ten fans of the scholar whose Id number is 595 (Table 2).

Table 2. Partial recommended score

Sender	Receiver	Res_score
95619	595	1.1137539924235118
105349	595	1.0146073995760714
95749	595	1.0173437605269569
519	595	1.2062965536397061
3592	595	1.1146058750097334
52105	595	1.0142073991231219
94729	595	1.0136073995760724
106637	595	1.0146045315141275
529	595	1.4622173232898676
95508	595	1.0846173465245337

Ranking the top 10 scholars with the final recommendation score in Top-N, the personalized recommendation list of integrated academic social network scholars under the teacher service platform is displayed, and the effect is shown in Fig. 7.

Fig. 7. Personalized recommendation display

5   Conclusions

Online academic social networking has increasingly become the mainstream way of communication for scholars. Making full use of the non-private data sharing concept and communication mode of social networks, a college faculty information management service platform centered on academic social networks has been constructed. Online trials have been carried out in many colleges and universities, and good feedback has been obtained. On this basis, combining the similarity and trust calculation models on teachers' academic social networks provides a new channel for friend recommendation for teachers' academic communication and cooperation, which significantly improves the accuracy of teacher recommendation and enriches academic exchanges and scientific research between teachers and scholars.

Of course, there are some shortcomings in the study. The influence of the time decay factor on the recommendation effect is not considered, and due to a large amount of related data such as teachers' profiles and papers, it is still a way of offline calculation. As a result, it is impossible to recommend friends for teachers immediately according to the personalized recommendation algorithm, a real-time recommendation will be a major research direction in the future.

Acknowledgements. This work was supported by National Natural Science Foundation of China under grant number U1811263, by National Natural Science Foundation of China under grant number 6177221.

References

1. Wan, H., Zhang, Y., Zhang, J., et al.: AMiner: search and mining of academic social networks. J. Data Intell. **1**(1), 58–76 (2019)

2. Zhang, M., Chen, W.: Optimised tags with time attenuation recommendation algorithm based on tripartite graphs network. Int. J. Comput. Sci. Eng. **21**(1), 30 (2020)
3. Seng, D., Liu, J., Zhang, X., et al.: Top-N recommendation based on mutual trust and influence. J. Int. J. Comput. Commun. Control **14**(4), 540–556 (2019)
4. Jeckmans, A., Tang, Q., Hartel, P., et al.: Poster: privacy-preserving profile similarity computation in online social networks. In: 18th Conference on Computer and Communications Computer and Communications Security (CCS), pp. 793–796. ACM, Chicago (2011)
5. Li, J., Xu, H.: Suggest what to tag: recommending more precise hashtags based on users' dynamic interests and streaming tweet content. J Knowl. Based Syst. **106**, 196–205 (2016)
6. Wang, H., Xia, H.: Collaborative filtering recommendation algorithm mixing LDA model and list-wise model. J. Comput. Sci. **46**(9), 216–222 (2019)
7. Takano, Y., et al.: Improving document similarity calculation using cosine-similarity graphs. In: Barolli, L., Takizawa, M., Xhafa, F., Enokido, T. (eds.) AINA 2019. AISC, vol. 926, pp. 512–522. Springer, Cham (2020). https://doi.org/10.1007/978-3-030-15032-7_43
8. Wang, Z., He, M., Du, Y.: Text similarity computing based on topic model LDA. J. Comput. Sci. **40**(12), 229–232 (2013)
9. Sun, S., Lin, X., Peng, B., et al.: A recommendation method for scholars based on trust and research interests. J. Comput. Digit. Eng. **047**(003), 608–615 (2019)
10. Alhijawi, B., Kilani, Y.: The recommender system: a survey. J. Int. J. Adv. Intell. Paradigms **15**(3), 229–251 (2020)
11. Dou, Y., Yang, H., Deng, X., et al.: A survey of collaborative filtering algorithms for social recommender systems. In: 12th International Conference on Semantics, Knowledge and Grids (SKG), pp. 40–46. IEEE, Los Alamitos (2016)
12. Chen, H., Wang, Z.: Summary of personalized recommendation algorithms. J. Enterp. Sci. Technol. Dev. (02), 56–57 (2019)
13. Abdi, M., Okeyo, G., Mwangi, R., et al.: Matrix factorization techniques for context-aware collaborative filtering recommender systems: a survey. J. Comput. Inf. Sci. **11**(2), 1–10 (2018)
14. Nguyen, T., Tran, D., Dam, G., Nguyen, M.: Estimating the similarity of social network users based on behaviors. Vietnam J. Comput. Sci. **5**(2), 165–175 (2018). https://doi.org/10.1007/s40595-018-0112-1
15. Zeng, J., Li, F., He, X., et al.: Fused collaborative filtering with user preference, geographical and social influence for point of interest recommendation. J Int. J. Web Serv. Res. **16**(4), 40–52 (2019)
16. Du, C., Li, C., Zheng, Y., et al.: Collaborative filtering with user-item co-autoregressive models. In: Thirty-Second Conference on Artificial Intelligence, pp. 2175–2182. AAAI, New Orleans (2018)
17. Zhang, Z., Liu, Y., Jin, Z., et al.: A dynamic trust based two-layer neighbor selection scheme towards online recommender systems. J. Neurocomput. **285**, 94–103 (2018)
18. Zheng, J., Wang, S., Li, D., et al.: Personalized recommendation based on hierarchical interest overlapping community. J. Inf. Sci. **47**, 55–75 (2019)
19. Liu, Z., Xiong, H., Liu, J., et al.: Recommendation algorithm fusing implicit similarity of users and trust. In: 21st International Conference on High Performance Computing and Communications (HPCC), pp. 2084–2092. IEEE, Zhangjiajie (2019)
20. Yuan, C., Bao, Z., et al.: Incorporating word attention with convolutional neural networks for abstractive summarization. J. World Wide Web. **23**(1), 267–287 (2020). https://doi.org/10.1007/s11280-019-00709-6
21. Yuan, C., et al.: Citation based collaborative summarization of scientific publications by a new sentence similarity measure. In: Romdhani, I., Shu, L., Takahiro, H., Zhou, Z., Gordon, T., Zeng, D. (eds.) Collaborative Computing: Networking, Applications and Worksharing. LNICSSITE, vol. 252, pp. 680–689. Springer, Cham (2018). https://doi.org/10.1007/978-3-030-00916-8_62

22. Zhang, X., Chen, X., Seng, D., et al.: A factored similarity model with trust and social influence for top-n recommendation. J. Int. J. Comput. Commun. Control **14**(4), 590–607 (2019)
23. Wang, J., Xu, W., Yan, W., et al.: Text similarity calculation method based on hybrid model of LDA and TF-IDF. In: 3rd International Conference on Computer Science and Artificial Intelligence, pp. 1–8, Beijing (2019)
24. Chen, S., Luo, B., Sun, Z.: Social friend recommendation algorithm based on trust of paths between mixed friends. J. Comput. Technol. Dev. **28**(02), 74–77 (2018)

Academic Paper Recommendation Method Combining Heterogeneous Network and Temporal Attributes

Weisheng Li, Chao Chang$^{(\boxtimes)}$, Chaobo He, Zhengyang Wu, Jiongsheng Guo, and Bo Peng

South China Normal University, Guangzhou 510631, China
{weishengli,changchao,wuzhengyang,johnsenGuo,
bpeng}@m.scnu.edu.cn

Abstract. In the case of information overload of academic papers, the demand for academic paper recommendation is increasing. Most of the existing paper recommendation methods only utilize scholar friendship or paper content information, and ignore the influence of temporal weight on research interest, and hence they are hard to obtain good recommendation quality. Aiming at this problem, the method HNTA for academic paper recommendation based on the combination of heterogeneous network and temporal attributes is proposed. HNTA firstly constructs a heterogeneous network composed of different types of entities to calculate the similarity between two papers, and then the temporal attribute is introduced into scholars' research interests which are divided into instant interests and continuous interests to calculate the similarity between scholars and papers. Finally, by weighting the above two similarities, the purpose of recommending papers to scholars is achieved. Overall, HNTA can not only comprehensively utilize both relationships of scholars and the content information of papers, but also it considers the impact of the temporal weight of scholars' research interests. By conducting comparative experiments on the data set of the real academic social network: SCHOLAT, the results show that HNTA performs better than traditional paper recommendation methods.

Keywords: Academic paper recommendation · Heterogeneous network · Temporal attributes · Academic social networks

1 Introduction

In recent years, with the rapid development of the scientific research field, academic papers are increasing exponentially. Scholars need to quickly obtain papers related to their research interests among thousands of academic papers. In this case, the social recommendation system [1, 2] stands out to help users alleviate the problem of information overload and recommend papers to users with their relevant interests. Therefore, the academic paper recommendation system has become an indispensable tool for scholars to find papers.

© Springer Nature Singapore Pte Ltd. 2021
Y. Sun et al. (Eds.): ChineseCSCW 2020, CCIS 1330, pp. 456–468, 2021.
https://doi.org/10.1007/978-981-16-2540-4_33

At present, the existing recommendation methods at home and abroad are mainly divided into three categories: recommendation based on content information [3–5]; recommendation based on Collaborative filtering [6–8]; recommendation based on hybrid approach [9–11]. At the same time, with the emergence of a large number of academic papers, a number of academic paper recommendation methods have been proposed.

In paper recommendations, Liu et al. [12] utilize keyword-driven and popularization consciousness, and then propose a paper recommendation algorithm based on undirected paper citation maps. Parvin et al. [13] propose a new collaborative filtering algorithm to predict the similarity rating of users. Manju et al. [14] construct a heterogeneous graph of papers and use the random walk method to alleviate the problem of cold start in the paper recommendation system. Pan et al. [15] combine the citation relationship between two papers and the contextual knowledge in the paper, and then propose a method based on heterogeneous graphs for academic papers recommendation. Meng et al. [16] use the coupling relationship to implement TOP-N recommendations of keywords. Catherine et al. [17] propose a recommendation method based on the knowledge graph. Guo et al. [18] calculate the similarity of Title-Abstract Attentive Semantics for recommending. Yue et al. [19] propose a listwise learning-to-rank recommendation method based on heterogeneous network analysis in social networks, and then construct a heterogeneous network and utilize the link of relationships on the meta path to perform list-level ranking for recommendation.

The above methods mainly considers the binary relationship between users and items and is more based on the scholar's basic information and the content of the paper itself. However, in the actual situation, the paper recommendation method is usually not only based on the content information associated with the scholars themselves, but it should also take into account the changes in scholars' research interests at different time periods. The research interest and research history of scholars represent the progress and progress of scholars. Research interest includes the decay and growth of interest, and the time factor will have a certain impact on the research interest of scholars [20]. At the same time, academic social networks usually contain rich and interrelated academic information features, which can be used as a heterogeneous network containing multiple entity types and relationship types [21], from which link relationships of scholars and papers can be more conveniently obtained. In this paper, we propose an academic paper recommendation method (HNTA) that combines heterogeneous network and temporal attributes.

The main work of this paper include:

1. Constructing a heterogeneous network through different types of entities in the papers.
2. Adding temporal attributes to scholars' research interest which is divided into instant interest and continuous interest.
3. Calculating the similarity between papers and papers and the similarity between scholars and papers through the above heterogeneous network and temporal attributes. By using the data set of the real academic social platform: SCHOLAT[1], our experiment verifies that HNTA is practical and effective.

---

[1] http://www.scholat.com/.

2 Methodology

2.1 Constructing Heterogeneous Network

Firstly, the recommended words are extracted. The text information in the original paper data set of SCHOLAT is subjected to word segmentation, and a list of stop words is used to filter the text information to remove irrelevant stop words and symbols. The TF-IDF algorithm [22] and the Information Gain algorithm [23] are used to segment the word After processing the data, the recommended words are obtained. During this process, the keywords in the papers is added to a custom dictionary to improve the precision in the segmentation of texts.

$$TF - IDF = TF_{ij} \times IDF_j = \frac{n_{ij}}{\sum_k n_{k,j}} \times \log \frac{|D|}{1 + |j; t_i \in d_j\}|} \tag{1}$$

$$IG(T) = H(C) - H(C \mid T) = -\sum_{i=1}^{m} p(c_i) \log p(c_i) +$$
$$p(t) \sum_{i=1}^{m} p(c_i \mid t) \log p(c_i \mid t) + p(\bar{t}) \sum_{i=1}^{m} p(c_i \mid \bar{t}) \log p(c_i \mid \bar{t}) \tag{2}$$

Suppose there is a collection of n papers, $P = \{P_1, P_2, \dots, P_n\}$. Each paper in the collection is represented by m different types of features, $F = \{F_1, F_2, \dots, F_m\}$. Each feature Fi consists of a set of feature sets, that is, $F_i = \{f_{1i}, f_{2i}, \dots, f_{ni}\}$. The heterogeneous network is a huge network graph with semantic relations, which consists of the relationship between entities and entities that exist in the paper. After extracting the feature, the basic information of author, title, abstract, keywords and other information are obtained, and then different types of entities are represented as interrelated and different-shaped nodes to construct a heterogeneous network based on scholars and papers (Fig. 1).

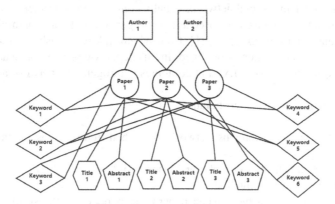

Fig. 1. Example of a heterogeneous network.

2.2 Similarity Calculation Based on Heterogeneous Network

Similarity of Pairwise Recommended Words. The extracted recommended words are constructed into a set of word vectors through the Word2Vec model [24], and then the word vectors are used to calculate the semantic similarity between word pairs: $W_{similaity}$. The calculating processes are shown in Eqs. 3 and 4,

$$W(X, Y) = \begin{cases} 0 & X, Y \notin \text{keywords} \\ W_{\text{similarity}} & X, Y \in \text{keywords} \end{cases} \tag{3}$$

$$W(\omega_i, \omega_j) = \alpha \times W_T(\omega_i, \omega_j) + \beta \times W_K(\omega_i, \omega_j) + \gamma \times W_A(\omega_i, \omega_j) \tag{4}$$

Where W_T, W_k, W_T respectively represent the similarity of the word pairs ω_i and ω_j in the thesis among the topics, keywords and abstracts, α, β, γ respectively represent the similarity of the topics, keywords and abstracts in the thesis.

Similarity of Pairwise Papers. The similarity between two papers can be measured according to the similarity between recommended words. At the same time, there are other similar papers in the papers and scholars' entities to jointly construct a similarity model of the papers. Then, the similarity between the recommended words is obtained by training the word vector: $S_w(p_1, p_2)$. The calculating processes are shown in Eqs. 5,

$$S_w(p_1, p_2) = \alpha \times W_T(p_1, p_2) + \beta \times W_K(p_1, p_2) + \gamma \times W_A(p_1, p_2) \tag{5}$$

Where $W_T(p_1, p_2)$ represents the similarity between the recommended phrases of the topics between the two papers, $W_K(p_1, p_2)$ represents the similarity between the recommended phrases of the keywords between the two papers, and $W_A(p_1, p_2)$ represents the similarity of the recommended words in the abstracts of the two papers. α, β, and γ respectively represent the proportion of attributes in similarities, at the same time, different types of papers have different parameter proportions.

2.3 Research Interest Similarity Based on Temporal Attributes

Interest in Continuous Research. Scholars' continuous research interest mainly refers to the relevant research interests that scholars continuously have research achievement in the process of research for a long time. In the time distribution of the data set of SCHOLAT, the time span of the paper is large, so a time decay function is used with a small time decay in the initial process: exponential time decay function. The calculating processes are shown in Eqs. 6,

$$W(t_y, t) = \frac{1}{1 + e^{\mu \times (t - t_y)}} \tag{6}$$

The sum of word vectors for sexual research interests is:

$$CV_{user} = \sum_{i=0}^{m} W(MAX(L_i(year))) \times V(L_i) \tag{7}$$

Where m is the total number of persistent keywords, L_i is the word vector of persistent keywords, W is the time decay function, V is the word vector of recommended words after training. Then, the similarity between scholars and scholars in continuing research interests is:

$$Csimilarity = \cos(CV_{user1}, CV_{user2}) = \frac{\sum_{i=1}^{n} CVuser_1 \times CV_{user2}}{\sqrt{\sum_{i=1}^{n}(CV_{user1})^2} \times \sqrt{\sum_{i=1}^{n} CV_{user2}}} \tag{8}$$

Interest in Instant Research. The scholar's instant research interest represents the research interest of scholars within a certain year. The key words of instant research are the concentrated expression of scholars' research interest in a certain year, which mainly represents the change of scholars' research interest.

$$R_T = R - R_c \tag{9}$$

$$IV_{user} = \Sigma_{i=0}^{k} W(R_T(i, year)) \times V(R_T(i)) \tag{10}$$

$$Isimilarity = \cos(TV_{user1}, TV_{user2}) = \frac{\sum_{i=1}^{n} IV_{user1} \times IV_{user2}}{\sqrt{\sum_{i=1}^{n}(IV_{user1})^2} \times \sqrt{\sum_{i=1}^{n} IV_{user2}}} \tag{11}$$

Where R represents the original recommended words extracted by the instant research interest, R_c represents the Continuous research keywords of scholars, R_T represents the instant research interests after removing the persistent keywords, k represents It is the length of R_T. IV_{user} is the sum of word vectors of scholars' instant research interest. *Isimilarity* is the similarity of the instant research interests between scholars and scholars. Finally, the similarity of research interests between scholars and scholars is:

$$Ssimilarity = \sigma \times Csimilarity + \tau \times Isimilarity \tag{12}$$

Where $\sigma + \tau = 1$.

2.4 HNTA Recommendation Model

In the method of calculating the similarity proposed above, the similarity is measured through two aspects as the final recommendation algorithm. Figure 2 shows the overall process diagram of the HNTA for Top-N recommendation.

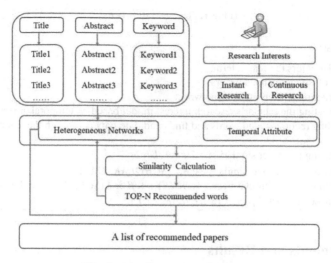

Fig. 2. The framework of HNTA.

Similarity Between Search Terms and Recommended Terms. By calculating the same keywords with the highest similarity to the current text as recommended words, where the similarity of keywords between different attributes is measured according to the similarity between different attributes. The calculation of semantic similarity between search words and candidate words is shown in Eqs. 13,

$$S(key1, key2) = \frac{\sum_{i=1}^{m} S_i(key1, key2)}{m} \qquad (13)$$

Where m is the number of attributes to be compared in this article, which mainly represent the abstract of the paper, keywords and topics.

Similarity between Scholars and Recommended Words. Calculate the similarity between recommended words and scholars by combining scholars' instant research interest and continuation research interest.

$$S_{sw}(SCHOLAR, CW) = \psi \times SCV_{SCHOLAR} + \omega \times SIV_{SCHOLAR} \qquad (14)$$

Where $SCV_{SCHOLAR}$ represents the similarity of scholars' continuous research interest, and $STV_{SCHOLAR}$ represents the scholar's instant research interest. ψ and ω respectively represent the proportion of each module. Finally, the recommended words between user-search words are constructed, and the comprehensive standard of the above two measures is used as the final TOP-N recommended standard:

$$S_{TOP-N}(SCHOLAR, KEY) = \alpha \times S(KEY) + \beta \times S(SCHOLAR) \qquad (15)$$

Finally, finding the corresponding papers of TOP-N recommended words in the heterogeneous network, so as to obtain a list of recommended papers for scholars. Table 1 shows the overall process of the HNTA.

Table 1. The overall process.

HNTA Process:
Input: a scholar enters the search term; Output: a list of Recommended papers; Step1. Constructing a heterogeneous network; Step2. Constructing the scholar's research interest model based on temporal attributes; Step3. Obtaining recommended words and find candidate papers through the heterogeneous network; Step4. Calculating the paper similarities: $S(key1, key2)$; Step5: Calculating the scholar similarities: $S_{sw}(SCHOLAR, CW)$; Step6: Weighting the two similarities: $\alpha \times S(KEY) + \beta \times S(SCHOLAR)$, and getting a list of top-n recommended paper: $S_{TOP-N}(SCHOLAR, KEY)$

3   Experiments and Results

3.1   Datasets

The algorithm proposed in this paper is recommended for academic papers in academic social networks, so we choose the real paper data set of the online academic information service platform-SCHOLAT for experiments. The data set mainly included scholars' basic information, academic paper titles, abstracts, and keywords from 34,518 papers after data preprocessing. Figure 3 shows a relationship diagram of a user's paper in the SCHOLAT data set.

3.2   Baseline Methods and Evaluation Metrics

In order to verify the recommended effect of HNTA method, the following experimental methods are used for comparative analysis. The three methods are:

(1) PWFC [25], which uses a co-authored network constructed by academic achievements among scholars and the theme of published papers, builds a three-layer paper recommendation model, constructs keyword vectors and classifies academic achievements, and then researches on scholars. The co-authored network has added a random walk model to extract relevant features for TOP-N recommendation.

(2) UPR [26], which mainly proposes to construct the recommendation model of scholars' papers according to the recent research results of users and the relationship between citations and citations between two papers. Among them, it mainly constructs the recent research interest model of scholars and the feature vector of candidate papers.

(3) CB [17], which mainly adopts the classical content-based recommendation model. Through the similarity of the research results published by scholars in the content to make recommendations related to TOP-N.

Fig. 3. The relationship diagram of the paper.

In this paper, Recall, Precission and F1-score value are used to evaluate the effect of TOP-N recommendation.

$$Recall = \frac{1}{N} \sum_{i=1}^{n} \frac{L_i}{L_n} \tag{16}$$

$$Precission = \frac{1}{N} \sum_{i=1}^{n} \frac{L_i}{R_i} \tag{17}$$

$$F1 - Score = \frac{2 \times Precission \times Recall}{Precission + Recall} \tag{18}$$

Where L_n represents all the research keywords that all users like among the recommended research keywords. N represents the number of samples recommended by test research interest, L_i represents the number of research keywords recommended in the sample, and R_i represents the total number of data recommended by TOP-N in the fourth sample.

3.3 Parameter Tuning

In the time attenuation function, we could find that scholars will have a great interest in a certain research in a period of time, and grow rapidly in a certain period of time.

Temporal weight parameter plays an important role in the calculation of similarity of research interest. In the data set, the time attenuation function have a large time span. Compared with exponential time attenuation function, linear time attenuation function, Logistic time attenuation function and Ebbinghaus time attenuation function, the exponential attenuation function with smaller attenuation is selected finally. After analyzing the temporal weight, it is found that when the attenuation factor is $\gamma = 0.3$, it will have the best effect. Figure 4 shows the temporal attribute weights. Besides, we set $\alpha = 0.4$, $\beta = 0.4$, $\gamma = 0.2$ in formula (4) and $\alpha = 0.3$, $\beta = 0.5$, $\gamma = 0.2$ in formula (5).

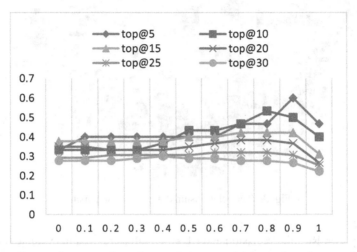

Fig. 4. According to the proportion of research interest, 0.1 is used as the step size, and the range is [0, 1]. The recommended results of different proportion are compared, and the time attenuation is exponential attenuation. In the selection of parameters, choose the one with higher recommendation Precision, that is, scholars' research interest accounts for a higher proportion: 0.9.

3.4 Comparisons and Analysis

From the similarity among abstracts, titles and keywords of the previously trained papers, the first m are selected for comparison. In this experiment, the experimental results are compared in terms of Recall, Precission and F1-Score, in which the TOP-N is 5, 10, 15, 20, 25, 30 respectively. Table 2 resports the relevant recommended words of a scholar on "friend recommendation". Table 3 is a list of papers recommended by the scholar after searching for the keyword "friend recommendation".

The following figures are comparative analysis of HNTA and other algorithms: PWFC, UPR, CB about Precission, Recall and F1-score.

As shown in Fig. 5 and Fig. 6, they respectively show the verification of the Precision, Reacll and F1-score of the relevant results of the SCHOLAT data set using the three algorithms. It could be found that, to a certain extent, the HNTA has good results in various indicators.

Table 2. The result of the recommended word of "friends recommendation".

Number	Recommended words	Comprehensive similarity score
1	Personalized recommendation	0.6772
2	Recommendation algorithm	0.6681
3	User similarity	0.6511
4	Friend relationship	0.6428
5	Link prediction	0.6412
6	Social network	0.6322
7	User interest	0.6333

Table 3. Scholar Tang Yong's recommended list of papers.

Num	Recommendation_Paper_title
1	Friend Recommendation in Social Network Using Nonnegative Matrix Factorization
2	Explicit and Implicit Feedback Based Collaborative Filtering Algorithm
3	Design of Learning Resource Recommendation Model Based on Interest Community
4	A Novel Hybrid Friends Recommendation Framework for Twitter
5	Multiple Criteria Recommendation Algorithm Based on Matrix Factorization and Random Forest

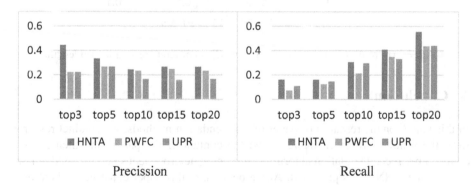

Fig. 5. Precision and recall results of HNTA in different recommended lengths.

As shown in Fig. 7, through the comparison between the HNTA algorithm and the CB algorithm, it is further proved that the HNTA algorithm will have better paper recommendation results when there is no citation relationship or the citation relationship network between scholars in the social network is not complete. The above experiments prove the feasibility and practicability of HNTA.

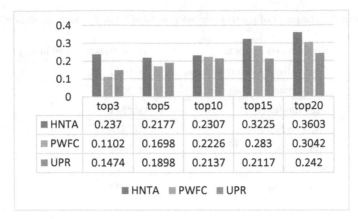

	top3	top5	top10	top15	top20
■ HNTA	0.237	0.2177	0.2307	0.3225	0.3603
■ PWFC	0.1102	0.1698	0.2226	0.283	0.3042
■ UPR	0.1474	0.1898	0.2137	0.2117	0.242

■ HNTA ■ PWFC ■ UPR

Fig. 6. F1-score comparions with different recommended length.

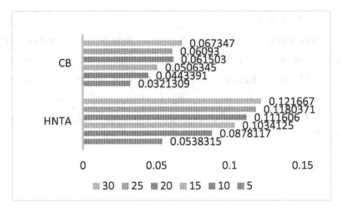

Fig. 7. The result of F1-score of HNTA and CB in different recommended length.

4 Conclusion

In this paper, on the research of paper recommendation methods, we conduct research from the scholar's research interests and the content of the paper, and then an academic paper recommendation algorithm combining heterogeneous network and temporal attributes (HNTA) is proposed. At the beginning, this paper constructs a heterogeneous network composed of different entities to calculate the similarity between papers. Then, the temporal attributes are introduced into the extraction of scholars' research interest in order to calculate the similarity between scholars and papers. Finally, by weighting the above two similarities, the purpose of recommending the papers to scholars is achieved. The data is extracted from the real paper data set in the academic platform: SCHOLAT, and the recommendation algorithm proposed in this paper was used for experimental verification. Precision, recall, and F1-score in this paper recommendation experiment were respectively carried out for the recommendation results. Experiment verifies that the HNTA method has a good recommendation effect. Since the paper contains a large amount of information that could be mined, our next step will be to consider adding trust between scholars and citation relationships between papers to further improve the accuracy of similarity in order to obtain better recommendation results.

References

1. Chun-Hua, T., Peter, B.: Exploring social recommendations with visual diversity-promoting interfaces. ACM Trans. Interact. Intell. Syst. **10**(1), 5:1–5:34 (2020)
2. Venugopal, K., Srikantaiah, K., Nimbhorkar, S.: Web Recommendations Systems, 1st edn. Springer, Heidelberg (2020). https://doi.org/10.1007/978-981-15-2513-1
3. Yiu-Kai, N.: CBRec: a book recommendation system for children using the matrix factorisation and content-based filtering approaches. IJBIDM **16**(2), 129–149 (2020)
4. Dimosthenis, B., Christos, T.: Promoting diversity in content based recommendation using feature weighting and LSH. J. Artif. Intell. Appl. Innov. **16**(1), 452–461 (2020)
5. Braja, G., Vahed, M., Babak, S., Nan, D., et al.: A content-based literature recommendation system for datasets to improve data reusability - A case study on Gene Expression Omnibus (GEO) datasets. J. Biomed. Informat. **104**, 103399 (2020)
6. Kadyanan, I., Dwidasmara, I., Mahendra, I.: A hybrid collaborative filtering recommendation algorithm: integrating content information and matrix factorisation. IJGUC **11**(3), 367–377 (2020)
7. Shunmei, M., Qianmu, L., Jing, Z., et al.: Temporal-aware and sparsity-tolerant hybrid collaborative recommendation method with privacy preservation. Concurr. Comput. Pract. Exp. **32**(2), 5447 (2020)
8. Depeng, D., Chuangxia, C., Wenhui, Y., et al.: A semantic-aware collaborative filtering recommendation method for emergency plans in response to meteorological hazards. Intell. Data Anal. **24**(3), 705–721 (2020)
9. Antonio, G., Maxim, N., Dheevatsa, M., et al.: Mixed dimension embeddings with application to memory-efficient recommendation systems. CoRR abs/1909.11810 (2019)
10. Kaya, B.: Hotel recommendation system by bipartite networks and link prediction. Inf. Sci. **46**(1), 53–63 (2019)
11. Liulan, Z., Jing, L., Weike, P., et al.: Sequence-aware factored mixed similarity model for next-item recommendation. In: BigComp, pp. 181–188 (2020)
12. Hanwen, L., Huaizhen, K., Chao, Y., et al.: Keywords-driven and popularity-aware paper recommendation based on undirected paper citation graph. Complexity **2020**, 2085638:1–2085638:15 (2020)
13. Hashem, P., Parham, M., Shahrokh, E.: TCFACO: trust-aware collaborative filtering method based on ant colony optimization. Expert Syst. Appl. **18**, 152–168 (2019)
14. Manju, G., Abhinaya, P., Hemalatha, M.R., et al.: Cold start problem alleviation in a research paper recommendation system using the random walk approach on a heterogeneous user-paper graph. IJIIT **16**(2), 24–48 (2020)
15. Pan, L., Dai, X., Huang, S., Chen, J.: Academic paper recommendation based on heterogeneous graph. In: Sun, M., Liu, Z., Zhang, M., Liu, Y. (eds.) Chinese Computational Linguistics and Natural Language Processing Based on Naturally Annotated Big Data. LNCS (LNAI), vol. 9427, pp. 381–392. Springer, Cham (2015). https://doi.org/10.1007/978-3-319-25816-4_31
16. Xiangfu, M., Longbing, C., Xiaoyan, Z., et al.: Top-k coupled keyword recommendation for relational keyword queries. Knowl. Inf. Syst. **50**(3), 883–916 (2017). https://doi.org/10.1007/s10115-016-0959-3

17. Rose, C., Kathryn, M., Maxine, E., et al.: Explainable entity-based recommendations with knowledge graphs. CoRR. abs/1707.05254 (2017)
18. Guibing, G., Bowei, C., Xiaoyan, Z., et al.: Leveraging title-abstract attentive semantics for paper recommendation. In: AAAI, pp. 67–74 (2020)
19. Feng, Y., Hangru, W., Xinyue,Z., et al.: Study of listwise learning-to-rank recommendation method based on heterogeneous network analysis in scientific social networks. Comput. Appl. Res. 1–7 (2020)
20. Yongbin, Q., Yujie, S., Xiao, W.: Interest mining method of weibo users based on text clustering and interest attenuation. Comput. Appl. Res. 36(05), 1469–1473 (2019)
21. Guojia, W., Bo, D., Shirui, Pa., et al.: Reinforcement learning based meta-path discovery in large-scale heterogeneous information networks. In: AAAI, pp. 6094–6101 (2020)
22. Wang, J., Xu, W., Yan, W., et al.: Text similarity calculation method based on hybrid model of LDA and TF-IDF. In: Computer Science and Artificial Intelligence, pp. 1–8 (2019)
23. Gonen, S., Roee, A., Irad, B.: A weighted information-gain measure for ordinal classification trees. Expert Syst. Appl. 152, 113375 (2020)
24. Martin, G.: word2vec, node2vec, graph2vec, X2vec: towards a theory of vector embeddings of structured data. In: PODS, pp. 1–16 (2020)
25. Lantian, G., Xiaoyan, C., Fei, H., et al.: Exploiting fine-grained co-authorship for personalized citation recommendation. J. IEEE Access 5, 12714–12725 (2017)
26. Kazunari, S., Min-Yen, K.: Scholarly paper recommendation via user's recent research interests. J. ACM. 29–38 (2010)

An Analysis on User Behaviors in Online Question and Answering Communities

Wanxin Wang, Shiming Li, Yuanxing Rao, Xiaoxue Shen, and Lu Jia(✉)

China Agricultural University, Beijing, China
s20193081366@cau.edu.cn

Abstract. Community-based Question and Answering, or CQA, has been extensively studied nowadays. Studies focus on the characteristics on users help us
understand the usage of the platforms, and pave the way for downstream tasks
such as churn prediction, expert recommendation and spam detection. Different
from previous works, we focus on analyzing the behavioral differences between
different types of users, take the feedback of the community into consideration,
and predict users' future behaviors by their historical behaviors and feedback of
the community. In this paper, we conduct a case study on Zhihu.com. We collect
data of questions, answers, comments and users from Zhihu.com, make a brief
analysis of the usage, divide the users into 3 groups named Takers, Devotees and
Participants and study the differences between each type by KS tests, and conduct
linear regressions to predict users' future behaviors using their historical behaviors and feedback from the community. We find significant differences between
each type of users, and find several positive correlations between users' historical
behaviors, feedback from the community and users' future behaviors. By the analysis of users' behaviors, our research provides a theoretical basis for the follow-up
studies.

Keywords: CQA · Community-based Question and Answering · User behaviors

1 Introduction

Community-based Question and Answering (CQA) platforms have been shed light on
nowadays for their huge potential value. Traditional Question and Answering (QA)
platforms have many shortages, for example: (i) many questions are asked in natural
languages, and they seek for people's personal opinions and advices [1] so that they
can just be asked by humans; (ii) The lack of social functions. Social functions can
contribute to answer perfecting or user loyalty to the platform. On the contrast, CQA
platforms provide a community pattern for users to contact to each other. This way,
they can make use of UGC (User-Generated Content), which is a concept rised up
with the concept of Web2.0, which advocates personalization as its main feature. Users
become publishers and spreaders of information contents, instead of just receivers [2].
Through participating in questioning, answering, commenting as well as voting, users
can solve their own problems with the help of the others, share their own knowledge and
experience, and further gain a sense of satisfaction and identity.

© Springer Nature Singapore Pte Ltd. 2021
Y. Sun et al. (Eds.): ChineseCSCW 2020, CCIS 1330, pp. 469–483, 2021.
https://doi.org/10.1007/978-981-16-2540-4_34

CQA platforms can be divided into 2 types: single-domain platforms, which focus on only one specific domain, such as Stack Overflow [3] and Stack Cooking [4]; and open-domain platforms, which focus on more than one domain on the same time, such as Quora [5], Yahoo!Answers [6] and Zhihu.com [7]. In this paper, we choose Zhihu.com, an open-domain CQA platform that is well-received in China, as our research object. Users ask all kinds of questions, ranging from daily life troubles to car manufacture. Instead of questions themselves, Zhihu.com focuses more on the answers, so that everyone's unique life experience can be displayed in public. Through observations, we found that users behave in different ways on the platform.

The predictions and motivations of the users' behaviors are studied in different ways [4, 8–12]. As innovation points, we consider the combination of users' historical behaviors and feedback on themselves from the other users as the input variable of the prediction.

To do a detailed research on users' behaviors, in this paper, we conduct a case study on Zhihu.com. We collect information from the platform within 35 days, which is consisted by more than 900 thousand questions, two million answers, four million comments and 1 million users. We will mainly focus on the questions below:

– What are the major types of users according to their behaviors? What are the differences between each type?
– What are the relationships between users' historical behaviors, feedback from the platform and users' future behaviors?

For the first question, we divide the users into 3 groups according to their asking and discussing behaviors, and conduct two-tailed Kolmogorov-Smirnov (KS) tests to test the differences between them from several aspects. For the second question, we extract several features from users' historical behaviors, following behaviors and feedback on their behaviors from others, and conduct linear regressions to predict users' future behaviors. Our analysis made several discoveries:

– Firstly, through dividing the users into 3 groups as takers, devotees, and participants, we observed significant differences between them from several aspects. In general, Participants perform the best in the three, whether in behavior aspects or feedback from others.
– Secondly, through linear regressions we found that, in the prediction of the number of questions, the number of historical questions and the number of answers obtained by historical questions are positively correlated to it; in the prediction of the number of answers, the number of users' historical answers, the number of users' followers and followees are positively correlated to it; on the prediction of the number of comments, the number of users' historical answers, the number of reviews, the number of followers and followees are positively correlated to them, and the number of comments received by users' questions had a positive impact.

The rest of this article is organized as follows. In Sect. 2, we give a brief review of the related works. In Sect. 3, we give a brief introduction of Zhihu.com and our dataset. In Sect. 4, we divide the users into 3 groups and study the differences between them. In

Sect. 5, we study the influence factors of users' future behaviors. Both in Sects. 4 and 5, we suggest the hypotheses first and verify them by experiments. Finally, we conclude our work in Sect. 6.

2 Related Works

CQA platforms are extensively studied before. Srba et al. [13] made a comprehensive research on studies of CQA platforms. As for information of users' behaviors, they are used onto several research fields:

Expert Finding. Wang [1] made a comprehensive survey on expert finding in CQA platforms. For example, prediction of the activity level of the expert users [12] and depiction of users' expertise [14], two typical research interests of expert finding, are studied by making use of users' behaviors. Question recommendation [4], which is also included in this domain, can also be studied through users' behaviors. Expert finding contributes to the effectiveness of knowledge sharing through recommending experts to questions or otherwise.

Spam or Abusive User Detection. Spam detection is early conducted on e-mails, but there are a certain number of spam or abusive users observed on CQA platforms. Kayes et al. [15] applied machine learning methods to the abusive user detection on Yahoo!Answers. Le et al. [16] found struggling students through detecting their behaviors to break the vicious cycle.

Churn Prediction. Dror [17] predicted the churn of new-comers by users' profile, rate of activity and interactions with others. They found that two important signals for churns are the amount of answers users provided, which is closely related to the motivations, and the recognition got from the community, which is measured in counts of best answers, amounts of thump-ups, and positive feedback from the askers. Inspired by it, we consider the feedback from the community as a feature used to predict the future behaviors of the users.

Best Answer Finding. Through analyzing users' voting behaviors [18] and users' profile [19], the quality of answers can be evaluated and, to go a step further, the best answer can be found or predicted.

Researches also focused on the motivations of users' behaviors. Instead of analyzing directly on the users' behaviors, different factors are concerned, including their national culture [9] and privacy concerns [11].

The researches mentioned above use user behaviors into different research interests, or detect the motivations of user behaviors. In this paper, we consider the combination of users' historical behaviors and feedback on them from the other users as the input variable of the prediction of users' future behaviors.

3   Dataset Introduction

We choose Zhihu.com as our objective for the case study, which is a typical open-domain CQA platform well-received in China. On the contrary of strategy of CQA platforms which select the best answer for each question, there's no best answer for any question, but there are always worthy answers, such as useful experience of experienced professionals, practical skills of specified domains, and list of books or articles shared for newcomers, etc. Every answer is regarded seriously as some kind of writings, and is protected by strict intellectual property protection measures. Based on such reasons, instead of short quires, users of Zhihu tend to create long articles as answers. A typical answer and its comments on Zhihu.com are shown in Fig. 1.

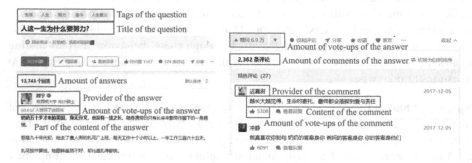

Fig. 1. An answer and its comments on Zhihu.com

We collect the questions asked from 2019/03/30 00:00:00 to 2019/05/03 23:59:59, 35 days total, along with their answers and comments provided among the same time interval, with a clawer. We collected a total of 912,287 questions, including 188,672 anonymous questions, 2,788,430 answers, including 250,892 anonymous answers, 4,533,118 comments of the answers, including 141,073 anonymous comments, and 1,987,372 users related to them, who have a normal status of login.

The serious polarization can be seen through the basic analysis. The CDF of answers to questions is shown in Fig. 2.

Among 513,116 questions which are answered, 36.82% of them has only 1 answer, while 94.62% of them have ten or less. But at the same time, some questions have more than 10,000 answers in total. The same circumstance occurs on the comments of answers.

To make a further analysis, we find that among all the answers, 21.82% of them are provided the same day as the questions are asked, while 83.17% of them take 10 days or less.

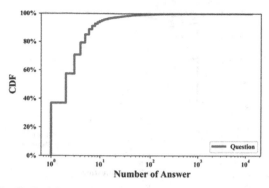

Fig. 2. CDF of the number of answers each question has collected

4   User Classification

We divide the users into 3 groups according to their behaviors.

– *Takers.* Users who only asked questions within the time interval of our experiment, or in other words, within the 35 days. They did not involve themselves in the discussions. There are 343,872 takers in total.
– *Devotees.* Users who only took part in the discussions, including answering questions and commenting, within the 35 days. They did not ask any questions. There are 1,433,381 devotees in total.
– *Participants.* Users who got involved in all the activities, including asking, answering and commenting, within the 35 days. In other words, within the time interval, participants asked at least 1 question and discussed with others for 1 time at least. There are 210,119 participants in total.

To detect the differences between the 3 kinds of users, first we make the several hypotheses.

– *Hypothesis 1.* Takers and Participants behave differently in question asking.
– *Hypothesis 2.* Devotees and Participants behave differently in discussing, including answering and commenting.
– *Hypothesis 3.* Takers, Devotees and Participants behave differently in following users and being followed by users.

To test the hypotheses, we subdivided the asking and discussing aspects from users' historical behaviors and feedback of the community, respectively. While the former includes the amount of questions, answers and comments user provided, the latter includes the answers for the users' questions, the comments for the users' answers, and the vote-ups gained from answers and comments.

For the asking aspect, CDFs are shown in the figures below (Figs. 3, 4 and 5).

An overwhelming majority of the takers and the participants asked 10 or less questions. Overall, the participants performed better than the takers in asking questions, and

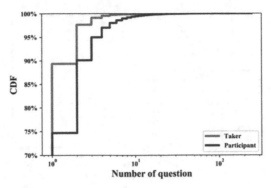

Fig. 3. CDFs of the number of questions raised by takers and by participants, respectively.

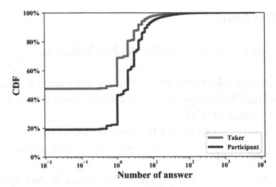

Fig. 4. CDFs of the average number of answers received by takers and by participants, respectively.

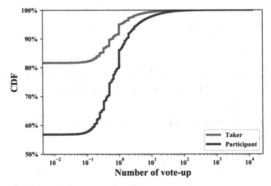

Fig. 5. CDFs of average vote-ups gained from answers received by takers and participants, respectively.

the degree of polarization was greater. As for the feedback of the community, most of the takers and participants gained at an average of 10 or less answers for their questions. Generally speaking, the participants performed better. In terms of the quality of the answers, indicated by the vote-ups they gained, the participants performed better as well.

To make a quantitative test, we conduct two-tailed Kolmogorov-Smirnov tests, and the results are shown below (Table 1).

Table 1. Results of KS tests of the takers and participants

Variable		P value
Questioning	Amount of questions	***
Feedbacks from the community	Amount of questions that got answers	***
	Amount of questions that got no answers	***
	Average amount of answers for questions	***
	Average amount of vote-ups got from answers for questions	***

[1]*: $P < 0.1$; **: $P < 0.01$; ***: $P < 0.001$

The results indicate the significant differences between the takers and the participants, both on behaviors of asking and feedback gained from the community. In that case, the Hypothesis 1 is supported.

Similarly, the performance on discussing aspects of the devotees and the participants are shown below (Table 2).

Table 2. Result of KS tests of the devotees and participants

Variable		P value
Answering	Amount of answers	***
Commenting	Amount of comments	***
Feedbacks from the community	Amount of answers which got comments	***
	Amount of answers which got no comments	***
	Average amount of comments for answers	***
	Average amount of vote-ups got from comments for answers	***
	Average amount of vote-ups for answers	***
	Average amount of vote-ups for comments	***

[1]*: $P < 0.1$; **: $P < 0.01$; ***: $P < 0.001$

The results indicate the significant differences between the devotees and the participants. In that case, the Hypothesis 2 is supported.

In terms of the aspect of following, which is indicated by the amount of followees and followers, the performances of the takers, devotees and participants are shown below (Table 3).

Table 3. Results of KS tests between 3 kinds of users

		P value
Number of followees	Takers & Devotees	***
	Takers & Participants	***
	Devotees & Participants	***
Number of followers	Takers & Devotees	***
	Takers & Participants	***
	Devotees & Participants	***

[1]*: $P < 0.1$; **: $P < 0.01$; ***: $P < 0.001$

The significant differences between 3 kinds of users are indicated from the results, so that the Hypothesis 3 is supported.

To make a brief summary of this section, we divide the users into 3 groups, and prove that there are significant differences between them from the aspects of questioning, discussing and following, including their historical behaviors and the feedback from the community.

5 User Behaviors Prediction

To make a better understanding of users' behaviors and the relationships between them, we conduct linear regressions to predict the future behaviors of users. Before start, we also need to develop several hypotheses.

- *Hypothesis 4*. The users' historical behaviors, including questioning and discussing, are positively correlated with their future behaviors.
- *Hypothesis 5*. The feedback on users' historical behaviors are positively correlated with their future behaviors.
- *Hypothesis 6*. The following behaviors are positively correlated with users' future behaviors.

For Hypothesis 5 and 6, we subdivide them into several detailed hypotheses.

- *Hypothesis 5.1.* The amount of answers gained by questions is positively correlated with users' future questioning behaviors.
- *Hypothesis 5.2.* The quality of answers gained by questions, indicated by the number of vote-ups of the answers, is positively correlated with users' future questioning behaviors.
- *Hypothesis 5.3.* The admiration of the answers that a user provided, indicated by the number of vote-ups of the answers, is positively correlated with the users' future answering behaviors.
- *Hypothesis 5.4.* The amount of comments gained by answers is positively correlated with users' future answering behaviors.
- *Hypothesis 5.5.* The quality of comments gained by answers, indicated by the number of vote-ups of the comments, is positively correlated with users' future answering behaviors.
- *Hypothesis 5.6.* The admiration of the comments that a user provided, indicated by the number of vote-ups of the comments, is positively correlated with the users' future commenting behaviors.
- *Hypothesis 5.7.* The amount of comments gained by answers is positively correlated with users' future commenting behaviors.
- *Hypothesis 5.8.* The quality of comments gained by answers, indicated by the number of vote-ups of the comments, is positively correlated with users' future commenting behaviors.
- *Hypothesis 5.9.* The admiration of the answers that a user provided, indicated by the number of vote-ups of the answers, is positively correlated with the users' future commenting behaviors.
- *Hypothesis 6.1.* The number of followees is positively correlated with users' future questioning behaviors.
- *Hypothesis 6.2.* The number of followees is positively correlated with users' future answering behaviors.
- *Hypothesis 6.3.* The number of followees is positively correlated with users' future commenting behaviors.
- *Hypothesis 6.4.* The number of followers is positively correlated with users' future questioning behaviors.
- *Hypothesis 6.5.* The number of followers is positively correlated with users' future answering behaviors.
- *Hypothesis 6.6.* The number of followers is positively correlated with users' future commenting behaviors (Tables 4, 5 and 6).

We conduct linear regressions on users' amount of questions, answers and comments on the 5th week, respectively. We use information among 4 weeks, including users' historical behaviors, feedback from the platform and information of followers and followees, as the independent variables. We randomly extracted 20% data as testing set, and the rest are treated as the training set. As for result, our model for questioning prediction gains $R^2 = 0.504$ and RMSE $= 0.689$; our model for answering prediction gains $R^2 = 0.511$ and RMSE $= 0.697$; our model for commenting prediction gains $R^2 = 265$ and RMSE $= 0.856$. All the VIF of variables are lower than 3, indicating that there's no serious multi-collinearity between them (Tables 7, 8 and 9).

Table 4. Statistical results of the variables of user historical behaviors

Variable		Value
Amount of questions that got answers (P1)	Min	0
	Mean	0.54
	Max	190
Amount of questions that got no answer (P2)	Min	0
	Mean	0.21
	Max	76
Amount of answers that got comments (P3)	Min	0
	Mean	2.72
	Max	183
Amount of answers that got no comment (P4)	Min	0
	Mean	7.43
	Max	240
Amount of comments (P5)	Min	0
	Mean	15.91
	Max	369

Table 5. Statistical results of the variables of feedback of the platform

Variable		Value
Average of answers got from questions (P6)	Min	0.00
	Mean	1.73
	Max	6,545.50
Average of comments got from answers (P7)	Min	0.00
	Mean	3.28
	Max	2,797.50
Average of vote-ups of answers (P8)	Min	0.00
	Mean	33.48
	Max	55,428.00
Average of vote-ups of answers got from questions (P9)	Min	0.00
	Mean	1.09
	Max	4,570.00
Average of vote-ups for comments (P10)	Min	0.00
	Mean	6.94
	Max	1,412.00
Average of vote-ups for comments got from answers (P11)	Min	0.00
	Mean	1.54
	Max	2,120.50

Table 6. Statistical results of the variables of followers and followees

Variable		Value
Followees (P12)	Min	0
	Mean	105.14
	Max	14,408
Followers (P13)	Min	0
	Mean	1721.40
	Max	372,668

Table 7. Result of linear regression on users' questioning behaviors

Variable	Parameter	Std. error	P value
P1	0.5280	0.004	<0.001
P2	0.2893	0.004	<0.001
P3	0.0152	0.005	<0.01
P4	0.0052	0.005	0.251
P5	−0.0069	0.004	0.077
P6	−0.0068	0.004	0.057
P7	−0.0052	0.005	0.305
P8	0.0012	0.005	0.804
P9	−0.0011	0.004	0.754
P10	−0.0058	0.004	0.106
P11	−0.0033	0.004	0.375
P12	0.0035	0.004	0.332
P13	−0.0016	0.004	0.672

Users' historical amount of questions, answers and comments are positively correlated with their future questioning, answering and commenting behaviors respectively, thus, the Hypothesis 4 is supported. Also, the amount of answers that got comments is positively related to uses' future questioning behaviors, and the amount of answers that got comments is positively related to users' future commenting behaviors. The amount of vote-ups of answers is positively related to users' future commenting behaviors, thus the Hypothesis 5.9 is supported. The amount of followees and followers are positively correlated to users' future answering behaviors, thus the Hypothesis 6.2 and 6.5 are supported. The amount of the vote-ups of answers, the amount of followees and followers are positively correlated with users' future commenting behaviors, thus the Hypothesis 6.3 and 6.6 are supported.

Table 8. Result of linear regression on users' answering behaviors

Variable	Parameter	Std. error	P value
P1	−0.0061	0.004	0.154
P2	−8.574e−06	0.004	0.998
P3	0.2672	0.005	<0.001
P4	0.5284	0.005	<0.001
P5	−0.0084	0.004	0.032
P6	−0.0042	0.004	0.242
P7	−0.0065	0.005	0.209
P8	−0.0014	0.005	0.789
P9	−0.0025	0.004	0.496
P10	0.0020	0.004	0.582
P11	0.0060	0.004	0.107
P12	0.0133	0.004	<0.001
P13	0.0180	0.004	<0.001

Table 9. Result of linear regression on users' commenting behaviors

Variable	Parameter	Std. error	P value
P1	−0.0024	0.005	0.625
P2	−0.0066	0.005	0.170
P3	0.0283	0.006	<0.001
P4	−0.0234	0.005	<0.001
P5	0.5801	0.005	<0.001
P6	−0.0031	0.005	0.496
P7	−0.0383	0.005	<0.001
P8	0.0368	0.005	<0.001
P9	0.0012	0.004	0.794
P10	0.0056	0.004	0.203
P11	0.0033	0.005	0.495
P12	0.0235	0.005	<0.001
P13	0.0272	0.005	<0.001

On the contrast, other hypotheses are not supported.

To make a brief conclusion of this section, we conduct linear regressions to predict users' future behaviors using the data of their historical behaviors, feedback from the community and their following behaviors, and find out several positive correlations between them.

6 Conclusion and Future Work

In this paper, we conduct a case study on Zhihu.com, an open-domain CQA platform. We give a brief introduction of the usage of the platform, divide the users into 3 groups and detect the differences between them, and make predictions of their future behaviors. We develop several hypotheses and the verification results are shown below (Table 10).

Table 10. Verification results on the hypotheses

Hypothesis	Result	Hypothesis	Result
H1	Supported	H5.7	Unsupported
H2	Supported	H5.8	Unsupported
H3	Supported	H5.9	Supported
H4	Supported	H6.1	Unsupported
H5.1	Unsupported	H6.2	Supported
H5.2	Unsupported	H6.3	Supported
H5.3	Unsupported	H6.4	Unsupported
H5.4	Unsupported	H6.5	Supported
H5.5	Unsupported	H6.6	Supported
H5.6	Unsupported		

We find it interesting that the feedback from the community doesn't have a wildly positive relevance on users' future answering behaviors, which is not in line with our expectations. For further studies, we will try to do more in-depth studies on them.

References

1. Xianzhi, W., Chaoran, H., Lina, Y., et al.: A survey on expert recommendation in community question answering. J. Comput. Sci. Technol. **33**(4), 625–653 (2018)
2. Lin, X., Li, L., Zhai, Y., Qi, J.: UGC quality evaluation based on user communities and contents. In: 2014 4th International Conference on Wireless Communications, Vehicular Technology, Information Theory and Aerospace & Electronic Systems (VITAE), pp. 1–5, Aalborg (2014). https://doi.org/10.1109/VITAE.2014.6934410
3. Reyyan, Y., Jamie, C.: Analyzing bias in CQA-based expert finding test sets. In: 37th international ACM SIGIR Conference on Research & Development in Information Retrieval (SIGIR 2014), pp. 967–970. Association for Computing Machinery, New York (2014). https://doi.org/10.1145/2600428.2609486

4. Grégoire, B., Paul, M., Yulan, H., Harith, A.: Predicting answering behaviour in online question answering communities. In: 26th ACM Conference on Hypertext & Social Media (HT 2015), pp. 201–210. Association for Computing Machinery, New York (2015). https://doi.org/10.1145/2700171.2791041

5. Haocheng, W., Wei, W., Ming, Z., Enhong, C., Lei, D., Heung-Yeung, S.: Improving search relevance for short queries in community question answering. In: 7th ACM International Conference on Web Search and Data Mining (WSDM 2014), pp. 43–52. Association for Computing Machinery, New York (2014). https://doi.org/10.1145/2556195.2556239

6. Yandong, L., Jiang, B., Eugene, A.: Predicting information seeker satisfaction in community question answering. In: 31st Annual International ACM SIGIR Conference on Research and Development in Information Retrieval (SIGIR 2008), pp. 483–490. Association for Computing Machinery, New York (2008). https://doi.org/10.1145/1390334.1390417

7. Xin, L., Yiqun, L., Rongjie, C., Shaoping, M.: Investigation of user search behavior while facing heterogeneous search services. In Proceedings of the Tenth ACM International Conference on Web Search and Data Mining (WSDM 2017), pp. 161–170. Association for Computing Machinery, New York (2017). https://doi.org/10.1145/3018661.3018673

8. Kang, J., Yu, Z., Liang, Y., Xie, J., Guo, B.: Characterizing collective knowledge sharing behaviors in social network. In: 2019 IEEE SmartWorld, Ubiquitous Intelligence & Computing, Advanced & Trusted Computing, Scalable Computing & Communications, Cloud & Big Data Computing, Internet of People and Smart City Innovation (SmartWorld/SCALCOM/UIC/ATC/CBDCom/IOP/SCI), pp. 869–876, Leicester, United Kingdom (2019). https://doi.org/10.1109/SmartWorld-UIC-ATC-SCALCOM-IOP-SCI.2019.00178

9. Imrul, K., Nicolas, K., Daniele, Q., Adriana, I., Francesco, B.: Cultures in community question answering. In: 26th ACM Conference on Hypertext & Social Media (HT 2015), pp. 175–184. Association for Computing Machinery, New York (2015). https://doi.org/10.1145/2700171.2791034

10. Sun, Y., Guo, B., Li, Z., Cheng, J., Wang, L., Yu, Z.: Leveraging user profiling in click-through rate prediction based on Zhihu data. In: 2019 2nd China Symposium on Cognitive Computing and Hybrid Intelligence (CCHI), pp. 131–136, Xi'an, China (2019)

11. Imrul, K., Nicolas, K., Francesco, B., Adriana, I.: Privacy concerns vs. user behavior in community question answering. In Proceedings of the 2015 IEEE/ACM International Conference on Advances in Social Networks Analysis and Mining 2015 (ASONAM 2015), pp. 681–688. Association for Computing Machinery, New York (2015). https://doi.org/10.1145/2808797.2809422

12. Liu, Z., Xia, Y., Liu, Q., He, Q., Zhang, C., Zimmermann, R.: Toward personalized activity level prediction in community question answering websites. ACM Trans. Multimedia Comput. Commun. Appl. 14(2s), 41:1–41:15 (2018). https://doi.org/10.1145/3187011

13. Ivan, S., Maria, B.: A comprehensive survey and classification of approaches for community question answering. ACM Trans. Web 10(3), 63 (2016). https://doi.org/10.1145/2934687. Article id 18

14. Baoguo, Y., Suresh, M.: Exploring user expertise and descriptive ability in community question answering. In: Proceedings of the 2014 IEEE/ACM International Conference on Advances in Social Networks Analysis and Mining (ASONAM 2014), pp. 320–327. IEEE Press, Beijing (2014). https://doi.org/10.1109/ASONAM.2014.6921604

15. Imrul, K., Nicolas, K., Daniele, Q., Adriana, I., Francesco, B.: The social world of content abusers in community question answering. In: 24th International Conference on World Wide Web (WWW 2015). International World Wide Web Conferences Steering Committee, pp. 570–580. Republic and Canton of Geneva, CHE (2015). https://doi.org/10.1145/2736277.2741674

16. Long, T.L., Chirag, S., Erik, C.: Bad users or bad content? Breaking the vicious cycle by finding struggling students in community question-answering. In: 2017 Conference on Conference Human Information Interaction and Retrieval (CHIIR 2017), pp. 165–174. Association for Computing Machinery, New York (2017). https://doi.org/10.1145/3020165.3020181
17. Gideon, D., Dan, P., Oleg, R., Idan, S: Churn prediction in new users of Yahoo! answers. In: 21st International Conference on World Wide Web (WWW 2012 Companion), pp. 829–834. Association for Computing Machinery, New York (2012). https://doi.org/10.1145/2187980.2188207
18. Chong, L., Eduarda, R., Gabriella, K., Nata, M., Aleksandar, I.: Model for voter scoring and best answer selection in community Q&A services. In: 2009 IEEE/WIC/ACM International Joint Conference on Web Intelligence and Intelligent Agent Technology - Volume 01 (WI-IAT 2009), pp. 116–123. IEEE Computer Society, USA (2009). https://doi.org/10.1109/WI-IAT.2009.23
19. Zhimin, Z., Man, L., Zhengyu, N., Yue, L.: Exploiting user profile information for answer ranking in cQA. In: 21st International Conference on World Wide Web (WWW 2012 Companion), pp. 767–774. Association for Computing Machinery, New York (2012). https://doi.org/10.1145/2187980.2188199

An Overlapping Local Community Detection Algorithm Based on Node Transitivity and Modularity Density

Xintong Huang[1,2] , Ling Wu[1,2] , and Kun Guo[1,2,3]($\boxtimes$)

[1] College of Mathematics and Computer Sciences, Fuzhou University, Fuzhou 350116, China
{wuling1985,gukn}@fzu.edu.cn
[2] Fujian Provincial Key Laboratory of Network Computing and Intelligent Information Processing, Fuzhou, China
[3] Key Laboratory of Spatial Data Mining and Information Sharing, Ministry of Education, Fuzhou 350116, China

Abstract. The local community detection (LCD) method can discover the local community structure in which the seed node is located. Compared with global community detection, local community detection is characterized by its low cost and high efficiency. However, most existing LCD methods only return a non-overlapping community. Individuals in the real world may participate in multiple communities, which can only be discovered by using overlapping local community detection methods. In this study, an overlapping local community detection algorithm based on modularity and node transitivity. First, the scope and structure information of the overlapping communities are obtained according to the node transitivity. Second, NMF is used to obtain the number of overlapping communities. Finally, the local modularity density based on edge weights is used to refine the detected local communities. The experimental results validate the high performance of our method to the other method in comparison.

Keywords: Complex network · Local community detection · Node transitivity · Local modularity

1 Introduction

Early community detection algorithms focused on mining the entire community structure of a network, which is called global community detection [1]. However, at some point, people only need to obtain a community structure where some nodes are located. Therefore, local community detection came into being. Since Clauset proposed the concept of local community detection [2], a series of excellent correlation algorithms have been proposed successively [3, 4]. These algorithms aim to find a single local community. But in reality, individuals in the real world usually belong to multiple groups. Therefore, overlapping local community detection has more practical significance.

For a given seed node, overlapping local community detection is dedicated to discovering overlapping communities where it is located. Most overlapping local community

© Springer Nature Singapore Pte Ltd. 2021
Y. Sun et al. (Eds.): ChineseCSCW 2020, CCIS 1330, pp. 484–498, 2021.
https://doi.org/10.1007/978-981-16-2540-4_35

detection algorithms obtain overlapping communities by extracting nodes or node sets similar to the seed nodes. However, there are deficiencies in subgraph extraction and the accuracy of overlapping local communities mining. In this study, an Overlapping Local Community Detection algorithm based on node Transitivity and Modularity density (OLCDTM) is proposed. The main contributions of OLCDTM include:

(1) We proposed node transitivity to find the scope and structure information of the overlapping communities which ensures that the correct number of overlapping communities can be found.
(2) A local modularity density based on edge weight is proposed to expand the overlapping local communities by considering the internal structure information and the weight of edges, which can precisely excavate the real structure of overlapping communities.
(3) Experimental comparison is made on real and artificial datasets. The results show the efficiency of the OLCDTM algorithm.

2   Related Work

2.1 Local Community Detection

There are many types of LCD algorithms. A classic LCD algorithm is based on local modularity. In 2005, Clauset et al. [2] first designed the local modularity R and the corresponding local community detection algorithm. In 2008, Luo et al. [4] designed another local modularity, namely M, and corresponding algorithm LWP. In 2018, Luo et al. [5] designed two algorithms based on R and M. In 2019, Meng ct al. [6] designed the FuzLhocd methods, FuzLhocd is based on higher-order information.

Another local community detection algorithms mainly detect the central node. Chen et al. [3] designed the classical method LMD, which is utilized by the centrality of the node. In 2020, Luo et al. [7] designed the LCDNN method, which is by using the concept of nearest nodes with greater centrality (NGC). Jian et al. [8] proposed CLOSE by using the local structure of networks.

He et al. [9] designed an algorithm by using spectral clustering, which divides communities by computing the Krylov subspace. Li [10] designed an LCD method which can be used for multi-layer networks.

2.2 Local Overlapping Community Detection

Up to now, the existing overlapping local community detection algorithms are relatively few. In 2017, Hollocou et al. [11] proposed an overlapping local community detection algorithm MULTICOM. MULTICOM embeds areas near seed nodes into low-dimensional vector Spaces to find overlapping communities. However, MULTICOM requires multiple parameters but also has low time efficiency and accuracy.

In 2019, Ni et al. [12] proposed a framework LOCD and implemented six different versions of traditional non-overlapping methods based on this framework. However, the accuracy of LOCD depends largely on the local community detection method used.

In 2018, Kamuhanda [13] designed the MLC method by using non-negative matrix factorization. MLC uses BFS mining subgraphs, the number of layers in BFS limits the subgraphs from obtaining complete information about the overlapping communities, which reduces the accuracy of community count. Meanwhile, MLC uses NMF to mine the community, it cannot detect the complete structure of the community. Based on the above, our method firstly naturally obtains the maximum range of subgraph by using the node transitivity. Then, we use the local modularity density based on edge weight to detect the structure and optimization of the overlapping communities.

3   The OLCDTM Algorithm

3.1   Basic Concepts

Given a complex network $G = (V, E)$, where V is the nodes of G, E is the edges of G. A local community structure C is given by Fig. 1. In Fig. 1, C can be specifically divided into a core node that has no contact with the outside and a boundary node that has contact with the outside. For the rest of the network, the details are unknown. The basic concepts used by OLCDTM are described below.

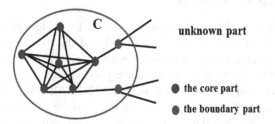

Fig. 1. Local community structure

Definition 1 (neighbors of node). The direct neighbors of node v, $\Gamma(v)$, are defined as follows:

$$\tau(v) = \{u \in V | (u, v) \in E\} \tag{1}$$

Correspondingly, $\Gamma(C)$, are defined as follows:

$$\tau(C) = \cup_{v \in C} \tau(v) - C \tag{2}$$

Definition 2 (transitivity of node). The transitivity of two nodes, $T(u, v)$, are defined by:

$$T(u, v) = \frac{\sum_{z \in (\tau(u) \cap \tau(v))} \Delta z}{\Delta u + \Delta v + 1} \tag{3}$$

where Δu refers to the number of triangles with node u as a vertex, and the side length of the triangle is 1.

Definition 3 (nonnegative matrix factorization). NMF transforms a matrix $X \in R^{m*n}$ into two non-negative matrices $V \in R^{m*c}$ and $H \in R^{c*n}$, such that $X \cong VH$. This proved to be NP-hard. Therefore, a minimum error formula based on the Frobenius norm is used, which is defined as follows:

$$V, H_{\geq 0}^{min} X - VH_F \tag{4}$$

where the Frobenius norm for the matrix X is defined as follows:

$$X_F = \sqrt[2]{\sum_{i=1}^{m} \sum_{j=1}^{n} |x_{ij}|^2} \tag{5}$$

Definition 4 (local modularity density based on edge weight). The local modularity density based on edge weight of C, $QW(C)$, is defined by:

$$QW(C) = \frac{d_{in}(C)}{d(C)} \cdot wp(C) - \left(\frac{d_{out}(C)}{d(C)} \right) \tag{6}$$

where $d_{in}(C)$ is the total degrees of each node in C. $d_{out}(C)$ is the number of connections between C and the rest of the network. $d(C)$ is the total degrees of C. $wp(C)$ is a density parameter based on edge weight:

$$wp(C) = \frac{\sum_{(u,v) \in E_C} \frac{d(u) \cdot d(v)}{2|E_C|}}{V_C \cdot (V_C - 1)} \tag{7}$$

The $wp(C)$ take into account the overall density of the community and avoids the problem of resolution limit. So $wp(C)$ replaces the number of edges with the probability that each edge is connected.

Definition 5 (node penalty factor). The penalty factor for v to be added to C, $r(C, v)$, is defined by:

$$r(C, v) = \frac{2 \cdot \sum_{i,j \in (\tau(v) \cap C)} \varepsilon_{ij}}{|\tau(v) \cap C| \cdot (|\tau(v) \cap C| - 1)} \tag{8}$$

where $\varepsilon_{ij} = 1$ if there exists an edge between i and j, otherwise $\varepsilon_{ij} = 0$. The penalty factor reflects the importance of node v to be added.

Definition 6 (incremental local modularity density based on edge weight). The local modularity density based on edge weight increment can be expressed as:

$$\Delta QW = QW(C') - QW(C) + r(v) \tag{9}$$

where C' is the set after node v is added to C.

Definition 7 (Jaccard coefficient). The Jaccard coefficient of two nodes, $J(u, v)$, is defined by:

$$J(u, v) = \frac{|\tau(u) \cap \tau(v)|}{|\tau(u) \cup \tau(v)|} \tag{10}$$

Definition 8 (community tightness). The tightness of community C, $\rho(C)$, is defined by:

$$\rho(C) = \frac{\sum_{i,j \in C} J(i,j) \cdot \alpha_{ij}}{\sum_{i \in B, j \in (\tau(i) - C)} J(i,j) \cdot \alpha_{ij}} \tag{11}$$

where $\alpha_{ij} = 1$ if there exists an edge between i and j, otherwise $\alpha_{ij} = 0$. B represents the boundary part of C.

Definition 9 (similarity between nodes and communities). The sim (v, C) is defined by:

$$sim(v, C) = \frac{|\tau(v) \cap \tau(C)|}{min(|\tau(v)|, |\tau(C)|)} \tag{12}$$

3.2 Algorithm Description

Framework of the Algorithm. The proposed OLCDTM algorithm based on node transitivity and modularity density is mainly divided into three stages: (1) finding core subgraph. (2) determining community number. (3) detecting overlapping local communities. The framework of OLCDTM is shown in Algorithm 1.

Algorithm 1 OLCDTM
___
Input: $G=(V, E)$, the seed node s.

Output: $C_1, C_2, ..., C_k$.
1 : C_{sub} =FindCoreSubgraph (G,s);
2 : k, ori= DetermineCommunityNumber (C_{sub});
3 :   **For each** k **Do**
4 : C_k =DetectOverlappingLocalCommunity(G, ori, k);
5 :   **End for**
6 :   return $C_1, C_2, ..., C_k$.

Finding Core Subgraph. First, the transitivity of the seed node with each of its neighbors is calculated. Then, OLCDTM selects the node with non-zero and the largest transitivity value in the neighbors as the transfer object. Second, OLCDTM marks the seed node and the selected transfer object, and add the transfer object to the core subgraph. Third, OLCDTM starts a new search from the transfer object. The function repeats the above steps until the transfer object and all its unmarked neighbors have a transitive value of 0. Finally, all nodes in the core subgraph and their neighbors are returned as the final core subgraph. The details of the finding core subgraph stage are given in Function 1.

Function 1 FindCoreSubgraph

Input: $G=(V, E)$, seed node s.
Output: Core subgraph C_{sub}.
1 : $N_{trans}=[s]$, $N_{labled}=\{s\}$, $i=0$, $C_{sub} = \varnothing$;
2 :   **While** True **Do**
3 : $v_{transnode} = N_{trans}[i]$;
4 : $V_c = \varnothing$,
5 : $V_{unlabled} = \varnothing$;
6 : $V_{unlabled} = \tau(v_{transnode}) - N_{labled}$;
7 :   **For each** node v in $V_{unlabled}$ **Do**
8 :   calculate $T(v_{transnode}, v)$ according to equation (3);
9 :   **If** $T(v_{transnode}, v) > 0$ **then**
10 : $V_c = V_c \cup \{v\}$;
11 :   **End if**
12 :   **End for**
13 :   **If** $V_c = \varnothing$ **Then**
14 :   break;
15 :   **End if**
16 : $V_{max}=\{v \mid v \in V_c \ \& \ max(T)= T(v_{transnode}, v)\}$;
17 :   **If** $len(V_{max}) > 1$ **Then**
18 :   **For each** v in V_{max} **Do**
19 :   calculate $d(v)$; //calculate the degree of node v
20 :   **End for**
21 : $V_{maxdeg}=\{v \mid v \in V_{max} \ \& \ max(d)= d(v)\}$;
22 : $V_{max} = V_{maxdeg}$;
23 :   **End if**
24 : N_{trans} .append(V_{max});
25 : $N_{labled} = N_{labled} \cup \{ V_{max} \}$;
26 : $i += 1$;
27 : **End while**
28 : $C_{sub} = C_{sub} \cup N_{trans}$;
29 : **For each** v in N_{trans} **Do**
30 : $C_{sub} = C_{sub} \cup \tau(v)$;
31 : **End for**
32 : return C_{sub} ;

Determining Community Number. The application of NMF in community detection algorithms usually exists in global community detection [0-0]. In 2018, Kamuhanda et al. first applied NMF to local community detection [0]. The algorithm decomposed the network's adjacency matrix A to the matrix $V \in R^{m*k}$ and $H \in R^{k*n}$ through NMF. The algorithm starts NMF from $k = 1$ until k reaches the maximum value n. NMF stops running when it finds an optimal value for k. The appropriate k value refers to that when $k = k + 1$, the matrix H output by NMF exists a whole row with an element value of 0 after post-processing.

Elements in H need to be normalized in each iteration:

$$Normalize(h_{ij}) = \frac{h_{ij}}{\sum_{x=1}^{k} h_{xj}} \tag{13}$$

After normalization, Eq. (14) is used to keep the centers of nodes.

$$Subtract(h_{ij}) = \begin{cases} 1 \ h_{ij} = 1 \\ 0 \ h_{ij} < 1 \end{cases} \tag{14}$$

The details of the determining community number stage are given in Function 2.

Function 2 DetermineCommunityNumber

Input: Subgraph C_{sub}, A: the adjacency matrix of C_{sub}.

Output: The number of overlapping community k, original members of each community ori.

```
 1 :   k = 1, i = 1, , ori= Ø;
 2 :   While k < n Do    //n is the size of the subgraph C_sub
 3 :       V,H=NMF(A,k);
 4 :       For each element h_ij in H Do
 5 :           h_ij = Normalize(h_ij) according to equation (13);
 6 :           h_ij = Subtract(h_ij) according to equation (14);
 7 :       End for
 8 :       If h_ij( j = 0,1,2,...,n) = 0 Then
 9 :           break;
10 :       Else
11 :           k += 1;
12 :       End if
13 : End while
14 : While i ≤ k Do
15 :       If h_ij( j = 0,1,2,...,n) = 1 Then
16 :           ori_i = ori_i ∪ { h_ij };
17 :       End if
18 :       i += 1;
19 : End while
20 :   return k, ori;
```

Detecting Overlapping Local Community. At this stage, k times community detection is performed, thereby obtaining k communities. First, the algorithm uses Eq. (11) to divide different moments. Then, community expansion is carried out through a constant optimization Eq. (9). When $\rho(C) > 1$, the community has been formed. The details of this stage are shown in Function 3.

3.3 Complexity Analysis

Suppose the local community has n nodes, m edges, each node connects d edges on average. Function 1 requires $O(pn^2 + d^2n^2)$ time. Function 2 requires $O(kr^2)$ time. Function 3 requires $O(k \ (qmn^2 + dn^2))$ time. In summary, OLCDTM's time complexity is $O(pn^2 + d^2n^2 + kr^2 + kqmn^2 + kdn^2)$. Since $p, d, k, r,$ and $q << n$, the time complexity of OLCDTM is $O(mn^2)$.

The space complexity of OLCDTM is $O(nd)$ because the neighboring nodes of the entire local community need to be stored in the worst case.

Function 3 DetectOverlappingLocalCommunity

Input: $G = (V, E)$, the number of overlapping community k, original members of each community *ori*.

Output: $C_1, C_2, ..., C_k$.

```
1 :   Initialize i=1;
2 :   While i ≤ k Do
3 :       Ci=orii;
4 :       calculate ρ( Ci ) according to equation (11);
5 :       While ρ( Ci ) < 1 Do
6 :           For each u∈τ(ori) Do
7 :               calculate ΔQW according to equation (9);
8 :           End for
9 :           QW max= {u | u∈τ(orii) & max( ΔQW )= ΔQW(u)};
10 :          Ci= Ci∪ { QW max };
11 :          calculate  ρ( Ci ) according to equation (11);
12 :      End while
13 :      While True Do
14 :          maxsim = 0;
15 :          For each node u ∈τ( Ci ) Do
16 :              calculate Δ ρ( Ci );
17 :              If Δ ρ( Ci ) ≥ 0 Then
18 :                  calculate sim (u, Ci ) according to equation (12);
19 :                  If sim (u, Ci ) > maxsim Then
20 :                      maxsim = sim (u, Ci )
21 :                  End if
22 :              Else
23 :                  continue;
24 :              End if
25 :          End for
26 :          If !(u∈τ( Ci ) & Δ ρ( Ci )≥0) Then
27 :              break;
28 :          End if
29 :          Ci = Ci∪ {u}; //u refers to the node with the maximum similarity
30 :      End While
31 : End While
32 : return C1, C2, ..., Ck.
```

---

4 Experiments

4.1 Datasets Description

Real Datasets. We selected 8 real datasets for experimental analysis. They are the Strike dataset [16], the Karate Club dataset [17], the Dolphin dataset [18], the Polbooks dataset [19], the Adjnoun dataset [20], the Football dataset [21], the email-Eu-core dataset [22] and the Amazon dataset [23]. The information of the 8 real datasets is given in Table 1.

Table 1. Description of real datasets

Network	Nodes	Edges	Average degree	Communities
Strike	24	75	6.25	3
Karate	34	78	4.59	2
Dolphin	62	159	5.13	2
Polbooks	105	441	8.40	3
Adjnoun	112	425	7.59	2
Football	115	616	10.66	13
email-Eu-core	1005	25571	50.89	42
Amazon	334863	925872	5.53	75149

Artificial Datasets. The LFR benchmark generation tool [24] is used to manufacture artificial datasets required for the experiment. The specific parameter settings of each network used in the experiment are given in Table 2.

Table 2. Description of the artificial datasets

Dataset	Parameter
D1	$N = 1k, \mu = 0.1$–$0.6, k = 10, maxk = 30, minc = 20, maxc = 100, on = 100, om = 3$
D2	$N = 1k, \mu = 0.2, k = 10, maxk = 30, minc = 20, maxc = 100, on = 100, om = 2$–$5$
D3	$N = 1k$–$4k, \mu = 0.2, k = 10, maxk = 30, minc = 20, maxc = 100, on = 100, om = 3$
D4	$N = 1k, \mu = 0.2, k = 10, maxk = 30, minc = maxc = 20$–$100, on = 100, om = 3$

4.2 Experimental Scheme

In the experiments, we selected 4 local community detection algorithms as baseline algorithms. Namely, Clauset [2], LWP [4], MLC [13] and MULTICOM [11]. The parameters of all algorithms are used as suggested by the authors.

4.3 Evaluation Metrics

F-Score [25] is selected as the evaluation index. The definition of *F-Score* is as follows:

$$F-Score = \frac{2 * p * r}{p + r} \tag{15}$$

$$r = \frac{|F \cap T|}{T} \tag{16}$$

$$p = \frac{|F \cap T|}{F} \tag{17}$$

where F is the size of the results detected by LCD methods, T is the size of the true community. And p represents precision, r represents recall. $F\text{-}Score \in [0, 1]$.

4.4 Experimental Results on Accuracy

Experimental Results on Real Datasets. We utilize various seed selection strategies for different real datasets. For the first six datasets in Table 1, specifically, each vertex is used as the seed node to calculate the $F\text{-}score$, then take the arithmetic mean of the experiment as the result. For the last two networks that have a large scale of nodes, we randomly select some nodes as seed nodes. The experimental result is given in Table 3.

Table 3. *F-Score* on real datasets

Network	Clauset	LWP	MLC	MULTICOM	OLCDTM
Strike	0.6246	0.8693	**0.8576**	0.7183	0.7596
Karate	0.6482	0.6837	0.8631	0.3938	**0.8936**
Dolphin	0.4158	0.4619	0.7152	0.6919	**0.8911**
Polbooks	0.4900	0.5338	0.7131	0.1770	**0.8155**
Adjnoun	0.2090	0.2364	0.5541	0.2625	**0.6707**
Football	0.7207	0.7221	0.3289	0.5352	**0.7360**
email-Eu-core	0.5419	0.5546	0.2824	0.1736	**0.6483**
Amazon	0.6087	0.6603	0.6954	0.2543	**0.7754**

In Table 3, OLCDTM achieves the highest value of F-score on almost all networks. A possible reason is that OLCDTM uses the triangle-based node transitivity to explore the maximum range of the overlapping information and uses the local modularity density based on edge weight to detect communities. As a result, OLCDTM can obtain higher experimental accuracy. The MLC algorithm obtains a second better performance and the results of Clauset and MULTICOM are poor. LWP gets the highest performance on the Strike network because the local modularity M used in LWP pays more attention to the compactness within the community, which is more suitable for the Strike network that the boundaries between communities are sparse. Due to the high density of the email-Eu-core network, the accuracy of all algorithms has been reduced to a certain extent, but OLCDTM still obtains the optimal value.

Experimental Results on Artificial Datasets

(1) The variation of mixing parameter μ
We first conduct experiments about precision on non-overlapping seed nodes to objectively describe the accuracy of each algorithm. For the six networks of the D1 dataset, we randomly select 200 non-overlapping nodes in each network as seed nodes. We divide these seed nodes into high degree groups and low degree groups to analyze the sensitivity of seed nodes. The results are given in Fig. 2.

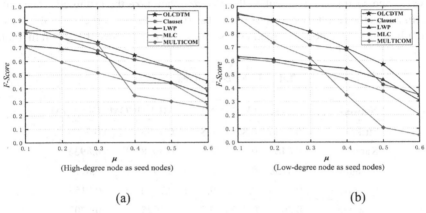

(a)   (b)

Fig. 2. *F-score* on non-overlapping nodes

In Fig. 2, the accuracy of each algorithm varies with different μ values. The larger the μ is, the smaller the *F-score* is. A significant reason is that the ability of the algorithm to correctly identify the community is gradually increased due to the community results become fuzzy when the value of μ is increasing. Furthermore, we observe that OLCDTM achieves the best results in both the high degree group and low degree group, and the MLC algorithm obtains the second better performance. A primary reason is that OLCDTM uses the local modularity density based on edge weight to detect community technique to effectively improve the robustness of the algorithm.

To effectively evaluate the accuracy of OLCDTM in overlapping local community detection, we select 200 overlapping nodes from the D1 dataset as seed nodes and compare with two overlapping local community detection algorithms MLC and MULTICOM. Since Clauset and LWP only return a single local community, they are not taken into account. The results are given in Fig. 3.

In Fig. 3, OLCDTM performed well in both high and low-degree groups. The *F-score* values of the three algorithms in the high group are lower than those in the low degree group. A possible reason is that the low degree overlapped nodes which are usually located at the junction of two communities are more suitable to explore multiple different overlapping communities by using them as seed nodes. However,

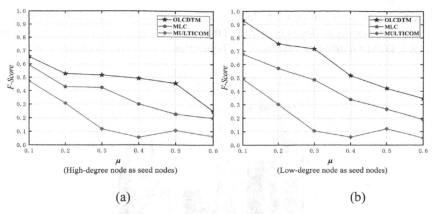

Fig. 3. *F-score* on overlapping nodes

the location of height nodes is more inclined to the community center, which makes it difficult to find overlapping communities.

(2) Different values of *om*

We use om to control the changes of overlapping communities of each secd node. In Fig. 4, the accuracy of OLCDTM, MLC, and MULTICOM does not change much with the increase of om. OLCDTM and MLC algorithm obtains better F-score in low degree group, which shows that the stage of finding core subgraph can effectively mine the real overlapping communities of seed nodes. Furthermore, the finding core subgraph stage based on node transitivity adopted by OLCDTM can provide more comprehensive real information. As a result, OLCDTM achieves higher accuracy than the other algorithms on the D2 dataset.

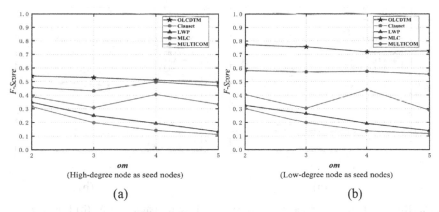

Fig. 4. *F-score* on parameter *om*

(3) Comparison with different seed nodes

We compare OLCDTM, MLC, and MULTICOM on different seed nodes. Namely, seed nodes in non-overlapping communities, seed nodes with om equal 3, and seed nodes with om equal 5. In Fig. 5, the performance of each algorithm on non-overlapping nodes is higher than that on overlapping nodes. The accuracy of OLCDTM on different seed nodes is better than that of the other two algorithms.

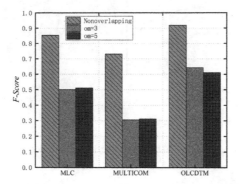

Fig. 5. *F-score* on different types of seed nodes

4.5 Experimental Results on Running Time

We randomly select 100 seed nodes from each network in both D3 and D4 groups, and then take the average running time as the experimental result. In Fig. 6(a), the running time does not change much with the increase of network scale. A primary reason is that the time consumed by the local community detection methods has nothing to do with the overall size of the network.

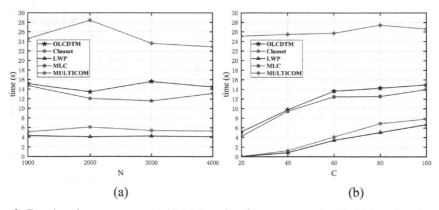

(a)   (b)

Fig. 6. Running time on parameter *N*. (a) Running time on parameter *N*. (b) Running time on different community sizes

In Fig. 6(b), MULTICOM takes the highest time. The difference in time cost between the OLCDTM algorithm and the MLC algorithm is not obvious because they both use

NMF to mine the number of overlapping communities and detect multiple overlapping communities simultaneously. Clauset and LWP have lower time cost because they only return a single local community.

5 Conclusions

This study proposes the OLCDTM method. On the one hand, the information of over-lapping communities is mined by nodes' transitivity to ensure that the exact number of overlapping communities is obtained. On the other hand, we use a new local modularity calculation equation to locate the overlapping local communities more precisely. In the future, we consider developing incremental strategies to adapt OLCDTM to dynamic networks.

Acknowledgements. This work is partly supported by the National Natural Science Foundation of China under Grant No. 61672159, No. 61672158, No. 62002063 and No. 61300104, the Fujian Collaborative Innovation Center for Big Data Applications in Governments, the Fujian Industry-Academy Cooperation Project under Grant No. 2017H6008 and No. 2018H6010, the Natural Science Foundation of Fujian Province under Grant No. 2019J01835 and No.2020J01230054, and Haixi Government Big Data Application Cooperative Innovation Center.

References

1. Gregory, S.F.: Finding overlapping communities in networks by label propagation. New J. Phys. **12**(10), 103018 (2010)
2. Clauset, A.F.: Finding local community structure in networks. Phys. Rev. E **72**(2), 026132 (2005)
3. Chen, Q.F., Wu, T.S., Fang, M.S.: Detecting local community structures in complex networks based on local degree central nodes. Phys. A: Stat. Mech. Appl. **392**(3), 529–537 (2013)
4. Luo, F.F., Wang, J.S., Promislow, E.T.: Exploring local community structures in large networks. In: 2006 IEEE/WIC/ACM International Conference on Web Intelligence, pp. 233–239. IEEE, Hong Kong (2006)
5. Luo, W.F., Zhang, D.S., Jiang, H.T.: Local community detection with the dynamic membership function. IEEE Trans. Fuzzy Syst. **26**(5), 3136–3150 (2018)
6. Meng, T.F., Cai, L.S., He, T.T.: Local higher-order community detection based on fuzzy membership functions. IEEE Access **7**, 128510–128525 (2019)
7. Luo, W.F., Lu, N.S., Ni, L.T.: Local community detection by the nearest nodes with greater centrality. Inf. Sci. **517**, 377–392 (2020)
8. Jian, Z.F., Ma, H.S., Huang, J.T.: CLOSE: local community detection by LOcal structure expansion in a complex network. In: 2019 International Conference on Technologies and Applications of Artificial Intelligence, pp. 1–6. IEEE, Taiwan (2019)
9. He, K.F., Shi, P.S., Bindel, D.T.: Krylov subspace approximation for local community detection in large networks. ACM Trans. Knowl. Discov. Data (TKDD) **13**(5), 1–30 (2019)
10. Li, X.F., Xu, G.S., Lian, W.T.: Multi-layer network local community detection based on influence relation. IEEE Access **7**, 89051–89062 (2019)
11. Hollocou, A.F., Bonald, T.S., Lelarge, M.T.: Multiple local community detection. ACM SIGMETRICS Perform. Eval. Rev. **45**(3), 76–83 (2018)

12. Ni, L.F., Luo, W.S., Zhu, W.T.: Local overlapping community detection. ACM Trans. Knowl. Discov. Data **14**(1), 1–25 (2019)
13. Kamuhanda, D.F., He, K.S.: A nonnegative matrix factorization approach for multiple local community detection. In: 2018 IEEE/ACM International Conference on Advances in Social Networks Analysis and Mining, pp. 642–649. IEEE, Barcelona (2018)
14. Binesh, N.F., Rezghi, M.S.: Fuzzy clustering in community detection based on nonnegative matrix factorization with two novel evaluation criteria. Appl. Soft Comput. **69**, 689–703 (2018)
15. Liu, X.F., Wang, W.S., He, D.T.: Semi-supervised community detection based on non-negative matrix factorization with node popularity. Inf. Sci. **381**, 304–321 (2017)
16. Dohleman, B.F.: Exploratory social network analysis with Pajek. Psychometrika **71**(3), 605 (2018)
17. Zachary, W.F.: An information flow model for conflict and fission in small groups. J. Anthropol. Res. **33**(4), 452–473 (1977)
18. Lusseau, D.F., Schneider, K.S., Boisseau, O.T.: The bottlenose dolphin community of Doubtful Sound features a large proportion of long-lasting associations. Behav. Ecol. Sociobiol. **54**(4), 396–405 (2003)
19. Political books network. http://www.orgnet.com. Accessed 28 June 2020
20. Rossi, R.F., Ahmed, N.S.: The network data repository with interactive graph analytics and visualization. In: Twenty-Ninth AAAI Conference on Artificial Intelligence, pp. 4292–4293. AAAI, Texas (2015)
21. Girvan, M.F., Newman, M.S.: Community structure in social and biological networks. Proc. Natl. Acad. Sci. **99**(12), 7821–7826 (2002)
22. Yin, H.F., Benson, A.S., Leskovec, J.T.: Local higher-order graph clustering. In: Proceedings of the 23rd ACM SIGKDD International Conference on Knowledge Discovery and Data Mining, pp. 555–564. ACM, Halifax (2017)
23. Yang, J.F., Leskovec, J.S.: Defining and evaluating network communities based on ground-truth. Knowl. Inf. Syst. **42**(1), 181–213 (2015)
24. Lancichinetti, A.F., Fortunato, S.S., Radicchi, F.T.: Benchmark graphs for testing community detection algorithms. Phys. Rev. E **78**(4), 046110 (2008)
25. Schütze, H.F., Manning, C.S., Raghavan, P.T.: Introduction to Information Retrieval. Cambridge University Press, Cambridge (2008)

SONAS: A System to Obtain Insights on Web APIs from Stack Overflow

Naixuan Wang, Jian Cao$^{(\boxtimes)}$, Qing Qi, Qi Gu, and Shiyou Qian

Department of Computer Science and Engineering, Shanghai Jiao Tong University,
Shanghai, China
{cao-jian,qi_ng616,guqi,qshiyou}@sjtu.edu.cn

Abstract. In recent years, we have witnessed the rapid development of types and quantities of Web APIs. However, it is challenging for users to select Web APIs that best match their requirements and to learn how to invoke a Web API correctly. Although Web API providers often publish documents that describe the functionalities of Web APIs and how to use them, users still have to collect information to acquire knowledge about the usage information of Web APIs. Stack Overflow, the largest programming-related question-and-answer (Q&A) website, has many posts about Web APIs. Therefore, we have designed and implemented a *S*ystem to *O*btain i*N*sights on Web *A*PIs from *S*tack Overflow (SONAS). SONAS collects questions related to Web APIs and classifies them into different categories using a deep learning model. The statistics on the numbers of different types of questions indicate the usage information of Web APIs. Furthermore, SONAS predicts the future usage trends of Web APIs, based on a long short-term memory model with multi-task learning. The experiments on a real-world dataset prove SONAS can provide useful insights on Web APIs.

Keywords: Web API · Stack overflow · Data mining · Text classification · Time series prediction

1 Introduction

In recent years, Web services on the Internet have grown rapidly [1] and Web APIs have become a major type of Web services [2]. Many organizations package their capabilities, resources or data into services and publish them on the Web in the form of Web APIs, causing an exponential increase both in the number of Web APIs and the functions they provide. For example, on ProgrammableWeb (https://www.programmableweb.com), more than 20,000 Web APIs have been published so far.

When a developer wants to use Web APIs, he has to consider many factors such as functionality, maintainability and efficiency. Web API providers often publish informative documents on their Web API to assist developers to learn about a Web API's functionalities. Moreover, semantic technologies can

© Springer Nature Singapore Pte Ltd. 2021
Y. Sun et al. (Eds.): ChineseCSCW 2020, CCIS 1330, pp. 499–514, 2021.
https://doi.org/10.1007/978-981-16-2540-4_36

be applied to describe the functions of Web APIs and classify them based on these documents'[3]. For example, the LDA model can be used to search for Web APIs with similar functions [4], and the semantic similarity can be calculated to help classify Web APIs [5]. Researchers have provided techniques to measure and predict the quality of service (QoS) of Web APIs based on historical QoS information [6, 7]. However, this information is not sufficient because developers want to obtain knowledge on the usage information of Web APIs, such as Web APIs' popularity.

There are many sources to mine insights on Web APIs, such as Alexa.com [8] and some technical forums. Stack Overflow (https://stackoverflow.com/) is the largest online technical forum for developers. There are many posts about Web APIs. Mining these questions and answers will help us obtain knowledge of these Web APIs. Unfortunately, there is no research on how to obtain insights into a specific Web API from Stack Overflow as yet.

It is very challenging to mine knowledge of Web APIs from Stack Overflow. Firstly, to collect and filter Web API-related questions and answers is not an easy task since Stack Overflow is not designed specifically for Web APIs. Secondly, the questions and answers are expressed in natural language, which should be processed by applying natural language processing techniques before mining. Thirdly, the topics covered by the posts are diverse which adds to the difficulty of extracting the knowledge we need. Therefore, we have designed and implemented a *S*ystem to *O*btain i*N*sights on Web *A*PIs from *S*tack Overflow (SONAS).

The main contributions of this paper are as follows: Firstly, we propose a data processing method based on positive and unlabeled learning to filter relevant questions. Secondly, we classify questions into *HOWTO* -type, *ERROR/BUG* -type and *OTHERS* -type using a TextCNN model. Thirdly, we design an LSTM model with multi-task leaning to predict the numbers of different types of questions, which can indicate the usage information of Web APIs. We also perform experiments to show that the techniques applied by SONAS efficiently obtain insights on Web APIs from Stack Overflow. To the best of our knowledge, SONAS is the first system that is able to mine knowledge on a specific Web API from Stack Overflow.

This paper is organized as follows. Section 2 introduces the related work. In Sect. 3, the framework of SONAS is presented. In Sect. 4, we describe the methods for Web API-related data collection and filtering. Section 5 reports how questions are classified. The mining and prediction approaches of Web APIs' usage information are introduced In Sect. 6. Section 7 describes the system implementation of SONAS and the experiment results. Section 8 summarizes the threats to validity of our work and Sect. 9 concludes the paper and makes suggestions for future work.

2   Related Work

There is a lot of research on how to obtain information on the functionality and performance of Web Services. In [2], Maleshkova et al. point out that the use of Web APIs requires a lot of manual work. Based on a manual inspection and analysis of the main body of public Web APIs, they provide detailed information about common description forms, output types and reusability levels, etc. In order to better describe Web APIs, semantic descriptions can be provided so that they are easier to browse and more meaningful for humans while also being machine-interpretable [3]. Moreover, based on the descriptions of Web APIs, they can be classified or clustered using machine learning algorithms. For example, in [4], a word embedding augmented LDA model is proposed to cluster web services. Yang et al. [9] propose a deep neural network model for services classification. In order to help developers select Web APIs, some researchers have proposed approaches for searching and ranking Web APIs. For example, in [10], a faceted approach is proposed for searching and ranking Web APIs that takes into consideration the API's attributes as found in their HTML descriptions.

In addition to the functional descriptions of Web Services, developers also want performance information. In recent years, many researchers have proposed approaches to measure and predict the QoS of Web services. For example, in [6], probabilistic graphical models (PGMs) are utilized to yield near-future time series predictions for QoS information based on the invocation records collected from Web services. Some researchers have proposed context-aware QoS prediction models. In [7], the users' geographical information, company affiliations and country affiliations of the services are modeled as contexts. Then an ensemble model is applied for context-aware QoS prediction.

Some researchers use information on the social websites to help developers use Web services. For example, Torres et al. [11] proposed a faster Web API discovery method using social information from ProgrammableWeb. In [12], a popular ranking framework that considers three factors of Web APIs is proposed, which is also based on the datasets from ProgrammableWeb. In [13], a time-aware linear model is proposed to predict the popularity of an API using the time series feature of APIs and API's self-features from ProgrammableWeb. Although ProgrammableWeb provides rich information about Web APIs and Mashups, it lacks information on the usage of Web APIs in practical applications.

Many researchers make use of the data on Stack Overflow to propose approaches or develop tools. For example, [14] reports on the prevalence and severity of API misuse on Stack Overflow. In [15], an approach to automatically augment API documentation with "insight sentences" from Stack Overflow is presented. In [16], a prototype plug-in that augments Stack Overflow with definitions and examples of API calls in the questions and answers is presented to help novice programmers.

In order to identify the concerns of client developers when using Web APIs, Venkatesh et al. undertook an empirical study on developer forums and Stack Overflow [17]. Our work also mines Web API-related data from Stack Overflow, however the difference is that we aim to obtain knowledge for each specific Web API whereas Venkatesh et al. investigated the problems of using Web APIs from a global view. They conduct an empirical study on 32 popular Web APIs from seven domains. Although their case study shows the results of each Web API, their findings focus on the global views of Web APIs. In [18], Rodrigucz et al. investigated Apache spark usage by mining Stack Overflow questions. We also mine Stack Overflow questions but our goal is to provide a general framework to obtain knowledge on different Web APIs, not for dedicated software.

3   The Framework of SONAS

SONAS consists of three modules. Figure 1 presents the overview of the three modules of SONAS.

- Data Collection and Filtering Module: This is responsible for collecting and selecting the relevant data from Stack Overflow on Web APIs. We do not know whether a question is related to a Web API or not. Therefore we try to find questions which possibly discuss a Web API and filter out questions which are not actually related to Web APIs.
- Question Type Classification Module: The filtered questions are classified into different types by this module.
- Usage Information Mining and Prediction of Web APIs Module: After the questions are classified, this module undertakes a statistical analysis on these questions and the usage information of a Web API can be measured in terms of the number of different types of questions. Predictions of the number of different types of questions also tell us the future trends in relation to a Web API. Different types of questions can indicate different metrics of the Web API.

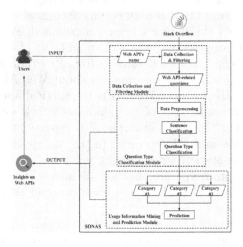

Fig. 1. Overview of SONAS

Users can input a Web API's name, and SONAS collects and filters the Web API-related data from Stack Overflow first, and then Question Type Classification Module and Usage Information Mining and Prediction of Web APIs

Module mines insights on the Web API, finally the results of these two modules are organized and output to users.

The Data Collection and Filtering Module uses positive and unlabeled learning (PUL) [19] to filter relevant data. After this step, we can obtain clean datasets on Web APIs. The Question Type Classification Module is based on a TextCNN model. Predictions of Web APIs' different types of questions relies on a long short-term memory (LSTM) with multi-task learning. The details are introduced in the following sections.

4   Web API-Related Data Collection and Filtering

4.1   Data Format

Stack Overflow makes its data publicly available in XML format. There are several XML documents such as posts.xml, comments.xml, users.xml, tags.xml, etc. [20]. In this paper, we focus on posts.xml which contains the text in the body of the posts. So far, there have been over 40 million posts in posts.xml, spanning 127 months, from August 2008 to February 2019 and there are two types of posts, questions and answers.

4.2   Data Collection and Filtering

Data Collection On Stack Overflow, every question-type post has 0 to 5 tags such as <html>, <android> etc. to show what this post is about [21]. Since tags can be added by any users in an arbitrary way, there are often quite a few tags for the same Web API. For example, there are over 10 tags for the YouTube API, such as <youtube-api>, <youtube> and <youtube-api-v3>. For all these tags, the size of the corresponding data is different and the relevance of the data to a Web API also differs.

For each Web API, in order to acquire the relevant data as completely as possible, in the first step, we combine a keyword search and a tag search to collect the data. The data containing the keywords of the Web API in the title or the text body and those containing the tags related to this Web API are all collected.

Spurious Relevant Data Removal. Some of the data obtained in the first step does not have tags related to the Web API or a keyword in its title, and the keywords only appear in the code segments or the HTML hyperlinks of the text body. We remove this spurious relevant data directly.

Irrelevant Data Removal Through PUL. If the keywords of a Web API only appear in the text body one or two times, we cannot easily judge whether this data is Web API-relevant or not. At the same time, different tags and the positive data may have different relevance. We need to filter out irrelevant data from the dataset obtained in the previous two steps.

If we consider the relevant data as positive samples and the irrelevant data as negative samples, then all the original data in the dataset represent unlabeled samples. We need to find all the positive samples from the unlabeled ones. Of the different tags for the same Web API, we can pick out the tag that is the mostly related to this Web API. For example, <youtube-api> is the tag most related to the YouTube API. We consider data with this tag as part of the positive samples. After some positive samples and a lot of unlabeled samples are obtained, we can use PUL to collect all the positive samples from the unlabeled samples.

PUL is a branch of semi-supervised learning. There are many techniques to solve PUL problems, such as PU bagging [22] and positive unlabeled random forests [23]. After comparing several algorithms, we finally choose a two-step approach [24] to solve this problem. The algorithm begins with a positive dataset P and an unlabeled dataset U. RN is used to store the reliable negative examples. In the first step, we view P as a positive dataset and U as a negative dataset, and the data is used to train a Naive Bayes Classifier. Then the data in U is predicted, and all the negative data is put into RN. In the second step, we build a SVM classifier and use data in P and RN to train the model. Let $Q = U - RN$, then we use the SVM classifier to predict the data in Q, and the negative data is put into RN. We repeat step two until $RN == U$ or all the data in Q is positive. Finally the algorithm returns the dataset Q, which contains all the positive data in U.

By using this algorithm, we can find much positive data from unlabeled dataset U when we only be provided a small positive dataset P, and it is acceptable that this algorithm may not collect all the positive data, which will not cause large errors in the results.

5   Question Type Classification for Web APIs

5.1   The Types of Questions About Web APIs

The questions posted on Stack Overflow are of different types. Studying the statistics on these types of questions on a Web API helps developers understand which aspect is most likely to cause problems for the users and it also indicates the usage information of this Web API in actual applications to some extent. So the classification of question types is meaningful.

In [25], sentences are divided into 5 categories, which are *ANOMALY*, *EXPLANATION, HOWTO, PROPERTY* and *OTHER*. By carefully examining the relevant data on Web APIs, we decide to study two types of questions, which are *HOWTO* and *ERROR/BUG*. *HOWTO* -type questions are generally caused by developers' lack of skills or the absence of technical documents. A typical example of this type question is *"How to print Twitter API requests reply?"*. For *HOWTO* -type questions, the words *how to, is it possible to* or *is there a way to* often appear in their titles or their text bodies. *ERROR/BUG* -type questions are usually caused by bugs in codes or abnormal return values in programs. Typical words or phrases appearing in this type of questions are *error, fail* or *bug*. According to the statistics, *HOWTO* and *ERROR/BUG* -type questions account for over 60% of all questions, so we decide to divide the questions into three categories, i.e., *HOWTO, ERROR/BUG* and *OTHERS*.

After analyzing the characteristics of the posts, we find that the types of questions can be distinguished from their titles and text bodies. In most cases, the title of a question is a single sentence or a few phrases, which is a summarization of the question. However, the title is often too short or is written in a casual format, making it very difficult to classify the question based only on its title. On the other hand, the text body of a question is usually composed of several sentences, and almost all developers can clearly express their questions in the text body. Therefore, we decide to use both the title and the text body of a question for type classification.

5.2   Question Type Classification Process

Data Preprocessing. Firstly, because there is much information in the text bodies that is irrelevant to question classification, such as the HTML tags like <p> and <code>, code segments and HTML hyperlinks, we delete these components to remove noises.

Secondly, we separate the long text bodies into short sentences. The lengths of the text bodies of questions vary greatly. The text body is often composed of multiple sentences with different meanings and expressions, while a single sentence usually contains a much simpler meaning and clearer expressions. We find that many sentences do not contribute to the classification of the question types, if we classify long text directly, it can easily cause large errors, so we separate the long text bodies into short sentences using Natural Language Toolkit (NLTK) [26] and classify them first. Because the titles of questions are usually very short and rarely contain more than one sentence, we treat each title as a sentence.

Sentence Classification. This problem can be viewed as a short text multi-classification problem, and we refer to TextCNN [27,28], a convolutional neural network (CNN) model. We can use several convolution kernels of different sizes to get more features.

Question Type Classification. After obtaining the classification results of all the sentences, we convert them to the results of the questions. We count the number of sentences in the different types as new features and use k-Nearest Neighbor (kNN) to solve this multi-classification problem.

6   Usage Information Mining and Prediction of Web APIs

6.1   Usage Information Mining and Prediction

After classifying the questions into types, we can count the number of different types of questions. The total number of all the three types of questions can indicate popularity of the Web API, it shows how many discussions there are about the Web API on Stack Overflow. And the number of $HOWTO$ -type questions tells us whether a Web API is easy to use, so it indicates usability of the Web API in a way. By counting the number of $ERROR/BUG$ -type questions, we know how often a developer encounters an error or bug, so we can learn the reliability of the Web API roughly.

In addition to mining the current usage information of Web APIs, developers also need to know the future trends so they can make a more rational choice. Since the number of the different types of questions per time unit (such as per month) can be organized into regular time series data, time series prediction approaches can be applied to forecast future trends in relation to Web APIs.

6.2 LSTM Models for Prediction

SONAS uses a type of recurrent neural network (RNN) [29], the long short-term memory (LSTM) model [30], to predict the number of questions of Web APIs. However, as there are internal relationships between the number of different types of questions, we can use multi-task learning to learn common features to perform multiple predictions at the same time.

Multi-task learning is a type of transfer learning algorithm with broad applications. Given m learning tasks, where all or part are related but not exactly the same, the goal of multi-task learning is to help improve the performance of each task using the knowledge contained in these m tasks [31,32]. Multi-task learning is based on shared representation, which learns multiple related tasks together to obtain better generalization performance [33].

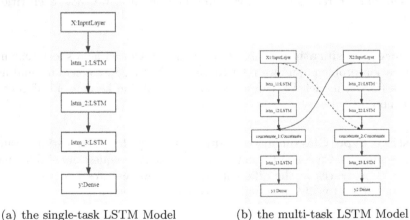

(a) the single-task LSTM Model   (b) the multi-task LSTM Model

Fig. 2. The Structures of the single-task LSTM Model and the multi-task LSTM Model

There are many forms of multi-tasking learning models. We set the number of tasks to 2 and the structures of the single-task model and multi-task learning model are shown in Figure 2(a) and 2(b) respectively. The multi-task learning model we design is a multiple-input multiple-output model, in which each task's output value is affected by both inputs at the same time. The model is trained using the weighted loss values of the two tasks.

7   System Implementation and Experiments

7.1   System Implementation

SONAS is built based on the Python Flask framework. It provides two functions in the form of Web APIs. The first function is to analyze the usage information of one Web API. When users input the name of a Web API, this name is passed to the system as a string. Then the data collection and filtering algorithm, the TextCNN model and the LSTM model are invoked step by step. This process only takes a few minutes. The results obtained in every step are displayed. The second function is a usage information comparison between two Web APIs. When users input two Web API names, each API's results are collected, and some results are compared.

7.2   Experiments

Dataset. We select 50 categories with the most Web APIs on Programmable, and sort the Web APIs of each category by followers. Then for each Web API, we check whether there is a corresponding Stack Overflow tag and make sure that the number of posts of this tag is greater than 200, bacause it is difficult to draw reliable conclusions based on too few posts. Finally, we collect 105 Web APIs from 11 different categories on ProgrammableWeb.

Data Processing Experiment. For any Web API, We assume that the tag distributions of the positive data and negative data are different so we can use tag information in the PUL algorithm.

In order to verify the correctness of the assumption, we first check whether the data with the most related tag is positive. We randomly pick 20 Web APIs, and for each one, we pick a tag which is most related to this Web API. Of these data, we randomly choose 20 posts to manually check and the average percent of positive samples is 94.5%, which means that it is reasonable to view this kind of data as positive samples.

Then we verify that any Web API's tag distributions of positive data and negative data are different, which is measured in terms of cosine similarity. For each Web API, we select the most related tag and regard questions with this tag as positive samples. We calculate the tag distribution similarities between positive samples and unlabeled samples, and the average value is 0.83. We also randomly separate the positive dataset into two parts and the average tag distribution similarity between them is 0.95. From this result we can conclude that

Table 1. Performance prediction results with LSTM models

Web API name	LSTM (Single target sequence)	LSTM (target sequence + driving sequences)	AR	MA	ARIMA	ES
Admob API	0.1660	**0.1503**	0.2007	0.2020	0.2035	0.2039
Amazon EC2 API	0.2119	**0.1786**	0.2235	0.2292	0.2236	0.2491
Amazon S3 API	0.2843	**0.2688**	0.3110	0.3181	0.2942	0.3301
Buffer API	0.1904	**0.1828**	0.2010	0.1989	0.1989	0.1984
Facebook API	0.0341	**0.0266**	0.0308	0.0310	0.0307	0.0315
Gmail API	0.1515	**0.1180**	0.2099	0.2119	0.1972	0.2118
Google App Engine API	0.0652	**0.0582**	0.0668	0.0673	0.0669	0.0677
Google Drive API	0.0926	**0.0782**	0.0846	0.0845	0.0794	0.0847
Google Maps Android API	0.0161	**0.0139**	0.0147	0.0151	0.0158	0.0161
Google Maps Places API	0.2834	**0.2822**	0.3360	0.3406	0.3356	0.3393
Google Maps API	0.1717	**0.1689**	0.1734	0.1728	0.1728	0.1729
Google Plus API	0.0084	**0.0081**	0.0089	0.0092	0.0089	0.0105
Heroku API	0.1475	**0.1201**	0.1396	0.1414	0.1308	0.1728
iTunes and iTunes Connect API	0.0310	0.0245	**0.0174**	0.0178	0.0178	0.0180
Payments Gateway API	0.2690	**0.2508**	0.2875	0.3040	0.2878	0.2969
Paypal API	0.0127	**0.0112**	0.0334	0.0339	0.0335	0.0347
SMS API	0.0831	**0.0757**	0.0832	0.0835	0.0838	0.0838
Twitter API	0.0441	**0.0419**	0.0575	0.0593	0.0587	0.0587
WooCommerce API	0.2280	**0.2255**	0.2583	0.2644	0.2391	0.2569
YouTube API	0.1255	**0.1222**	0.1296	0.1278	0.1278	0.1285

* The bold font indicates the best result, and the underline indicates the second best result.

** The traditional models we use are autoregressive model (AR), moving average model (MA), autoregressive integrated moving average model (ARIMA) and exponential smoothing (ES).

positive samples are very similar whereas unlabeled datasets which include negative samples have different tag distributions. Therefore, it is feasible to use tag information to find positive data in an unlabeled dataset.

After data processing, the positive dataset and negative dataset are obtained for the selected Web APIs. The average similarity of tag distributions in questions for the positive datasets and negative datasets is 0.76, while it is 0.97 for positive dataset. From the results it can be found that after data processing, the similarity between the positive datasets and the negative datasets has dropped significantly, while the similarity between the two separate positive datasets is still very high, which means most of the selected data from the unlabeled datasets are positive. To evaluate the results, for each Web API, we also manually check some data, the average precision is 92.0% and the average recall is 84.4%. This result proves the data processing algorithm is effective.

Question Classification Experiment. We label 5000 sentences for the sentence classification and 200 questions for question type classification. We use the 90% of the data to train the model and the remaining data to test the performance of the model.

The accuracy of the TextCNN classification model on the test set is 94.8%, which proves our labeling and experiment settings are effective. The prediction accuracy of the kNN algorithm is 99.6%, which means by using the classification results of sentences, we can accurately identify the question type.

As an example, we select Twitter API to see the question classification results. For Twitter API, most questions are the *HOWTO* -type, while fewest questions are the *ERROR/BUG* -type. The results indicates that the biggest problem of using Twitter API is not to fix the errors or bugs, but the way to implement some functions or use some methods.

Experiment to Predict Future Trends for Web APIs. We collect monthly data on 20 Web APIs from August 2008 to February 2019. For each Web API, we get three time series, which is the number of questions per month, the number of *HOWTO* -type questions per month and the number of *ERROR/BUG* -type questions per month. We set the input step to 4 and the output step to 1. To compare the prediction performance, we first use the time series of the number of total questions per month as an univariate input, and then we use the other two series as driving series to help improve the results of the first series. We use mean squared error (MSE) to measure the performance and compare this with traditional models (See Table 1). For each prediction task, we set 5 random number seeds and train the model 5 times and calculate the average result. From the table, we can see that if we only use one sequence as input to learn an LSTM model, sometimes the result can be worse than using a traditional model. However, if we use two sequences related to the input sequence as driving sequences, the results will be better.

As an example, we predict the number of questions on the Gmail API for the next 4 months, which is shown in Figure 3. We can see the number of questions will decrease a little in the next few months, and there is an interesting relationship between question number and the API versions.

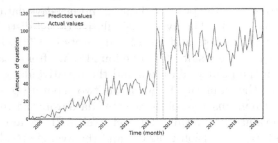

Fig. 3. The Predicted number of questions per month for Gmail API

MTL&LSTM Experiment Results Analysis. For each Web API, we use the number of *ERROR/BUG* and *HOWTO* -type questions as task 1 and task 2. The results are compared with single-task models, as shown in Table 2. We can see that there are only three combinations that do not work well, so we can draw the conclusion that using the LSTM model with multi-task learning improves the prediction performance.

8   Threats to Validity

Our approach relies on problem type classification. We divide question types into three categories because HOWTO and ERROR/BUG-type questions account for over 60% of all questions. It is possible to divide questions into more types. However, it will lead to problems of insufficient data samples for some minor types and performance degradation problems. Compared with some other approaches [34], our classification approach is relatively simple. Fortunately, our approach can get accurate results (accuracy 99.6%).

The prediction relating to one version of the Web API may not be applicable to the prediction on the next version. Unfortunately, for most questions that are related to Web APIs, their tags do not contain the Web API version information. Currently, we obtain the version update events from other data sources and simply display them on the statistical curves of problem numbers. The influences of Web API version updates have not been modeled in our model. The good news is in most cases changes between consecutive Web API versions are not significant so that the prediction on one version is applicable to the prediction on the next version in these cases.

Currently, SONAS only provides general information about a Web API's usage. It can not obtain precise and quantitative performance information of Web APIs because there is no enough quantitative performance data reported by the posts on Stack Overflow. However, general usage information is still helpful for developers.

Table 2. Multi-task learning & LSTM model results

Web API name	Single-task LSTM	Multi-task LSTM	Web API name	Single-task LSTM	Multi-task LSTM
Admob API	0.1691	**0.1648**	Google Maps API	0.1354	**0.0879**
	0.0495	**0.0493**		0.1913	**0.1862**
Amazon EC2 API	0.1492	**0.1487**	Google Plus API	0.0114	**0.0076**
	0.1436	**0.1381**		0.0105	**0.0087**
Amazon S3 API	0.1717	**0.1401**	Heroku API	0.1863	**0.1849**
	0.1907	0.2036		0.1387	**0.1296**
Buffer API	0.1324	**0.1273**	iTunes and iTunes Connect API	0.0097	**0.0086**
	0.0974	0.0988		0.0699	**0.0644**
Facebook API	0.0658	**0.0588**	Payments Gateway API	0.2522	**0.2464**
	0.0305	**0.0283**		0.2423	**0.2395**
Gmail API	0.2458	**0.1947**	Paypal API	0.0368	**0.0351**
	0.2123	**0.1821**		0.0436	**0.0346**
Google App Engine API	0.0814	**0.0734**	SMS API	**0.0216**	0.0218
	0.0326	**0.0308**		**0.0873**	0.0889
Google Drive API	0.1315	**0.1069**	Twitter API	0.0294	**0.0261**
	0.0860	**0.0822**		0.0398	**0.0314**
Google Maps Android API	0.0393	**0.0392**	WooCommerce API	0.2252	**0.2160**
	0.0174	**0.0153**		0.2800	**0.2778**
Google Maps Places API	0.2403	**0.2331**	YouTube API	0.1281	**0.1237**
	0.1958	**0.1870**		0.1404	**0.1311**

* The bold font indicates the best result.
** For each Web API, the first row is the result of $ERROR/BUG$ -type questions and the second row is the result of $HOWTO$ -type questions.

9   Conclusions and Future Work

In this paper, we introduce SONAS, a system to obtain insights on Web APIs from Stack Overflow. Although there are many questions and answers that are relevant to Web APIs on Stack Overflow, it is a difficult task to filter data that actually discusses Web API-related issues. Therefore, SONAS uses positive and unlabeled learning to select Web API-related questions. Then the questions are classified based on a TextCNN model. The numbers of different types of questions roughly indicate the popularity, reliability and usability of the Web APIs. Therefore, SONAS predicts the number of different types of questions using an LSTM model with multi-task learning. Experiments on a set of Web APIs prove SONAS can give insights to developers.

Currently, SONAS only divides the questions into three types. If we can divide the questions into more types, the measurement will be more comprehensive. In the current version of SONAS, only questions relevant to Web APIs are processed. However, the answers to these questions are also of great value. In the future, we will mine the answers and questions at the same time. Last but not least, to combine the data from Stack Overflow and other sources to mine knowledge of Web APIs is also a very promising research topic.

Acknowledgments. This work is partially supported by National Key Research and Development Plan (No. 2018YFB1003800) and China National Science Foundation (Granted Number 62072301).

References

1. Bouguettaya, A., et al.: A service computing manifesto: the next 10 years. Commun. ACM **60**(4), 64–72 (2017)
2. Maleshkova, M., Pedrinaci, C., Domingue, J.: Investigating web APIs on the world wide web. In: Eighth IEEE European Conference on Web Services 2010, pp. 107–114. IEEE (2010)
3. Cheron, A., Bourcier, J., Barais, O., Michel, A.: Comparison matrices of semantic RESTful APIs technologies. In: Bakaev, M., Frasincar, F., Ko, I.-Y. (eds.) ICWE 2019. LNCS, vol. 11496, pp. 425–440. Springer, Cham (2019). https://doi.org/10.1007/978-3-030-19274-7_30
4. Shi, M., Liu, J., Zhou, D., Tang, M., Cao, B.: WE-LDA: a word embeddings augmented LDA model for web services clustering. In: IEEE International Conference on Web Services (icws), 2017, pp. 9–16. IEEE (2017)
5. Maleshkova, M., Zilka, L., Knoth, P., Pedrinaci, C.: Cross-lingual web API classification and annotation. In: Proceedings of the 2nd International Conference on Multilingual Semantic Web, vol. 775, pp. 1–12. CEUR-WS. org (2011)
6. Wang, H., Wang, L., Yu, Q., Zheng, Z., Bouguettaya, A., Lyu, M.R.: Online reliability prediction via motifs-based dynamic Bayesian networks for service-oriented systems. IEEE Trans. Softw. Eng. **43**(6), 556–579 (2016)
7. Xu, Y., Yin, J., Deng, S., Xiong, N.N., Huang, J.: Context-aware QoS prediction for web service recommendation and selection. Exp. Syst. Appl. **53**, 75–86 (2016)

8. Neumann, A., Laranjeiro, N., Bernardino, J.: An analysis of public REST web service APIs. IEEE Trans. Serv. Comput. 1 (2018, early access). https://doi.org/10.1109/TSC.2018.2847344

9. Yang, Y., Qamar, N., Liu, P., Grolinger, K., Wang, W., Li, Z., Liao, Z.: Servenet: a deep neural network for web service classification. arXiv preprint arXiv:1806.05437, 2018

10. Gomadam, K., Ranabahu, A., Nagarajan, M., Sheth, A.P., Verma, K.: A faceted classification based approach to search and rank web APIs. In: 2008 IEEE International Conference on Web Services, pp. 177–184. IEEE (2008)

11. Torres, R., Tapia, B., et al.: Improving web API discovery by leveraging social information. In: 2011 IEEE International Conference on Web Services, pp. 744–745. IEEE (2011)

12. Li, C., Zhang, R., Huai, J., Sun, H.: A novel approach for API recommendation in mashup development. In: 2014 IEEE International Conference on Web Service, pp. 289–296. IEEE (2014)

13. Wan, Y., Chen, L., Wu, J., Yu, Q.: Time-aware API popularity prediction via heterogeneous features. In: 2015 IEEE International Conference on Web Services, pp. 424–431. IEEE (2015)

14. Zhang, T., Upadhyaya, G., Reinhardt, A., Rajan, H., Kim, M.: Are code examples on an online q&a forum reliable?: A study of API misuse on stack overflow. In: 2018 IEEE/ACM 40th International Conference on Software Engineering (ICSE), pp. 886–896. IEEE (2018)

15. Treude, C., Robillard, M.P.: Augmenting API documentation with insights from stack overflow. In: 2016 IEEE/ACM 38th International Conference on Software Engineering (ICSE), pp. 392–403. IEEE (2016)

16. Venigalla, A.S.M., Lakkundi, C.S., Agrahari, V., Chimalakonda, S.: Stackdoc-a stack overflow plug-in for novice programmers that integrates q&a with API examples. In: 2019 IEEE 19th International Conference on Advanced Learning Technologies (ICALT), vol. 2161, pp. 247–251. IEEE (2019)

17. Venkatesh, P.K., Wang, S., Zhang, F., Zou, Y., Hassan, A.E.: What do client developers concern when using web APIs? An empirical study on developer forums and stack overflow. In: 2016 IEEE International Conference on Web Services (ICWS), pp. 131–138. IEEE (2016)

18. Rodríguez, L.J., Wang, X., Kuang, J.: Insights on apache spark usage by mining stack overflow questions. In: IEEE International Congress on Big Data (BigData Congress) 2018, pp. 219–223. IEEE (2018)

19. Elkan, C., Noto, K.: Learning classifiers from only positive and unlabeled data. In: Proceedings of the 14th ACM SIGKDD International Conference on Knowledge Discovery and Data Mining, pp. 213–220 (2008)

20. Barua, A., Thomas, S.W., Hassan, A.E.: What are developers talking about? An analysis of topics and trends in stack overflow. Empiric. Softw. Eng. 19(3), 619–654 (2014)

21. González, J.R.C., Romero, J.J.F., Guerrero, M.G., Calderón, F.: Multi-class multi-tag classifier system for stackoverflow questions. In: IEEE International Autumn Meeting on Power, Electronics and Computing (ROPEC), 2015, pp. 1–6. IEEE (2015)

22. Mordelet, F., Vert, J.-P.: A bagging SVM to learn from positive and unlabeled examples. Pattern Recognit. Lett. 37, 201–209 (2014)

23. Li, C., Hua, X.-L.: Towards positive unlabeled learning for parallel data mining: a random forest framework. In: Luo, X., Yu, J.X., Li, Z. (eds.) ADMA 2014. LNCS (LNAI), vol. 8933, pp. 573–587. Springer, Cham (2014). https://doi.org/10.1007/978-3-319-14717-8_45

24. Kaboutari, A., Bagherzadeh, J., Kheradmand, F.: An evaluation of two-step techniques for positive-unlabeled learning in text classification. Int. J. Comput. Appl. Technol. Res. **3**, 592–594 (2014)
25. Sandor, A., Lagos, N., Vo, N.P.A., Brun, C.: Identifying user issues and request types in forum question posts based on discourse analysis. In: Proceedings of the 25th International Conference Companion on World Wide Web, pp. 685–691 (2016)
26. Bird, S.: NLTK: The natural language toolkit, pp. 69–72 (2006)
27. Kim, Y.: Convolutional neural networks for sentence classification, arXiv preprint arXiv:1408.5882 (2014)
28. Zhang, Y., Wallace, B.: A sensitivity analysis of (and practitioners' guide to) convolutional neural networks for sentence classification, arXiv preprint arXiv:1510.03820 (2015)
29. Zaremba, W., Sutskever, I., Vinyals, O.: Recurrent neural network regularization, arXiv preprint arXiv:1409.2329 (2014)
30. Greff, K., Srivastava, R.K., Koutník, J., Steunebrink, B.R., Schmidhuber, J.: LSTM: a search space odyssey. IEEE Trans. Neural Netw. Learn. Syst. **28**(10), 2222–2232 (2016)
31. Zhang, Y., Yang, Q.: An overview of multi-task learning. Natl. Sci. Rev. **5**(1), 30–43 (2018)
32. Zhang, Y., Yang, Q.: A survey on multi-task learning, arXiv preprint arXiv:1707.08114 (2017)
33. Caruana, R.: Multitask learning. Mach. Learn. **28**(1), 41–75 (1997)
34. Ahasanuzzaman, Md, Asaduzzaman, M., Roy, C.K., Schneider, K.A.: CAPS: a supervised technique for classifying stack overflow posts concerning API issues. Empiric. Softw. Eng. **25**(2), 1493–1532 (2019). https://doi.org/10.1007/s10664-019-09743-4

Anatomy of Networks Through Matrix Characteristics of Core/Periphery

Chenxiang Luo, Cui Chen, and Wu Chen$^{(\boxtimes)}$

Southwest University, Chongqing 400700, China
{luolatazhu8,cc1026}@email.swu.edu.cn, chenwu@swu.edu.cn

Abstract. In the traditional network structure analysis research, it mainly analyzes the characteristics of individuals in the network or the way of connection. However, these studies do not reflect the general phenomenon of social networks, for example, each organization usually includes individuals with different characteristics. Only by fully understanding the characteristics and the original hierarchical structure of the network, can we improve the network security and the management of the network. Therefore, we aim to analyze the overall characteristics of the network and dissect the relationship between the layers of various networks as well as the relationship between individuals in each layer without destroying the network structure. From a new point of view, the network is analyzed by using the characteristics of network matrix structure and some properties of core/periphery structure. We analyze several real networks and verifies the intensity of the core/periphery structure relationship in the network.

Keywords: Layer · Core/periphery structure · Intensity

1 Introduction

Research on the structure of social networks is very important in many fields [1,15], which not only helps to understand the structure and operation of the network, but also plays a key role in the design and implementation of some strategies of social groups [8,11]. With the continuous development of society, some characteristics of social networks will change accordingly. The research on the network structure has never stopped. Although the network structure keeps changing with time, the characteristics of the relationship between nodes have always been regular.

Regarding the general laws of the network, first, there are generally hierarchical relationships in the network; second, the relationship between the layers, or the nodes within each layer may also be very different, then the connections between these nodes will be very different (Such as collaborative relationship between management and employees), and this relationship conforms to the core-periphery (CP) structural characteristics. In recent years, stratification has attracted scholars research interest [4,12]. These studies have almost divided the

© Springer Nature Singapore Pte Ltd. 2021
Y. Sun et al. (Eds.): ChineseCSCW 2020, CCIS 1330, pp. 515–529, 2021.
https://doi.org/10.1007/978-981-16-2540-4_37

network into several layers based on a certain feature, and have not explored the relationship between nodes in the layer and the relationship between layers after layering. We will further reveal the strong intra-layer CP relationship and inter-layer CP relationship in the network on the basis of layering. There are several problems that need to be explored

- Given any network, how to quickly determine the type of its structure?
- How to analyze the hierarchical relationship of the structure in different networks?
- How to detect the core/periphery relationship within and between layers of networks with different structures?

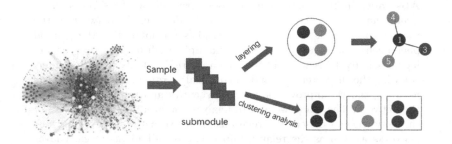

Fig. 1. Contrasting clustering with layering. Notes: the same color represents the same feature

It is worth mentioning that although many clustering methods are used for data mining, cluster analysis is a multivariate statistical method for classifying research objects according to certain characteristics. It does not care about the causal relationship between variables. Therefore, when exploring the core/periphery relationship of each part, layering is better than cluster analysis. As shown in Fig. 1, the core and periphery nodes can obviously appear on the same layer. We also verified later in the experimental part.

There are many studies on core/periphery structure, and almost all of them divide a network into the core and the periphery. In fact, multilayered core/periphery structure may exist between and within the network layer, we will reveal it. In response to the above three problems, given an arbitrary network, first, we quickly identify the type of network structure, and at the same time find that any network is in line with the three types we proposed. Secondly, different matrix analysis methods are used to analyze the above three types of structures. Finally, there is also a strong core/periphery relationship within each layer parsed, and the strength of the relationship within the layer is verified using the characteristics of the CP structure.

2   Related Work

The research on the network structure ranges from local to global. There are local research on the clustering coefficient and correlation of the network system structure [2,16], and global research on the group or community structure of the network [6]. And community testing and similar community testing to detect CP structure [5,10]. These studies have provided great help for many scholars in the back, but they did not consider both local and global at the same time. We conduct research from the perspective of the combination of global and local network structures. There is also a separate inquiry into the issue of network centers. And the super hubs of social networks tend to avoid collaboration and form a potentially competitive subgroup at the core of the network [7,18,20]. The competitive relationship exists not only at the same layer, but also at different layers. We have verified the CP relationship at different layers. Of course, the hierarchical issues on social networks have also attracted a lot of attention. The layering of k-shell [12], the layering of cells in the human body [9], and the multi-layer network inferred by the data collected by the human microbiome project [21], and the use of k-core or k-truss to disrupt the network layer by layer [23,24]. These are not the same as using the network's matrix features and CP structure to analyze the network locally and globally at the same time. These studies decompose the network. The purpose of this paper is to analyze the relationship between the layers. Therefore, their method may help solve some of the problems raised, but it cannot be applied to solve the whole problem.

3   Preliminaries

In this section, we present some basic notation and concepts that will be used throughout the paper.

3.1   Basic Core-Periphery Structure Notation

Let $G = (V, E)$ denote a social network, where $V = \{v_1, v_2, \cdots, v_n, \}$ is the set of n nodes and $E \in V \times V$ is the set of edges. The number of nodes in the network is n, which can be expressed as a matrix A. The matrix A is represented as a measurement matrix, and the elements inside are represented by a_{ij}, $A = (a_{ij})_{n \times n}$.

As a notion of ideal core/periphery structure, Borgatti and Everett defined [3]: The lower right corner of the matrix is all 0, and the rest are all 1. In the article, we use B to represent an ideal core/periphery matrix, and the elements inside are represented by b_{ij}, $B = (b_{ij})_{n \times n}$. When $i \neq j$, $b_{ij} = \begin{cases} 0 & if\ i \geq k, j \geq k \\ 1 & otherwise \end{cases}$, where k is the number of nodes in the core. Because the link relationship of agent itself is not considered, when $i = j$, the value of b_{ij} is set to null, and adjacency matrix can be represented as Fig. 2. To make the notation more readable, we will often denote two arbitrary nodes by u and v and the i-th node is denoted by u_i or v_i.

Fix a matrix B, if the core nodes are $u_1, u_2, ..., u_k$, the periphery nodes are $u_{k+1}, u_{k+2}, ..., u_n$, then it can be concluded that the subset in $\{u_1, u_2, \cdots, u_k\}$ is the core of $\{u_{k+1}, u_{k+2}, \cdots, u_n\}$, etc. Note that for convenience of calculation,

$$B=$$

	u_1	u_2	u_3	u_4	u_5	u_6	u_7	u_8	u_9	u_{10}
u_1		1	1	1	1	1	1	1	1	1
u_2	1		1	1	1	1	1	1	1	1
u_3	1	1		1	1	1	1	1	1	1
u_4	1	1	1		1	1	1	1	1	1
u_5	1	1	1	1		0	0	0	0	0
u_6	1	1	1	1	0		0	0	0	0
u_7	1	1	1	1	0	0		0	0	0
u_8	1	1	1	1	0	0	0		0	0
u_9	1	1	1	1	0	0	0	0		0
u_{10}	1	1	1	1	0	0	0	0	0	

Fig. 2. Example of idealized core/periphery structure matrix B

when $i = j$ in matrix A, a_{ij} is null, corresponding to the matrix (Fig. 2), which will not affect our experimental results. Measurement matrix A is not necessarily a matrix of nodes in the whole network, but a matrix of some nodes in the network. To detect the strength of core/periphery relationship, these nodes need to be compared with matrix B, so the number of rows and columns of matrix B must be the same as that of matrix A.

3.2 Measurement Criteria

In this section we will talk about a measurement method used to measure the results of subsequent experiments. An arbitrary network G or a matrix A of an organizational structure, the closer the matrix A is to B, the more obvious the core and the periphery, indicating that it conforms to the core/periphery structure. For matrix A and matrix B, if the number of equal values of a_{ij} and b_{ij} is more, A and B tend to be equal. When the values of elements in matrix A and matrix B are completely equal, then A is the matrix of core/periphery structure network in ideal state. We use the similarity p to measure the final result of their comparison. The similarity value is between 0 and 1, so we do a normalization process, dividing the number of equal elements in two matrices by the total number of elements:

$$p = \frac{\sum\limits_{i=1}^{n} \sum\limits_{j=1}^{n} a_{ij} \odot b_{ij}}{n*n - n}, a_{ij} \in A, b_{ij} \in B \tag{1}$$

where $\odot$ represents XNOR. We assume that the total number of nodes is n, excluding the diagonal elements of the matrix, then the number of elements in the matrix is $n*n - n$. $\sum\limits_{i=1}^{n} \sum\limits_{j=1}^{n} a_{ij} \odot b_{ij}$ can be seen in the matrix as the number of a_{ij} equal to 0 in the lower right corner and the number of other parts equal to 1. We assume that the number of network nodes is n and the number of core nodes is $m - 1$, so the intensity of core/periphery relationship can be expressed as follows:

$$p = \frac{\sum\limits_{i=m}^{n} \sum\limits_{j=m}^{n} a_{ij} \odot b_{ij} \odot 0 + \left(\sum\limits_{i=1}^{n} \sum\limits_{j=1}^{m-1} a_{ij} \odot b_{ij} \odot 1 + \sum\limits_{i=1}^{m-1} \sum\limits_{j=1}^{n} a_{ij} \odot b_{ij} \odot 1 \right)}{n*n - n} \tag{2}$$

In the equation, $a_{ij} \in A, b_{ij} \in B$. For the convenience of later description, we express the value of $\sum\limits_{i=1}^{n} \sum\limits_{j=1}^{m-1} a_{ij} \odot b_{ij} \odot 1 + \sum\limits_{i=1}^{m-1} \sum\limits_{j=1}^{n} a_{ij} \odot b_{ij} \odot 1$ as x, and the value of $\sum\limits_{i=m}^{n} \sum\limits_{j=m}^{n} a_{ij} \odot b_{ij} \odot 0$ as y, which is equivalent to $p = (x+y)/(n*n-n)$.

4   Problem Statement and Its Theoretical Analysis

Definition 1. *Let* $G = (V, E)$ *denote a network, the connection relationship between nodes is represented by matrix A, $A = (a_{ij})_{n \times n}$. We use $deg(u)$ to represent the degree of u.*

$$F(u) = \{v : v \in V \& deg(v) > deg(u)\} \tag{3}$$

In the sorted matrix A in Fig. 3, $F(u_4) = \{u_5, u_2\}$.

Definition 2. *Let* $s = \begin{cases} 0 & if \ <u, v> \in E \\ 1 & otherwise \end{cases}$, $sum_s(v)$ *is the number of nodes whose rows and columns are greater than that of node v and whose state is s. k and l represent the row and column of node v, respectively.*

$$sum_1(v) = \sum_{i=k+1}^{n} a_{ik} + \sum_{j=l+1}^{n} a_{lj} \tag{4}$$

$$sum_0(v) = 2(n-k) - sum_1(v) \tag{5}$$

$$C(u) = \{v : v \in F(u) \& sum_1(v) \geq sum_0(v)\} \tag{6}$$

The values of $sum_0(v)$ and $sum_1(v)$ for each node are illustrated in Fig. 3(b), and Fig. 4 is an example of calculating $F(u)$ and $C(u)$.

As mentioned in the measurement criteria, it can be seen from matrix B that 0 converges in the lower right corner and 1 converges in other parts. That is to say, the more 0 in the lower right corner, the better, and the more 1 in other parts, the better.

At the beginning, the matrix is irregular, so it is necessary to store and exchange node positions. Try to make more rows of 1 up and more rows of 0 down, so we will sort the nodes according to degree. The purpose is to make it easy to observe and calculate the next step by letting the number of rows with a concession value of 1 go up and the number of rows with a concession value of 0 go down (Fig. 3(a)).

In Sect. 3.2, it has been mentioned that the intensity of core/periphery relationship is related to the number of 0,1. When comparing matrices A and B, the number of 0 or 1 of the first row and the first column is calculated, so the elements of the first row or the first column should not be calculated again when calculating the second column or the second column. For example, (Fig. 3(b)) a_{1j} and a_{i1} should not be calculated when calculating $sum_1(u_2)$ or $sum_0(u_2)$.

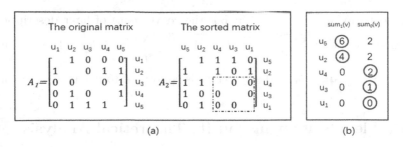

Fig. 3. Example of matrix sorting. (b) The number of 0 or 1 after each node excludes $F(u)$.

$$
\begin{array}{cccccc}
 & u_1 & u_2 & u_3 & u_4 & u_5 \\
A_3 = & \begin{bmatrix} 1 & 0 & 1 & 1 \\ 1 & & 0 & 0 & 0 \\ 0 & 0 & & 1 & 0 \\ 1 & 0 & 1 & & 0 \\ 1 & 0 & 0 & 0 & \end{bmatrix} & & & & \begin{matrix} u_1 \\ u_2 \\ u_3 \\ u_4 \\ u_5 \end{matrix}
\end{array}
$$

$F(u_4)=\{u_1,u_2,u_3\}$

$C(u_4)=\{u_1, u_3\}$

Fig. 4. Example of calculating $F(V), C(V)$ after matrix sorting.

Theorem 1. *In a sorted matrix, if $sum_1(v) \geq sum_0(v)$, the node is suitable as the core rather than the periphery.*

Proof. In Eq. 2, p is affected by two parts, x and y. If a node v satisfies $sum_1(v) \geq sum_0(v)$, we can know $sum_1(v)/(n*n-n) \geq sum_0(v)/(n*n-n)$ according to this condition. Adding $sum_0(v)$ to y is not as good as adding $sum_1(v)$ to x to make p higher. Nodes in the core can make the measurement matrix A closer to the matrix B, so node v is suitable as the core rather than the periphery.

Example 1. In Fig. 3, each node is judged step by step. As shown in Fig. 3(b), the $sum_0(v)$ and $sum_1(v)$ corresponding to u_5 are 6 and 2 respectively. After excluding $F(v)$ from nodes u_2, u_4, u_3 and u_1, the number of 0 and 1 corresponding to the row and column is calculated as shown in Fig. 3(b). $sum_1(v)$ of u_2 and u_5 is greater than $sum_0(v)$, so they are more suitable as core nodes. On the contrary, when u_4, u_3 and u_1 are calculated, $sum_0(v)$ is greater than $sum_1(v)$, so they are more suitable as periphery nodes. It can be seen that the bottom right corner of the sorted matrix A_2 is all 0, and the rest 1 is more. The highest accuracy of A_2 in accordance with the core/periphery structure is $(6+4+2+1+0)/(5\times5-5) = 65\%$. The core nodes are u_5, u_2, and the periphery nodes are u_4, u_3, u_1.

In fact, from Example 1, we can find a rule can be proposed that we can directly search for the closest or equal positions of $sum_0(v)$ and $sum_1(v)$, and compare them with $sum_0(v)$ and $sum_1(v)$ in the following line, respectively. In case of inversion, it is actually the demarcation point between the core and the periphery (the demarcation point of example 1 is u_4). In this way, we can search this demarcation point directly, which greatly reduces the computational complexity. The demarcation point we mentioned above is equivalent to the demarcation point between the core and the periphery of a layer. In the following discussion, the demarcation point we will mention is also the demarcation point between layers of the network. For example (Fig. 4), the demarcation point within the layer is u_2, u_4, and the demarcation point between layers is u_3. The demarcation points between layers satisfy $sum_1(u_3) \geq sum_0(u_3)$, and the upper row($u_1, u_3$) of the demarcation points(u_2, u_4) in the layer satisfies $sum_1(u_1 \ or \ u_3) \geq sum_0(u_1 \ or \ u_3)$.

Theorem 2. *In the sorted matrix, each demarcation point u and the following(behind) nodes can only be used as periphery nodes of nodes ($C(v)$) satisfying $sum_1(v) \geq sum_0(v)$ in $F(u)$, $\forall v \in V$.*

Proof. In the sorted matrix, the number of 1 decreases gradually. In $C(u)$, $\forall v \in C(u), sum_0(v) \leq sum_1(v)$. And because any node v' below (behind) the demarcation point satisfies $sum_0(v') \leq sum_1(v)$ and $sum_1(v') \leq sum_1(v)$, each demarcation point and the nodes below can only be used as the periphery nodes of $C(v)$.

The problem of demarcation points has been solved. In this section, we will analyze different networks. The following will give an example of how to determine the network type, the layering method for the three types of network structure, and the structural relationship between the nodes in the layer and the nodes in each layer. We use IN to express the relationship between layers, and OUT to express the relationship within each layer. Several types of matrices are sorted according to the above method. In order to observe the structural features conveniently, the diagonal elements in matrix A are not neglected in figure for the time being and are presented in the figure. And they are not considered in calculation, which will not affect the results. Three types will be discussed below, i and j represent the rows and columns of matrix A, respectively.

Type 1: When $i = 1, j = 1$, $sum_0(v) \leq sum_1(v)$. Under this condition, it is defined as type 1, then the first node is the core and a core node of the whole network. Next, all demarcation points are searched. A boundary line appears at the position where the reversal occurs after $sum_0(v)$ and $sum_1(v)$ comparison. The first node below the boundary line is the demarcation point. The upper left corner of a square with a solid line is the demarcation point (Fig. 5). If there are two demarcation points, the demarcation points are two nodes u_1 and u_2 corresponding to a_{kk} and a_{qq} lines. Because when $i = k, j = k$, $sum_0(u_1) \leq sum_1(u_1)$. The same reason, when $i = q, j = q$, $sum_0(u_2) \leq sum_1(u_2)$, which shows that u_1 is the core of the node in the middle solid square box. The same u_2 is the core of the node in the smallest solid wire frame.

The upper and lower layers of each demarcation line can correspondingly be divided into two layers of the network. Draw three layers of the network (Fig. 5). The first layer contains the nodes between the dashed line and the middle solid

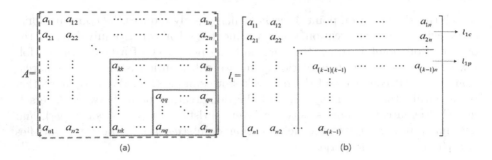

Fig. 5. Example of layering of type 1 . (b) Layer l_1 is the first layer of matrix A .

frame. In this layer, $sum_0(v)$ and $sum_1(v)$ in each line is compared by the above method. When $sum_0(v)$ is greater than $sum_1(v)$, it is the dividing line between the core and the periphery in the first layer. Because the first row above the demarcation point must be $sum_0(v)$ greater than $sum_1(v)$, the corresponding node of that row must be the periphery node of the upper layer. It is concluded that each of the three layers above has its own core and periphery. We use l_i for layer i. The first layer l_1 has been drawn in the Fig. 5(b). The solid line represents the boundary between the core and periphery of layer l_1. The core and periphery of the first layer are expressed as l_{1c} and l_{1p}, respectively, and other layers are defined by analogy. Layer i is the core of layer j with $l_i \leftarrow l_j$, and layer i is the core of layer j and layer k with $l_i \leftarrow l_j + l_k$. Figure 5(a) shows the relationship between layers and Fig. 5(b) shows the relationship between the inner core/periphery structure of each layer. As for the network hierarchy that appears above, it can be expressed as: $L = l_1 + l_2 + \cdots + l_n$
The relationship within each layer satisfies(IN):

$$l_{1c} \leftarrow l_{1p}, l_{2c} \leftarrow l_{2p}, \cdots, l_{nc} \leftarrow l_{np} \tag{7}$$

The relationship between type 1 layers satisfies(OUT):

$$\begin{cases} l_{1c} \leftarrow l_{1p} + l_2 + l_3 + \cdots + l_n \\ l_{2c} \leftarrow l_{2p} + l_3 + l_4 + \cdots + l_n \\ \quad\quad \cdots \\ l_{(n-1)c} \leftarrow l_n \end{cases} \tag{8}$$

L sees the entire network as a whole layer. Each relationship can be compared with matrix B to calculate the accuracy and reflect the strength of the relationship. This network hierarchy presents water wave shape. This kind of network is multi-layer CP structure, similar to onion structure, which is different from the two-layer CP network proposed before [17].

Type 2: When $i = 1, j = 1$, $sum_0(v) > sum_1(v)$, then the first node is not suitable for the core of the whole network, so it is necessary to search for the

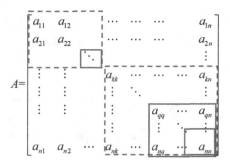

Fig. 6. Example of layering of type 2

existence of demarcation points. If there is a demarcation point, this is defined as type 2. Then search the first demarcation point from top to bottom, and divide the whole network into two main parts, which are defined as *upper plate* and *lower plate*. They are separated from the first demarcation point (as shown in the Fig. 6). The line above the row corresponding to a_{kk} is the dividing line. At this time, the network is mainly composed of two parts, and the two parts are analyzed separately. Figure 6 shows that both parts conform to type 1. In real life, there are two main parts of a network, and each has its own management system. It is worth noting the difference between type 1 and type 2. Type 1 is to treat the network as an onion-like whole, while type 2 is to divide the network into two parts. Any one of the two parts may conform to type 1, type 2 and type 3. The figure shows that two small parts correspond to type 1. Note that if there is no demarcation point on the upper or lower part, or the plate is relatively large, you can choose to search for it in the same way as type 3. About type 2, as for the network layer that appears above (Fig. 6), it can be expressed as: $L = l_1 + l_2$
The intra-layer relationship(IN) is:

$$l_{1c} \leftarrow l_{1p}, l_{2c} \leftarrow l_{2p}, \cdots, l_{nc} \leftarrow l_{np} \tag{9}$$

The interlayer relationship(OUT) of the upper plate(l_1) is

$$\begin{cases} l_{11c} \leftarrow l_{11p} + l_{12} + l_{13} + \cdots + l_{1n} \\ l_{12c} \leftarrow l_{12p} + l_{13} + l_{14} + \cdots + l_{1n} \\ \qquad \cdots \\ \qquad l_{1(n-1)c} \leftarrow l_{1n} \end{cases} \tag{10}$$

The interlayer(OUT) relationship of the lower plate(l_2) is

$$\begin{cases} l_{21c} \leftarrow l_{21p} + l_{22} + l_{23} + \cdots + l_{2n} \\ l_{22c} \leftarrow l_{22p} + l_{23} + l_{24} + \cdots + l_{2n} \\ \qquad \cdots \\ \qquad l_{2(n-1)c} \leftarrow l_{2n} \end{cases} \tag{11}$$

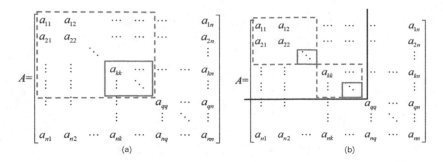

Fig. 7. Example of layering of type 3, (a) type 3 converted to type1, and (b) type 3 converted to type 2.

Type 3: When $i = 1, j = 1$, $sum_0(v) > sum_1(v)$, then the first node is not suitable for the core of the whole network. And if there is no demarcation point in search, it is defined as type 3. Then this time is more special, generally because the network is larger. The core of the network is hidden, and many new agents may join the network at any time. Almost all new agents are on the periphery of the network, with a high centrifugal rate [14,22]. There are few network cores and too many periphery nodes. We define these nodes as *libertarian*, represented by the symbol α. Their addition affects the network matrix, does not establish strong links with the core, and the peripheral links of the matrix are sparse. The more *libertarian*, the more sparse the network is. For example, in scale-free networks, *libertarian* is particularly common [13]. And satisfy the relationship:

$$L = l_1 + l_2 + \cdots + l_n + \alpha \tag{12}$$

At this point, we need to find the largest major part of the network that excludes *libertarian*. We exclude these nodes in order from the right to the left of the matrix. When we exclude the first one, we judge whether the matrix satisfies when $i = 1, 1 < j < n - 1, sum_0(v) \leq sum_1(v)$. The number of elements of *libertarian* is expressed by t. If not, then judge in turn whether it satisfies: when $t = 2, 3, \cdots, n$ and $i = t, t < j < n - t$, $sum_0(v) \leq sum_1(v)$.

When the demarcation point appears in the middle of the matrix, it stops. Apart from *libertarian*, the rest goes back to type 2 (Fig. 7(b)). Or when the $sum_0(v)$ of the first line is less than or equal to $sum_1(v)$. At this point, excluding *libertarian*, the rest goes back to the type 1 (Fig. 7(a)).

For an arbitrary network, the type is first determined and then calculated in three ways. Our experiments(NTDL) show that the first two types almost exist in smaller networks, while the last one generally exists in larger networks. We can solve any complex network according to the above three types, analyze its overall and internal relations, and help us improve network security or increase cooperation by intervening in the network.

5 Experimental Results and Discussions

Table 1. Statistics of datasets

Dataset	monkey	football	bitcoinalpha	GrQc	bitcoinotc	HepTh	Facebook	Orkut
Vertices	35	115	3783	5242	5881	12008	134833	3072441
Edges	69	613	14124	14496	21492	118521	1380293	117185083
max deg	12	12	511	81	795	491	1469	33313

Datasets: 8 real-life network networks are deployed in our experiments. The original data of all are from http://snap.stanford.edu/ and http://konect.uni-koblenz.de/. Table 1 shows statistics of 8 datasets which are listed in increasing order of their edge numbers.

In the experimental part, the three algorithms are first compared with each other. In Fig. 8, the Random algorithm and the CPDG algorithm are used for comparison. The Random algorithm selects the same network as our proposed algorithm (NTDL), but it uses a random extraction method when selecting nodes as the core and periphery. Compared with random algorithm, in order to verify that the proposed algorithm can well test the core/periphery relationship between nodes in the network. In addition, the NTDL and CPDG algorithms both sort the nodes according to their degree. The difference between the two algorithms is that the NTDL algorithm uses the method of finding the demarcation point in Sect. 4. The CPDG algorithm is to divide core nodes and the periphery nodes equally. It can be seen from the bar graph that the proposed algorithm has greater advantages than Random and CPDG.

In Fig. 8, the line graph tests the strength of the relationship between OUT and IN in the network(intensity). According to different networks, it is divided into three different types. The monkey network has only 35 nodes, a very small network, which was detected as type 1 through experiments; the football network is also relatively small, which is detected as type 2; the other networks are larger, which is type 3. It can be concluded from the line graph that many nodes in the network have a strong core/periphery relationship.

In addition, we used cluster analysis as a comparison. We adopt k-means++ clustering method based on network embedding features. Figure 9 shows the intensity of CP relationships in clusters and between clusters after clustering through 8 networks. Figure 9(a) shows the situation within the cluster. Since clustering generally classifies nodes with similar features into one cluster, the CP relationship in the cluster is not obvious. It can also be seen from the experiment that the abscissa represents the cluster. For example, the monkey network is divided into 2 clusters (corresponding to the layer we divided). Figure 9(b) is to avoid the bad results caused by the similarity of the nodes in a cluster. Let two different clusters construct the core and the periphery separately to see if they can produce different effects. However, experiments have shown that there

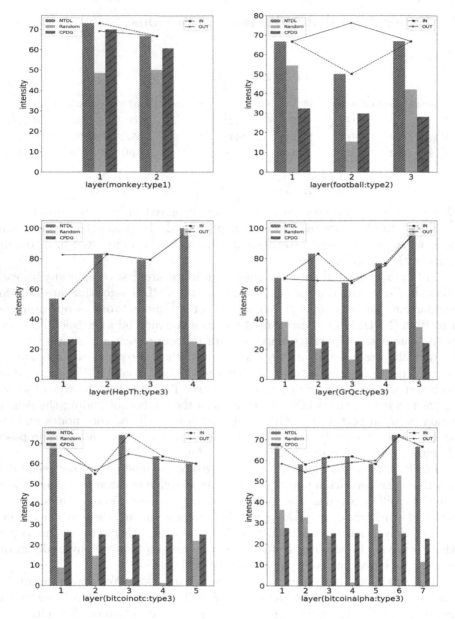

Fig. 8. The relationship strength among network layers(OUT) and within network layers(IN), and the comparison among three algorithms.

is no good effect, and the intensity does not exceed 50%. The abscissa represents the combination of different clusters after network clustering, for example, the football network is divided into 3 clusters, then there are $A(3,2) = 6$ arrangements. The networks including GrQc, bitcoinotc, bitcoinalpha, Facebook and

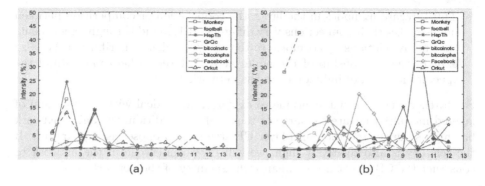

Fig. 9. Clustering algorithm explores core/periphery relationship

Orkut have $A(5,2)$, $A(5,2)$, $A(7,2)$, $A(10,2)$ and $A(13,2)$ cases, respectively. We extracted 12 cases and presented them in the figure. It is further proved that when the clustering algorithm is studying the strength of CP relationship, the algorithm we proposed is more effective than it.

6   Conclusion

We propose and study the problems and methods of comprehensive analysis of the network from a new perspective. The purpose is to find out the ubiquitous cooperative relationship and distribution of nodes in the network. Experimental research shows that the real network generally has a certain hierarchical structure. There may be a relationship between network layers. The collaboration between nodes within the layer also shows strong core/periphery characteristics. The method proposed in this paper is feasible on small or large networks and has scalability.

In short, we make three main contributions:

– First, we use the structural characteristics of the network matrix and the connection of nodes to determine the type of network structure. For example, a multi-layer CP network similar to the onion structure has high security [19]; there is also a scale-free network that is familiar to everyone, and the core is good in concealment, but once attacked, the entire network is easy to collapse [13]. You can analyze the structure and surrounding conditions of the core, make certain adjustments to the network, and improve the security of the network. Through matrix analysis, we classify the network into three types and explain each type.
– Second, we analyze the layer of the network. Although every network is constantly changing, the algorithm proposed is applicable to any stage of the network, which helps to constantly detect the characteristics of the network structure, facilitate management and make appropriate adjustments.

- Third, regarding nodes in the layer, the cooperation or competition between nodes makes their connections very different. It is found through experiments that many relationships conform to the core/periphery distribution. We also explain the relationship of the nodes at each layer to help understand the specific situation of individuals in the network.

For further study, we will continue to explore how to deal with the problem of layered modules that are too large or too small. In addition, the onion network has high security, and the multi-layer CP structure proposed is similar to it. If the security is also relatively high, we will try to intervene in the network to construct the CP structure to enhance the security of the network.

References

1. Albert, R., Barabási, A.L.: Statistical mechanics of complex networks. Rev. Mod. Phys. **74**(1), 47 (2002)
2. Boccaletti, S., Latora, V., Moreno, Y., Chavez, M., Hwang, D.U.: Complex networks: structure and dynamics. Phys. Rep. **424**(4–5), 175–308 (2006)
3. Borgatti, S.P., Everett, M.G.: Models of core/periphery structures. Soc. Netw. **21**(4), 375–395 (2000)
4. Clauset, A., Moore, C., Newman, M.E.: Hierarchical structure and the prediction of missing links in networks. Nature **453**(7191), 98–101 (2008)
5. Diani, M., Wasserman, S., Faust, K.: Social Network Analysis: Methods And Applications, p. 825. Cambridge, Cambridge University Press (1994). Italian Political Science Review/Rivista Italiana di Scienza Politica **25**(3), 582–584 (1995)
6. Fortunato, S.: Community detection in graphs. Phys. Rep. **486**(3–5), 75–174 (2010)
7. Goyal, S.: Connections: An Introduction to the Economics of Networks. Princeton University Press, Princeton (2012)
8. Haggerty, K.L., Griffith, J., McGuire, J., Molnar, B.: Elder mistreatment and social network composition: an exploratory study. Soc. Netw. **59**, 23–30 (2019)
9. Ingram, D.A., et al.: Identification of a novel hierarchy of endothelial progenitor cells using human peripheral and umbilical cord blood. Blood **104**(9), 2752–2760 (2004)
10. Kovács, I.A., Palotai, R., Szalay, M.S., Csermely, P.: Community landscapes: an integrative approach to determine overlapping network module hierarchy, identify key nodes and predict network dynamics. PloS One **5**(9), e12528 (2010)
11. Krogan, N.J., et al.: Global landscape of protein complexes in the yeast saccharomyces cerevisiae. Nature **440**(7084), 637–643 (2006)
12. Lahav, N., Ksherim, B., Ben-Simon, E., Maron-Katz, A., Cohen, R., Havlin, S.: K-shell decomposition reveals hierarchical cortical organization of the human brain. New J. Phys. **18**(8), 083013 (2016)
13. Lewis, T.G.: Network Science: Theory and Applications. Wiley, Hoboken (2011)
14. Moskvina, A., Liu, J.: How to build your network? A structural analysis. arXiv preprint arXiv:1605.03644 (2016)
15. Newman, M.E.: The structure and function of complex networks. SIAM Rev. **45**(2), 167–256 (2003)
16. Newman, M.E.: Complex systems: A survey. arXiv preprint arXiv:1112.1440 (2011)
17. Rombach, M.P., Porter, M.A., Fowler, J.H., Mucha, P.J.: Core-periphery structure in networks. SIAM J. Appl. Math. **74**(1), 167–190 (2014)

18. Saavedra, S., Malmgren, R.D., Switanek, N., Uzzi, B.: Foraging under conditions of short-term exploitative competition: the case of stock traders. Proc. R. Soc. B: Biol. Sci. **280**(1755), 20122901 (2013)
19. Schneider, C.M., Moreira, A.A., Andrade, J.S., Havlin, S., Herrmann, H.J.: Mitigation of malicious attacks on networks. Proc. Natl. Acad. Sci. **108**(10), 3838–3841 (2011)
20. Smilkov, D., Kocarev, L.: Rich-club and page-club coefficients for directed graphs. Phys. A **389**(11), 2290–2299 (2010)
21. Stanley, N., Shai, S., Taylor, D., Mucha, P.J.: Clustering network layers with the strata multilayer stochastic block model. IEEE Trans. Netw. Sci. Eng. **3**(2), 95–105 (2016)
22. Yan, B., Liu, Y., Liu, J., Cai, Y., Su, H., Zheng, H.: From the periphery to the center: information brokerage in an evolving network. arXiv preprint arXiv:1805.00751 (2018)
23. Zhang, F., Li, C., Zhang, Y., Qin, L., Zhang, W.: Finding critical users in social communities: the collapsed core and truss problems. IEEE Trans. Knowl. Data Eng. **32**(1), 78–91 (2018)
24. Zhang, F., Zhang, Y., Qin, L., Zhang, W., Lin, X.: Finding critical users for social network engagement: the collapsed k-core problem. In: Proceedings of the Thirty-First AAAI Conference on Artificial Intelligence, pp. 245–251 (2017)

A Light Transfer Model for Chinese Named Entity Recognition for Specialty Domain

Jiaqi Wu[1], Tianyuan Liu[1,3], Yuqing Sun[1,2(✉)], and Bin Gong[1,2]

[1] School of Software, Shandong University, Jinan, China
`oofelvis@163.com, zodiacg@foxmail.com`
[2] Engineering Research Center of Digital Media Technology, Shandong University, Jinan, China
`{sun_yuqing,gb}@sdu.edu.cn`
[3] School of Computer Science and Technology, Shandong University, Jinan, China

Abstract. Named entity recognition (NER) for specialty domain is a challenging task since the labels are specific and there are not sufficient labelled data for training. In this paper, we propose a simple but effective method, named Light Transfer NER model (LTN), to tackle this problem. Different with most traditional methods that fine tune the network or reconstruct its probing layer, we design an additional part over a general NER network for new labels in the specific task. By this way, on the one hand, we can reuse the knowledge learned in the general NER task as much as possible, from the granular elements for combining inputs, to higher level embedding of outputs. On the other hand, the model can be easily adapted to the domain specific NER task without reconstruction. We also adopt the linear combination on each dimension of input feature vectors instead of using vector concatenation, which reduces about half parameters in the forward levels of network and makes the transfer light. We compare our model with other state-of-the-art NER models on real datasets against different quantity of labelled data. The experimental results show that our model is consistently superior than baseline methods on both effectiveness and efficiency, especially in case of low-resource data for specialty domain.

Keywords: Named entity recognition · Light transfer model · Specialty domain

1 Introduction

Name entity recognition (NER) refers to the recognition of entities with specific meanings in the text, such as person, location, organization etc. It is a fundamental task in natural language processing that provides useful information on constructing knowledge graph, syntactic parsing or information retrieval.

There are lots of research works on the general NER task, One of the representative kind of methods focus on handcrafted rules, which need a lot of

© Springer Nature Singapore Pte Ltd. 2021
Y. Sun et al. (Eds.): ChineseCSCW 2020, CCIS 1330, pp. 530–541, 2021.
https://doi.org/10.1007/978-981-16-2540-4_38

linguistic knowledge, and the rules of different languages or domains are not the same. Kim [10] proposed to use Brill rule and Hanisch et al. [5] proposed a system named ProMiner to recognize entities in biomedical text. Build handcrafted rules is laborious, and not portable. Traditional machine learning models like maximum entropy Markov models (MEMMs) [16] and conditional random fields (CRF) [11] have a good performance in NER task. But these methods often require feature engineering, which is troublesome. Compared to these models, the neural models handle the NER task more effectively in an end-to-end mode. The use of neural network structure to solve the problem of NER can be traced back to 2003, when Hammerton [4] first proposed a NER model based on LSTM. In 2011, Collobert et al. [2] proposed a multilayer neural network to tackle the NER task, where they use the language model to compute word embeddings and apply multiple tasks to train the model. Then, Huang et al. [7] proposed the neural architecture BiLSTM-CRF for NER, where BiLSTM is used to extract sequence representations and CRF is for decoding tags. Similarly, Ma and Hovy [15] adopt both word level and character level features to feed the proposed LSTM-CNN-CRF model. In recent years, language model pretraining methods such as BERT [3] and ELMO [18] achieved state-of-the-art performance in many NLP tasks including NER. Li et al. [12] proposed a BERT-MRC architecture for NER which recognitions entities through machine reading comprehension task and then Li et al. [13] introduce dice loss to improve the performance of BERT-MRC.

Although most of these methods achieve good performances, they usually require a lot of labelled data for training the neural network. Considering the general NER task in their works, i.e. the entities are recognized by common sense, the data can be shared across different methods.

There is an increasing need to recognize specific *entities* in specialty domain. For example, in the field of Biology, the entities that are considered for amino acids, such as Glycine, Glutamic acid, Lysine, rather than person names as the general NER task. In another specialty example of procuratorate, the extracted elements are domain specific. The considered entities are the crime suspect, procurator, court, and etc. Although the purpose of NER task for specialty domain is similar with the general NER task, the label set are completely different. Thus the pretrained neural network for the general NER task can not be directly applied to this specific task. Furthermore, there are usually not sufficient labelled data in specialty domain to train a specific neural network.

To address this challenging problem for low-resource NER in specialty domain, some works adopt the transfer learning. Wang et al. [19] propose a label-aware double transfer learning framework, where both the Bi-LSTM feature representations and the CRF parameters are transferred to the specific NER task. Lin and Lu [14] adapt the top layers of existing NER neural architecture to solve the specific task. Recently, Jia et al. [8] consider the language model as the companied task with NER model and adopt multi-task learning to tackle the NER task for a new specialty domain. As expected, these methods benefit a lot from the general NER task by transfer learning. But they are not appropriate for the Chinese NER task since they do not take advantage of the characteristics of Chinese components. The most related work is the lattice-structured LSTM model for Chinese NER, proposed by Zhang et al. [20]. It encodes a sequence of input characters as well as all potential words that match a lexicon and achieves the best results on their provided specialty domain datasets. However, in order to encode lexicon effectively, the lattice-structured LSTM model is designed with a very complex cell structure. Motivated by this work, we explore the effective use of character level and word level information of Chinese. But different with their method, we design the linear combination on each dimension of input feature vectors rather than the concatenation operation, which reduces about half parameters in the forward levels of network and makes the transfer light. Besides, our transfer part is delicately constructed. Our contribution are summarized as follows.

Firstly, to tackle the problem of NER in specialty domain, we propose a simple but effective Light Transfer NER model (LTN) that consists of the general part (LTN-G) and transfer part (LTN-T), as illustrated in Fig. 1. LTN-G is based on the combination of BiLSTM and CRF networks, which is pre-trained by a general NER dataset. For each token of the input sequence, we make full use of Chinese characteristics, including the pretrained embeddings of characters and words, as well as the information on part of speech (POS) tags. The POS embeddings are randomly initialized and updated with the network. For the light weight purpose, we combine the character and word embeddings in a linear way on each dimension instead of the concatenation operation on the whole vector as traditional methods, which can integrate the semantics and at the same time reduce the number of parameters in the forward network.

Secondly, in the transfer part, we try to reuse the knowledge learned in the general part as much as possible, from the granular elements for combining inputs, to a higher level embeddings of sentence. Thus, we take into account the hidden states of the final level of LTN-G as the inputs to the transfer part since these high-level sequences contain rich semantics and latent syntax knowledge. Similarly, the outputs of general tag sequence are embedded and are fed to the transfer part also. For the light purpose, we adopt the gated recurrent unit (GRU) in this part since it has fewer parameters compared to LSTM.

Finally, we perform experiments on real datasets and compare with other state-of-the-art methods. The results on two Chinese specialty domain datasets show that our model outperforms other baseline methods in the Chinese NER task on both effectiveness and efficiency. Even with a small quantity of labelled data, our model still has an acceptable result. Besides, it is worth mentioning that there is no need to change either the data tags or the CRF layer of LTN-G in the transfer process.

2   The Light Transfer Model

2.1   General Part

As illustrated in Fig. 1, the general part of our model LTN is based on the BiLSTM-CRF [6,7] structure where the BiLSTM encodes context information better than LSTM and the CRF has proved to be effective in NER task. It is pretrained on the general NER datasets and is fine-tuned for the specific NER task.

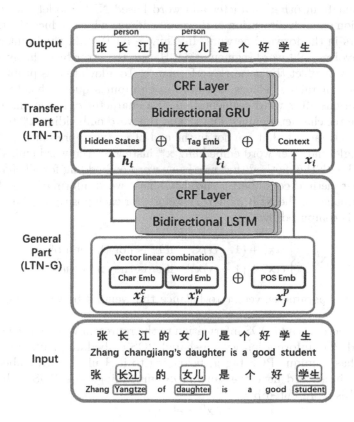

Fig. 1. The light transfer NER model (LTN)

Let $\mathbf{x} = c_1, ..., c_n$ be a Chinese character sequence of input sentence, and $\mathbf{y} = y_1, ..., y_n$ be the ground truth NER tag sequence for $\mathbf{x}$. Each character c_i is mapped to vector $\mathbf{x}_i^c$ using a pretrained character embedding lookup table:

$$\mathbf{x}_i^c = \mathbf{e}^c(c_i). \tag{1}$$

After word segmentation, the input sentence can be seen as a word sequence $\mathbf{x}' = w_1, ..., w_m$. Each word w_j is mapped to vector $\mathbf{x}_j^w$ with pretrained word embeddings:

$$\mathbf{x}_j^w = \mathbf{e}^w(w_j). \tag{2}$$

If a word consists of only a single character after word segmentation, it is also regarded as a word that is pretrained with other words, denoted by $\mathbf{x}_i^w = \mathbf{e}^w(c_i)$. Each word embedding has the same dimensions with a character embedding. For clarity, we adopt $\mathbf{e}^w$ to represent the pretrained word embedding lookup table and $\mathbf{e}^c$ to represent the one for characters. The footmark i represents the ith character in $\mathbf{x}$ and j represents the jth word in $\mathbf{x}'$.

Different from other character and word based NER models, we use linear combination on each dimension of their embeddings which reduce the number of parameters in the forward network. In Chinese NER task, the input sequences usually involves gold word segmentation. Hence, the error brought by word segmentation will affect the network performance. To alleviate this problem, in our model, each character belongs to a word in an input sequence has two different representations after word segmentation. If a character c_i is the end of a word w_j, we use the character embedding $\mathbf{x}_i^c$ and the word embedding $\mathbf{x}_j^w$ to represent it. For the characters that are not in the end of w_j, we represent the characters with a single character word embedding $\mathbf{x}_i^w$ instead of the word embedding $\mathbf{x}_j^w$. Compared with the method that uses same word embedding for all characters in a word, our method could better encode Chinese word information and improve the performance of the NER model. Therefore, for each position c_i, the combined vector $\mathbf{x}_i^l$ is computed by:

$$\mathbf{x}_i^l = \begin{cases} \boldsymbol{\alpha}\mathbf{x}_i^c + (1 - \boldsymbol{\alpha})\mathbf{x}_i^w & \text{if } c_i \text{ is not the end of } w_j \\ \boldsymbol{\alpha}\mathbf{x}_i^c + (1 - \boldsymbol{\alpha})\mathbf{x}_j^w & \text{if } c_i \text{ is the end of } w_j \end{cases} \tag{3}$$

where $\boldsymbol{\alpha}$ is a parameter vector to balance the weights between character and word.

Furthemore, we add POS information to the representation of each character. We embed POS information with a shared POS embedding lookup table. Each word w_j has a unique POS tag p_j belonging to it. Unlike word embedding, all characters in a word have the same POS embedding. The POS embedding of each word is represented by:

$$\mathbf{x}_j^p = \mathbf{e}^p(p_j). \tag{4}$$

Here, $\mathbf{e}^p$ is the POS embedding lookup table which is randomly initialized at the beginning of the training and updates during the training.

For the input sequence $\mathbf{x} = c_1, ..., c_n$, the final input representation at cell i is:

$$\mathbf{x}_i = \mathbf{x}_i^l \oplus \mathbf{x}_i^p \tag{5}$$

where $\oplus$ represents concatenation. We apply a standard bi-directional LSTM for $\mathbf{x}$ to learn the context, where each cell accepts $\mathbf{x}_i$ and computes it with the previous hidden state for the current state $\mathbf{h}_i$:

$$\overrightarrow{\mathbf{h}}_i = LSTM(\overrightarrow{\mathbf{h}}_{i-1}, \mathbf{x}_i) \tag{6}$$

$$\overleftarrow{\mathbf{h}}_i = LSTM(\overleftarrow{\mathbf{h}}_{i+1}, \mathbf{x}_i) \tag{7}$$

$$\mathbf{h}_i = \overrightarrow{\mathbf{h}}_i \oplus \overleftarrow{\mathbf{h}}_i \tag{8}$$

The standard CRF model is used as the output layer for LTN-G. The output probability $p(\mathbf{y}|\mathbf{x})$ is computed over label sequence $\mathbf{y} = y_1, ..., y_n$:

$$p(\mathbf{y}|\mathbf{x}) = \frac{exp\{\sum_i (\mathbf{w}_{CRF}^{y_i} \mathbf{h}_i + b_{CRF}^{(y_{i-1}, y_i)})\}}{\sum_{y'} exp\{\sum_i (\mathbf{w}_{CRF}^{y_i'} \mathbf{h}_i + b_{CRF}^{(y_{i-1}', y_i')})\}} \tag{9}$$

where y' represents a possible label sequence, and $\mathbf{w}_{CRF}^{y_i}$ is a model parameter specific to y_i, and $b_{CRF}^{(y_{i-1}, y_i)}$ is a bias specific to y_{i-1} and y_i. We use the first-order $Viterbi$ algorithm to find the label sequence with a highest score. Given a set of labelled training data G from general domain, the sentence-level negative log-likelihood loss with L2 regularization is used to train the general part model:

$$Loss = - \sum_{(\mathbf{x},\mathbf{y}) \in G} log(p(\mathbf{y}|\mathbf{x})) + \lambda \|\Theta\|^2 \tag{10}$$

where λ is the L2 regularization parameter and Θ represents the parameter set.

2.2   Transfer Part

The transfer part LTN-T is based on BiGRU-CRF structure since GRU [1] has fewer parameters and is more efficient than LSTM. We takes the output tag sequence and hidden states of LTN-G as the inputs to the transfer part. For an input sequence $\mathbf{x} = c_1, ..., c_n$, let $\mathbf{l} = l_1, ..., l_n$ denote its output tag sequence of by LTN-G. Each tag l_i is mapped to an embedding $\mathbf{x}_i^t$ represented as:

$$\mathbf{x}_i^t = \mathbf{e}^t(l_i). \tag{11}$$

Here $\mathbf{e}^t$ represents a shared tag embedding lookup table. So the input sequence $\mathbf{t} = t_1, ..., t_n$ of LTN-T is computed by:

$$t_i = \mathbf{x}_i \oplus \mathbf{x}_i^t \oplus \mathbf{h}_i \tag{12}$$

where $\mathbf{h}_i$ represents output of general part BiLSTM at step i and $\mathbf{x}_i$ is the input representation from the general part. We apply a standard bi-directional GRU layer for the input T, where the computation at each cell i is as follows.

$$\overrightarrow{\mathbf{h}}_i = GRU(\overrightarrow{\mathbf{h}}_{i-1}, t_i) \tag{13}$$

$$\overleftarrow{\mathbf{h}}_i = GRU(\overleftarrow{\mathbf{h}}_{i+1}, t_i) \tag{14}$$

$$\mathbf{h}_i = \overrightarrow{\mathbf{h}}_i \oplus \overleftarrow{\mathbf{h}}_i \tag{15}$$

Similar with the general part, we apply the standard CRF as the output of transfer part and use the first-order $Viterbi$ algorithm to compute the highest scored label sequence. The loss function is the same with Eq. 10.

2.3   Training Process

We first train the general part LTN-G with the labelled data for the general NER task. After it performs well in the general task, it can be reused for NER tasks in different domains. When we train the network for a specific NER task, there is no need to change the tag set of the CRF layer in LTN-G. The new labels for the specific task are only applied to the CRF layer of the transfer part. The whole process is presented in Algorithm 1. When applying it to different specialty domains, we can reuse the parameters of the general part, which makes transfer model light.

Algorithm 1. Transfer learning

Require: General domain data G and tag set T^G; Specialty domain data S and tag set T^S
Ensure: Transfered model for target domain
 1: **General part training:**
 2: **while** training steps not end **do**
 3:     split training data into minibatches $G_m \subset G$:
 4:     $\Delta h^{LSTM}, \Delta w^{CRF}, \Delta \alpha... \leftarrow train(G_m)$
 5:     update general part parameters Θ
 6: **end while**
 7: load the parameters Θ of LTN-G
 8: **Whole model training:**
 9: **while** training steps not end **do**
10:     split training data into minibatches $S_m \subset S$:
11:     $\Delta h^{GRU}, \Delta h^{LSTM}, \Delta w^{CRF}, \Delta \alpha... \leftarrow train(S_m)$
12:     update LTN parameters
13: **end while**

---

3   Experiments

3.1   Datasets and Experiment Settings

We adopt three chinese datasets in different domains, **MSRA** NER [9], **Weibo** NER [17] and **Resume** NER [20], the statistics of which are shown in Table 1.

MSRA [9]: It is a news domain dataset which contains three types of entities: PER (person), LOC (location), and ORG (organization). **MSRA** is used as shared task on SIGNAN backoff 2006. Since this dataset contains sufficient data and is widely adopted in the Chinese NER research works, we select it to pretrain LTN and other comparison models for the general NER task.

Weibo [17]: It is a low-source social media domain dataset from Sina Weibo which contains four named entities: PER.NAM (person), LOC.NAM (location), ORG.NAM (organization) and GPE.NAM(geo-political). It also contains four nominal entities:PER.N0M (person), LOC.N0M (location), ORG.N0M (organization) and GPE.N0M(geo-political). There are five types of entities different from **MSRA**.

Resume [20]: It is a resume domain dataset annotated by Zhang et al. [20]. It contains eight types of entities: NAME (name), LOC (location), ORG (organization), CONT (Nationality), RACE (race), TITLE (title), EDU (education) and PRO (profession). There are five types of entities different from **MSRA**.

Comparatively, both **Weibo** and **Resume** contain different entities. Thus, they are used for the specific NER task. It is worth mentioning that the labelled entities in **Weibo** are more sparse than **Resume**.

We implement the algorithm in Pytorch. The sizes of embeddings of character, word and part-of-speech are set 50. Dropout rate is set 0.5. We use stochastic gradient descent(SGD) for optimization and the learning rate is set to 0.01.To ensure the fairness of the experiment, we use the same pre-trained embedding to initialize the embedding layer of all models.

Table 1. Datasets statistics

Dataset	Type	Train	Dev	Test	Entity/Sentence
MSRA	Sentence	46.4k	–	4.4k	1.58
	Char	2169.9k	–	172.6k	
	Entity	73.3k	–	4.3k	
Resume	Sentence	3.8k	0.46k	0.48k	3.54
	Char	124.1k	13.9k	15.1k	
	Entity	13.438k	1.63k	1.497k	
Weibo	Sentence	1.4k	0.27k	0.27k	1.35
	Char	73.8k	14.5k	14.8k	
	Entity	1.885k	0.414k	0.389k	

3.2 Comparison Methods and Metrics

The comparison models in this paper are as follows, where the precision (P), recall (R) and F1-score (F1) are adopted as the evaluation metrics. CharLSTM was selected as the baseline model to test the improvement brought by semantic information embedding, such as POS and word embedding.

1. **CharLSTM:** the BiLSTM-CRF structure network with the character embeddings as input. To compare the effects of transfer learning, we introduce a variant **CharLSTM-T** that is pretrained for the general NER and transferred for the specific NER by replacing the CRF layer of **CharLSTM**.

Table 2. Comparison results on **Weibo**

Models	Precision	Recall	F1
CharLSTM	58.77	48.55	53.17
Lattice	67.61	51.93	58.74
LTN-GC	**69.16**	51.45	59.00
LTN-G	68.47	**51.93**	**59.07**
La-DTL*	–	–	57.74
CharLSTM-T	62.35	48.79	54.74
Lattice-T	67.69	53.14	59.54
LTN	**69.45**	**58.21**	**63.34**

*indicates the result reported in the corresponding reference.

Table 3. Comparison results on **Resume**

Models	Precision	Recall	F1
CharLSTM	92.69	92.64	92.67
Lattice	**94.81**	94.11	94.46
LTN-GC	94.45	94.05	94.25
LTN-G	94.37	**94.60**	**94.49**
CharLSTM-T	93.89	94.36	94.12
Lattice-T	94.67	94.85	94.76
LTN	**94.87**	**95.40**	**95.14**

2. Lattice [20]: the state-of-the-art Chinese NER model that achieves the best results on **Weibo** and **Resume**. To compare the effects of transfer learning, we also introduce a variant **Lattice-T** that is transferred for the specific NER on its CRF layer.
3. La-DTL [19]: a transfer learning model for cross-specialty NER which conducts both feature representation transfer and parameter transfer with label-aware constraints.
4. LTN and Variants: LTN is our transfer model. The general part of LTN named LTN-G and the transfer part of LTN named LTN-T. In addition, to verify the effect of linear combination between character embedding and word embedding in the general part, we design the model **LTN-GC** to replace the linear combination with concatenation operation.

3.3 Experiment Results

Effectiveness. We first verify the performance of our model comparing with other models and present the results in Table 2 and 3, respectively. The models in the upper part of each table are trained directly on the dataset for specialty domain without any transfer learning, while the lower part contains the models with transfer learning. The models in the lower part of each table are first pretrained on **MSRA**, and then transfered to **Weibo** and **Resume**.

On both datasets, our model **LTN** achieves the best results on F1-score. It is worth mentioning that the result of our model general part (LTN-G) on two datasets is similar with Lattice without transfer learning. Comparing the variants of our model, LTN-G and LTN-GC achieve similar results on both two datasets. This illustrates that the linear combination of the input vectors in LTN-G reduces the parameters of network without loss of accuracy.

The results in the lower part of Table 2 and 3 show that our transfer model outperforms other methods. This illustrates that the adoption of contextual information learned in the general domain benefits the transfer process, which is better than simply adapting pre-training weights to a specialty domain. On **Weibo**, our model LTN also performs better than La-DTL. Although La-DTL uses more sophisticated techniques for the transfer learning approach, it does not leverage the semantic information of the general domain more effectively than LTN.

Comparing the results on two datasets, we can see that the performance on **Resume** is much better that **Weibo**. This is because **Resume** contain more labelled data than **Weibo** such that models can be trained well without transfer learning. Our model **LTN** improves a lot comparing with baseline methods, which convinces the performance of transfer learning and illustrates the effectiveness of reusing knowledge from the general task.

Table 4. F1-score against different data size on **Weibo**

Model	10% data	25% data	50% data	100% data
LTN	45.02	51.41	62.57	63.34
Lattice	28.39	39.45	53.96	58.74
CharLSTM	22.71	32.48	48.16	53.17

Table 5. Model processing speed on **Weibo**

Model	Train (sent/sec)	Test (sent/sec)
LTN	19.37	67.87
Lattice	8.10	24.32
CharLSTM	23.98	135.45

Robustness. Then we quantify the robustness of models under insufficient data, we conducted the experiments against different proportion of data on **Weibo**. **Lattice** and **CharLSTM** are trained directly on **Weibo** without transfer learning. Table 4 shows the F1-score of models trained with **Weibo** of different dataset size, namely 10%, 25%, 50%, and 100%, respectively.

The results in Table 4 show our model performs much better than **Lattice** and **CharLSTM** on all cases. Specially, even in the case of 10%, our model still achieves an acceptable result. This convinces that the performance benefits from transfer learning, while **Lattice** and **CharLSTM** model are affected much by the dataset size. It is worth mentioning that our model needs only half of the data to achieve the equivalent result of Lattice.

The results on the robustness of LTN illustrate that the proposed transfer learning based method can be adapted to a new area without much labelled data. This attributes to that the knowledge that are learned in the general NER are reused in the domain specific NER task.

Efficiency. Since a light model should process data fast in either training or testing, we compare the proposed model with others on this metric. Table 5 shows the processing speed in terms of sentences per minute at the same environment. The results on both training and test processes show that our model **LTN** is much faster than **Lattice**. **Lattice** is designed in order to better use lexicon information and the complicated network structure leads to the slow speed of training and testing. Although the **CharLSTM** model with the simplest network structure is faster than our model, together considering its performance, our model is light and with the best performance.

4 Conclusion

We propose a light and effective transfer model for the NER task for specialty domain. By designing an additional part over a general NER network, we reuse the knowledge learned in the general NER task as much as possible, from the granular elements for combining inputs, to a higher level embeddings of outputs. At the same time the model can be easily adapted to the domain specific NER task without reconstruction. We compare our model with related works on real datasets. The experimental results show that our model is consistently superior than baseline methods on both effectiveness and efficiency, especially in case of low-resource data for specialty domain.

Acknowledgement. This work was supported by the National Key R&D Program of China (2018YFC0831401), the Key R&D Program of Shandong Province (2019JZZY010107), the National Natural Science Foundation of China (91646119), the Major Project of NSF Shandong Province (ZR2018ZB0420), and the Key R&D Program of Shandong province (2017GGX10114). The scientific calculations in this paper have been done on the HPC Cloud Platform of Shandong University.

References

1. Chung, J., Gülçehre, Ç., Cho, K., Bengio, Y.: Empirical evaluation of gated recurrent neural networks on sequence modeling. CoRR abs/1412.3555 (2014)
2. Collobert, R., Weston, J., Bottou, L., Karlen, M., Kavukcuoglu, K., Kuksa, P.: Natural language processing (almost) from scratch. JMLR **12**, 2493–2537 (2011)
3. Devlin, J., Chang, M., Lee, K., Toutanova, K.: BERT: pre-training of deep bidirectional transformers for language understanding. CoRR abs/1810.04805 (2018)
4. Hammerton, J.: Named entity recognition with long short-term memory. In: Proceedings of the Seventh Conference on Natural Language Learning at HLT-NAACL 2003, CONLL 2003, vol. 4, pp. 172–175. Association for Computational Linguistics, USA (2003)
5. Hanisch, D., Fundel, K., Mevissen, H.T., Zimmer, R., Fluck, J.: Prominer: rule-based protein and gene entity recognition. BMC Bioinform. **6**, S14 (2005)
6. Hochreiter, S., Schmidhuber, J.: Long short-term memory. Neural Comput. **9**(8), 1735–1780 (1997)
7. Huang, Z., Xu, W., Yu, K.: Bidirectional LSTM-CRF models for sequence tagging. CoRR abs/1508.01991 (2015)
8. Jia, C., Liang, X., Zhang, Y.: Cross-domain NER using cross-domain language modeling. In: ACL, pp. 2464–2474 (2019)
9. Jin, G., Chen, X.: The fourth international Chinese language processing bakeoff: Chinese word segmentation, named entity recognition and Chinese POS tagging. In: Proceedings of the Sixth SIGHAN Workshop on Chinese Language Processing (2008)
10. Kim, J.H., Woodl, P.: A rule-based named entity recognition system for speech input. In: ICSLP (2000)
11. Lafferty, J.D., McCallum, A., Pereira, F.C.N.: Conditional random fields: probabilistic models for segmenting and labeling sequence data. In: ICML, pp. 282–289 (2001)
12. Li, X., Feng, J., Meng, Y., Han, Q., Wu, F., Li, J.: A unified MRC framework for named entity recognition. In: Proceedings of the 58th Annual Meeting of the Association for Computational Linguistics, pp. 5849–5859. Association for Computational Linguistics (2020)
13. Li, X., Sun, X., Meng, Y., Liang, J., Wu, F., Li, J.: Dice loss for data-imbalanced NLP tasks. In: Proceedings of the 58th Annual Meeting of the Association for Computational Linguistics, pp. 465–476. Association for Computational Linguistics (2020)
14. Lin, B.Y., Lu, W.: Neural adaptation layers for cross-domain named entity recognition. In: EMNLP, pp. 2012–2022 (2018)
15. Ma, X., Hovy, E.: End-to-end sequence labeling via bi-directional LSTM-CNNS-CRF (2016). arXiv preprint arXiv:1603.01354
16. McCallum, A., Freitag, D., Pereira, F.C.N.: Maximum entropy Markov models for information extraction and segmentation. In: ICML, pp. 591–598 (2000)
17. Peng, N., Dredze, M.: Named entity recognition for Chinese social media with jointly trained embeddings. In: EMNLP, pp. 548–554 (2015)
18. Peters, M.E., et al.: Deep contextualized word representations. CoRR abs/1802.05365 (2018)
19. Wang, Z., et al.: Label-aware double transfer learning for cross-specialty medical named entity recognition. In: NAACL, pp. 1–15 (2018)
20. Zhang, Y., Yang, J.: Chinese NER using lattice LSTM. In: ACL, pp. 1554–1564 (2018)

Outlier Detection for Sensor Data Streams Based on Maximum Frequent and Minimum Rare Patterns

Xiaochen Shi[1], Saihua Cai[2,3], and Ruizhi Sun[1,4]($\boxtimes$)

[1] College of Information and Electrical Engineering, China Agricultural University, Beijing 100083, China
sunruizhi@cau.edu.cn
[2] School of Computer Science and Communication Engineering, Jiangsu University, Zhenjiang 212013, China
caisaih@ujs.edu.cn
[3] Jiangsu Key Laboratory of Security Technology for Industrial Cyberspace, Jiangsu University, Zhenjiang 212013, Jiangsu, China
[4] Scientific Research Base for Integrated Technologies of Precision Agriculture (Animal Husbandry), The Ministry of Agriculture, Beijing 100083, China

Abstract. Identifying outliers in data is an essential assignment in data mining. It aims to find the data that is significantly different from other data in the data streams, and scholars have proposed many outlier detection methods in recent years. Because pattern-based outlier recognition technology clearly illustrates the causes of anomalies, they have received extensive attention. However, the existing pattern-based outlier identification techniques main use frequent patterns or rare patterns with perform the outlier identification operations, which will affect the precision from outlier identification. To improve the exactness of outlier detection on data streams, in this paper, the maximum frequent patterns and minimum rare patterns are combined together and regarded as a whole as the pattern used in the process of outlier detection, and then it redefines the deviation index, so as to perfectly measure the abnormal degrees of outlier in the data streams. Dependent upon those search patterns and characterized deviation index, Novel maximum frequent and minimum rare pattern-based outlier detection method are put forward, that is, specific MFaMRP-OD, which can faultlessly recognize those possibility outliers.

Keywords: Outlier detection · Maximum frequent patterns · Minimum rare patterns · Data streams · Deviation index

1 Introduction

Identifying outliers in data is an essential assignment in data mining. It aims to identify outliers (aka abnormal data instances) that are significantly different from other data instances. Compared with normal data, some important information is often hidden behind the outliers. For example, the sales number of luxury goods is very few in the

© Springer Nature Singapore Pte Ltd. 2021
Y. Sun et al. (Eds.): ChineseCSCW 2020, CCIS 1330, pp. 542–557, 2021.
https://doi.org/10.1007/978-981-16-2540-4_39

commodity transactions, thus, compared with the daily necessities, the luxury goods transaction records are more like the outliers. However, since the transaction of luxury goods not only can increase the transaction amount of the mall, but also can reflect the purchase level and consumption habits of the users, which helps the mall to make the targeted adjustments. Thus, the transaction of luxury goods should be attracted more attention than that of daily necessities. Outlier detection technology has been generally utilized within fraud detection [1], network intrusion detection [2], network traffic detection [3], weather prediction [4] and other fields.

With the development of communication technology, a large amount of dynamic data transmitted in the form of streams are collected from the cameras, sensors and Internet, they are called data streams. Compared with static data, data streams are continuous, rapidly generated and unlimited amount of data instances. These characteristics of data streams put forward the following more stringent requirements for outlier detection: 1) Data stream will produce data continuously, which makes it impossible to store all the data information completely in memory. Thus, it is essential to identify those possibility outliers starting with information streams then afterward filtering them for best once [5]; 2) The quickly and continuously arrival of the data streams require the detection algorithm has fast response capabilities [6]; 3) The fluctuation and dynamic change of the data instances in the data streams requires the detection algorithm has a good adaptive ability [7]. However, for static data sets, a larger part of the existing outlier recognition would be recommended. They performed the outlier identification operations through examining those datasets numerous times. In addition, in the algorithm of outlier detection using distance and density, the distance between each data instance is used to figure the deviation level of transactions. However, the limitation of the distance will make the distance-based algorithms use the numerical value to represent the classification characteristics of the data in the process of outlier calculation, so as to judge the deviation degree. This operation will affect the precision from outlier identification by the numerical changes. To dispose of this trouble, scholars have proposed outlier detection algorithms based on pattern mining [9–13] to search for the outliers, in which the frequent patterns are used to better identify the potential outliers in the data streams and what's more, demonstrate those outliers sensibly.

The algorithm for detecting outliers by pattern is consisted of pattern mining and outlier identification, and the transactions that contain more frequent patterns are judged as outliers. According to the above judgment principles, the running process of frequent pattern-based outlier detection algorithms can be summarized as follows: (1) Mining frequent patterns with occurrence times exceed or equal to the user-defined minimum occurrence frequency; (2) designing the corresponding deviation index based on the mined frequent patterns; (3) Ascertain those deviation levels of transaction dependent upon designed outlier deviation indexes., thus accurately identifying the potential outliers. In addition, to decrease those scale about incessant utilized patterns within those outlier identification stage, the outlier detection algorithm based on maximum frequent patterns [11] and the outlier detection algorithm based on closed frequent patterns [12] are proposed. Compared with the frequent patterns, the rare patterns are more in line with the definition of the outliers, thus, scholars also put forward the algorithm to detect outliers in data streams by using the least rare pattern [13].

Because the data distribution in the data stream has uncertain characteristics, those frequent pattern based outlier identification calculations or those base rare pattern-based outlier identification calculations can't adjust with known data conveyance situations. For example, in the traffic data streams, the data distribution during the commuting time period is very dense, while the data distribution during other time periods is relatively sparse. Therefore, the changes of the data streams result the following problems no matter which type of pattern-based outlier detection method is selected: (1) When processing the dense data streams, the use of outlier recognition calculation based on rare pattern will cause low recognition accuracy because of the small scale of the mined rare patterns; (2) When preparing these poor data streams, the use of outlier recognition calculation based on frequent pattern will lay a foundation for the problem of low recognition accuracy, because there are indeed several frequent examples that could be mined in the design mining stage. In addition, the detection accuracy is also affected by the differences of minimum support threshold. For example, it has the possibility that cannot mine any frequent patterns when the minimum support threshold is set too large, which will affect the performance of the frequent pattern-based outlier detection algorithms. The inverse may be that it shows the likelihood that can't mine any extraordinary designs when those base backing edge will be situated as well small, which will decrease the identification exactness from rare pattern-based outlier identification calculations. Therefore, neither of these two algorithms can obtain good detection results when dealing with data distributions that are not good at.

The judging guideline about outlier identification dependent upon design mining may be that the frequent characteristics would be included for the transaction, the more improbable it may be to make confirmed concerning an outlier. Conversely, the more rare pattern features contained in a transaction, the greater the probability that the transaction will be judged as an outlier. It can be seen that frequent patterns and rare patterns work together on the process of outlier detection. Frequent patterns have a negative effect on outliers, and infrequent patterns have a positive effect on outliers. In the data stream, a single consideration of one of the outlier characteristics may affect the outlier detection rate due to the uncertainty of the data stream distribution. To further enhance the adaptability of the outlier detection algorithm to uncertain data distribution, we consider both the frequent patterns and rare patterns on the transform of outlier identification, so as to effectively deal with data streams with different data distribution. So as with quicken those transforming velocity about outlier detection, we diminish the number for designs utilized patterns within the outlier identification transform eventually, perusing utilizing the maximum frequent patterns and the minimum rare pattern to displace the frequent patterns and the rare patterns individually. Furthermore, we put forward an algorithm (MFaMRP-OD) to detect outliers by using the maximum frequent pattern and the minimum infrequent pattern on faultlessly recognizing the possibility outliers from data streams.

The primary center points as takes after:

(1) We put forward to consider both frequent patterns and rare patterns in the process of identifying outliers. Under the mined maximum frequent patterns and minimum rare patterns, we outline an successful deviation indicator MFaMRP, which can a chance to be used to assess the deviation level of the transactions.
(2) Under the designed deviation indicator MFaMRP and the excavated maximum frequent patterns and minimum rare patterns, we raise an outlier detection algorithm MFaMRP-OD should faultlessly recognize those possibility outliers in the data streams.
(3) We perform a considerable measure of investigations on a true dataset, and the test outcomes exhibit that the suggested MFaMRP-OD algorithm might faultlessly recognize the possibility outliers.

Whatever remains of this paper will be orchestrated similarly as takes after. The second portion discusses the previous research. The third portion gives some basic definitions. In the fourth portion, we first propose the maximum frequent pattern mining algorithm and the minimum rare pattern mining algorithm, and then propose the outlier detection algorithm. The fifth portion gives the test outcomes on the real datasets. The sixth portion concludes the full content.

2 Previous Research

Outliers are an important task and have received widespread attention. In order to accurately detect outliers, many methods have been proposed in the past, which can be summarized as: statistics-based outlier detection method [14], distance-based outlier detection method [15], density-based outlier detection method [16], clustering-based outlier detection method [17] and pattern-based outlier detection algorithms [18–20]. Compared to different sorts of outlier detection methods that use accurate numerical values to identify outliers, the pattern-based outlier detection method uses the frequency of data as the basis for outlier determination. The accuracy of outlier detection will not fluctuate strongly with the change of data streams. Therefore, this method has received extensive attention. Combined data mining and outlier detection for the first time, He et al. put forward an outlier detection algorithm FindFPOF [18] in light of the mining of frequent patterns, which enables users in the business field to directly use association rules to achieve outlier detection. However, this method is very sensitive to the setting of the parameters, thus, it may dig out expansive number of frequent patterns which may be could reasonably be expected on bring about excessively yearn running time. In response to the over problems, Zhang et al. found that the long running time is caused by the repeated calculation of frequent patterns, and proposed an improved method LFP [19]. LFP uses the downward closure of frequent items to effectively solve the problem of long running time, but the input parameters are still very sensitive which will have a greater impact on the detection accuracy. Lin et al. proposed an outlier detection algorithm OODFP [11] that uses the maximum frequent pattern for the data stream of time series. The detection accuracy and time consumption of this method are lower than

FindFPOF, but when the input parameters are small, the time advantage of this method is greater small. In addition, closed frequent patterns are a lossless compression of frequent patterns, thus Hao et al. used closed frequent patterns instead of frequent items to tackle the issue of long time-consuming identify outliers phase, and proposed the FCI-Outlier outlier detection method [12]. Its computational cost is lower than FindFPOF, and with the greater the degree of support, the detection accuracy can reach 100% at some times. Since rare patterns are more in line with the definition of outliers and are helpful for detecting outliers, Hemalatha et al. first proposed the MIFPOD [13] outlier detection algorithm starting with the viewpoint of the base rare pattern, and the identification precision about this technique is superior to FindFPOF, whose running time may be shorter. Cai et al. proposed an outlier detection method MRI-AD [20] for rare patterns for sensor data streams. This method guarantees detection accuracy while ensuring a small running time and memory.

In short, the design of outlier index only considers a single influencing factor, and does not consider frequent patterns and rare patterns together as abnormal factors. Aiming at the problem of uncertain data stream distribution, this paper adds frequent patterns and rare patterns together as abnormal factors into the design of outlier indicators to achieve the goal of improving the accuracy of outlier detection.

3 Basic Definitions

In this portion, we basically present fundamental standpoints of the data streams, pattern, frequent pattern, rare pattern, maximum frequent pattern, and minimum rare pattern.

Definition 1: Data streams: Assume that the transactions $T_1, T_2, T_3,..., T_n, T_{(n+1)}$ are arriving continuously in a certain order, and then the data streams D is represented as $D = \{T_1, T_2, T_3,..., T_n, T_{(n+1)}\}$, where T_n is the transaction that arrives at n-th time.

As data continues to arrive, we use the sliding window (SW) to partition the latest arrival data and form the index of the sliding window according to the division order. The volume of the sliding window ($|SW|$) indicates the amount of the most recent transactions that the algorithm can handle at one time.

Definition 2: Pattern: Pattern $P = \{P_1, P_2, P_3,..., P_n, P_{(n+1)}\}$ represents the collection of all items appearing in the latest sliding window transaction. The pattern $P_k = \{P_1, P_2, P_3, ..., P_m\} \subset P$ indicates that P_k is a subset of P, and P_k is called a k-pattern when $|P_k| = k$, where the value of k is ranged in $[1, m]$.

Definition 3: Support: the pattern P_m is supported by the number of occurrences in the sliding window, which is recorded as $sup(P_m)$.

Definition 4: Frequent pattern (FP): If the support of pattern P_m is not less than the predefined minimum support threshold (min_sup), then the P_m is called frequent pattern, it can be expressed as $FP = \{P_m \mid sup(P_m) \geq min\_sup\}$.

Definition 5: Maximum frequent pattern *(MFP)*: Assume that all supersets of frequent pattern P_m are marked as P_k. If the support of all patterns in the set P_k is less than the predefined minimum support threshold, then P_m is the maximum frequent pattern, it can be expressed as $MFP = \{P_m \mid \forall x \in P_k, sup(x) < min\_sup \wedge (P_m \subseteq x) \wedge (sup(P_m) \geq min\_sup)\}$.

Definition 6: Rare pattern *(RP)*: If the support about pattern P_m may be short of the predefined minimum support threshold, P_m is called a rare pattern, it can be expressed as $RP = \{P_m \mid sup(Pm) < min\_sup\}$.

Definition 7: Minimum rare pattern (MRP): If all subsets of the rare pattern P_m are frequent patterns, then P_m is called a minimum rare pattern, it can be expressed as $MRP = \{P_m \mid sup(Pm) < min\_sup \wedge \{\forall y \subseteq P_m, sup(y) \geq min\_sup\}\}$.

4 Outlier Detection Algorithm

For the proposed maximum frequent and minimum rare pattern-based outlier detection algorithm MFaMRP-OD. It identifies potential outliers through two stages, first, search all the required patterns appearing in the data, and then use the required patterns to set outlier indicators to identify outliers. Thus, we detailed describe the proposed MFaMRP-OD algorithm with two sections, and then use an example to show the specific operations of outlier detection.

4.1 Pattern Mining Stage

At that stage of search all patterns, the traditional methods continuously scanned the datasets and performed the pattern mining operations from $(k-1)$-pattern to k-pattern in an iterative manner, so the time complexity is very high. However, when the item being processed may be a data stream that continuously generates data, in the customary technique, the data instance may not have the opportunity to be saved quickly, or reach the speed alternately equal to the data stream generation speed. Because the MRI-Mine algorithm [17] uses a data structure that combines a hash table and a two-dimensional matrix, it may be more effective to filter those transactions in the data stream to store data information by scanning once. In addition, the support calculation operation of the extended patterns can be quickly completed through the intersection operation in the matrix list, thus, the time consumption is relatively short. Based on the advantages of the MRI-Mine algorithm, we use the idea of the MRI-Mine algorithm to mine the maximum frequent patterns and minimum rare patterns.

However, the MRI-Mine algorithm is recommended for mining the minimum rare patterns, but they can not completely search out the maximum frequent patterns. Therefore, some methods are needed to mine the maximum frequent patterns simultaneously. In the process of searching for the desired pattern, we use the anti-monotone property [15] to whether the frequent pattern meets the condition of the maximum frequent pattern, thus making the algorithm can simultaneously mine the minimum rare patterns and maximum frequent patterns. The improved MRI-Mine method mines the corresponding patterns through the following steps.

Step 1: Use a two-dimensional hash list and matrix to store the information of each data instance occurrence, where the hash table stores the support of each data instance. If the data instance exists in the transaction, then the corresponding position is written as 1, otherwise, the corresponding position is written as 0.

Step 2: Erase extraordinary designs whose support will be less *min_sup* in the hash table, same time the pattern designs whose support will be not short of what *min_sup* will be spared with performing those further "pattern extension" operations. In addition, the support of each pattern in the hash table is taken out to determine the support of the 2- pattern position that supports the expansion based on the rare patterns required by the minimum support, and the frequent 2-pattern is also used to perform those further "pattern expansion" operations.

Step 3: Haphazardly select two (k-1) patterns for those same prefix (the length of prefix is (k-2)) from the hash table to generate k-patterns (k > 2). And then, the corresponding support values of the patterns in the two-dimensional matrix are taken out to perform the union operation, thus Ascertaining those support of the pattern with length k which is recombined by two k-1 patterns, and the support of the recombined pattern with length k is put away in the hash table. If the support value of the k-pattern is not less than *min_sup*, it is a frequent pattern, and then the maximum frequent pattern is detected through the anti-monotone property; Assuming that its support will be under *min_sup*, afterward perform the base minimum rare identification operation. If any subset of extended k-patterns is rare pattern, it should be deleted from the hash table, otherwise, it is a minimum rare pattern. Repeat the above operations until all maximum frequent patterns and minimum rare patterns are discovered.

And then, we use the data stream that is shown in Table 1 to describe the improved MRI-Mine method, where |SW| is set on 5, and the *min_sup* will be situated on 3, and the particular procedure of the design mining operations is shown in Fig. 1.

Step 1: Display all the information of the data instance in the data streams, it is shown in Fig. 1(a). The hash table records all patterns in the sliding window and their corresponding appearing times, while the matrix shows the specific distribution of the pattern in different transactions.

Step 2: In Fig. 1 (b), frequent 1-pattern with support not under than *min_sup* is shown. In view those support from item {D} is under *min_sup*, it is a minimum rare pattern and should be deleted from the matrix and hash table. And then, the result of 2-patterns generated by randomly connecting the frequent 1-patterns is shown in Fig. 1(c), where the rare 2-patterns are deleted and the frequent 2-patterns are saved in the matrix, it is shown in Fig. 1(d). Because the subsets of 2-pattern {AC} ({A} and {C}) are frequent patterns, thus, pattern {AC} is a minimum rare pattern.

Step 3: Presentation those consequences about 3-patterns that is presented stretched by those frequent 2-patterns, furthermore it is demonstrated over Fig. 1(e). Following the support of the 3-patterns is judged. The frequent patterns are {ABE} and {BCE}, and they are shown in Fig. 1(f), while the rare patterns are {ABC} and {ACE}. Because the subset {AC} of {ACE} and {ABC} is a rare pattern, thus, these two patterns are not rare patterns, they should be deleted directly. And then, the remaining two frequent 3-patterns are extended to 4-pattern {ABCE}, and it will be indicated previously Fig. 1(g). Because the support for {ABCE} is 2, it is a rare pattern. Because the subset of {ABCE} (including: {ABC}, {AC} and {ACE}) are rare patterns, therefore, example {ABCE} may not be the minimum rare pattern. Finally, the minimum rare patterns are {D} and {AC}, and the maximum frequent patterns are {ABE} and {BCE}.

Table 1. Illustration from data mining

TID	Transactions
T_1	A, D, B, E
T_2	A,B,D
T_3	A,E,C,B
$T4$	B,C,E
T_5	A,B,E,C
...	....

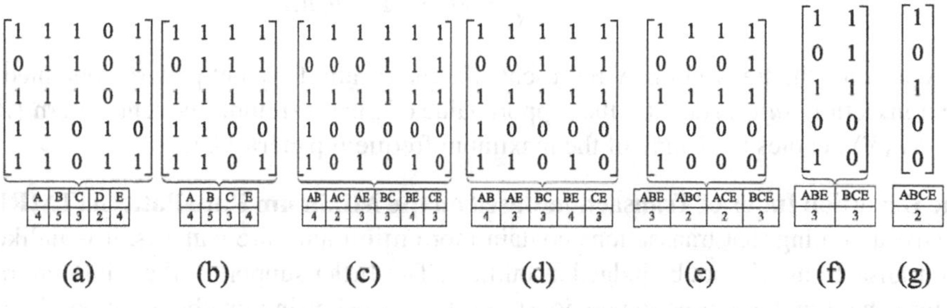

$$
\begin{bmatrix} 1 & 1 & 1 & 0 & 1 \\ 0 & 1 & 1 & 0 & 1 \\ 1 & 1 & 1 & 0 & 1 \\ 1 & 1 & 0 & 1 & 0 \\ 1 & 1 & 0 & 1 & 1 \end{bmatrix}
\begin{bmatrix} 1 & 1 & 1 & 1 \\ 0 & 1 & 1 & 1 \\ 1 & 1 & 1 & 1 \\ 1 & 1 & 0 & 0 \\ 1 & 1 & 0 & 1 \end{bmatrix}
\begin{bmatrix} 1 & 1 & 1 & 1 & 1 & 1 \\ 0 & 0 & 0 & 1 & 1 & 1 \\ 1 & 1 & 1 & 1 & 1 & 1 \\ 1 & 0 & 0 & 0 & 0 & 0 \\ 1 & 0 & 1 & 0 & 1 & 0 \end{bmatrix}
\begin{bmatrix} 1 & 1 & 1 & 1 & 1 \\ 0 & 0 & 1 & 1 & 1 \\ 1 & 1 & 1 & 1 & 1 \\ 1 & 0 & 0 & 0 & 0 \\ 1 & 1 & 0 & 1 & 0 \end{bmatrix}
\begin{bmatrix} 1 & 1 & 1 & 1 \\ 0 & 0 & 0 & 1 \\ 1 & 1 & 1 & 1 \\ 0 & 0 & 0 & 0 \\ 1 & 0 & 0 & 0 \end{bmatrix}
\begin{bmatrix} 1 & 1 \\ 0 & 1 \\ 1 & 1 \\ 0 & 0 \\ 1 & 0 \end{bmatrix}
\begin{bmatrix} 1 \\ 0 \\ 1 \\ 0 \\ 0 \end{bmatrix}
$$

A	B	C	D	E
4	5	3	2	4

A	B	C	E
4	5	3	4

AB	AC	AE	BC	BE	CE
4	2	3	3	4	3

AB	AE	BC	BE	CE
4	3	3	4	3

ABE	ABC	ACE	BCE
3	2	2	3

ABE	BCE
3	3

ABCE
2

(a)   (b)   (c)   (d)   (e)   (f)   (g)

Fig. 1. Display of specific pattern mining operations

4.2 Outlier Detection Stage

In the outlier identification stage, the fundamental errand is to design the outlier deviation index to assess those outlier levels of transactions. From the point of view of these excavated maximum frequent patterns and the minimum rare patterns, it will be fundamental to consider the specificity of the two designs when planning deviation index. The deviation index may be supportive, and will perfectly identify the abnormal values that could be reasonably expected.

Unique in relation to the existing the maximum frequent patterns outlier detection algorithms and the minimum rare patterns outlier detection algorithms, it may be fundamental to consider those impact of both the maximum frequent patterns and the minimum rare patterns on the judgment of the deviation level of each transaction at the same time. Therefore, in order to accurately identify the possibility of outliers, we designed the index of accompanying deviation.

The Deviation Index of Transaction in Light of the Maximum Frequent Pattern Design. (*TMFP*): The more frequent pattern is included over a transaction, the more improbable the transaction is on making an outlier, and the subset of the maximum frequent pattern is frequent pattern. In other words, the transaction is more averse to being an outlier if it holds a greater amount maximum frequent patterns. Under certain circumstances, the longer the maximum frequent pattern, the more subset of the maximum frequent pattern. Therefore, for two transactions containing the same number of maximum frequent patterns, the transaction containing the longer maximum frequent pattern length is more likely to be normal, that is, the length of the maximum frequent pattern is negatively related to the deviation degree of the transaction. On the other hand, the support of the mined maximum frequent pattern indicates the number of occurrences of the pattern. Thus, the support degree of the maximum frequent pattern is the greater, the transaction containing the maximum frequent pattern has the more common characteristics, and the degree of transaction outlier is lower, that is, the support of the maximum frequent pattern plays an opposite role to the judgment of transaction outlier degree. In summary, the deviation index TMFP of the maximum frequent patterns is designed as follows:

$$TMFP(T_i) = \sum_{X \subseteq T_i} \frac{1}{sup(X) \times len(X)} \tag{1}$$

T_i represents the transaction, X represents the maximum frequent pattern contained in the transaction, $sup(X)$ denotes the support value of the maximum frequent pattern $\{X\}$, and $len(X)$ denotes the length of the maximum frequent pattern $\{X\}$.

The Deviation Index of Transaction Based on the Minimum Rare Pattern (TMRP): Firstly, assuming that transactions contain more minimum rare patterns, it is unlikely that transactions will not be judged as outliers. The littler support of the minimum rare patterns held in the transaction demonstrates that the showing up abnormal possible of the example is substantially higher. Thus, transaction should be a chance to be an outlier through holds those minimum rare patterns for little support. For the held minimum rare patterns in the transaction, those shorter lengths imply that there would be more additional extraordinary possibilities that the more rare patterns can have a chance come into being through this minimum rare pattern, so the transaction that contains the shorter minimum rare pattern will not object as abnormal. Through the above analysis, the deviation index of the transaction based on the minimum rare pattern is designed as follows:

$$TMRP(T_i) = \left(\sum_{Y \subseteq T_i} \frac{min\_sup - sup(Y)}{len(Y)} \right) \times \frac{\sum\limits_{Y \subseteq T_i} sup(Y)}{|MRP|} \times (count_{Z \in MRP, len(Z)=1}(Z) + 1)$$

(2)

$|MRP|$ represents the number of the minimum rare patterns, $len(Y)$ represents the length of the minimum rare pattern $\{Y\}$, $\{Z\}$ represents the minimum rare pattern with a length of 1, $count(Z)$ represents the number of rare pattern $\{Z\}$ that is contained in the transaction.

Transaction Deviation Index (MFaMRP): For a transaction in the data streams, if it contains less frequent patterns or more rare patterns, then this transaction will be judged as an outlier. Since the maximum frequent patterns and the minimum rare patterns successfully correspond to the frequent patterns and rare patterns respectively, the TMFP and TMRP could make use of those two index to viably assess those deviation level of the transaction. The maximum frequent pattern is positively correlated to the determination of the outlier, and the minimum rare pattern is negatively correlated with the measurement of abnormal degree. In light of the above analysis, those transaction deviation index MFaMRP is outlined as takes after:

$$MFaMRP(T_i) = TMFP(T_i) - TMRP(T_i)$$

(3)

According to the designed deviation index, the degree of deviation is assessing, and then those transactions are sort as stated by the diminishing deviation degrees, that is, a transaction with a greater deviation degree will be judged as the final outlier. It should be noted that if a transaction does not contain any maximum frequent pattern, then the *TMFP* is directly set to 0. If a transaction does not contain any minimum rare pattern, then the *TMRP* value is directly set to 0. The recommended outlier detection algorithm based maximum frequent and minimum rare pattern MFaMRP-OD is demonstrated in in Algorithm 1.

Algorithim1:MFaMRP-OD

Input: Data streams, $|SW|$, *min_sup*

Output: *Outliers in descending order*

01.Mine *MRPs* and *MFPs*

02.$TMFP(T_i)=0$, $TMRP(T_i)=0$, $MFaMRP(T_i)=0$

03.**for each** T_i in SW **do**

04. **if** *MFP* $\{X\}$ in T_i **then**

05. $$TMFP(T_i) = \sum_{X \subseteq T_i} \frac{1}{sup(X) \times len(X)}$$

06. **else**

07. $TMFP(T_i)=0$

08. **end if**

09. **if** *MRFP* $\{Y\}$ and $\{Z\}$ in T_i **then**

10. $$TMRP(T_i) = (\sum_{Y \subseteq T_i} \frac{min\_sup - sup(Y)}{len(Y)}) \times \frac{\sum_{Y \subseteq T_i} sup(Y)}{|MRP|} \times (count_{Z \in MRP, len(Z)=1}(Z)+1)$$

11. **else**

12. $TMRP(T_i)=0$

13. **end if**

14. $MFaMRP(T_i) = TMFP(T_i) - TMRP(T_i)$

15.**end for**

16.sort T_i according to their decreasing *MFaMRP* values

17.top k transactions with large *MFaMRP* values→ Outliers

18.return *Outliers*

19.slide the window

20.return 01

And then, we use the above example to demonstrate the specific operation of the MFaMRP-OD algorithm, where the mined maximum frequent patterns are {ABE} and {BCE}, and the mined minimum rare patterns are {D} and {AC}.

Step 1: Calculate the TMFP for each transaction.
$TMFP(T_1) = 0.11; TMFP(T_2) = 0; TMFP(T_3) = 0.22; TMFP(T_4) = 0.11; TMFP(T_5)$
$= 0.22$.
Step 2: Calculate the TMRP for each transaction.
$TMRP(T_1) = 3; TMRF(T_2) = 3; TMRF(T_3) = 1.5; TMRF(T_4) = 0; TMRF(T_5) = 1.5$.
Step 3: Calculate the MFaMRP for each transaction.
$MFaMRP(T_1) = -2.89, MFaMRP(T_2) = -3, MFaMRP(T_3) = -1.28, MFaMRP(T_4)$
$= 0.11, MFaMRP(T_5) = -1.28$.

Therefore, the probability of a transaction that will be determined as an outlier in the descending order is: T_4, T_5, T_3, T_1, T_2.

5 Experiments

When testing the effectiveness of the MFaMRP-OD method, we use the related algorithms similar to the algorithm strategy as control experiments, including method of detecting outliers utilizing frequent patterns: FindFPOF and LFP, method FCI-Outline for detecting outliers utilizing closed frequent patterns, and method of detecting outliers utilizing minimum rare patterns: MRI-AD and MIFPOD. All analysis and comparison will be performed on the machine running Windows10 framework with inter(R) Core(TM)I5-5200U 2. 2GHz processor, the execution environment will be Python 3. 7. The dataset utilized within those analysis is the Wisconsin breast tumor dataset (WBCD) from UCI machine Taking in library. The size of the dataset is 699, including 458 benign cases and 241 malignant cases, where the malignant are recorded as outliers, and each transaction has 9 integer attributes. In the experiments, two assessment indices about "Precision" and "Recall" would be used to assess the identification effectiveness of the MFaMRP-OD algorithm. Precision will be basically attained toward ascertaining the proportion of the amount of outliers detected by those algorithms to the real outliers in the current window. The recall will be basically attained toward ascertaining the proportion of the amount for outliers distinguished by those algorithms calculation in the top n bits for data, the place n speaks to the real amount for outliers in the current window. Those examinations are tried under separate |SW| and corresponding min_sup, the place the |SW| will be set with 20, 30 and 40, the corresponding minimum support value setting will always follow the ratio of the minimum support value to |SW|, which is 40%, 50% and 60%, regardless of the size of |SW|. The test effects might make seen from Fig. 2 to Fig. 4. In the figures, those X-axis speaks to the particular amount of the sliding window, and the Y-axis speaks to the identification effectiveness.

It can have a chance to be seen from Fig. 2 that at the extent from required sliding window may be situated should 20, those "Precision" and "Recall" of the suggested MFaMRP-OD algorithm realize the most elevated to the majority sliding windows. In any case of the min_sup may be situated on 8, 10 or 12, these two evaluation indexes often can arrive at with 100% inside the window. Compared with the method of comparison, the identification effectiveness of the MIFPOD method is practically near the recommended MFaMRP-OD method. Although the identification effectiveness of the MIFPOD method may be generally superior other result, these two assessment indexes would be much higher when the identification effectiveness of the MIFPOD technique would be compared with the MFaMRP-OD method. Compared with the method of comparison, the identification effectiveness of FCI-Outlier and MRI-AD methods are at middle of the level done a large portion cases, the "Precision" and "Recall" from these two strategies could compass with 100% in uncommon cases. The identification effectiveness of the FindFPOF and LFP techniques would be moderately poor. At the same time the identification effectiveness of LFP may be much more terrible, a direct result the blue line clinched alongside Fig. 2 would be fundamentally brought down over the green line.

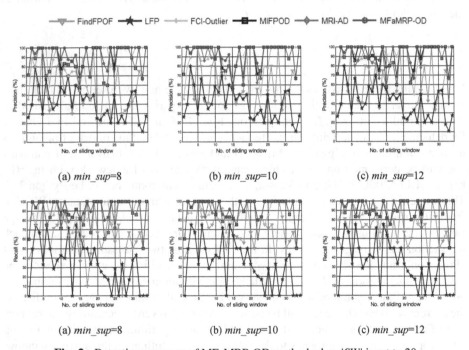

Fig. 2. Detection accuracy of MFaMRP-OD method when |SW| is set to 20

The sliding window will be situated to 30 and the size of *min_sup* may be increasing, the lines for diverse shades on Fig. 3 have fewer intersections, but the presented of "Precision" and "Recall" by MFaMRP-OD technique are more effective. Done the vast majority cases, the identification precision of MFaMRP-OD may be at present higher over other algorithms and can even considerably range to 100% under the same conditions. The recall rates from LFP are still positioned to keep going to a minimum. These phenomena substantiate those conclusions we got starting with Fig. 2.

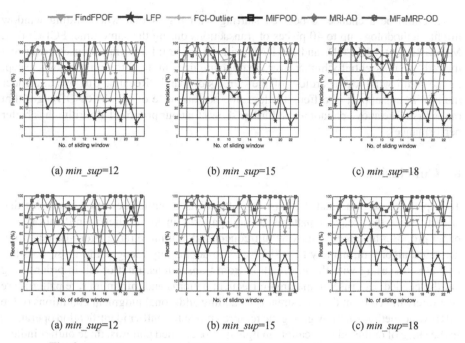

(a) *min_sup*=12   (b) *min_sup*=15   (c) *min_sup*=18

(a) *min_sup*=12   (b) *min_sup*=15   (c) *min_sup*=18

Fig. 3. Detection accuracy of MFaMRP-OD method when |*SW*| is set to 30

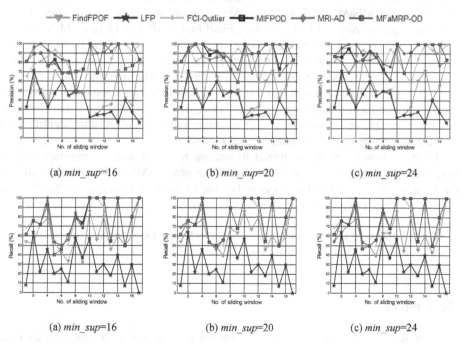

(a) *min_sup*=16   (b) *min_sup*=20   (c) *min_sup*=24

(a) *min_sup*=16   (b) *min_sup*=20   (c) *min_sup*=24

Fig. 4. Detection accuracy of MFaMRP-OD method when |*SW*| is set to 40

It could be a chance to be seen starting with Fig. 4 that when sliding window might methodology up to 40 pieces of transactions during the same time, FCI-Outlier, MFaMRP-OD, MIFPOD, and MRI-AD will the more information focuses cover and the red line is essentially higher over the opposite lines in the different windows except to the ninth window, regardless of the "Precision" and "Recall". But overall, under the minimum support of different values, the performance of MFaMRP-OD algorithm ranks first. The above phenomenon further shows that our proposed algorithm has better performance.

6 Conclusions

To explore those issue that the existing pattern-based outlier detection strategies that is just utilize the frequent patterns or rare patterns in the outlier identification, this paper proposes an effective maximum frequent and minimum rare pattern-based outlier detection algorithm, namely MFaMRP-OD, to adequately recognizing those possibility outliers starting with the information from transactions. First of all, in order to dig out the needed patterns, the maximum frequent patterns and minimum rare patterns are accurately mined from the data streams by adding additional judgment conditions of the MRI-Mine method, thereby giving the required mode for outlier identification operation. In the stage of outlier identification, in light of those mined patterns, three outlier indicators are needed to realize the abnormal sorting of each transaction, so as to promote the recognition accuracy in this way. Those test effects on the open dataset WBCD hint at that those suggested MFaMRP-OD technique can attain superior identification effectiveness on slide windows over that from claiming FindFPOF, LFP, FCI-Outline, MRI-AD and MIFPOD techniques under distinctive sizes about sliding window and unique min_sup in the corresponding window. In addition, the identification effectiveness of MFaMRP-OD method might scope to 100% over sliding windows. The suggested MFaMRP-OD method might explore the issue that frequent pattern-based outlier identification can't acquire outlier identification effectiveness under those vast min_sup values, and the minimum rare pattern-based outlier identification routines can't acquire identification effectiveness under the substantial min_sup values.

In the next work plan, we intend to arrange to coordinate the existing pattern-based outlier detection algorithm and the recommended MFaMRP-OD algorithm in the framework of visualization, so as to be able to more intuitively show the reasons why transactions are judged as outliers to the users.

Acknowledgment. This research was funded in part by National Key Research and Development Program of China, grant number2016YFB05001805.

References

1. Kim, K., Choi, Y., Park, J.: Pricing fraud detection in online shopping malls using a finite mixture model. Electron. Commer. Res. Appl. **12**(3), 195–207 (2013)

2. Mamalakis, G., Diou, C., Symeonidis, A., Georgiadis, L.: Of daemons and men: reducing false positive rate in intrusion detection systems with file system footprint analysis. Neural Comput. Appl. **31**(11), 7755–7767 (2018). https://doi.org/10.1007/s00521-018-3550-x

3. Kim, M.S., Kong, H.J., Hong, S.C., Chung, S.H., Hong, J.W.: A flow-based method for abnormal network traffic detection. In: 2004 IEEE/IFIP Network Operations and Management Symposium (IEEE Cat. No.04CH37507), pp. 599–612. IEEE, USA (2004)

4. Sideris, I.V., Foresti, L., Nerini, D., Germann, U.: NowPrecip: localized precipitation nowcasting in the complex terrain of Switzerland. Q. J. R. Meteorol. Soc. **146**(729), 1768–1800 (2020)

5. Tang, X., Li, G., Chen, G.: Fast detecting outliers over online data streams. In: 2009 International Conference on Information Engineering and Computer Science, pp. 1–4. IEEE, USA (2009)

6. Said, A.M., Dominic, P.D.D., Faye, I.: Data stream outlier detection approach based on frequent pattern mining technique. Int. J. Bus. Inf. Syst. **20**(1), 55–70 (2015)

7. Toshniwal, D., Yadav, S.: Adaptive outlier detection in streaming time series. In: Proceedings of International Conference on Asia Agriculture and Animal (ICAAA 2011), vol. 13, pp. 186–191. Springer, Heidelberg (2011)

8. Yuan, J., Wang, Z., Sun, Y., Zhang, W., Jiang, J.: An effective pattern-based Bayesian classifier for evolving data stream. Neurocomputing **295**, 17–28 (2018)

9. Cai, S., Li, L., Li, S., Sun, R., Yuan, G.: An efficient approach for outlier detection from uncertain data streams based on maximal frequent patterns. Expert Syst. Appl. **160**, 1–17 (2020)

10. Cai, S., Li, S., Yuan, G., Hao, S., Sun, R.: MiFI-Outlier: minimal infrequent itemset-based outlier detection approach on uncertain data stream. Knowl.-Based Syst. **191**, 1–22 (2020)

11. Feng, L., Wang, L., Jin, B.: Research on maximal frequent pattern outlier factor for online high dimensional time-series outlier detection. J. Converg. Inf. Technol. **5**(10), 66–71 (2010)

12. Hao, S., Cai, S., Sun, R., Li, S.: FCI-Outlier: an efficient frequent closed itemset-based outlier detecting approach on data stream. In: CCF Conference on Computer Supported Cooperative Work and Social Computing, pp. 371–385. Springer, Heidelberg (2018)

13. Sweetlin Hemalatha, C., Vaidehi, V., Lakshmi, R.: Minimal infrequent pattern based approach for mining outliers in data streams. Expert Syst. Appl. **42**, 1998–2012 (2015)

14. Aggarwal, C.C., Yu, P.S.: Outlier detection for high dimensional data. Proc. ACM Sigmod May Santa Barbara **30**(2), 37–46 (2001)

15. Kontaki, M., Gounaris, A., Papadopoulos, A.N., Tsichlas, K., Manolopoulos, Y.: Efficient and flexible algorithms for monitoring distance-based outliers over data streams. Inf. Syst. **55**(C), 37–53 (2016)

16. Zhang, L., Lin, J., Karim, R.: Adaptive kernel density-based anomaly detection for nonlinear systems. Knowl.-Based Syst. **139**, 50–63 (2018)

17. Elahi, M., Li, K., Nisar, W., Lv, X., Wang, H.: Efficient clustering-based outlier detection algorithm for dynamic data stream. In: 5th International Conference on Fuzzy Systems and Knowledge Discovery, Shandong, pp. 298–304. IEEE, USA (2008)

18. He, Z., Xu, X., Huang, Z., Deng, S.: FP-Outlier: Frequent pattern based outlier detection. Computer Science and Information Systems **2**(1), 103–118 (2005)

19. Zhang, W., Wu, J., Yu, J.: An improved method of outlier detection based on frequent pattern. In: WASE International Conference on Information Engineering (ICIE), pp. 3–6. IEEE, USA (2010)

20. Cai, S., Sun, R., Mu, H., Shi, X., Yuan, G.: A minimum rare-itemset-based anomaly detection method and its application on sensor data stream. In: Sun, Y., Lu, T., Yu, Z., Fan, H., Gao, L. (eds.) ChineseCSCW 2019. CCIS, vol. 1042, pp. 116–130. Springer, Singapore (2019). https://doi.org/10.1007/978-981-15-1377-0_9

Embedding Based Personalized New Paper Recommendation

Yi Xie[1,2], Shaoqing Wang[2], Wei Pan[1], Huaibin Tang[3], and Yuqing Sun[1,4](✉)

[1] School of Software, Shandong University, Jinan, China
heilongjiangxieyi@163.com, 530834115@qq.com, sun_yuqing@sdu.edu.cn
[2] School of Computer Science and Technology, Shandong University, Jinan, China
492358046@qq.com
[3] School of Microelectronics, Shandong University, Jinan, China
tanghuaibin@sdu.edu.cn
[4] Engineering Research Center of Digital Media Technology, Shandong University,
Jinan, China

Abstract. It is meaningful for researchers to find the interested and high quality new papers. We propose the Joint Text and Influence Embedding recommendation model (JTIE) to consider both the paper quality and the content correlation. We train a paper embedding based on its core elements: contents, authors and publication venues. The quality of a new paper is evaluated based on the author authority and the venue reputation. The citation relationships between papers are considered asymmetric such that they can reflect the user's consideration on the intrinsic influence of a paper. We learn user interests by one's historical references or a set of query keywords. Finally, papers are recommended according to the relatedness between user interests and paper embeddings. We perform experiments against three real-world datasets. The results show that our model outperforms baseline methods on both the personalized recommendation and the query keywords based retrieval.

Keywords: Academic paper · Recommendation · Embedding

1 Introduction

There is a large number of academic papers published every year. To continuously have creative ideas, researchers are interested in the state-of-the-art theory and technology. It is important to find the high quality new papers. Personalized paper recommendation has become a major technique for helping researchers handle huge amounts of papers. To improve user experience, it is essential that the recommendation model predicts users' preferences on papers and provides explainable results. To satisfy the above requirements, there are some challenges, such as the assessment of a new paper quality and the inherent correlation between a user interest and the paper content.

© Springer Nature Singapore Pte Ltd. 2021
Y. Sun et al. (Eds.): ChineseCSCW 2020, CCIS 1330, pp. 558–570, 2021.
https://doi.org/10.1007/978-981-16-2540-4_40

The recent attention on explainability has lead to the development of a series of explainable recommendation models. A fundamental question explainable recommendation aims to answer is how we balance accuracy and explainability. The purpose of this paper is to illustrate how to effectively optimize accuracy and explainability in a joint and unified framework. The key idea is that fully exploiting the correlations between the recommendation task and the explanation task potentially enables both tasks to be better off than when they are considered separately.

To evaluate a paper quality, most of existing works consider paper citations, including the citation amount and the quality of them [17,24,29]. The more citation a paper receives, the better quality it has. However, such metric is not applicable for a new paper quality evaluation, because there is less citation. Many works based on their publication texts and calculate the content correlation between the profiles and papers [12,14,23]. Generally, representative publications and references well illustrate one's research interests. But these content-based methods neither model the user's preference on paper quality nor consider the inherent semantic connection between user interests and papers for recommendation.

To solve the problems, we propose a Joint Text and Influence Embedding (JTIE) based recommendation method for providing related new papers. Given an academic corpus, which includes paper contents and citation relationships, our model learns the latent influence of new papers and recommends the most related papers to researchers. Based on the contents and the citation relationships, we model the semantics and influence of the elements related to papers, such as the authors and the venues. We embedding paper contents, authors and venues as vectors in the same latent space. Then the representation of a paper is fused on the basis of the former elements. Paper citation reflects the preferences of authors's consideration on the intrinsic influence of a paper, which consists the author authority and the venue reputation. The idea behind is that a paper and its references always share some common features, the embedding of them should be close in the latent space. We model the influence as the asymmetric probabilistic propagation based on citation relationships.

When recommending new papers to a user, we first consider the user interests. Research interests could be learned from user published papers and references. We utilize these information to predict his interests and judge if he could be interested in a new paper. Since a user always follow the academic works or authors in a specific area, we profile a user interest from the papers he ever cited, and recommend the most similar new papers to him. We also consider a general query requirement for a researcher in the form of query words. We perform experiments on three real-world datasets and the results illustrate that our method performs better than the existing methods.

The rest of this paper is organized as follows. First, we introduce related works about embedding and paper recommendation systems in Sect. 2. Then we present definitions and propose our models in Sect. 3, detail our recommendation strategies in Sect. 4, and show our experiments and evaluations in Sect. 5. Finally, we draw our conclusions in Sect. 6.

2 Related Work

Content-Based Embedding. Content-based Embedding is high related with our work, such as word embedding models, which are very helpful means to embed items into lower-dimension latent vectors seeing that the relationships between neighbor nodes remain close in the latent space, such as ISOMAP [28] and Laplacian eigenmap [3]. Word2vec is a method of word embedding, of which the most frequently mentioned two models are CBOW and Skip-gram. Proposed by Mikolov et al. [18], the vectors learned by these models have been utilized to find semantically similar words and proved their great performance.

Most of the current methods focus on the relatedness between a user interest and paper contents. A widely adopted method with paper recommendation is Content Based Filtering (CBF). CBF calculates similarity among items and recommends similar items to target users. Chakraborty et al. [4] propose a diversified citation recommendation system for scientific queries, which considers the semantically correlated articles. Sugiyama et al. proposed to use TF-IDF values of keywords as elements in a paper's vector and represent an author with his publications and citations [24]. However, this method does not consider the semantic ambiguity on user chosen words. However, these paper recommendation methods based on explicit factors of similar items or users without learning various kinds of relationships in the Citation Network.

Structure-Based Embedding. Many network embedding methods were inspired by word embedding. LINE [26] and DeepWalk [20] are network embedding models which be can applied to large-scale networks. They are designed to preserve the proximity between vertices and keep the structure of the network. Supervised PTE were then proposed as an expanding method for the heterogeneous networks but it is not specific for a bibliometric task [25]. Embedding has also been used in research of author identification [5]. However, they only consider homogeneous networks where all the nodes are of the same type and cannot scale to a heterogeneous network containing various kinds of nodes and relations such as author-write-paper and author-published in-venue. To recommend related high-quality works to researchers, some works use network structure and relationships between items and users to promote the accuracy of recommendation. Network-based recommender systems [2,9,10] are also related closely with our work. However, their network refers to social network, which cannot be widely applied to networks that do not contain such explicit social relationships, such as "following" or "mutual following" relationships. Besides, they only leverage the network structure without considering the semantics of the textual items such as papers or books.

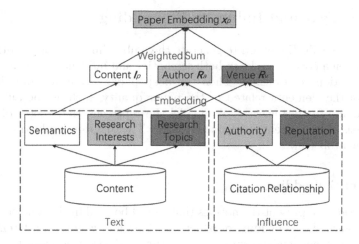

Fig. 1. The joint text and influence embedding model

Paper Recommendation. Another related personalised paper recommendation method is CF based systems. Sugiyama et al. [23] propose a comprehensive evaluation of scholarly paper recommendation by an adaptive neighbor selection method in a collaborative filtering framework. Matrix Factorization (MF) is a model-based Collaborative Filtering and was used in recommender system in previous works [7,14,16]. It considers the latent semantics of users and items. For example, Zhang et al. [32] use a low-dimensional linear model to describe the user rating matrix in a recommendation system. They present a hybrid approach, which learns an additive linear combination of canonical to represent each user's rating profile. However, the dataset in this problem contains millions of authors and papers. Even if we extract hundreds of authors amongst them, the matrix could be extremely sparse. Besides, MF is a CF method so it also could not solve new paper recommendation problem.

To solve the above problem, some works explore the use of deep neural networks to learn the interaction function from data. However, there is little related work on employing deep neural networks for recommendation in contrast to the vast amount of models on MF methods. He et al. [11] present a Neural network-based Collaborative Filtering architecture to model latent features of users and items in recommendation task. Yates et al. [30] adopt the PageRank method to quantify the authority of an author by the citation network. Xie et al. [29] consider the new paper recommendation problem by learning the inherent relativity of word usages from an academic dataset, and computing the potential influence of a new paper by contents, publication venue and author reputation. There are some other academic recommendation, which also seem related with our work, such as experts ranking [1,19,21], partner recommendation [6,31], hot research topic prediction [22] and etc.

3 Joint Text and Influence Embedding

We propose the JTIE method to generate the embedding of papers, authors and venues. Given a corpus, we learn both semantics and influence from the contents and the citation relationship. Then the deep paper semantics and the latent features of the elements related to the paper quality, such as the authors and the venue, are further learned. We model the representation vector of a paper on the basis of its content, authors and venue The diagram is illustrated in Fig. 1 in a down-top view.

3.1 Item Embedding

In this section, we present the notions that would be used in the following discussion, and learn the embedding of papers, authors and venues in a latent space.

For a given academic corpus, we construct a citation network $G = (P, E)$, where $=P$ denotes the paper set, $E = \{(p,r)|p, r \in P, \ p \ cites \ r\}$ denotes the citation relationship between papers. We definite the keyword set, the author set and the venue set of the academic corpus as K, A and V, respectively. Each paper p is accompanied with an attribute set (K_p, a_p, v_p), in which $K_p \subseteq K$ denotes the set of keywords mentioned in p, a_p means the author of p and v_p means the publication venue of p.

At first, we model the representation of a paper based on the learned embedding of its elements. The contents and the quality are two main concerns when users refer papers, thus these are key points when we model the representation vector for a paper. With respect to the paper contents, keywords are direct features in depicting paper contents, which are characteristic words extracted from the paper. Let $\mathbf{I}_p$ represents content vector of p, which is the mean value of all keyword vectors extracted from this paper. $\mathbf{I}_p$ is calculated as $\mathbf{I}_p = \frac{\sum_{k_i \in K_p} \mathbf{I}_{k_i}}{|K_p|}$, where $\mathbf{I}_{k_i}$ denote the vector representation of keyword k_i.

With respect to the paper quality, we consider both the authority of the author and the reputation of the venue. The author of a paper reflects its authority and its research directions. A paper is more likely to be a good work if its author has published many high citation papers. And an author usually focus on a few research directions, which will reflect on his papers. Also, the venue of a paper can reflect both its reputation and research topics. If a paper is published in top conference, it tends to be more influential in a certain research field. Besides, each conference concentrates on the topics within a specific direction, so papers published in similar venues share similar research contents. Let $\mathbf{x}_p$ denotes the representation vector of p, which is calculated by a weighted sum of its content vector, author embedding and venue embedding: $\mathbf{x}_p = \alpha \mathbf{I}_p + \beta \mathbf{R}_{a_p} + \gamma \mathbf{R}_{v_p}$. $\mathbf{R}_{a_p}$ and $\mathbf{R}_{v_p}$ denote the author and the venue embedding vector for paper p, respectively. $\alpha, \beta, \gamma \in [0, 1]$ are parameter for balancing the weight of keywords, author and venue, $\alpha + \beta + \gamma = 1$.

Based on the paper embedding, we consider the asymmetric relationships between papers. We define a relevance score between a paper and its reference, which is expected to be high if there is citation relationship between them. Let pair (p, r) means two papers $p, r \in P$. The relevance score $g(p, r)$ denotes the extent of p citing r $g(p, r)$ should reflect three features: the semantic relatedness between two papers, the research interest of a_p and v_p, the influence of a_r and v_r. We adapt the dot product of $\mathbf{x}_p$ and $\mathbf{x}_r$ to model their semantic relevance score. The reference relationship is unidirectional and asymmetric, so we add local bias vectors $\mathbf{M_1}$ and $\mathbf{M_2}$, which shift $\mathbf{x}_p$ and $\mathbf{x}_r$ due to their different semantics in the path. Then the relevance score between the two papers is defined as

$$g(p, r) = \mathbf{I}_p^T \mathbf{I}_r + \mathbf{M_1}^T \mathbf{x}_p + \mathbf{M_2}^T \mathbf{x}_r \tag{1}$$

We adopt soft-max function to model the probability that there exists citation relationship between pair (p, r), which is denoted as $Pr(p|r)$.

$$Pr(p|r) = \frac{\exp(g(p, r))}{\sum_{p' \in P} \exp(g(p', r))} \tag{2}$$

3.2 Objective Function

After embedding the items into a latent space, we formulate the loss function as

$$\mathbb{L} = - \sum_{(p,r) \in G} \log Pr(p|r) \tag{3}$$

Then we have the objective function as follow:

$$\arg \min_{\theta} \mathbb{L} = \arg \min_{\theta} - \sum_{(p,r) \in G} \log \frac{\exp(g(p, r))}{\sum_{p' \in P} \exp(g(p', r))} \tag{4}$$

where θ is the parameter to be learned. Since the time cost to re-calculate the normalization part in Eq. 2 is intolerable, we adopt negative sampling as in NCE [8] which solves the problem of estimation of un-normalized data. Let $Pr^+(p|r) = Pr(p|r)$ denotes the Probability Distribution Function(PDF) of every positive sample $(p, r) \in E$, $Pr^-(p|r)$ denotes the artificial noise distribution of each negative sample $(p, r) \notin E$, and G' denote the set of all negative samples. The loss function can be re-written as follows

$$\mathbb{L} = - \sum_{(p,r) \in G} \log \sigma(\log Pr^+(p|r)) - \sum_{(p',r') \in G'} \log(1 - \sigma(\log Pr^-(p'|r'))) \tag{5}$$

Stochastic gradient descent method is adopted to update the parameters $\mathbf{R}_a$, $\mathbf{R}_v$, $\mathbf{M_1}$ and $\mathbf{M_2}$. There might be millions of pairs in the form of (p, r), so storage for all these pairs is intolerable. Sampling must be applied to solve this problem. A straightforward and efficient way of sampling is to sample nodes following the sequence of the tuple, i.e., to r, p, and a one by one. Papers with more citations are more likely to be sampled as the first node of the sampled tuple, i.e., r in our instance. So we pre-compute the citations of all papers and then sample r according to distributions of numbers of their citations. Once r has been sampled, positive sampling and negative sampling are performed next.

4 Personalized New Paper Recommendation

4.1 Personalized Recommendation Based on User Interests

We recommend related new papers to users based on their research interests. Deep research interests could be learned from user publications and references. We utilize these information to predict their interests and judge if they could be interested in the new paper. Let U denote the set of users who have publications. Each user in $a \in U$ has a reference set RE_a, i.e., all papers he referenced before the publish time of the new paper. The representation of user interest is calculated by the mean of reference vectors as $\mu_a = \frac{\sum_{r \in RE_a} \mathbf{x_r}}{|RE_a|}$, where $\mathbf{x_r}$ is representation vector of a paper $r \in RE_a$. With the above metric, we calculate the similarity between new papers and researcher's interests. Given a researcher, we recommend the most similar new papers to him.

4.2 Personalized Recommendation Based on Keyword Query

We consider a general query requirement for recommendation. Users could provide a set of query keywords, denoted by $Q \subset K$. We solve the problem in two aspects: contents relevance and venue reputation. We model the user query as the mean of keyword representation vector, which is calculated as $\mu_q = \frac{\sum_{k \in Q} \mathbf{I_k}}{|Q|}$, where $|Q|$ denotes the number of keywords in Q.

Then we calculate the similarity between μ_q and the content vectors of new papers, choose the most related ones from the user as candidates. Besides user interests, it is better to recommend papers with high reputation and trusts, so we choose top k papers published in top conferences from the candidate set as our results. This recommending method could also be applied to new researchers who have not published any paper yet.

5 Experiments and Analysis

5.1 Datasets

We adopt three datasets to verify our model. The first dataset is the citation network of DBLP downloaded from AMiner [27]. It contains information about title, abstract, venue, author, year of publication, reference information about a paper. The second dataset is crawled from *Scopus* website dataset[1]. Each paper contains a title, authors, an abstract, keywords, citation information and discipline labels.

Another dataset is the internal patent database released by United States Patent and Trademark Office (PT for short)[2], which includes the information on all published patents at the PT. Each patent contains the ownership (referred to as authors in this paper), mark characteristics, classification, prosecution events,

---

[1] https://www.scopus.com/.
[2] https://bulkdata.USPTO.gov.

Table 1. Dataset statistics

Dataset	#papers/patents	#authors	#keywords	#venues
AMiner	3056388	1752401	354693	11397
Scopus	1304907	482602	127630	7653
PT	182260	73974	-	-

references, renewal and maintenance history. The patent dataset does not contain the venues and keywords information, so we consider the impact of the authors.

We adopt the Seg-phrase model [15] to extract the keywords, which iterates the process of phrase segmentation and keyword extraction to obtain the most representative phrases. We combine the phrases (two or more words) extracted by Seg-phrase and unigrams with relatively higher tf-idf values, and take this as the keyword set. Stopwords are removed from the keyword set.

After cleaning, the statistics of these datasets are shown in Table 1. As a matter of convenience, papers and patents are named collectively as papers.

5.2 Baseline Methods and Metrics

We compare our model with the following methods.

- *WNMF-5* [32]: This is the weighted nonnegative matrix decomposition method, where each entry in Matrix is set 1 if the researcher has cited the paper, and 0 otherwise. The feature number is set to 5.
- *WNMF-10*: The same method with the above with the feature number being 10.
- *NBCF* [23]: A neighborhood-based Collaborative Filtering algorithm, which is a comprehensive evaluation of scholarly paper recommendation by an adaptive neighbor selection method in a collaborative filtering framework.
- *MLP* [11]: It uses Multi-Layer Perceptron to learn the non-linear interaction function of embeddings from data.

All the parameters of baseline methods are empirically set to the optimal values.

For each user, we prepare k candidate papers/patents. Each candidate set contains 1 really cited paper/patent at least. The candidate papers are ranked according to the relatedness between user interests and paper vectors. We dopt $nDCG@k$ as a measurement [13]. $nDCG@k$ is often used to measure effectiveness of web search engine algorithms. $nDCG@k$ is calculated as $nDCG@k = \frac{DCG@k}{IDCG}$, $DCG@k = \sum_{i=1}^{k} \frac{rel_i}{log_2(i+1)}$, where $rel_i = 5$ if the i-th paper is really cited by the researcher, otherwise $rel_i = 0$. $IDCG = \sum_{i=1}^{|Cite|} \frac{5}{log_2(i+1)}$ is the ideal discounted cumulative gain, in which $|Cite|$ means the number of papers that are really cited by the researcher among the candidate papers.

Table 2. The experiments on the personalised new paper recommendation

nDCG@k	Aminer			Scopus			PT		
	k = 20	k = 30	k = 50	k = 20	k = 30	k = 50	k = 20	k = 30	k = 50
WNMF-5	0.7914	0.7852	0.7256	0.7797	0.7686	0.6819	0.6797	0.6586	0.5419
WNMF-10	0.8265	0.7892	0.7316	0.7889	0.7725	0.7052	0.6789	0.6625	0.5652
NBCF	0.8331	0.7994	0.7322	0.7932	0.7856	0.7272	0.6932	0.6756	0.6272
MLP	0.8391	0.8011	0.7649	0.8263	0.8201	0.7305	0.7063	0.6801	0.6504
Our Method	**0.8693**	**0.8512**	**0.8053**	**0.8309**	**0.8257**	**0.7399**	**0.7309**	**0.7057**	**0.6899**

Table 3. Query based recommendation compared with MLP

User Query	Our method	MLP
J. Han Information Network Knowledge Discoverry Data Mining	1. DeepTutor: An Effective, Online Intelligent Tutoring System That Promotes Deep Learning 2. Quantifying Robustness of Trust Systems against Collusive Unfair Rating Attacks Using Information Theory 3. Deep Learning Architecture with Dynamically Programmed Layers for Brain Connectome Prediction 4. Evaluating the statistical significance of biclusters 5. A Gaussian Process Latent Variable Model for BRDF Inference	1. Convergence properties of general network selection games 2. Quantifying the Targeting Performance Benefit of Electrostatic Haptic Feedback on Touchscreens 3. The Role of Environmental Predicta- bilityand Costs in Relying on Automation 4. Reflective Informatics: Conceptual Dimensions for Designing Technologies of Reflection 5. When hybrid cloud meets flash crowd: Towards cost-effective service provisioning
D. Patterson High Performance Computing Computer Architecture Parallel	1.Markov Mixed Membership Models 2.The effect of head mounted display weight and locomotion method on the perceived naturalness of virtual walking speeds 3.Influence at Scale: Distributed Computation of Complex Contagion in Networks	1.Convergence properties of general network selection games 2.Quantifying the Targeting Performance Benefit of Electrostatic Haptic Feedback on Touchscreens 3.The Role of Environmental Predicta- bilityand Costs in Relying on Automation

5.3   Evaluation on Personalized New Paper Recommendation

Recommendation Based on User Publications and References. To evaluate the performance of our recommendation, we compare our model with other methods on new research papers in this subsection. We randomly select 300, 100 and 50 researchers in Aminer, Scopus and PT datasets, respectively, to verify the performance of our model. The authors must satisfy the following conditions:

- Have published at least 5 papers and have cited at least 5 papers before year Y;
- Have cited at least 1 paper after year Y;
- The above academic papers should contain titles, authors, venues and abstracts, while the patents should contain titles, abstracts and authors.

The dataset is separated into two parts according to their published years. Papers published before year Y are used for training and after year Y are used

Table 4. The comparison between methods with different number of representative papers (#rp)

Average nDCG	Aminer			Scopus		
	#rp = 1	#rp = 3	#rp = 5	#rp = 1	#rp = 3	#rp = 5
WNMF-5	0.664	0.754	0.770	0.633	0.706	0.765
WNMF-10	0.680	0.760	0.790	0.659	0.715	0.761
NBCF	0.690	0.769	0.821	0.676	0.721	0.782
MLP	0.759	0.853	0.871	0.681	0.747	0.805
Our Method	**0.771**	**0.861**	**0.874**	**0.706**	**0.752**	**0.828**

Table 5. The comparison between methods with different ratios between samples (positive samples: negative samples)

Average nDCG	Aminer			Scopus		
	10:1	1:1	1:10	10:1	1:1	1:10
WNMF-5	0.679	0.754	0.730	0.633	0.726	0.693
WNMF-10	0.702	0.761	0.753	0.659	0.732	0.706
NBCF	0.720	0.775	0.753	0.676	0.742	0.731
MLP	0.743	0.821	0.770	0.680	0.775	0.734
Our Method	**0.791**	**0.869**	**0.780**	**0.708**	**0.801**	**0.754**

for testing. Then for each user, we prepare k candidate papers, each candidate paper set must satisfy the following conditions:

- Each candidate set contains 1 really cited paper at least;
- All candidates should be published after year Y, the papers contain titles, authors, venues and abstracts, while the patents should contain titles, abstracts and authors;
- The authors of these candidates should have been embedded during the training process;
- The candidates should be textually similar with the papers that cited by the user before year Y. The cosine similarity is adopted to calculate the semantic similarity between candidates and really cited papers.

Year Y is set 2014, then we rank the candidate papers for new paper recommendation.

The evaluation results on the three datasets are shown in Table 2. We can see that our method is very helpful in improving the performance of new paper recommendation.

Recommendation Based on Keyword Query. Our method could recommend for users who provide their queries. However, we do not have comparing metrics in aspect of contents relevance and we do not have their publication information so we cannot examine through their reference. Therefore, we test if our search results of some queries could retrieve content-relevant papers with higher reputation. We rank all the papers in a given area and see if papers retrieved by our method are more relevant to the query and querist. The results in Table 3 shows that our recommendations are more likely to the querists.

Evaluation on Parameters Settings. To quantify the influence by different parameter settings, we perform the above experiments with different settings on Aminer and Scopus. As the results shown in Table 4, the performance of our proposed personalized cross domain paper recommendation increases with a increasing number of papers. It is easy to understand that our recommendation method better grasp an author's requirement when she has more publications. Table 5 shows that our model performs best when the ratio between positive and negative samples is 1 : 1.

6   Conclusion

In this paper, we propose the Joint Text and Influence Embedding method for personalised recommendation on new papers. A paper is represented by the vectors of content, authors and published venue in the same latent space and the objective function is designed for the consistency of content semantics and paper influences in the citation network. Then we adopt the stochastic gradient descent method for optimization. A new paper is evaluated by its contents, authors' authority and the reputation of the publication venue. A user interest is learned either by one's historical references or a set of query keywords. Then we recommend the top-k related new papers. The results of the experiments show our method outperforms other methods.

Acknowledgement. This work was supported by the National Key R&D Program of China (2018YFC0831401), the Key R&D Program of Shandong Province (2019JZZY010107), the National Natural Science Foundation of China (91646119), the Major Project of NSF Shandong Province (ZR2018ZB0420), and the Key R&D Program of Shandong province (2017GGX10114). The scientific calculations in this paper have been done on the HPC Cloud Platform of Shandong University.

References

1. Asiwal, K., Suresh, B.K., Reddy, G.R.M.: Analysis of academic research networks to find collaboration partners. In: Stephanidis, C. (ed.) HCI 2016. CCIS, vol. 618, pp. 8–14. Springer, Cham (2016). https://doi.org/10.1007/978-3-319-40542-1_2
2. Bedi, P., Kaur, H., Marwaha, S.: Trust based recommender system for semantic web. IJCAI **7**, 2677–2682 (2007)
3. Belkin, M., Niyogi, P.: Laplacian eigenmaps and spectral techniques for embedding and clustering. NIPS **14**, 585–591 (2001)

4. Chakraborty, T., Modani, N., Narayanam, R., Nagar, S.: Discern: a diversified citation recommendation system for scientific queries. In: 2015 IEEE 31st International Conference on Data Engineering (ICDE), pp. 555–566. IEEE (2015)
5. Chen, T., Sun, Y.: Task-guided and path-augmented heterogeneous network embedding for author identification (2016)
6. Chikhaoui, B., Chiazzaro, M., Wang, S.: A new granger causal model for influence evolution in dynamic social networks: the case of DBLP. In: AAAI, pp. 51–57 (2015)
7. Gu, Q., Zhou, J., Ding, C.: Collaborative filtering: weighted nonnegative matrix factorization incorporating user and item graphs. In: Proceedings of the 2010 SIAM International Conference on Data Mining, pp. 199–210. SIAM (2010)
8. Gutmann, M.U., Hyvärinen, A.: Noise-contrastive estimation of unnormalized statistical models, with applications to natural image statistics. J. Mach. Learn. Res. 13(Feb), 307–361 (2012)
9. Guy, I., et al.: Personalized recommendation of social software items based on social relations. In: Proceedings of the Third ACM Conference on Recommender Systems, pp. 53–60. ACM (2009)
10. He, J., Chu, W.W.: A social network-based recommender system (SNRS). In: Memon, N., Xu, J., Hicks, D., Chen, H. (eds.) Data Mining for Social Network Data. Annals of Information Systems, vol. 12. Springer, Boston (2010). https://doi.org/10.1007/978-1-4419-6287-4_4
11. He, X., Liao, L., Zhang, H., Nie, L., Hu, X., Chua, T.S.: Neural collaborative filtering. In: Proceedings of the 26th International Conference on World Wide Web, pp. 173–182. International World Wide Web Conferences Steering Committee (2017)
12. Hu, Y., Koren, Y., Volinsky, C.: Collaborative filtering for implicit feedback datasets. In: Eighth IEEE International Conference on Data Mining. ICDM 2008, pp. 263–272. IEEE (2008)
13. Järvelin, K., Kekäläinen, J.: IR evaluation methods for retrieving highly relevant documents. In: Proceedings of the 23rd Annual International ACM SIGIR Conference on Research and Development in Information Retrieval, pp. 41–48. ACM (2000)
14. Koren, Y., Bell, R., Volinsky, C.: Matrix factorization techniques for recommender systems. Computer 42(8) (2009)
15. Liu, J., Shang, J., Wang, C., Ren, X., Han, J.: Mining quality phrases from massive text corpora. In: Proceedings of the 2015 ACM SIGMOD International Conference on Management of Data, pp. 1729–1744. ACM (2015)
16. Ma, H., Yang, H., Lyu, M.R., King, I.: Sorec: social recommendation using probabilistic matrix factorization. In: Proceedings of the 17th ACM Conference on Information and Knowledge Management, pp. 931–940. ACM (2008)
17. McNee, S.M., et al.: On the recommending of citations for research papers. In: Proceedings of the 2002 ACM Conference on Computer Supported Cooperative Work, pp. 116–125. ACM (2002)
18. Mikolov, T., Sutskever, I., Chen, K., Corrado, G.S., Dean, J.: Distributed representations of words and phrases and their compositionality. In: Advances in Neural Information Processing Systems, pp. 3111–3119 (2013)
19. Moreira, C., Calado, P., Martins, B.: Learning to rank academic experts in the DBLP dataset. Expert. Syst. 32(4), 477–493 (2015)
20. Perozzi, B., Al-Rfou, R., Skiena, S.: Deepwalk: online learning of social representations. In: Proceedings of the 20th ACM SIGKDD International Conference on Knowledge Discovery and Data Mining, pp. 701–710. ACM (2014)

21. Priem, J., Piwowar, H., Hemminger, B.: Altmetrics in the wild: Using social media to explore scholarly impact (2012)
22. Song, M., Heo, G., Kim, S.: Analyzing topic evolution in bioinformatics: investigation of dynamics of the field with conference data in DBLP. Scientometrics **101**(1), 397–428 (2014)
23. Sugiyama, K., Kan, M.: A comprehensive evaluation of scholarly paper recommendation using potential citation papers. Int. J. Digit. Libr. **16**(2), 91–109 (2015)
24. Sugiyama, K., Kan, M.Y.: Exploiting potential citation papers in scholarly paper recommendation. In: Proceedings of the 13th ACM/IEEE-CS Joint Conference on Digital Libraries, pp. 153–162. ACM (2013)
25. Tang, J., Qu, M., Mei, Q.: PTE: predictive text embedding through large-scale heterogeneous text networks. In: Proceedings of the 21th ACM SIGKDD International Conference on Knowledge Discovery and Data Mining, pp. 1165–1174. ACM (2015)
26. Tang, J., Qu, M., Wang, M., Zhang, M., Yan, J., Mei, Q.: Line: large-scale information network embedding. In: Proceedings of the 24th International Conference on World Wide Web, pp. 1067–1077. ACM (2015)
27. Tang, J., Zhang, J., Yao, L., Li, J., Zhang, L., Su, Z.: Arnetminer: extraction and mining of academic social networks. In: Proceedings of the 14th ACM SIGKDD International Conference on Knowledge Discovery and Data Mining, pp. 990–998. ACM (2008)
28. Tenenbaum, J.B., De Silva, V., Langford, J.C.: A global geometric framework for nonlinear dimensionality reduction. Science **290**(5500), 2319–2323 (2000)
29. Xie, Y., Sun, Y., Shen, L.: Predicating paper influence in academic network. In: 2016 IEEE 20th International Conference on Computer Supported Cooperative Work in Design (CSCWD), pp. 539–544. IEEE (2016)
30. Yates, E., Dixon, L.: Pagerank as a method to rank biomedical literature by importance. Source Code Biol. Med. **10**(1), 16 (2015)
31. Zhang, A., Bhardwaj, A., Karger, D.: Confer: a conference recommendation and meetup tool. In: Proceedings of the 19th ACM Conference on Computer Supported Cooperative Work and Social Computing Companion, pp. 118–121. ACM (2016)
32. Zhang, S., Wang, W., Ford, J., Makedon, F.: Learning from incomplete ratings using non-negative matrix factorization. In: Proceedings of the 2006 SIAM International Conference on Data Mining, pp. 549–553. SIAM (2006)

Short Papers

Scholar Knowledge Graph and Its Application in Scholar Encyclopedia

Kun Hu, Jianguo Li^(✉), Wande Chen, Chengzhe Yuan, Qiang Xu, and Yong Tang

South China Normal University, Guangzhou 510631, China
{lkome,jianguoli,chenwande,qiangxu,ytang}@m.scnu.edu.cn

Abstract. Knowledge graphs (KGs) are developing rapidly and has a wide range of applications in various fields. But there have been relatively few Chinese encyclopedias with the scholar knowledge graph so far. Therefore, we rely on the academic social network-SCHOLAT to complete the construction of scholar KG. The purpose of this scholar KG is to solve the problems that the information of scholars is not updated in time and scattered in the organizations' websites. This paper introduces in detail the key technologies for constructing scholar KG, such as knowledge extraction, organization and management of scholar KG. At the same time, we take advantages of KG in semantic expression into account. We apply the scholar KG to the scholar encyclopedia system. Based on the scholar KG, we completed the scholar recommendation on the Scholar Encyclopedia. On the other hand, the scholar KG will also provide the basis for more intelligent applications of SCHOLAT in the future.

Keywords: Knowledge graph · Artificial intelligence · Named-entity recognition · Scholar recommendation

1 Introduction

Nowadays the Internet has entered the era of big data. Massive amounts of data provide good conditions for scientific research scholars, allowing them to create greater academic value for scientific research. But on the other hand, the massive amount of data itself can easily lead to information overload, so it takes a lot of time and energy to extract useful information from it. At the same time, searching for the information of countless scholars themselves is also a relatively expensive task. So it is necessary to know how to use computer technology to manage the large amount of information data that has accumulated on the Internet. At the same time, we must also consider the accuracy and effectiveness of network information, because this is not only related to the actual experience of network users, but also has been a difficult problem studied by experts and scholars and industry. KGs can solve this problem because it has natural advantages in organizing and understanding the massive information on the Internet. Therefore, in order to realistically reflect the things in the objective world and reduce the cost of Internet data acquisition, Tim Berners-Lee believes that the Web should not only be a link between web pages, and proposed the concept of the Semantic Web in 1998. That

© Springer Nature Singapore Pte Ltd. 2021
Y. Sun et al. (Eds.): ChineseCSCW 2020, CCIS 1330, pp. 573–582, 2021.
https://doi.org/10.1007/978-981-16-2540-4_41

is, a web page is not just a web page, but an entity in the objective world, such as people, institutions, and places. At the same time, it is believed that the links between web pages represent the relationship between entities, such as birthplace, founder, and location [1]. Google proposed the concepts of KGs in 2012. The original purpose was to achieve a more intelligent search method than the existing search engines to improve the search experience and improve the search quality [2]. The essence of KG is a knowledge base of the semantic web, which is built on a directed graph. The nodes of the graph represent entities, corresponding to the web pages in the semantic web. The edges represent the relationships between entities and correspond to the links between web pages in the semantic web. Major technology companies have proposed that KG is the foundation of artificial intelligence. The reason is that KG can provide knowledge reserves for machines to think and understand problems like humans.

In this paper, we mainly rely on the data of SCHOLAT[1] to construct scholar KG. At the same time, in order to make up for the shortcomings of the number of entities extracted from the scholar network, we have also integrated the knowledge of Chinese academic knowledge fields from different sources on the Internet, such as general domain knowledge base Baidu Encyclopedia, Ownthink[2], Baidu academic etc. After extracting a large amount of relevant knowledge from these different data sources, we begin to construct scholar KG. And after the construction of scholar KG is completed, the scholar KG is applied to the Scholar Encyclopedia knowledge website for efficient query and scholar recommendation. At the same time, it also provides a professional scholar KG to ordinary users, which reduces the cost of acquiring scholar information, and provides knowledge reserves for other KGs applications, such as semantic understanding technology, intelligent search technology, human-computer dialogue Technology etc.

2   Related Work

KG is the current research hotspot and its application is very extensive. At present, KGs that are relatively well-known abroad include DBpedia, Yago, Babelnet, etc. DBpedia [3] uses an automated extraction framework to extract knowledge information in various languages from Wikipedia sites, with more than 30 million entities and hundreds of millions of RDF triples. Yago [4] integrates data from Wikipedia, WordNet[3], and GeoNames to add time and space attribute descriptions to knowledge items.

There are well-known KGs such as Baidu Zhixin, Sogou Zhicub, Zhishi.me, XLore and so on in China. Among them, Zhishi.me developed by Shanghai Jiao Tong University has the following knowledge sources: Baidu Encyclopedia, Chinese Wikipedia, Interactive Encyclopedia, etc. There are currently 14,307,056 entities extracted from Baidu Encyclopedia, the number of entities extracted from Interactive Encyclopedia is up to about 5,551,163, and 903,462 entities extracted from Chinese Wikipedia are integrated. In addition, Tsinghua University has developed XLore [5]. XLore's main sources of knowledge are the following: Baidu Encyclopedia, Chinese and English Wikipedia,

---

[1] https://www.scholat.com/.

[2] https://www.ownthink.com/knowledge.html.

[3] https://wordnet.princeton.edu/.

Interactive Encyclopedia, etc. A great feature is the support of bilingual Chinese and English. So far it has contained about 10856042 examples, 663740 concepts and 56449 attributes.

However, Chinese encyclopedias that use scholar KG are still rare. The domestic typical scholar knowledge libraries include CNKI scholar library and Tsinghua University scholar library. Among them, there are many scholar entities included in the CNKI scholar library. But the information of scholars in the knowledge library is often not updated in time, so it is not accurate enough to some extent. And on the other hand, the scholar's entity has more redundancy. However, the scope of scholars at Tsinghua University is relatively limited, and only scholars from Tsinghua University are included. In order to solve the problem that the scholar information is not updated in time and the scholar information is scattered in the organization's website. We will build a KG for the scholar field based on the real scholar data of SCHOLAT and integrate the scholar data of other Chinese encyclopedias. Finally, the constructed scholar KG is applied in the Scholar Encyclopedia system.

3   The Construction of Scholar Knowledge Graph

When constructing domain KGs, we refer to the general KGs construction technology to a large extent. Taking the domain specificity of the scholar KG into account, there is also a certain professional specificity in the construction of the scholar KG. So the first step in constructing scholar KG is to use Named Entity Recognition (NER) [6, 7] to extract knowledge from unstructured data on the Internet [8], including named-entity extraction and relationship extraction. Among them, the source of structured data is mainly from SCHOLAT [10]. The semi-structured data and unstructured data are mainly from Baidu Encyclopedia, Baidu Academic and Ownthink. Then we perform data fusion and entity alignment on the knowledge from different data sources. The second step is to construct knowledge ontology, organize and apply scholar knowledge. The technical process of scholar KG construction is shown in Fig. 1.

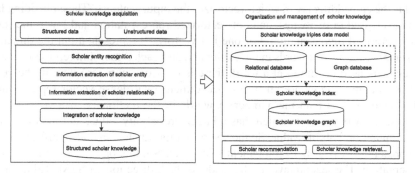

Fig. 1. The technical process of scholar KG construction

3.1 Knowledge Extraction

The first step in the processes of constructing KG is knowledge extraction. The main task of knowledge extraction is to extract useful information from the massive heterogeneous data of the Internet, and use the knowledge extracted by knowledge extraction technology as the knowledge unit for constructing scholar KG. Knowledge extraction is a technology that automatically extracts structured information such as entities, relationships, and entity attributes from unstructured data [12].

3.2 Named Entity Recognition

As an important technology of natural language processing (NLP), NER is mainly used to identify entities in text data, such as names of people, places, and organizations [11]. In a specific field, there will be different vocabularies as the entities in the field. For example, in the academic field, the name of a paper can be defined as an entity and the name of a project can be defined as an entity. In the field of e-commerce, the name of a product can be defined as an entity. The names of the entities are infinite, flexible and ambiguous. There is a large amount of triple knowledge in KGs. And entities are an important part of triple knowledge. Therefore, NER technology plays an indispensable role when constructing and expanding KG.

Generally, NER is mainly divided into role labeling framework and sequence labeling framework. NER based on role labeling includes perceptron sequence labeling, conditional random field-based sequence labeling [13], Hidden Markov Model sequence labeling [14] and so on. Both the role tagging framework and the sequence tagging framework rely strongly on manually annotated data sets. With the development of deep learning technology, Huang Z et al. [7, 15] proposed a BiLSTM-CRF model based on recurrent neural network to solve the problem that NER tasks rely heavily on hand-made features and specific domain knowledge. And the model does not depend on specific languages and external dictionary resources. Therefore, this paper uses the BiLSTM-CRF model for NER.

The structure of BiLSTM-CRF model is shown in Fig. 2. The first layer of the model is the embedding layer, which vectorizes the input sentence. The embedding layer can be expressed in two ways, one-hot vectorized representation and low-dimensional dense vectorized representation. For one-hot vectorized representation, the dimension represents the number of words in the dictionary, and each dimension represents a word. If a vector is used to represent a word, there is only one non-zero item in the vector, and the remaining items are all 0. For low-dimensional dense vector representation, its dimension is less than the number of words in the dictionary. If a vector is used to represent a word, a certain item has a larger probability value, and the other items have a smaller probability value. So the vector in this way has spatial significance. One-hot representation is easy to implement with code. So, this paper uses one-hot vector representation. The second layer of the model is the bidirectional LSTM [16] layer. The multidimensional vector of the sentence is used as the input of the BiLSTM layer at each time step. For a sentence $X = (x_1, x_2, ..., x_n)$ contains N words, and each word can be mapped to an n-dimensional vector, where n is one- hot vector dimension. LSTM

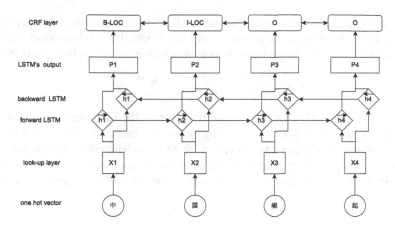

Fig. 2. The structure of BiLSTM-CRF model

is implemented using the following (1)–(5) formulas below, where ' $*$' denotes the convolution operator and ' $\odot$', as before, denotes the Hadamard product:

$$f_t = \sigma\left(W_{xf} * x_t + W_{hf} * h_{t-1} + W_{cf} \odot c_{t-1} + b_f\right) \tag{1}$$

$$i_t = \sigma\left(W_{xi} * x_t + W_{hi} * h_{t-1} + W_{ci} \odot c_{t-1} + b_i\right) \tag{2}$$

$$c_t = f_t \odot c_{t-1} + i_t \odot \tanh(W_{xc}x_t + W_{hc}h_{t-1} + b_c) \tag{3}$$

$$o_t = \sigma\left(W_{xo}x_t + W_{ho}h_{t-1} + W_{co}c_t + b_o\right) \tag{4}$$

$$h_t = o_t \odot \tanh(c_t) \tag{5}$$

where σ represents the logistic sigmoid function, and o, i and f are respectively the output gate, input gate, forget gate. W_{hi} represents the hidden-input gate matrix and W_{xo} represents the input-output gate matrix.

Combine the hidden state sequences output by the forward and reverse LSTMs to obtain a complete hidden state sequence. Then a linear layer is connected, and the hidden state vector is mapped from n-dimensional to k-dimensional, thereby obtaining the sentence feature matrix M, where k is the number of labels in the BIO label set. Each dimension in the feature matrix M can be regarded as the score of a certain label. If we directly perform Softmax multi-classification, it is equivalent to performing independent k classification for each position. In this way, the information that has been marked cannot be used, and illegal label sequences cannot be avoided. For example, the location entity must start with B-LOC, but Softmax may get illegal results such as I-LOC or I-LOC, so we need to change the Softmax result Access to the third layer to solve the problem of illegal labels. The third layer of the model is the CRF [17] layer. Assuming the sentence length is w, the input of the CRF layer is a (w + 2)*k-dimensional matrix, which represents the transfer score from the first label to the kth label. When performing label prediction, we can use the previously marked labels. The final output of the model is the result obtained by the probability scoring function, as shown in formula (6).

$$S(X, Y) = \sum_{i=0}^{n} A_{y_i y_{i+1}} + \sum_{i=1}^{n} P_{i,y_i} \tag{6}$$

Where $X = (x_1, x_2, ..., x_n)$ is the word vector of the input sentences after word segmentation, and $Y = (y_1, y_2, ..., y_n)$ represents the predicted labels. After training, the accuracy rate on the test data set can reach 88.62%. The entity recognition model supports direct input of sentences and returns all the person, location, and organization entities in the sentence. This interface can be provided to the Scholar Encyclopedia website.

3.3 Knowledge Graph Representation

Knowledge graph representation refers to modeling KG, so that knowledge calculation and reasoning can be conveniently carried out. The Semantic Web is based on the resource description framework RDF. In RDF, knowledge always appears in the form of <subject, predicate, object> triples. The subject is an individual. A predicate is an attribute, and an attribute can connect two individuals, or connect an entity and an instance of a data type. The object can be an individual or an instance of a data type. RDFS (RDF Schema) provides a simple description of classes and attributes, thereby providing modeling primitives for RDF data.

RDF and RDFS can provide simple semantic representation, but their expressive power is still weak. For example, it is impossible to prove whether two classes or two individuals are equivalent (for example, it cannot be stated that "Shuren Zhou" and "Xun Lu" are the same person). For this reason, W3C proposed OWL (Web Ontology Language)[4] language extension RDFS, which is a recommended language for representing ontology on the Semantic Web.

RDFS and OWL use discrete symbols to represent KGs, which can effectively represent data in a structured manner. However, discrete symbols cannot express corresponding semantic information and perform semantic calculations in the computer. Therefore, it cannot be well applied to downstream tasks, such as intelligent answer and semantic search. In order to solve this problem, the concept of knowledge embedding is proposed, which is to map entities and relations to a low-dimensional continuous vector space. This kind of research on mapping the entities and relationships in KG to the continuous vector space is called Knowledge Graph Embedding [9] (KGE) and Representation Learning. Wang Q et al. [9] introduces the related methods and research progress of KGE.

The idea of KGE is to learn and train the model through the method of supervised machine learning. During the training process, the model can learn the information of the semantic layer. The main idea of the classic TransE [18] model is to vectorize the triples. If a triple <h, r, t> holds, the formula $h + r \approx t$ holds.

---

[4] https://www.w3.org/OWL/.

4 Scholar Recommendation Application Based on Scholar Knowledge Graph

Scholar KG has stored tens of thousands of relationship attributes, such as colleague relationship, alumni relationship, paper cooperator relationship, project cooperator relationship, etc. In the graph database, we can perform efficient data query according to the depth of the relationship. Therefore, we use the scholar KG to recommend scholars. This paper mainly uses content-based scholar recommendations. We recommend scholars by calculating the similarity of atlas entities and the relationship between entities. The recommended process is shown in the Fig. 3.

When a user visits a scholar's information homepage, the page document is loaded first, and then an asynchronous request is sent through AJAX technology. Finally, the relevant page returns the packaged recommendation list and loads it into the page. After receiving the request in the background, call relevant background methods to find the thesis collaborators, project collaborators, patent collaborators, scholars in the same work unit and scholars in the same research field. When acquiring entities are in the background method, query the scholar KG using the query statement Cypher of Neo4j database. Compared with the relational database, because the multi-table Cartesian product operation is avoided, the query efficiency will be significantly improved. And the home page of scholars is shown in the Fig. 4. The following describes the specific implementation of the recommendation algorithm.

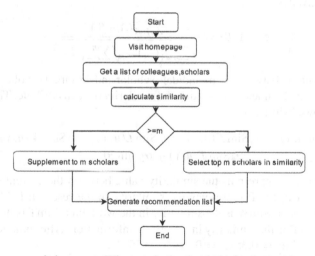

Fig. 3. The recommended process. Where m is the threshold for the number of recommended scholars. If the number of candidate scholars is more than m, the m scholars with the highest scores are selected. If the number of candidate scholars is less than m, the corresponding number of scholars will be supplemented according to the number of homepage visits.

Calculate the similarity between the attributes of the current homepage entity and all the scholars in the recommendation list. The main attributes that need to be compared are character tags, research fields, work units, job titles, and personal profile information

Fig. 4. The home page of scholars

in plain text. Before the calculation of similar paragraphs, these text data need to be processed, mainly including word segmentation and stop words. Then compare with the current homepage scholar, calculate the cosine similarity of each attribute separately. The formula for calculating the cosine similarity is shown in (6). A and B represents two vectors respectively.

$$\cos(A, B) = \frac{\sum_{i=1}^{n}(A_i \cdot \beta_i)}{\sqrt{\sum_{i=1}^{n}(A_i)^2} \cdot \sqrt{\sum_{i=1}^{n}(B_i)^2}} \tag{7}$$

For the similarity between each candidate scholar and the target scholar, we obtain the scholar's similarity by linearly weighting the similarity of each attribute. The calculation formula is shown below (7).

$$\begin{aligned} Sim(u, v) = {}&\alpha_1 Sim\,F(u, v) + \alpha_2 Sim\,L(u, v) + \alpha_3 Sim\,W(u, v) + \\ &\alpha_4 Sim\,T(u, v) + \alpha_5 Sim\,R(u, v) \end{aligned} \tag{8}$$

Where $Sim(u, v)$ represents the similarity value between the current scholar u and the target scholar u, SimF represents the similarity in the research field, SimL is the similarity in the label, SimW is the similarity in the work unit, SimT is the similarity in the title, and SimR is the similarity in the profile information. After practice, we choose $\alpha_1 = 0.1\ \alpha_2 = 0.1\ \alpha_3 = 0.35\ \alpha_4 = 0.1\ \alpha_5 = 0.35$.

Finally, we use the list of similarity Top5 scholars. If less than 5 scholars are obtained in the recommendation model, the corresponding number of scholars will be randomly selected from scholars who are in Top 20 clicks.

5   Conclusions

In this paper, we construct scholar KG based on SCHOLAT data and other open KGs data. In addition, some key technologies of constructing scholar KG are introduced, including knowledge extraction, knowledge ontology construction, KG storage and visualization. Based on this scholar KG, we have realized the scholar recommendation function in Scholar Encyclopedia. Because the scholar KG we constructed is mainly based on the real data of the SCHOLAT, and filters out those scholars who have not updated the profile for a long time. Therefore, the scholar information included in the scholar KG is authentic and time-sensitive. At the same time, the scholar KG also lays a solid foundation for more intelligent applications of Scholar Encyclopedia and SCHOLAT.

Although scholar KG has so many advantages, it is not mature enough at present. For example, the number of entities and relationships in scholar KG is still relatively limited. The relationship between entities needs to be optimized. The entity types of KG can be more abundant. In the next work, we will spend more time to expand the number of entity relationships and provide a more complete knowledge reserve for more intelligent applications in SCHOLAT. At the same time, we will continue to optimize the structure of scholar KG. And in order to make scholar KG more rich and more diverse, we will continue to expand the types of entities.

Acknowledgements. This work was supported by National Natural Science Foundation of China under grant number 61772211, by National Natural Science Foundation of China under grant number U1811263.

References

1. Berners-Lee, T., Hendler, J., Lassila, O.: The semantic web. Sci. Am. **284**(5), 34–43 (2001)
2. Vincent, J.: Apple boasts about sales; Google boasts about how good its AI is [EB/OL], 4 October 2016. Accessed 1 May 2018. https://www.theverge.com/2016/10/4/13122406/goo gle-phone-event-stats
3. Bizer, C., Lehmann, J., Kobilarov, G., et al.: DBpedia-A crystallization point for the web of data. J. Web Semant. **7**(3), 154–165 (2009)
4. Suchanek, F.M., Kasneci, G., Weikum, G.: Yago: a core of semantic knowledge. In: Proceedings of the 16th International Conference on World Wide Web, pp. 697–706 (2007)
5. Wang, Z., Li, J., Wang, Z., et al.: XLore: a large-scale English-Chinese bilingual knowledge graph. In: Proceedings of the ISWC 2013 Posters & Demonstrations Track, pp. 121–124 (2013)
6. Mohit, B.: Named entity recognition. In: Zitouni, I. (ed.) Natural Language Processing of Semitic Languages, pp. 221–245. Springer, Berlin (2014)
7. Lample, G., Ballesteros, M., Subramanian, S., et al.: Neural architectures for named entity recognition. In: Proceedings of the 2016 Conference of the North American Chapter of the Association for Computational Linguistics, pp. 260–270 (2016)
8. Zheng, S., Wang, F., Bao, H., et al.: Joint extraction of entities and relations based on a novel tagging scheme. In: Proceedings of the 55th Annual Meeting of the Association for Computational Linguistics, pp. 121–124 (2017)
9. Wang, Q., Mao, Z., Wang, B., et al.: Knowledge graph embedding: a survey of approaches and applications. IEEE Trans. Knowl. Data Eng. **29**(12), 2724–2743 (2017)
10. Tang, J., Zhang, J., Yao, L.M., et al.: ArnetMiner: extraction and mining of academic social networks. In: Proceedings of the 14th ACM SIGKDD International Conference on Knowledge Discovery and Data Mining, pp. 990–998. ACM, New York (2018)

11. Ling, X., Weld, D.S.: Fine-grained entity recognition. In: Proceedings of the 26th AAAI Conference on Artificial Intelligence, pp. 94–100. AAAI, Palo Alto (2012)
12. Cowie, J., Lehnert, W.: Information extraction. Commun. ACM **39**(1), 80–91 (1996)
13. Lafferty, J., McCallum, A., Pereira, F.: Conditional random fields: probabilistic models for segmenting and labeling sequence data. In: Proceedings of International Conference on Machine Learning (ICML), pp. 282–289 (2001
14. Collier, N., Nobata, C., Tsujii, J.: Extracting the names of genes and gene products with a Hidden Markov model. In: Proceedings of the 18th Conference on Computational Linguistics, pp. 201–207. Association for Computational Linguistics, Stroudsburg (2000)
15. Huang, Z., Xu, W., Yu, K.: Bidirectional LSTM-CRF Models for Sequence Tagging. CoRR, 2015, abs/1508.01991. https://arxiv.org/abs/1508.01991.
16. Hochreiter, S., Schmidhuber, J.: Long short-term memory. Neural Comput. **9**(8), 1735–1780 (1997)
17. Lin, B.Y., Xu, F., Luo, Z., et al.: Multichannel BiLSTM-CRF model for emerging named entity recognition in social media. In: Proceedings of the 3rd Workshop on Noisy UserGenerated Text, pp. 160–165. Association for Computational Linguistics, Stroudsburg (2017)
18. Bordes, A., Usunier, N., Garcia-Duran, A., et al.: Translating embeddings for modeling multirelational data. In: Proceedings of Advances in Neural Information Processing Systems 26: 27th Annual Conference on Neural Information Processing Systems, pp. 2787–2795 (2013)

MPNet: A Multiprocess Convolutional Neural Network for Animal Classification

Bin Jiang$^{(\boxtimes)}$ ⓘ, Wei Huang ⓘ, Yun Huang ⓘ, Chao Yang ⓘ, and Fangqiang Xu ⓘ

College of Computer Science and Electronic Engineering, Hunan University, Lushan Road (S), Yuelu District, Changsha, China
{jiangbin,hwei,hy1122,yangchaoedu,xfq0204}@hnu.edu.cn

Abstract. Recent deep learning-based approaches have achieved remarkable performance in the animal-image classification field. However, previous deep learning-based approaches consume large amounts of computational resources, thus is not suitable for deployment under resource-constrained environments. To address this problem, we propose a novel Multiprocess Convolutional Network (MPNet). Specifically, this network contains two subnetworks. The first one employs a convolutional network to extract abstract semantic features from a horizontal viewpoint. To make full use of semantic information, we design the other subnetwork to extract features from a vertical viewpoint. Then we calculate the gram matrix of these feature maps by element-wise multiplication. Meanwhile, we adopt weight sharing strategy to reduce model parameters. Experiments on the Animals with Attributes (AWA) dataset has demonstrated that our proposed approach achieves 87.54% top-1 accuracy with 33.57MB parameters. Compared with the other state-of-the-art approaches, our model saves more computation cost and yields higher accuracy.

Keywords: Animal classification · Multiprocess convolutional network · Weight sharing

1 Introduction

The animal classification is a hot research task in the field of computer vision. The task aims at recognizing the animal from a picture of the natural scenes, which can be utilized in many fields such as the wild animals researches, the protection of endangered animals, and the animal husbandry [23]. The core challenge of animal classification is to maintain balance in recognition accuracy, execution speed, and memory usage.

Traditional works [2,6] utilize Bags-of-Visual-Words (BVW) techniques to tackle with the classification problems, most of traditional approaches calculate Scale-Invariant Feature Transform (SIFT) [17] features, then employ k-means cluster SIFT features to obtain Word Frequency Vector (WFV). Many

© Springer Nature Singapore Pte Ltd. 2021
Y. Sun et al. (Eds.): ChineseCSCW 2020, CCIS 1330, pp. 583–593, 2021.
https://doi.org/10.1007/978-981-16-2540-4_42

approaches combine the WFV with machine learning algorithms to classify animal images. Though achieving a fast speed, most of methods obtain an unsatisfied performance in recognition accuracy. Because the features of traditional methods are manually extracted, these low-level features contain a lot of noise information.

Compare with traditional methods, deep learning methods [3,7,22], can extract high-level features yet yield promising accuracy. Chen et al. [5] build a lightweight convolutional network that only contains one convolutional layer and two fully connected layers. However, the features generated by only one convolutional layer still contain a lot of noise information, leading to unsatisfied recognition accuracy. Yuan et al. [28] compare MobleNet-Beta [28] with ten widely-used models include ResNet50 [10], ResNet101, MobileNet [11], MobileNet V2 [19], DenseNet121 [12] and so on. They demonstrate that the ResNets [28] yield extremely high accuracy. The residual network can extract sufficient the high-level semantic information by many convolutional layers. Besides, information propagated directly from the shallow convolutional layer to the deep convolutional layer by utilizing the skip connection structure, which avoids vanishing gradient problem. However, ResNets have too many convolutional layers, resulting in huge amounts of computational cost and time-consuming inference, which is not feasible to the application in devices with limited memory resources. The MobileNet [11] proposes depthwise separable convolution to reduces model parameters. Compared to the ResNets, the MobileNet achieves a good balance between accuracy and efficiency.

To make full use of the semantic information, we investigate the process of the human recognize object. From a physiological view, Tanenbau and Freeman [21] have demonstrated that the system of human vision consists of two paths. Concretely, Tanenbau et al. [21] believe the fact: When people observe an image, people first observe the position of the object, then observe the appearance of the object. From a cognitive view, our proposed approach is similar to the three-views drawing in mathematics. We can reshape the general appearance of the object by observing three-views drawing. This process motivates us to design a model for achieving an obvious performance improvement.

Based on the above analysis, we propose a Multiprocess Convolutional Network (MPNet), which recognizes the animal categories as the process of the human cognitive. Initially, we design two convolutional networks called horizontal subnetwork and vertical subnetwork, respectively. Each subnetwork can generate the same size of the feature maps. To save memory usage, we adopt weight sharing measure. Thereafter, we reshape the size of subnetwork-2 feature maps. Then we calculate the gram matrix of all the network feature maps to fuse semantic information.

Similar to most current networks [27], we divide the MPNet into two stages. The first stage, which consists of two parallel convolutional layers, is designed to collecte semantic information. The second stage, which contains a fully connected layer and an output layer, is presented to recognize the animal categories. We utilize an average cross entropy loss function to optimize the MPNet.

In this paper, we carry out experiments on the Animals with Attributes (AWA) dataset to evaluate the effectives of our method. The MPNet achieves 87.54% top-1 accuracy and has only 33.57MB parameters, which outperforms MobileNet V2 [19] by 3.4% top-1 accuracy and only 20.42% the number of parameters of ResNet101 [28]. To more comprehensive evaluate our model, we compare some metric including the parameter of network and the Float-point operations (Flops) with other popular models. Consequently, the consumption of memory is greatly reduced, and incorporating the multiviewpoints features improves recognization accuracy. In addition, we employ the Receiver Operating Characteristic (ROC) curve to evaluate the performance of MPNet.

In summary, our main contributions are listed as follows:

- A novel trade-off model called Multiprocess Convolutional Network for recognizing animal image. Our proposed network can extract semantic information from both horizontal and vertical views.
- We adopt weight sharing measure to reduce the number of parameters and utilize the gram matrix of feature maps to improve model accuracy.
- To verify our model, we conduct a series of experiments to compare several models in terms of computational efficiency and effectiveness. In particular, our model outperforms MobileNet V2 [19] by 3.4% top-1 accuracy, and the usage of memory is 7 times smaller than the ResNets [28].

The remainder of this paper is organized as follows. Section 2 will introduce current methods related to animal classification and discuss the advantages and disadvantages of these works. Section 3 will describe the details of the proposed model. Section 4 will conduct experiments and discuss the results. Finally, Sect. 5 will draw a conclusion and have a outlook on the future work.

2   Related Works

In summary, previous animal image classification methods can generally be divided into two groups: Non-learning classifying methods and Learning classifying methods.

Non-learning methods [18, 21, 22] extract abstract semantic information by Word Frequency Vector (WFT), then combine with other traditional machine learning algorithms to recognize images. Manohar *et al.* [18] propose an animal recognition system with supervised and unsupervised approaches. They firstly extract Gabor features, then utilize Linear Discriminate Analysis (LDA) [13] approach to reduce the dimension of the feature maps. In addition, Manohar *et al.* [18] reduce the dimension of the feature via adopting Principle Component Analysis (PCA) [25], then utilize k-means [1] algorithm to divide features. However, these traditional methods only extract low-level semantic information. The results of classification hardly can satisfy the actual demands.

Learning approaches mainly utilize deep learning techniques [4, 13, 28]. These approaches can extract more high-level semantic information. Yuan *et al.* [28]

propose a novel network, called MobileNet-Beta, they firstly pretrain the convolutional neural network with SISURF feature maps, then train the network again on original dataset. The MobileNet-Beta model obtains a more better result compared with MobileNet [11]. To make full use of these abstract information, Lin *et al.* [16] propose a bilinear model for fine-grained categorization, which consists of two feature extractors. Nevertheless, the bilinear model occupies a substantial amount of computation resources. Huang *et al.* [12] establish the connection between different layers, which reduces the number of parameter. Wang *et al.* [24] utilize a bank of convolutional filters to enhance the mid-level representation learning. However, these methods do not consider the local information in the image. Although these methods have achieved better recognition accuracy, the number of network parameters is larger. Comparing with these approaches, our approach not only achieves promising accuracy but also requires less memory, which is more feasible for being applied to the restricted hardware devices.

3   Multiprocess Convolutional Neural Network

In this section, we will introduce the proposed Multiprocess Convolutional Neural Network (MPNet), which consists of two subnetworks, called subnetwork-1 and subnetwork-2, respectively. The overview framework of our model as shown in Fig. 1. Let I_{gt} be the ground truth images and I_{in} be the input images of the network. At first, we transform I_{gt} into I_{in}, then feed I_{in} into subnetwork-1. At the same time, we feed I_{in} into subnetwork-2, then reshape the size of the feature maps of subnetwork-2. Finally, we combine the feature maps of subnetwork-1 and subnetwork-2, then feed them into the fully connected layer.

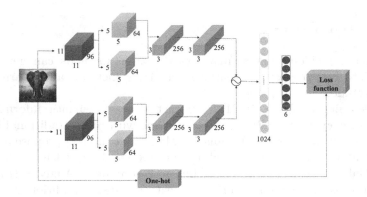

Fig. 1. The architecture of our model. We utilize the double network to discriminate animal image

3.1   The Structure of the Multiprocess Convolutional Neural Network

The input I_{in} of MPNet is a $3 \times 256 \times 256$ image. The subnetwork-1 consists of the four convolutional layers. For the first two convolutional layers, the kernels of convolution are set to 11×11 and 5×5 for enlarging respective fields and avoiding too much information loss. The kernel of the last two convolutional layers is set to 3×3, which not only increases the learning ability of the network but also reduces the number of parameters. Meanwhile, the other subnetwork setting adopts weight sharing measure. Both subnetworks are able to be trained simultaneously in this way.

3.2   The Calculate of the Multiprocess Convolutional Neural Network

In this subsection, we further discuss the detail of MPNet, we assume that the output feature maps of subnetwork-1 are F_1 and the subnetwork-2 are F_2. Because the dimensions of output features are four dimensions and we only focus on the third and the fourth dimension. Let $F_1^{[height,width]}$ represent the third and the fourth dimension of the F_1, the $F_2^{[height,width]}$ be the third and the fourth dimension of the F_2, the formula for $F_1^{[height,width]}$ and $F_2^{[height,width]}$ as follow:

$$F_1^{[height,width]} = [\alpha_1, \alpha_2, \ldots, \alpha_n]^T, \tag{1}$$

$$F_2^{[height,width]} = [\beta_1, \beta_2, \ldots, \beta_n]^T, \tag{2}$$

where α, β both are n dimensional row vectors, n is the number of columns. Then we calculate the gram matrix of these feature maps by element-wise multiplication to integrate F_1 with F_2. The process of calculating the gram matrix can be presented as follow:

$$Result = F_1^{[height,width]} [F_2^{[height,width]}]^T, \tag{3}$$

specifically,

$$Result = \begin{bmatrix} \alpha_1\beta_1^T & \alpha_1\beta_2^T & \cdots & \alpha_1\beta_{n-1}^T & \alpha_1\beta_n^T \\ \alpha_2\beta_1^T & \alpha_2\beta_2^T & \cdots & \alpha_2\beta_{n-1}^T & \alpha_2\beta_n^T \\ \alpha_3\beta_1^T & \alpha_3\beta_2^T & \cdots & \alpha_3\beta_{n-1}^T & \alpha_3\beta_n^T \\ \vdots & \vdots & \ddots & \vdots & \vdots \\ \alpha_n\beta_1^T & \alpha_n\beta_2^T & \cdots & \alpha_n\beta_{n-1}^T & \alpha_n\beta_n^T \end{bmatrix}, \tag{4}$$

note that the value of F_1 is equal to the value of F_2 by utilizing weight sharing measure. Therefore, the product of $(\alpha_i \cdot \beta_j^T)$ is equivalent to $(\alpha_i \cdot \alpha_j^T)$. Finally, the result of integrating F_1 with F_2 is called gram matrix in mathematics.

In the training process, the back propagation of the above transformation operation is shown in Fig. 2.

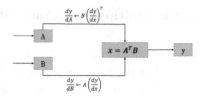

Fig. 2. An overview diagram of back propagation of multiprocesses mixed processing

Intuitively, the kind of fusion can be regarded as the combination of different semantic information under the same features of the image, which has been demonstrated in [8,15]. From a cognitive viewpoint, we describe this fusion process as presented in Fig. 3.

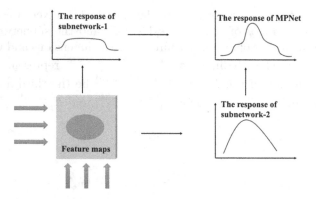

Fig. 3. The Multiprocesses Network semantic analysis

We observe feature maps from horizontal view obtain the features of subnetwork-1, and obtain the features of subnetwork-2 from vertical view. By calculating the gram matrix of subnetworks features, we can obtain multi-viewpoint features, which takes full advantage of extracting abstract semantic information.

We first handle the original image with the one-hot encoding method. Meanwhile, we use the average cross entropy loss function to train our model. Let y_{true} be the true label, y_{pred} represent the value of predicting and L represent average cross entropy loss function L is defined as:

$$L = -\frac{1}{M}\sum_{x} y_{true}(x)\log(y_{pred}(x))$$
$$+(1 - y_{true}(x))\log(1 - y_{pred}(x)),$$
(5)

where M is the number of mini-batch size and x is the input image. With the increase of the gap between the values of y_{pred} and y_{true}, the value of L becomes larger. What is more, this growth is similar to exponential growth, which causes the greater "penalty" for the current model. The advantage of L that the model tends to make the y_{pred} closer to the y_{true} and makes convergence speed faster.

4   Experiments

4.1   Experimental Dataset

We evaluate our model on the Animals with Attributes (AWA) dataset [26]. Before feeding image into the network, we do pixel mean processing to image, which not only removes the average intensity of the image but also removes common parts and highlight individual differences.

4.2   Training Strategy

We implement all experiments with mini-batch size 64 and 150 epochs. Our model is optimized by the Adam algorithm with a learning rate of 0.0001.

4.3   Evaluation Metrics

In this work, we employ four metrics to evaluate our model. Firstly, the accuracy of the recognition indicates how well the true value matches predicted value, which is significant to evaluate the learning ability of the model. The formula is as follows:

$$acc = \frac{N_1}{N},\tag{6}$$

where acc is the accuracy of recognition, N_1 is the number of correctly recognized sample, N is the total number of samples. Secondly, network parameters represent the number of parameters learned by the network, which reflects how much memory the model occupies, contains the number of parameters of the convolutional layer and fully connected layer. Thirdly, the Floating-point operations per second (Flops) presents the computational complexity of the network, which benefits to analyze the running time of the network. Finally, to evaluate the generalization performance of the model, we draw Receiver Operating Characteristic (ROC) curves and computer Area Under ROC curve (AUC).

4.4   Experiment Result

We conduct a series of experiments to verify the effective of our model, including the Multiprocess Convolutional Neural Network (MPNet) and the Single Process Convolutional Network (SPNet) are compared in terms of accuracy, the influence of depth on model performance, and compare the MPNet with other state-of-the-art models.

Comparison the Multiprocess Convolutional Neural Network with the Single Process Convolutional Network: The Single Process Convolutional Network (SPNet) is a standard convolutional network. For fairness, the experiments of SPNet and MPNet are carried out on the same conditions. Table 1 shows the recognition accuracy of the two networks in the test dataset.

Table 1. Comparison of evaluation metrics

Network	Accuracy	Model Size(MB)
SPNet6	75.90%	6.44
MPNet6	87.45%	33.54

In the test datasets, we can see from Table 1 that the accuracy of the MPNet outperforms the SPNet by 11.55% top-1 accuracy, whereas the MPNet6 increases four times parameters compared with the SPNet.

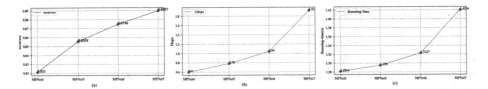

Fig. 4. a: The accuracy of different depth networks b: The flops of different depth networks c: The running time of different depth networks

Influence of Depth on the Multiprocess Convolutional Neural Network: Previous works [9,20] have demonstrated that deeper neural networks are beneficial to object classification on large data size. Therefore, we investigate the effect of different depth on model performance. All the evaluation matrix performance presented in Table 2. The depth of each network: from two convolutional layers in the MPNet4 to five convolutional layers in the MPNet7.

Figure 4 shows that the evaluation matrix of different models. According to Fig. 4, we find out that model achieves higher accuracy as the depth of network increases. However, large amounts of learnable parameters in the networks are required to be optimized, which results in a lower inference speed.

The Analysis of the ROC Curve of the Multiprocess Convolutional Neural Network: After the above analysis, to achieve a better trade-off between accuracy and efficiency, we choose MPNet6 as our base model. Figure 5 presents the ROC curves of the MPNet6. As can be seen from Fig. 5, the average value of the AUC of the MPNet6 is 0.9799, which demonstrates that our network has obtained better generalization ability.

Table 2. Comparison of evaluation metrics of different depth networks

Network	Depth	Accuracy	Model Size(MB)	Running time(s)	Flops
MPNet4	4	82.10%	8.40	1.28	0.6G
MPNet5	5	85.74%	33.57	1.29	0.78G
MPNet6	6	87.45%	33.57	1.32	1.04G
MPNet7	7	88.97%	33.58	1.42	1.92G

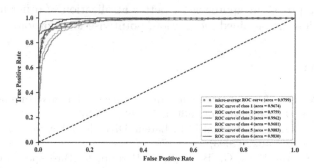

Fig. 5. The ROC curve of the multiprocesses network

Comparison of the Multiprocess Convolutional Neural Network and Other Convolutional Neural Networks: We compare the MPNet with six models including AlexNet [14], ResNet50, ResNet101, ResNet152, MobileNet V1, and MobileNet V2 on the AWA dataset. The results are presented in the following Table 3.

As shown in Table 3. Although the ResNet152 achieve 96.32% top-1 accuracy, this model increases 7 times parameters compared with our model. In addition, our model outperforms MobileNet V2 [19] by 3.4% top-1 accuracy whereas the MobileNet V2 occupies less memory. For a comprehensive comparison, our model yields better performance than other models and more suitable for the use of limited hardware conditions.

Table 3. Comparison of evaluation metrics of different networks

Network	Accuracy	Model Size(MB)
ResNet50	95.95%	99.77
ResNet101	96.60%	164.38
ResNet151	96.32%	224.94
MobileNet	82.53%	23.39
MobileNet V2	84.12%	22.12
AlexNet	76.9%	60.9
MPNet6	87.54%	33.57

5 Conclusion and Future Work

In this paper, we present a Multiprocess Convolutional Neural Network (MPNet), which contains two subnetworks. The first one extracts semantic information from a horizontal viewpoint, the other extracts semantic information from a vertical viewpoint. Moreover, we employ weight sharing measure to reduce memory consumption. A series of experiments have demonstrated the effectiveness of our methods in the field of animal recognition. In the future, we will further improve our approach for better application in other fields, such as object detection, fine-grained recognition, pointwise localization.

References

1. Ahmad, A., Dey, L.: A k-mean clustering algorithm for mixed numeric and categorical data. Data Knowl. Eng. **63**(2), 503–527 (2007)
2. Blei, D.M., Ng, A.Y., Jordan, M.I.: Latent Dirichlet allocation. J. Mach. Learn. Res. **3**(Jan), 993–1022 (2003)
3. Branson, S., Van Horn, G., Belongie, S., Perona, P.: Bird species categorization using pose normalized deep convolutional nets. arXiv preprint arXiv:1406.2952 (2014)
4. Cao, Z., Principe, J.C., Ouyang, B., Dalgleish, F., Vuorenkoski, A.: Marine animal classification using combined CNN and hand-designed image features. In: OCEANS 2015-MTS/IEEE Washington, pp. 1–6. IEEE (2015)
5. Chen, G., Han, T.X., He, Z., Kays, R., Forrester, T.: Deep convolutional neural network based species recognition for wild animal monitoring. In: 2014 IEEE International Conference on Image Processing (ICIP), pp. 858–862. IEEE (2014)
6. Csurka, G., Dance, C., Fan, L., Willamowski, J., Bray, C.: Visual categorization with bags of keypoints. In: Workshop on Statistical Learning in Computer Vision, ECCV, vol. 1, pp. 1–2. Prague (2004)
7. Farabet, C., Couprie, C., Najman, L., LeCun, Y.: Learning hierarchical features for scene labeling. IEEE Trans. Pattern Anal. Mach. Intell. **35**(8), 1915–1929 (2012)
8. Gatys, L.A., Ecker, A.S., Bethge, M.: Image style transfer using convolutional neural networks. In: Proceedings of the IEEE Conference on Computer Vision and Pattern Recognition, pp. 2414–2423 (2016)
9. He, K., Sun, J.: Convolutional neural networks at constrained time cost. In: Proceedings of the IEEE Conference on Computer Vision and Pattern Recognition, pp. 5353–5360 (2015)
10. He, K., Zhang, X., Ren, S., Sun, J.: Deep residual learning for image recognition. In: Proceedings of the IEEE Conference on Computer Vision and Pattern Recognition, pp. 770–778 (2016)
11. Howard, A.G., et al.: Mobilenets: Efficient convolutional neural networks for mobile vision applications. arXiv preprint arXiv:1704.04861 (2017)
12. Huang, G., Liu, Z., Van Der Maaten, L., Weinberger, K.Q.: Densely connected convolutional networks. In: Proceedings of the IEEE Conference on Computer Vision and Pattern Recognition, pp. 4700–4708 (2017)
13. Kong, S., Fowlkes, C.: Low-rank bilinear pooling for fine-grained classification. In: Proceedings of the IEEE Conference on Computer Vision and Pattern Recognition, pp. 365–374 (2017)

14. Krizhevsky, A., Sutskever, I., Hinton, G.E.: Imagenet classification with deep convolutional neural networks. In: Advances in Neural Information Processing Systems, pp. 1097–1105 (2012)
15. Li, Y., Wang, N., Liu, J., Hou, X.: Demystifying neural style transfer. arXiv preprint arXiv:1701.01036 (2017)
16. Lin, T.Y., RoyChowdhury, A., Maji, S.: Bilinear CNN models for fine-grained visual recognition. In: Proceedings of the IEEE International Conference on Computer Vision, pp. 1449–1457 (2015)
17. Lowe, D.G.: Distinctive image features from scale-invariant keypoints. Int. J. Comput. Vision **60**(2), 91–110 (2004)
18. Manohar, N., Kumar, Y.S., Kumar, G.H.: Supervised and unsupervised learning in animal classification. In: 2016 International Conference on Advances in Computing, Communications and Informatics (ICACCI), pp. 156–161. IEEE (2016)
19. Sandler, M., Howard, A., Zhu, M., Zhmoginov, A., Chen, L.C.: Mobilenetv 2: Inverted residuals and linear bottlenecks. In: Proceedings of the IEEE Conference on Computer Vision and Pattern Recognition, pp. 4510–4520 (2018)
20. Simonyan, K., Zisserman, A.: Very deep convolutional networks for large-scale image recognition. arXiv preprint arXiv:1409.1556 (2014)
21. Tenenbaum, J.B., Freeman, W.T.: Separating style and content with bilinear models. Neural Comput. **12**(6), 1247–1283 (2000)
22. Teng, J., Zhang, D., Lee, D.J., Chou, Y.: Recognition of Chinese food using convolutional neural network. Multimedia Tools Appl. **78**(9), 11155–11172 (2019)
23. Veech, J.: A comparison of landscapes occupied by increasing and decreasing populations of grassland birds. Conservation Biol.: J. Soc. Conservation Biol. **20**, 1422–1432 (2006)
24. Wang, Y., Morariu, V.I., Davis, L.S.: Learning a discriminative filter bank within a CNN for fine-grained recognition. In: Proceedings of the IEEE Conference on Computer Vision and Pattern Recognition, pp. 4148–4157 (2018)
25. Wold, S., Esbensen, K., Geladi, P.: Principal component analysis. Chemom. Intell. Lab. Syst. **2**(1–3), 37–52 (1987)
26. Xian, Y., Lampert, C.H., Schiele, B., Akata, Z.: Zero-shot learning—a comprehensive evaluation of the good, the bad and the ugly. IEEE Trans. Pattern Anal. Mach. Intell. **41**(9), 2251–2265 (2018)
27. Yu, C., Zhao, X., Zheng, Q., Zhang, P., You, X.: Hierarchical bilinear pooling for fine-grained visual recognition. In: Ferrari, V., Hebert, M., Sminchisescu, C., Weiss, Y. (eds.) ECCV 2018. LNCS, vol. 11220, pp. 595–610. Springer, Cham (2018). https://doi.org/10.1007/978-3-030-01270-0_35
28. Yuan, D.: Research on animal recognition algorithm based on convolutional neural network. Master's thesis, South China University of Technology (2018)

An Analysis of the Streamer Behaviors in Social Live Video Streaming

Yuanxing Rao, Wanxin Wang, Xiaoxue Shen, and Lu Jia[✉]

China Agricultural University, Beijing, China
{ryx,Ljia}@cau.edu.cn

Abstract. Social live video streaming systems, for example Facebook Live, Youtube Live, and Twitch.tv, are currently very popular both in industry and in academia. In this article, we provide a detailed analysis of broadcasting patterns in social live streaming based on our publicly available (anonymized) dataset that contains behavioral information on over 5,819 streamers. We found that streamers have distinctive streaming patterns, which we summarize into 5 groups, including professional streamers, regular streamers, casual streamers, hit-and-run streamers and lost professional streamers. We investigate the behavioral differences among them, namely the amount of donations streamers received and the number of danmaku messages streamers received. To alleviate the influence of unrealistic online situation, for our measurement, we analyze the factors that potentially influence the amount of donations streamers received except professional streamers. Among other results, we find that *the number of person who have donated free gifts*, is positively correlated with the amount of donations the streamers received for regular streamers. In contrast, for other three types of streamers, we conject that there are much audience who maybe prefer to donate free gifts rather than valuable gifts. Our findings and discussions are insightful for improving social live video streaming services.

Keywords: Broadcasting pattern · Behavioral difference · Donation

1 Introduction

Social live video streaming has become very popular in both industry and academia with the flexibility and interactivity of their novel services, they have become a serious challenge to traditional TVs [1] and have made great impacts in our daily lives, in related industries [2], and in our culture [3]. These platforms allow users to interact with each other in many different ways. Beyond entertainment, they join to socialize and to communicate [4–6], to learn and to cooperate [6, 7], and to seek out experiences that are different from their own lives [6]. In return, broadcasters satisfy their social needs [6], gain (sometimes world-level) appreciations, and receive tangible rewards for their effort and some even build their career and make livings [8] through user donations.

With the proliferation of social live video streaming systems, in recent years plenty of research has been carried out, ranging from quantitatively revealing the basic characteristics of these systems [9–14] and optimizing the resource allocation to meet the

© Springer Nature Singapore Pte Ltd. 2021
Y. Sun et al. (Eds.): ChineseCSCW 2020, CCIS 1330, pp. 594–604, 2021.
https://doi.org/10.1007/978-981-16-2540-4_43

dynamic and heterogeneous demands from the huge user base [15–18], to qualitatively analyzing user engagement and arguing user motivations to watch and to broadcast [4–6]. However, understandings on streamers' broadcasting patterns and exploring the impact of the amount of donations the streamers received are rather limited. In this article, we use Douyu [19], a popular social live video steaming system in China, as an example and we analyze broadcasting patterns in social live video streaming systems. Besides traditional streaming features, Douyu displays user donations and the user chats during streaming, making it possible to collect and to analyze the relationship between the broadcasting patterns and the amount of gifts streamers received.

Following the above motivations, the research questions we intend to answer are as follows:

- *What are the behavioral differences between the professional streamers, regular streamers, casual streamers, hit-and-run streamers and lost professional streamers?*
- *What are the factors that influence the amount of gifts streamers received, and do they have similar effects for the latter four types of streamers?*
- To answer the first question, we conducted one tailed Kolmogorov-Smirnov test. For the second question, we build ridge regression models for the regular streamers, casual streamers, hit-and-run streamers and lost professional streamers respectively, to identify factors that influence their receiving gifts behavior. We consider a variety of features and particularly, we intend to examine whether the maximum number of consecutive online days could be an influence. We have the following findings:
- Firstly, we observe significant behavioral differences between the above mentioned five broadcaster groups, in terms of the amount of user donations and the number of danmaku messages streamers received.
- Secondly, from streamer's perspective, we find that the number of cases that streamer was offline the day before but online in the next day is negatively correlated with how many valuable gifts they received for most streamers.
- Thirdly, from donor's perspective, the number of danmaku messages is positively correlated with the amount of user donations for streamers. But the number of users who send danmaku messages plays a positive role only for casual and lost professional streamers.

2 Related Work

Characterizing Social Live Video Streaming Systems. With the rise of social live video streaming systems, in recent years a number of studies were carried out for characterizing these systems. Kaytoue et al. proposed models for predicting video popularity in Twitch [11]. Deng et al. studied the workload of the Twitch streams and found that it is affected significantly by game tournaments [9]. Jia et al. compared Twitch with other systems and investigated their repositories and user activities [10], Raman et al. investigated Facebook Live and found that many attributes of Facebook Live deviate from both the concepts of live and broadcast [12]. Siekkinen et al. [13] and Wang et al. [14] analyzed the Quality of Service in Periscope including playback smoothness, latency, video quality, and the energy consumption.

User Donation. User donation is a rather unexplored area in social live video streaming. Lee et al. [20] analyzed 12 videos of live streams and found that viewers were motivated by the reciprocal acts of streamers. Based on surveys and interviews, Lu et al. [6] investigated on the motivations and the experiences of the viewers and provided insight into the role user donation and fan group-chat play in engaging users. Closet to our work, Wang et al. [21] and Jia et al. [22] carried out large-scale analyses and provided preliminary characterizations of user donations in online social video systems. In contrast, our analysis is based on a dataset containing all user activities for five months, which allows us to reveal both the short-term and long-term dynamics of user donations and streamer churn.

3 Hypotheses Development

3.1 Behavioral Differences Between Different Streamers

For our analysis, we have only considered streamers who have streamed in prior three months, and even last two months, 5,819 streamers in total. There is typically an imbalance in between streamers who broadcast online. A small number of streamers can guarantee live broadcast every week, and the average number of live broadcast is 6 days or more per week, and even the number of cases that streamer was offline the day before but online in the next day is not more than once every two weeks, we will here refer to as *professional streamers*. Approximately about 10% of the streamers' total live broadcast days are lower than that of professional streamers, but they basically guarantee an average live broadcast of more than 5 days per week, we call it *regular streamers*. Some streamers with a live broadcast lifetime which less than 60 days in 5 months, and the total live broadcast days that are less than 30 days, are believed to be *hit-and-run streamers*. A fraction of streamers who had a continuous broadcast of more than one month in the early stage, and then suddenly disappeared, will be described as *lost professional streamers*. And those other streamers that weren't mentioned above, here referred to as *casual streamers*.

Hypothesis 1. Those streamers behave differences in the amount of user donations streamers received and the number of danmaku messages streamers received.

3.2 Impact of the Amount of Donations Streamers Received

While most previous works on user donations analysis have considered features related to the number of danmaku messages and the number of free gifts received, we believe that (e.g., the number of cases that streamer was offline the day before but online in the next day) also impact user willingness to donate valuable gift, as users who have recognized that streamers often be offline or uncertain might not be encouraged to donate as well. We make the following hypothesis:

Hypothesis 2. The number of cases that streamer was offline the day before but online in the next day is negatively correlated with the amount of user donations.

Hypothesis 3. The number of online sessions is positively correlated with the amount of user donations.

Hypothesis 4. The online time length is positively correlated with the amount of user donations.

Hypothesis 5. The minimum number of consecutive offline days is negatively correlated with the amount of user donations.

Hypothesis 6. The maximum number of consecutive offline days is negatively correlated with the amount of user donations.

Hypothesis 7. The minimum number of consecutive online days is positively correlated with the amount of user donations.

Hypothesis 8. The maximum number of consecutive online days is positively correlated with the amount of user donations.

Hypothesis 9. The number of danmaku messages is positively correlated with the amount of user donations.

Hypothesis 10. The number of users who have sent danmaku messages is positively correlated with the amount of user donations.

Hypothesis 11. The number of free gifts is positively correlated with the amount of user donations.

Hypothesis 12. The number of users who have donated free gifts is positively correlated with the amount of user donations.

4   Methodology and the Douyu Dataset

In this section, we give a brief introduction of the ecosystem of Douyu, and we introduce our measurement methodology and the dataset used throughout this article.

4.1   An Overview of Douyu

Douyu.com is a prominent crowdsourced live video broadcasting website in China where any registered user can set up a channel and broadcast live videos on the Internet. The success of any streaming communities lies in the broadcasters actively providing the contents and as an effective encouragement, many popular streaming communities such as YouTube Live and Twitch have adopted user donation mechanisms [9–14]. However, in these enterprise giants the behaviors of streamer and audience remain a secret and it is not clear what pattern exist among streamers about online living-style nor what are the behavioral differences between different streamers with different patterns. It's critical to collect and to reveal the cluster of streamers and their behavioral difference that needs real effort to be established.

Table1. Douyu donation dataset.

# streamers	5,819
# streaming sessions	392,099
Session length aggregate/mean	2,043,750/351.22 h
# danmaku messages aggregate/mean	81,853,985/14,067
Amount of user donations aggregate/mean	206,913,644/35,558.28 USD

Table2. Statistics on the broadcaster activity and popularity.

	Mean	Median	Max	Description
Alldays(p1)	58.56	46	154	The number of days online
Lifespan(p2)	129.68	137	154	Lifetime
#Conversion (p3)	11.48	10	43	The number of cases that streamer was offline the day before but online in the next day
Times(p4)	67.38	50	389	The number of online sessions
Length(p5)	351.22	192	3,644	The online time length
Ter0min(p6)	1.81	1	57	Minimum number of consecutive offline days
Ter0max(p7)	39.95	37	148	Maximum number of consecutive offline days
Ter1min(p8)	2.81	1	154	Minimum number of consecutive online days
Ter1max(p9)	20.19	10	154	Maximum number of consecutive online days
Value(z0)	35,558.28	3.71	45,136,797	The amount of donations streamers received
danmaku(z1)	98,466.73	12,046	77,186,496	The number of danmaku messages streamer received
danmakuU(z2)	19,932.63	7,348	5,535,982	The number of person who have sent danmaku messages
GiftNMR(z3)	215,973.2	1,825	346,021,290	The number of free gifts streamer received
GiftNMU(z4)	2,739.32	752	1,503,456	The number of person who have donated free gifts

4.2 Measurement Methodology and the Dataset

Our data collection is carried out based on Douyu's RESTful APIs, which started in April, 2019 and we have considered only broadcasters of the game League of Legends (LOL). For our analysis, we repeatedly crawled and recorded their streaming activities every ten minutes. Meanwhile, we collected all the danmaku messages and donation information for the streamers in real time through Douyu's danmaku API. In total, we get 5,819 valid streamers who have streamed at least once from April 8th to July 8th, 2019, and have streamed some sessions in the last two month periods (from July 8th to September 8th, 2019). In this way, we get a comprehensive view for these broadcasters, including (i) the time and the duration of their streams, (ii) the time and the users who have sent danmaku and the messages of the danmaku, and (iii) the time and the users who have made donations and the value of the donations. Overall, the 5,819 broadcasters in our analysis have streamed 392,099 sessions in the five month periods, with an average session length of 5.21 h. In total, they have accumulated 206,913,644 US dollar worth of donations. The basic statistics of our dataset and of the streamers are shown in Table 1 and Table 2, respectively.

5 Results

In this section, we demonstrate the basic characteristics of streamers with different broadcasting pattern in Douyu, and test whether different types of streamers behave differently on the amount of donations received and the number of danmaku messages they received. In addition, from both the streamers' and the donors' perspectives, we decide to explore the factors that how to impact the amount of donations streamers received.

5.1 Descriptive Analysis

In this article, we observe that there are different broadcast modes among streamers. Some streamers broadcast almost every day, some of them disappear after a month or so, and some even have a high frequency in the early stage but almost disappear in the later stage. In order to have a deeper communication on the number of danmaku messages streamers received and the amount of donations the streamers received, the influencing factors of it are analyzed more clearly, and the streamers are divided into 470 professional streamers, 614 regular streamers, 4,441 casual streamers, 113 hit-and-run streamers and 181 lost professional streamers, respectively, based on the characteristics of live broadcasting in Sect. 3.1. The statistical results of key variables of them are described in Table 3.

Table 3. Statistics results of the variables.

Variable		Professional	Regular	Casual	Hit-and-run	Lost professional
The number of days online	Mean	145.14	123.41	40.89	8.27	78.57
	Median	146.00	124.00	33.00	5.00	78.00
	Max	154.00	147.00	130.00	29.00	121.00
Lifetime	Mean	153.58	149.14	127.25	41.00	116.69
	Median	154.00	153.00	132.00	44.00	114.00
	Max	154.00	154.00	154.00	59.00	154.00
The number of cases that streamer was offline the day before but online in the next day	Mean	5.31	12.38	12.29	4.16	6.72
	Median	5.00	13.00	11.00	4.00	6.00
	Max	11.00	23.00	44.00	13.00	16.00
The number of online sessions	Mean	170.49	146.21	46.00	9.06	93.43
	Median	161.00	140.00	35.00	6.00	90.00
	Max	389.00	332.00	236.00	39.00	209.00
The online time length	Mean	1,246.18	816.61	191.68	34.53	560.70
	Median	1,109.25	765.25	122.00	16.33	506.67
	Max	3,643.83	3,336.00	2,390.83	321.17	1,961.33
The minimum number of consecutive offline days	Mean	1.03	1.14	1.83	8.73	1.45
	Median	1.00	1.00	1.00	2.00	1.00
	Max	7.00	20.00	48.00	57.00	33.00
The maximum number of consecutive offline days	Mean	3.03	10.14	46.79	71.90	49.28
	Median	2.00	8.00	44.00	71.00	47.00
	Max	12.00	34.00	148.00	90.00	111.00
The minimum number of consecutive online days	Mean	19.55	3.14	1.05	1.18	2.59
	Median	3.00	1.00	1.00	1.00	1.00
	Max	154.00	140.00	16.00	10.00	121.00
The maximum number of consecutive online days	Mean	76.01	41.74	10.68	3.64	45.60
	Median	68.00	37.00	7.00	2.00	41.00
	Max	154.00	140.00	91.00	26.00	121.00

5.2 Verification of Behavioral Differences Between Streamers

We conducted one tailed Kolmogorov-Smirnov (KS) test to test whether different types of streamers behave differently on two aspects, namely the amount of donations received and the number of danmaku messages they received. The results are shown in Table 4, from which we observe significant behavioral differences. Hypotheses 1 is supported.

5.3 Exploration on the Factors in Donations Received

In this section, to alleviate the influence of unrealistic online situation, for our measurement, we examine the factors that potentially influence the amount of donations streamers received except professional streamers (Hypotheses 2–12), we propose ridge regression models. As for factors, we mainly consider two aspects, including *streaming activity* and *streamer popularity* as shown in Table 3. The results of the regression model for four types of streamers are reported in Table 5.

Table 4. Results of the one tailed KS test.

Variable	P_value	
	#Danmaku messages	Donation received
Professional VS Regular	***	***
Professional VS Casual	***	***
Professional VS hit-and-run	***	***
Professional VS Lost professional	***	***
Regular VS Casual	***	***
Regular VS hit-and-run	***	***
Regular VS Lost professional	***	***
Casual VS hit-and-run	***	***
Casual VS Lost professional	***	***
hit-and-run VS Lost professional	***	***

1 *:p < 0.05; **:p < 0.01; ***:p < 0.001.

Impact of the Amount of Donations for Casual Streamers, Hit-and-Run Streamers and Lost Professional Streamers. As for the casual streamers, their live broadcasting time may not be regular enough to accumulate a large number of loyal fans. Therefore, although the live broadcasting time is long, they receive few gifts (parameter $= -0.35$). However, from the perspective of danmaku (parameter $= 0.11$) and free gifts (parameter $= 0.1$), it still meets our expectations. By contrast, the number of person who have donated free gifts shows negative effect on the amount of donations streamers received. We conject that the viewers maybe prefer to donate free gifts rather than valuable donations, so H2, H3, H5, H6, H8, H9, H10 and H11 are supported for casual streamers.

For only broadcast on the early stage of the streamer (hit-and-run streamers), the experimental results show that the gifts they receive mainly affected in danmaku (parameter $= 0.06$) and free gifts (parameter $= 0.04$). When it comes to the lost professional streamers, it's worth noting that the minimum number of consecutive offline days (parameter $= -147.9$) is negatively correlated with the amount of donations. We guess that the audience is still regretting the sudden departure of the streamers and hoping for his early return, so H5 is supported. Now we summarize hypotheses testing results in Table 6.

Table 5. Results of ridge regression for different streamers.

Variable	Regular R^2: 0.99	Casual R^2: 0.99	hit-and-run R^2: 0.97	Lost professional R^2: 0.79
The number of cases that streamer was offline the day before but online in the next day	2115.3	−10.64	−7.14	−20.74
The number of online sessions	122.07	8.26	1.65	0.91
The online time length	−0.05	−0.35	0.01	−0.08
The minimum number of consecutive offline days	921.23	−21.52	0.3	−147.9
The maximum number of consecutive offline days	500.66	−4.67	1.43	−16.56
The minimum number of consecutive online days	−158.2	−243.2	−12.08	37.07
The maximum number of consecutive online days	184.26	37.78	1.66	−21.23
The number of danmaku messages streamers received	0.03	0.11	0.06	0.08
The number of users who have sent danmaku messages	−0.07	0.24	−0.08	0.01
The number of free gifts streamers received	0.07	0.1	0.04	0.05
The number of person who have donated free gifts	9.62	−1.61	−0.32	−0.85

Table 6. Hypotheses testing results.

Hypotheses	Regular	Casual	hit-and-run	Lost professional
H1	Supported	Supported	Supported	Supported
H2	Unsupported	Supported	Supported	Supported
H3	Supported	Supported	Supported	Supported
H4	Unsupported	Unsupported	supported	Unsupported
H5	Unsupported	Supported	Unsupported	Supported
H6	Unsupported	Supported	Unsupported	Supported
H7	Unsupported	Unsupported	Unsupported	Supported
H8	Supported	Supported	Supported	Unsupported
H9	Supported	Supported	Supported	Supported
H10	Unsupported	Supported	Unsupported	Supported
H11	Supported	Supported	Supported	Supported
H12	Supported	Unsupported	Unsupported	Unsupported

6   Conclusion and Future Work

In this work, we provided an in-depth analysis on user donation for four types of streamers in LOL. We revealed the behavioral differences between different streamers, in term of the amount of user donations and the number of danmaku messages they received. We found that professional streamers receive relatively more messages and gifts, the regular streamers next, and the third is lost professional streamers. We further discussed the factors that influence the amount of gifts streamers received, and observed that the number of cases that streamer was offline the day before but online in the next day is negatively correlated with how many valuable gifts they received for streamers, except regular streamers, and the minimum number of consecutive offline days is negatively correlated with the amount of user donations for casual and lost professional streamers. On the other hand, from the user's perspective, the number of danmaku messages is positively correlated with the amount of user donations for streamers. But the number of users who have sent danmaku messages only play positive role for casual and lost professional streamers. As for the number of free gifts, it is positively correlated with the amount of user donations for streamers. In contrast, the number of users who have donated free gifts is positively correlated with the amount of user donations only for regular streamers. Our research provide valuable information for maintaining of social live streaming systems.

References

1. Spilker, H.S., Ask, K., Hansen, M.: The new practices and infrastructures of participation: how the popularity of Twitch.tv challenges old and new ideas about television viewing. Inf. Commun. Soc. **23**(4), 605–620 (2018)
2. Johnson, M., Woodcock, J.: The impacts of live streaming and Twitch.tv on the video game industry. Media Cult. Soc. **41**, 5 (2019)
3. Lu, Z., Annett, M., Fan, M., Wigdor, D.: Streaming and engaging with intangible cultural heritage through livestreaming. In: Proceedings of the SIGCHI Conference on Human Factors in Computing Systems (CHI 2019) (2019)
4. Haimson, O.L., Tang, J.C.: What makes live events engaging on facebook live, periscope, and snapchat. In: Proceedings of the SIGCHI Conference on Human Factors in Computing Systems (CHI 2017) (2017)
5. Hu, M., Zhang, M., Wang, Y.: Why do audiences choose to keep watching on live video streaming platforms? An explanation of dual identification framework. Comput. Hum. Behav. **75**, 594–606 (2017)
6. Lu, Z., Xia, H., Heo, S., Wigdor, D.: You watch, you give, and you engage: a study of live streaming practices in China. In: Proceedings of the SIGCHI Conference on Human Factors in Computing Systems (CHI 2018) (2018)
7. Faas,T., Dombrowski, L., Young, A., Miller, A.D.: Watch me code: programming mentorship communities on Twitch.tv. In: Proceedings of the ACM 2018 Conference on Computer Supported Cooperative Work (CSCW 2018) (2018)
8. Johnson, M., Woodcock, J.: It's like the gold rush: the lives and careers of professional video game streamers on Twitch.tv. Inf. Commun. Soc. **22**(3), 336–351 (2017)
9. Deng, J., Cuadrado, F., Tyson, G., Uhlig, S.: Behind the game: exploring the twitch streaming platform. In: Network and System Support for Games (NetGames 2015) (2015)

10. Jia, A.L., Shen, S., Epema, D.H., Iosup, A.: When game becomes life: the creators and spectators of online game replays and live streaming. ACM Trans. Multimed. Comput. Commun. Appl. **12**, 4 (2016)
11. Pires,K., Simon, G.: youtube live and twitch: a tour of user-generated live streaming systems. In: Multimedia Systems Conference (MMSys 2015) (2015)
12. Raman, A., Tyson, G., Sastry, N.: Facebook (A)Live? Are live social broadcasts really broadcasts? In: Proceeding of the 26th International World Wide Web Conference (WWW 2017) (2017)
13. Siekkinen, M., Masala, E., Kamarainen, T.: The first look at quality of mobile live streaming experience: the case of periscope. In: Proceedings of the 16th ACM SIGCOMM Conference on Internet Measurement Conference (IMC 2016) (2016)
14. Wang, B., Zhang, X., Wang, G., Zheng, H., Zhao, B.Y.: Anatomy of a personalized livestreaming system. In: Proceedings of the 16th ACM SIGCOMM Conference on Internet Measurement Conference (IMC 2016) (2016)
15. He, Q., Liu, J., Wang, C., Li, B.: Coping with heterogeneous video contributors and viewers in crowdsourced live streaming: a cloud-based approach. IEEE Trans. Multimed. **18**(5), 916–928 (2016)
16. Ma, M., Zhang, L., Liu, J., Wang, Z., Pang, H., Sun, L., Li, W., Hou, G., Chu, K.: Characterizing user behaviors in mobile personal livecast: towards an edge computing-assisted paradigm. ACM Trans. Multimed. Comput. Commun. Appl. **14**, 3 (2018)
17. Ray, D., Kosaian, J., Rashmi, K.V., Seshan, S.: Vantage: optimizing video upload for time-shi ed viewing of social live streams. In: Proceeding of ACM Special Interest Group on Data Communication (SIGCOMM 2019) (2019)
18. Zhang, C., Liu, J., Wang, H.: Cloud-assisted crowdsourced livecast. ACM Trans. Multimed. Comput. Commun. Appl. **13**, 3 (2017)
19. Douyu (2019). https://www.douyu.com. Accessed 01 July 2020
20. Lee, Y.C., King, J.T., Yen, C.H., Fu, W.T., Chiu, P.T.: Tip Me!: tipping is changing social interactions on live streams in China. In: Proceedings of the SIGCHI Conference on Human Factors in Computing Systems (CHI 2018 Extended Abstracts) (2018)
21. Wang, X., Tian, Y., Lan, R., Yang, W., Zhang, X.: Beyond the watching: understanding viewer interactions in crowdsourced live video broadcasting services. IEEE Trans. Circ. Syst. Video Technol. **29**(11), 3454–3468 (2019). Early access
22. Jia, A.L., Shen, X., Shen, S., Fu, Y., Peng, L., Xu, J.: User donations in a user generated video system. In: Proceeding of the 26th International World Wide Web Conference (WWW 2019 Companion, the Web Conference) (2019)
23. Ding, Y., Du, Y., Hu, Y., Liu, Z., Wang, L., Ross, K.W., Ghose, A.: Broadcast yourself: understanding youtube uploaders. In: Proceedings of the 5th Internet Measurement Conference (IMC 2011) (2011)

Research on the Method of Emotion Contagion in Virtual Space Based on SIRS

Heng Liu[1,2], Xiao Hong[1,2], Dianjie Lu[1,2(✉)], Guijuan Zhang[1,2], and Hong Liu[1,2]

[1] School of Information Science and Engineering, Shandong Normal University,
Jinan 250358, China
[2] Shandong Provincial Key Laboratory for Novel Distributed Computer Software Technology,
Jinan 250358, China

Abstract. There is emotional contagion in virtual space, and the process of emotional contagion is also affected by emotional recurrence. However, the existing emotional contagion model does not consider the impact of individual emotional recurrence on emotional contagion in the virtual space. Therefore, the emotional contagion process cannot be simulated. The paper proposes a method of recurrent emotional contagion in virtual space. First, we construct an individual emotional recurrence contagion model based on SIRS (Susceptible-Infected-Recovered-Susceptible), we propose emotional recurrence contagion rules, and consider the impact of emotional recurrence in the virtual space on the emotional contagion process; secondly, we construct a virtual space emotional recurrence contagion model (REC) and solve it to simulate the process of emotional recurrence contagion. Third, we build a WS small-world network based on REC simulation. Provide REC (Recurrent emotional contagion model) simulation process in WS small-world network to verify the rationality of the REC. Finally, the REC is simulated to show the simulation effect of emotional recurrence contagion.

Keywords: Emotional recurrence contagion · SIRS model · Virtual cyberspace · The small-world network

1 Introduction

After Hatfield [1] put forward the definition of emotional contagion, a large number of researchers were attracted by emotional contagion, and thus a large number of emotional contagion models were produced. Currently, emotional contagion models can be divided into two categories, emotional contagion models in physical space and emotional contagion models in virtual space. The emotional contagion model in physical space is mainly to study the emotional contagion of individuals in real life, such as when disasters such as earthquakes and fires occur, people's emotional changes. The emotional contagion model in virtual space is mainly to study the emotional contagion of users in the network, such as tracking Facebook updates to study large-scale emotional contagion in social networks [2].

On the basis of emotional contagion, more and more researchers have begun to study the influence of factors such as age, gender [3], and personality [4] on the process of

© Springer Nature Singapore Pte Ltd. 2021
Y. Sun et al. (Eds.): ChineseCSCW 2020, CCIS 1330, pp. 605–615, 2021.
https://doi.org/10.1007/978-981-16-2540-4_44

emotional contagion, but they have ignored the factor that emotions will recur. Therefore, the emotional contagion process cannot be genuinely simulated. In the context of virtual space, understanding the impact of personal emotional recurrence on the emotional transmission process is a challenging issue.

The emotional contagion process in the virtual space will be affected by the recurrence of individual emotions. Common infectious disease models include SIS [5] (Susceptible-Infecte2d-Susceptible) model, SIR [6] (Susceptible-Infected-Recovered) model and SIRS model [7]. After considering several models, in order to reproduce the emotional recurrence contagion, we refer to the classic SIRS model.

This paper uses the classic infectious disease model (SIRS model) to model emotions, and proposes a method of emotional recurrence contagion in virtual space. We first proposed a method of recurrent emotional contagion in virtual space; Second, construct and solve the emotional recurrence contagion model. Finally, the model is simulated to show the simulation effect of emotional recurrence contagion. The research content and innovations of this article are as follows:

(1) Methods of recurrent emotional contagion in virtual space

Aiming at the problem that the process of emotional contagion cannot be truly simulated, a method of recurrent emotional contagion in virtual space is proposed Methods. First, we construct an individual emotional recurrence contagion model, propose the emotional recurrence contagion rules, and consider the impact of recurring emotions on the emotional contagion process in the virtual space; Second, we construct and solve the emotional recurrence contagion model (REC) in the virtual space.

(2) Simulation on WS small-world network based on emotion recurrence model

We construct a WS small-world network based on REC simulation, and discuss the state transition of nodes in the small-world network. Provide the REC simulation process in WS small-world network to verify the rationality of the REC.

(3) Visualization of emotional contagion process

We simulate REC to verify and visualize the theoretical analysis results. This method can simulate the process of emotional recurrence contagion. The experimental results show that the emotional recurrence contagion process described by the above method is in good agreement with reality.

2 Related Work

After the emotional contagion was proposed, many researchers began to study the emotional contagion in physical space or the emotional contagion in virtual space. The emotional contagion model in virtual space main targets the emotional contagion of social platform users [8]. R. Fan et al. [9] proposed a model to simulate emotional contagion in online networks. Kramer et al. [10] conducted a large-scale experiment on Facebook.

As researchers deepen their research on emotional contagions, many researchers have added other factors such as personality, age, gender, and groupings to the emotional contagion process. Researchers usually divide the factors that affect emotions into two categories: social factors and personal factors. Social factors are mainly factor such as environment, status, closeness of relationship, and trust. Personal factors are mainly influencing factors such as personality, gender, and age. Guy et al. [10] modeled the behavior of heterogeneous people based on the theory of personality traits. Lhommet et al. [4] proposed a model based on relationship. Lu et al. [11] proposed a personality model to simulate heterogeneous traffic behavior.

In recent years, emotional contagions are no longer limited to negative emotional contagions, and positive emotional contagions have also begun to be widely studied. Faroqi et al. [11] designed a system to investigate how security personnel alleviate the fear of the crowd. The evacuation simulator built by Cai et al. [12] showed that safety officials have a more substantial contagion effect, which can slow down the speed of evacuees. Zhang et al. [13] proposed a positive emotion contagion based on the Internet of Things (IoT-PEC) method to evacuate people, and solve congestion and confusion during crowd evacuation by appropriately deploying security personnel and maximizing the impact of positive emotions. However, most researchers have not considered the characteristic of emotion recurrence. In this paper, we considered the importance of emotional recurrence in the process of emotional contagion in virtual space.

3 Emotional Recurrence Contagion in Virtual Cyberspace

In order to simulate the process of emotional recurrence contagion, this section first constructs an individual emotional recurrence contagion model, and then proposes the emotional recurrence contagion rules. The following two subsections will introduce these contents separately.

3.1 Emotional Contagion Model

In the individual emotional recurrence contagion model, this article considers the emotional recurrence factors in the virtual space. These factors will be introduced in two parts: recurrent emotions and contagion space.

Recurrent emotional contagions exist in virtual spaces. Therefore, In this section, we set a triples $S(i) = (State\ (i,\ t),\ \mu_V,\ \delta_V)$ to represent the attributes of individual i in virtual space. To quantify, $State(i,\ t)$ represents the state of individual i at any time t, μ_V represents the cure rate of infected persons, and δ_V represents the recovery rate of temporarily immunized persons.

This section divides the individual states into the following three categories: susceptible persons in virtual space, infected persons in virtual space, and temporarily immune persons in virtual space. This section defines $State\ (i,\ t) = (XS,\ XI,\ XR)$, XS means susceptible persons in virtual space, XI means infected persons in physical space, and XR means temporarily immune persons in virtual space.

3.2 Emotional Contagion Rules

Emotional contagion will occur in virtual space, and emotional recurrence will also have an impact on the crowd's emotional contagion process. Therefore, this section proposes the rules of emotional recurrent contagion to simulate emotional recurrent contagion. The probability of a susceptible individual being infected as an infected individual is β_V, the probability of an infected individual being cured as a temporarily immunized individual is μ_V, and the probability of a temporary immunized individual returning to a susceptible individual is δ. This section proposes the following emotional contagion rules.

(1) The susceptible person XS in the virtual space can be infected by the virtual space infected person XI with probability to become the virtual space infected person XI.
(2) The infected person XI in the virtual space can be cured by the virtual space temporarily immunized XR with probability to become the virtual space temporarily immunized XR.
(3) XR who is temporarily immune to virtual space has a probability of reverting to virtual space susceptible XS.

4   REC Construction and Solution

This section first constructs REC to simulate the process of contagion. This section first uses the mean field theory to establish the REC model, it then uses the finite difference method to numerically solve the REC.

4.1 The Mean Field Equations for the REC

In this section, the average field theory is used to derive the evolution process of REC, and the average field equation of REC is obtained as follows. As shown in Table 1, this section defines the parameters of the REC average field equation.

Table 1. Parameter values in the contagion space

Parameters	Description
N	The size of the crowd
$XS(t)$	At time t, the proportion of susceptible persons
$XI(t)$	At time t, the proportion of infected persons
$XR(t)$	At time t, the proportion of temporarily immunized

The framework of the REC model is shown in Fig. 1. The solution of the REC satisfy the following constraints:

$$XS(t) + XR(t) + XR(t) = 1 \tag{1}$$

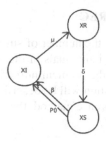

Fig. 1. REC model frame diagram. Among them, *XS* is a susceptible individual in the virtual space, *XI* is an infected individual in the virtual space, and *XR* is a temporarily immune individual in the virtual space. β_V is the contagion rate, μ_V is the cure rate, δ is the recovery rate, and P_0 is the factors that causes emotional relapse due to factors other than emotion.

The change in the number of individuals infected person *XI* in unit time Δt can be expressed as:

$$N[X^I(t + \Delta t) - X^I(t)] = \beta_V X^S(t)NX^I(t)\Delta t - \mu_V NX^I(t)\Delta t + P_0 X^S(t)N\Delta t \quad (2)$$

$\beta_V X^S(t)NX^I(t)\Delta t$ indicates the number of susceptible individuals infected, $\mu_V NX^I(t)\Delta t$ indicates the number of infected individuals cured, $P_0 X^S(t)N\Delta t$ represents the number of susceptible individuals in the virtual space infected by factors other than those infected.

The change in the number of individuals susceptible person *XS* in unit time Δt can be expressed as:

$$N[X^S(t + \Delta t) - X^S(t)] = -\beta_V X^S(t)NX^I(t)\Delta t - P_0 X^S(t)N\Delta t + \delta X^R(t)N\Delta t \quad (3)$$

$\beta_V X^S(t)NX^I(t)\Delta t$ indicates the number of susceptible individuals infected, $P_0 X^S(t)N\Delta t$ indicates the number of susceptible individuals infected, $\delta X^R(t)N\Delta t$ represents the number of temporarily immunized individuals restored to susceptible individuals in the virtual space.

The change in the number of individuals in XR of the temporarily immunized person in the virtual space within the unit time Δt can be expressed as:

$$N[X^R(t + \Delta t) - X^R(t)] = \mu_V NX^I(t)\Delta t - \delta X^R(t)N\Delta t \quad (4)$$

$\mu_V NX^I(t)\Delta t$ indicates the number of infected individuals in the virtual space cured by temporary immunizations, $\delta X^R(t)N\Delta t$ indicates the number of temporarily immunized persons in the virtual space restored to susceptible.

The average field equation of REC obtained from the above is:

$$\frac{dX^I(t)}{dt} = \beta_V X^S(t)X^I(t) - \mu_V X^I(t) + P_0 X^S(t) \quad (5)$$

$$\frac{dX^S(t)}{dt} = -\beta_V X^S(t)X^I(t) + \delta X^R(t) - P_0 X^S(t) \quad (6)$$

$$\frac{dX^R(t)}{dt} = \mu_V X^I(t) - \delta X^R(t) \quad (7)$$

4.2 Numerical Solution of the REC

In order to explore the changes in the number of susceptible individuals, infected individuals, and temporarily immune individuals in the REC model over time, we use the finite difference method to solve the problem numerically. The numerical solutions of different individuals in the population will be described in detail below.

At time t, using the finite difference method, the mean field equation of *XI(t)* can be expressed as:

$$X^I(t + \Delta t) = [\beta_V X^S(t) X^I(t) - \mu_V X^I(t) + P_0 X^S(t)]\Delta t + X^I(t) \tag{8}$$

At time t, using the finite difference method, the mean field equation of *XS(t)* can be expressed as:

$$X^S(t + \Delta t) = [-\beta_V X^S(t) X^I(t) - P_0 X^S(t) + \delta X^R(t)]\Delta t + X^S(t) \tag{9}$$

At time t, using the finite difference method, the mean field equation of *XR(t)* can be expressed as:

$$X^R(t + \Delta t) = [\mu_V X^I(t) - \delta X^R(t)]\Delta t + X^R(t) \tag{10}$$

Given the initial conditions of individuals in different states in the population at time $t = 0$, the proportion of individuals at time can be obtained according to the above equation.

4.3 Simulation for the REC

This section will introduce REC simulation to describe how emotions recur in a virtual space scene.

This paper considers the impact of emotional recurrence on the emotional contagion process. Therefore, this section builds a WS small-world network for REC simulation. This section uses $G(V, E)$ to represent the WS small-world network. Finally, in order to reflect the emotional recurrence, we add a small probability P_0, that is, other factors that can cause emotional recurrence.

This section uses the roulette method to explore the state transition of nodes in the recurrent WS small-world network. This section analyzes the state transition of any node in the recurrent WS small-world network at the time interval [t, t + 1]. First, calculate the state transition probability of each node, and then determine the state of the node at the next moment based on this probability.

5 Experiments and Analysis

This section is divided into three parts: the first part is the numerical simulation of REC, the second part is the REC simulation in the cyclic WS small-world network, and the rationality of the model is verified, and the third part is the visualization of the emotional transmission process.

5.1 Numerical Simulation of the REC

REC is numerically simulated to explore the changing trends of different groups in the virtual space and analyze the thresholds of some important parameters. Set $\mu_V = 0.1$, $\delta = 0.05$, $P_0 = 10^{-7}$, $XI = 100$, $XS = 4 \times 10^4$, $XR = 0$ as the default value of the experiment in this section.

Threshold Analysis of β. Perform threshold analysis on the contagion rate of virtual space. In Fig. 2, when $\beta_V = 3 \times 10^{-6}$, obvious emotional recurrence can be observed. When $\beta_V = 5 \times 10^{-6}$, no recurrence of emotions. When $\beta_V = 10^{-7}$, significant emotional recurrence. Therefore, the virtual space contagion rate has a threshold, threshold at $[10^{-7}, 3 \times 10^{-6}]$.

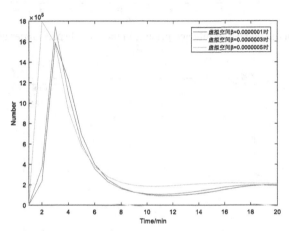

Fig. 2. Changes in the number of infected persons in the virtual space under different contagion rates.

Threshold Analysis of μ. Analysis of the cure rate μ_V of Virtual space, by observing Fig. 3, it is found that when $\mu_V = 0.001$, the Mood relapses disappear. When $\mu_V > 0.001$, it can show normal recurrence. So there is a threshold.

Threshold Analysis of δ. Observing Fig. 4, it is found that when the recovery Rate $\delta = 0.15$, the number of contagions gradually stabilized, and a small-scale emotional recurrence occurred. When the recovery rate $\delta = 0.01$, the number of susceptible decreases first and then gradually increases, while the number of contagions gradually decreases, and there is no recurrence. When $\delta = 0.05$, emotional recurrence occurs, so there is a threshold for the recovery rate. The threshold is (0.01, 0.15).

5.2 Simulation of the REC on Recurrent WS Small-World Network

In this section, a recurrent WS small-world network with 5000 nodes is used for REC simulation to further analyze the rationality of REC. Each iteration of the simulation

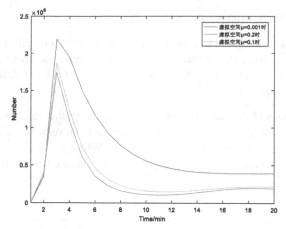

Fig. 3. Changes in the number of infected persons in the virtual space under different cure rates.

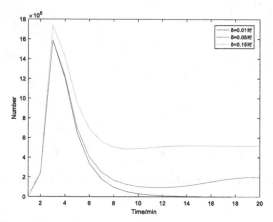

Fig. 4. Changes in the number of infected persons in the virtual space under different recovery rates.

experiment is 1000 times. Set $\mu_V = 0.3$, $\delta = 0.1$ in this experiment. Through 50 experiments and 1000 iterations, M_V, which is the proportion of infected individuals in the virtual space, is obtained. Observing Fig. 5, it is found that the numerical simulation curve and the network simulation curve have a high degree of fit. Therefore, REC has good stability.

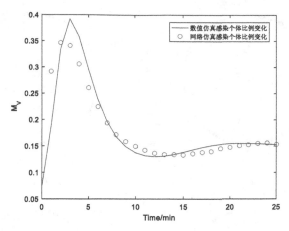

Fig. 5. Simulation of REC on WS small-world Network

5.3 Visualization of the Emotional Contagion Process

We visualized the results of the recurrence of emotions in the virtual space.

Visualization of Emotional Recurrence Contagion. The experiment explores the emotional contagion of individuals in the virtual space through the recurrent WS small-world network. In this experiment, the nodes are divided into three categories, namely susceptible nodes, infected nodes and temporary immune nodes. In this experiment, N = 100, k = 4, μ = 0.3, δ = 0.1 and β are determined by the number of infected nodes in neighbor nodes. Figure 6(a)–(d) shows the individual emotional state in the virtual space at different times, showing the entire process of emotional recurrence contagion in the virtual space. The red mark indicates the randomly selected infected node in the recurrent WS small-world network. Blue markers indicate susceptible individuals, and green markers indicate temporarily immunized individuals. The yellow marks indicate re-infected individuals. The experimental result shows the state transition of the node.

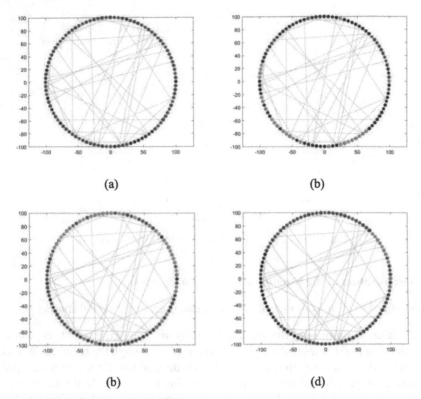

(a)   (b)

(b)   (d)

Fig. 6. Process of emotional recurrence contagion (Color figure online)

6 Conclusions

The paper proposes a method of recurrent emotional contagion in a virtual space based on SIRS. First, we construct an individual emotional recurrence contagion model, propose emotional recurrence contagion rules, and consider the impact of emotional recurrence in the virtual space on the emotional contagion process; Second, we construct and solve the emotional recurrence contagion model (REC) in the virtual space to simulate the emotional recurrence contagion process more realistically. Third, we build a WS small-world network based on REC simulation. Provide the REC simulation process in the WS small-world network to verify the rationality of the model. Finally, REC is simulated to show the simulation effect of emotional recurrence contagion. The method can simulate the emotional contagion process more realistically, and can guide crowd evacuation.

References

1. Hatfield, E., Cacioppo, J.T., Rapson, R.L.: Emotional contagion. Curr. Direct. Psychol. Sci. **2**(3), 96–100 (1993)
2. Kramer, A.D.I.: The spread of emotion via Facebook. In: Proceedings of the SIGCHI Conference on Human Factors in Computing Systems, pp. 767–770. ACM (2012)

3. Cao, M., Zhang, G., Wang, M., et al.: A method of emotion contagion for crowd evacuation. Phys. A: Stat. Mech. Appl. **483**, 250–258 (2017)
4. Lhommet, M., Lourdeaux, D., Barthès, J.P.: Never alone in the crowd: a microscopic crowd model based on emotional contagion. In: Proceedings of the 2011 IEEE/WIC/ACM International Conferences on Web Intelligence and Intelligent Agent Technology, vol. 02, pp. 89–92. IEEE Computer Society (2011)
5. Ma, K., Li, W., Guo, Q., et al.: Information spreading in complex networks with participation of independent spreaders. Phys. A: Stat. Mech. Appl. **492**, 21–27 (2018)
6. Nowzari, C., Preciado, V.M., Pappas, G.J.: Analysis and control of epidemics: a survey of spreading processes on complex networks. IEEE Control Syst. **36**(1), 26–46 (2016)
7. Zheng, M., Wang, C., Zhou, J., et al.: Non-periodic outbreaks of recurrent epidemics and its network modelling. Sci. Rep. **5**, 16010 (2015)
8. Kramer, A.D.I., Guillory, J.E., Hancock, J.T.: Experimental evidence of massive-scale emotional contagion through social networks. Proc. Natl. Acad. Sci. **111**(24), 8788–8790 (2014)
9. Guy, S.J., Kim, S., Lin, M.C., Manocha, D.: Simulating heterogeneous crowd behaviors using Personality Trait Theory. In: Proceedings of SCA 2011: ACM SIGGRAPH/Eurographics Symposium on Computer Animation, pp. 43–52 (2011)
10. Lu, X., Wang, Z., Xu, M., et al.: A personality model for animating heterogeneous traffic behaviors. Comput. Anim. Virt. Worlds **25**(3–4), 363–373 (2014)
11. Faroqi, H., Mesgari, M.: Agent-based crowd simulation considering emotion contagion for emergency evacuation problem. ISPRS - Int. Arch. Photogram. Remote Sens. Spat. Inf. Sci. **40**, 193–196 (2015)
12. Tsai, J., Fridman, N., Bowring, E., et al. ESCAPES - Evacuation simulation with children, authorities, parents, emotions, and social comparison. In: 10th International Conference on Autonomous Agents and Multiagent Systems (AAMAS 2011), Taipei, Taiwan, 2–6 May 2011, vol. 1–3. International Foundation for Autonomous Agents and Multiagent Systems (2011)
13. Zhang, G., Lu, D., Liu, H.: IoT-based positive emotional contagion for crowd evacuation. IEEE Internet Things J. (2020). https://doi.org/10.1109/JIOT.2020.3009715

A Comprehensive Evaluation Method for Container Auto-Scaling Algorithms on Cloud

Jingxuan Xie, Shubo Zhang, Maolin Pan, and Yang Yu[✉]

School of Data and Computer Science, Sun Yat-sen University, Guangzhou, China
{xiejx9,zhangshb6}@mail2.sysu.edu.cn,
{panml,yuy}@mail.sysu.edu.cn

Abstract. Container technology becomes increasingly popular in public cloud platforms. Although many Container Auto-Scaling Algorithms (CASAs) have been proposed recently, there is still a lack of standardized frameworks to evaluate them comprehensively. This paper proposes a comprehensive evaluation method for CASAs. We firstly proposed a set of CASA evaluation metrics considering the requirements of cloud provider and user, and then designed a test data set based on real-world system load traces and 6 workload patterns. Experiments on some representative CASAs are conducted to demonstrate the effectiveness of the proposed evaluation method. Our research can provide cloud providers, operators and users with more comprehensive and systematic information about CASAs.

Keywords: Cloud computing · Container auto-scaling algorithm · Evaluation metrics · Workloads · Service level agreement

1 Introduction

Cloud computing is used to achieve efficient, convenient, and on-demand access to the configured computing resources (network, servers, storage, applications, services). With the establishment of the CNCF organization, especially the maturity of PaaS cloud platforms such as Kubernetes, container technology becomes increasingly popular in public cloud platforms. Since the container provides computing resources under the PaaS platform, the Container Auto-Scaling Mechanism becomes an important infrastructure for the container cloud platform.

At present, there are many challenges on CASA evaluation. The most of all are the diversity of user's requirements and the complexity of varieties of application scenarios. In recent years, although many CASAs have been proposed [8,16,19,20], , there is still no unified evaluation standard. At present, there is a lack of regulatory frameworks to comprehensively evaluate CASA, which increase the difficulty for users to select appropriate CASA for their applications. First, some metrics and measurement methods of the metrics system [10,27] built

© Springer Nature Singapore Pte Ltd. 2021
Y. Sun et al. (Eds.): ChineseCSCW 2020, CCIS 1330, pp. 616–627, 2021.
https://doi.org/10.1007/978-981-16-2540-4_45

on the IaaS platform are not suitable for container clouds. Secondly, some of the literature on CASA [8,16,19,20] used their own metrics lacking a uniform definition. Another problem of evaluating CASA is the lack of a comprehensive load test data set to cover a variety of application scenarios, such as literature [2,8,16,19–21].

Our goal is to propose a standard to comprehensively evaluate the applicable scenarios and metrics of the CASA. The main jobs include:

- Our paper attempts to unify and clarify the metrics from the literature and we propose a container cloud metrics system for the evaluation of auto-scaling algorithms.
- Our paper proposes how to construct load test data sets, then designs a test data set based on real-world system load traces and 6 workload patterns to test the CASA.
- Our paper uses multiple metrics to characterize the compared algorithms, covering different aspects of performance. And we adopts radar diagrams to allow easy visual comparison between algorithms.

The rest of the paper is organized as follows. Section 2 summarizes the related works. Section 3 describes the system architecture. Section 4 proposes a CASAs metrics system. Section 5 shows details on the workloads. Section 6 describes the experiments' setting and results. Section 7 presents the conclusions.

2   Related Work

2.1   Cloud Service Evaluation Metrics

Cloud service evaluation metrics focuses on evaluation from the perspective of cloud providers or end-users. Feng et al. in [10] collected key metrics such as resources, cost, and QoS from different providers to evaluate different cloud services. Literature [22,27] has proposed cloud service frameworks for evaluating specific performance and quality of service attributes. Garg et al. in [11] proposed the SMICloud framework and mechanism to measure service quality and rank cloud services provided by different cloud providers.

2.2   CASA Evaluation Metrics

Almost all CASA experiments focuses on performance, resource, or other evaluations. For example, Kan et al. in [19] gives a metrics of instance scaling with load. Literature [20] gave similar metrics, and it also introduced elasticity. Chang et al. in [8] gives the number of instance scaling and resource utilization. Horovitz et al. in [16] proposed the experimental evaluation metrics including resource utilization, response time, and SLA violation.

2.3 Workload in Auto-Scaling Algorithm

The auto-scaling algorithm literature usually uses real loads or self-constructed loads for experimental evaluation. Literature [2] used the page request data of German Wikipedia. Chang et al. in [8] constructed a rising-smooth load scenario. Load scenarios similar to sine waves have been constructed in [16,21], which are highly predictive. Kan et al. in [19] constructed a rising-smooth-falling load scenario and an up-and-down oscillating load scenario and used the real load dataset of FIFA WorldCup98.

3 System Design

3.1 System Overview

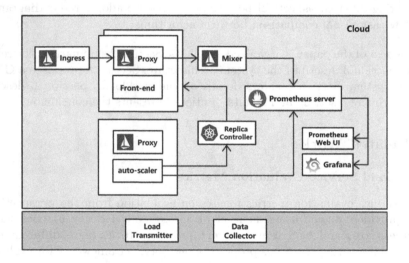

Fig. 1. System architecture

Figure 1 shows the overall structure of our framework. The microservice application and CASAs used in our experiments were deployed on the cloud platform supporting Istio service mesh in the form of Docker images. Mixer collects telemetry data from various proxies and stores the data in a time series database, such as Prometheus used in our system.

In the experiment, we use the load transmitter to send requests and access our application through the ingress gateway. The CASA we deploy in the pod obtains the current service performance metrics from the time series database, and make scaling decisions based on algorithmic strategies. Then perform inward or outward scaling of services through Replica Controller. Finally, we collect and analyze experimental data through the data collector.

3.2 Definition of Collected Data

Prometheus provides a functional expression language PromQL, which enables users to find and aggregate time series data in real time. The data can be displayed in the form of graphs, such as Grafana. Our data collector queries the metrics through PromQL syntax as follows:

- **Response time** t. Response time is a commonly used QoS metrics, and it is most intuitive to measure service quality and SLA satisfaction.
- **Request success number** N_s/**failure number** N_f. The success and failure of the request number are used to measure the reliability of a service.
- **Service available time** T_a/**unavailable time** T_u. The availability and unavailability of service time are measure of service availability.
- **Supply pods** n. The number of supply pods reflects the changes in service instances that currently apply CASA.

4 Metrics System

ISO 9126 software quality model [18] is an international standard for the evaluation of software quality. The standard defines six quality characteristics, namely functionality, reliability, usability, efficiency, maintainability, and portability. Further, the Cloud Services Measurement Initiative Consortium (CSMIC) has proposed the Service Measurement Index (SMI) attributes [24] based on the international standard in 2011. Based on the above, this paper adopts some SMI attributes, such as reliability, availability, service stability, and elasticity. At the same time, combined with the characteristics of container clouds and existing literature results, we propose a metrics system for container-based service auto-scaling algorithm, as shown in Fig. 2.

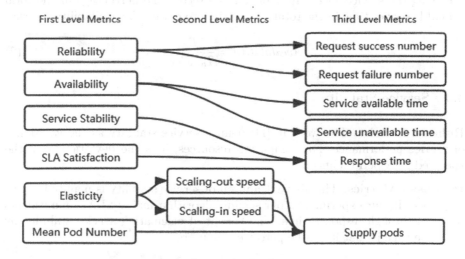

Fig. 2. Hierarchical relationship between metrics

4.1 Reliability

Related Metrics. The reliability defined in [11] is the ability of a service to maintain normal operation without failure in a given time and condition. The reliability is calculated based on the mean time to failure promised and the number of failed users. The service reliability defined in [3] is the ability of the system to provide acceptable services, which is essentially the part where the service request is successfully served.

Proposed Metrics. Reliability refers to the probability of a service being successfully completed in a certain period of time and under certain conditions. In our experiments, if the CASA cannot effectively complete the instance expansion, it will cause a large number of user requests to fail. In order to evaluate the reliability of the CASA, we define this metrics as:

$$Reliability = \frac{N_s}{N_s + N_f} \tag{1}$$

4.2 Availability

Related Metrics. A general availability definition is given by checking the status code of the web server response in [13]. In [3,26], it was considered that the service interruption event caused a long time user request to fail and retry, the service was considered unavailable. The work in [11] defines availability as the percentage of time a user can access the service.

Proposed Metrics. Availability is the percentage of time the system is available. The definition in [3,26] combines downtime, which is more suitable for the VM environment. Our experiments are closer to the definition of [11] above. By accessing the service successfully or not, we record the percentage of the total available service time in the total service time to obtain the availability metrics:

$$Availability = \frac{T_a}{T_a + T_u} \tag{2}$$

4.3 Service Stability

Related Metrics. The work in [11] defines service stability as the variability of service performance. For computing resources, it is the deviation from the specified SLA performance.

Proposed Metrics. The above definition of service stability is mainly for performance. In our experiments, our scaling algorithm makes SLA commitments to users through response time. The benefit of calculating service stability by response time is that it can be perceived by users.

$$Service\,Stability = \frac{1}{\sqrt{\frac{(t_1-\bar{T})^2+\cdots+(t_i-\bar{T})^2}{i}}} \tag{3}$$

where t_i is the response time at time i, and $\bar{t}$ is the average response time within a service time T.

4.4 Elasticity

Related Metrics. The definition of elasticity in [14] refers to the degree to which the system adapts to workload changes by automatically providing or removing resources. The work in [2,15] also use the above definition of elasticity, the difference is the way of metrics aggregation. The work in [23] defines elasticity from the perspective of service consumers.

Proposed Metrics. Through investigation, we found that the definition of elasticity in [14] has been widely accepted and used for service evaluation in cloud computing. We modified part of the elasticity measurement and propose a new elasticity evaluation metrics to measure the elasticity of services after applying different CASAs. We define the deployment speed of containers as:

$$S = \frac{\Delta n}{\Delta T} \tag{4}$$

Where Δn represents the number of container changes, ΔT represents the time period of container changes. It calculates the number of containers that can be deployed within a scaling period of time. We define the elasticity as the deployment speed of containers within a scaling period of time, and divide the elasticity into scaling-out elasticity E_o and scaling-in elasticity E_i:

$$S_o = \frac{\Delta n_o}{\Delta T_o} \tag{5}$$

$$E_o = \frac{S_o}{\Delta T_o} \tag{6}$$

Where S_o is the scaling-out speed of container, Δn_o is the number of scaling-out container changes, ΔT_o is the time period of scaling-out container changes. Same goes for E_i. Therefore, the elasticity in this paper is:

$$Elasticity = \omega_o \times E_o + \omega_i \times E_i \tag{7}$$

$$with\ \omega_o, \omega_i \in [0,1],\ \omega_o + \omega_i = 1$$

4.5 SLA Satisfaction

Related Metrics. The SLA defined in [25] requires that the cloud system allocate resources not less than the requirements of cloud-based applications. The work in [29] defines the SLA violation rate as the number of SLA violations divided by the number of service requests from the application in each deployment scenario. The work in [1] calculates the violation rate by setting the SLA threshold for violation of response time, also [12].

Proposed Metrics. The SLA satisfaction metrics in this paper is measured by response time. We define the SLA satisfaction metrics as the ratio of the response time that meets the maximum allowable response time to the overall response time. Within the measurement time T:

$$SLA\,Satisfaction = 1 - \frac{\sum_{i=1}^{T} \max(t_i - t_{sla}, 0)}{\sum_{i=1}^{T} t_i} \tag{8}$$

where t_i is the response time at time i, and t_{sla} is the maximum allowable response time we set in advance.

5 Workload Structure

In order to make our test more practical and reflect different load scenarios, we use the real-world traffic statistics data sets as the basis for simulating load scenarios. For example Wikipedia [28], which is more popular as the load of the CASA, and it has been used in [2,19] and other papers. According to the data, combined with the load scenario division in [7,17], the load feature classification of this paper is shown in Fig. 3:

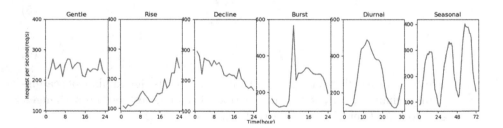

Fig. 3. Workload characteristics

- **Gentle.** This scenario describes where the load variation range is not large, and usually there is only a small amplitude fluctuation.
- **Rise.** This scenario describes where the load monotonically rises within a certain period of time, such as a web application activity being promoted.

- **Decline.** This scenario describes where the load decreases monotonically within a certain period of time, such as users terminating a session activity.
- **Burst.** This scenario describes where a peak load suddenly occurs at a certain time and then returns to normal, such as some product browsing pages of on-time activities.
- **Diurnal.** This scenario describes scenes with obvious day and night load changes.
- **Seasonal.** This scenario describes a pattern that recurs periodically.

Based on the classification of load characteristics, we extract multiple pieces of load data from the real-world traffic statistics data sets, and compress each piece of data into an hourly experimental sample.

6 Experimental Evaluation

6.1 Experiment Settings

The CASA we used for evaluation was deployed on a cloud platform supporting Istio and Kubernetes. There are five 4 core 8G nodes and two 8 core 16G nodes in our Kubernetes cluster. The operating system is CentOS7. In this experimental environment, we deployed a modified application Bookinfo [5] as a test application. Our experiment stipulates that only the number of instances should be scaled, so we set a fixed resource quota for the instance with CPU of 0.25 cores and memory of 512 MB.

The auto-scaling algorithms we used for evaluation are: the static threshold algorithm and Q-learning algorithm mentioned in [9], the SARSA algorithm mentioned in [4], and the ARIMA algorithm mentioned in [6]. Among them, static threshold and Q-learning are reactive algorithms, and SARSA and ARIMA are proactive algorithms. We set the same SLA of response time for these four algorithms as $t_{sla} = 250$ ms.

6.2 Results Analysis

We calculated the relative metrics vector of each metrics in the four algorithms, and compare the differences between different CASAs in the form of radar diagrams. The smaller the mean pod number, the better our expectations of the algorithm, the rest is opposite. We classify CASAs under different load scenarios, and get the comparative performance of the four algorithms under six load scenarios, as shown in Fig. 4 and Table 1.

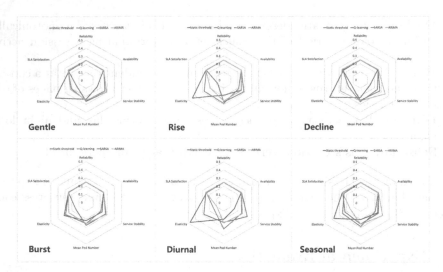

Fig. 4. Comparison of four auto-scaling algorithms in different scenarios

Table 1. The data of algorithms in different scenarios

Scenario	Algorithm	Reliability	Availability	Service Sta.	Mean Pod Num.	Elasticity	SLA Sat.
Gentle	Static threshold	99.55%	97.44%	0.01514	**3.04**	**42.16%**	93.36%
	Q-learning	**100.00%**	**100.00%**	**0.02744**	3.37	12.99%	**99.78%**
	SARSA	**100.00%**	**100.00%**	0.02531	3.29	19.94%	99.36%
	ARIMA	99.97%	**100.00%**	0.02727	3.18	21.46%	98.95%
Rise	Static threshold	**100.00%**	**100.00%**	0.02355	**2.29**	**61.07%**	99.40%
	Q-learning	**100.00%**	**100.00%**	**0.02721**	3.33	11.61%	**99.99%**
	SARSA	99.94%	99.86%	0.01999	3.54	37.84%	97.56%
	ARIMA	**100.00%**	**100.00%**	0.02391	3.16	26.25%	99.89%
Decline	Static threshold	99.52%	96.64%	0.00822	**3.04**	**31.78%**	91.35%
	Q-learning	99.67%	99.10%	0.01673	3.51	9.26%	96.29%
	SARSA	99.64%	99.02%	0.01465	3.73	12.27%	94.84%
	ARIMA	**100.00%**	**100.00%**	**0.02132**	3.57	20.60%	**99.18%**
Burst	Static threshold	99.68%	98.13%	0.01534	**2.65**	**23.89%**	95.75%
	Q-learning	99.77%	99.27%	0.01863	3.33	11.37%	**97.23%**
	SARSA	99.39%	98.41%	0.01418	3.08	22.14%	94.39%
	ARIMA	**99.86%**	**99.55%**	**0.01944**	3.07	18.70%	**97.23%**
Diurnal	Static threshold	99.57%	97.38%	0.01040	**3.15**	**15.96%**	91.92%
	Q-learning	99.78%	99.29%	**0.02337**	3.74	9.43%	**97.22%**
	SARSA	99.37%	97.59%	0.01773	5.01	4.73%	92.32%
	ARIMA	**99.80%**	**99.48%**	0.01859	3.86	3.22%	96.20%
Seasonal	Static threshold	**100.00%**	**100.00%**	0.01788	**2.88**	**11.47%**	98.97%
	Q-learning	99.91%	99.59%	0.01864	3.66	6.75%	97.17%
	SARSA	99.35%	98.14%	0.01541	4.10	7.79%	90.16%
	ARIMA	**100.00%**	**100.00%**	**0.02304**	3.69	4.49%	**99.40%**

Comprehensively, the ARIMA algorithm has relatively good reliability and availability. The static threshold algorithm has less instance cost and higher elasticity. The Q-learning algorithm and ARIMA algorithm are always the first or second in service stability and SLA satisfaction. We can know that ARIMA has better SLA satisfaction in dealing with seasonal load scenario, which because the seasonal load scenario can be followed regularly. The improvement of the SLA satisfaction is related to the mean pod number. But the more the mean pod number is not always able to get a higher SLA satisfaction, such as SARSA algorithm. This is due to the scaling characteristic of SARSA itself.

In general, we can draw the conclusion that the same CASA acts on different workload patterns may have different performance and different CASA acts on the same workload pattern, performance metrics have their own advantages and disadvantages.

The above algorithm case study shows that the method we proposed is effective. Therefore, cloud providers, operators, and users can use it to analyze different CASA from multi-dimensional metrics, and they should more carefully combine their own service workload patterns to choose more suitable algorithm.

7   Conclusion and Future Work

This paper proposes a comprehensive evaluation method for CASAs, which is dedicated to comprehensively and systematically evaluating the applicable scenarios and various metrics of CASAs. This paper proposes a container cloud-based metrics system and builds a load data set for testing the CASA. Experiments on some representative CASAs are conducted to demonstrate the effectiveness of the proposed evaluation method.

Our research might provide important help for algorithm designers, users, and operators to fully understand, monitor, and evaluate the service quality and capabilities of the container cloud. In the future, we will further explore the evaluation metrics system of the CASA on complex services to make the method proposed in this article more applicable.

Acknowledgments. This work is Supported by the National Key Research and Development Program of China under Grant No. 2017YFB0202201; the National Natural Science Foundation of China (NSFC) under Grant No. 61972427; the NSFC-Guangdong Joint Fund Project under Grant No. U1911205; the Research Foundation of Science and Technology Plan Project in Guangdong Province under Grant No. 2020A0505100030.

References

1. Abdullah, M., Iqbal, W., Erradi, A.: Unsupervised learning approach for web application auto-decomposition into microservices. J. Syst. Softw. **151**, 243–257 (2019)
2. Bauer, A., Lesch, V., Versluis, L., Ilyushkin, A., Herbst, N., Kounev, S.: Chamulteon: coordinated auto-scaling of micro-services. In: 2019 IEEE 39th International Conference on Distributed Computing Systems (ICDCS), pp. 2015–2025. IEEE (2019)
3. Bauer, E., Adams, R.: Reliability and Availability of Cloud Computing. Wiley, Hoboken (2012)

4. Benifa, J.B., Dejey, D.: RLPAS: reinforcement learning-based proactive auto-scaler for resource provisioning in cloud environment. Mob. Netw. Appl. **24**(4), 1348–1363 (2019). https://doi.org/10.1007/s11036-018-0996-0

5. Bookinfo. https://github.com/istio/istio/tree/master/samples/bookinfo

6. Calheiros, R.N., Masoumi, E., Ranjan, R., Buyya, R.: Workload prediction using ARIMA model and its impact on cloud applications' QoS. IEEE Trans. Cloud Comput. **3**(4), 449–458 (2014)

7. Calzarossa, M.C., Massari, L., Tabash, M.I., Tessera, D.: Cloud autoscaling for HTTP/2 workloads. In: 2017 3rd International Conference of Cloud Computing Technologies and Applications (CloudTech), pp. 1–6. IEEE (2017)

8. Chang, C.C., Yang, S.R., Yeh, E.H., Lin, P., Jeng, J.Y.: A kubernetes-based monitoring platform for dynamic cloud resource provisioning. In: GLOBECOM 2017–2017 IEEE Global Communications Conference, pp. 1–6. IEEE (2017)

9. Dutreilh, X., Moreau, A., Malenfant, J., Rivierre, N., Truck, I.: From data center resource allocation to control theory and back. In: 2010 IEEE 3rd International Conference on Cloud Computing, pp. 410–417. IEEE (2010)

10. Feng, D., Wu, Z., Zhang, Z., Fu, J.: On the conceptualization of elastic service evaluation in cloud computing. J. Inf. Technol. Res. (JITR) **12**(1), 36–48 (2019)

11. Garg, S.K., Versteeg, S., Buyya, R.: A framework for ranking of cloud computing services. Future Gener. Comput. Syst. **29**(4), 1012–1023 (2013)

12. Ghobaei-Arani, M., Souri, A., Baker, T., Hussien, A.: ControCity: an autonomous approach for controlling elasticity using buffer management in cloud computing environment. IEEE Access **7**, 106912–106924 (2019)

13. Herbst, N., et al.: Ready for rain? A view from SPEC research on the future of cloud metrics. arXiv preprint arXiv:1604.03470 (2016)

14. Herbst, N.R., Kounev, S., Reussner, R.: Elasticity in cloud computing: what it is, and what it is not. In: Proceedings of the 10th International Conference on Autonomic Computing, ICAC 2013, pp. 23–27 (2013)

15. Herbst, N.R., Kounev, S., Weber, A., Groenda, H.: BUNGEE: an elasticity benchmark for self-adaptive IaaS cloud environments. In: 2015 IEEE/ACM 10th International Symposium on Software Engineering for Adaptive and Self-Managing Systems, pp. 46–56. IEEE (2015)

16. Horovitz, S., Arian, Y.: Efficient cloud auto-scaling with SLA objective using Q-learning. In: 2018 IEEE 6th International Conference on Future Internet of Things and Cloud (FiCloud), pp. 85–92. IEEE (2018)

17. Iqbal, W., Erradi, A., Mahmood, A.: Dynamic workload patterns prediction for proactive auto-scaling of web applications. J. Netw. Comput. Appl. **124**, 94–107 (2018)

18. ISO: IEC 9126 software engineering—product quality—part 1: Quality model. Geneva: International Organization for Standardization (2001)

19. Kan, C.: DoCloud: an elastic cloud platform for web applications based on docker. In: 2016 18th International Conference on Advanced Communication Technology (ICACT), pp. 478–483. IEEE (2016)

20. Klinaku, F., Frank, M., Becker, S.: Caus: an elasticity controller for a containerized microservice. In: Companion of the 2018 ACM/SPEC International Conference on Performance Engineering, pp. 93–98 (2018)

21. Lolos, K., Konstantinou, I., Kantere, V., Koziris, N.: Elastic management of cloud applications using adaptive reinforcement learning. In: 2017 IEEE International Conference on Big Data (Big Data), pp. 203–212. IEEE (2017)

22. Saravanan, K., Kantham, M.L.: An enhanced QoS architecture based framework for ranking of cloud services. Int. J. Eng. Trends Technol. (IJETT) 4(4), 1022–1031 (2013)
23. Schulz, F.: Elasticity in service level agreements. In: 2013 IEEE International Conference on Systems, Man, and Cybernetics, pp. 4092–4097. IEEE (2013)
24. Siegel, J., Perdue, J.: Cloud services measures for global use: the service measurement index (SMI). In: 2012 Annual SRII Global Conference, pp. 411–415. IEEE (2012)
25. Tran, D., Tran, N., Nguyen, G., Nguyen, B.M.: A proactive cloud scaling model based on fuzzy time series and SLA awareness. Proc. Comput. Sci. 108, 365–374 (2017)
26. Tripathi, A., Pathak, I., Vidyarthi, D.P.: Integration of analytic network process with service measurement index framework for cloud service provider selection. Concurr. Comput.: Pract. Exp. 29(12), e4144 (2017)
27. Upadhyay, N.: Managing cloud service evaluation and selection. Proc. Comput. Sci. 122, 1061–1068 (2017)
28. Wiki Source. https://dumps.wikimedia.org/other/pagecounts-raw/2016/
29. Zheng, T., Zheng, X., Zhang, Y., Deng, Y., Dong, E., Zhang, R., Liu, X.: SmartVM: a SLA-aware microservice deployment framework. World Wide Web 22(1), 275–293 (2019). https://doi.org/10.1007/s11280-018-0562-5

Well Begun Is Half Done: How First Respondeners Affect Issue Resolution Process in Open Source Software Development?

Yiran Wang and Jian Cao[✉]

Department of Computer Science and Engineering, Shanghai Jiaotong University, Shanghai 200240, China
{wangyiran33,cao-jian}@sjtu.edu.cn

Abstract. Addressing issue reports is an integral part of open source software (OSS) projects. Although several studies have attempted to discover the factors that affect issue resolution, few pay attention to first responders, who is the first to post their comments after an issue report is published. We are interested at how first responders affect issue resolution process for OSS projects. Therefore, we extract the data from Github to perform our empirical study. By obtaining their identity types and speech acts, we analyze the impact of first responders with different identity types on the efficiency of issue resolution based on three metrics and find that identified users especially collaborators make first response can improve the efficiency of issue resolution. Furthermore, we make use of the identity type information of the first responders to forecast the issue lifetime and the results show that this information can also improve the accuracy for short-term prediction. It also verifies that first responders have a direct influence on the issue resolution process.

Keywords: Issue resolution · First responder · Issue resolution efficiency · Issue lifetime prediction

1 Introduction

Issue registering, tracking and resolution are very important in open source software (OSS) projects [3]. Anyone can submit an issue through an issue tracking system on open source hosting platforms such as GitHub and Google Code. However, tackling these issues is burdensome for developers and it is usual that these issue resolution tasks are delayed or even completely ignored. Previous studies show that problems of issue overstocking may arise over time [10] and a considerable portion of issues in GitHub are pending for months or even over a year [7]. Therefore, effective strategies should be implemented to improve the

© Springer Nature Singapore Pte Ltd. 2021
Y. Sun et al. (Eds.): ChineseCSCW 2020, CCIS 1330, pp. 628–638, 2021.
https://doi.org/10.1007/978-981-16-2540-4_46

issue resolution process such as prioritizing issues to be resolved and allocating resources for each issue clearly in practice and more importantly, we should understand the influential factors behind successful issue resolution processes.

As a collaboration process, issue resolution consists of multiple actions that occur in order and involves several persons. Just as the old saying goes, "Well begun is half done", first responder tends to play a decisive role in the process of issue resolution. It can be assumed if the first responder leads issue resolution well, then other stakeholders will be willing to participate in the resolution process and the efficiency of issue resolution will increase. On the contrary, if the first responder misleads the direction of resolution or increase the complexity of the problem, the whole process will slow down, which may result in a pended issue.

There are various types of roles involved in OSS projects. People of different identity types have different responsibilities and authority in projects and all of them may be involved in issue resolution. We want to study the response patterns of people with different identity types as the first responders and their first responses' impact on the efficiency of issue resolution. At the same time, whether the identity type information of first responders is useful for issue lifetime prediction is also our concern. Through this research, we are looking forward to forming some suggestions to increase the efficiency of issue resolution.

There are different types of issues and they may have very different microprocess patterns. In this paper, we focus on the issues relating to the maintenance activities. Moreover, the ISO/IEC 14764 standard [11] defines four types of maintenance activities spanning the different motivations that software engineers have while undertaking changes to an existing software system. Issues can be mapped to these maintenance activities through a classification model. In this paper, we only focus on issues relating to corrective maintenance, which corresponds to bugs and accounts for nearly half of all issues.

Therefore, in this paper, we aim at answering the following research questions:

RQ1: What is the majority identity types of first responders and what are the common speech acts of their first responses?

RQ2: How do the first responders with different identity types affect issue resolution?

RQ3: Is first responders' identity type information useful for issue lifetime prediction?

In order to answer these questions, firstly, we collect and preprocess data from Github to extract first responders of issues and construct models to extract their speech acts in first responses. The distributions and characteristics of identity types and their common speech acts are analyzed. Then, we analyze the impact of first responders with different identity types on the efficiency of issue resolution based on three metrics. Finally, we try to utilize the identity type information of first responders as an additional feature for issue lifetime predictions.

2   Related Work

In previous work, several factors are discovered that affect issue lifetime. Murgia et al. [11] find that the issue resolution time depends on the type of maintenance performed, where corrective and perfective maintenance are generally faster than adaptive and preventive maintenance. Destefanis et al. [4] discover that the level of politeness in the communication process among developers has an effect on the time required to fix issues. Liao et al. [10] find an association between issue commits and the lifetime of a project and build a three-segment model and define three phases, which provides new knowledge about issues. [16] examines factors affecting bug fixing time along three different dimensions and compares them using descriptive logistic regression models. In our work, we explore the impact of first responder's identity type on issue resolution time, which has never been explored before.

Bug-fixing time prediction and issue lifetime prediction have been received significant attentions in recent years. Weiss et al. [14] predict the time spent on fixing an issue based on the average time of its similar issues. Al-Zubaidi et al. [2] use multi-objective search-based approach to estimate issue resolution time which makes estimation models accurate and simple simultaneously. Giger et al. [5] and Rees-Jones et al. [13] present decision-tree-based models to predict bug fix time while Panjer et al. [12] utilize logistic regression models. Kikas [8] predict issue resolution time based on random forest models using dynamic and contextual features. Zhang [17] present a kNN-based classification model for predicting whether a bug will get a slow or a fast fix. In this work, we also construct issue lifetime prediction models but our emphasis is to show an improvement in the predictors' performance when putting the identity type of first responder into features.

3   Dataset and Methodology

3.1   Dataset

In this study, we collect issue data using public APIs from GitHub and only closed issues updated at or after January 1, 2017 are included. The ten popular projects used in this study vary in domain, scale and programming language and they all provide dynamic platforms for bug reporting, discussing and fixing.

Since we focus on issues relating to corrective maintenance, issue reports must be classified first. We rely on labels applied to each issue report in GitHub to identify their maintenance type. Finally, 4863 issues relating to corrective maintenance are collected and included in the dataset. We do some important preprocessing before extracting first responders of an issue. Usually people take for granted that first responder is the first one who leaves a comment under an issue. However, we find that about 15% of first commenters discovered using this method are issue creators themselves and what they do in their first comment is usually information supplement. When issue creators become their issues' first commenters, they often add what is supposed to be clarified in the body of an issue and we can not regard them as the first response. The solution for this problem is to filter issue creators' comments until other users participate in and we regard the first comment not by the issue creator as the first response and the commenter as the first responder.

1. Bug Identification
   1a. Reproduction-needed code Request (RR)
   1b. Reproduction-needed code Offer (RO)
   1c. Reproduction Confirmation (RC)

2. Problem Definition and Solution Development
   2a. Problem Analysis, Solution Development and Progress Reporting (SD)
   2b. Problem Critique and Solution Evaluation (PC)

3. Non period
   3a. Pinging Others and Progress Inquiring (PI)
   3b. Duplication Statement (DS)
   3c. Non-task (NT)

Fig. 1. A classification standard for speech acts in the process of issue resolution

3.2 Speech Act Classification

In order to answer RQ1, we need to do speech act classification for the first response. This relies on our previous work and since it is not our emphasis, we introduce our classification method briefly here.

After reading a lot of issues, we proposed a classification standard for speech acts in the issue resolution process based on the dialog behavior of the comments in the issue Fig. 1. According to the stage of the speech acts of comments, they can be divided into three categories: Bug Identification, Problem Definition and Solution Development and Non-period. The first two categories correspond to the two stages of issue resolution, and the third category is speech acts that occur at any time during the issue resolution process.

We would like to automatically classify every comment in the issue resolution process, so we need to construct a machine learning model. First, we performed the labeling task and manually labeled 1915 comments as a training data set. In feature selection, we select 1–3 n-grams with word frequency ≥ 1, use chi2 method to select top250 features, then manually add some important keywords which we find in the manual labeling stage, and finally add some non-verbal features. Through comparison, the model adopts the best-performing random forest method, and the final performance metrics Micro-F1 (Accuracy) reaches 0.703.

3.3   Issue Lifetime Prediction

To answer RQ3, we construct prediction models to show an improvement in the predictors' performance when calculating the identity types of first responders into features.

Feature Selection. The performance of prediction models relies on features that are properly selected. Our feature engineering is based on the work of [8] and [13]. Furthermore, we also add the following new features to the features we chose based on previous work: *CodeIncluded* for whether the body of an issue includes code or not, *CleanedTitleLength* for the number of words in the issue body with markdown parsed and tags removed and *CreatorAuthority* for whether the creator has authoritative identity type in the project.

The selected features can be divided into three classes, i.e., **Issue features**, **Issue creator features** and **Project features**. Issue features describe the contents of an issue and its related events. Issue creator features reflect the characteristics of the author of an issue, which relates to issue contents and quality. The resolution processes of issues are also be affected by their projects and project features reflects their issue resolution statuses and activity levels.

Model Training. We train prediction models based on observation point at 1 day, we collect all issues that were first responded in 1 day in our dataset to predict whether they will be closed with different prediction horizons. Naturally, the prediction horizon should end after the observation point. The prediction horizons are chosen to match calendric periods (week, fortnight, month, quarter, semester and year), which leads to six prediction horizons (7, 14, 30, 90, 180 and 365 days). Since we have one observation point, for each prediction horizon, we should train one model.

Evaluation Metrics. Although the prediction task is a binary classification problem in our study, we still utilize the macro-averaged F1-score and micro-averaged F1-score to evaluate the classifiers because correctly predicting the fact that an issue can be closed in a given prediction horizon or not is equally important so that traditional metrics for binary-classification are not sufficient. Instead, macro-averaged F1-score and micro-averaged F1-score provide overall insights into the performance of models for this problem.

4   Experiments and Results

4.1   Identity Type and Speech Act of First Responders

Overall Identity Type Distribution of First Responders. As a collaboration process, issue resolution involves all kinds of people with different identity types. There are 5 types of identities in GitHub: *Owner, Collaborator, Contributor, Member and None*, the classification standard is project users' authority

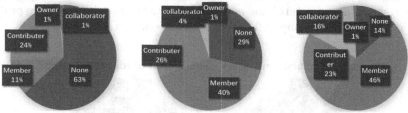

Fig. 2. The distribution of identity types of issue creators, all responders and first responders

and contribution in the project. Their specific definitions described by GitHub [1] are as follows:

- *Owner:* The person/s who has administrative ownership over the organization or repository.
- *Collaborator:* The person who has Read, Write, or Admin permissions to one or more repositories in an organization.
- *Contributor:* The person who has contributed something to the project.
- *Member:* The person who uses the project and might be active in conversations or expresses his opinion on the project's direction.
- *None:* Other users.

Roughly speaking, we name 'None' as ordinary users and others as identified users.

We first make a statistics of identity type distribution in issue creators, all responders and first responders, respectively. The distribution pie chart is shown as Fig. 2.

From the distribution of identity types of issue creators we can easily find that most (63%) issue creators are ordinary users without special identity type which indicates that *submitting issue reports is a low barrier operation* that any GitHub user can submit an issue report while coming across an issue. Of course, experienced users like contributors, collaborators and owners can raise an issue for others' help when they can't handle it alone, which reflects that issue resolution is a collaboration process that needs collective intelligence.

However, for a user to participate in the discussion of issue resolution, the threshold is greatly increased. The majority (71%) of responders are at least active users in the project and the proportion of ordinary users has dropped significantly. This is in line with our common sense that *only senior users with prior knowledge can participate in problem solving well.*

As for first responders, the proportion of ordinary users is further reduced to 14%. *Starting issue resolution requires higher professionalism and authority.* Comparing distribution of identity types of all responders and first responders, another major difference is the proportion of collaborators, which increases 12%

in first responders. Project collaborators are more accustomed to actively accepting the responsibility to start issue resolution due to their authority in the project and this identity type group will be our focus in our further study.

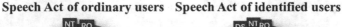

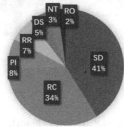 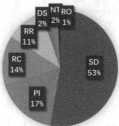

Fig. 3. The distribution of speech acts of ordinary users and identified users

Common Speech Acts of First Response. We do speech act classification for the first response of ordinary users and identified users respectively and Fig. 3 shows the result. The most significant distinction between ordinary users and identified users lies in the difference of proportion of RC (Reproduction Confirmation), SD (Problem Analysis and Solution Development) and PI (Progress Inquiring and Pinging Others).

For identified users, their first responses' speech acts have a 20% decline in RC while have a 12% rise in SD and a 9% rise in PI compared with ordinary users. This reflects the diversity of function between ordinary and identified users' first responses. For ordinary users, they tend to prove the bug can be reproduced like 'I have come across the same issue!', then the issue created will more likely to be proved as a bug and others will join in the resolution communication. For identified users, based on their knowledge and their responsibility, they are more customized to push forward the resolution process by assigning related people or directly trying problem analysis and solution development.

To answer RQ1 in general, first responders are identified users in majority(86%) and collaborators are especially accustomed to starting the issue resolution communication. Ordinary users are likely to confirm the bug while identified users do more assignment or development during their first response.

4.2   Impacts on Issue Resolution with Different First Responders

In this section, we will show the difference in efficiency of issue resolution when users with different identity types make first responses. We propose three metrics to characterize efficiency of issue resolution:

– *First Response Hours:* The interval from the creation time of an issue to the time that the first response is made.

Table 1. The efficiency of issue resolution for first responders with different identity types

Metrics		Ordinary	Identified				
		None	Overall	Member	Contri.	Collab.	Owner
First response hours	Mean	996	402	481	366	242	**166***
	Median	47	5	8	9	**0***	5
Average response hours	Mean	949	767	842	757	**386***	4392
	Median	214	102	138	167	**11***	3382
Issue lifetime	Mean	4296	3547	3927	3486	**1783***	19095
	Median	982	572	758	797	**188***	22781

- *Average Response Hours:* The average interval between any two adjacent responses.
- *Issue lifetime:* The interval from the creation time of an issue to the time that the issue is closed.

These three metrics measure the efficiency of issue resolution from three perspectives, i.e., start stage, middle stage and whole process. Table 1 presents the results.

First we analyze the result from a macro perspective, which is the difference between ordinary or identified users. Compared to issues with ordinary first responder, those with identified first responder achieve better efficiency in all dimensions, i.e., first response hours, average response hours and issue lifetime are all lower than those of unidentified first responder. This may bring us a heuristic conclusion: **To encourage identified users to make first response can improve the efficiency of issue resolution.**

Then we focus on the detailed diversity among issues with different kinds of identified first responders. It is obvious that issues with collaborators as first responders perform best in most metrics and the gap is really huge. It should be pointed out that the median first response hour for collaborators is 0 which means half and more of collaborators can make first response for an issue within 1 h, which is very impressive. Except for high 'starting speed', when first responders are collaborators, the efficiency of the issue resolution process and final issue resolution time will also be sharply reduced. We attribute this tremendous advance to collaborators' professionalism, influence and authority:

- Collaborators' experience and knowledge allow them to make fast and effective first response to identify the issue or assign it to related developers, which provides a good beginning for issue resolution.
- Collaborators usually have certain influence and authority in one project, which may motivate other project users to join in the process of issue resolution.

Unexpectedly, issues whose first responder is project owner himself suffer from extremely high average response hours and issue lifetime. We conduct a

Table 2. Prediction Performances of models for different observation points and prediction horizons

Prediction horizon days	macro-F1		micro-F1		AUC	
	initial	*+identity type*	*initial*	*+identity type*	*initial*	*+identity type*
Observation at 1 day after issue is opened						
7	0.528	**0.562***	0.740	**0.777***	0.531	**0.559***
14	0.575	**0.619***	0.664	**0.696***	0.574	**0.618***
30	0.653	**0.669***	0.657	**0.673***	0.653	**0.669***
90	0.696	0.694	0.717	0.716	0.692	0.691
180	0.741	0.741	0.802	0.804	0.726	0.725
365	0.781	0.783	0.893	0.894	0.747	0.750

sample survey and find that the cause is these issues are mostly open and difficult issues remaining multiple rounds of dialogue, which determines that they cannot be closed in a short time.

To answer RQ2, When identified users, especially collaborators make first response, the efficiency of issue resolution will be improved.

4.3   Identity Type of First Responder on Issue Lifetime Prediction

In order to answer RQ3, we construct models with first responder's identity type information in contrast to models without identity type information to show that this information is beneficial to lifetime prediction.

We analyze model performance for different observation points and compare initial models (initial) and models with first responder's identity type information (+identity type). Table 2 shows the result.

It is not hard to see an improvement when taking first responder's identity type information into account for short-term prediction. The reason may be considering their identity type information is like getting first responder's help to make a decision. First responder's identity type information can be regarded as a prior probability. At an early period when other features can barely provide information, adding prior probability is of great help for classifiers.

5   Threads to Validity

We present various threats to validity of our study according to the guideline reported in [15]. Any conclusions made from this work must be considered with the following issues in mind.

Threats to internal validity are concerned with factors that may affect a dependent variable and were not considered in the study. In this study, we examine the influence on issue resolution efficiency of first responder's identity type in various projects. We assume that properties of issues in each project are uniformly distributed so that we can build a correlation between efficiency of issue resolution and first responder's identity type. However, the heterogeneity of issues in each project cannot be avoided. In addition, there are several relationships between some issues in a project, i.e., they are not independent. These will inevitably affect our results to some extent.

Also, issue misclassification can happen [6,9], which may have impact on our study.

Threats to external validity are concerned with the generality of our results. In our study, we use issues from 10 projects that are representatives of the open source domain which have different backgrounds, development practices and goals. To improve generality, we can extend our study to more representative projects.

6 Conclusion

In this paper we try to study the impact of identity type of the first responder on issue resolution. Based on the issues extracted from 10 distinctive and representative projects in the open source domain, we collect data and do preprocessing to extract first responders of these issues and extract their common speech acts in the first response. We find most of first responders are at least active users in a project. The distributions and characteristics of each identity type and their common speech acts are analyzed. Then, we analyze the impact of first responders with different identity types on efficiency of issue resolution based on three metrics we proposed. We find that allowing identified users especially collaborators to make first response can improve the efficiency of issue resolution. After this, we construct models for issue lifetime prediction in GitHub projects for different calendric periods with identity type of first responders in order to demonstrate value within first responder's identity type information. The results show that models with first responder's identity type achieve better accuracy for issue lifetime prediction for short-term prediction.

In this study we only focus on issues relating to corrective maintenance. However, there are other types of issues. Although the number of these issues is smaller, they are also of great significance in open-source software project maintenance. Specifically, perfective maintenance aims to enhance performance, adaptive maintenance deals with feature requests and preventive maintenance involves actions to be taken to avoid future bugs. One possible direction for future work is to study the impact of first responder's identity type on resolution time of these types of issues.

Acknowledgments. This work is partially supported by National Key Research and Development Plan (No. 2018YFB1003800) and China National Science Foundation (Granted Number 62072301).

References

1. How to contribute to open source? [EB/OL]. https://opensource.guide/how-to-contribute

2. Al-Zubaidi, W.H.A., Dam, H.K., Ghose, A., Li, X.: Multi-objective search-based approach to estimate issue resolution time. In: Proceedings of the 13th International Conference on Predictive Models and Data Analytics in Software Engineering, pp. 53–62. ACM (2017)
3. Bertram, D., Voida, A., Greenberg, S., Walker, R.: Communication, collaboration, and bugs: the social nature of issue tracking in small, collocated teams. In: Proceedings of the 2010 ACM Conference on Computer Supported Cooperative Work, pp. 291–300. ACM (2010)
4. Destefanis, G., Ortu, M., Counsell, S., Swift, S., Marchesi, M., Tonelli, R.: Software development: do good manners matter? PeerJ Computer Science 2, 1–35 (2016)
5. Giger, E., Pinzger, M., Gall, H.: Predicting the fix time of bugs. In: Proceedings of the 2nd International Workshop on Recommendation Systems for Software Engineering, pp. 52–56. ACM (2010)
6. Herzig, K., Just, S., Zeller, A.: It's not a bug, it's a feature: how misclassification impacts bug prediction. In: Proceedings of the 2013 International Conference on Software Engineering, pp. 392–401. IEEE Press (2013)
7. Kikas, R., Dumas, M., Pfahl, D.: Issue dynamics in Github projects. In: Abrahamsson, P., Corral, L., Oivo, M., Russo, B. (eds.) PROFES 2015. LNCS, vol. 9459, pp. 295–310. Springer, Cham (2015). https://doi.org/10.1007/978-3-319-26844-6_22
8. Kikas, R., Dumas, M., Pfahl, D.: Using dynamic and contextual features to predict issue lifetime in Github projects. In: Proceedings of the 13th International Conference on Mining Software Repositories, pp. 291–302. ACM (2016)
9. Kochhar, P.S., Le, T.D.B., Lo, D.: It's not a bug, it's a feature: does misclassification affect bug localization? In: Proceedings of the 11th Working Conference on Mining Software Repositories, pp. 296–299. ACM (2014)
10. Liao, Z., He, D., Chen, Z., Fan, X., Zhang, Y., Liu, S.: Exploring the characteristics of issue-related behaviors in Github using visualization techniques. IEEE Access 6, 24003–24015 (2018)
11. Murgia, A., Concas, G., Tonelli, R., Ortu, M., Demeyer, S., Marchesi, M.: On the influence of maintenance activity types on the issue resolution time. In: Proceedings of the 10th International Conference on Predictive Models in Software Engineering, pp. 12–21. ACM (2014)
12. Panjer, L.D.: Predicting eclipse bug lifetimes. In: Proceedings of the Fourth International Workshop on Mining Software Repositories, p. 29. IEEE Computer Society (2007)
13. Rees-Jones, M., Martin, M., Menzies, T.: Better predictors for issue lifetime. arXiv preprint arXiv:1702.07735 (2017)
14. Weiss, C., Premraj, R., Zimmermann, T., Zeller, A.: How long will it take to fix this bug? In: Fourth International Workshop on Mining Software Repositories (MSR 2007: ICSE Workshops 2007), p. 1. IEEE (2007)
15. Yu, C.H., Ohlund, B.: Threats to validity of research design (2010). Accessed 12 Jan 2012
16. Zhang, F., Khomh, F., Zou, Y., Hassan, A.E.: An empirical study on factors impacting bug fixing time. In: 2012 19th Working Conference on Reverse Engineering, pp. 225–234. IEEE (2012)
17. Zhang, H., Gong, L., Versteeg, S.: Predicting bug-fixing time: an empirical study of commercial software projects. In: Proceedings of the 2013 International Conference on Software Engineering, pp. 1042–1051. IEEE Press (2013)

The Design and Implementation of KMeans Based on Unified Batch and Streaming Processing

Hao Chen[1,2], Kun Guo[1,2,3], Yuzhong Chen[1,2,3]($\boxtimes$), and Hong Guo[1]

[1] College of Mathematics and Computer Science, Fuzhou University, Fuzhou 350108, China
`{gukn,yzchen,guohong}@fzu.edu.cn`
[2] Fujian Provincial Key Laboratory of Network Computing and Intelligent Information Processing, Fuzhou, China
[3] Key Laboratory of Spatial Data Mining and Information Sharing, Ministry of Education, Fuzhou 350108, China

Abstract. Clustering algorithms aim at gathering similar data points from a dataset in an unsupervised manner. Although the batch clustering algorithms have relatively high accuracy, they cannot make use of the dynamic clustering results efficiently. The requirement of using the whole dataset in calculation results in the problems of resource waste and high time cost. On the contrary, incremental clustering only needs to update the varied part of a model upon the arrival of new data, which makes it unnecessary to recluster the whole data all the time. The feature is very suitable for the streaming data process, but it decreases the accuracy of the algorithms and cannot satisfy the low latency requirement of real-time data processing. In response to this problem, the paper proposes a novel unified batch and streaming clustering model (UBSCM) based on streaming computation, which includes a streaming cluster feature updating mechanism (SCFUM). The Flink framework is used to implement a new streaming KMeans algorithm based on UBSCM (KMeansUBSP). The experiments on the real-world datasets validate that the new streaming KMeans algorithm is effective in clustering the batch and streaming data in a unified manner.

Keywords: Unified batch and streaming · KMeans · Streaming cluster update

1 Introduction

With the development of electronic information technology and the popularization of the mobile Internet of Things. Real-time data processing becomes an important issue. As an unsupervised learning algorithm, which is widely used in clustering similar data to mine the hidden value of data.

According to different processing scenarios, clustering algorithms can be roughly divided into three categories, including batch clustering algorithms, incremental clustering algorithms, and streaming clustering algorithms [1–3]. The batch clustering algorithm mainly solves batch data clustering. However, unable to get all the data at once due

© Springer Nature Singapore Pte Ltd. 2021
Y. Sun et al. (Eds.): ChineseCSCW 2020, CCIS 1330, pp. 639–649, 2021.
https://doi.org/10.1007/978-981-16-2540-4_47

to the limitation of memory size and disk space. Thus, incremental clustering appears. The incremental clustering algorithm can update the later data incrementally, which solves the defect that the traditional clustering algorithm requires full computation. With the development of digital informatization, timeliness has become a very important label for big data. In this way, many streaming clustering algorithms have emerged to adapt to the data that is continuously generated in real-time. Typical representatives include CluStream [4] based on hierarchy and DenStream [5] based on grid density. However, the streaming clustering model cannot support the cluster analysis of batch data effectively. To meet the increasing demand for stream data processing, a large number of computing engines have emerged [9]. The implementation of clustering algorithms includes mainstream parallel computing frameworks based on Apache Hadoop [6], Apache Spark [7], Apache Flink [8], etc. Among them, Hadoop is suitable for a large amount of static data processing, but its frequent IO read and write cannot satisfy low-latency streaming data processing. Spark pioneered memory computing by storing data in memory for iterative computation. Spark Streaming is a discrete stream model that uses micro-batch processing, not real streaming. Flink is a more novel streaming computing engine that provides features such as stateful streaming processing, event time semantics, exactly one-time processing, and backpressure control. In terms of the design principle, Flink is more suitable for lower processing delay and more complex streaming scenarios than Spark.

Because of the different application scenarios of batch data clustering and streaming data clustering, the Flink streaming computing engine is used in this article to treat batch data as bounded streaming data processing. Combining batch clustering and streaming clustering algorithms, the realization mechanism and scheme of batch-stream unified KMeans are proposed. The main contributions of this paper are as follows:

(1) Propose a unified batch and stream clustering model based on stream computing. The cluster center is initialized in the first time window, and the cluster is continuously updated in subsequent windows. The algorithm satisfies the high precision of batch clustering and the low latency of streaming clustering.
(2) Propose a cluster feature update mechanism based on streaming iteration. Create cluster features as an iterative stream for the update. Using this mechanism can realize a streamlined iterative clustering algorithm.
(3) Implement the KMeans based on unified batch and stream processing using Flink. The unified batch and stream clustering model and the cluster feature update mechanism based are used to implement the batch-stream unified KMeans algorithm. It is verified through experiments that the algorithm meets the requirements of batch data and streaming data, and also has better performance in accuracy and time.

2   Related Work

Related work mainly aims at improving the design principles and implementation schemes of clustering algorithms.

In terms of clustering design, the earliest traditional KMeans clustering algorithm was proposed by Macqueen in 1967 [10]. KMeans clustering algorithm's idea is simple and widely used. However, the obvious disadvantage of this algorithm is that its clustering accuracy depends on the selection of the initialization center. Davi et al. Proposed kmeans++ algorithm [11]. This method uses the point as far away as possible in the data set as the cluster center for initialization, which aims at solving the problem of difficulty in initialization. Unfortunately, the aforementioned clustering algorithm cannot dynamically add data. Because of this defect, Pham et al. proposed incremental based on objective optimization function K-means incremental clustering algorithm [12]. In reference [7], an improved initialization method of incremental KMeans is proposed. However, incremental KMeans can't support streaming data generated by high speed and low delay. In reference [4], a two-stage data stream clustering CluStream is proposed. Although the algorithm can meet the requirements of stream clustering, it cannot find outliers. Reference [13] proposed a new data stream clustering algorithm DenStream. Introduce potential microclusters and outlier microclusters to optimize the shortcomings of CluStream that cannot find outliers. However, these two stream clustering methods cannot meet the requirements of high precision in the case of batch data.

In the implementation of parallel clustering and streaming clustering algorithms, mahout [14], as a machine learning algorithm library based on MapReduce [15] in the Hadoop computing framework, encapsulates a large number of data mining algorithms, including clustering algorithms such as KMeans, canopy [16], fuzzy KMeans [17], etc. However, the Hadoop computing framework will produce a large number of disk IO read and write operations in the calculation. So it cannot support the implementation of streaming clustering. In order to speed up the calculation, the Spark computing framework saves the data in memory for calculation. Spark ML algorithm library is a perfect machine learning algorithm library in the ecosystem. The clustering algorithm includes traditional KMeans, Gaussian mixture clustering GMM, and label propagation LDA [18]. Streaming KMeans [19] algorithm is implemented by Spark streaming. Flink regards batch data as a special case of streaming, providing more natural streaming. Alibaba proposed the machine learning algorithm platform Alink [20], which supplemented the deficiencies of Flink in machine learning algorithms. However, it does not implement a streaming generation model algorithm. The unified clustering model proposed in this paper combines batch clustering rules with flow clustering rules. A unified batch and stream KMeans clustering algorithm is implemented to solve different clustering scenarios.

3 Preliminary

3.1 MapReduce Computing Model

The MapReduce model and its open-source implementation of Hadoop have been widely adopted by industry and academia, attributable to a simple and powerful programming model, which hides the complexity of parallel task execution and fault tolerance from users [15]. The computing framework is a parallel computing framework for large-scale data, which is widely used with low cost, high reliability, and high scalability. Among them, the Mahout machine learning algorithm library is based on the Hadoop MapReduce computing framework. The core idea of this framework is to divide and conquer, using Map and Reduce to process data. In a nutshell, the data is divided into numerous small chunks, which are then processed by the Map function, then aggregated by using the Reduce function. Unfortunately, the intermediate results generated by the calculation need to be stored in the HDFS distributed file system of Hadoop, which seriously affects the calculation performance.

3.2 DAG Model

Spark and Flink introduced the Directed Acyclic Graph (DAG) model based on the MapReduce computing model. At a higher level, both engines implement a driver that describes the application's high-level control flow, which relies on two main parallel programming abstractions [21]. The DAG model provides support for streaming data processing. Spark caches intermediate data results and high-frequency usage in memory, storing data in a format called resilient distributed dataset (RDD). This mode is very suitable for a large number of iterative calculations in machine learning. Spark Streaming uses "micro-batch" to process streaming data, unifying the streaming in batch processing. Flink engine has attracted much attention in the field of stream computing in recent years. It regards batch data as a special case of streaming data. What's more, it provides exactly one-time semantics, stateful computing, streaming iteration, and three temporal semantics to satisfy various streaming application scenarios. These advanced streaming features provide higher fault tolerance and lower latency.

4 Unified Batch and Streaming KMeans Algorithm

4.1 Unified Batch and Stream Clustering Model

Different from other clustering algorithms, USCM supports both batch data and stream data clustering. Read multi-source data from the file system, message queue, network, etc. using the Flink streaming computing framework. Then use the window function for clustering update. The processing structure of the unified batch and stream clustering model is shown in Fig. 1.

In the model, you can set the time window size and the number of initialization iterations. You can also increase the number of initialization iterations through parameters, resulting in higher accuracy. The Cluster Center is initialized for the first window data. When the subsequent window data arrives, the algorithm broadcasts the iterative stream of cluster features to all the nodes in the Keyby() partition and updates them. To guarantee the consistency of data processing, the cluster characteristics are saved as states. Finally, the updated cluster characteristics are sent back to the iterative stream. If the time window is large enough, the first window can hold all the data. In this way, both batch and stream data clustering can be satisfied.

4.2 Streaming Cluster Feature Update Mechanism

SCFUM designs the cluster feature into an iterative stream. The updated cluster features are returned to the iterative stream to rebroadcast. For the partition-based clustering

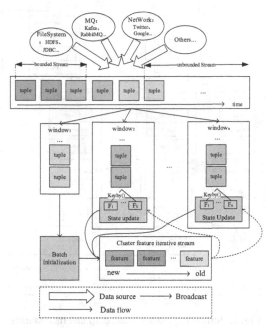

Fig. 1. The processing structure of unified batch and stream clustering model

algorithm, the cluster feature is the cluster center. The update rule of the streaming KMeans in the window is shown in the following formula.

$$c_{t+1} = \frac{(c_t \times n_t \times a) + (x_t \times m_t)}{n_t + m_t} \tag{1}$$

$$n_{t+1} = n_t \times a + m_t \tag{2}$$

Where c is the cluster center; n is the number of points corresponding to the cluster center; a is the decay factor used to define the influence of past points in the new cluster; x is the cluster center generated by the window data point; m is the number of points corresponding to the window data generated cluster center. The specific update process of the streaming cluster feature update mechanism is shown in Fig. 2.

As shown in Fig. 2, first of all, the cluster feature update mechanism reads incoming streaming data and partitions the data randomly. In seconds, the broadcast iterated stream of cluster features is read in the partition, and the data points are divided according to the acquired cluster features. Then divide the different partitions and use formula (1) and formula (2) to recalculate each partition to obtain new cluster features. Finally, the new cluster feature is passed back to the cluster feature stream.

4.3 The KMeans Based on Unified Batch and Stream Processing

The implementation of the KMeansUBSP algorithm uses UBSCM and SCFUM. The overall flow of the algorithm is as follows. First of all, the program will continuously

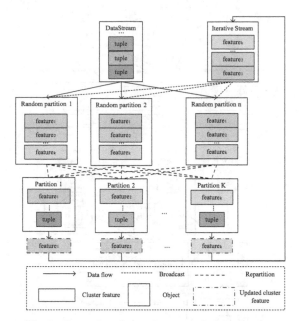

Fig. 2. Streaming cluster feature update mechanism

read multi-source data provided by Flink source, including the file system, distributed file system, Kafka, Socket, etc. In seconds, the read data is Data pre-processing in real-time, and the preprocessing includes filtering outliers and missing values. A time window partition for scrolling by the given window size. Then call the flow clustering method to update the cluster center. Finally, the continuous output produces the result of cluster partition and output. The implementation process of the streaming clustering method is as follows. First, read the window data from the outside. Determine whether it is the first window. If it is, use KMeans++ to initialize the cluster center for the window data. Then update the cluster iteratively. Finally, the cluster center is created as an iterative stream for broadcasting. If not, then directly read the iterative stream of the previous window update and broadcast it. The window data and the cluster center are merged, and the data points are divided into different partitions according to the nearest cluster. Then divide the partition into a window, and update the cluster in the partition window according to formulas (1) and (2). Finally, the updated clusters are passed back to the stream and the clustering results are returned. The specific implementation of the KMeansUBSP is shown in Algorithm 1.

Algorithm 1. KMeansUBSP
Input: The point DataStream ds, the number of cluster centers k, window size winS, the number of initial iterations iterN, decay factor decay.
Output: The point division result rResult.
1.   ds ← ds.filter(null);
2.   dataWin ← ds.createWindow(winS); // Partition window.
3.   isFirst ← TRUE;
4.   **IF**(isFirst)**THEN** // Determine if it is the first window.
5.   isFirst ← FALSE;
6.   centers ← InitCenter(dataWin,k,winS,iterN); // Use KMeans++ to initialize the cluster center.
7.   cIteStream ← centers.iterate(); // Create a central iteration stream.
8.   **END IF**
9.   **WHILE** (dataWin.hasData) **DO** // Window has data.
10.   conStream ← ds.connect(cIteStream.broadcast()); // Use the Flink operator to connect the two streams and broadcast the cluster center to all parallel nodes.
11.   pResult ← FindClosest(dataWin,conStream); // Find the nearest cluster center based on Euclidean distance.
12.   cUpdate ← UpdateCenter(pResult,decay); // Update the cluster center according to formulas (1) and (2).
13.   cIteStream ← cUpdat; // Pass the updated cluster back to the iteration stream.
14.   **END WHILE**
15.   return pResult;

5   Experiments

5.1   Experimental Environment

In the experiment, a total of 5 physical machine servers were used to build Spark and Flink clusters. Each server has a total CPU core number of 20, total memory of 256G, and a hard disk capacity of $2 \times 300G$.

5.2   Experimental Datasets

The real datasets of UCI [22] are used to verify the method proposed in this paper. Some information about the data set is given in Table 1 below.

Table 1. UCI datasets

Data sets	Clusters	Dimension	Size
Seeds	3	7	210
Wine	3	13	178
HTRU2	2	9	17898
Occupancy detection	2	7	20560

5.3   Evaluation Metrics

A widely used evaluation index "Macro F1" [23] is selected to verify the performance of the KMeansUBSP algorithm. The macro_F1 is the weighted harmonic average of the macro precision rate (macro_P) and macro recall rate (macro_R), which is often used to evaluate the quality of the model. The formula is as follows:

$$macro\_P = \frac{1}{n} \sum_{i=1}^{n} Pi \tag{3}$$

$$macro\_R = \frac{1}{n} \sum_{i=1}^{n} Ri \tag{4}$$

$$macro\_F1 = \frac{2 \times macro\_P \times macro\_R}{macro\_P + macro\_R} \tag{5}$$

5.4   Precision Experiment

Batch Precision Experiment. In this experiment, batch datasets of Seeds, Wine, and HTRU2 were transferred to Spark Streaming-KMeans, Spark KMeans, Alink KMeans, and KMeansUBSP for calculation and comparison of macro_F1 values.

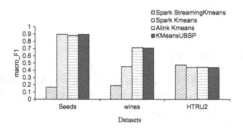

Fig. 3. Batch accuracy results

As shown in Fig. 3, Spark StreamingKMeans has a low macro value, around 0.2 in Seeds and wines, because it only provides random initialization and specifies cluster center initialization, and does not support iterative computation. For HTRU2 data, the high value of the macro is due to the prediction error caused by the big difference between positive and negative examples, and the defect that KMeans Algorithm cannot detect outliers effectively. The KMeansUBSP algorithm proposed in this paper introduces the distance probability initialization of KMeans ++ to the first window and provides iterative calculation so that the value of macro_F1 in Seeds and wines is higher than the Spark Streaming algorithm by about 0.6 and 0.4. Spark KMeans uses the default random initialization method but provides iterative calculations, which is lower than the KMeansUBSP algorithm by about 0.15 in wines. Alink KMeans uses the initialization of KMeansll by default, which is equivalent to the accuracy of KMeansUBSP. In summary, the KMeansUBSP algorithm has higher accuracy results for batch data clustering.

Streaming Precision Experiment. This experiment uses the 7-day data from February 4, 2015, to February 10, 2015, in the time series data set Occupancy Detection, and the incremental experiment is performed according to data blocks divided into days. This experiment uses Spark Stream-ingKMeans and KMeansUBSP to compare real-time macro_F1 values.

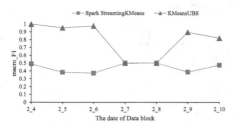 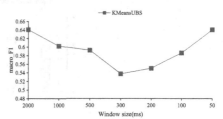

Fig. 4. Streaming accuracy results   **Fig. 5.** Time relationship results

As shown in Fig. 4, in a data block clustering on February 4th, the value of macro_F1 of the KMeansUBSP algorithm is close to 1, while the value of Spark StreamKMeans is only close to 0.5. This is because of the different initialization methods and iterative updates of the two algorithms. In the second and third windows, the macro_F1 value of the KMeansUBSP algorithm is higher than Spark Stream-ingKMeans by about 0.6. This is because the KMeansUBSP algorithm saves the previous clustering results through the Flink state mechanism and extracts them for use in the next clustering. In the data blocks on February 7th and February 8th, the macro_F1 of the two algorithms is the same and only 0.5. This is because the data is too scattered and the KMeans algorithm cannot identify outliers. In the clustering of data blocks on February 9, the macro_F1 value of the KMeansUBSP algorithm began to rise again, exceeding 0.5 of the StreamingKMeans algorithm. This is because KMeansUBSP saves the updated cluster center and can continue to be used in the future to produce higher accuracy. We set windows of different sizes to read the Occupancy Detection data set in Kafka for clustering, for exploring the relationship between the macro_F1 score and the window size of the KMeansUSBP algorithm. Use the macro_F1 score to score the results. The result is shown in Fig. 5. As the window decreases, the value of macro_F1 first decreases and then increases. This is because when the window is greater than 300ms, the accuracy is mainly affected by the amount of data. When it is less than 300ms, the accuracy is mainly affected by the number of updates.

5.5 Running Time Experiments

This experiment uses data from the time series datasets Occupancy Detection for 5 days from February 5, 2015, to February 9, 2015. Kafka is used to continuously send datasets to the corresponding topics, and the algorithm continues to consume the corresponding topics until all the data is processed.

As shown in Fig. 6, when the window size is less than 500 ms, the KMeansUBSP algorithm proposed in this paper is faster than Spark StreamingKMeans. The execution time of the program is 1 s and 2 s faster in the window size of 300 ms and 200 ms, respectively. This is because KMeansUBSP uses the Flink streaming engine, which has a smaller granular delay. When the window is greater than 500 ms and reaches 2000 ms, the processing time of the Spark StreamingKMeans algorithm is 2 s faster than the KMeansUBSP algorithm. This is because KMeansUBSP iterates on the initial clusters and the delay increases as the data in the time window increases.

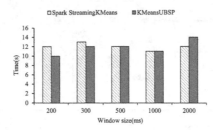

Fig. 6. Streaming runtime results

6 Conclusions

This paper proposes a batch stream unified clustering model based on stream processing, the model uses the cluster feature update mechanism based on the iterative stream. Considering the difference between batch clustering and streaming clustering, the cluster features are created into iterative stream processing, and the cluster features stored in the state are constantly updated. Then transfer the updated clusters to the stream, which can effectively handle both batch and stream clustering scenarios. Finally, achieve the effect of a unified batch and stream. The KMeansUBSP implemented in this paper belongs to a partition-based clustering algorithm. To make the proposed unified batch and stream model more general. Next step, the UBSCM and SCFUM proposed in this article will be combined to implement other types of clustering algorithms, such as density-based, hierarchical-based, grid-based, etc.

Acknowledgements. This work is partly supported by the National Natural Science Foundation of China under Grant No. 61672159. No. 61672158, No. 62002063 and No. 61300104, the Fujian Collaborative Innovation Center for Big Data Applications in Governments, the Fujian Industry-Academy Cooperation Project under Grant No. 2017H6008 and No. 2018H6010, the Natural Science Foundation of Fujian Province under Grant No. 2019J01835 and No. 2020J01230054, and Haixi Government Big Data Application Cooperative Innovation Center.

References

1. Saxena, A., Prasad, M., Gupta, A., et al.: A review of clustering techniques and developments. Neurocomputing **267**, 664–681 (2017)
2. Nguyen, H.-L., Woon, Y.-K., Ng, W.-K.: A survey on data stream clustering and classification. Knowl. Inf. Syst. **45**(3), 535–569 (2014). https://doi.org/10.1007/s10115-014-0808-1

3. Young, S., Arel, I., Karnowski, T.P., Rose, D.: A fast and stable incremental clustering algorithm. In: 2010 Seventh International Conference on Information Technology: New Generations, pp. 204–209 (2010)
4. Aggarwal, C.C., Philip, S.Y., Han, J., Wang, J.: A framework for clustering evolving data streams. In: Proceedings 2003 VLDB Conference, pp. 81–92 (2003)
5. Cao, F., Estert, M., Qian, W., Zhou, A.: Density-based clustering over an evolving data stream with noise. In: Proceedings of the 2006 SIAM International Conference on Data Mining, pp. 328–339 (2006)
6. Apache Hadoop. https://hadoop.apache.org/. Accessed 04 June 2020
7. Apache Spark. https://Spark.apache.org/. Accessed 05 June 2020
8. Apache Flink. https://Flink.apache.org/. Accessed 09 June 2020
9. Chintapalli, S., Dagit, D., Evans, B., et al.: Benchmarking streaming computation engines: storm, Flink and spark streaming. In: 2016 IEEE International Parallel and Distributed Processing Symposium Workshops, pp. 1789–1792 (2016)
10. MacQueen, J.: Some methods for classification and analysis of multivariate observations. In: Proceedings of the Fifth Berkeley Symposium on Mathematical Statistics and Probability, vol. 1, no. 14, pp. 281–297 (1967)
11. Bagirov, A.M., Ugon, J., Webb, D.: Fast modified global k-means algorithm for incremental cluster construction. Pattern Recogn. **44**(4), 866–876 (2011)
12. Pham, D.T., Dimov, S.S., Nguyen, C.D.: An incremental K-means algorithm. Proc. Inst. Mech. Eng. J. Mech. Eng. Sci. **218**(7), 783–795 (2004)
13. Cao, F., Estert, M., Qian, W., et al.: Density-based clustering over an evolving data stream with noise. In: Proceedings of the 2006 SIAM International Conference on Data Mining, pp. 328–339 (2006)
14. Apache Mahout. https://mahout.apache.org. Accessed 05 June 2020
15. Dean, J., Ghemawat, S.: MapReduce: Simplified data processing on large clusters. Commun. ACM **51**(1), 107–113 (2008)
16. McCallum, A., Nigam, K., Ungar, L.H.: Efficient clustering of high-dimensional data sets with application to reference matching. In: Proceedings of the Sixth ACM SIGKDD International Conference on Knowledge Discovery and Data Mining, pp. 169–178 (2000)
17. Bezdek, J.C.: Pattern recognition with fuzzy objective function algorithms. Springer, Boston (2013). https://doi.org/10.1007/978-1-4757-0450-1
18. Wei, X., Croft, W.B.: LDA-based document models for ad-hoc retrieval. In: Proceedings of the 29th Annual International ACM SIGIR Conference on Research and Development in Information Retrieval, pp. 178–185 (2006)
19. Spark StreamingKMeans. https://Spark.apache.org/docs/latest/ml-guide.html. Accessed 08 June 2020
20. Alibaba Alink. https://github.com/alibaba/Alink. Accessed 08 June 2020
21. Marcu, O.C., Costan, A., Antoniu, G., Pérez-Hernández, M.S.: Spark versus Flink: understanding performance in big data analytics frameworks. In: 2016 IEEE International Conference on Cluster Computing, pp. 433–442 (2016)
22. UCI Machine Learning. https://archive.ics.uci.edu/ml/index.php. Accessed 07 May 2020
23. Zhou, Z.-H.: Machine Learning, 2nd edn. Tsinghua University Press, Beijing (2016)

Image Super-Resolution Using Deformable Convolutional Network

Chang Li, Lunke Fei$^{(\boxtimes)}$, Jianyang Qin, Dongning Liu, Shaohua Teng, and Wei Zhang

School of Computer Science and Technology, Guangdong University of Technology,
Guangzhou, China
{liudn,shteng,weizhang}@gdut.edu.cn

Abstract. Social media network is inseparable from image recognition, and image super-resolution (SR) reconstruction plays an important role in image recognition. The changes of scale and geometry are rarely considered in the image super-resolution reconstruction based on deep learning over the years, we introduce a super-resolution reconstruction network based on deformable convolutional network. We replace the ordinary convolution with the deformable convolution to pretend the geometric deformation and extract abundant local features. The image super-resolution reconstruction is usually based on the conventional convolutional neural network (CNN). Most CNN-based SR models do not utilize the features of the original low resolution (LR) image as much as possible, resulting in lower performance. After introducing the idea of deformable convolution, though the complexity is increased, the recognition accuracy is obviously raised.

Keywords: Deformable convolution · Super resolution · Residual learning

1 Introduction

Single image Super-Resolution (SISR) [5] is commonly used for some mapping relationship between LR and SR. SISR has been widely and frequently studied in the field of computer vision. From surveillance images to medical images, the details of the images required are increasing. Conventional SISR research methods contain interpolation-based and reconstruction-based [7]. Sparse-representations and bicubic interpolation are familiar interpolation methods [1, 9, 14, 15].

SRCNN [3, 4] is the first algorithm that applies deep learning to solve SISR problem, and completes image reconstruction with three ordinary convolutional layers. Due to the slow convergence speed of SRCNN during training, FSRCNN [2] improved on the basis of SRCNN and achieved better reconstruction performance by changing the number and dimension of convolutional layers. VDSR [11] uses residual learning to perform image reconstruction and adds a skip connection to the excessively deep network, which increases the convergence speed of the network. DRCN [18] brings recursion into the study of images and sets a skip connection for each convolutional layer, which deepens the structure of the network and increases the receptive field of the network, thus improving the performance. In the method like SRCNN, the LR image needs to be input

© Springer Nature Singapore Pte Ltd. 2021
Y. Sun et al. (Eds.): ChineseCSCW 2020, CCIS 1330, pp. 650–660, 2021.
https://doi.org/10.1007/978-981-16-2540-4_48

into the network by up-sampling interpolation to obtain the same size as the HR image, which means that convolution operation needs to be carried out at a higher resolution, resulting in the increase of computational complexity. ESPCN [12] proposed an efficient method to extract features directly from the LR image size and calculate the HR image. ResNet is local residual learning for chain mode. VDSR is global residual learning. DRCN is global residual learning + single weight recursive learning + multi-objective optimization. DRRN [6] is the local residuals learning + global residuals learning + recursive learning with multiple weights.

We all know that conventional convolution, no matter how you add it up, the basic results are rectangular, but there are many limitations in modeling irregular objects in real life. Therefore, we hope to find a new convolution method that can overcome the current limitation and apply it to SISR research. Deformable convolution [10] is different from ordinary convolutional layer, and we consider migrating it to SISR and design a network model based on deformable convolution. The contribution of the new network is as follows:

Strong ability to model the deformation and scale of objects. Deformable convolution adopts to the scale of the object. The receptive field is much larger than the general convolution because of the deviation.

2 Related Works

Super-Resolution Convolutional Neural Network (SRCNN) [3] which is produced by Dong et al. first attempted to establish a link between deep learning and image super-resolution reconstruction. The proposed idea aimed at associating LR image to HR image in an end-to-end mapping. It is significantly better than the conventional non-DL method. Along with taking a deconvolution layer to upscale, Fast Super-Resolution Convolutional Neural Networks (FSRCNN) [2] changes characteristic dimension and shares the mapping layers, thus the original LR image can be directly input. Accurate Image Super-Resolution Using Very Deep Convolutional Networks (VDSR) [11] stacks more convolutional layers to deepen the network. The accompanying vanishing/exploding gradients problems can be eased with residual learning. Efficient Sub-Pixel Convolutional Neural Network (ESPCN) [12] proposed by Shi et al. introduces a sub-pixel convolution layer. For LR image, only one up-sampling operation is performed at the end of the network. It also reduces computational complexity when you get HR image.

The super-resolution reconstruction methods mentioned above are all based on standard convolution operation. In essence, CNN does not have rotation invariance or scale invariance, CNN training requires training samples to contain enough geometric changes for learning and fitting. However, there is a lack of strategic basis for judging deformation for standard convolution. Assuming that the direction of geometric transformation is fixed and known, we can use prior knowledge to expand data and design algorithm. However, when this assumption is not established, the learning will lose the generalization ability, making it difficult to continue. Feature maps are all computed with the same convolution even if the positions of the same layer's feature map may correspond to the objects with different shapes. Feature map is obtained by the same convolution kernel, so it is hard to deal with multiple deformations simultaneously [13].

The conventional methods are mainly performed by scaling, rotating and flipping to get better learning performance. However, the operation will expand the number of samples. Correspondingly, the computing time of the network will be longer [14, 17].

These limitations come from the fixed geometry of CNN module. To improve the ability of model geometric transformation, Dai et al. proposed the idea of deformable convolution. Different from the convolutional operation in the traditional dot product operation with kernel translation on the image, deformable convolutional layer improves the processing capacity for different changes of geometric images without increasing the number of training samples. It adds a 2D offset to the regular grid sampling position in standard convolution. The offset is learned by the additional convolutional layer from the preceding feature diagram. The size of the DCN kernels and position can be adjusted according to the current need to identify the dynamic image content, we can see it visually that different location of Convolution kernels will sampling point location based on image content adaptive changes, so as to accommodate different shape and size of objects such as geometric deformation. As illustrated in Fig. 1. To sum up what has been mentioned above, we propose Image Super-Resolution using Deformable Convolutional Network (SRDCN). We will detail our SRDCN in next section.

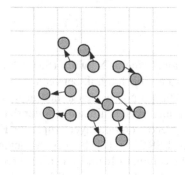

Fig. 1. Image of sampling positions in 3 × 3 conventional and deformable convolution.

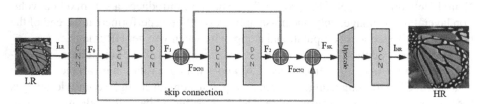

Fig. 2. Our proposed network architecture.

3   Proposed Network

3.1   The Structure of SRDCN+

Figure 2 shows our proposed SRDCN+, which mainly contains the following structures. feature extraction layer, double deformable convolution layers, feature fusion layer and up-sampling layer. The details are as follows. For the first standard convolutional layer operating on LR, we set it as a filter size 5×5 for appropriate receptive field. The kernel sizes of the deformable convolutional layers are all set at 3×3. At the beginning of the network, LR image is taken as input, and corresponding features were extracted through a standard convolution operation. The selection of convolution kernel should increase the receptive field as much as possible. We will define I_{LR} to represent the input, from this we can get

$$F_0 = G_{CNN}(I_{LR}) \tag{1}$$

where G_{CNN} indicates convolution operation. F_0 is the output after the convolution of the first layer and serves as the input to the next layer and skip connection. The obtained feature map generates F_1 after two deformable convolutional layers.

$$F_1 = G_{D1}(F_0) \tag{2}$$

where G_{D1} denotes the calculation with the same deformable convolutional kernel twice. F_0 is fused with F_1 using skip connection, can be obtained by

$$F_{DCN1} = F_0 + F_1 \tag{3}$$

where $+$ denotes fusion operation. Similar to what we did before, we can figure out

$$F_2 = G_{D2}(F_{DCN1}) \tag{4}$$

$$F_{DCN2} = F_{DCN1} + F_2 \tag{5}$$

$$F_{SK} = F_{DCN2} + F_0 \tag{6}$$

where G_{D2} denotes the convolution operation of the second double deformable convolutional layers. The reconstructed part of the network consists of a sub-pixel convolutional layer and a deformable convolutional layer. The features extracted from the previous network are finally reconstructed. The final output I_{SR} is

$$I_{SR} = G_{REC}(F_{SK}) \tag{7}$$

where G_{REC} denotes the computation part of the last reconstruction.

Deformable convolutional layer works as follows. We all know that the standard convolutional network for convolutional kernel migration is sliding in order on $\Re$.

$$\Re = \{(-1, -1), (-1, 0), \ldots, (0, 1), (1, 1)\} \tag{8}$$

where $\mathfrak{R}$ denotes the size and expansion of the receptive field, here we define $\mathfrak{R}$ as a 3×3 convolutional kernel. For each position l_0 on the output feature map y, we have

$$y(l_0) = \sum_{l_i \in \mathfrak{R}} w(l_i) \cdot x(l_0 + l_i) \qquad (9)$$

where l_0 denotes the central point, l_i ($i = 1, 2, \cdots, 9$) represents each position in $\mathfrak{R}$. As the offset progresses, convolutional computation is completed on the regular grid. And deformable convolution is adding an offset to $x(l_0 + l_i)$, the Eq. (9) is converted to

$$y(l_0) = \sum_{l_i \in \mathfrak{R}} w(l_i) \cdot x(l_0 + l_i + \Delta l) \qquad (10)$$

where Δl denotes an additional offset. However, Δl is usually expressed as a decimal. In order to facilitate the computation, processing it into an integer form by bilinear interpolation. The result calculated by Eq. (10) is the pixel value of the image.

Deformable convolutional layer produces an offset field of 2N at each position of the feature map rather than the convolution kernel. N denotes the number of channels which is the same as the feature map. Then the offset is corresponding to each position of the original feature map, and deformable convolution can be operated on the original feature map [10].

3.2 SRCNN with DCN

SRCNN mainly contains three layers of network structure, namely Patch extraction and representation, Non-linear mapping and Reconstruction [4]. The process of SRCNN is as follows:

Firstly, the LR image is expanded to the specified size computed by bicubic interpolation (for example, to 2, 3 and 4 times). After being up-sampled to the target size, the output is the next input to the next layer of the network.

LR images were input into a three-layer convolutional neural network, in one of the experimental Settings, Y channel in the YCrCb color space was reconstructed in the form of (conv1 + relu1) - (conv2 + Relu2) - (conv3)). The first layer convolution: convolution kernel size 9×9(F1 $\times$ F1), number of convolution kernels 64(N1), and output 64 feature graphs; The second layer of convolution: size of convolution kernel 1 $\times$ 1(F2 $\times$ F2), number of convolution kernel 32(N2), output 32 feature diagrams; The third layer of convolution: kernel size is set as 5×5(F3 $\times$ F3), the number of kernel is 1(N3), and the result of one feature map is the last reconstruction of high-resolution image [3, 4].

We exchange the standard convolutional layer of SRCNN's extraction layer into a deformable convolutional layer. When extracting features from the first-layer network, the batch size can be set lower because of the offset, and the convolution kernel size remains unchanged. The network increases the processing capacity of different geometric image changes, so the number of training samples does not need to be enlarged, and the training time will be longer.

4 Experiment

In this section, we will investigate whether deformable convolution performs better than traditional convolution operation in image super-resolution reconstruction. First of all, the comparison experiment between SRCNN under the SRDCN method proposed in this paper and the original SRCNN method is conducted to prove that deformable convolution can improve the image reconstruction effect.

Secondly, we study the effect of residual-learning network on reconstruction under deformable convolution structure.

Thirdly, the SRDCN method proposed in this paper is compared with Bicubic, A+ [8], SRCNN, SRCNN+, and others, and the idea of deforming convolution has a better performance in improving the effect of image reconstruction.

4.1 Setting

Training Dataset. 291-images dataset is widely utilized in the field of SR reconstruction. To ensure consistency with the previous experimental conditions, 291 images which contains 91 images provided by Yang et al. and 200 images from Berkeley Segmentation Dataset are taken as training data. Beyond that, the expansion of dataset is used by rotating the training set by 90°, 180° and 270°. The enlargement of training dataset leads to more accurate results.

Testing Dataset. For benchmark, we follow SRCNN. Following SRCNN and SRDCN, we use the Set5, Set14, and BSD100 dataset for testing. PSNR [16] and SSIM are used to evaluate the SR results on the Y channel of the transformed YCbCr space.

The convolutional kernel size of the first layer is 5×5, and the kernel size of other deformable convolutional layers are set as 3×3. Batch size is commonly set as 32, the total number of epoch is 30, MSE is used as the loss function. The learning rate is set at 0.001, and it is cut down by 10% after per 10 generations. The training takes roughly three days on a GTX 1060 6G GPU [19, 20].

4.2 SRDCN and SRDCN +

To illustrate whether the presence of deformable convolution layer influence on image reconstruction results, we replaced all the deformable convolutional layers in the network with the standard convolutional layers. Name the network containing deformable convolution as SRDCN+ and the network composed of standard convolutional layers as SRDCN. Then compared the two algorithms on Set5 dataset for super-resolution reconstruction. As can be seen from Fig. 3, when the network based on standard convolution and the network consists of deformable convolution layers are in the case of amplification order $\times 2$, $\times 3$ and $\times 4$, the PSNR change diagram of reconstructed images is shown. The effect of the method based on deformable convolutional layer is better than that based on standard convolutional layer. We can know that deformable convolution plays a certain role in image super resolution reconstruction.

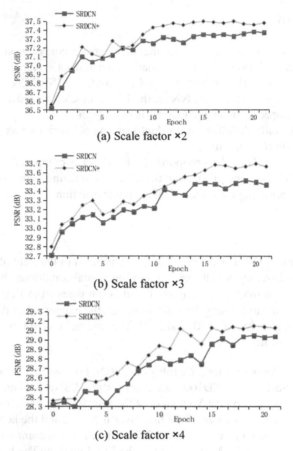

Fig. 3. Convergence performance curve of SRDCN and SRDCN+ with different scales.

4.3 SRDCN-S and SRDCN+

Remove all skip connections in the SRDCN+ network, and the conditions of other layers are consistent with SRDCN+. After training, the two networks were compared on Set5. As shown in Fig. 4, under the up-scaling scale factor ×4, the performance of the deformable convolutional network based on residual learning is significantly better than that the one without residual learning. The result shows that the network using residual learning has a significantly faster convergence rate, which indicates the residual learning still takes an important position in the deformable convolutional network.

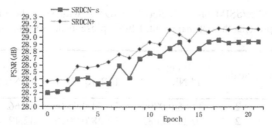

Fig. 4. Convergence performance curve of SRDCN-s and SRDCN+ with up-scaling scale ×4. which SRDCN-s denotes a network without residual learning.

4.4 SRCNN with DCN

The standard convolution of feature extraction layer is replaced by deformable convolution. In order to explain the role of the deformable convolutional layer, we replace the standard convolution of the feature extraction layer in SRCNN with the deformable convolutional layer and name it SRCNN+. The original algorithm was compared with the new algorithm in Set5 data set for super-resolution reconstruction. As shown in Fig. 5, when the feature extraction network based on standard convolution and the feature extraction network based on deformable convolution are in the case of amplification order ×4, The variation trend of PSNR from the reconstructed images is shown in Fig. 5. With the increase of the epoch, the value of PSNR increased with it and finally tended to be stable. At the same time, the effect of the method based on deformable convolutional layer is better than that based on standard convolutional layer. We can know that deformable convolution plays a certain role in image super resolution reconstruction.

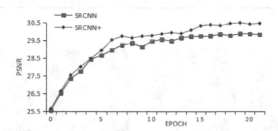

Fig. 5. Convergence performance curve of SRCNN and SRCNN+ with up-scaling scale ×4.

In order to make it more acceptable and clearly see the reconstruction performance of comparing our method with others, the numerical of reconstruction consequences corresponding to other methods is given in Table 1. And Fig. 6 shows the comparison of results between SRDCN+ and other 6 methods when the up-sampling scale is 4.

Table 1. Average PSNR/SSIM on Set5, Set14, BSD100 with scale factor ×2, ×3 and ×4 respectively. Red represents the optimal result. Blue represents the second optimal result.

DataSet	Scale	Bicubic PSNR/SSIM	A+[8] PSNR/SSIM	SRCNN[3] PSNR/SSIM
Set5	2	33.66/0.9299	36.54/0.9544	36.66/0.9542
Set14	2	30.24/0.8682	32.28/0.9056	32.42/0.9063
BSD100	2	29.56/0.8431	31.21/0 8663	31.36/0.8879
Set5	3	30.39/0.8682	32.58/0.9088	32.75/0.9090
Set14	3	27.55/0.7742	29.13/0.8188	29.28/0.8209
BSD100	3	27.21/0.7385	31.21/0.8863	28.41/0.7863
Set5	4	28.42/0.8104	30.28/0.8603	30.48/0.8628
Set14	4	26.00/0.7027	27.32/0.7491	27.49/0.7503
BSD100	4	25.96/0.6675	26.82/0.7087	26.90/0.7101

DataSet	Scale	SRCNN+ PSNR/SSIM	SRDCN PSNR/SSIM	SRDCN+ PSNR/SSIM	VDSR[11] PSNR/SSIM
Set5	2	36.72/0.9533	36.95/0.9548	36.99/0.9521	37.53/0.9587
Set14	2	33.55/0.9035	33.64/0.9125	33.69/0.8945	33.03/0.9124
BSD100	2	31.65/0.8741	31.70/0.8846	31.75/0.8671	31.90/0.8960
Set5	3	33.15/0.9110	33.24/0.9014	32.95/0.9128	33.66/0.9213
Set14	3	29.23/0.8309	29.25/0.8247	29.80/0.8296	29.77/0.8314
BSD100	3	28.22/0.7877	28.36/0.7647	28.40/0.7560	28.82/0.7976
Set5	4	30.64/0.8809	30.86/0.8987	30.44/0.8324	31.35/0.8838
Set14	4	27.56/0.7634	27.63/0.7525	27.69/0.7469	28.01/0.7674
BSD100	4	27.02/0.7169	27.04/0.7156	27.23/0.7167	27.29/0.7251

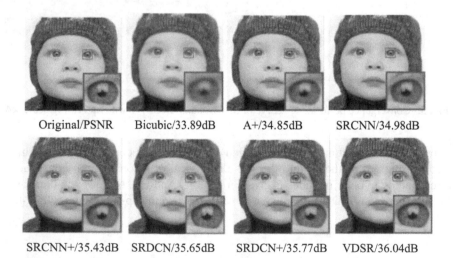

Original/PSNR   Bicubic/33.89dB   A+/34.85dB   SRCNN/34.98dB

SRCNN+/35.43dB   SRDCN/35.65dB   SRDCN+/35.77dB   VDSR/36.04dB

Fig. 6. "Baby" image from Set5 with a scale factor ×4.

5 Conclusion

When conducting research on SISR, a huge amount of images that have undergone complex geometric deformation in reality will be treated. Conventional convolutional operation has certain limitations, while the processing ability of deformable convolution to image geometric transformation is significant. In paper, a deformable convolutional network is proposed for image super-resolution reconstruction. In addition to replacing part of standard convolutions with deformable convolutions, the application of residual learning improves the efficiency of operation. Finally, a sub-pixel convolution layer is added in the network for the final up-sampling. By comparing our experimental results with previous methods, PSNR value and SSIM value are improved to some extent, which proves that deforming convolution plays a certain role in image super-resolution reconstruction. What needs to be studied next is the extra computing overhead generated when the deformable convolution layer stacks into a deeper network, and whether there is a more reasonable combination of traditional convolution and deformable convolution, which is more efficient and effective in network training.

Acknowledgements. This paper was supported in part by the National Natural Science Foundation of China under Grant 61702110, and Grant 61972102, in part by the Natural Science Foundation of Guangdong Province under Grant 2019A1515011811, in part by the Research and Development Program of Guangdong Province under Grant 2020B010166006, in part by the Guangzhou Science and technology plan project under Grant 202002030110.

References

1. Bengio, Y., Goodfellow, I.J., Courville, A.: Deep learning. https://www.deeplearningbook.org
2. Schulter, S., Leistner, C., Bischof, H.: Fast and accurate image upscaling with super-resolution forests. In: IEEE Conference on Computer Vision and Pattern Recognition, pp. 3791–3799. IEEE Press, New York (2015). https://doi.org/10.1109/CVPR.2015.7299003
3. Dong, C., Loy, C., He, K., Tang, X.: Learning a deep convolutional network for image super-resolution. In: Fleet, D., Pajdla, T., Schiele, B., Tuytelaars, T. (eds.) ECCV 2014. LNCS, vol. 8692, pp. 184–199. Springer, Cham (2014). https://doi.org/10.1007/978-3-319-10593-2_13
4. Dong, C., Loy, C.C., He, K.M., Tang, X.O.: Image super-resolution using deep convolutional networks. IEEE Trans. Pattern Anal. Mach. Intell. **38**(2), 295–307 (2015). https://doi.org/10.1109/TPAMI.2015.2439281
5. Huang, J.-B., Singh, A., Ahuja, N.: Single image super resolution from transformed self-exemplars. In: IEEE Conference on Computer Vision and Pattern Recognition, pp. 5197–5206. IEEE Press, New York (2015). https://doi.org/10.1109/CVPR.2015.7299156
6. Tai, Y., Yang, J., Liu, X.: Image super-resolution via deep recursive residual network. In: 30th IEEE/CVF Conference on Computer Vision and Pattern Recognition, pp. 2790–2798. IEEE Press, New York (2017). https://doi.org/10.1109/CVPR.2017.298
7. Timofte, R., De Smet, V., Van Gool, L.: Anchored neighborhood regression for fast example based super-resolution. In: IEEE International Conference on Computer Vision, pp. 1920–1927. IEEE Press, New York (2013). https://doi.org/10.1109/ICCV.2013.241
8. Timofte, R., De Smet, V., Van Gool, L.: A+: Adjusted Anchored Neighborhood Regression for Fast Super-Resolution. In: Cremers, D., Reid, I., Saito, H., Yang, M.-H. (eds.) ACCV 2014. LNCS, vol. 9006, pp. 111–126. Springer, Cham (2015). https://doi.org/10.1007/978-3-319-16817-3_8

9. Yang, J., Wright, J., Huang, T.S., Ma, Y.: Image super-resolution via sparse representation. IEEE Trans. Image Process. 2861–2873 (2010). https://doi.org/10.1109/TIP.2010.2050625

10. Dai, J., Qi, H., Xiong, Y., Li, Y., Zhang, G., Hu, H., Wei, Y.: Deformable convolutional networks. In: IEEE International Conference on Computer Vision, pp. 2380–7504. IEEE Press, New York (2017). https://doi.org/10.1109/ICCV.2017.89

11. Kim, J., Kwon Lee, J., Mu Lee, K.: Accurate image super-resolution using very deep convolutional networks. In: IEEE Conference on Computer Vision and Pattern Recognition, pp. 1063–6919. IEEE Press, New York (2016). https://doi.org/10.1109/CVPR.2016.182

12. Shi, W.Z., et al.: Real-time single image and video super-resolution using an efficient subpixel convolutional neural network. In: IEEE Conference on Computer Vision and Pattern Recognition, pp. 1874–1883. IEEE Press, New York (2016). https://doi.org/10.1109/CVPR.2016.207

13. Bevilacqua, M., Roumy, A., Guillemot, C., AlberiMore, M.L.: Low-complexity single-image super-resolution based on nonnegative neighbor embedding. In: 23rd British Machine Vision Conference. BMVA Press, Guildford (2012). https://doi.org/10.5244/C.26.135

14. Fei, L., Zhang, B., Zhang, L., Jia, W., Wen, Jie, Wu, J.: Learning compact multifeature codes for palmprint recognition from a single training image per palm. IEEE Trans. Multimedia, 1–13 (2020). https://doi.org/10.1109/TMM.2020.3019701

15. Martin, D., Fowlkes, C., Tal, D., Malik, J.: A database of human segmented natural images and its application to evaluating segmentation algorithms and measuring ecological statistics. In: 8th IEEE International Conference on Computer Vision, pp. 416–423. IEEE Press, Los Alamitos (2001). https://doi.org/10.1109/iccv.2001.937655

16. Wang, Z., Bovik, A.C., Sheikh, H.R., Simoncelli, E.P.: Image quality assessment: from error visibility to structural similarity. IEEE Trans. Images Process. 13, 600–612 (2004). https://doi.org/10.1109/TIP.2003.819861

17. Zeyde, R., Elad, M., Protter, M.: On single image scale-up using sparse-representations. In: Boissonnat, J.-D., et al. (eds.) Curves and Surfaces 2010. LNCS, vol. 6920, pp. 711–730. Springer, Heidelberg (2012). https://doi.org/10.1007/978-3-642-27413-8_47

18. Kim, J., Lee, J.K., Lee, K.M.: Deeply-recursive convolutional network for image super-resolution. In: IEEE Conference on Computer Vision and Pattern Recognition, pp. 1637–1645. IEEE Press, New York (2016). https://doi.org/10.1109/CVPR.2016.181

19. Bengio, Y., Simard, P., Frasconi, P.: Learning long-term dependencies with gradient descent is difficult. IEEE Trans. Neural Netw. 5, 157–166 (1994). https://doi.org/10.1109/72.279181

20. Hui, T.-W., Loy, C., Tang, X.: Depth map super-resolution by deep multi-scale guidance. In: Leibe, B., Matas, J., Sebe, N., Welling, M. (eds.) ECCV 2016. LNCS, vol. 9907, pp. 353–369. Springer, Cham (2016). https://doi.org/10.1007/978-3-319-46487-9_22

Research on Cognitive Effects of Internet Advertisements Based on Eye Tracking

Yanru Chen[1,2], Yushi Jiang[2], Miao Miao[2], Bofu Feng[2], Liangxiong Wei[1],
Yanbing Yang[1], Bing Guo[1], Youling Lin[1(✉)], and Liangyin Chen[1(✉)]

[1] Sichuan University, Chengdu 610065, China
{yanruchen,linyouling}@stu.scu.edu.cn, weilx_scu@qq.com,
{yangyanbing,guobing,chenliangyin}@scu.edu.cn
[2] Southwest Jiaotong University, Chengdu 610031, China
jysnjrf@sohu.com, miaomiao@swjtu.edu.cn, 2656058839@qq.com

Abstract. Thanks to eye tracking systems, movement data of eyes can be obtained when audiences gaze at advertisements. Eye tracking emerges as an effective means to obtain cognitive effects. Advanced techniques generally require customized equipment and systems. Meanwhile, acquisition of experimental data costs too much. It greatly limits the application scope of eye tracking systems. In this paper, our work consists of two parts. For the first part, we study the cognitive effect of ads through tracking the eyes of participants taking part in our experiments. In the other part, due to the limitation of the eye tracking, we propose a new eye tracking system based on edge computing to extend the application range. The experiment results show that products' involvement and sexual appeal have an impact on the cognitive effect. Furthermore, by taking advantage of the new eye tracking system, we believe it can make psychological research easier.

Keywords: Edge computing · Eye tracking technology · Internet advertisements · Product involvement · Sexual appeal Ads

1 Introduction

Factors such as color, size, and background of online advertisements may have influences on cognitive effect [9]. However, the cognitive effects of sexual appeal ads have not been sufficiently studied to derive a theoretical framework [3]. Moreover, conclusions from different scholars are even conflicted with one another. On the one hand, some studies [5] show that sexual appeal elements could be brilliant sales strategies. On the other hand, some studies present that ads with sexual themes will damage product cognition and brand memory. The different conclusions of various researches are caused by different ads stimulus materials, research objects, experiment processes and experimental methods. Some studies show that the degree of involvement of products also affects people's cognition.

With the help of eye tracker, the eye movement data of target audience can be gained easily. It has been widely used in advertising psychology, cognitive linguistics and other

© Springer Nature Singapore Pte Ltd. 2021
Y. Sun et al. (Eds.): ChineseCSCW 2020, CCIS 1330, pp. 661–669, 2021.
https://doi.org/10.1007/978-981-16-2540-4_49

researches [4, 10]. Despite eye trackers are the most common method to obtain eye movement data, there are still many limitations. Firstly, due to the price of the eye tracker, it is impossible to equip every subject with an eye tracker. As a result, gathering all subjects together is the only reasonable manner. Therefore, it may lead to a small group of subjects, e.g. number, kinds of occupation and age range. Secondly, the acquisition of experiment data costs too much. Thirdly, data cannot be updated in a timely manner. Finally, as a result of the above limitations, few advertisements can be used for testing.

These limitations motivate us to propose a new edge computing eye tracking system (ECET). The core idea of ECET is to establish a system by taking full use of already existing image sensors and edge computing resources. It can automatically obtain a huge amount of eyes movement data from many different people. ECET make us not only analyze eye movement data, but also conveniently acquire and analyze other data, i.e., facial expression data, ECG or brain wave data.

In the meantime, we study the effects of ad cognition for different involvement of products, different appeal expressions of ads, and different genders of participants. We control the multiple factors, for example, layout, color, size, complexity, and cognitive load of Internet ads. Finally, based on the results of the experiment, we put forward reasonable suggestions on how to match various elements of Internet ads. We hope that advertising design strategies are valuable for advertising designers.

The rest of the article is structured as follows. Section 2 mainly introduces the related work. In Sect. 3, we provide the experiments and the result. In Sect. 4, we propose our new eye tracking system. Finally, Sect. 5 is the conclusion part.

2 Related Work

With recent technological innovations, existing research about advertising psychology, cognitive linguistics, et al. generally based on eye tracking [2, 6, 13]. With the help of eye tracker, pupil diameter [12] and micro saccades can be captured, which are indicators of cognitive load [7]. Eye tracking studies show that people view pictures or read text in different ways [1].

Marta et al. [1] studied the differences in the way of looking at pictures and texts between verbalizers and visualizers. The results show that the verbalizers need to spend more time observing the text, while the visualizers need to spend more time observing the picture.

Krejtz et al. [7] confirm the work of micro saccade magnitude. Moreover, they extend this word by regarding micro saccade metrics as pupillometric measures. And this is the first research to study the sensitivity and reliability of micro saccadic measures and task-evoked pupillary of cognitive load.

Many researchers doubt the authenticity of the questionnaires. Therefore, Chauliac et al. [8] use eye tracking technology to study the relationship between the real mental activities and the results of the questionnaire.

Wang et al. [11] use both eye tracking and questionnaire to investigate the relationship between consumer attention, purchase intention and the background complexity of a product picture. The experimental results show that the more complicated the background of the product image is, the more attention can be attracted to the consumers. However, consumers pay the most attention to the background.

3 Experiments

3.1 Experiment Design

In this work, we designed two groups of experiments. In experiment 1, without involving cognitive load, we used two-factor repeated experiments to study the cognitive effects. Based on cognitive load, by introducing a third factor in experiment 2, we used three-factor repeated experiments to study the cognitive effects. To decrease the participants' cognitive processing capability, time pressure was applied to each participant.

Through a poster on our campus, we recruited more than 100 candidates. We randomly chose 10 males and 10 females as participants who were aged between 21 and 26. In experiment 2, The procedure of selecting participants was the same as that in experiment 1, and there were 26 participants. As for experimental materials, from the Internet, we chose ads with sexual appeal and non-sexual appeal as the basic materials. In experiment 2, with the significance test, we choose 4 products' picture (iPhone, hardware appliances, lady's bags, men's razors) for experiment.

Then, in experiment 1, we grouped 200 pictures of advertisement (include 40 filter ads) into 4 categories. In experiment 2, by introducing sexual appeal elements or not, we categorized 68 ad pictures (include 20 filter ad pictures) into 8 groups.

3.2 Results and Analyses of Experiment 1

Logo and product pictures in ads are defined as the interest area. After analyzing the participants' eye movement data, we got averaged values of the participants' fixation duration (FD) and fixation times (FT) towards different interest areas in ads. Details are illustrated in Table 1 and Table 2, in which L means interest area of the logo, P means interest area of product picture, L&P means interest area of both logo and product picture. G1–G4 means Group 1–Group 4 in experiment 1, respectively.

Table 1. Men's average FD and FT towards logo and product picture with different groups.

Men	G1	G2	G3	G4	G1 + G2	G3 + G4
FD for L (ms)	578.30	774.54	2721.40	2289.08	1352.84	5010.48
FD for P (ms)	1148.70	1396.00	2910.20	2242.70	2544.70	5152.90
FD for L & P (ms)	1727.00	2170.54	5631.60	4531.78	3897.54	10163.38
FT for L (times)	1.98	2.65	8.71	6.57	4.63	15.28
FT for P (times)	5.03	7.84	18.61	12.21	12.87	30.82
FT for L & P (times)	7.01	10.49	27.32	18.78	17.50	46.10

Table 1 and Table 2 show the difference in fixation duration and times in different sexual appeal background. For the sexual appeal and non-sexual appeal backgrounds, we use a one-factor variance method to process the participants' fixation duration and times towards interest area respectively. Besides, as shown in Table 3, without cognitive

Table 2. Women's average FD and FT towards logo and product picture with different groups.

Women	G1	G2	G3	G4	G1 + G2	G3 + G4
FD for L (ms)	1213.30	1450.90	2678.40	2103.34	2664.20	4781.74
FD for P (ms)	2259.90	2134.60	2460.20	2595.50	4394.50	5055.70
FD for L & P (ms)	3473.20	3585.50	5138.60	4698.84	7058.70	9837.44
FT for L (times)	4.76	5.57	9.27	7.85	10.33	17.12
FT for P (times)	10.07	11.98	15.25	14.13	22.05	29.38
FT for L & P (times)	14.83	17.55	24.52	21.98	32.38	46.50

load, people are significantly different in fixation duration and times for the Region of Interest in ads when the background of ads are of different appeal expressions.

Table 3. Data analysis results 1 for experiment 1.

Item	Variance method	FD for logo	FD for product	FT for logo	FT for product
Homogeneity of variance	Levene	0.733	1.037	0.625	1.502
	P	$0.637 > 0.050$	$0.367 > 0.050$	$0.746 > 0.050$	$0.144 > 0.050$
Confidence interval: a = 95%	F(3, 796)	4.644	8.381	3.220	7.825
	P	$0.015 < 0.050$	$0.035 < 0.050$	$0.023 < 0.050$	$0.041 < 0.050$

We use independent samples to perform T-test to further analyze whether there are differences between the fixation duration and times towards interest area of Logo and Product respectively. As shown in Table 4, without cognitive load, there is no significant difference between men and women in fixation duration and times of ROI (Region of Interest) when the backgrounds are expressions of non-sexual appeal. Without cognitive load, men take significantly less fixation duration and times in ROI than women, and take significantly more fixation duration and times in the background pictures of ads than women when the ads use expressions of sexual appeal.

Table 4. Data analysis results 2 for experiment 1.

Variance method	FD for logo	FD for product	FT for logo	FT for product
Test	2.881	3.074	2.244	2.266
F's P	$0.008 < 0.050$	$0.004 < 0.050$	$0.032 < 0.050$	$0.036 < 0.050$

3.3 Results and Analyses of Experiment 2

The experiment 2 is similar to experiment 1, we analyze and summarize eye movement index data from different ads. The results are illustrated in Table 5 and Table 6, in which L, P, and L&P mean the same as those in experiment 1, and G1–G8 mean Group 1–Group 8 in experiment 2, respectively.

Table 5. Men's average FD and FT towards interest area of logo and product picture.

Men	G1	G2	G3	G4	G5	G6	G7	G8
FD for L (ms)	1318.60	1671.20	622.75	1199.80	1062.85	1295.25	1204.40	1682.30
FD for P (ms)	2271.25	2514.60	1207.20	1480.20	1528.90	1899.95	1983.20	2352.30
FD for L & P (ms)	3589.85	4185.80	1829.95	2680.00	2591.75	3195.20	3187.60	4034.60
FT for L (times)	6.76	8.56	3.19	6.15	5.44	6.63	6.17	8.62
FT for P (times)	11.03	12.88	6.18	7.58	7.83	9.73	10.16	12.05
FT for L & P (times)	17.79	21.44	9.37	13.73	13.27	16.36	16.33	20.67

Table 6. Women's average FD and FT towards interest area of logo and product picture.

Women	G1	G2	G3	G4	G5	G6	G7	G8
FD for L (ms)	1109.75	1470.30	503.30	843.85	1354.70	1135.85	844.54	1154.65
FD for P (ms)	1720.90	2102.05	1259.00	1058.60	1856.25	2570.20	1428.75	1809.00
FD for L & P (ms)	2830.65	3572.40	1762.30	1902.50	3210.95	3706.05	2273.29	2963.65
FT for L (times)	5.68	7.53	2.25	4.32	9.94	5.82	4.33	5.95
FT for P (times)	8.82	10.77	6.45	5.42	9.51	13.17	7.32	9.26
FT for L & P (times)	14.50	18.30	8.70	9.74	19.45	18.99	11.65	15.21

We use MANOVA (multivariate analysis of variance) to analyze the eye movement data in Table 5 and Table 6. On the one hand, two tables indicate that even under the presence of cognitive load, the attention which the participants paid to the interest area in sexual appeal ads is still obviously less than that in non-sexual appeal ads. It suggests that the participants have paid more attention to the sexual appeal background elements in the ads. As shown in Table 7, with cognitive load, the participants' fixation duration and fixation times of ROI in sexual appeal ads are significantly less than those in non-sexual appeal ads. On the other hand, two tables indicate that under the presence of cognitive load, there are significant differences in the fixation duration and fixation times for the participants paying attention to the interest areas in different product involvement. Specifically, the female participants' fixation duration and fixation times with high product involvement for females are much more than that with low product involvement for females, and the male participants' fixation duration and fixation times with high product involvement for males are also much more than that with low involvement for males. Moreover, the interaction effects for different product involvement combined with different advertising appeal methods are not significant. With cognitive load, there is no significant difference in the fixation duration and times for male and female participants towards interest areas of different advertising appeal methods.

Table 7. Data analysis results for experiment 2.

Variance method	For different appeal ads		For different involvement		For different genders of the participants	
	FD	FT	FD	FT	FD	FT
F(3, 312)	31.336	21.199	23.886	28.370	1.299	1.201
P	0.000 < 0.050	0.000 < 0.050	0.000 < 0.050	0.000 < 0.050	0.361 > 0.050	0.276 > 0.050

4 Proposed New Eye Tracking System

In our experiments, we study the influence of appeal expressions and product involvement for cognitive effects. This work derives the following conclusions. Firstly, people's eye movement traces and attention degree are significantly affected by whether to adopt expressions of sexual appeal or not. The participants pay more attention to sexual appeal ads than non-sexual appeal ads obviously. Secondly, without cognitive load, males tend to be more attracted by sexual appeal ads. There is no significant difference in attraction of sexual appeal ads between males and females. Thirdly, product involvement greatly affects participants' eye movement trace and watching behavior. It is recommended that, for products with high involvement, advertisers should use expressions of sexual appeal. At the same time, for products with low involvement, expressions of sexual appeal should not be used.

This work has fully considered every factor that may affect the experiment processes or results. However, flaws, insufficiencies, and limitations of the work may be unavoidable. First, in our experiments, we do not consider the impacts of webpage types, ad positions, ad size, and ad presentation, etc. Second, the presentation time of experimental materials is settled which means every participant has the same time to look through all ads. However, in reality, different people have different interests in different ads. Third, the participants in this work are merely selected from students. In the future, with enlarging sample size, participants will be chosen from different ages and different cultural levels to investigate their attitude differences over sexual appeal ads. Fourth, it takes a lot of time and cost to collect experimental data. Finally, data cannot be updated in a timely manner and few advertisements can be used for testing.

These limitations motivate us to propose a new edge computing eye tracking system (ECET). The core idea of ECET is to establish a system by taking full use of already existing image sensors and edge computing resources. It can automatically obtain a huge amount of eye movement data from many different people.

Nowadays, the Internet of Things (IoT) has played a key role in our lives [14, 15]. A vast array of devices is being interconnected via IoT techniques. At the same time, a large amount of data will be produced. In cloud computing, data will be sent to centralized servers, and will be sent back when finish computation. Therefore, this process greatly increases the cost of data transmission. So, edge computing is regarded as one important technology to address the problem. The peak of traffic can be effectively alleviated because of the locations of edge computing devices. In addition, the demand for bandwidth of centralized servers can be significantly mitigated while the transmission latency being reduced during data computing or storage in IoT [16, 17].

With the quick deployment of edge computing resources, most IoT devices with limited computing resources or limited power can offload their computing tasks to edge clouds, hence improve their computing performance [26]. In addition, there still are many computing resources, data, services in traditional Cloud. So, edge cloud and center cloud must work cooperatively. On current trends of quick development of IoT and edge computing, we propose ECET to preferably study the advertising psychology in this paper. The core idea is to reuse numerous existing resources, including image sensors and edge computing resources, to collect mass data used in the researches of advertising psychology. Specifically, image sensors can be the camera of a user's mobile phone or the camera of a monitoring system in a supermarket and the computing resources can be from edge computing resources. As a result, huge related data can be collected automatically. The architecture diagram of ECET is given in Fig. 1. We can see that image processing services, including image data analysis, pupil feature extraction, pupil coordinate superposition, and other related services are segregated from image sensors and located in the edge computing center. The algorithms library and offline data involved in these services can be pulled or updated from remote Cloud. The extensible architecture can provide huge eye tracker data with high performance with the help of a powerful edge and remote cloud computing resources.

With the help of ECET, huge eye tracker data can be obtained through the Internet. For example, we can develop an Android application which can automatically guide users to make the experiments. Specifically, when one user opens this application, it will present some ad pictures that are downloaded from the data set stored in the edge cloud or remote cloud. These ad pictures are collected from the Internet using automated crawler tools and stored in the remote cloud in advance. When one user first requests one picture, it will be pulled from the remote cloud and stored in edge cloud meanwhile. After that, users will be pulled the picture from edge cloud. When the user watches the picture, the application will continually obtain images of users with the help of the front camera of the user's mobile phone. Then, these images will be sent to edge cloud. In edge cloud, these images will be processed by image data analysis service, pupil feature extraction service and pupil coordinate superposition service. The algorithm library involved in these services can be pulled or updated from the remote cloud. After processing in edge cloud, eye tracker data can be obtained. All eye tracker data distributed in many different edge clouds will be sent to the remote cloud and these data can be analyzed by special researchers or computer programs. In order to encourage users, the application can reward them some electronic cash. Another example is presented below. With the help of an existing camera of a monitoring system in an elevator, when one user watching the ad picture posted in the wall of the elevator, the image data can be obtained and then be processed in the edge cloud.

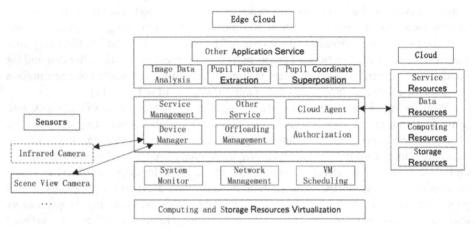

Fig. 1. The architecture of ECET.

ECET makes us not only analyze eye tracker data butss also conveniently acquire and analyze other data, i.e., facial expression data obtained from image sensors, ECG (electrocardiogram) or brainwave data collected from wearable devices.

One example of the application of ECET-based advertising psychology study is that, in online web pages, we can recommend different ad pictures or ad videos in real time based on the real-time analysis results about the eye tracker data or other related sensing data of the users.

5 Conclusions

In this article, we conduct two sets of experiments using eye tracking system. We conclude that for the products with different involvements or different genders, different sexual appeal ads should be considered. Meanwhile, considering the limitations of experiments, we propose a new edge computing eye tracking system (ECET). It can take full use of already existing image sensors and edge computing resources to automatically obtain a huge amount of eye tracking data.

References

1. Marta, K., Tim, H., Gun-Brit, T., et al.: Visualizers versus verbalizers: effects of cognitive style on learning with texts and pictures–an eye-tracking study. Comput. Hum. Behav. **68**, 170–179 (2017)
2. Ahonniska-Assa, J., Polack, O., Saraf, E., et al.: Assessing cognitive functioning in females with Rett syndrome by eye-tracking methodology. Eur. J. Paediatr. Neurol. **22**(1), 39–45 (2018)
3. Hairong, L., Ang, L., Shuguang, Z.: Internet advertising strategy of multinationals in China. Int. J. Advertising **28**(1), 125–146 (2015)
4. Eckstein, M.K., Guerra-Carrillo, B., Singley, A.T.M., et al.: Beyond eye gaze: what else can eyetracking reveal about cognition and cognitive development? Dev. Cognit. Neurosci. **25**, 69–91 (2017)
5. Jessica, S., George, E.B., Michael, A.B.: The effects of sexual and non-sexual advertising appeals and information level on cognitive processing and communication effectiveness. J. Advertising **19**, 14–22 (2013)
6. Dean, B., Ginnell, L., Ledsham, V., et al.: Eye-tracking for longitudinal assessment of social cognition in children born preterm. J. Child Psychol. Psychiatry (2020)
7. Krejtz, K., Duchowski, A.T., Niedzielska, A., et al.: Eye tracking cognitive load using pupil diameter and microsaccades with fixed gaze. PloS One **13**(9), e0203629 (2018)
8. Chauliac, M., Catrysse, L., Gijbels, D., et al.: It is all in the surv-eye: can eye tracking data shed light on the internal consistency in self-report questionnaires on cognitive processing strategies? Frontline Learn. Res. **8**(3), 26–39 (2020)
9. Michael, L., Kimberly, A.W., Hoegg, J.: Customer relationship stage and the use of picture-dominant versus textdominant advertising. J. Retail. **89**, 263–343 (2013)
10. Kessler, S.H., Zillich, A.F.: Searching online for information about vaccination: assessing the influence of user-specific cognitive factors using eye-tracking. Health Commun. **34**(10), 1150–1158 (2019)
11. Qiuzhen, W., Da, M., Hanyue, C., et al.: Effects of background complexity on consumer visual processing: An eye-tracking study. J. Bus. Res. **111**, 270–280 (2020)
12. Dong, L., Chao, C.: Eye gaze estimation method based on pupil center cornea reflection technique. Comput. Digit. Eng. **1**, 158–161 (2016)
13. Stevenson, M.P., Dewhurst, R., Schilhab, T., et al.: Cognitive restoration in children following exposure to nature: evidence from the attention network task and mobile eye tracking. Frontiers Psychol. **10**, 42 (2019)
14. Cheng, Z., Hailiang, Z., Shuiguang, D.: A density-based offloading strategy for IoT devices in edge computing systems. IEEE Access **6**, 73520–73530 (2018)
15. Wei, Y., Fan, L., Xiaofei, H., William, G.H., et al.: A survey on the edge computing for the Internet of Things. IEEE Access **6**, 6900–6919 (2018)
16. Jiping, J., Lingyu, C., Xuemin, H., Jianghong, S.: A heuristic algorithm for optimal facility placement in mobile edge networks. KSII Trans. Internet Inf. Syst. **11**, 3329–3350 (2017)
17. Weisong, S., Jie, C., Quan, Z., Youhuizi, L., Lanyu, X.: Edge computing: vision and challenges. IEEE Internet Things J. **3**, 637–646 (2016)

A Short Text Classification Method Based on Combining Label Information and Self-attention Graph Convolutional Neural Network

Hongbin Wang[1,2], Gaorong Luo[1,2], and Runxin Li[1,2(✉)]

[1] Faculty of Information Engineering and Automation, Kunming University of Science and Technology, Kunming 650500, China
[2] Yunnan Key Laboratory of Artificial Intelligence, Kunming University of Science and Technology, Kunming 650500, China

Abstract. Deep learning classification model has achieved good results in many text classification tasks. However, these methods generally consider single text feature, and it is difficult to represent short text. To this end, this paper proposes a text classification method based on combines label information and self-attention graph convolutional neural network. This method introduces an attention mechanism based on label information to obtain more accurate text representation, and it extracts the global features of multiple texts by self-attention graph convolution network for text classification. The experimental results based on R8 and MR show that compared with the latest model, our model increases the F1 values and accuracy by 2.58% and by 2.02% on the MR data; and increases the F1 values and accuracy by 3.52% and by 2.52% on R8 dataset.

Keywords: Text classification · Graph convolutional neural network · Label vector · Deep learning · Self-Attention

1 Introduction

Text classification is an important task in natural language processing. And text representation is the intermediate link of text classification and an important content of text classification. Classification methods based on machine learning obtain text features through statistical methods, and their generalization ability is poor. But the neural network algorithm based on word vector takes word vector as input, and then researchers integrate information such as entities and entity positions in the text into text representation. These neural network algorithms include convolution neural network [1], recurrent neural network [2], etc. Deep learning classification methods only extract features from a single text and do not extract global features from multiple texts. Then Henaff [3] proposed graph convolution neural network, which can extract global features and improve the text classification results.

In this paper, we propose a short text classification method based on combining label information and self-attention graph convolutional neural network. Self-attention graph

Y. Sun et al. (Eds.): ChineseCSCW 2020, CCIS 1330, pp. 670–677, 2021.
https://doi.org/10.1007/978-981-16-2540-4_50

convolution network can learn common features between nodes and nodes. In addition, in order to obtain more accurate text representation, an attention mechanism based on label information is used to integrate label semantic information into text representation. In summary, the contributions of this paper are as follows:

(1) In order to obtain more accurate text representation, a short text representation method combining label vectors is proposed, which makes full use of the rich semantic information contained in label;
(2) Compared with the previous method based on graph convolution neural network, this paper proposes a new method of constructing graphs;
(3) Experiments show that the model in this paper can achieve better performance than the latest methods.

2 Related Work

The development of text classification mainly has three stages: text classification based on traditional machine learning, deep learning text classification and graph convolution neural network text classification.

Text classification based on traditional machine learning mainly focus on feature engineering and classification algorithms. The implementation of that method is divided into two steps: firstly, feature is extracted by machine statistical algorithms such as document frequency method (DF) [4], information gain method (IG), and then feature is classified by machine learning algorithms such as Naive Bayesian algorithm and support vector machine (SVM) [5]. But the machine statistical algorithm for feature extraction needs to make appropriate choices for different dataset, and its generalization ability is weak.

The deep learning text classification method is based on word vector [6], then convolution neural network (CNN) [1] or recurrent neural network (RNN) [7] is used to extract features, and researcher uses the two networks to extract different features [8] respectively. Finally, the features are input into the classifier to obtain classification result. These methods do not require feature engineering, but only extract the features of a single text and do not extract global features from multiple texts.

Graph convolution neural network (GCN) for text classification was proposed by Velikovi [9]. Velikovii uses PMI algorithm to construct graphs, and then uses GCN to extract global features. However, the model did not highlight important nodes, and then the researcher [10] has adopted attention mechanism to solve this problem.

CNN has strong ability to extract features for a single text, while GCN has strong ability to fuse global features, so the two methods can be fused to improve the performance of classification model.

3 Method

In order to extract rich label semantic information and global features from multiple texts, we propose a short text classification method combining label information and self-attention graph convolutional neural network. Firstly, the attention mechanism based on label information is used to obtain text representation, and then the attention mechanism based on GCN is used to extract global features. Finally, the features are input into the classifier to obtain the classification results. Framework of the proposed model shows as Fig. 1.

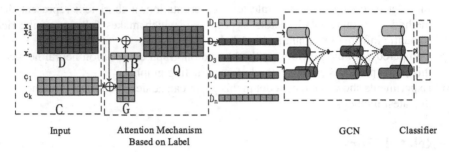

Input   Attention Mechanism   GCN   Classifier
Based on Label

Fig. 1. Framework of the proposed model. Input includes text sequences X = {x₁, x₂...xₙ} and categories sequences C = {c₁, c₂...cₖ}. ⊕ is an operation between matrices, Q is a text representation, then the text representation is obtained through the average word vector and graph convolutional neural network, which is input into the classifier to obtain the classification results.

Attention mechanism based on label. In order to use of label information, an attention mechanism based on label information is adopted.

$$G = D \times C^T \tag{1}$$

$$\beta_i = f(G_{i-r:i+r}w_1 + b_1) \tag{2}$$

$$N_i = \max -pooling(\beta_i) \tag{3}$$

Where $D = \{x_1^d, x_2^d, x_3^d...x_n^d\} \in R^{n \times d}$, $C = \{c_1^d, c_2^d, c_3^d...c_n^d\} \in R^{k \times d}$, $\times$ is matrix point multiplication. Then feature is extracted by CNN, f is a activation function. And the text representation is obtained by averaging word vectors:

$$M = soft \max(N) \tag{4}$$

$$V = \{x_1 \otimes M_1; x_2 \otimes M_2; x_3 \otimes M_3...x_n \otimes M_n\} \tag{5}$$

$$D = \frac{1}{n}\sum_{i=1}^{n} V_i \tag{6}$$

Graph convolutional neural network (GCN). We use GCN to extract global features. Firstly, we need to construct adjacency matrices. $A \in R^{N \times N}$ Formally, A is defined as formula (7):

$$A_{ij} = \begin{bmatrix} 1, & TF\_IDF(i,j) > t \\ 1, & Co(i,j) > n \\ 0, & other \end{bmatrix} \tag{7}$$

Where $TF\_IDF(i, j)$ is the value of frequency–inverse document frequency between sentence i and sentence j. $Co(i, j)$ is number of Co-occurrence Words between sentence i and sentence j. The value of t and n are obtained by experiment. Then the text representation $D_i^{''}$ is obtained by the following equation:

$$e_{i,j} = a[w_i h_i; w_j h_j] \tag{8}$$

$$D_i^{''} = \sigma(\sum_{j \in C} a_{i,j} w_{i,j} D_j) \tag{9}$$

Where $h_i \in R^d$ and $h_j \in R^d$ are text feature of node i and node j. C is the set of nodes connected to node i. We perform k-head attention mechanism on (9), and σ is the activation function.

$$D_i^{''} = \sigma(\frac{1}{k} \sum_{h=1}^{k} \sum_{j \in C} a_{i,j}^h w_{i,j}^h D_j) \tag{10}$$

Classifier. Finally, the features are input into the classifier and the parameters are updated by the cross entropy loss function.

4 Evaluation and Analysis

4.1 Dataset

Because the previous research work was based on R8 of Reuters 21578 and Movie Review (MR), we ran our experiments on R8 and MR. R8 has 8 categories, and was split to 5,485 training and 2,189 test documents. MR has 5,331 positive and 5,331 negative reviews. We divide the training set and the test set according to the ratio of 7:3.

4.2 Parameter Setting

For our model, we use 300 dimensional GloVe words embedding and set the window size as 5. We also tuned other parameters and set the learning rate as 0.005, dropout rate as 0.5, apply K = 8 output attention heads. We trained our model for a maximum of 400 epochs using Adam and stop training if the validation loss does not decrease for 10 consecutive epochs. In Table 1, Co denotes the degree of co-occurrence in two sentences. It includes two steps. Firstly, count the words that appear together in the two sentences, and then calculate the proportion of the words that appear together. For example, Co-0.29 means that if the co-occurrence degree is greater than 0.29, an edge is constructed. Tf denotes the similarity between two sentences calculated by TF_IDF. The following Table 1 gives the selection process of t and n. Specifically, we removed stop words are defined in NLTK [11] and low frequency words appearing less than 5 times for R8.

Table 1. Experimental result of selection t and n.

Dataset	R8		MR	
Evaluation	Accuracy	F1	Accuracy	F1
Co-0.29	0.8950	0.8726	0.5656	0.5968
Co_0.30	0.9100	0.8737	0.5606	0.5231
CO-0.35	**0.9275**	**0.9052**	0.6187	0.6134
CO-0.4	0.9125	0.8936	0.6388	0.6460
CO-0.45	0.8950	0.8878	0.6421	0.6200
CO-0.5	0.9025	0.8746	0.6515	0.6523
CO-0.55	0.8925	0.8819	**0.6818**	**0.6810**
Co_0.60	0.8850	0.8786	0.5959	0.6672
Tf_0.035	0.9075	0.6581	0.6464	0.6445
Tf_0.04	0.9075	0.6494	0.6414	0.6433
Tf_0.045	0.9225	0.6764	0.6288	0.6091
Tf_0.05	0.9200	0.8091	0.6616	0.6552
Tf_0.55	0.9125	0.4097	0.6565	0.6767

In Table 1, the method of constructing adjacency matrix with co-occurrence words is generally better than the method of constructing adjacency matrix with TF_IDF values. The best set of results in the method of constructing adjacency matrix with co-occurrence words is selected as the results of this paper, and the corresponding co-occurrence words value is selected as n value. That is, n value of R8 and MR are 0.35 and 0.55 respectively.

4.3 Baselines

We compare our model with multiple text classification methods as follows:

CNN [1]: a CNN used in text classification. We input pre-trained word embedding to CNN. Word vector does not participate in training.
CNN [13]: a CNN used in text classification. We input character embedding to CNN. Character embedding participates in training.
LSTM: a bi-directional LSTM used in text classification. We input pre-trained word embedding to Bi-LSTM.
LEAM [12]: label-embedding attentive model, which embeds the words and labels in the same joint space for text classification. It utilizes label descriptions.
SWEM [14]: simple word embedding model, which employs simple pooling strategies operated over word embedding.
Text-GCN [9]: PMI and TF_IDF algorithms are used to construct graphs, and graph convolution neural network is used to extract features. We removed stop words defined in NLTK and low frequency words appearing less than 5 times for R8.

4.4 Test Performance

In this paper, two groups of comparative tests were carried out. One group is ablation test (see Table 2). The other group is a comparison experiment with the baseline model (see Table 3). Notes: "(-) no-label" represents the label information is not used, and the word vector as model input, "(-) no-graph" represents graph convolution neural network is not used, "(-no-attention)" represents self-attention mechanism is not used.

Table 2. Experimental results of ablation test.

Dataset	R8		MR	
Evaluation	Accuracy	F1	Accuracy	F1
(-)no-label	0.5025	0.5500	0.4621	0.4511
(-)no-graph	0.9175	0.7041	0.6414	0.6407
(-)no-attention	0.8750	0.6063	0.6565	0.6752
Our method	**0.9275**	**0.9052**	**0.6818**	**0.6810**

In Table 2, when the label vector is not used, the accuracy and F1 are greatly reduced, which demonstrates the label vector is a very important part of the model. When self-attention and graph convolution neural network is not used, the accuracy and F1 both decrease to different degrees on the two datasets. It's proved that self-attention mechanism and graph convolution neural network are both essential parts of the model.

Table 3. Comparison of experimental results between the model and baselines.

Dataset	R8		MR	
Evaluation	Accuracy	F1	Accuracy	F1
CNN [1]	0.8650	0.8500	0.5152	0.5200
CNN [13]	0.8950	0.8600	0.6700	0.6700
Text-GCN [9]	0.9050	0.8700	0.6450	0.6515
LSTM	0.8225	0.8200	0.4621	0.2900
SWEM [14]	0.9000	0.7600	0.3950	0.2400
LEAM [12]	0.9100	0.7600	0.6667	0.4848
Our method (Co)	**0.9275**	**0.9052**	**0.6818**	**0.6810**

In Table 3, in the CNN model, the character vector-based method is superior to the word vector-based method. Compared with LSTM-based model, CNN has better result, which shows that CNN is better than LSTM in extracting features on short text. On R8 dataset, SWEM and LEAM perform quite well, weaker than Text-GCN, which demonstrates the effectiveness of simple pooling methods and label embedding. However, there is a big difference on MR dataset, which indicates that the graph convolution neural network can aggregate global features in short texts. The effect of our model is better than Text-GCN, which proves that the self-attention graph convolution network can pay attention to important node information.

5 Conclusion and Future Work

Because short text contains relatively little semantic information, it is difficult to accurately represent the text and the traditional convolution neural network cannot fuse the features of multiple texts, we have proposed a text classification method based on combines label information and self-attention graph convolutional neural network. Firstly, the text vector matrix and label vector matrix are obtained according to the pre-trained word vector, and the text representation is obtained according to the text information and label information. Then, a graph is constructed according to the co-occurrence words in the text. Finally, global features are extracted through graph convolution, and the classification result is obtained. The model in this paper achieved better results than the baselines model on R8 dataset and MR dataset, which shows that the method is effective.

In the future work, we plan to explore the application effect of the model in other natural language processing, such as question answering, fine-grained emotion analysis, etc. At the same time, the application of this model in knowledge graph is explored.

Acknowledgments. This work was supported by the National Natural Science Foundation of China under Grant 61966020.

References

1. Kim, Y.: Convolutional neural networks for sentence classification. EprintArxiv (2014)
2. Chung, J., Gulcehre, C., Cho, K., Bengio, Y.: Empirical evaluation of gated recur-rent neural networks on sequence modeling. arXiv preprint arXiv (2016)
3. Henaff, M., Bruna, J., Le Cun, Y.: Deep convolutional networks on graph-structured data. arXiv preprint arXiv:1506.05163 (2015)
4. Joachims, T.: Text categorization with support vector machines: learning with many relevant features. In: Nédellec, C., Rouveirol, C. (eds.) Machine Learning: ECML-98. LNCS, vol. 1398, pp. 137–142. Springer, Berlin (1998). https://doi.org/10.1007/BFb0026683
5. Mladenic, D., Grobelnik, M.: Feature selection for unbalanced class distribution and Naive Bayes. In: 16th International Conference on Machine Learning, Slovenia, Bled, pp. 258–267 (1999)
6. Pennington, J., Socher, R., Manning, C.: Glove: global vectors for word representation. In: 2014 Conference on Empirical Methods in Natural Language Processing (EMNLP), pp. 1532–1543 (2014)
7. Luo, Y.: Recurrent neural networks for classifying relations in clinical notes. J. Biomed. Inform. **72**, 85–95 (2017)
8. Zhang, Y., Lu, W., Ou, W., Zhang, G., Zhang, X., Cheng, J., Zhang, W.: Chinese medical question answer selection via hybrid models based on CNN and GRU. Multimedia Tools Appl. **79**(21–22), 14751–14776 (2019). https://doi.org/10.1007/s11042-019-7240-1

9. Yao, L., Mao C.: Graph convolutional networks for text classification. In: the AAAI Conference on Artificial Intelligence, vol. 33, pp. 7370–7377 (2019)
10. Veličković, P., Cucurull, G., Casanova, A., Romero, A., Lio, P., Bengio, Y.: Graph attention networks. arXiv preprint arXiv:1710.10903 (2017)
11. Natural Language Toolkit Homepage. https://www.nltk.org/. Accessed 14 Oct 2020
12. Wang, G., et al.: Joint em-bedding of words and labels for text classification. arXiv preprint arXiv:1805.04174 (2018)
13. Zhang, X., Zhao, J., Le Cun, Y.: Character-level convolutional networks for text classification. In Advances in Neural Information Processing Systems, pp. 649–657 (2015)
14. Shen, D., et al.: Baseline needs more love: on simple word-embedding-based models and associated pooling mechanisms. arXiv preprint arXiv:1805.09843 (2018)

Community-Based Propagation of Important Nodes in the Blockchain Network

Xin Li$^{(\boxtimes)}$, Weidong Zheng, and Hao Liao

College of Computer Science and Software Engineering, Shenzhen University,
Shenzhen 518060, China
1910273046@email.szu.edu.cn

Abstract. With the rapid development of Blockchain technology, this new technology is being widely used in finance, public services, and other fields. In recent years, frequent security problems in Blockchain's application have brought huge losses to relevant industries, so its security has been widely discussed. Among them, most security incidents occurred in the field of digital currency. If we can effectively identify the communities and important nodes in the currency transaction network, and strengthen the protection measures for these nodes, it will be beneficial to improve digital currency transactions' security. This paper combines the community detection algorithm Infomap and the node influence algorithm IMM, and proposes an important node ranking method based on the propagation of influence in the community, named CIIN. Using real data from Ethereum currency transactions, we ranked important nodes in the currency transaction network. The experimental results show that the community based on ranking method CIIN can effectively extract the most vital exchange or individual account in the Blockchain currency transaction records.

Keywords: Blockchain currency transaction · Important nodes · Community detection

1 Introduction

With the popularization of digitalization and informatization in today's society, people pay more attention to their privacy and property security [1]. Blockchain technology has developed rapidly in recent years due to its unique advantages such as nontamperability and transparency, and various industries are paying more and more attention to Blockchain technology. Because of the advantages of Blockchain technology in avoiding information leakage, reducing transaction time, reducing fraud and cybercrime risks, and observing real-time transaction

Supported by: National Natural Science Foundation of China (Grant Nos. 71471118 and 71871145), Guangdong Province Natural Science Foundation (Grant Nos. 2019A1515011173 and 2019A1515011064).

© Springer Nature Singapore Pte Ltd. 2021
Y. Sun et al. (Eds.): ChineseCSCW 2020, CCIS 1330, pp. 678–686, 2021.
https://doi.org/10.1007/978-981-16-2540-4_51

information [2], it has attracted the attention of financial giants. Many financial institutions and Blockchain companies, including Citibank and HSBC, have agreed to develop and apply Blockchain technology in the financial industry.

Many scenes in life can be regarded as a complex system, and these systems can be studied by complex network modeling. Using the theories and methods in complex networks can analyze the trend of problems more clearly. Previous research on central nodes of complex networks is mostly based on standard networks such as aviation networks and social networks, and a few types of research are based on transaction data networks. Transaction in the Blockchain network can be regarded as one of the complex network models, and methods such as node influence in the research of complex networks can also be applied to the Blockchain network. In the process of complex network research, important node detection and community detection are more essential directions.

In recent years, many digital currencies have flourished in Blockchain technology. However, due to lack of supervision and other reasons, many digital currency platforms have successively closed down, causing tens of thousands of investors to lose their money. This paper aims to explore the influential nodes of the community in the Blockchain network. The expectation is that these nodes can be supervised to ensure that investors in Blockchain digital currency transactions receive more security protection.

2   Related Work

Blockchain's concept was proposed in 2008, which is a new technology that combines cryptography, computer science, and other fields. This new technology has the advantages of centrality, openness, and non-tampering. Traditional transactions are generally operated on a trusted third party, while transactions in the Blockchain apply peer-to-peer technology (P2P) [4,5]. Bitcoin is the original product of Blockchain technology, and then various types of digital currencies have been derived. For example, Ethereum, Ripple, Litecoin, Libra, etc. The continuous innovation of digital currency also indicates the vigorous development of Blockchain technology [2].

In a complex network, some closely connected points form a community, and they are sparsely connected to points in other communities. The purpose of the community detection algorithm is to dig out communities that meet certain similar conditions. The types of algorithms currently used for community detection include hierarchical clustering, modularity-based optimization, tag-based propagation, and dynamics-based methods [6,7]. The GN algorithm [8], a classic community mining algorithm, is based on hierarchical clustering. The algorithm uses edge betweenness as the measurement method. Because they need a standard to evaluate the community partition results, Newman et al. [9] proposed the concept of modularity. Raghavan et al. [10] proposed in 2007 that the LPA algorithm is based on an algorithm based on tag propagation. Therefore, its community partition speed is breakneck, suitable for networks with a large amount of data. In 2002, van Dongen first proposed the clustering algorithm MCL based on Markov random walk theory [20]. In 2008, Rosvall et al. [10] proposed the Infomap algorithm to find an effective community structure through mapping equations.

The evaluation of node influence in complex networks has important practical significance. Identifying nodes with greater influence can target a node's characteristics and activities to make correct decisions and deployments. Node importance evaluation can be expanded from the local attributes, global attributes, location, and random walk of the network [11,14]. Kitsak et al. [12] in 2010 proposed the k-shell, which considers the position of nodes in the network. K-core decomposition's central idea is to recursively find the node with the largest k-shell value, which is the most influential node. The node importance algorithm based on the random walk is mainly used in the field of information search. The most classic algorithms are PageRank algorithm [13] and HITS algorithm [15]. In addition, Lempel et al. [16] proposed the SALSA algorithm. Rafiei et al. [17] proposed the Reputation algorithm.

3 Method

The goal of our model is to identify the community structure and find important nodes in the Blockchain currency transaction network. As shown in Fig. 1, the community Partition method is responsible for the Blockchain transaction network's community partition. After getting the divided community, we use the Node Influence Ranking method to find the node ranking in the community.

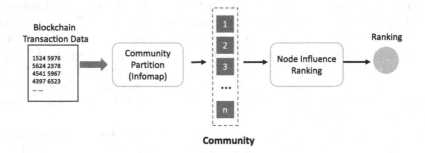

Fig. 1. The framework of CIIN.

3.1 Community Partition

Community partition uses the Infomap algorithm [10], which is a community partition algorithm based on the random walk proposed by Rosvall. This algorithm uses the coded form to record the results of random walks and perform clustering. Infomap adopts a two-level coding structure [17]. Two-level encoding can describe paths with fewer bits than first-level encoding. The first code is the community number of the node, and the second code is the community number of the node. Each community has a start and exit code that determines whether the migration path leaves the current community. The process of two-level coding can be illustrated with the example in Fig. 2.

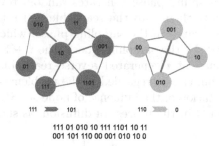

Fig. 2. In the figure, the starting code of the blue community is 111, and the exit code is 101. the starting code of the orange community is 110, and the exit code is 0.

Use information entropy formula to calculate the shortest code length of the community. Where p_α is defined as the probability of occurrence of node α, and $p_{i\frown}$ is defined as the probability of occurrence of community i.

$$L(Q) = -\sum_i \frac{q_{i\frown}}{q_\frown} \log \frac{q_{i\frown}}{q_\frown} \tag{1}$$

$$L\left(P^i\right) = -\frac{q_{i\frown}}{p_{i\smile}} \log \frac{q_{i\frown}}{p_{i\smile}} - \sum_{\alpha \in i} \frac{q_i}{p_{i\smile}} \log \frac{p_\alpha}{p_{i\smile}} \tag{2}$$

$$K(M) = q_\frown L(Q) + \sum_i p_{i\smile} L\left(P^i\right) \tag{3}$$

Equation (1) calculates the shortest average coding length of the community. Equation (2) calculates the shortest average coding length of node i. Equation (3) gets the total shortest average code length of the community.

After getting the preliminary results of community partition, we use the concept Q of modularity proposed by Newman to judge the results. The range of Q value is between $[-0.5, 1)$. When the Q value is between $[0.3, 0.7]$, the result of community partition is relatively accurate. Otherwise, repeat the steps of community partition for nodes with poor partitioning results. The calculation of Q value is shown in Eq. (4):

$$Q = \frac{1}{2m} \sum_{ij} (A_{ij} - P_{ij}) \partial(C_i, C_j) \tag{4}$$

where A is the adjacency matrix, and m is the number of edges. P_{ij} represents the expected number of edges of nodes i and j, and k_i is the degree value of vertex i. When two nodes i and j are in the same community, $\partial(C_i, C_j)$ equal to 1, otherwise equal to 0.

3.2   Node Influence Ranking

The calculation of node influence in the community uses the Influence Maximization via Mar-tingales(IMM) algorithm [18], which is a node influence algorithm calculated by martingale. The algorithm is divided into two parts: sampling and node selection. The sampling phase generates random reverse reachable (RR) sets iteratively and puts them into the reachable (R) set. The R set is a set of random RR sets generated in the sampling phase, which stops until certain conditions are satisfied. In the sampling phase, the $i + 1$ RR set will only be generated when the current i generated reverse reachable set meets the stop condition. The algorithm can be regarded as the problem of solving OPT. OPT is the maximum expectation of the influence of node-set S_k^*, $I(S)$ is the influence of node-set S on graph G in the process of diffusion, as shown in the Eq. (5):

$$E(I(S)) = \frac{n}{\theta} \cdot E\left[\sum_{i=1}^{\theta} x_i\right] \qquad (5)$$

where θ represents the number of random RR sets in the R set. Add the random RR set to the R set until the number of RR sets in the R set equals θ. Finally, the R set is returned as the result of the sampling phase.

4   Experiment

4.1   Data and Baseline

Data from the etheric transaction data of EthereumG1 and EthereumG2 is available on the X-Block website[1]. The information of the data is shown in Table 1.

Table 1. Basic information of the data set.

	Nodes	Edges
EthereumG1	3832	225714
EthereumG2	10628	222876

We use PageRank as our baseline. PageRank is widely used to evaluate the importance of academic papers and other fields, and has a great influence on the influence of research nodes.

---

[1] http://xblock.pro/.

4.2   Results and Analysis

First, the Infomap used to divide the communities of EthereumG1 and EthereumG2, and the results are shown in Table 2 We can see that the number of small and medium-sized communities is relatively large.

Table 2. Number of nodes in the community.

	Above 2	10–50	50–200	200–500	Above 500
EthereumG1	510	23	5	2	1
EthereumG2	827	29	7	4	4

From the Q value of the community partition results, we see that the results of the EthereumG1 community partition at more than 50,100,200 nodes are excellent, but the community partition results at more than 500 nodes are not ideal. The results of the community partition of EthereumG2 are excellent, as shown in Table 3.

Table 3. Q value of community partition.

	Above 50	Above 100	Above 200	Above 500
EthereumG1	0.42	0.45	0.41	−1.0E-14
EthereumG2	0.72	0.73	0.74	0.68

This paper mainly studies larger communities' community behaviors and selects community 6 in EthereumG2 with a relatively uniform number of nodes and edges for analysis. The top 10 nodes of the two algorithms are shown in Fig. 3.

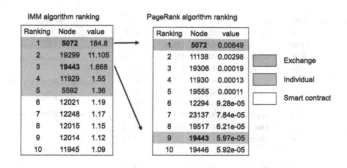

Fig. 3. IMM and PageRank algorithms for node ranking.

684   X. Li et al.

It is found that except for the same node in the first place, the rest of the ranking has only one common node and belongs to different rankings. From the IMM value and PageRank value, we can see that the first node's value is far from the values of the other ranking nodes. This is also why the two rankings are the same except for the first place, but the other rankings are quite different.

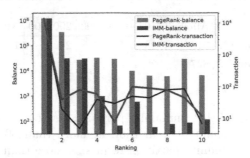

Fig. 4. IMM and PageRank ranking nodes. -balance represents the highest account balance of the account, -transaction represents the average number of daily transactions of the account.

Rank the account balance and the number of transactions of all nodes in Community 6, as shown in Fig. 4. The bar chart represents the balance of the account, and the line chart represents the number of transactions of the account. It can be seen that the ranking of IMM results is biased towards the nodes with more transactions, while the ranking results of PageRank are biased towards the nodes with more transaction amounts.

To compare the effects of different algorithms, we counted the transaction amount, the number of transactions, and the number of transaction objects of the nodes in Community 6 and ranked them respectively. The information of the top 10 nodes is shown in Fig. 5.

Ranking	IMM	PageRank	N1	N2	N3
1	5072	5072	5072	5072	5072
2	19299	11328	11138	12336	12003
3	19443	19306	12336	12272	12336
4	11929	11930	11946	11971	3142
5	5592	19555	11944	12060	4389
6	12021	12294	11953	12048	3804
7	12248	23137	11930	2380	2380
8	12015	19517	19299	11956	2551
9	12014	19443	19443	11975	11956
10	11945	19446	12272	12029	19424

Exchange

Individual

Smart contract

N1:Amount of the transaction

N2:Number of transactions

N3:Number of trading partners

Fig. 5. Nodes of different ranking strategies.

It can see from Fig. 5. It is difficult to find a valuable personal account by sorting an attribute value. In contrast, the IMM algorithm has a high and concentrated ranking of influential nodes. It can be seen that the algorithm to find the ranking strategy suitable for the data set has a better ranking effect than simple attribute ranking. Because the algorithm does not only consider a particular attribute, it also considers the probability, connection, and other factors.

In the trading community, the most important node is the number one exchange account, which participates in many smart contracts, and these contract accounts are linked to them. In the PageRank algorithm, each node has a corresponding score, and it is associated with the node that has a link relationship. However, the IMM algorithm is judged by the propagation scope and breadth of node influence. Generally speaking, the influence of individual nodes is more significant than that of smart contract nodes, so the IMM algorithm's ranking strategy is more suitable for Blockchain transaction networks.

5 Conclusion

In this paper, we propose a method for ranking community nodes' influence in the Blockchain transaction network. The results show that the CIIN method can be applied to not only the traditional model but also the Blockchain transaction model. In the blockchain transaction network, our method can more effectively identify important nodes in the community comparing with PageRank and provide a foundation for the next step of important nodes protection. As further improvement, we can optimize the CIIN method to improve nodes raking accuracy and robustness through introducing more different factors.

References

1. Liu, A.D., Du, X.H., Wang, N., et al.: Research progress of blockchain technology and its application in information security. J. Softw. **29**(7), 2092–2115 (2018)
2. Underwood, S.: Blockchain beyond bitcoin. Commun. ACM **59**(11), 15–17 (2016)
3. He, P., Yu, G., Zhang, Y.F., et al.: Survey on blockchain technology and its application prospect. Comput. Sci. **44**(04), 23–29 (2017)
4. Swan, M.: Blockchain: Blueprint for a New Economy. O'Reilly Media, Inc. (2015)
5. Zheng, Z., Xie, S., Dai, H., et al.: An overview of blockchain technology: architecture, consensus, and future trends. In: 2017 IEEE International Congress on Big Data (BigData Congress), pp. 557–564. IEEE (2017)
6. Liu, D.Y., Jin, D., He, D.X.: Community mining in complex networks. Comput. Res. Dev. **50**(10), 2140–2154 (2013)
7. Chen, X.Q., Shen, H.W.: Community structure of complex networks. Complex Syst. Complexity Sci. **08**(1), 57–70 (2011)
8. Girvan, M., Newman, M.E.J.: Community structure in social and biological networks. Proc. Natl. Acad. Sci. U.S.A. **99**(12), 7821–7826 (2002)
9. Nandini, R.U., Réka, A., Soundar, K.: Near linear time algorithm to detect community structures in large-scale networks. Phys. Rev. E Stat. Nonlinear Soft Matter Phys. 76 (2007)

10. Rosvall, M., Bergstrom, C.T.: Maps of random walks on complex networks reveal community structure. Proc. Natl. Acad. Sci. U.S.A. **105**(4), 1118–1123 (2008)
11. Xindong, W., Yi, L., Lei, L.: Influence analysis of online social networks. J. Comput. **37**(4), 735–752 (2014)
12. Kitsak, M., Gallos, L.K., Havlin, S., et al.: Identification of influential spreaders in complex networks. Nat. Phys. **6**(11), 888–893 (2010)
13. Brin, S., Page, L.: The anatomy of a large-scale hypertextual web search engine (1998)
14. Chen, W.: Research on influence diffusion in social networks. Big Data Res. 2015031
15. Kleinberg, J.M.: Authoritative sources in a hyperlinked environment. J. ACM (JACM) **46**(5), 604–632 (1999)
16. Lempel, R., Moran, S.: The stochastic approach for link-structure analysis (SALSA) and the TKC effect. Comput. Netw. **33**(1–6), 387–401 (2000)
17. Rosvall, M., Bergstrom, C.T.: Multilevel compression of random walks on networks reveals hierarchical organization in large integrated systems. PloS One **6**(4), e18209 (2011)
18. Tang, Y., Shi, Y., Xiao, X.: Influence maximization in near-linear time: a martingale approach. In: Proceedings of the ACM SIGMOD International Conference on Management of Data, pp. 1539–1554 (2015)
19. Borgs, C., Brautbar, M., Chayes, J., et al.: Maximizing social influence in nearly optimal time. In: Proceedings of the Twenty-Fifth Annual ACM-SIAM Symposium on Discrete Algorithms. Society for Industrial and Applied Mathematics, pp. 946–957 (2014)
20. Enright, A.J., Van Dongen, S., Ouzounis, C.A.: An efficient algorithm for large-scale detection of protein families. Nucleic Acids Res. **30**(7), 1575–1584 (2002)

ScholatAna: Big Data-Based Academic Social Network User Behavior Preference System

Wenjie Ma, Ronghua Lin$^{(\boxtimes)}$, Jianguo Li, Chengjie Mao, Qing Xu, and Angjian Wen

School of Computer Science of South China Normal University, Guangzhou 510631, Guangdong, China

`{wjma,rhlin,jianguoli,maochj,qingXu,2018022669}@m.scnu.edu.cn`

Abstract. In recent years, the rapid development of academic social networks has greatly promoted academic exchanges and scientific research collaborations among users. At the same time, with the various social behaviors of many users, massive log text data are accumulated on academic social networks. In this regard, in order to extract the available information from the massive log text data, this paper takes the case of implicit interactive information generated by scholars on SCHOLAT (https://www.scholat.com) as a research, and analyzes the user behavior in the past year based on the user's spatiotemporal behavior characteristics and preference behavior characteristics. Thus, ScholatAna, a framework based on big data technology for Academic Social Networking (ASN) is proposed. Considering that the information generated by users is up to tens of millions of chaotic log files. Therefore, this paper combines distributed computing methods, uses Hadoop ecosystem technology to extract available data, and uses TF-IDF and social collaborative filtering algorithms to perform faster and more accurate statistics and analysis of data. The experimental results are demonstrated and evaluated by using visual analysis techniques. To a certain extent, these results revealing the user's behavior trends and regulars in the domain of academic social, which affects the development of scientific research.

Keywords: Academic social network · Log text big data · Distributed computing · User behavior analysis · SCHOLAT

1 Introduction

With the rapid development of social networks, more and more users are no longer satisfied with traditional social networks. Therefore, a series of domain-specific networks focused on a certain field have been refined to be popular among users. In this case, the academic social network (ASN), a new type of social network focused on providing scholars with academic exchanges and scientific research cooperation, has attracted more and more scholars. Since ASN surmounts the limitations of time and space, it facilitates academic sharing and research collaborations. Due to the frequent interactions between users and ASN, it accumulates massive structured and unstructured data every day. These rapidly growing data provide valuable resources for the development of social networks.

© Springer Nature Singapore Pte Ltd. 2021
Y. Sun et al. (Eds.): ChineseCSCW 2020, CCIS 1330, pp. 687–696, 2021.
https://doi.org/10.1007/978-981-16-2540-4_52

At present, the well-known academic social networking platforms mainly include Academia, ResearchGate, Mendeley, etc. They focus on different disciplines and have their own strengths. In China, since 2008, academic social networks such as Scholat.com and ScholarMate have begun to appear one after another, providing massive information on user behavior [1]. At present, most of the analysis of this information is concentrated in the existing structural databases, such as mining analysis based on personal attribute information and behavior information of the user relation database only [2], or behavior analysis in the mass social field [3]. However, "contactless" social behaviors such as user's login behavior [4], user's search behavior [5] and user browsing and access link behavior [6] account for more than 80% of human social interaction, which means that it is more meaningful to analyze the user behavior by using massive implicit data [7].

Therefore, for the log text big data generated after the behavior of all academic users, we can analyze the behavior of academic users to in-depth the operation of academic social networks and the hot trend of academic information. That is, how to filter, calculate and analyze from the massive text big data. It has become a major challenge for text data analysis in today's big data era to perform user behavior operations and determine user social behaviors.

The main goal of this work is to design and implement ScholatAna, which is a domain-specific ASN-based big data technology framework to identify the behavior patterns of ASN users and explain them in an intuitive, visual, and easy-to-understand manner. The workflow of ScholatAna is mainly as follows.

1. Data preprocessing, through researching the data extracted from ASN and storing it in the big data platform.
2. Data analysis, using Spark distributed computing, and using TF-IDF [8] and collaborative filtering [9–11] and other algorithms to analyze user search keyword popularity and recommend popular academic courses, academic teams, information content, and predict academic research hotspots.
3. Demonstrate the results based on appropriate visual analysis techniques and tools.

In the following, we first we review related work and discuss the motivation in Sect. 2. We introduce the workflow and implementation of our system in detail in Sect. 3. We then evaluate our system with extensive experiments in Sect. 4. Finally, we conclude our work in Sect. 5.

2 Related Work

Many scholars have conducted mining and analysis on user behaviors with the massive data information in social networks such as Facebook, Twitter, Sina Weibo, and QQ. Lu Q et al. [12] took e-mail as an example to analyze the information such as the keywords, subject, and event content of the e-mail. Guo Z et al. [13] took Sina Weibo as an example to monitor the spammers in popular topics by observing the user's posting behavior. Vithayathil J et al. [14] applied media richness theoryand social network analysis to explore the relationship between social media and shopping preferences based on survey data from physical supermarkets and online stores.

However, with the application of big data, when using traditional technology for user analysis and calculation of massive data, no matter whether it is in the storage or computing capacity of the computer, bottlenecks are encountered. At this time, distributed computing technology is introduced into big data processing, thus breaking through the original limitations have realized the load balancing of big data calculations by multiple computing nodes and completed the effective high-speed processing of data. Therefore, more research institutions and related universities in the world began to use distributed computing to carry out user behaviors on big data. Analyze research projects.

Magdalena Cantabella et al. [15] used Hadoop MapReduce big data processing framework and association rule technology to analyze a large amount of data in the learning management system, and found the trend and deficiencies of students using the learning management system. Liu J et al. [16] used the distributed computing method of the behavior change model TTM to filter the multi-label classification of keywords in Twitter tweets, extracted alcohol-related tweets and analyzed the relationship between user network communication, accident behavior, and drinking. Lamia Karim et al. [17] apply the Spark-Hadoop ecosystem to analyze pedestrian road traffic risks, propose a data flow model for drivers and pedestrians, and analyze pedestrian road traffic safety; the well-known website Facebook also uses distributed computing methods, through pulling Starting from the Hadoop cluster nodes of the entire website, collecting the HTTP request information of users clicking and visiting the website, and calculating the user visit behavior in a distributed manner to recommend advertisements. In addition, AOL, IBM, Amazon and many other IT giants are also using Hadoop.

In these social network big data user behavior analysis studies, most of them are aimed at the general public [18], and there is no user behavior analysis for specific groups. The data is derived from the data recorded in the database by user operations and network requests. It is not comprehensive enough to do user behavior analysis from such explicitly visible data, and it is impossible to mine the more underlying behavior rules and trends of users. This paper focuses on the specific groups of scholar users in academic social networks, extracting and mining massive log text data generated by academic social networks for user behavior analysis, these logs record the academic behavior of academic users, browsing access behavior and academic social The direction contains rich behavior trends and laws, and analyzing these hidden log user behavior data is a powerful supplement and further exploration of the explicitly visible database data.

3   System Design

This system model mainly includes the preprocessing of big data in the back-end log text of the scholar network, the user behavior analysis of the scholar network scholar and the visual processing of the user behavior data. Massive log text data does not exist in an orderly and structured form; therefore, the original text data needs to be collected, filtered, analyzed, and calculated through a series of operations such as data preprocessing to be transferred into relational database data. Then, use TF-IDF and collaborative filtering algorithms to analyze and process the relational database data, and display the data in the form of charts according to the front-end visualization technology. Figure 1 is a system flowchart.

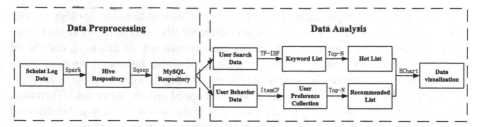

Fig. 1. System flow chart

3.1 Data Preprocessing Design

For the realization of the processing of log text big data, this paper combines the content of the Hadoop ecosystem to collect, extract, analyze and calculate the log big data in a distributed and efficient parallel system, so as to extract the valid data available for use.

First, collect the log text data through the Flume log collection system to extract the most original and unprocessed text data; save the extracted data files to the HDFS distributed file system; then use the Spark analysis engine to Scan and analyze the data files, analyze the required data information in units of behavior, and continuously insert the data information into the Hive data table. Finally, use the Sqoop data transfer tool to transfer the Hive warehouse data table into relational database data. Log text data preprocessing logic is shown in Fig. 2.

In the design of log text data processing, considering the large number of log files generated by the three servers Tomcat and containing massive data information, a distributed environment is used to collect, filter, analyze, and calculate log text data. This system A series of technologies in the Hadoop ecosystem are mainly used to complete the distributed processing of log text data. The use of the Hadoop ecosystem to complete the processing of log text data mainly includes four parts, which are the construction of a Hadoop cluster, the collection of log text data, the calculation of log text data, and the migration of calculation result data.

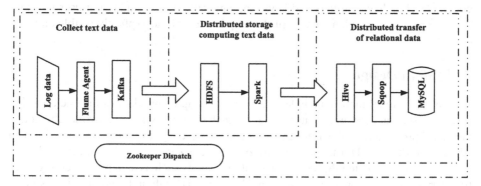

Fig. 2. Log text data preprocessing logic architecture diagram

3.2 Data Analysis

- User search behavior analysis

  According to the user's search situation, the user search content text is obtained in different search sections (scholar search, news search, course search, team search), and each user search content is recorded as a text record, and the user search content list is obtained; then Use the TF-IDF algorithm to calculate the corresponding important user search keyword list in different search sections. The keyword list filters the corresponding user search content list, and excludes the search content that is irrelevant or unimportant to the search section, so as to filter out the matching search The search keyword popularity list of the section content analyzes the user's search popularity behavior. For the implementation of the TF-IDF algorithm, the user search content extracted by the Hadoop distributed system is used as the data raw material, and the importance level of keywords in different search sectors is determined by the value of the product of TF and IDF. The definition is as follows:

$$tf_{ij} = \frac{n_{ij}}{\sum_k n_{k,j}} \tag{1}$$

$$idf_i = \log \frac{|D|}{|\{j : t_i \in d_j\}|} \tag{2}$$

$$TF - IDF_{if} = \frac{n_{ij}}{\sum_k n_{k,j}} \times \log \frac{|D|}{|\{j : t_i \in d_j\}|} \tag{3}$$

  Among them, in the definition (1), tf_{ij} represents the frequency of occurrence of a search keyword in the text, n_{ij} represents the number of times a search keyword appears in a document, $\sum_k n_{k,j}$ represents the total number of user searches in a document; in the definition (2), idf_i represents the importance measure of a search keyword, $|D|$ represents the total number of documents in the data raw material, $|\{j : t_i \in d_j\}|$ represents t_i The total number of documents searched for keywords. In the definition (3), $TF - IDF_{if}$ indicates the importance level of a search keyword, that is, the product value of tf_{ij} and idf_i, the larger the value, the higher the importance level. According to the above definition, the keyword list calculated by $TF - IDF$ is used to filter the user search content list, and finally the filtered list results are sorted in descending order, and the top ten search keywords are taken out, that is, the search behavior of all users is analyzed.

- User preference behavior analysis

  Extract academic user browsing access information from the relational database, including academic user access link time and academic user access link address. User preference behavior analysis first scans and filters user access link addresses to extract link addresses related to academic teams, academic courses, and academic information; then, in three different link addresses, according to the academic team, academic courses, and academic information The ID field further subdivides different academic teams, academic courses, and academic information; and then uses the item-based collaborative filtering algorithm to calculate and obtain and recommend the academic teams, academic courses, and academic information preferred by scholar users within a certain period of time; Finally, use Echarts technology to visualize the data.

Item-based collaborative filtering algorithms are used to calculate academic courses, academic teams, and information content that users may be interested in or like to make recommendations, thereby analyzing user preference behaviors. Using the user browsing and access behavior data extracted by the Hadoop distributed system as data materials, the data is collected and calculated according to the historical user data, and the similarity between academic courses, academic teams, and information content is obtained, and finally based on the similarity and The user's historical browsing access behavior data generates a corresponding recommendation list for the user [19], and uses the Top-N analysis method to recommend the first five courses and team data, and the first six information data, which are defined as follows:

$$w_{i,j} = \frac{|N(i) \cap N(j)|}{\sqrt{|N(i)\|N(j)|}} \tag{4}$$

$$p_{u,j} = \sum\nolimits_{i \in N(u) \cap S(i,K}} w_{i,j} r_{u,i} \tag{5}$$

Taking the calculation of user preference for academic courses as an example, in the definition (4), $w_{i,j}$ represents the similarity between courses browsed by the user, and $|N(i)|$ is the course i visited by the user The number of times, $|N(j)|$ is the number of times the user visits the browsing course j, $|N(i) \cap N(j)|$ is the number of times the user visits the browsing course i and course j at the same time Similarity, if the value of the similarity between the two courses is higher, it means that the two courses may be liked by more users. In the definition (5), $p_{u,j}$ represents the interest and likeness of all users u to course j, $N(u)$ is the set of courses that users like, and $S(i, K)$ is the most similar to course i Of K courses, $w_{i,j}$ is the similarity between course i and course j that the user browses and visits as described above, $r_{u,i}$ is the interest of all users u in course i, here simplified $r_{u,i}$ takes the value 1. According to the above definition, the recommended list is sorted according to the calculated value, and the courses ranked higher in the recommended list are more likely to be historical courses that users visit and browse similar courses of interest or like, that is, analyze the user's preferred course behavior.

4   Experiment Analysis

4.1   Experimental Environment

This system uses five physical machines to build a distributed system environment. The Hadoop distributed environment is shown in Table 1. In addition, Spring Boot + EChart + LayUI is used to build the user behavior analysis system of Scholat.

4.2   Experimental Results and Analysis

In all user search behavior analysis interfaces, the relationship between all users' different times and search times is drawn in the form of a histogram. It is analyzed that the users

Table 1.

Host name	IP	Running process
Hadoop01	192.168.195.131	NameNode Master
Hadoop02	192.168.195.132	NameNode Master
Hadoop03	192.168.195.133	DataNode QuorumPeerMain Worker
Hadoop04	192.168.195.134	DataNode QuorumPeerMain Worker
Hadoop05	192.168.195.135	DataNode QuorumPeerMain Worker

of the social network scholar network using the PC terminal still occupy the majority, see Fig. 4 for details. In general, there is a downward trend from April to June, which is related to the learning and work needs of users in the social network and the courses and teams. During the semester, more users will publish relevant academic and academic information, monthly summaries and plans, etc., so users search more frequently and more frequently.

Query all users' search keyword popularity, the front-end interface initiates a request, and the background business logic first sends the corresponding search section data request according to different search sections; the UserSearchResitory data operation business class specifies the search section content after receiving the request, and the Service business class based on the search content Classify and use the TF-IDF algorithm to calculate the keywords of all users' search content in the current search section; filter the top ten search terms according to the keywords, encapsulate the data as a collection object and return to the front end; finally the front end uses Echats The data visualization technology will analyze and display the popularity of all user search keywords in the front end. Taking the course search section as an example, all users search keyword popularity analysis as shown in Table 2 (Fig. 3).

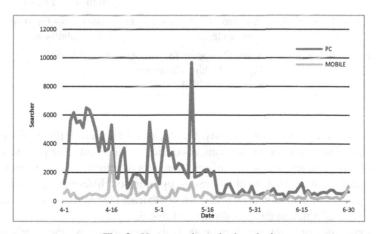

Fig. 3. User search method analysis

Table 2. User search hot ranking

Serial number	Course name	Searches
1	Discrete mathematics	1787
2	Computer application foundation	1766
3	Computer network	1747
4	Circuit analysis	957
5	Operating system	709
6	Software engineering	658
7	Data structure	634
8	UI interface design	577
9	Accounting Professional Ethics	568
10	Database	549

Table 3. User preference behavior analysis

Hot courses recommendation	Hot team recommendation	Hot information recommendation
C language programming	Social Network Data Intelligence Lab	IEEE CSCWD 2020 Conference Call for Papers
Basic Computer Science	Network and Security Lab	Welcome to Scholar Network (Welcome to SCHOLAT)
Database principle	School of Computer Science, South China Normal University 2016	Supplementary Provisions of the School of Software on the Training and Introduction of High-level Talents
Computer Basic Experiment	Professor Shu Lei's scientific research team	Instructions for using the course platform
Computer network training	Youth Computer Technology Forum of China Computer Society YOCSEF	PAAP 2019

In user preference behavior analysis, the back-end business logic performs corresponding data classification calculations according to different user preferences, and extracts relevant courses, teams and information data for business logic processing and cumulative calculations, so as to obtain users' opinions on different courses and teams. And the degree of information preference, finally extract the top five courses and teams, and the top six information data, as shown in Table 3.

The above user preference behavior analysis interface draws the user's preferred courses, teams, and information in the form of graphics. Analyze the situation that the social network scholar network users pay attention to in the browsing and access operation behavior, and understand the user's behavior process when using the social network scholar network, that is, what courses, teams, and information the user stays longer, read And more operations. Taking the preference team as an example, users pay more attention to and visit the "Social Network Data Intelligence Lab (SCHOLAT LAB)/Guangdong Service Computing Engineering Research Center" to understand the specific situation of user preferences.

5   Conclusion

In the era of social networking, with the massive increase in semi-structured and unstructured data, the value obtained from the computational analysis and mining of textual big data is increasing. In front of massive amounts of data, using a distributed environment to extract, mine, and calculate available valid data not only saves the time and cost of calculation, but also improves the efficiency of data calculation. Therefore, this paper uses the Hadoop distributed ecosystem technology to extract and calculate the log text big data of the social network scholar network, and designs and develops an available user behavior analysis system to analyze the behavior trends and laws of the scholar network users. This system uses a series of Hadoop distributed ecosystem technologies to extract the large text data of the scholar's log text, and uses TF-IDF algorithm and collaborative filtering technology and Spring Boot, Echats and other system development technologies to achieve an analysis of user behavior and demonstrated system. The main contributions of this paper include:

- With the social network scholar network as the background, consult social network related theories, text big data processing related technical documentation, study its log data processing method and analyze the social behavior of scholar network users.
- Build a distributed environment of the Hadoop ecosystem, import logs recorded by all servers, and analyze and calculate in the distributed environment, thereby generating data raw materials as the data source for the analysis of this system.
- According to the user behavior data extracted from the log text data, the functional modules of this system are designed through demand analysis, combined with related theoretical algorithms, to complete the analysis of background logical user behavior data.
- According to the data results obtained after the analysis of background logical data, design different functional modules, display the front-end visual interface, develop and complete the system.

Acknowledgement. This work was supported by National Natural Science Foundation of China under grant number U1811263, by National Natural Science Foundation of China under grant number 61772211.

References

1. Zhao, Y., Li, L.: Review and reflection on the status quo of academic social network research at domestic and abroad. Inf. Doc. Serv. **11**(6), 41–47 (2016)

2. Huang, T., Elghafari, A., Relia, K., et al.: High-resolution temporal representations of alcohol and tobacco behaviors from social media data. PACM **1**(1), 1–26 (2017)
3. Al Hasan Haldar, N., Li, J., Reynolds, M., Sellis, T., Yu, J.: Location prediction in large-scale social networks: an in-depth benchmarking study. VLDB J. **28**(5), 623–648 (2019). https://doi.org/10.1007/s00778-019-00553-0
4. Du, F., Plaisant, C., Spring, N., et al.: EventAction: visual analytics for temporal event sequence recommendation. In: 2016 IEEE Conference on Visual Analytics Science and Technology (VAST), pp. 2–4. IEEE (2016)
5. Yin, H., Cui, B., Chen, L., et al.: A temporal context-aware model for user behavior modeling in social media systems. In: Proceedings of the 2014 on SIGMOD'14 PhD Symposium, pp. 1543–1554. ACM, Snowbird (2014)
6. Kravi, E.: Understanding user behavior from online traces. In: Proceedings of the 2016 on SIGMOD'16 PhD Symposium, pp. 27–31. ACM, San Francisco (2016)
7. Navar-Gill, A.: Knowing the audience in the information age: big data and social media in the US television industry. In: Companion of the 2017 ACM Conference on Computer Supported Cooperative Work and Social Computing, CSCW, Portland, pp. 89–92 (2017)
8. Huang, C., Yin, J., Hou, F.: A text similarity measurement combining word semantic information with TF—IDF method. Chin. J. Comput. **34**(5), 856–864 (2011)
9. Shi, Y., Larson, M., Hanjalic, A.: Collaborative filtering beyond the user-item matrix: a survey of the state of the art and future challenges. ACM Comput. Surv. **47**(1), 1–45 (2014)
10. Chen, Z., Wang, Y., Zhang, S., et al.: Differentially private user-based collaborative filtering recommendation based on K-means clustering. arXiv preprint arXiv:1812.01782 (2018)
11. Shao, Y., Xie, Y.: Research on cold-start problem of collaborative filtering algorithm. Comput. Syst. Appl. **28**(2), 246–252 (2019)
12. Lu, Q., Zhang, Q., Luo, X., Fang, F.: An email visualization system based on event analysis. In: Sun, Y., Lu, T., Yu, Z., Fan, H., Gao, L. (eds.) ChineseCSCW 2019. CCIS, vol. 1042, pp. 658–669. Springer, Singapore (2019). https://doi.org/10.1007/978-981-15-1377-0_51
13. Guo, Z., Liu, S., Wang, Y., Wang, L., Pan, L., Wu, L.: Detect cooperative hyping among VIP users and spammers in Sina Weibo. In: Sun, Y., Lu, T., Xie, X., Gao, L., Fan, H. (eds.) ChineseCSCW 2018. CCIS, vol. 917, pp. 241–256. Springer, Singapore (2019). https://doi.org/10.1007/978-981-13-3044-5_18
14. Vithayathil, J., Dadgar, M., Osiri, J.: Social media use and consumer shopping preferences. Int. J. Inf. Manag. **53**(6), 1–13 (2020)
15. Cantabella, M., Martinez-Espana, R., Ayuso, B., et al.: Analysis of student behavior in learning management systems through a Big Data framework. Future Gener. Comput. Syst. **90**(3), 262–272 (2019)
16. Liu, J., Weitzman, E., Chunara, R., et al.: Assessing behavior stage progression from social media data. In: CSCW 2017, pp. 1320–1333. ACM, Portland (2017)
17. Karim, L., Boulmakoul, A., Mandar, M., et al.: A new pedestrians' intuitionistic fuzzy risk exposure indicator and big data trajectories analytics on Spark-Hadoop ecosystem. In: The 11th International Conference on Ambient Systems, Networks and Technologies (ANT), pp. 137–144 (2020)
18. Meo, P., Ferrara, E., Abel, F., et al.: Analyzing user behavior across social sharing environments. ACM Trans. Intell. Syst. Technol. **5**(1), 14:1–14:31 (2014)
19. Zhang, H., Huang, T., Lv, Z., Liu, S., Zhou, Z.: MCRS: A course recommendation system for MOOCs. Multimedia Tools Appl. **77**(6), 7051–7069 (2017). https://doi.org/10.1007/s11042-017-4620-2

Pedestrian Detection Algorithm in SSD Infrared Image Based on Transfer Learning

Jing Feng[1] and Zhiwen Wang[2]

[1] South China Institute of Software Engineering, GU, Guangzhou 510990, Guangdong, China
[2] Guangxi University of Science and Technology, LiuZhou 545006, Guangxi, China

Abstract. Because of difficulty in feature extraction of infrared pedestrian images, the traditional methods of object detection usually make use of the labor to obtain pedestrian features, which suffer from the low-accuracy problem. With the rapid development of machine vision, deep learning has gradually become a research hotspot and a mainstream method for many pattern recognition and object detection problems. In this paper, aiming at the defects of deep convolutional neural network, such as the high cost on training time and slow convergence, a new algorithm of SSD infrared image pedestrian detection based on transfer learning is proposed, which adopts a transfer learning method and the Adam optimization algorithm to accelerate network convergence. For the experiments, we augmented the OUS thermal infrared pedestrian dataset and our solution enjoys a higher mAP of 94.8% on the test dataset. After experimental demonstration, our proposed method has the characteristics of fast convergence, high detection accuracy and short detection time.

Keywords: Deep learning · Transfer learning · SSD · Pedestrian detection

1 Introduction

At present, pedestrian detection technology at night mainly uses visible light images, lidar, infrared images and other technologies. In recent years, with the rapid development of artificial intelligence, deep learning algorithms have gradually matured, and have successfully solved many current pattern recognition problems (including object detection problems) [1]. Compared with traditional target detection methods, it does not require experts to set a specific feature [2]. A large number of samples can automatically extract features through deep learning algorithm models, and then use the extracted features to train the classifier [3–6]. Although there are not many achievements in using deep learning algorithms in infrared pedestrian detection, from the perspective of development trend, deep learning will become the mainstream of infrared pedestrian detection algorithms and many target detection fields [7]. The pedestrian detection technology based on deep learning infrared images has great market application potential.

© Springer Nature Singapore Pte Ltd. 2021
Y. Sun et al. (Eds.): ChineseCSCW 2020, CCIS 1330, pp. 697–704, 2021.
https://doi.org/10.1007/978-981-16-2540-4_53

2 SSD Model Introduction

2.1 SSD Network Structure

The SSD algorithm is composed of the basic network (VGG-16), and then several volume layers [8–10]. Figure 1 shows the SSD network.

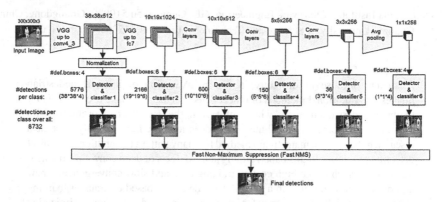

Fig. 1. SSD network

2.2 SSD Network Loss Function

The loss function of the SSD network is weighted sum of location loss (LOC) and confidence loss (CONF) [11], which can be calculated by Eq. (1).

$$L(x, c, l, g) = \frac{1}{N}(L_{conf}(x, c) + \alpha L_{loc}(x, l, g)) \tag{1}$$

In Eq. (1), N is the number of positive samples of the a priori box, c is the category confidence prediction value, l is the position prediction value of the bounding box corresponding to the a priori box, g is the position parameter of the true bounding box, and α is weight coefficient.

The LOC is defined by Eq. (2) with Smooth L1 loss.

$$L_{loc}(x, l, g) = \sum_{i \in Pos}^{N} \sum_{m \in \{cx, cy, w, h\}} x_{ij}^k smooth_{L1}(l_i^m - \hat{g}_j^m) \tag{2}$$

For the CONF, softmax loss is used to calculate it with Eq. (3).

$$L_{conf}(x, c) = -\sum_{i \in Pos}^{N} x_{ij}^p \log(\hat{c}_i^p) - \sum_{i \in Neg} \log(\hat{c}_i^0) \text{ where } \hat{c}_i^p = \frac{\exp(\hat{c}_i^p)}{\sum_p \exp(\hat{c}_i^p)} \tag{3}$$

In Eq. (3), $x_{ij}^p \in \{1, 0\}$ is a index. When $x_{ij}^p = 1$, it means that i-th predicted bounding box coincides with j-th ground bounding box, and category is p. Higher the probability prediction of p, the smaller the loss, and the probability is generated by softmax.

3 Key Detection Technology

3.1 Transfer Learning

Transfer learning is a machine learning method, which transfers knowledge from the source domain to the target domain to improve the learning effect of the target domain [12, 13]. Through the data features and model parameters learned in the previous field, the same function can be realized in the new field [14]. Transfer learning has many advantages, such as smaller training data requirements; good classification capabilities on non-training data; more robust training process, making training more stable. And it can accelerate training and improve depth Learn the performance of the model to improve network performance [15].

3.2 Main Implementation Steps

This paper uses transfer learning to detect infrared pedestrian images. That is to say, the network parameters are initialized based on the network weights that SSD has trained and iterated 100000 times on PASCAL VOC dataset. The main steps are:

1. Use PASCAL VOC dataset to train SSD network and save the weight;
2. After expand the OUS dataset, we divide the training set and test set into 8:2, and transform them into.tfrecord format file;
3. Use the training set to fine tune the pre-trained SSD network. Firstly, the pre-trained weight of PASCAL VOC dataset is converted to the weight on the hot infrared pedestrian dataset of OUS. Then take layer that directly involved output layer layer. Relearn their network weights;
4. Input the test set into the trained model, and adjust the super parameters according to the results.

The experimental flow is shown in Fig. 2. The black arrow represents the training process, red represents the verification process, and blue represents the test process.

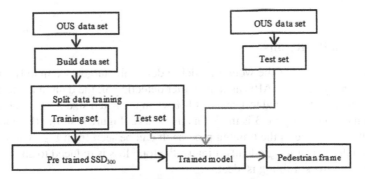

Fig. 2. SSD Transfer learning process

4 Experimental Results

4.1 Experiment Configuration and Super Parameter Setting

The hardware and software configuration of this experiment is shown in Table 1.

Table 1. Experimental configuration

Name	Related configuration
CPU	Inter Core i7
Deep learning framework	Tensorflow-1.13.0rc2
data processing	Python 3.5, Open CV

4.2 Dataset

Paper use the OUS hot infrared pedestrian dataset. As there are only 284 images in OUS dataset, in order to achieve better results, this chapter uses data enhancement method to expand the dataset. Then the data set is divided according to the ratio of training set and test set of 8:2. Table 2 shows the data set values used in this paper.

Table 2. Specific data set values

Data set	Number of images	Number of pedestrians
Training set	2258	5568
Test set	565	1336
total	2823	6904

4.3 Experimental Results

To verify the accuracy of the weight model to detect the target, we mainly look at the average accuracy mean (mAP) calculated after detecting all the data in the test set.

The accuracy of the final test set is 94.8% when the batch size is 16 and the number of iterations is 9000. Figure 3 is mAP function curve. Figure 4 shows the convergence curve of the loss value in the training process. It can be seen from Fig. 3 and 4 that the network model converges gradually. Finally, the total loss is reduced to about 1.94, and the result of network training is ideal.

Data sets used in this paper consists of the original the OUS thermal infrared pedestrian dataset and the expanded image based on it. Figure 5 and Fig. 6 are partial image detection results.

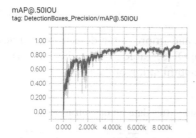

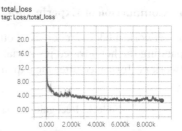

Fig. 3. mAP function curve   **Fig. 4.** Loss function curve

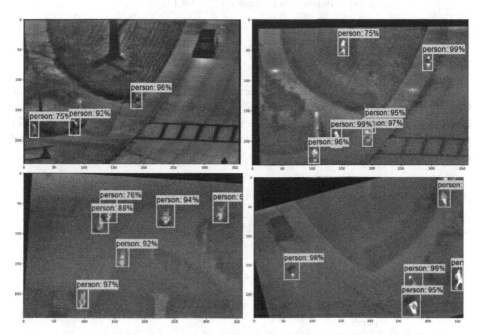

Fig. 5. Pedestrian detection results in infrared image

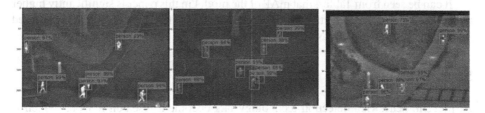

Fig. 6. Other image detection results

4.4   Comparison of Experimental Results

The network model can achieve better results and improve the detection accuracy by expanding the hot infrared pedestrian dataset of OUS. Table 3 shows the comparison of network performance after the expansion and optimization of data set.

Table 3. Recognition results before and after data set optimization

Data set processing method	mAP (%)
Before data expansion	64.17
After data expansion	90.23
After data optimization	94.80

In this paper, the mAP, recall and detection time are used as the evaluation criteria. Considering the two requirements of accuracy and completeness, the overall performance of the algorithm is measured, and the detection performance of different algorithms for pedestrian target is evaluated more comprehensively. Table 4 shows the mAP, R and detection time of single image of different models in this dataset.

Table 4. Recognition results of different network models

Model	mAP (%)	R (%)	Test time (ms)
Faster-Rcnn	79.1	74.35	2374
YOLO	75.25	69.79	1653
SSD	77.86	72.33	3428
Methods of this paper	94.8	85.94	530

It can be seen from Table 4 that mAP of the model in this paper is significantly higher than other models, and detection time is greatly reduced. The detection effect of infrared image is better than fast RCNN, Yolo and SSD. Its mAP can reach 94.8%.

5   Conclusion

At present, pedestrian detection technology has become a research hotspot, which is widely used in vehicle assistant driving, driverless vehicle, intelligent video monitoring, intelligent transportation and other aspects. In this article, a new pedestrian detection algorithm based on transfer learning is proposed by using deep learning method, and SSD is used in network weight on Pascal VOC dataset initializes the new SSD network. The weight of the layer directly related to the output layer will be learned again and again through the infrared image data set, and a reasonable model of data fitting is obtained, which is more suitable for infrared image pedestrian detection. Input the test set into the trained model, adjust the super parameters according to the results. Experiments show that this algorithm improves the network mAP, reduces network training time and iteration times, and speeds up the network convergence. Through the transfer learning, the complexity and running time of the network are reduced.

Acknowledgements. This work was supported in part by the National Natural Science Foundation of China under Grant (6192007, 61462008, 61751213, 61866004), the Key projects of Guangxi Natural Science Foundation of China under Grant (2018GXNSFDA294001, 2018GX-NSFDA281009), the Natural Science Foundation of Guangxi of China under Grant (2018GXNS-FAA294050, 2017GXNSFAA198365), 2015 Innovation Team Project of Guangxi University of Science and Technology under Grant (gxkjdx201504), Research Fund Guangxi Key Lab of Multi-source Information Mining&Security under Gant (MIMS19-0, Innovation Project for College Students of Guangxi University of Science and Technology (GKYC201708), Scientific Research and Technology Development Project of Liuzhou under Grant (2016C050205).

References

1. Olmeda, D., de la Escalera, A., Armingol, J.M.: Contrast invariant features for human detection in far infrared images. In: Intelligent Vehicles Symposium, pp. 117–122. IEEE (2012)
2. Kim, D.S., Kim, M., Kim, B.S., et al.: Histograms of local intensity differences for pedestrian classification in far-infrared images. Electron. Lett. **49**(4), 258–260 (2013)
3. Ren, S., He, K., Girshick, R., et al.: Faster R-CNN: towards real-time object detection with region proposal networks. In: Advances in Neural Information Processing Systems 28, pp. 1137–1149 (2015)
4. David, R., André, M., Pedro, M., et al.: A real-time deep learning pedestrian detector for robot navigation. In: Evalu 2017 IEEE International Conference on Autonomous Robot System and Competitions, pp. 165–171 (2017)
5. Yandong, L., Zongbo, H., Hang, L.: A survey of convolutional neural networks. Comput. Appl. **36**(9), 2508–2515 (2016)
6. Song, W., Shumin, F.: SSD (single shot multibox detector) research and improvement of target detection algorithm. Ind. Control Comput. **32**(4), 103–105 (2019)
7. Xudong, X., Liqian, M.: Control chart recognition based on transfer learning and convolution neural network. Comput. Appl. **38**(S2), 290–295 (2018)
8. Girshick, R.B., Donahue, J., Darrell, T., et al.: Rich feature hierarchies for accurate object detection and semantic segmentation. In: Computer Vision and Pattern Recognition, pp. 580–587 (2014)
9. Kaiming, H., Xiangyu, Z., Shaoqing, R., et al.: Spatial pyramid pooling in deep convolutional networks for visual recognition. In: 2014 European Conference on Computer Vision, pp. 1904–1916 (2014)
10. Simon, M., Rodner, E.: Neural activation constellations: unsupervised part model discovery with convolutional networks. In: 2015 IEEE International Conference on Computer Vision and Pattern Recognition, pp. 1143–1151 (2015)
11. Liu, W., et al.: SSD: single shot multibox detector. In: Leibe, B., Matas, J., Sebe, N., Welling, M. (eds.) ECCV 2016. LNCS, vol. 9905, pp. 21–37. Springer, Cham (2016). https://doi.org/10.1007/978-3-319-46448-0_2

12. Yong, L.T., Ping, L., Xiao, G.W., et al.: Deep learning strong parts for pedestrian detection. In: 2015 IEEE International Conference on Computer Vision, pp. 1904–1912 (2015)
13. Gowdham, P., Binsu, K., Sudha, N., et al.: Obstacle detection and classification using deep learning for tracking in high-speed autonomous driving. In: 2017 IEEE Region 10 Symposium (TENS YMP), pp. 1–6 (2017)
14. Jing, F., Zhiwen, W., Min, Z., Xinliang, C.: Flower recognition based on transfer learning and adam deep learning optimization algorithm. In: Proceedings of the 2019 International Conference on Robotics, Intelligent Control and Artificial Intelligence, pp. 598–604 (2019)
15. Xudong, X., Liqian, M.: Control chart recognition based on transfer learning and convolutional neural network. Comput. Appl. **38**(S2), 290–295 (2018)

How to Make Smart Contract Smarter

Shanxuan Chen, Jia Zhu$^{(\boxtimes)}$, Zhihao Lin, Jin Huang, and Yong Tang

School of Computer Science of South China Normal University,
Guangzhou 510631, Guangdong, China
{shanxuanchen,jzhu}@m.scnu.edu.cn

Abstract. The smart contract is a self-executing code that is managed by blockchain nodes, providing coordination and enforcement framework for agreements between network participants. However, smart contracts are not particularly "smart" since the virtual machine (VM) does not support the running environment of machine learning. To make the smart contract smarter, we propose a decentralized blockchain oracle framework to support smart contract training the machine learning model. In view of malicious nodes which may attack the process of training, we propose a consensus algorithm to prevent the malicious attack from malicious nodes. At the end of this paper, we do an experiment on two datasets: MNIST and CIFAR10. The result shows that our framework can prevent malicious attack efficiently and keep high accuracy. With our proposed framework, smart contracts have an ability to train or call machine learning model, making a smart contracts "smarter".

Keywords: Blockchain · Blockchain Oracle · Consensus mechanism

1 Introduction

The smart contract is a self-executing code that is managed by blockchain nodes, providing coordination and enforcement framework for agreements between network participants. Leading AI textbooks define the field as the study of "smart": any device that perceives its environment and takes actions that maximise its chance of successfully achieving its goals. Unfortunately, smart contracts are not particularly "smart", since the smart contract can only be as smart as the people coding it, all the information available at the time of coding must be taken into consideration.

To make the smart contract smarter, we propose a decentralised blockchain oracle framework, in which the smart contract has an ability to call the machine learning algorithm and train the machine learning model. There are two challenges to this framework. First, smart contracts cannot access off-chain data or service. Second, ethereum virtual machine, interpreting and executing smart contracts, do not support any machine learning algorithm model.

For the first challenge, inspired by blockchain oracle technology, which can help smart contracts access external information, we introduce the blockchain oracle framework to obtain external information from the outside. It is worth

© Springer Nature Singapore Pte Ltd. 2021
Y. Sun et al. (Eds.): ChineseCSCW 2020, CCIS 1330, pp. 705–713, 2021.
https://doi.org/10.1007/978-981-16-2540-4_54

mentioning that blockchain oracles are vital in the blockchain ecosystem because they expand the scope of smart contracts. Without blockchain oracle, the use of smart contracts will be very limited because they can only access data from their own network [13].

For the second challenge, we expend some machine learning components in the oracle node to train machine learning model, like tensorflow and pytorch. Since the proposed framework is decentralised, some participants may be not rigorously following the protocol, in which the malicious output is well crafted to maximise its effect on the network. In consideration of some malicious participants, we propose a consensus mechanism to prevent the attack during the process of model training.

In this paper, we present an overview of the decentralized blockchain oracle framework to bridge the smart contract and machine learning model. Furthermore, we present a consensus algorithm to prevent the malicious attack and keep model high accuracy. At the end of this paper, we analyse the Byzantine fault tolerance ability of this consensus by constructing a experiment on MNIST and CIFAR10.

2   Framework

In this section, we introduce a decentralised intelligent oracle framework. Decentralized intelligent oracle framework is a third-party service that provides smart contracts with machine learning components. It connects the blockchain and the outside services, like intelligent service. First, we describe the overview of the framework. Then, we introduce the consensus mechanism between oracle nodes during training.

2.1   Overview

As illustrated in Fig. 1, we provide decentralized oracle services to the blockchain for the validity and accuracy of the data. By deploying multiple oracle nodes monitoring with each other, the decentralized oracle guarantees reliable data that can be provided to the smart contract [13].

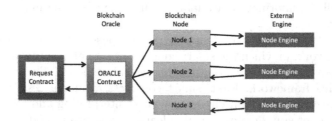

Fig. 1. Decentralized intelligent Oracle framework

When the requester contract sends an on-chain transaction to an oracle contract, the oracle contract tracks the balances from the requester, gets training data from the blockchain, and emits an event to notify blockchain node. After receiving the signal from oracle contracts, blockchain node would create a task to node engine. Node engine is an environment with TensorFlow or PyTorch within a docker container. Like the distributed machine learning, we distribute the machine learning workload across multiple node engine. In view of malicious nodes which may attack the process of training, we design a consensus mechanism to tolerant some attacks from malicious nodes, and we will talk about in Sect. 2.2.

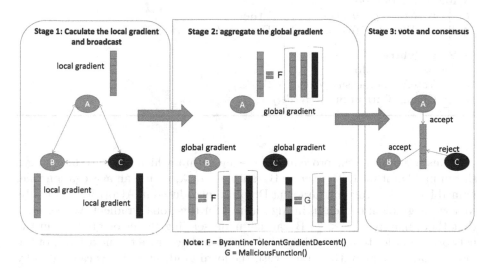

Note: F = ByzantineTolerantGradientDescent()
G = MaliciousFunction()

Fig. 2. An overview of the algorithm includes three important stages. The Byzantine node (the black node) not only can send the malicious local gradient, but also can construct a malicious global gradient by calling the malicious function G.

2.2   Consensus Mechanism

In view of malicious nodes which may attack the process of training, it is necessary to design a secure consensus mechanism to prevent the malicious attack. One of the famous attacks is data poison which at-tacks compromise the training data to change the model's behaviour at inference time. The other famous attack is the backdoor attack which changes the model's behaviour only on specific attacker-chosen inputs, without impacting its performance on the main task.

As illustrated in Fig. 2, we show a scenario to expose the process of the Byzantine attack. At stage 1, when each node calculates and broadcasts the local gradient, Byzantine nodes send a malicious local gradient to the trust nodes. After stage 1, each node needs to aggregate the gradients from each node. It is a chance for Byzantine nodes to propose a malicious global gradient and affect the consensus.

Algorithm 1: Main Algorithm

---

1 **Stage One:**
2 **for** $i \leftarrow 1$ **to** n **do**
3   Caculate gradient vi and BrocastGradient($Node_i$)

4 **Stage Two:**
5 **for** $i \leftarrow 1$ **to** n **do**
6   **if** *Receive gradients* $(v_1, \ldots, v_n)$ **then**
7   ByzantineTolerantGradientDescent($(v_1, \ldots, v_n)$)

8 **Stage Three:**
9 **for** $i \leftarrow 1$ **to** n **do**
10   VoteAndConsensus($Node_i$)
11   Change Current State ($Node_i$)

Thus, we propose the procedure of the algorithm, which includes three stages shown in Algorithm 1. To tolerate Byzantine attack, we use an aggregation rule, named Byzantine Tolerant Gradient Descent, introduced in Algorithm 3. Finally, we use the gradient consensus in Algorithm 4 for the global gradient. We consider that there may be $f(f < \frac{n}{3})$ Byzantine node sending the error gradient in the network. In order to update the global model and generate the next block of the model chain, the algorithm starts with the local gradient which is calculated by each node, and we will show in the next section.

Broadcast Gradient. Before the execution of the Byzantine Agreement algorithm between N nodes, each node calculates the local gradient first with local data set depend on the current state, and then broadcast the local gradient to each other. Due to the unsafe network in the real world, there are two possible hacks that may happen in Byzantine nodes such as the malicious dataset attack and the malicious gradient attack. Mistrusting of the local gradient, we propose a consensus algorithm to prevent the Byzantine attack in the next subsection.

Byzantine Tolerant Gradient Descent. We define basic requirements on aggregation consensus. We consider that there are n trust nodes and f of them are malicious. Each node exchanges the vector lg and output a vector F. When each node agrees on the aggregation consensus result, we consider the vector F is the optimal aggregation result for the steepest direction of the cost function.

After stage one (Broadcast Gradient), each node has all local gradient results $V = (lg_1, \ldots, lg_N)$. We design an aggregation function F to the aggregation rule

---

Algorithm 2: Broadcast Gradient

Input: the local minibach size B, the dataset $\mathcal{P} = (P_1, P_2, \ldots, P_N)$, the number of local epochs E, the learning rate η, and the current state w_t

1   **for** $i \leftarrow 1$ **to** n **do**
2      $\mathcal{B} \leftarrow$ (split P_i into batches of size B)
3   **for** *batch* $b \in \mathcal{B}$ **do**
4        $lg \leftarrow W_t - \eta \nabla \ell(W_t; b)$
5   multicast lg to other nodes
6   wait for $(lg_1, \ldots, lg_{i-1}, lg_{i+1}, \ldots, lg_N)$

---

of all nodes. During the round t, Each node updated the parameter vector using the following equation.

$$w_{t+1} = w_t - \gamma_t \cdot F\left(lg_1^t, \ldots, lg_N^t\right) \tag{1}$$

Algorithm 3: Byzantine Tolerant Gradient Descent

Input: gradient vector array $(v_1, \ldots, v_n)$, Number of nodes N, Number of Byzantine node f

1   **for** $i \leftarrow 1$ **to** n **do**
2      $lg \leftarrow W_t - \eta \nabla \ell(W_t; b)$
3   multicast lg to other nodes
4   wait for $(lg_1, \ldots, lg_{i-1}, lg_{i+1}, \ldots, lg_N)$
5   **for** $i \leftarrow 1$ **to** n **do**
6      $s(i) = \sum_{(i,j) \in V_i} \|v_i - v_j\|^2$
7    $MinI = \{i | s(i) \in min_{k \subset N} s(k)\}$
8    $F =$ the mean of v_i in $\{v_i \mid i \in MinI\}$

We introduce an algorithm, named Byzantine Tolerant Gradient Descent which satisfies the Byzantine resilience condition and the definition of consensus shown in Algorithm 2. As illustrated in Fig. 3, first, we compute euclidean distances between all gradients in each round. Then, we sum up the $n - f - 2$ closest distances and picks the gradient with the lowest sum as the global gradient for the next round. Geometrically illustrated as Fig. 3, v_i belongs to the ball centered at g with radius r, then the scalar product is bounded below by a term involving $\sin \alpha = r / \|g\|$.

Vote and Consensus. Each node has a global gradient after running Byzantine Tolerant Gradient Descent algorithm, and each correct node has the same global gradient because the Byzantine Tolerant Gradient Descent algorithm F is deterministic. Considering the Byzantine node's existence, each correct node

in the network also needs to communicate with each other to certify whether
the global gradient is correct or not.

Algorithm 4: Vote And Consensus

---
1 **for** $i \leftarrow 1$ ***to*** n **do**
2      $Node_i$ multicast F to other nodes for voting F
3      $Node_i$ wait for the vote of other's node
4      $F' \leftarrow$ the gradient F whose number of vote $> n - f$
5      $Node_i$ update the next state by $W_{t+1} = W_t - \gamma_t * F'$

As shown in Algorithm 4, the consensus between nodes depends on the number of the same message. Because we assume that there are $n - f$ normal nodes
in the network, the global gradient which gets $n - f$ replies seems correct.

3 Experimental Evaluation

In this section, to evaluate our proposed framework efficiently, we experiment
with attacking the proposed framework, with and without the presence of the
defences, and provide an experiment for validating the applicability of the framework by attacking real-world networks.

Attack Model. We propose an assumption on the behaviour of the Byzantine nodes - the attacker will interfere with the process with the mere desire of
obstructing the server from reaching good accuracy by random gradients. Given
current model weight W_t, Byzantine attacker random select a value R to infect
the gradient lg_b.

$$lg_b \leftarrow W_t - \eta \nabla \ell(W_t; b) * R \qquad (2)$$

Experiment Detail. Following the definition described mentioned above, we
consider simple architecture on the first two datasets: MNIST and CIFAR10. To
strength our claims, we also experimented on the modern WideResNet architecture on CIFAR100. We assume that there are $n = 15$ normal blockchain oracle
nodes, out of which $m = 12$ were corrupted and non-omniscient.

As illustrated as Fig. 3, graph A is the accuracy result of MNIST and graph C
is the accuracy result of CIFAR10, and graph B is the error rate of MNIST and
graph D is the error rate of CIFAR10. Comparing with the graph B and graph
D, with the influence of Byzantine attack, these two graphs show that malicious
blockchain oracle nodes could increase the rate of error, and it is fragile to train
the machine learning model in the decentralized blockchain network. Against
the simple attack model, we use our proposed consensus mechanism to defend
the Byzantine attack. After training, the accuracy of the attack down to 30%,
while the accuracy of our proposed method keeps 83%, means that our proposed
method can keep high accuracy, and prevent attack successfully.

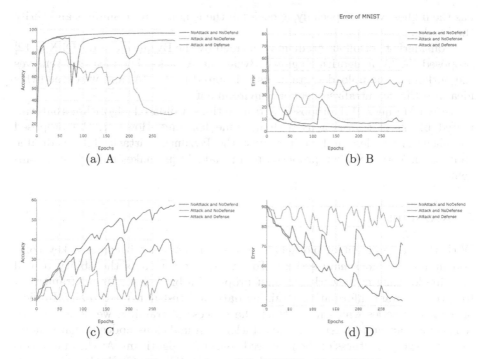

(a) A   (b) B

(c) C   (d) D

Fig. 3. Experiment result in MNIST and CIFAR10

4 Related Work

In this section, we review the related research for consensus mechanism about machine learning and blockchain.

Chen, Xuhui [7] proposed a privacy-preserving learning system named LearningChain. LearningChain is a decentralized blockchain system which can learn the general model without a trusted central server. What's more, they proposed an aggregation algorithm so that their system can resist potential Byzantine attacks. Similar to the aggregation algorithm mentioned above, we propose a consensus mechanism to against Byzantine attack.

Pawlak, M [9] proposed a supervised and non-remote internet voting system based on blockchain, called ABVS. ABVS combines voting system with reinforcement learning, that shows how to integrate the intelligent node and blockchain network. However, the proposed system is nearly centralized and do not use a consensus mechanism between intelligent nodes.

Rana, Shehla [8] proposed a Byzantine fault-tolerant consensus mechanism, named iPath, by constructing a conflict graph to discover the reliable communication between a pair of nodes and disagreement.

Li, Kejiao [10] proposed a new consensus mechanism, named proof of vote (POV), for the consortium blockchain. In this consensus mechanism, nodes vote to the block which they consider it is valid. Finally, they generate a block which has the highest vote. Especially, agencies in the league vote to summit and verify the blocks.

Considering poisonous gradients uploaded by the Byzantine node, Qi Xia [12] proposed FABA, a algorithm against Byzantine Attacks. The idea that removes the outliers of the uploaded gradients is inspired by FABA, and we extend this idea into the decentralized network environment.

Bravo-Marquez [11] proposed a transactions validated algorithm that validated blocks of transactions with machine learning. However, the proposed consensus is trustless and can't against the Byzantine attack. The Auditable Blockchain Voting System proposed by Pawlak, M [9] makes blockchain intelligent.

5   Conclusion

With the development of blockchain, the smart contract becomes a key component for the blockchain system. To give smart contracts the ability to call machine learning frameworks, we first proposed a blockchain Oracle framework to access relevant information (training data and test data) externally. In view of malicious nodes which may attack the process of training, we design a consensus mechanism to tolerant some attacks from malicious nodes. What's more, we prove the correctness of the proposed consensus algorithm. At the end of this paper, we do an experiment on two datasets: MNIST and CIFAR10. The result shows that our framework can prevent malicious attack effectively and keep high accuracy.

Acknowledgements. This work was supported by the National Natural Science Foundation of China (No. 62077015, No. U1811263 and No. 61772211), the Key-Area Research and Development Program of Guangdong Province, China (No. 2018B010109002) and the Science and Technology Project of Guangzhou Municipality, China (No. 201904010393), as well as the Guangzhou Key Laboratory of Big Data and Intelligent Education (No. 201905010009).

References

1. Nakamoto, S.: Bitcoin: a peer-to-peer electronic cash system (2008). https://bitcoin.org/bitcoin.pdf
2. Dinh, T.T.A., Liu, R., Zhang, M., Chen, G., Ooi, B.C., Wang, J.: Untangling blockchain: a data processing view of blockchain systems. IEEE Trans. Knowl. Data Eng. **30**, 1366–1385 (2018)
3. Yeow, K., Gani, A., Ahmad, R.W., Rodrigues, J.J.P.C., Ko, K.: Decentralized consensus for edge-centric internet of things: a review, taxonomy, and research issues, pp. 1513–1524, June 2017
4. Castro, M., Liskov, B.: Practical Byzantine fault tolerance, pp. 173–186 (1999)
5. Ongaro, D., Ousterhout, J.: In search of an understandable consensus algorithm. In: 2014 USENIX Annual Technical Conference (USENIXATC 2014), pp. 305–319 (2014)

6. Divyakant, M.: Issues and challenges with blockchain: a survey. Int. J. Comput. Sci. Eng. **6**, 488–491 (2018)
7. Chen, X., Ji, J., Luo, C., Liao, W., Pan, L.: When machine learning meets blockchain: a decentralized, privacy-preserving and secure design. In: 2018 IEEE International Conference on Big Data (Big Data), pp. 1178–1187 (2018)
8. Rana, S.S., Vaidya, N.H.: iPath: intelligent and optimal path selection for Byzantine fault tolerant communication. In: 2015 IEEE Conference on Computer Communications (INFOCOM), pp. 1095–1103 (2015)
9. Pawlak, M., Poniszewska-Marańda, A., Guziur, J.: Intelligent agents in a blockchain-based electronic voting system. In: Yin, H., Camacho, D., Novais, P., Tallón-Ballesteros, A.J. (eds.) IDEAL 2018. LNCS, vol. 11314, pp. 586–593. Springer, Cham (2018). https://doi.org/10.1007/978-3-030-03493-1_61
10. Li, K., Li, H., Hou, H., Li, K., Chen, Y.: Proof of vote: a high-performance consensus protocol based on vote mechanism & consortium blockchain. In: 2017 IEEE 19th International Conference on High Performance Computing and Communications, pp. 466–473 (2017)
11. Bravo-Marquez, F., Reeves, S., Ugarte, M.: Proof-of-learning: a blockchain consensus mechanism based on machine learning competitions. In: 2019 IEEE International Conference on Decentralized Applications and Infrastructures (DAPPCON), pp. 119–124 (2019)
12. Xia, Q., Tao, Z., Hao, Z., Li, Q.: FABA: an algorithm for fast aggregation against byzantine attacks in distributed neural networks. In: IJCAI 2019, pp. 4824–4830 (2019)
13. Beniiche, A.: A Study of Blockchain Oracles. arXiv 2004.07140 (2020)

Population Mobility Driven COVID-19 Analysis in Shenzhen

Ziqiang Wu[1], Xin Li[1], Juyang Cao[1], Zhanning Zhang[2], Xiaozhi Lin[2], Yunkun Zhu[2], Lijun Ma[3], and Hao Liao[1(✉)]

[1] College of Computer Science and Software Engineering, Shenzhen University, Shenzhen 518060, China
[2] Technical Reconnaissance Detachment, Shenzhen Public Security Bureau, Shenzhen 518000, China
[3] College of Management, Shenzhen University, Shenzhen 518060, China

Abstract. In the beginning of 2020, the COVID-19 epidemic broke out and spread all over the world in just a few months. COVID-19 spread analysis has attracted considerable research efforts in many areas including the impact of population mobility on the epidemic development. However, most studies do not use real data on population mobility, or choose an overly wide range of objects. This paper studies the COVID-19 epidemic in Shenzhen from January 26 to February 16, focusing on the impact of population mobility on the epidemic development. Combined with the population mobility data, we propose the Source-SEIR model. We estimated that the basic reproduction number of SARS-CoV-2 is 2.61. The experiment results show that the combination of population mobility data is helpful to the evaluation of the epidemic development, and the restrictions on population mobility in Shenzhen have played a role in curbing the deterioration of COVID-19 epidemic. Without restrictions on population mobility, there will be more than 600 confirmed cases of COVID-19 in Shenzhen.

Keywords: COVID-19 · Population mobility · Reproduction number · SEIR model · Epidemic analysis

1 Introduction

Global transport and communication networks enable information, ideas and infectious diseases to spread far faster than historically possible [1]. With the COVID-19 spreading, lots of scholars around the world began to focus on the research of COVID-19. The basic reproduction number of SARS-CoV-2 is one

Supported by: National Natural Science Foundation of China (Grant Nos. 61803266, 61703281, 91846301, 71790615, 71471118 and 71871145), Guangdong Province Natural Science Foundation (Grant Nos. 2019A1515011173 and 2019A1515011064), Shenzhen Fundamental Research-general project (Grant Nos. JCYJ20190808162601658 and JCYJ2018030512462-8810).

© Springer Nature Singapore Pte Ltd. 2021
Y. Sun et al. (Eds.): ChineseCSCW 2020, CCIS 1330, pp. 714–721, 2021.
https://doi.org/10.1007/978-981-16-2540-4_55

of the topic in these work [2,3]. Most studies show that the basic reproduction number of SARS-CoV-2 is about 3 [4,5]. It is higher than the basic reproduction number of SARS in 2002 [6], which means that SARS-CoV-2 is more contagious. SARS-CoV-2 has high concealment and high spread, which is the main reason of COVID-19 outbreak [4]. The high concealment is reflected in the low detection rate of COVID-19 patients. Bommer C et al. [7] calculated 40 countries with the worst epidemic, and concluded that the detection rate of COVID-19 was only 6%. It also leads to the high dissemination of SARS-CoV-2. In addition, de Arruda GF et al. [8] proposed that the high spread of SARS-CoV-2 is related to the recovery rate of COVID-19. Although the spread is much higher than that of SARS, the death rate is much lower. Kissler SM et al. [9] found that SARS-CoV-2 is more similar to OC43 and HKU1, which also have high spread and low death rate. Verity R et al. [10] estimated that the death rate of COVID-19 in China is 1.38%.

On January 23, Wuhan was locked down, and various cities began to implement prevention measures. Studies have found that prevention before the outbreak played an important role [11]. Du Z et al. [12] found that the implementation of social distance control in China has slowed down the epidemic development of COVID-19. Tian H et al. [13] pointed out the city closure of Wuhan is closely related to the postponement of the COVID-19 outbreak in other cities. The conclusions of studies [14] show that Chinese efforts in epidemic prevention have achieved results.

The success of epidemic prevention in China has made more scholars focus on the spread and the prevention of COVID-19. Cui P et al. [15] found that the large-scale spread of the epidemic is caused by the mutual promotion of ordinary and close social contact. Experts generally agree that social contact is an important way for SARS-CoV-2 to spread, proposing models base on contact to analyze the COVID-19 epidemic [16,17]. Some epidemic model studies based on population mobility have emerged [3,11].

However, some of the population mobility data used in these studies are simulated or not accurate. In addition, the wide scope of the study make it difficult for the experiment results to reflect the real situation in a certain area and are easily affected by areas with severe outbreaks. In this paper, we use the real population mobility data of Shenzhen, combined with the local epidemic data from January 26 to February 16, to study the impact of population mobility on the epidemic development. According to the spreading dynamics, we propose **Source-SEIR model** based on population mobility.

2   Data and Methods

2.1   Data Sources and Experiment Setup

After the outbreak of COVID-19, the National Health Commission released daily the epidemic situation of all provinces and cities, including data on confirmed cases, recovery cases and death cases. On January 19, 2020, the first confirmed case in Guangdong occurred in Shenzhen. To prevent the COVID-19 epidemic,

Shenzhen Public Security Bureau has strengthened the screening of migrant population, especially those from Wuhan and Hubei. The Information Center of Shenzhen Public Security Bureau updates daily the number of floating people from Hubei and Wuhan, as well as the inflow and outflow population of Shenzhen. We used it as the real-time data of population mobility in the experiment. As shown in Fig. 1, the positive number indicates an inflow of population, while the negative indicates an outflow of population.

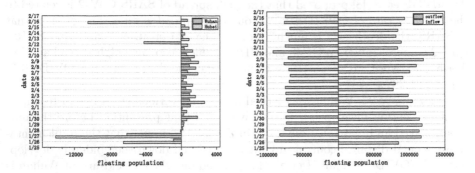

Fig. 1. Population mobility data in Shenzhen.

We assume the generation interval of SARS-CoV-2 T_G is 8.4 days [6]. The median latent period of COVID-19 P_L is 3.0 days, estimated by Nanshan Zhong from 1099 COVID-19 patients [18]. By Eq. 1, the infection period of COVID-19 P_I is 5.4 days.

On December 8, officials reported the first case of unexplained pneumonia, which was later named as COVID-19. We assume it as the first onset day of COVID-19, which is considered to be the beginning of the whole epidemic [2].

$$P_I = T_G - P_L \tag{1}$$

2.2 Reproduction Number

The reproduction number R, refers to the average number of people infected before an infected person recovers.

At the beginning of the epidemic, the virus spread freely in a population of all susceptible people, with R_0 representing the basic reproduction number. R_0 is one of the important conditions to judge whether an epidemic breaks out. When $R_0 > 1$, the epidemic will break out. When $R_0 < 1$, it will die out [19]. R_0 can be used to indicate the infectious ability of the virus. The outbreak of COVID-19 shows that R_0 of SARS-CoV-2 must be greater than 1.

In the middle and later stages of the epidemic, the reproduction number under the influence of various factors is called the effective reproduction number, expressed by R_t. R_t is calculated according to the epidemic situation of a certain period, and the choice of period has an impact on the calculation. R_t can reflect the real-time status of the epidemic. When $R_t < 1$, it indicates that the epidemic is receding.

The reproduction number R can be calculated as follows [20]:

$$R = (1 + rP_L)(1 + rP_I), \tag{2}$$

$$b(t) = b(t - \Delta t)e^{r\Delta t}, \tag{3}$$

where P_L and P_I indicate the latent period and infection period of SARS-CoV-2 respectively, $b(t)$ indicates the number of new confirmed cases on day t, r indicates the growth rate of new confirmed cases, $r > max(-\frac{1}{P_L}, -\frac{1}{P_I})$. r at the beginning of the epidemic, can be used to calculate R_0, while in the middle and later stage can be used to calculate R_t.

2.3   Source-SEIR Model

As COVID-19 patients have an latent period, we use SEIR model to analyze the COVID-19. In the SEIR model, the population was divided into four groups: Susceptibles(S), Exposed(E), Infected(I) and Recovered(R). The meaning of recovered in the model is those who are no longer involved in the process of infection. Therefore, in addition to recovery cases, it also includes death cases.

SEIR model can not only predict the epidemic situation in Shenzhen more accurately, but also estimate the number of latent patients. By means of fitting, the proportion of infections from Wuhan and Hubei who inflow and outflow from Shenzhen can be estimated. we put forward the **Source-SEIR model** combined with the data of population mobility in Shenzhen [3]. The differential equations of the model are as follows:

$$\frac{\mathrm{d}S(t)}{\mathrm{d}t} = -\frac{S(t)}{N(t)} \cdot \frac{R}{P_I}I(t) + L_{sin}(t) - L_{sout}(t), \tag{4}$$

$$\frac{\mathrm{d}E(t)}{\mathrm{d}t} = \frac{S(t)}{N(t)} \cdot \frac{R}{P_I}I(t) - \frac{E(t)}{P_L} + k_1 \cdot L_{wein}(t) + k_2 \cdot L_{hein}(t), \tag{5}$$

$$\frac{\mathrm{d}I(t)}{\mathrm{d}t} = \frac{E(t)}{P_L} - \frac{I(t)}{P_I}, \tag{6}$$

$$\frac{\mathrm{d}R(t)}{\mathrm{d}t} = \frac{I(t)}{P_I}, \tag{7}$$

$$\frac{\mathrm{d}N(t)}{\mathrm{d}t} = L_{sin}(t) - L_{sout}(t) + k_1 \cdot L_{wein}(t) + k_2 \cdot L_{hein}(t), \tag{8}$$

where $S(t)$, $E(t)$, $I(t)$, $R(t)$ and $N(t)$ indicate the number of individuals in susceptibles, exposed, infected , recovered and total population on day t, respectively. R indicates the reproduction number. $L_{sin}(t)$ and $L_{sout}(t)$ represent the inflow and outflow population in Shenzhen on day t, respectively. $L_{wein}(t)$ and $L_{hein}(t)$ represent the population of Wuhan and Hubei that flowed into Shenzhen on day t, respectively. k_1 and k_2 indicate the proportion of latent patients which deduce according to the epidemic data of Shenzhen. P_L and P_I respectively indicate latent period and infection period.

In the model, the population inflow and outflow of Shenzhen mainly affect the susceptibles and the total population in Shenzhen, while the number of people from Hubei and Wuhan mainly affect the exposed.

3   Results Analysis

3.1   Basic Reproduction Number and Effective Reproduction Number

On January 23, Wuhan implemented the city closure. According to $P_L = 3.0$, as of January 26th, it is free spread stage of SARS-CoV-2. We chose to calculate the basic reproduction number R_0 on January 26(1462 newly confirmed patients), so we obtained $R_0 = 2.61$. It is similar to the $R_0(3)$ in other studies. The error may be due to the detection rate of confirmed patients was considered in other studies.

We calculated the reproduction number at the beginning of the epidemic in Shenzhen and found that it was much greater than 2.61. The reason is that the new cases in Shenzhen include not only the number of newly infections, but also a large number of imported cases. This part of imported cases does not belong to local spread and is not within the calculation range of R_0.

When calculating R_0, a larger range such as country or global should be selected, with little or no impact on imported cases. According to early reproduction number in Shenzhen and the R_0 in the whole country, we can analyze the proportion of imported cases among the new cases in the early period of the COVID-19 epidemic in Shenzhen.

In February, the COVID-19 epidemic in Shenzhen was gradually brought under control, and the number of imported cases decreased greatly. The calculated reproduction number can be regarded as R_t. As of February 4, R_t of Shenzhen has been reduced to less than 1. On February 10, Shenzhen ushered in the return peak of corporate rework. R_t of Shenzhen rose again, but remained below 0.5, which shows that the prevention measures in Shenzhen were effective.

3.2   Impact Analysis of Population Mobility

The COVID-19 epidemic in Shenzhen is divided into: climbing period, prevention period, and resume work prevention period. The epidemic was studied in stages according to the Source-SEIR model, and the simulation result as shown in Fig. 2. We estimate that during the climbing period, about 60% of people from Wuhan and 10% of people from Hubei to Shenzhen are latent patients. At this period, the epidemic spread most rapidly without prevention. As a result of a large number of imported cases, the reproduction number in Shenzhen is much greater than 2.61, which provides us estimate the latent patients from Wuhan and Hubei.

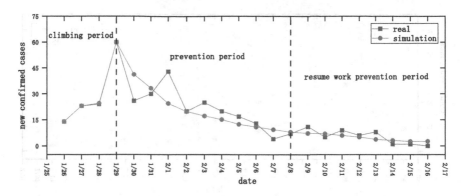

Fig. 2. New confirmed cases in Shenzhen.

In the prevention period, health prevention measures to curb the epidemic development of COVID-19. The reproduction number in local propagation is lower than the calculated results. However, due to the impact of population mobility, there are still latent patients inflow, resulting in a small number of new cases. We estimate that 0.8% of the people from Wuhan and 0.1% from Hubei to Shenzhen are latent patients.

In the resume work prevention period, the reason for the increase in R_t is that a large number of people getting back to work led to an increase in the population base in Shenzhen with an increase in the probability of new infection. We estimate that the proportion of latent patients from Wuhan and Hubei to Shenzhen has dropped to about 0.1% and 0.01% respectively.

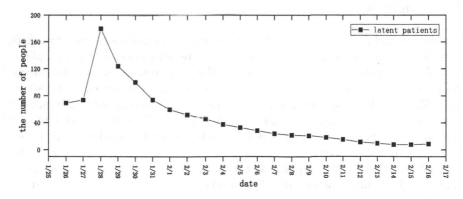

Fig. 3. Estimation of the number of latent patients in Shenzhen.

We estimated the number of latent patients through Source-SEIR model(See Fig. 3). The number of latent patients dropped to single digits after February 13, which shows that the local spread of the COVID-19 epidemic in Shenzhen is coming to an end soon.

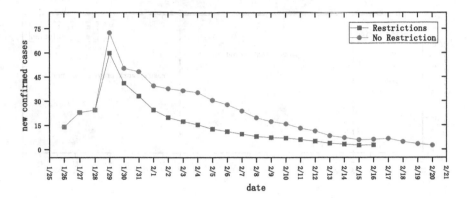

Fig. 4. Impact of population mobility restrictions on new confirmed cases in Shenzhen.

We also combined the population mobility data of Shenzhen in the same period in 2019(after the second day of the Spring Festival). Assuming that except for the restrictions on population movement, other health prevention measures in Shenzhen(such as wearing masks, strict security checks, etc.) remain unchanged, the simulation results are as shown in Fig. 4. It shows that the number of infections increased, and the date of new confirmed cases falling to single digits postponed. We estimate that if Shenzhen does not impose restrictions on population mobility and does not postpone resumption of work, the total number of COVID-19 infections will rise to more than 600.

4   Conclusion

Population mobility data can make epidemic assessment more realistic. Especially for infectious diseases with latent period, population mobility data plays a crucial role in predicting viral carriers [21]. Population base changes are affected significantly in prediction of epidemic development in local areas. Such as mega cities like Shenzhen, even after implementing restrictions on population mobility, millions of people continue to flow daily(See Fig. 1). Shenzhen, a city with 20 million people, has outstanding performance in COVID-19 epidemic prevention. Most experts are only qualitatively explained by results, while this paper illustrates quantitatively the important factors of effective prevention strategies in Shenzhen from empirical and model perspectives: population mobility restriction and effective isolation measures.

Population mobility driven COVID-19 analysis in Shenzhen is not only a proof of the impact of population mobility on epidemic development but also an improvement on SEIR model. However, there are still some shortcomings which need to be improved in future such as considering detection rate of infections, self-healing of latent patients, etc.

References

1. Moore, S, Rogers, T.: Predicting the speed of epidemics spreading in networks. Phys. Rev. Lett. **124**(6), 068301 (2020)
2. Zhou, T., Liu, Q., Yang, Z., et al.: Preliminary prediction of the basic reproduction number of the novel coronavirus 2019-nCoV. J. Evid. Based Med. **13**(1), 3–7 (2020)
3. Wu, J.T., Leung, K., Leung, G.M.: Nowcasting and forecasting the potential domestic and international spread of the 2019-nCoV outbreak originating in Wuhan, China: a modelling study. Lancet **395**(10225), 689–697 (2020)
4. Hao, X., Cheng, S., Wu, D., et al.: Reconstruction of the full transmission dynamics of COVID-19 in Wuhan. Nature **584**(7821), 420–424 (2020)
5. Cao, Z., Zhang, Q., Lu, X., et al.: Incorporating human movement data to improve epidemiological estimates for 2019-nCoV. medRxiv (2020)
6. Lipsitch, M., Cohen, T., Cooper, B., et al.: Transmission dynamics and control of severe acute respiratory syndrome. Science **300**, 1966–1970 (2003)
7. Bommer, C., Vollmer, S.: Average detection rate of SARS-CoV-2 infections is estimated around six percent. Lancet Infect Dis (2020)
8. de Arruda, G.F., Petri, G., Rodrigues, F.A., et al.: Impact of the distribution of recovery rates on disease spreading in complex networks. Phys. Rev. Res. **2**(1), 013046 (2020)
9. Kissler, S.M., Tedijanto, C., Goldstein, E., et al.: Projecting the transmission dynamics of SARS-CoV-2 through the postpandemic period. Science **368**(6493), 860–868 (2020)
10. Verity, R., Okell, L.C., Dorigatti, I., et al.: Estimates of the severity of coronavirus disease 2019: a model-based analysis. Lancet Infectious Dis. (2020)
11. Jia, J., Lu, X., Yuan, Y., et al.: Population flow drives spatio-temporal distribution of COVID-19 in China. Nature 1–5 (2020)
12. Du, Z., Xu, X., Wang, L., et al.: Effects of proactive social distancing on COVID-19 outbreaks in 58 cities, China. Emerg. Infect. Dis. **26**(9), 2267 (2020)
13. Tian, H., Liu, Y., Li, Y., et al.: An investigation of transmission control measures during the first 50 days of the COVID-19 epidemic in China. Science **368**(6491), 638–642 (2020)
14. Ali, S.T., Wang, L., Lau, E.H.Y., et al.: Serial interval of SARS-CoV-2 was shortened over time by nonpharmaceutical interventions. Science **369**(6507), 1106–1109 (2020)
15. Cui, P., Wang, W., Cai, S., et al.: Close and ordinary social contacts: how important are they in promoting large-scale contagion? Phys. Rev. E **98**(5), 052311 (2018)
16. Koher, A., Lentz, H.H.K., Gleeson, J.P., et al.: Contact-based model for epidemic spreading on temporal networks. Phys. Rev. X **9**(3), 031017 (2019)
17. Liu, Y., Gu, Z., Xia, S., et al.: What are the underlying transmission patterns of Covid-19 outbreak? An age-specific social contact characterization. EClinicalMedicine 100354 (2020)
18. Guan, W., Ni, Z., Hu, Y., et al.: Clinical characteristics of 2019 novel coronavirus infection in China. MedRxiv (2020)
19. Adnerson, R.M., Anderson, B., May, R.M.: Infectious Diseases of Humans: Dynamics and Control. Oxford University Press, Oxford (1992)
20. Wallinga, J., Lipsitch, M.: How generation intervals shape the relationship between growth rates and reproductive numbers. Proc. Roy. Soc. B: Biol. Sci. **274**(1609), 599–604 (2007)
21. Xu, X., Wen, C., Zhang, G., et al.: The geographical destination distribution and effect of outflow population of Wuhan when the outbreak of the 2019-nCoV Pneumonia. J. Univ. Electron. Sci. Technol. China **49**, 1–6 (2020)